UNITEXT for Physics

UNITEXT for Physics series publishes textbooks in physics and astronomy, characterized by a didactic style and comprehensiveness. The books are addressed to upper-undergraduate and graduate students, but also to scientists and researchers as important resources for their education, knowledge, and teaching.

Fabrizio Tafuri • Carmine Antonio Perroni •
Giulio De Filippis

Understanding Quantum Many-Body Theory

Fundamental Concepts and Methods

Springer

Fabrizio Tafuri
Department of Physics 'E. Pancini'
University of Naples Federico II
Naples, Italy

Carmine Antonio Perroni
Department of Physics 'E. Pancini'
University of Naples Federico II
Naples, Italy

Giulio De Filippis
Department of Physics 'E. Pancini'
University of Naples Federico II
Naples, Italy

ISSN 2198-7882 ISSN 2198-7890 (electronic)
UNITEXT for Physics
ISBN 978-3-032-17848-0 ISBN 978-3-032-17849-7 (eBook)
https://doi.org/10.1007/978-3-032-17849-7

This Springer imprint is published by the registered company Springer Nature Switzerland AG
The registered company address is: Gewerbestrasse 11, 6330 Cham, Switzerland

F.T. dedicates this book to his parents.
C.A.P. dedicates this book to his wife.
G.D.F. dedicates this book to his father.

Preface

This book originates from the lecture notes prepared by Fabrizio Tafuri, who at the time was a master's student attending the *Quantum Many-Body Course* at the Department of Physics of the *University of Naples Federico II* in 2018–2019. The course itself has been offered regularly since 2009 by Professors Carmine Antonio Perroni and Giulio De Filippis. Building upon those notes, the authors have substantially expanded and refined the material: calculations and derivations have been developed in greater detail, additional topics have been incorporated, and the overall exposition has been reorganized to provide a coherent and self-contained treatment. As a result, the present volume goes well beyond the original lecture notes, offering an accessible yet rigorous resource for MSc Physics students and a solid foundation for further study in quantum many-body physics.

The book is carefully organized in a formal and systematic style, with definitions, theorems, and proofs clearly delineated. This structure serves both pedagogical and practical purposes: it allows readers to approach the material at multiple levels. One may focus on the statements and their physical implications, tracing applications and examples, or delve into the detailed proofs and derivations to gain a deeper understanding of the underlying formalism. By combining rigor with clarity, the book provides a framework that is both educational and versatile, enabling students and researchers alike to navigate the material according to their interests and needs, whether for structured study, self-directed learning, or as a reference for advanced investigations.

After a brief review of the principles of quantum mechanics, the second quantization formalism is introduced in Part I. It enormously simplifies the treatment of many interacting particles: it allows one to avoid treating directly many-particle wavefunctions. Any system turns out to be described in terms of the field operators that replace the one-particle wavefunction in the Schrödinger representation. Field operators incorporate the Bose-Einstein and Fermi-Dirac statistics and allow to prove that any bosonic or fermionic system of identical particles can be described in terms of quanta.

The next step in Part II is the description of a solid by using the second quantization formalism. The student is gradually introduced to the full Hamiltonian:

(1) the starting point is the band theory, based on the assumption that nuclei/ions are fixed at the lattice positions; (2) later, the lattice vibrations around the equilibrium positions are taken into account, leading to the concept of normal modes in classical mechanics: they are quantized, demonstrating that oscillations in the classical theory are described in quantum mechanics in terms of non-interacting bosons, i.e., phonons, shedding further light on the wave-particle duality; (3) finally, the electron-phonon interaction is included in the Hamiltonian: it plays a crucial role in describing the onset of superconductivity, in metallic compounds with low critical temperature, and transport properties of metals.

The next important steps in Part III are the introduction of the retarded many-body Green's functions and the formulation of Feynman perturbation theory, based on the well-known Feynman diagrams. The linear response theory allows us to highlight the relevance of the Green's functions: the response of a system to a weak external perturbation can be deduced by investigating the fluctuations at thermal equilibrium in the absence of the perturbation. By merging linear response theory and retarded Green's functions, one can obtain the dielectric function, the magnetic susceptibility, and the optical absorption, i.e., all quantities experimentally accessible.

In Part IV, the book contains: (1) a simple description of the Bose-Einstein condensation both in the absence and in the presence of interactions, by using canonical transformations; (2) the quantization of the electromagnetic field, that leads to photons. The book ends with Part V providing a toolkit of group theory and symmetries in physics and Part VI dedicated to appendices.

The book also contains a simple model, describing an impurity interacting with phonons, exactly solved through unitary transformations in Chap. 6. The aim is to introduce the student to the most used approaches in the many-body theory, emphasizing that, for describing a system of interacting particles, one can choose among many and different techniques, each of them shedding light on specific features of the physical system under investigation.

Finally, a particular mention deserves the concept of quasiparticle, a brilliant idea proposed by Landau in 1941. Quasiparticles represent low-energy long-lived excitations that behave as weakly interacting particles even if the system contains many strongly interacting bodies. In the framework of quantum many-body theory and second quantization, the concept of a quasiparticle provides a powerful tool to describe interacting systems in terms of effective independent entities. A quasiparticle is not an elementary particle in the strict sense, but rather a collective excitation of the system that behaves as if it were a particle with well-defined energy, momentum, and, in some cases, effective mass or charge. In some cases, they are collective excitations of the original interacting particles, in other cases they coincide with the original particles dressed by the interaction either with the other particles or other fields. In the book, there are a lot of applications of this crucial physical concept. For instance, interacting electrons in a solid can be described in terms of dressed electrons (see Chap. 3), whose properties differ from those of free electrons due to interactions with other electrons and with the lattice. Similarly, lattice vibrations give rise to phonons (see Chap. 4), while collective oscillations of the electronic density are described as plasmons (see Chap. 10).

We hope that this book will encourage students to appreciate quantum many-body physics and help them grasp advanced approaches commonly used in the study of strongly interacting systems.

Naples, Italy
January 30, 2026

Fabrizio Tafuri
Carmine Antonio Perroni
Giulio De Filippis

Acknowledgments We express our sincere gratitude to our families for their constant support, patience, and understanding throughout the process of research and writing that led to this book.

We are grateful to Professor Patrizia Vitale for her valuable comments and suggestions on Chaps. 14 and 15, the Toolkit of Group Theory and Symmetries in Physics.

We also acknowledge the scientific and academic influence of Professors Vincenzo Marigliano Ramaglia and Gianpaolo Zucchelli. Their contributions to teaching and research played a significant role in shaping the scholarly environment in which we received our training. Although they are no longer active members of the academic community, their intellectual legacy continues through the work of successive generations of researchers.

The present book grows out of this scientific tradition, building upon established lines of research while presenting the authors' own contributions and perspectives.

Competing Interests The authors have no competing interests to declare that are relevant to the content of this manuscript.

General References

The development of this book has been guided by a number of foundational texts in quantum many-body theory, which have influenced both its overall structure and the viewpoint adopted throughout. While the exposition is intended to be fully self-contained, the following references offer the interested reader a wider perspective and additional insights that extend beyond the scope of these pages.

1. Abrikosov, A.A., Gorkov, L.P., Dzyaloshinski, I.E.: *Methods of Quantum Field Theory in Statistical Physics*. Dover Publications, New York, 1975. ISBN 978-0486632285.
2. Fetter, A.L., Walecka, J.D.: *Quantum Theory of Many-Particle Systems*. McGraw-Hill, New York, 1971.
3. Mahan, G.D.: *Many-Particle Physics*, 3rd ed. Kluwer Academic/Plenum Publishers, New York, 2000. ISBN 0-306-45423-7.
4. Nolting, W.: *Theoretical Physics 9—Fundamentals of Many-Body Physics*, 2nd ed. Springer, Cham, 2018. ISBN 978-3-319-75794-6.
5. Bruus, H., Flensberg, K.: *Many-Body Quantum Theory in Condensed Matter Physics: An Introduction*. Oxford Graduate Texts, Oxford University Press, Oxford, 2004. ISBN 978-0-19-856633-5.
6. Zagoskin, A.: *Quantum Theory of Many-Body Systems: Techniques and Applications*. Springer, New York, 1998. ISBN 978-0-387-98275-5.
7. Fabrizio, M.: *A Course in Quantum Many-Body Theory: From Conventional Fermi Liquids to Strongly Correlated Systems*. Springer, Cham, 2023. ISBN 978-3-031-11496-2.
8. Thouless, D.J.: *The Quantum Mechanics of Many-Body Systems*, 2nd ed. Dover Publications, Mineola, 2014 (1st ed. 1961). ISBN 978-0-486-49344-8.
9. Dickhoff, W.H., Van Neck, D.: *Many-Body Theory Exposed! Propagator Description of Quantum Mechanics in Many-Body Systems*, 2nd ed. World Scientific, Singapore, 2008. ISBN 978-981-281-380-4.
10. March, N.H., Young, W.H., Sampanthar, S.: *The Many-Body Problem in Quantum Mechanics*. Dover Publications, New York, 1995 (originally published 1967). ISBN 978-0-486-68812-7.

11. Schultz, T.D.: *Quantum Field Theory and the Many-Body Problem.* Academic Press, New York, 1964. ISBN 978-0126326509.
12. Mattuck, R.D.: *A Guide to Feynman Diagrams in the Many-Body Problem*, 2nd ed. Dover Publications, New York, 1992 (1st ed. 1967). ISBN 978-0-486-41938-7.
13. Peliti, L.: *Statistical Mechanics in a Nutshell.* Princeton University Press, Princeton, 2011. ISBN 978-0-691-14519-8.
14. Madelung, O.: *Introduction to Solid State Theory.* Springer Series in Solid-State Sciences, Volume 2. Springer, Berlin, 1996 (1st ed. 1978). ISBN 978-3-540-60443-3.
15. Cohen-Tannoudji, C., Dupont-Roc, J., Grynberg, G.: *Photons and Atoms: Introduction to Quantum Electrodynamics.* Wiley-Interscience, New York, 1989. ISBN 0-471-62181-0.
16. Essler, F.H.L., Frahm, H., Göhmann, F., Klümper, A., Korepin, V.E.: *The One-Dimensional Hubbard Model.* Cambridge University Press, Cambridge, 2005. ISBN 978-0-521-86018-8.
17. Born, M., Fock, V.A.: Beweis des Adiabatensatzes. *Z. Phys. A* 51, 165–180 (1928).
18. Kato, T.: On the Adiabatic Theorem of Quantum Mechanics. *J. Phys. Soc. Jpn.* 5, 435–439 (1950).
19. Altland, A., Simons, B.D.: *Condensed Matter Field Theory*, 2nd ed. Cambridge University Press, Cambridge, 2010. ISBN 978-0521769754.
20. Negele, J.W., Orland, H.: *Quantum Many-Particle Systems.* Addison-Wesley, Reading, MA, 1988. ISBN 978-0201125931.
21. Coleman, P.: *Introduction to Many-Body Physics.* Cambridge University Press, Cambridge, 2015. ISBN 978-0521864886.
22. Nagaosa, N.: *Quantum Field Theory in Condensed Matter Physics.* Springer, Berlin, 2010. ISBN 978-3642084850.
23. Tsvelik, A.M.: *Quantum Field Theory in Condensed Matter Physics*, 2nd ed. Cambridge University Press, Cambridge, 2007. ISBN 978-0521529808.
24. Dupuis, N.: *Field Theory of Condensed Matter and Ultracold Gases—Volume 1.* World Scientific, 2023. ISBN 978-1-80061-390-4.
25. Cornwell, J.F.: *Group Theory in Physics: An Introduction*, Vol. 1. Academic Press, London, 1997. ISBN 978-0121898001.

Contents

List of Symbols

This book covers topics ranging from classical mechanics to quantum mechanics in second quantization. As a consequence, many mathematical symbols can take on different meanings depending on the specific context in which they are used.

To reduce notational complexity and improve readability, we adopt the convention that only fundamental operators (such as position and momentum operators, as well as unitary or antiunitary symmetry operators, e.g., Wigner operators) are explicitly marked with a hat symbol ˆ. Generic interaction operators, including the Hamiltonian in first quantization, are instead left without a hat, while their corresponding second-quantized forms are consistently denoted with one. Less standard operators or scalar quantities are also left without a hat, unless otherwise specified.

This notational choice is carefully designed. For instance, the number operator in first quantization is denoted with a hat, in contrast to the number of particles, which is a scalar quantity. This distinction is intentional, as it avoids ambiguities in situations where the number operator and the number of particles appear side by side or may be interchanged depending on the context. In addition, creation and annihilation operators are likewise written without a hat, reflecting their pervasive use throughout the book.

Moreover, the List of Symbols provides the principal interpretation of the most important symbols used throughout the book. However, since some symbols may carry different meanings depending on the context, their specific usage will always be clarified directly in the book. This approach allows us to maintain a compact notation without compromising the formal precision required for the theoretical development.

$\mathbb{1}$	Identity operator or matrix; its precise meaning will be clear from the context.
0	Zero scalar or null operator; its precise meaning will be clear from the context.
$\delta_{i,j}$	Kronecker delta, equal to 1 if $i = j$, and 0 otherwise.

$\delta(\mathbf{r}-\mathbf{r}'), \ldots$	Generic Dirac delta distribution; its precise dimensionality (e.g., 1D, 3D) will be clear from the context.
$\bar{a}$	Generic n-dimensional vector. This notation is used in Chaps. 14 and 15.
$\mathbf{a}, \mathbf{A}, \mathbf{b}, \mathbf{B}, \ldots$	Generic three-dimensional vectors; their precise meaning will be specified case by case.
$\lvert\mathbf{a}\rvert, a, \lvert\mathbf{b}\rvert, b, \ldots$	Modulus of a three-dimensional vector. Depending on the context, the modulus of a vector (e.g., $\mathbf{a}$) may be denoted either by absolute value symbols ($\lvert\mathbf{a}\rvert$) or by the corresponding non-bold letter (a), to simplify notation. This dual notation also reflects the fact that in some cases it is important to emphasize the vectorial nature of the quantity, such as in the case of eigenvalues of the kinetic energy operator, which depend on the modulus squared of a three-dimensional wavevector, e.g., $\mathcal{E}_{\mathbf{k}} = \frac{\hbar^2\mathbf{k}^2}{2m}$. Additionally, it should be noted that a non-bold symbol (e.g., a) may sometimes indicate a scalar quantity that is not necessarily positive (e.g., a component of a vector), rather than the strictly positive modulus of a vector. In any case, the intended meaning is clear from the context.
H_1, H_o, H_s, H	One-particle Hilbert spaces (orbital and spin, orbital, spin, and generic).
H_N, H_N^A, H_N^S	N-particle Hilbert spaces (generic, antisymmetric for fermions, symmetric for bosons).
$\mathcal{L}^2(X)$	Space of square-integrable complex functions on a domain X.
$\psi, \varphi, \Psi, \Phi, \xi, \zeta, \ldots$	One-particle general wave vectors (configurational, spin, or other internal degrees of freedom).
$\mathbf{r}$	Spatial position.
$\hat{\mathbf{r}}$	Position operator.
∇	Nabla operator used to denote gradient, divergence, curl, and Laplacian (e.g., ∇f, $\nabla\cdot\mathbf{V}$, $\nabla\times\mathbf{V}$, $\nabla^2 f$).
$\hbar$	Reduced Planck constant; sets the fundamental scale of quantum effects ($\hbar \approx 1.055\times10^{-34}$ J s).
$\mathbf{p}$	Momentum.
$\hat{\mathbf{p}}$	Momentum operator, defined in position representation as $\hat{\mathbf{p}} = -i\hbar\nabla$.
$\mathbf{k}, \mathbf{q}$	These symbols may refer to different quantities (e.g., wave vector, crystal momentum, reciprocal–lattice vectors, Brillouin–zone vectors, etc.); its precise meaning will be specified case by case.

	Throughout this book, we will generally refer to this symbol as the momentum, as it is related to the physical momentum by a factor of $\hbar$.
$\mathbf{k}_F$	Fermi momentum.
$\hat{S}_i$ $(i = 1, 2, 3$ or $x, y, z)$	Components of the spin operator in first quantization.
$\hat{\mathbf{S}}^2$	Total spin–squared operator, with eigenvalues $S(S+1)$.
S	Spin quantum number of the particle; fixed integer or half-integer.
s	Spin projection index, taking values in $\{-S, -S+1, \ldots, S-1, S\}$.
σ, σ', σ_i $(i = 1, 2, 3$ or $x, y, z)$, $\boldsymbol{\sigma}$	These symbols may refer to different quantities (e.g., spin basis label (spin projection quantum number), eigenvalues of $\hat{S}_z$, Pauli matrices, permutation operators, optical conductivity, etc.); its precise meaning will be specified case by case.
$f_n, \eta_\sigma, \varphi_{n,\sigma}, \ldots$	Generic one-particle basis functions associated with configurational, spin, composite, or other internal degrees of freedom.
$\chi_\sigma(s)$	Component of the eigenfunction of the spin operator $\hat{S}_z$ with eigenvalue σ in the spin basis labeled by s.
α	This symbol may refer to different quantities (e.g., generical quantum state label, the quantum state label $(\mathbf{k}, \sigma)$, fugacity, etc.); its precise meaning will be specified case by case.
x	This symbols may refer to different quantities (compact notation $x = (\mathbf{r}, s)$, combining spatial position $\mathbf{r}$ and spin s for one particle, etc.); its precise meaning will be specified case by case.
ε_{ijk}	Levi-Civita tensor (fully antisymmetric symbol) with $\varepsilon_{123} = +1$; used to express cross products and angular-momentum commutation relations.
$\mathcal{H}$	First-quantized Hamiltonian.
$\mathcal{E}, \mathcal{E}_\alpha, \mathcal{E}_{n,\mathbf{k},\sigma}, \ldots$	Energies.
$\mathcal{E}_F$	Fermi energy.
μ	This symbols may refer to different quantities (chemical potential, atomic species index, etc.); its precise meaning will be specified case by case.
q_e	Electronic charge, $q_e = -1.602 \cdot 10^{-19}$C.

m	This symbol may refer to different quantities (e.g., particle mass, electronic band index, etc.); its precise meaning will be specified case by case.
m_e	Electron mass, $m_e = 9.109 \cdot 10^{-31}$ kg. We will use the symbol m_e (rather than m) whenever the discussion refers specifically to electrons rather than to generic particles.
μ_B	Bohr magneton, defined as $\mu_B = \frac{\hbar \lvert q_e \rvert}{2 m_e c}$.
T, T_E, T_D, T_F	Absolute temperature (in Kelvin), Einstein temperature, Debye temperature, Fermi temperature.
k_B	Boltzmann constant, $k_B = 1.380649 \cdot 10^{-23} \frac{J}{K}$.
β	Beta parameter, defined as $\beta = \frac{1}{k_B T}$.
t	Real time.
$U(t), U(t, t')$	Evolution operator in real-time formalism. $U(t)$ denotes evolution from a fixed initial time, while $U(t, t')$ gives the evolution between two arbitrary times. Their explicit form depends on the chosen picture (Schrödinger, Heisenberg, or interaction picture).
$S(t, t')$	Time-evolution operator in the interaction picture.
τ	This symbols may refer to different quantities (real time in the imaginary-time formalism, number of molecules in a solid, etc.); its precise meaning will be specified case by case.
$U(\tau), U(\tau, \tau')$	Evolution operator in imaginary-time formalism. $U(\tau)$ denotes evolution from a fixed initial imaginary time, while $U(\tau, \tau')$ gives the evolution between two arbitrary imaginary times. It is typically defined in the thermal interaction picture.
$S(\tau, \tau')$	Time-evolution operator in the thermal interaction picture.
$\Theta(t)$	Heaviside step function, equal to 1 for $t > 0$ and 0 for $t < 0$.
$a, A, b, B, \ldots,$ $a_\alpha, A_\alpha, b_\alpha, B_\alpha, \ldots,$ $a^\dagger, A^\dagger, b^\dagger, B^\dagger, \ldots,$ $a_\alpha^\dagger, A_\alpha^\dagger, b_\alpha^\dagger, B_\alpha^\dagger, \ldots$	These symbols may denote different objects depending on the context (generic operators, annihilation/creation operators (with arbitrary labels α), etc.); their precise meaning will be specified in context.
N	This symbols may refer to different quantities (number of particles, number of unit cells in a

	three-dimensional crystal, etc.); its precise meaning will be specified case by case.
$\hat{N}$, $\hat{N}_\alpha$, ...	Number operator.
$G(x,t,x',t')$, $G^{(r)}(x,t,x',t')$, $G^{(r)}_{AB}(t,t')$, $G^{(m)}_{AB}(\tau,\tau')$ G, $G^{(r)}$, $G^{(m)}$, ...	Green's functions in real or imaginary time, including retarded $G^{(r)}$ and Matsubara (thermal) $G^{(m)}$ types. The arguments may denote one-particle coordinates $(\mathbf{r}, t)$ or, in many-body contexts, generalized coordinates $x = (\mathbf{r}, s)$ incorporating spin or other internal degrees of freedom.
Z	Partition function (canonical or grand-canonical depending on context); its precise meaning will be specified case by case.
$\langle \ldots \rangle$	This symbol may refer to different quantities (thermal average, expectation value in a quantum state, statistical average, etc.); its precise meaning will be specified case by case.
ρ, $\rho(\mathbf{r})$, ...	Symbol that may denote different quantities (density matrix, reduced density matrix, particle density (fluctuation) operator in first quantization, density states, etc.); its precise interpretation will be specified case by case.
$g(x,x')$, $g(\mathbf{r},s,\mathbf{r}',s')$, $g(\mathbf{r},\sigma,\mathbf{r}',\sigma')$, ...	Pair correlation function.
ε	This symbol may refer to different quantities (e.g., statistical index: $\varepsilon = +1$ for bosons, $\varepsilon = -1$ for fermions, and it is used to express quantities that depend on particle statistics in a unified formalism; arbitrarily small constant, etc.); its precise meaning will be specified case by case.
$n_{-1}(\mathcal{E})$, $n_{-1}(\mathcal{E}_\alpha)$, ...	Fermi-Dirac statistics (or distribution).
$n_1(\mathcal{E})$, $n_1(\mathcal{E}_\alpha)$, ...	Bose-Einstein statistics (or distribution).
$n_\varepsilon(\mathcal{E})$, $n_\varepsilon(\mathcal{E}_\alpha)$, ...	Unified Bose-Einstein or Fermi-Dirac statistics (or distribution).
P_σ, P_{ij}	Permutation operators (general permutation; transposition of particles i and j).
$P^{(A,N)}$, $P^{(S,N)}$	Antisymmetrization and symmetrization operators for N particles.
$\mathcal{F}$, $\mathcal{F}^A$, $\mathcal{F}^S$	Fock space; antisymmetric (fermionic) and symmetric (bosonic) Fock spaces.
$\hat{\psi}$, $\hat{\psi}^\dagger$, $\hat{\psi}(x)$, $\hat{\psi}^\dagger(x)$, ...	Fields operator.

$V(x)$, V_N, $\hat{V}$	One-body interaction function; one-body operator in first quantization; one-body operator in second quantization.
$\hat{\mathcal{H}}$	Second-quantized Hamiltonian.
$\hat{T}_\tau$	Wick time-ordering operator.
$\hat{T}_D$	Dyson time-ordering operator.
$\hat{\rho}(\mathbf{r})$, $\hat{\rho}_\mathbf{q}$	Particle-density operator (and its Fourier components) in second quantization.
$U(x, y)$, U_N, $\hat{U}$	Two-body interaction function; two-body operator in first quantization; two-body operator in second quantization.
$\mathbf{R}$, $\mathbf{R}_\mathbf{n}$, $\mathbf{R}_{\mathbf{n},\mu,i}$, ...	Alternative notations for the lattice vector identifying the elementary cell of a direct Bravais lattice. Here $\mathbf{n} = (n_1, n_2, n_3)$ is the integer triplet labeling the cell, μ denotes the atomic species within the cell, and i labels one of the three spatial directions.
$\hat{T}_\mathbf{n}$	Lattice spatial translation operator by vector $\mathbf{R}_\mathbf{n}$.
$\varphi_{n,\mathbf{k}}(\mathbf{r})$, $U_{n,\mathbf{k}}(\mathbf{r})$	Bloch functions and periodic Bloch functions (cell-periodic functions).
$\omega_n(\mathbf{r}, \mathbf{R})$	Wannier functions.
$\varphi_n^{\mathrm{at}}(\mathbf{r})$	Atomic orbital functions.
$\mathbf{v}_{n,\mathbf{k}}$	Group velocity of the Bloch waves.
$\mathbf{S}$, $\mathbf{S}_\mathbf{n}$, $\mathbf{S}_{\mathbf{n},\mu,i}$, ...	Alternative notations for the displacement vector from the equilibrium position of atoms in a direct Bravais lattice. Here $\mathbf{n} = (n_1, n_2, n_3)$ labels the unit cell, μ denotes the atomic species within that cell, and i specifies one of the three spatial directions. In both the classical and quantum descriptions, the same symbol is used for the quantity and for its corresponding operator; the meaning is always clear from the context.
$W^{\mathbf{n},\mu,i}_{\mathbf{m},\nu,k}$, $D^{\mathbf{n},\mu,i}_{\mathbf{m},\nu,k}$	Force constants coupling the displacement of atom μ in direction i of cell $\mathbf{n}$ with the displacement of atom ν in direction k of cell $\mathbf{m}$.
$\boldsymbol{\varepsilon}^{(s)}_\mu(\mathbf{q})$	Polarization vector of atom μ for the s-th normal mode with wavevector $\mathbf{q}$, specifying the direction and relative amplitude of its displacement within the unit cell.
$A_{A,B}(\omega)$, $A_{A,B}(z)$, $A_\mathbf{q}(\omega)$, ...	Spectral function (frequency- or complex frequency-dependent), encoding the dynamical response associated with operators A and B, or with a mode labeled by $\mathbf{q}$.

$\Sigma(z)$, $\Sigma(x, \tau, x', 0)$, ... $\Sigma^*(z)$, $\Sigma^*(x, \tau, x', 0)$ ...	Proper (irreducible) self-energy; improper (reducible) self-energy (indicated with an asterisk).
$D^{(0)}_{\mathbf{q}_1}(\tau_1, \tau_2)$, $D_{\mathbf{q}_1}(\tau_1, \tau_2)$, ...	Phonon propagator.
$\Pi^{(0)}_{\mathbf{q}}(\tau)$, $\Pi_{\mathbf{q}}(\tau)$, $\Pi(\mathbf{r}, t, \mathbf{r}', t')$, $\Pi_{\mathbf{q}}(\omega)$, ...	Polarization propagator.
E, **D**, **H**, **B**	Electric field, electric displacement field, magnetic field (auxiliary field), and magnetic induction, respectively.
$\mathbf{J}$, $\mathbf{J}(\mathbf{r}, t)$, ...	This symbol may denote different quantities (current density, total angular momentum, etc.); the meaning will be clear from the context.
$\epsilon(\omega)$, $\varepsilon_{\mathbf{q}}(\omega)$, ...	Dielectric function.
ω_p	Plasma frequency.
s_i ($i = 1, 2, 3$ or x, y, z)	Components of the spin densities in first quantization.
$\hat{M}_i(\mathbf{r})$ ($i = 1, 2, 3$ or x, y, z), $\hat{M}_i(\mathbf{q})$ ($i = 1, 2, 3$ or x, y, z)	Magnetization density.
λ_T	De Broglie thermal wavelength.
$\Pi_{\mathbf{q},s}$	Canonical (conjugate) momentum associated with the Lagrangian variable $A^*_{\mathbf{q},s}$, the Fourier component of the vector potential.
$T(3)$, $SO(3)$, $E(3)$	Translation, rotation, and Euclidean groups, respectively.
$SU(2)$	Special unitary group of degree 2.
$\hat{R}$, $\hat{R}_{\mathbf{u}}$, ...	Rotation operator.
$\hat{\Omega}$, $\hat{\Omega}(R)$, $\hat{\Omega}(R, \mathbf{a})$, ...	Wigner operators representing group elements (rotations R, translations **a**, etc.).
$\hat{\Theta}$	Time-reversal operator.
$\hat{T}(\mathbf{a})$	Spatial translation operator by vector **a**.

Part I
Foundations of Quantum Many-Body Theory

Chapter 1
Quantum Mechanics Review

Abstract This chapter offers a compact yet rigorous overview of the fundamental concepts of quantum mechanics, essential for the topics developed in the following sections. It begins with the one-particle formalism, including spin as an intrinsic degree of freedom, followed by the introduction of plane waves as momentum eigenstates. An abstract framework based on creation and annihilation operators is then developed, within which coherent states are analysed, key in contexts such as the quantum harmonic oscillator and quantum field theory. The adiabatic theorem is presented next, describing the slow evolution of quantum systems under gradually varying external parameters. The chapter introduces the main pictures of time evolution, Schrödinger, Heisenberg, and interaction, followed by the one-particle Green's function, which captures the time evolution of quantum states while preserving essential properties like causality and unitarity. The final part draws on tools from statistical mechanics, beginning with the density matrix formalism, which extends the description of quantum states to include mixed states. Basic statistical notions, such as the Maxwell-Boltzmann distribution, are also recalled. Altogether, this summary provides a solid theoretical foundation and a concise reference for what follows.

1.1 One Particle Formalism

The physical states describing one quantum particle with spin, confined in a volume V, are represented by elements of a Hilbert space H_1 defined as the tensor product (see Appendix A for a simple finite-dimensional example of a tensor product) as follows

$$\begin{aligned} H_1 &= H_{\text{orbital}} \otimes H_{\text{spin}} \equiv \\ &\equiv H_o \otimes H_s, \end{aligned} \tag{1.1}$$

F. Tafuri et al., *Understanding Quantum Many-Body Theory*, UNITEXT for Physics, https://doi.org/10.1007/978-3-032-17849-7_1

i.e., it is the tensor product of the Hilbert space of the orbital degrees of freedom H_o and the finite Hilbert space H_s of the spin degree of freedom. A generic one particle state can then be written as a tensor product

$$\psi = \psi_o \otimes \psi_s, \tag{1.2}$$

where $\psi_o \in H_o$ encodes the orbital degrees of freedom and $\psi_s \in H_s$ encodes the spin degrees of freedom. Let $\mathcal{L}^2(V)$ denote the space of square-integrable complex functions on the domain V, where V is, in general, a three-dimensional spatial region. Then, for any $\Psi, \Phi \in \mathcal{L}^2(V)$, the scalar product in H_o is defined as

$$(\Psi, \Phi)_{H_o} = \int_V d^3\mathbf{r}\Psi^*(\mathbf{r})\Phi(\mathbf{r}). \tag{1.3}$$

This scalar product satisfies, among other properties, the Hermitian conjugation relation

$$(\Psi, \Phi)_{H_o} = (\Phi, \Psi)^*_{H_o}, \tag{1.4}$$

which expresses the conjugate symmetry of the inner product and is one of the defining axioms of a Hilbert space. By assumption, H_o is separable, that is, there exists a basis, i.e., a numerable, orthonormal and complete set of vectors $\{f_n\}$, which satisfy the orthonormality property

$$(f_n, f_{n'})_{H_o} = \int_V d^3\mathbf{r} f_n^*(\mathbf{r}) f_{n'}(\mathbf{r}) = \\ = \delta_{n,n'}, \tag{1.5}$$

and the completeness property

$$\sum_n f_n^*(\mathbf{r}) f_n(\mathbf{r}') = \delta\left(\mathbf{r} - \mathbf{r}'\right), \tag{1.6}$$

and the Hermitian conjugation property

$$(f_n, f_{n'})_{H_o} = (f_{n'}, f_n)^*_{H_o}. \tag{1.7}$$

A possible orthonormal basis $\{f_n(\mathbf{r})\}$ in H_o can be chosen as the eigenfunctions of the Hamiltonian. For instance, for a free particle in a finite volume, plane waves (see Sect. 1.1.2) form a natural basis since they are eigenfunctions of the kinetic energy operator and correspond to definite momentum states.

In many contexts, it is convenient to adopt the so-called Dirac notation, in which state vectors are denoted by symbols such as $|\psi\rangle$, while their duals (linear functionals) are represented by $\langle\psi|$. The inner product between states $|\Psi\rangle$, $|\Phi\rangle$, is written as $\langle\Psi|\Phi\rangle$, and Eq. 1.4 becomes

$$\langle\Psi|\Phi\rangle = \langle\Phi|\Psi\rangle^* . \tag{1.8}$$

The basis functions $f_n(\mathbf{r})$ can also be represented in the so-called Dirac notation as state vectors $|f_n\rangle$. By definition, in the position representation one has

$$\langle\mathbf{r}|f_n\rangle = f_n(\mathbf{r}), \tag{1.9}$$

and therefore

$$\langle f_n|\mathbf{r}\rangle = f_n^*(\mathbf{r}). \tag{1.10}$$

Often, for simplicity, one writes just $|n\rangle$, referring to the principal quantum number associated with the orbital state, without explicitly specifying the function in coordinate space. In this notation, the orthonormality and completeness properties, i.e., Eqs. 1.5 and 1.6, can be written as

$$\langle f_n|f_{n'}\rangle = \delta_{n,n'}, \tag{1.11}$$

$$\sum_n |f_n\rangle\langle f_n| = \mathbb{1}, \tag{1.12}$$

or equivalently,

$$\langle n|n'\rangle = \delta_{n,n'}, \tag{1.13}$$

$$\sum_n |n\rangle\langle n| = \mathbb{1}, \tag{1.14}$$

respectively. This notation is purely formal and does not imply a specific choice of representation: states cannot always be described as functions, but the notation is particularly useful for expressing general relations in a compact form, regardless of the basis used. For example, in the position representation, a state $|\psi\rangle \in H_o$ is described by the wavefunction $\psi(\mathbf{r})$, defined as

$$\begin{aligned}\psi(\mathbf{r}) &= \langle\mathbf{r}|\psi\rangle = \\ &= \sum_n c_n f_n(\mathbf{r}),\end{aligned} \tag{1.15}$$

where $|\mathbf{r}\rangle$ is a generalized eigenstate of the position operator $\hat{\mathbf{r}}$. As in the case of basis functions, we have

$$\begin{aligned}\psi^*(\mathbf{r}) &= \langle\mathbf{r}|\psi\rangle^* = \\ &= \langle\psi|\mathbf{r}\rangle =\end{aligned}$$

$$= \sum_n c_n^* f_n^*(\mathbf{r}). \tag{1.16}$$

In position representation, the orthonormality and completeness relations, i.e., Eqs. 1.5 and 1.6, take the form

$$\langle \mathbf{r} | \mathbf{r}' \rangle = \delta(\mathbf{r} - \mathbf{r}'), \tag{1.17}$$

$$\int_V d^3\mathbf{r} \, |\mathbf{r}\rangle\langle\mathbf{r}| = \mathbb{1}. \tag{1.18}$$

For discrete indices n, orthonormality is expressed with the Kronecker delta (see Eqs. 1.5 and 1.11), whereas for continuous indices, such as position $\mathbf{r}$, it is expressed with the Dirac delta (see Eq. 1.13). In complete analogy with the position representation, the orthogonality and completeness relations in the momentum representation take the form

$$\langle \mathbf{p} | \mathbf{p}' \rangle = \delta(\mathbf{p} - \mathbf{p}'), \tag{1.19}$$

$$\int d^3\mathbf{p} \, |\mathbf{p}\rangle\langle\mathbf{p}| = \mathbb{1}, \tag{1.20}$$

where the states $|\mathbf{p}\rangle$ are the generalized eigenstates of the momentum operator $\hat{\mathbf{p}}$.

Regarding the spin degrees of freedom, the spin Hilbert space $H_s = \mathbb{C}^{2S+1}$ is a complex vector space of dimension $2S + 1$, associated with a fixed spin quantum number S, which may be either integer or half-integer, depending on the nature of the particle (see Chap. 2). For any $\xi, \zeta \in H_s$, the scalar product in H_s is defined as

$$(\xi, \zeta)_{H_s} = \sum_{s=-S}^{S} \xi^*(s)\zeta(s). \tag{1.21}$$

Analogously to the orbital case, the scalar product in the spin space satisfies the Hermitian conjugation property, namely

$$(\xi, \zeta)_{H_s} = (\zeta, \xi)^*_{H_s} . \tag{1.22}$$

A convenient choice is to introduce an orthonormal basis $\{\eta_\sigma\}$, with $\sigma \in \{-S, -S+1, \ldots, S-1, S\}$: the vectors $\eta_\sigma \in H_s$, called spinors, are finite-dimensional complex vectors that encode the internal spin state of the particle. Such vectors form an orthonormal basis of H_s, satisfying the orthonormality relation

$$(\eta_\sigma, \eta_{\sigma'})_{H_s} = \sum_{s=-S}^{S} \eta^*_\sigma(s) \, \eta_{\sigma'}(s) =$$

$$= \delta_{\sigma,\sigma'}, \tag{1.23}$$

and the completeness relation

$$\sum_{\sigma=-S}^{S} \eta_\sigma^*(s)\, \eta_\sigma(s') = \delta_{s,s'}, \tag{1.24}$$

and the Hermitian conjugation property

$$(\eta_\sigma, \eta_{\sigma'})_{H_s} = (\eta_{\sigma'}, \eta_\sigma)^*_{H_s} \,. \tag{1.25}$$

As in the orbital case, this basis is not unique. A natural and widely used choice is given by the simultaneous eigenvectors of the operators $\hat{\mathbf{S}}^2$ and $\hat{S}_z$, which form an orthonormal and complete basis of H_s. These vectors, denoted by $\{\chi_\sigma\}$, satisfy the eigenvalue equations

$$\hat{\mathbf{S}}^2 \chi_\sigma = \hbar^2 S(S+1)\, \chi_\sigma, \tag{1.26}$$

$$\hat{S}_z \chi_\sigma = \hbar\sigma\, \chi_\sigma. \tag{1.27}$$

In particular, for the canonical basis vectors χ_σ, the components are given by

$$\chi_\sigma(s) = \delta_{\sigma,s}, \tag{1.28}$$

that is, each vector χ_σ in the basis of simultaneous eigenvectors of $\hat{\mathbf{S}}^2$ and $\hat{S}_z$ has all components equal to zero except for the one corresponding to its label σ, which is equal to one. Thanks to Eq. 1.28, we observe that, when spinors are written as column vectors in the basis of simultaneous eigenstates of $\hat{\mathbf{S}}^2$ and $\hat{S}_z$, the index s no longer appears explicitly. Indeed, each spinor $\chi_\sigma(s)$ is a Kronecker delta, so that all components vanish except for the one corresponding to $s = \sigma$. In other words, the index s is fixed by the structure of the basis, while it is σ that uniquely labels the spinor: the position of the single nonvanishing component is determined precisely by the value of σ.

Analogously to the orbital case, it is convenient to adopt the Dirac notation for the spin degrees of freedom, in which state vectors are denoted by symbols such as $|\eta\rangle$, while their duals are represented by $\langle\eta|$. In Dirac notation the orthonormality and completeness relations for the spin states, i.e., Eqs. 1.23 and 1.24, read

$$\langle\eta_\sigma|\eta_{\sigma'}\rangle = \delta_{\sigma,\sigma'}, \tag{1.29}$$

$$\sum_{\sigma=-S}^{S} |\eta_\sigma\rangle\langle\eta_\sigma| = \mathbb{1}, \tag{1.30}$$

or equivalently,

$$\langle\sigma|\sigma'\rangle = \delta_{\sigma,\sigma'}, \tag{1.31}$$

$$\sum_{\sigma=-S}^{S} |\sigma\rangle\langle\sigma| = \mathbb{1}, \tag{1.32}$$

Again, this notation is particularly useful for expressing general relations in a compact form, regardless of the basis used. For example, in the spin representation, a state $|\psi\rangle \in H_s$ is described by the spin wavefunction $\psi(s)$, defined as

$$\begin{aligned}\psi(s) = \langle s|\psi\rangle &= \\ &= \sum_{\sigma=-S}^{S} c_\sigma \eta_\sigma(s),\end{aligned} \tag{1.33}$$

where $|s\rangle$ is an eigenstate of the spin operator $\hat{S}_z$. In this representation, the orthonormality and completeness relations, i.e., Eqs. 1.23 and 1.24, take the form

$$\langle s|s'\rangle = \delta_{s,s'}, \tag{1.34}$$

$$\sum_{s=-S}^{S} |s\rangle\langle s| = \mathbb{1}. \tag{1.35}$$

Ultimately note that in the spin representation, and in the basis $|\sigma\rangle$ of simultaneous eigenvectors of $\hat{\mathbf{S}}^2$ and $\hat{S}_z$, Eq. 1.28 can be obtained as follows

$$\begin{aligned}\chi_\sigma(s) \equiv \langle s|\sigma\rangle &= \\ &= \delta_{s,\sigma},\end{aligned} \tag{1.36}$$

respectively.

It is important to stress the difference between the orbital Hilbert space H_o and the spin Hilbert space H_s. The space H_o is infinite-dimensional, and scalar products or completeness relations involve integrals over the continuous spatial variable $\mathbf{r}$. In contrast, H_s has finite dimension $(2S + 1)$, and sums over the discrete indices s or σ are ordinary finite sums, not integrals. This reflects the fundamental distinction between the continuous nature of spatial degrees of freedom and the discrete nature of spin degrees of freedom. Now, having discussed the orbital and spin Hilbert spaces separately, we now turn to the full one-particle Hilbert space 1.1, which we rewrite as

$$H_1 = \mathcal{L}^2(V) \otimes \mathbb{C}^{2S+1}, \tag{1.37}$$

and show that a basis for a particle with spin is given by the tensor product of a basis of H_o with a basis of H_s, i.e.,

$$\begin{aligned}\left\{\varphi_{n,\sigma}(\mathbf{r},s)\right\}_{n,\sigma} &\equiv \{f_n(\mathbf{r})\otimes\eta_\sigma(s)\}_{n,\sigma} \equiv \\ &\equiv \{f_n(\mathbf{r})\eta_\sigma(s)\}_{n,\sigma}\,. \end{aligned} \tag{1.38}$$

Indeed, the scalar product in H_1 is defined as

$$(\phi,\psi)_{H_1} = \sum_{s=-S}^{S}\int_V d^3\mathbf{r}\phi^*(\mathbf{r},s)\psi(\mathbf{r},s), \tag{1.39}$$

whose form is derived from the scalar products defined in the spaces H_o and H_s. Such a scalar product satisfies the orthonormality property

$$\begin{aligned}(\varphi_{n,\sigma},\varphi_{n',\sigma'})_{H_1} &= \sum_{s=-S}^{S}\int_V d^3\mathbf{r} f_n^*(\mathbf{r})f_{n'}(\mathbf{r})\eta_\sigma^*(s)\eta_{\sigma'}(s) = \\ &= (f_n,f_{n'})_{H_o}\,(\eta_\sigma,\eta_{\sigma'})_{H_s} = \\ &= \delta_{n,n'}\delta_{\sigma,\sigma'}, \end{aligned} \tag{1.40}$$

and completeness property

$$\begin{aligned}\sum_{n,\sigma}\psi^\dagger(\mathbf{r},s)\psi(\mathbf{r}',s') &= \sum_n f_n^*(\mathbf{r})f_n(\mathbf{r}')\sum_\sigma \eta_\sigma^*(s)\eta_\sigma(s') = \\ &= \delta\left(\mathbf{r}-\mathbf{r}'\right)\delta_{s,s'}, \end{aligned} \tag{1.41}$$

that is, $\left\{\varphi_{n,\sigma}(\mathbf{r},s)\right\}_{n,\sigma}$ is a basis. Moreover, such basis satisfies the Hermitian conjugation property

$$(\varphi_{n,\sigma},\varphi_{n',\sigma'})_{H_1} = (\varphi_{n',\sigma'},\varphi_{n,\sigma})^*_{H_1}\,. \tag{1.42}$$

Throughout this text, unless otherwise specified, we shall use the following compact notation

$$(\mathbf{r},s)\equiv x, \tag{1.43}$$

for spatial and spin variables, and

$$(n,\sigma)\equiv\alpha, \tag{1.44}$$

or

$$(\mathbf{k}, \sigma) \equiv \alpha, \tag{1.45}$$

depending on the context. Moreover, we write

$$\sum_{s=-S}^{S} \int_V d^3\mathbf{r} \equiv \int dx, \tag{1.46}$$

$$\delta(x - x') = \delta\left(\mathbf{r} - \mathbf{r}'\right) \delta_{s,s'}, \tag{1.47}$$

which, in turn, implies

$$\psi(x) = \sum_\alpha c_\alpha \varphi_\alpha(x), \tag{1.48}$$

$$\varphi_\alpha(x) = f_n(\mathbf{r}) \eta_\sigma(s), \tag{1.49}$$

$$(\varphi_\alpha, \varphi_{\alpha'}) = \delta_{\alpha,\alpha'}, \tag{1.50}$$

$$\sum_\alpha \varphi_\alpha(x)\, \varphi_\alpha^*(x') = \delta(x - x'), \tag{1.51}$$

$$\int dx \Psi^\dagger(x)\Phi(x) \equiv \sum_{s=-S}^{S} \int_V d^3\mathbf{r} \Psi^\dagger(\mathbf{r}, s)\Phi(\mathbf{r}, s), \tag{1.52}$$

for each state vector in the space H_1.

1.1.1 Spin-$\frac{1}{2}$ Particles and Pauli Matrices

Having introduced the spinors for a generic spin S, we now specialize to the case of spin-$\frac{1}{2}$ particles, such as electrons. This allows us to explicitly construct the eigenvectors of the spin operators and to introduce the Pauli matrices, which provide a convenient and widely used representation of spin-$\frac{1}{2}$ operators. We consider the spin operators in the basis formed by the simultaneous eigenvectors of $\hat{\mathbf{S}}^2$ and $\hat{S}_z$, in which they are conveniently represented by the Pauli matrices

$$\sigma_x \equiv \sigma_1 = \begin{pmatrix} 0 & 1 \\ 1 & 0 \end{pmatrix}, \tag{1.53}$$

$$\sigma_y \equiv \sigma_2 = \begin{pmatrix} 0 & -i \\ i & 0 \end{pmatrix}, \tag{1.54}$$

$$\sigma_z \equiv \sigma_3 = \begin{pmatrix} 1 & 0 \\ 0 & -1 \end{pmatrix}. \tag{1.55}$$

These are three Hermitian, traceless matrices (see Appendix B) that play a central role in the description of intrinsic angular momentum for spin-$\frac{1}{2}$ particles, as well as in the representation of the Lie algebra (see Chap. 14). Their fundamental properties are

- Hermitian: $\sigma_i^\dagger = \sigma_i$;
- Traceless: $\mathrm{Tr}(\sigma_i) = 0$;
- Determinant: $\det(\sigma_i) = -1$;
- Square: $\sigma_i^2 = \mathbb{1}$;
- Commutators:

$$[\sigma_a, \sigma_b] = 2i\varepsilon_{abc}\sigma_c, \; a, b, c = 1, 2, 3; \tag{1.56}$$

- Anticommutators:

$$\{\sigma_a, \sigma_b\} = 2\delta_{ab}\mathbb{1}, \; a, b = 1, 2, 3; \tag{1.57}$$

- Componentwise product:

$$\sigma_a\sigma_b = \delta_{ab}\,\mathbb{1} + i\sum_{c=1}^{3} \varepsilon_{abc}\,\sigma_c, \quad a, b = 1, 2, 3; \tag{1.58}$$

- Pauli identity:

$$(\boldsymbol{\sigma} \cdot \mathbf{v}_1)(\boldsymbol{\sigma} \cdot \mathbf{v}_2) = (\mathbf{v}_1 \cdot \mathbf{v}_2)\mathbb{1} + i\,\boldsymbol{\sigma} \cdot (\mathbf{v}_1 \times \mathbf{v}_2), \; \forall\, \mathbf{v}_1, \mathbf{v}_2. \tag{1.59}$$

The Pauli matrices are used to represent spin-$\frac{1}{2}$ operators $\left\{\hat{S}_i\right\}_{i=x,y,z}$ as follows

$$\hat{S}_i = \frac{\hbar}{2}\sigma_i, \; i = x, y, z. \tag{1.60}$$

As a particular case, we now consider the eigenvalue equations for the spin eigenfunctions of a spin-$\frac{1}{2}$ particle, such as the electron. Let $\chi_\sigma(s)$ denote the components of the spinors in the $\hat{\mathbf{S}}^2$ and $\hat{S}_z$ basis (see Eqs. 1.26 and 1.27), one gets

$$\chi_{-\frac{1}{2}}(s) = \begin{pmatrix} 0 \\ 1 \end{pmatrix}, \tag{1.61}$$

$$\chi_{+\frac{1}{2}}(s) = \begin{pmatrix} 1 \\ 0 \end{pmatrix}, \tag{1.62}$$

these results can be compactly expressed as

$$\hat{S}_z \chi_\sigma(s) = \hbar\sigma\, \chi_\sigma(s),\ \sigma = \pm\frac{1}{2}, \tag{1.63}$$

Considering the action of the operator $\hat{S}_x$, we have

$$\begin{aligned} \hat{S}_x \begin{pmatrix} 1 \\ 0 \end{pmatrix} &= \frac{\hbar}{2} \begin{pmatrix} 0 & 1 \\ 1 & 0 \end{pmatrix} \begin{pmatrix} 1 \\ 0 \end{pmatrix} = \\ &= \frac{\hbar}{2} \begin{pmatrix} 0 \\ 1 \end{pmatrix}, \end{aligned} \tag{1.64}$$

$$\begin{aligned} \hat{S}_x \begin{pmatrix} 0 \\ 1 \end{pmatrix} &= \frac{\hbar}{2} \begin{pmatrix} 0 & 1 \\ 1 & 0 \end{pmatrix} \begin{pmatrix} 0 \\ 1 \end{pmatrix} = \\ &= \frac{\hbar}{2} \begin{pmatrix} 1 \\ 0 \end{pmatrix}, \end{aligned} \tag{1.65}$$

these results can be compactly expressed as

$$\hat{S}_x \chi_\sigma(s) = \hbar\, |\sigma|\, \chi_{-\sigma}(s),\ \sigma = \pm\frac{1}{2}, \tag{1.66}$$

and similarly

$$\begin{aligned} \hat{S}_y \begin{pmatrix} 1 \\ 0 \end{pmatrix} &= \frac{\hbar}{2} \begin{pmatrix} 0 & -i \\ i & 0 \end{pmatrix} \begin{pmatrix} 1 \\ 0 \end{pmatrix} = \\ &= \frac{\hbar}{2} \begin{pmatrix} 0 \\ i \end{pmatrix} = \\ &= i\frac{\hbar}{2} \begin{pmatrix} 0 \\ 1 \end{pmatrix}, \end{aligned} \tag{1.67}$$

$$\hat{S}_y \begin{pmatrix} 0 \\ 1 \end{pmatrix} = \frac{\hbar}{2} \begin{pmatrix} 0 & -i \\ i & 0 \end{pmatrix} \begin{pmatrix} 0 \\ 1 \end{pmatrix} =$$

$$= \frac{\hbar}{2}\begin{pmatrix} -i \\ 0 \end{pmatrix} =$$

$$= -i\frac{\hbar}{2}\begin{pmatrix} 1 \\ 0 \end{pmatrix}, \tag{1.68}$$

these results can be compactly expressed as

$$\hat{S}_y \chi_\sigma(s) = i\hbar\sigma\chi_{-\sigma}(s), \; \sigma = \pm\frac{1}{2}. \tag{1.69}$$

1.1.2 Eigenfunctions of a Free Particle with Spin in a Finite Volume

As anticipated in Sect. 1.1, among the eigenstates of the orbital Hilbert space it is particularly convenient to select those of the one-particle Hamiltonian. We now make this choice explicit by constructing the corresponding basis, which will be used frequently throughout the text. Specifically, we consider momentum eigenstates subject to periodic boundary conditions. To fix ideas, let us take a free particle with spin S confined in a rectangular box of volume $V = L_x L_y L_z$. In the spinless case, the one-particle Hamiltonian reduces to the kinetic operator, i.e.,

$$\mathcal{H} = \frac{\hat{\mathbf{p}}^2}{2m}, \tag{1.70}$$

where $\hat{\mathbf{p}} = -i\hbar\nabla$ is the momentum operator and m is the particle mass. The eigenvalue problem

$$\mathcal{H}\varphi(\mathbf{r}) = \mathcal{E}\varphi(\mathbf{r}), \tag{1.71}$$

together with the periodic boundary conditions

$$\varphi(\mathbf{r} + L_i\mathbf{e}_i) = \varphi(\mathbf{r}), \; i = x, y, z, \tag{1.72}$$

yields the allowed spatial wavefunctions. To solve this problem explicitly, we take as ansatz a plane wave of the form

$$\varphi_\mathbf{k}(\mathbf{r}) = A\, e^{i\mathbf{k}\cdot\mathbf{r}}, \tag{1.73}$$

where A is a normalization constant and $\mathbf{k}$ is a wavevector to be determined. Acting with the Hamiltonian on this ansatz shows that it is an eigenfunction with energy

$$\mathcal{E}_{\mathbf{k}} = \frac{\hbar^2 \mathbf{k}^2}{2m}. \tag{1.74}$$

The periodic boundary conditions require that

$$e^{ik_i L_i} = 1, \; i = x, y, z, \tag{1.75}$$

which implies that each component of the wavevector is quantized, i.e.,

$$k_i = \frac{2\pi n_i}{L_i}, \; i = x, y, z, \; n_i \in \mathbb{Z}. \tag{1.76}$$

Therefore, the allowed spatial wavefunctions are plane waves with discrete, quantized momenta. After normalizing over the volume $V = L_x L_y L_z$, the eigenfunctions become the well-known spatial plane waves (see Appendix C)

$$\begin{aligned} \varphi_{\mathbf{k}}(\mathbf{r}) &= \langle \mathbf{r} | \mathbf{k} \rangle = \\ &= \frac{1}{\sqrt{V}} e^{i\mathbf{k}\cdot\mathbf{r}}. \end{aligned} \tag{1.77}$$

These functions form a basis of the orbital Hilbert space $H_o = \mathcal{L}^2(V)$, satisfying the ortonormality property

$$\int_V d^3\mathbf{r}\, \varphi^*_{\mathbf{k}}(\mathbf{r}) \varphi_{\mathbf{k}'}(\mathbf{r}) = \delta_{\mathbf{k},\mathbf{k}'}, \tag{1.78}$$

and the completeness property

$$\sum_{\mathbf{k}} \varphi^*_{\mathbf{k}}(\mathbf{r}) \varphi_{\mathbf{k}}(\mathbf{r}') = \delta(\mathbf{r} - \mathbf{r}'). \tag{1.79}$$

In the thermodynamic limit (see Eq. C.11), the completeness 1.79 is replaced by

$$\frac{V}{(2\pi)^3} \int d^3\mathbf{k}\, \phi^*_{\mathbf{k}}(\mathbf{r}) \phi_{\mathbf{k}}(\mathbf{r}') = \delta(\mathbf{r} - \mathbf{r}'). \tag{1.80}$$

If we also introduce the spin degrees of freedom, with the associated Hilbert space $H_s = \mathbb{C}^{2S+1}$, then a natural basis for the one-particle Hilbert space H_1 defined in Eq. 1.37 is given by tensor products of plane-wave states with spin eigenfunctions. Since the Hamiltonian $\mathcal{H}$ acts trivially on spin, the eigenfunctions of the full Hamiltonian take precisely this tensor-product form

$$\begin{aligned} \psi_{\mathbf{k},\sigma}(\mathbf{r}, s) &= \langle \mathbf{r}, s | \mathbf{k}, \sigma \rangle = \\ &= \varphi_{\mathbf{k}}(\mathbf{r}) \chi_\sigma(s) = \end{aligned}$$

$$= \frac{1}{\sqrt{V}} e^{i\mathbf{k}\cdot\mathbf{r}} \chi_\sigma(s), \tag{1.81}$$

where $\chi_\sigma(s)$ denote the components of the common eigenstates of $\hat{\mathbf{S}}^2$ and $\hat{S}_z$ (see Eqs. 1.26 and 1.27). The momentum-spin eigenfunctions satisfy the orthonormality relation

$$\sum_{s=-S}^{S} \int_V d^3\mathbf{r} \psi^*_{\mathbf{k},\sigma}(\mathbf{r}, s)\psi_{\mathbf{k}',\sigma'}(\mathbf{r}, s) = \delta_{\mathbf{k},\mathbf{k}'}\delta_{\sigma,\sigma'}, \tag{1.82}$$

and the completeness relation

$$\sum_{\mathbf{k}} \sum_{\sigma=-S}^{S} \psi^*_{\mathbf{k},\sigma}(\mathbf{r}, s)\psi_{\mathbf{k},\sigma}(\mathbf{r}', s') = \delta(\mathbf{r} - \mathbf{r}')\delta_{s,s'}. \tag{1.83}$$

Again, in the thermodynamic limit, the completeness 1.83 is replaced by

$$\sum_{\sigma=-S}^{S} \frac{V}{(2\pi)^3} \int d^3\mathbf{k} \psi^*_{\mathbf{k},\sigma}(\mathbf{r}, s)\psi_{\mathbf{k},\sigma}(\mathbf{r}', s') = \delta(\mathbf{r} - \mathbf{r}')\delta_{s,s'}. \tag{1.84}$$

Plane waves naturally emerge as the spatial eigenfunctions associated with a free particle when periodic boundary conditions are imposed. This requirement automatically leads to the discretization of the corresponding momentum eigenvalues. The spin degrees of freedom, being unaffected by the kinetic energy operator, simply multiply these spatial solutions and add a finite-dimensional internal space structure to the state vector. Thus, the state of a free particle with spin in a volume V is described by a discrete set of plane wave momenta combined with spin eigenstates, forming a complete orthonormal basis of the Hilbert space H_1.

Remark 1.1 (Plane-Wave Tensor Spin Basis) The plane-wave tensor spin basis defined in Eq. 1.81 is particularly useful in quantum many-body theory, where it forms the starting point for both numerical methods and analytical approaches based on perturbation theory. Throughout this text, we will make extensive use of this basis to formulate and analyze interacting systems in second quantization.

1.2 Quantum Mechanics Pictures

In quantum mechanics, dynamical pictures (also referred to as pictures or representations) are different but mathematically equivalent frameworks used to describe the time evolution of a quantum system. The most well-known are the Schrödinger and Heisenberg pictures, which apply to a general Hamiltonian. However, the interaction and thermal interaction pictures become particularly useful when the Hamiltonian $\mathcal{H}$ can be decomposed into the sum of a diagonalizable term $\mathcal{H}_0$ and a perturbative term $\mathcal{H}_I$, i.e.,

$$\mathcal{H} = \mathcal{H}_0 + \mathcal{H}_I. \tag{1.85}$$

This situation typically arises when the system involves some form of interaction, such as that of one particle with an external field or environment, or interactions among many particles. These representations allow one to treat the interaction term as a perturbation evolving over a solvable background defined by the diagonalizable part of the Hamiltonian. This separation proves especially useful in quantum field theory and quantum many-body physics, where exact solutions are seldom available and perturbative methods become essential for practical calculations.

1.2.1 Schrödinger Picture

In the Schrödinger picture, the state vectors of a quantum system depend on time, while the operators representing observables do not explicitly. The time evolution of a state vector $|\psi(t)\rangle$ is governed by the Schrödinger equation

$$i\hbar\frac{\partial}{\partial t}|\psi(t)\rangle = \mathcal{H}|\psi(t)\rangle, \tag{1.86}$$

where $\mathcal{H}$ is the Hamiltonian operator, which represents the total energy of the system. For systems where energy is conserved, the Hamiltonian $\mathcal{H}$ does not depend explicitly on time. In this case, the solution to the Schrödinger equation (see Eq. 1.86), which describes the evolution of the state from an initial time t' to a later time t, is given by

$$|\psi(t)\rangle = e^{-i\frac{\mathcal{H}}{\hbar}(t-t')}|\psi(t')\rangle. \tag{1.87}$$

More generally, let $U(t, t')$ be the unitary time-evolution operator, defined for $t \geq t'$, such that

$$|\psi(t)\rangle = U(t, t')|\psi(t')\rangle, \tag{1.88}$$

where the state $|\psi(t)\rangle$ is a solution of the Schrödinger equation.

Theorem 1.1 (Properties of the Time-Evolution Operator in the Schrödinger Picture) *In the Schrödinger picture, for a system with a time-independent Hamiltonian $\mathcal{H}$, the time-evolution operator $U(t, t')$ that governs the evolution of a state vector $|\psi(t)\rangle$ is characterized by the following properties:*

1. *Schrödinger equation. The operator $U(t, t')$ obeys the operator form of the Schrödinger equation (see Eq. 1.86):*

$$i\hbar\frac{\partial}{\partial t}U(t,t') = \mathcal{H}U(t,t'). \tag{1.89}$$

2. *Initial condition. At equal times, the time-evolution operator reduces to the identity:*

$$\lim_{t\to t'} U(t,t') = \mathbb{1}. \tag{1.90}$$

3. *Group property. The composition of time evolutions satisfies the group relation (see Chap. 14):*

$$U(t,t'') = U(t,t')U(t',t''), \forall\, t, t', t''. \tag{1.91}$$

4. *Unitarity. The time-evolution operator is unitary (see Eqs. I.2 and I.3):*

$$U^\dagger(t,t') = U^{-1}(t,t'). \tag{1.92}$$

Proof

1. By definition (see Eq. 1.88), we differentiate $|\psi(t)\rangle$ with respect to time t:

$$\begin{aligned}\frac{\partial}{\partial t}|\psi(t)\rangle &= \frac{\partial}{\partial t}\left(U(t,t')|\psi(t')\rangle\right)\\ &= \left(\frac{\partial}{\partial t}U(t,t')\right)|\psi(t')\rangle.\end{aligned}$$

From the Schrödinger equation, we also have

$$\begin{aligned}\frac{\partial}{\partial t}|\psi(t)\rangle &= -\frac{i}{\hbar}\mathcal{H}|\psi(t)\rangle =\\ &= -\frac{i}{\hbar}\mathcal{H}U(t,t')|\psi(t')\rangle.\end{aligned} \tag{1.93}$$

Equating the two expressions and noting that this must hold for an arbitrary state $|\psi(t')\rangle$, we conclude that

$$\frac{\partial}{\partial t}U(t,t') = -\frac{i}{\hbar}\mathcal{H}U(t,t').$$

Multiplying both sides by $i\hbar$ gives the desired result.

2. The time-evolution operator must transform a state from t' to t. If we evaluate the operator at $t = t'$, the state should not change, meaning the operator must be the

identity. The formal solution for a time-independent Hamiltonian is $U(t,t') = e^{-i\frac{\mathcal{H}}{\hbar}(t-t')}$.

$$\begin{aligned}\lim_{t\to t'} U(t,t') &= \lim_{t\to t'} e^{-i\frac{\mathcal{H}}{\hbar}(t-t')} = \\ &= e^{-i\frac{\mathcal{H}}{\hbar}(t'-t')} = \\ &= \mathbb{1}. \end{aligned} \tag{1.94}$$

3. The state at time t can be reached from the state at time t'' by evolving first to an intermediate time t' and then to t. We can write this as (see Eq. 1.88)

$$|\psi(t)\rangle = U(t,t')|\psi(t')\rangle, \tag{1.95}$$

$$|\psi(t')\rangle = U(t',t'')|\psi(t'')\rangle. \tag{1.96}$$

Substituting the second equation into the first gives

$$\begin{aligned}|\psi(t)\rangle &= U(t,t')\left(U(t',t'')|\psi(t'')\rangle\right) = \\ &= \left(U(t,t')U(t',t'')\right)|\psi(t'')\rangle. \end{aligned} \tag{1.97}$$

By the definition of the time-evolution operator, we also have

$$|\psi(t)\rangle = U(t,t'')|\psi(t'')\rangle. \tag{1.98}$$

Since this must hold for any state $|\psi(t'')\rangle$, the group relation immediately follows.

4. We need to show that $U^\dagger(t,t') = U^{-1}(t,t')$. Starting from the definition, $U(t,t') = e^{-i\frac{\mathcal{H}}{\hbar}(t-t')}$. The adjoint of this operator is:

$$\begin{aligned}U^\dagger(t,t') &= \left(e^{-i\frac{\mathcal{H}}{\hbar}(t-t')}\right)^\dagger = \\ &= e^{\left(-i\frac{\mathcal{H}}{\hbar}(t-t')\right)^\dagger}. \end{aligned} \tag{1.99}$$

Since $\mathcal{H}$ is self-adjoint, we have

$$\begin{aligned}\left(-i\frac{\mathcal{H}}{\hbar}(t-t')\right)^\dagger &= (-i)^*\frac{\mathcal{H}^\dagger}{\hbar}(t-t') = \\ &= i\frac{\mathcal{H}}{\hbar}(t-t'). \end{aligned} \tag{1.100}$$

Thus, $U^\dagger(t,t') = e^{i\frac{\mathcal{H}}{\hbar}(t-t')}$. The inverse of the operator is given by

$$\begin{aligned} U^{-1}(t,t') &= \left(e^{-i\frac{\mathcal{H}}{\hbar}(t-t')}\right)^{-1} = \\ &= e^{-(-i\frac{\mathcal{H}}{\hbar}(t-t'))} = \\ &= e^{i\frac{\mathcal{H}}{\hbar}(t-t')}, \end{aligned} \tag{1.101}$$

then U is unitary.

□

Even when the Hamiltonian $\mathcal{H}(t)$ depends explicitly on time, the fundamental properties of Theorem 1.1, such as unitarity and the semigroup relation, still hold. However, the time-evolution operator can no longer be expressed in the simple exponential form $e^{-i\frac{\mathcal{H}}{\hbar}(t-t')}$. In this case, its expression is given by a more complex form (see Eq. 2.202).

1.2.2 Heisenberg Picture

In the Heisenberg picture, the time dependence of a quantum system is transferred from the state vectors to the operators. That is, the state vectors remain time-independent, while the operators evolve in time according to the Heisenberg equation of motion. The time-evolved operator $A(t)$ is given by

$$A(t) = e^{i\frac{\mathcal{H}}{\hbar}t} A e^{-i\frac{\mathcal{H}}{\hbar}t}, \tag{1.102}$$

where $\mathcal{H}$ is the Hamiltonian of the system. This transformation ensures that the expectation values of observables remain the same as in the Schrödinger picture.

Now, let us assume that the Hamiltonian $\mathcal{H}$ is time-independent. By differentiating Eq. 1.102 with respect to t, we obtain

$$\begin{aligned} \frac{dA(t)}{dt} &= \frac{d}{dt}\left(e^{i\frac{\mathcal{H}}{\hbar}t}\right) A\, e^{-i\frac{\mathcal{H}}{\hbar}t} + e^{i\frac{\mathcal{H}}{\hbar}t} A \frac{d}{dt}\left(e^{-i\frac{\mathcal{H}}{\hbar}t}\right) = \\ &= \frac{i}{\hbar}\mathcal{H} e^{i\frac{\mathcal{H}}{\hbar}t} A e^{-i\frac{\mathcal{H}}{\hbar}t} - \frac{i}{\hbar} e^{i\frac{\mathcal{H}}{\hbar}t} A e^{-i\frac{\mathcal{H}}{\hbar}t}\mathcal{H} = \\ &= \frac{i}{\hbar}\left(\mathcal{H}A(t) - A(t)\mathcal{H}\right) = \\ &= \frac{i}{\hbar}\left[\mathcal{H}, A(t)\right]. \end{aligned} \tag{1.103}$$

The compact form of the equation of motion in the Heisenberg picture is therefore given by

$$\frac{dA(t)}{dt} = \frac{i}{\hbar}\left[\mathcal{H}, A(t)\right]. \tag{1.104}$$

Note that Eq. 1.104 can be equivalently expressed in a form more convenient for calculations. Indeed, from Eq. H.15, it follows that

$$\begin{aligned}\frac{dA(t)}{dt} &= \frac{i}{\hbar}\left[\mathcal{H}, e^{i\frac{\mathcal{H}}{\hbar}t}Ae^{-i\frac{\mathcal{H}}{\hbar}t}\right] = \\ &= \frac{i}{\hbar}\left(\mathcal{H}e^{i\frac{\mathcal{H}}{\hbar}t}Ae^{-i\frac{\mathcal{H}}{\hbar}t} - e^{i\frac{\mathcal{H}}{\hbar}t}Ae^{-i\frac{\mathcal{H}}{\hbar}t}\mathcal{H}\right) = \\ &= \frac{i}{\hbar}\left(e^{i\frac{\mathcal{H}}{\hbar}t}\mathcal{H}Ae^{-i\frac{\mathcal{H}}{\hbar}t} - e^{i\frac{\mathcal{H}}{\hbar}t}A\mathcal{H}e^{-i\frac{\mathcal{H}}{\hbar}t}\right),\end{aligned} \tag{1.105}$$

that is,

$$\frac{dA(t)}{dt} = \frac{i}{\hbar}e^{i\frac{\mathcal{H}}{\hbar}t}\left[\mathcal{H}, A\right]e^{-i\frac{\mathcal{H}}{\hbar}t}. \tag{1.106}$$

1.2.3 Interaction Picture

In the interaction (or Dirac's) picture both operators and states depend explicitly on time and the dynamical evolution of the operator A involve a unitary transformation (see Eqs. I.2 and I.3) related to $\mathcal{H}_0$ as follows

$$A^{(0)}(t) = e^{i\frac{\mathcal{H}_0}{\hbar}t}Ae^{-i\frac{\mathcal{H}_0}{\hbar}t}, \tag{1.107}$$

where the superscript 0 emphasizes that the evolution of the operators in the interaction representation is with respect to the Hamiltonian $\mathcal{H}_0$; on the other hand, the time evolution of the states is given by

$$\begin{aligned}|\psi(t)\rangle &= e^{i\frac{\mathcal{H}_0}{\hbar}t}e^{-i\frac{\mathcal{H}}{\hbar}t}\psi(0) \equiv \\ &\equiv U(t)|\psi(0)\rangle,\end{aligned} \tag{1.108}$$

where we have defined the unitary transformation

$$U(t) = e^{i\frac{\mathcal{H}_0}{\hbar}t}e^{-i\frac{\mathcal{H}}{\hbar}t}. \tag{1.109}$$

Note that

$$
\begin{aligned}
\langle A(t)\rangle &= \langle\psi(t)|A|\psi(t)\rangle = \\
&= \left\langle\psi(0)\left|e^{i\frac{\mathcal{H}}{\hbar}t}e^{-i\frac{\mathcal{H}_0}{\hbar}t}e^{i\frac{\mathcal{H}_0}{\hbar}t}Ae^{-i\frac{\mathcal{H}_0}{\hbar}t}e^{i\frac{\mathcal{H}_0}{\hbar}t}e^{-i\frac{\mathcal{H}}{\hbar}t}\right|\psi(0)\right\rangle = \\
&= \left\langle\psi(0)\left|e^{i\frac{\mathcal{H}}{\hbar}t}Ae^{-i\frac{\mathcal{H}}{\hbar}t}\right|\psi(0)\right\rangle,
\end{aligned}
\tag{1.110}
$$

as it must be.

Theorem 1.2 *The unitary transformation $U(t)$ associated with the picture representation 1.107 satisfies the relation*

$$
\frac{\partial U(t)}{\partial t} = -\frac{i}{\hbar}\mathcal{H}_I^{(0)}(t)U(t),
\tag{1.111}
$$

with

$$
U(0) = \mathbb{1}.
\tag{1.112}
$$

Proof We have

$$
\begin{aligned}
\frac{\partial U}{\partial t} &= \frac{i}{\hbar}\mathcal{H}_0 e^{i\frac{\mathcal{H}_0}{\hbar}t}e^{-i\frac{\mathcal{H}}{\hbar}t} - \frac{i}{\hbar}e^{i\frac{\mathcal{H}_0}{\hbar}t}\mathcal{H}e^{-i\frac{\mathcal{H}}{\hbar}t} = \\
&= \frac{i}{\hbar}e^{i\frac{\mathcal{H}_0}{\hbar}t}\mathcal{H}_0 e^{-i\frac{\mathcal{H}}{\hbar}t} - \frac{i}{\hbar}e^{i\frac{\mathcal{H}_0}{\hbar}t}\mathcal{H}e^{-i\frac{\mathcal{H}}{\hbar}t} = \\
&= \frac{i}{\hbar}e^{i\frac{\mathcal{H}_0}{\hbar}t}\left(\mathcal{H}_0 - \mathcal{H}\right)e^{-i\frac{\mathcal{H}}{\hbar}t} = \\
&= -\frac{i}{\hbar}e^{i\frac{\mathcal{H}_0}{\hbar}t}\mathcal{H}_I e^{-i\frac{\mathcal{H}}{\hbar}t} = \\
&= -\frac{i}{\hbar}e^{i\frac{\mathcal{H}_0}{\hbar}t}\mathcal{H}_I e^{-i\frac{\mathcal{H}_0}{\hbar}t}e^{i\frac{\mathcal{H}_0}{\hbar}t}e^{-i\frac{\mathcal{H}}{\hbar}t} = \\
&= -\frac{i}{\hbar}\mathcal{H}_I^{(0)}(t)U(t),
\end{aligned}
\tag{1.113}
$$

where we have used Eq. H.15 to obtain $\left[\mathcal{H}_0, e^{i\frac{\mathcal{H}_0}{\hbar}t}\right] = 0$, after which we have used the identity $\mathbb{1} = e^{-i\frac{\mathcal{H}_0}{\hbar}t}e^{i\frac{\mathcal{H}_0}{\hbar}t}$ to express $\mathcal{H}_I$ in the interaction picture with respect to $\mathcal{H}_0$. □

Together with the transformation $U(t)$, we define the matrix $S(t, t')$ such that

$$
|\psi(t)\rangle = S(t, t')|\psi(t')\rangle.
\tag{1.114}
$$

From

$$|\psi(t)\rangle = U(t)|\psi(0)\rangle, \tag{1.115}$$

$$\begin{aligned}|\psi(t)\rangle &= S(t,t')|\psi(t')\rangle = \\ &= S(t,t')U(t')|\psi(0)\rangle,\end{aligned} \tag{1.116}$$

we get

$$S(t,t')U(t') = U(t),\ \forall\ |\psi(0)\rangle, \tag{1.117}$$

which implies

$$\begin{aligned}S(t,t') &= U(t)U^{-1}(t') \equiv \\ &\equiv U(t)U^{\dagger}(t').\end{aligned} \tag{1.118}$$

By definition it follows

$$S(t,0) = U(t), \tag{1.119}$$

$$\begin{aligned}S(0,t) &= U^{-1}(t) = \\ &= U^{\dagger}(t).\end{aligned} \tag{1.120}$$

Theorem 1.3 *The evolution matrix $S(t,t')$ in Eq. 1.114 associated with the picture representation 1.107 satisfies the following general properties:*

1.

$$S(t,t) = \mathbb{1}, \tag{1.121}$$

which is the initial condition for the matrix $S(t,t')$;

2.

$$S(t,t')S(t',t'') = S(t,t''); \tag{1.122}$$

3.

$$S^{\dagger}(t,t') = U(t')U^{\dagger}(t). \tag{1.123}$$

Proof From the definition (see Eq. 1.114), it follows

1.

$$\begin{aligned} S(t,t) &= U(t)U^{\dagger}(t) = \\ &= \mathbb{1}; \end{aligned} \tag{1.124}$$

2.

$$\begin{aligned} S(t,t')S(t',t'') &= U(t)U^{\dagger}(t')U(t')U^{\dagger}(t'') = \\ &= U(t)U^{\dagger}(t'') \equiv \\ &\equiv S(t,t''); \end{aligned} \tag{1.125}$$

3.

$$\begin{aligned} S^{\dagger}(t,t') &= \left(U(t)U^{\dagger}(t')\right)^{\dagger} = \\ &= U^{\dagger\dagger}(t')U^{\dagger}(t) = \\ &= U(t')U^{\dagger}(t). \end{aligned} \tag{1.126}$$

□

Finally, note that matrix $S(t,t')$ can be written as a function of the Hamiltonian $\mathcal{H}$ as

$$\begin{aligned} S(t,t') &= U(t)U^{\dagger}(t') = \\ &= e^{i\frac{\mathcal{H}_0}{\hbar}t}e^{-i\frac{\mathcal{H}}{\hbar}t}\left(e^{i\frac{\mathcal{H}_0}{\hbar}t'}e^{-i\frac{\mathcal{H}}{\hbar}t'}\right)^{\dagger} = \\ &= e^{i\frac{\mathcal{H}_0}{\hbar}t}e^{-i\frac{\mathcal{H}}{\hbar}t}e^{i\frac{\mathcal{H}}{\hbar}t'}e^{-i\frac{\mathcal{H}_0}{\hbar}t'} = \\ &= e^{i\frac{\mathcal{H}_0}{\hbar}t}e^{-i\frac{\mathcal{H}}{\hbar}(t-t')}e^{-i\frac{\mathcal{H}_0}{\hbar}t'}. \end{aligned} \tag{1.127}$$

1.3 The One-Dimensional Quantum Harmonic Oscillator

In this section, we rigorously define and study the so-called coherent states, which are special quantum states of the harmonic oscillator with classical-like behavior. The quantum harmonic oscillator plays a central role in quantum mechanics because it is exactly solvable and provides an accurate approximation of many physical systems near the minimum of their potential energy. Consider a one-dimensional quantum harmonic oscillator with Hamiltonian

$$\mathcal{H} = \frac{\hat{p}^2}{2m} + \frac{1}{2}m\omega^2\hat{x}^2, \tag{1.128}$$

where $\hat{x}$ and $\hat{p}$ are the position and momentum operators, respectively. Now we introduce the annihilation and creation operators, respectively, defined by

$$a = \sqrt{\frac{m\omega}{2\hbar}}\left(\hat{x} + i\frac{1}{m\omega}\hat{p}\right), \tag{1.129}$$

$$a^\dagger = \sqrt{\frac{m\omega}{2\hbar}}\left(\hat{x} - i\frac{1}{m\omega}\hat{p}\right). \tag{1.130}$$

The annihilation and creation operators, that is, Eqs. 1.129 and 1.130, can be inverted to express $\hat{x}$ and $\hat{p}$ as

$$\hat{x} = \sqrt{\frac{\hbar}{2m\omega}}\left(a + a^\dagger\right), \tag{1.131}$$

$$\hat{p} = i\sqrt{\frac{\hbar m\omega}{2}}\left(a^\dagger - a\right). \tag{1.132}$$

The position and momentum operators $\hat{x}$ and $\hat{p}$ satisfy the canonical commutation relation

$$[\hat{x}, \hat{p}] = i\hbar\mathbb{1}, \tag{1.133}$$

which constitutes a fundamental postulate of quantum mechanics and underlies the algebraic structure of the quantum harmonic oscillator.

Theorem 1.4 *The annihilation and creation operators defined as in the one-dimensional harmonic oscillator, that is, Eqs. 1.129 and 1.130, satisfy the following commutation relation*

$$\left[a, a^\dagger\right] = \mathbb{1}. \tag{1.134}$$

Proof The above relations are an immediate consequence of the canonical commutation relation 1.133. Specifically,

$$\begin{aligned}\left[a, a^\dagger\right] = &\frac{m\omega}{2\hbar}\left[\hat{x}, \hat{x}\right] - i\frac{m\omega}{2\hbar}\frac{1}{m\omega}\left[\hat{x}, \hat{p}\right] + i\frac{m\omega}{2\hbar}\frac{1}{m\omega}\left[\hat{p}, \hat{x}\right] \\ &+ i(-i)\frac{m\omega}{2\hbar}\frac{1}{m^2\omega^2}\left[\hat{p}, \hat{p}\right] = \\ &= -i\frac{1}{2\hbar}\left[\hat{x}, \hat{p}\right] + i\frac{1}{2\hbar}\left[\hat{p}, \hat{x}\right] =\end{aligned}$$

$$= -i\frac{1}{2\hbar}\left[\hat{x}, \hat{p}\right] - i\frac{1}{2\hbar}\left[\hat{x}, \hat{p}\right] =$$

$$= -i\frac{1}{\hbar}\left[\hat{x}, \hat{p}\right], \tag{1.135}$$

which is Eq. 1.134. □

Obviously, the following also hold:

$$[a, a] = 0, \tag{1.136}$$

$$\left[a^{\dagger}, a^{\dagger}\right] = 0. \tag{1.137}$$

Remark 1.2 (Bosonic Interpretation of the Quantum Harmonic Oscillator) Theorem 1.4 allows one to interpret the quantum harmonic oscillator (see Sect. 1.3) as the simplest realization of a bosonic mode. This correspondence becomes evident by comparing Eqs. 1.134 and D.1, 1.136 and D.2, 1.137 and D.3: in each pair, the first equation refers to the harmonic oscillator, while the second expresses the analogous structure in terms of bosonic particle operators. This perspective emphasizes the role of the harmonic oscillator as a prototype for bosonic excitations (see, for instance, Chaps. 4 and 13).

Thanks to operators defined in 1.129 and 1.130, Hamiltonian 1.128 can be written as

$$\mathcal{H} = \hbar\omega\left(a^{\dagger}a + \frac{1}{2}\right), \tag{1.138}$$

Having established the algebraic structure of the creation and annihilation operators, we now construct the Hilbert space of the one-dimensional quantum harmonic oscillator explicitly. By defining the number operator as follows

$$\hat{N} = a^{\dagger}a, \tag{1.139}$$

the Hamiltonian 1.138 can be written as

$$\mathcal{H} = \hbar\omega\left(\hat{N} + \tfrac{1}{2}\right), \tag{1.140}$$

which shows that determining the eigenvalues and eigenvectors of $\mathcal{H}$ is equivalent to determining those of $\hat{N}$. Observe that, by definition, the number operator $\hat{N}$ is self-adjoint, in particular, its eigenvalues are real.

First, we observe that from the Jacobi identity (see Eq. F.1) and from Eqs. 1.134, 1.136 and 1.137, we obtain

$$\begin{aligned}\left[a, \hat{N}\right] = \left[a, a^\dagger a\right] = \\ &= \left[a, a^\dagger\right] a + a^\dagger [a, a] = \\ &= a,\end{aligned} \tag{1.141}$$

$$\begin{aligned}\left[a^\dagger, \hat{N}\right] = \left[a^\dagger, a^\dagger a\right] = \\ &= \left[a^\dagger, a^\dagger\right] a + a^\dagger \left[a^\dagger, a\right] = \\ &= -a^\dagger.\end{aligned} \tag{1.142}$$

Now, we write the eigenvalue equation for the number operator, namely

$$a^\dagger a|v\rangle = \lambda|v\rangle, \tag{1.143}$$

where λ is the eigenvalue and $|v\rangle$ the corresponding eigenvector. From Eqs. 1.134, 1.136, 1.137, 1.141 and 1.143, one finds

$$\begin{aligned}\hat{N}(a|v\rangle) &= (\hat{N}a)|v\rangle = \\ &= (a\hat{N} - a)|v\rangle = \\ &= a(\hat{N}|v\rangle) - a|v\rangle = \\ &= a(\lambda|v\rangle) - a|v\rangle = \\ &= (\lambda - 1)(a|v\rangle),\end{aligned} \tag{1.144}$$

$$\begin{aligned}\hat{N}(a^\dagger|v\rangle) &= a^\dagger a(a^\dagger|v\rangle) = \\ &= a^\dagger (aa^\dagger)|v\rangle = \\ &= a^\dagger (a^\dagger a + 1)|v\rangle = \\ &= a^\dagger (\lambda + 1)|v\rangle = \\ &= (\lambda + 1)(a^\dagger|v\rangle),\end{aligned} \tag{1.145}$$

that is, $a|v\rangle$ and $a^\dagger|v\rangle$ are eigenvectors of the number operator with eigenvalues $\lambda - 1$ and $\lambda + 1$, respectively. By iteration, one can show that, in general,

$$\hat{N}a^N|v\rangle = (\lambda - N)a^N|v\rangle, \tag{1.146}$$

$$\hat{N}a^{\dagger N}|v\rangle = (\lambda + N)a^{\dagger N}|v\rangle. \tag{1.147}$$

From

$$\langle v|a^\dagger a v\rangle = \lambda\, |v|^2 = \\ = |av|^2\,, \tag{1.148}$$

given that v is, by assumption, an eigenvector,

$$|v| \neq 0, \tag{1.149}$$

it follows

$$\lambda \geq 0, \tag{1.150}$$

$$\lambda = 0 \Leftrightarrow |av| = 0. \tag{1.151}$$

Moreover, from

$$\langle v|aa^\dagger v\rangle = \left|a^\dagger v\right|^2 = \\ = \langle v|\left(\mathbb{1} + a^\dagger a\right) v\rangle = \\ = (\lambda + 1)\, |v|^2\,, \tag{1.152}$$

combining with Eqs. 1.149 and 1.150, we get

$$\left|a^\dagger v\right| \neq 0. \tag{1.153}$$

From Eq. 1.146, since the spectrum is bounded from below (see Eq. 1.150), it follows

$$\exists N : a^{N+1} v = 0,\ a^N v \neq 0, \tag{1.154}$$

then

$$0 = a^{N+1} = \\ = aa^N v = \\ = a(\lambda - N)v, \tag{1.155}$$

and from Eq. 1.151, it follows

$$\lambda = 0,\ N = 0, \tag{1.156}$$

or

$$\lambda = N. \tag{1.157}$$

We have found that the eigenvalues of the number operator are the non-negative integers, and therefore the spectrum of the harmonic oscillator 1.138 has the structure

$$\mathcal{E}_N = \hbar\omega\left(N + \frac{1}{2}\right), \quad N = 0, 1, \dots \tag{1.158}$$

Before proceeding, we state a general result that will be useful later.

Theorem 1.5 *Let a and $a^\dagger$ be annihilation and creation operators that satisfy Eqs. 1.134, 1.136 and 1.137, then*

$$\left[a, \left(a^\dagger\right)^N\right] = N\left(a^\dagger\right)^{N-1}, \quad \forall\, N \in \mathbb{N}. \tag{1.159}$$

Proof We proceed by induction. For $N = 1$, the left-hand side coincides with Eq. 1.134, so the statement holds. Assume it holds for $N - 1$, i.e.,

$$\left[a, \left(a^\dagger\right)^{N-1}\right] = (N-1)\left(a^\dagger\right)^{N-2}, \tag{1.160}$$

and we check the validity for N. From Jacobi identity F.1, one finds

$$\begin{aligned}
\left[a, \left(a^\dagger\right)^N\right] = \left[a, a^\dagger\left(a^\dagger\right)^{N-1}\right] &= \\
= \left[a, a^\dagger\right]\left(a^\dagger\right)^{N-1} + a^\dagger\left[a, \left(a^\dagger\right)^{N-1}\right] &= \\
= \left(a^\dagger\right)^{N-1} + a^\dagger\left[a, \left(a^\dagger\right)^{N-1}\right] &= \\
= \left(a^\dagger\right)^{N-1} + a^\dagger(N-1)\left(a^\dagger\right)^{N-2} &= \\
= N\left(a^\dagger\right)^{N-1}, &
\end{aligned} \tag{1.161}$$

then Eq. 1.159 is proven. □

Now, to determine the eigenvectors, we start from the property

$$a|v_0\rangle = 0, \tag{1.162}$$

corresponding to the eigenvalue 0 (see Eq. 1.151). The eigenvector corresponding to $N > 0$ can then be obtained from

$$a^{\dagger N}|v_0\rangle. \tag{1.163}$$

From Eq. 1.159, one finds

$$\begin{aligned}
\left\langle a^{\dagger N} v_0 | a^{\dagger N} v_0 \right\rangle &= \left\langle v_0 | a^N a^{\dagger N} v_0 \right\rangle = \\
&= \left\langle v_0 | a^{N-1} a a^{\dagger N} v_0 \right\rangle = \\
&= \left\langle v_0 | a^{N-1} \left(a^{\dagger N} a + N \left(a^\dagger \right)^{N-1} \right) v_0 \right\rangle = \\
&= \left\langle v_0 | a^{N-1} a^{\dagger N} a v_0 \right\rangle + N \left\langle v_0 | a^{N-1} \left(a^\dagger \right)^{N-1} v_0 \right\rangle = \\
&= N \left\langle v_0 | a^{N-1} \left(a^\dagger \right)^{N-1} v_0 \right\rangle,
\end{aligned} \tag{1.164}$$

where in the last step we have used Eq. 1.162. By iterating, we obtain

$$\left\langle a^{\dagger N} v_0 | a^{\dagger N} v_0 \right\rangle = N! \langle v_0 | v_0 \rangle. \tag{1.165}$$

After normalizing v_0, the normalized eigenvectors for $N \geq 1$ are given by

$$v_N = \frac{a^{\dagger N} v_0}{\sqrt{N!}}. \tag{1.166}$$

Let us show that v_N is an eigenvector of the number operator:

$$\begin{aligned}
a^\dagger a\, v_N &= a^\dagger a \frac{a^{\dagger N} v_0}{\sqrt{N!}} = \\
&= \frac{a^\dagger \left(a a^{\dagger N} v_0 \right)}{\sqrt{N!}} = \\
&= \frac{a^\dagger \left(a^{\dagger N} a + N (a^\dagger)^{N-1} \right) v_0}{\sqrt{N!}} = \\
&= \frac{a^\dagger \left(N (a^\dagger)^{N-1} v_0 \right)}{\sqrt{N!}} = \\
&= \frac{N a^{\dagger N} v_0}{\sqrt{N!}} = \\
&= N v_N.
\end{aligned} \tag{1.167}$$

We also note that from Eq. 1.159, it also follows that

$$\begin{aligned} a v_N &= \frac{a a^{\dagger N} v_0}{\sqrt{N!}} = \\ &= \frac{a^{\dagger N} a v_0}{\sqrt{N!}} + N \frac{\left(a^{\dagger}\right)^{N-1} v_0}{\sqrt{N!}} = \\ &= \sqrt{N} \frac{\left(a^{\dagger}\right)^{N-1} v_0}{\sqrt{(N-1)!}}, \end{aligned} \tag{1.168}$$

then

$$a v_N = \sqrt{N} v_{N-1}. \tag{1.169}$$

Similarly, one can show that

$$a^{\dagger} v_N = \sqrt{N+1} v_{N+1}. \tag{1.170}$$

Finally, if we denote by $|N\rangle$ the normalized eigenstates of the number operator $\hat{N}$, namely

$$|N\rangle \equiv v_N = \frac{a^{\dagger N}|0\rangle}{\sqrt{N!}}, \tag{1.171}$$

then the main properties established so far for the annihilation, creation, and number operators, that is, Eqs. 1.167, 1.169, 1.162 and 1.170, respectively, can be summarized as follows.

- The number operator $\hat{N}$ satisfies

$$\hat{N}|N\rangle = N|N\rangle, \ \forall \, N \geq 0. \tag{1.172}$$

- The annihilation operator a acts as

$$a|N\rangle = \sqrt{N}|N-1\rangle, \ \forall \, N \geq 1, \tag{1.173}$$

$$a|0\rangle = 0. \tag{1.174}$$

- The creation operator $a^{\dagger}$ acts as

$$a^{\dagger}|N\rangle = \sqrt{N+1}|N+1\rangle, \ \forall \, N \geq 0. \tag{1.175}$$

In the one-dimensional case, the operators $a^{\dagger}$ and a act as creation and annihilation operators for quanta of energy, respectively. Each application of $a^{\dagger}$ increases the oscillator's energy by one quantum $\hbar\omega$, while a lowers it by the same amount, provided the system is not in the ground state. Physically, the integer N counts the number of energy quanta $\hbar\omega$ present in the oscillator. In other words, the number states $|N\rangle$ are discrete energy eigenstates, and the number operator $\hat{N}$ directly measures the number of energy quanta in each state.

We now aim to calculate the time evolution of the dimensionless observables. First, starting from Eq. 1.131 and 1.132, we introduce the dimensionless operators

$$\hat{X} = \sqrt{\frac{m\omega}{\hbar}}\hat{x}, \tag{1.176}$$

$$\hat{P} = \frac{\hat{p}}{\sqrt{\hbar m\omega}}. \tag{1.177}$$

Equations 1.129 and 1.130, become

$$a = \frac{1}{\sqrt{2}}\left(\hat{X} + i\hat{P}\right), \tag{1.178}$$

$$a^{\dagger} = \frac{1}{\sqrt{2}}\left(\hat{X} - i\hat{P}\right), \tag{1.179}$$

and Eq. 1.133 can be written as

$$\left[\hat{X}, \hat{P}\right] = i\mathbb{1}. \tag{1.180}$$

In the Schrödinger representation, we use Eq. 1.162 with a given by Eq. 1.178, namely

$$\begin{aligned} 0 &= \left(\hat{X} + i\hat{P}\right) v_0 = \\ &= \left[X + i\left(-i\frac{\partial}{\partial X}\right)\right] v_0, \end{aligned} \tag{1.181}$$

it follows that the state v_0 is a Gaussian function. At this point, we calculate the expectation values of the dimensionless position and momentum observables for an arbitrary eigenvector. We have

$$\begin{aligned} \langle v_N|\hat{X}|v_N\rangle &= \frac{1}{\sqrt{2}}\langle v_N|(a + a^{\dagger})|v_N\rangle = \\ &= \frac{1}{\sqrt{2}}\left(\sqrt{N}\langle v_N|v_{N-1}\rangle + \sqrt{N+1}\langle v_N|v_{N+1}\rangle\right) = \\ &= 0 \end{aligned} \tag{1.182}$$

indeed, eigenvectors corresponding to different eigenvalues are orthogonal. Similarly, for the dimensionless momentum operator, we have

$$\begin{aligned}\langle v_N|\hat{P}|v_N\rangle &= \frac{i}{\sqrt{2}}\langle v_N|(a^\dagger - a)|v_N\rangle = \\ &= \frac{i}{\sqrt{2}}\left(\sqrt{N+1}\langle v_N|v_{N+1}\rangle - \sqrt{N}\langle v_N|v_{N-1}\rangle\right) = \\ &= 0. \end{aligned} \tag{1.183}$$

Concerning the expectation values of the dimensionless quadratic operators, we have

$$\begin{aligned}\langle v_N|\hat{X}^2|v_N\rangle &= \frac{1}{2}\langle v_N|(a + a^\dagger)(a + a^\dagger)|v_N\rangle = \\ &= \frac{1}{2}\langle v_N|aa + a^\dagger a^\dagger + aa^\dagger + a^\dagger a|v_N\rangle = \\ &= \frac{1}{2}\langle v_N|aa^\dagger + a^\dagger a|v_N\rangle = \\ &= \frac{1}{2}\langle v_N|\mathbb{1} + 2a^\dagger a|v_N\rangle = \\ &= \frac{1}{2}(2N + 1), \end{aligned} \tag{1.184}$$

$$\begin{aligned}\langle v_N|\hat{P}^2|v_N\rangle &= -\frac{1}{2}\langle v_N|(a^\dagger - a)(a^\dagger - a)|v_N\rangle = \\ &= -\frac{1}{2}\langle v_N|a^\dagger a^\dagger - a^\dagger a - aa^\dagger + aa|v_N\rangle = \\ &= \frac{1}{2}\langle v_N|aa^\dagger + a^\dagger a|v_N\rangle = \\ &= \frac{1}{2}\langle v_N|\mathbb{1} + 2a^\dagger a|v_N\rangle = \\ &= \frac{1}{2}(2N + 1), \end{aligned} \tag{1.185}$$

then

$$\Delta X \Delta P = \frac{1}{2}(2N + 1), \tag{1.186}$$

$$\Delta x \Delta p = \frac{\hbar}{2}(2N + 1), \tag{1.187}$$

Let us return to the time evolution of the dimensionless operators. Since the Hamiltonian of the harmonic motion 1.138 is time-independent, we use Eq. 1.106 and compute the time-evolved dimensionless annihilation and creation operators (see Eqs. 1.178 and 1.179). We obtain

$$\begin{aligned}\frac{da(t)}{dt} &= \frac{i}{\hbar}e^{i\frac{\mathcal{H}}{\hbar}t}\left[\mathcal{H}, a\right]e^{-i\frac{\mathcal{H}}{\hbar}t} = \\ &= \frac{i}{\hbar}e^{i\frac{\mathcal{H}}{\hbar}t}\left[\hbar\omega\left(a^{\dagger}a+\frac{1}{2}\right), a\right]e^{-i\frac{\mathcal{H}}{\hbar}t} = \\ &= i\omega e^{i\frac{\mathcal{H}}{\hbar}t}\left[a^{\dagger}a, a\right]e^{-i\frac{\mathcal{H}}{\hbar}t} = \\ &= -i\omega e^{i\frac{\mathcal{H}}{\hbar}t}ae^{-i\frac{\mathcal{H}}{\hbar}t} = \\ &= -i\omega a(t),\end{aligned} \tag{1.188}$$

where we have employed Eq. 1.141. The solution is

$$a(t) = e^{-i\omega t}a. \tag{1.189}$$

Similarly, from Eqs. 1.106 and 1.142, one finds

$$\frac{da^{\dagger}(t)}{dt} = i\omega a^{\dagger}, \tag{1.190}$$

$$a^{\dagger}(t) = e^{i\omega t}a^{\dagger}. \tag{1.191}$$

The expectation values of the dimensionless observables are then given by

$$\begin{aligned}\langle \upsilon_N|\hat{X}(t)|\upsilon_N\rangle &= \frac{1}{\sqrt{2}}\langle \upsilon_N|(a(t)+a^{\dagger}(t))|\upsilon_N\rangle = \\ &= \frac{1}{\sqrt{2}}\left(\sqrt{N}\langle \upsilon_N|\upsilon_{N-1}\rangle e^{-i\omega t}+\sqrt{N+1}\langle \upsilon_N|\upsilon_{N+1}\rangle e^{i\omega t}\right) = \\ &= 0\end{aligned} \tag{1.192}$$

$$\begin{aligned}\langle \upsilon_N|\hat{P}(t)|\upsilon_N\rangle &= \frac{i}{\sqrt{2}}\langle \upsilon_N|(a^{\dagger}(t)-a(t))|\upsilon_N\rangle = \\ &= \frac{i}{\sqrt{2}}\left(\sqrt{N+1}\langle \upsilon_N|\upsilon_{N+1}\rangle e^{i\omega t}-\sqrt{N}\langle \upsilon_N|\upsilon_{N-1}\rangle e^{-i\omega t}\right) = \\ &= 0.\end{aligned} \tag{1.193}$$

and the results coincide with the stationary case, namely Eqs. 1.182 and 1.183, respectively. In contrast with the classical description of the harmonic oscillator, where position and momentum oscillate sinusoidally in time according to cosine and sine functions, the expectation values of these operators in the stationary states $|v_N\rangle$ are zero. To recover states that exhibit a behavior more closely resembling the classical motion of the oscillator, we must consider a different class of eigenstates. These are the so-called coherent states, which we now proceed to introduce and study in detail.

1.3.1 Coherent States

Coherent states are special quantum states of the harmonic oscillator that most closely resemble classical oscillatory behavior. These states play a central role in the description of bosonic many-body systems.

Definition 1.1 (Coherent State) A coherent state $|\alpha\rangle$ of the quantum harmonic oscillator is defined as an eigenstate of the annihilation operator a corresponding to the (generally complex) eigenvalue $\alpha \in \mathbb{C}$:

$$a|\alpha\rangle = \alpha|\alpha\rangle, \quad \alpha \in \mathbb{C}. \tag{1.194}$$

Since the annihilation operator a is not self-adjoint, its eigenvalues are not constrained to be real but can take any complex value.

Coherent states are not eigenstates of the Hamiltonian (unless $\alpha = 0$), but they have important semiclassical properties. We have

Theorem 1.6 (Construction of Coherent States) *A coherent state $|\alpha\rangle$ can be written as a superposition of number states $|N\rangle$, i.e.,*

$$|\alpha\rangle = e^{-\frac{|\alpha|^2}{2}} \sum_{N=0}^{\infty} \frac{\alpha^N}{\sqrt{N!}}|N\rangle. \tag{1.195}$$

Proof We will now derive this result by explicitly constructing the coherent state from the properties of the annihilation operator. To construct the coherent state, we begin by assuming that $|\alpha\rangle$ can be expressed as

$$|\alpha\rangle = \sum_{N=0}^{\infty} c_N|N\rangle. \tag{1.196}$$

We apply the annihilation operator a to this expression, using Eq. 1.173, then

$$a|\alpha\rangle = \sum_{N=1}^{\infty} c_N \sqrt{N} |N-1\rangle. \tag{1.197}$$

Next, we relabel the sum by setting $N' = N - 1$, so we get

$$a|\alpha\rangle = \sum_{N'=0}^{\infty} c_{N'+1} \sqrt{N'+1} |N'\rangle. \tag{1.198}$$

From Eq. 1.194, we also have

$$a|\alpha\rangle = \alpha \sum_{N'=0}^{\infty} c_{N'} |N'\rangle. \tag{1.199}$$

By equating the coefficients of $|N'\rangle$ on both sides, we derive the recurrence relation

$$c_{N'+1} \sqrt{N'+1} = \alpha c_{N'}. \tag{1.200}$$

This simplifies to

$$c_{N'+1} = \frac{\alpha}{\sqrt{N'+1}} c_{N'}. \tag{1.201}$$

By iterating this relation starting from c_0, we obtain

$$c_N = \frac{\alpha^N}{\sqrt{N!}} c_0. \tag{1.202}$$

Finally, we normalize $|\alpha\rangle$ by requiring that $\langle \alpha | \alpha \rangle = 1$, which gives

$$1 = |c_0|^2 \sum_{N=0}^{\infty} \frac{|\alpha|^{2N}}{N!} =$$

$$= |c_0|^2 e^{|\alpha|^2}, \tag{1.203}$$

$$c_0 = e^{-\frac{|\alpha|^2}{2}}, \tag{1.204}$$

which leads to the thesis. □

We now present two equivalent ways of expressing a coherent state.

Theorem 1.7 (Coherent State in the Exponential Creation Operator Form) *A coherent state $|\alpha\rangle$ can be expressed as*

$$|\alpha\rangle = e^{-\frac{|\alpha|^2}{2}} e^{\alpha a^\dagger} |0\rangle, \tag{1.205}$$

where $|0\rangle$ is the vacuum state (see Eqs. 1.162 and 1.174).

Proof By definition, a coherent state is an eigenstate of the annihilation operator. We propose a solution of the form

$$|\alpha\rangle = f(\alpha) e^{\alpha a^\dagger} |0\rangle, \tag{1.206}$$

where $f(\alpha)$ is a scalar to be determined, which we take as our Ansatz. Using the conjugation formula H.5, we can write

$$e^{-\alpha a^\dagger} a\, e^{\alpha a^\dagger} = a + \left[\alpha a^\dagger, a\right] + \frac{1}{2!}\left[\alpha a^\dagger, \left[\alpha a^\dagger, a\right]\right] + \ldots \tag{1.207}$$

From the commutation relation 1.134, in particular $\left[a^\dagger, a\right] = -\mathbb{1}$, all higher nested commutators vanish, and we immediately obtain

$$e^{-\alpha a^\dagger} a\, e^{\alpha a^\dagger} = a + \alpha. \tag{1.208}$$

Multiplying both sides by $e^{\alpha a^\dagger}$ on the left gives

$$a\, e^{\alpha a^\dagger} = e^{\alpha a^\dagger}(a + \alpha). \tag{1.209}$$

Now, we apply a to Eq. 1.206 and we get

$$\begin{aligned} a|\alpha\rangle &= f(\alpha) a\, e^{\alpha a^\dagger} |0\rangle = \\ &= f(\alpha) e^{\alpha a^\dagger} (a + \alpha)|0\rangle = \\ &= f(\alpha) e^{\alpha a^\dagger} (\alpha|0\rangle) = \\ &= \alpha|\alpha\rangle, \end{aligned} \tag{1.210}$$

where we used $a|0\rangle = 0$. This shows that our Ansatz satisfies the eigenvalue equation. Now, to fix $f(\alpha)$, we require that the coherent state $|\alpha\rangle$ be normalized, that is,

$$\begin{aligned} 1 &= \langle\alpha|\alpha\rangle = \\ &= |f(\alpha)|^2 \left\langle 0 \left| e^{\alpha^* a} e^{\alpha a^\dagger} \right| 0 \right\rangle = \\ &= |f(\alpha)|^2 e^{|\alpha|^2}, \end{aligned} \tag{1.211}$$

where we used $e^{\alpha^* a} e^{\alpha a^\dagger} = e^{\alpha a^\dagger} e^{\alpha^* a} e^{|\alpha|^2}$ and Eq. 1.174. Solving for $f(\alpha)$, we finally obtain

$$f(\alpha) = e^{-\frac{|\alpha|^2}{2}}. \tag{1.212}$$

□

Theorem 1.8 (Coherent State via the Displacement Operator) *A coherent state* $|\alpha\rangle$ *can be expressed as*

$$|\alpha\rangle = D(\alpha)|0\rangle, \tag{1.213}$$

where

$$D(\alpha) = e^{\alpha a^\dagger - \alpha^* a} \tag{1.214}$$

is called the displacement operator and $|0\rangle$ *is the vacuum state (see Eqs. 1.162 and 1.174).*

Proof Let us define

$$S(\alpha) = \alpha a^\dagger - \alpha^* a. \tag{1.215}$$

Now, we want to rewrite the displacement operator $D(\alpha) = e^{S(\alpha)}$ in a form that separates creation and annihilation operators. We set

$$X = \alpha a^\dagger, \tag{1.216}$$

$$Y = -\alpha^* a, \tag{1.217}$$

so that $D(\alpha) = e^{X+Y}$. From the commutation relation 1.134, the commutator of X and Y is given by

$$\begin{aligned}
[X, Y] &= \left[\alpha a^\dagger, -\alpha^* a\right] = \\
&= -\alpha\alpha^* \left[a^\dagger, a\right] = \\
&= -|\alpha|^2(-\mathbb{1}) = \\
&= |\alpha|^2\, \mathbb{1},
\end{aligned} \tag{1.218}$$

then we can apply Glauber's formula (see Eq. H.18) and we get

$$D(\alpha) = e^{\alpha a^\dagger - \alpha^* a} =$$

$$= e^{-\frac{|\alpha|^2}{2}} e^{\alpha a^\dagger} e^{-\alpha^* a}. \tag{1.219}$$

Applying this operator to the vacuum gives

$$\begin{aligned} D(\alpha)|0\rangle &= e^{-\frac{|\alpha|^2}{2}} e^{\alpha a^\dagger} e^{-\alpha^* a}|0\rangle = \\ &= e^{-\frac{|\alpha|^2}{2}} e^{\alpha a^\dagger}|0\rangle = \\ &= |\alpha\rangle, \end{aligned} \tag{1.220}$$

which is exactly the standard coherent state form obtained in Theorem 1.7. □

The representations of a coherent state, Eqs. 1.205 and 1.213, are mathematically equivalent, but each has its own advantages depending on the context. The exponential creation-operator form is particularly convenient for explicit calculations of expectation values and expansions in the basis, as it directly shows how the state is built from the vacuum using repeated applications of $a^\dagger$. The displacement-operator form is more natural when studying the action of unitary transformations (see Appendix I), quantum dynamics, or phase-space properties, since it highlights the coherent state as a unitary displacement of the vacuum.

We now compute the expectation value of the number operator in a coherent state, showing that it corresponds to the squared modulus of the coherent amplitude.

Theorem 1.9 *For a coherent state $|\alpha\rangle$, the expectation value of the number operator 1.139 is*

$$\left\langle \hat{N} \right\rangle = |\alpha|^2. \tag{1.221}$$

Proof From the expansion 1.195, we have

$$\begin{aligned} \left\langle \hat{N} \right\rangle &= \left\langle \alpha \left| a^\dagger a \right| \alpha \right\rangle = \\ &= \sum_{N=0}^{\infty} N\, |\langle N|\alpha\rangle|^2. \end{aligned} \tag{1.222}$$

Using

$$\begin{aligned} \langle N|\alpha\rangle &= \left\langle N \left| e^{-\frac{|\alpha|^2}{2}} \sum_{N'=0}^{\infty} \frac{\alpha^{N'}}{\sqrt{N'!}} \right| N' \right\rangle = \\ &= e^{-\frac{|\alpha|^2}{2}} \sum_{N'=0}^{\infty} \frac{\alpha^{N'}}{\sqrt{N'!}} \langle N|N'\rangle = \end{aligned}$$

$$
\begin{aligned}
&= e^{-\frac{|\alpha|^2}{2}} \sum_{N'=0}^{\infty} \frac{\alpha^{N'}}{\sqrt{N'!}} \delta_{N,N'} = \\
&= e^{-\frac{|\alpha|^2}{2}} \frac{\alpha^N}{\sqrt{N!}},
\end{aligned} \tag{1.223}
$$

it follows that

$$
\left\langle \hat{N} \right\rangle = e^{-|\alpha|^2} \sum_{N=0}^{\infty} N \frac{|\alpha|^{2N}}{N!}. \tag{1.224}
$$

The term $N = 0$ is zero, so the sum can start from $N = 1$, i.e.,

$$
\begin{aligned}
\left\langle \hat{N} \right\rangle &= e^{-|\alpha|^2} \sum_{N=1}^{\infty} N \frac{|\alpha|^{2N}}{N!} = \\
&= e^{-|\alpha|^2} \sum_{N=1}^{\infty} \frac{|\alpha|^{2N}}{(N-1)!} = \\
&= e^{-|\alpha|^2} |\alpha|^2 \sum_{N=1}^{\infty} \frac{|\alpha|^{2(N-1)}}{(N-1)!}.
\end{aligned} \tag{1.225}
$$

Setting $N' = N - 1$ gives

$$
\begin{aligned}
\left\langle \hat{N} \right\rangle &= e^{-|\alpha|^2} |\alpha|^2 \sum_{N'=0}^{\infty} \frac{|\alpha|^{2N'}}{N'!} = \\
&= e^{-|\alpha|^2} |\alpha|^2 e^{|\alpha|^2},
\end{aligned} \tag{1.226}
$$

which is the thesis. □

Let us examine the time evolution of an initial coherent state $|\alpha\rangle$. According to the Schrödinger equation (see Eq. 1.86), the state at time t is given by

$$
|\psi(t)\rangle = e^{-\frac{i}{\hbar}\mathcal{H}t} |\alpha\rangle. \tag{1.227}
$$

In the Heisenberg picture (see Eq. 1.102), the annihilation operator evolves according to Eq. 1.189, then

$$
\begin{aligned}
a|\psi(t)\rangle &= a\, e^{-\frac{i}{\hbar}\mathcal{H}t} |\alpha\rangle = \\
&= e^{-\frac{i}{\hbar}\mathcal{H}t} \left(e^{\frac{i}{\hbar}\mathcal{H}t} a e^{-\frac{i}{\hbar}\mathcal{H}t} \right) |\alpha\rangle =
\end{aligned}
$$

$$
\begin{aligned}
&= e^{-\frac{i}{\hbar}\mathcal{H}t}\, a(t)\,|\alpha\rangle = \\
&= e^{-\frac{i}{\hbar}\mathcal{H}t}\, (ae^{-i\omega t})|\alpha\rangle = \\
&= e^{-i\omega t}\, e^{-\frac{i}{\hbar}\mathcal{H}t}\, a|\alpha\rangle = \\
&= e^{-i\omega t}\, \alpha\, |\psi(t)\rangle.
\end{aligned} \tag{1.228}
$$

Therefore, $|\psi(t)\rangle$ is still an eigenstate of a with eigenvalue

$$\alpha(t) = \alpha e^{-i\omega t}, \tag{1.229}$$

that is,

$$a|\psi(t)\rangle = \alpha(t)|\psi(t)\rangle. \tag{1.230}$$

Hence, a coherent state preserves its coherence under time evolution, with its complex amplitude rotating in phase space at the oscillator frequency ω. Equivalently, one can write

$$e^{-\frac{i}{\hbar}\mathcal{H}t}|\alpha\rangle = c(t)|\alpha(t)\rangle, \tag{1.231}$$

where $c(t)$ is a time-dependent phase factor, whose explicit form is not essential here, and satisfies

$$|c(t)|^2 = 1. \tag{1.232}$$

Ultimately, for a coherent state $|\alpha\rangle$, we can explicitly compute the expectation values of the position and momentum operators.

Theorem 1.10 *Let $\hat{x}$ and $\hat{p}$ be the position and momentum operators of the one-dimensional quantum harmonic oscillator (see Eqs. 1.131 and 1.132), and let*

$$|\psi(t)\rangle = e^{-\frac{i}{\hbar}\mathcal{H}t}|\alpha\rangle \tag{1.233}$$

be the time-evolved coherent state. Then the time-dependent expectation values and uncertainties satisfy

$$\langle\hat{x}\rangle(t) = \sqrt{\frac{2\hbar}{m\omega}}\mathrm{Re}\ \alpha(t), \tag{1.234}$$

$$\langle\hat{p}\rangle(t) = \sqrt{2\hbar m\omega}\ \mathrm{Im}\ \alpha(t), \tag{1.235}$$

and

$$\Delta x(t)\Delta p(t) = \frac{\hbar}{2}, \ \forall t, \tag{1.236}$$

so that coherent states remain minimum-uncertainty states at all times and their expectation values oscillate sinusoidally in time, in complete analogy with the classical motion of a harmonic oscillator.

Proof Since the time-evolved state is a coherent state with eigenvalue $\alpha(t)$, we have

$$a|\psi(t)\rangle = \alpha(t)|\psi(t)\rangle, \tag{1.237}$$

and therefore

$$\langle a\rangle(t) = \alpha(t), \tag{1.238}$$

$$\langle a^\dagger\rangle(t) = \alpha^*(t), \tag{1.239}$$

and more generally

$$\left\langle a^N\right\rangle(t) = \alpha^N(t), \tag{1.240}$$

$$\left\langle a^{\dagger N}\right\rangle(t) = \alpha^{*N}(t). \tag{1.241}$$

Then

$$\begin{aligned}\langle\hat{x}\rangle(t) &= \sqrt{\frac{\hbar}{2m\omega}}\left(\langle a\rangle(t) + \langle a^\dagger\rangle(t)\right) = \\ &= \sqrt{\frac{\hbar}{2m\omega}}\left(\alpha(t) + \alpha^*(t)\right) = \\ &= \sqrt{\frac{2\hbar}{m\omega}}\mathrm{Re}\,\alpha(t),\end{aligned} \tag{1.242}$$

$$\begin{aligned}\langle\hat{p}\rangle(t) &= i\sqrt{\frac{\hbar m\omega}{2}}\left(\langle a^\dagger\rangle(t) - \langle a\rangle(t)\right) = \\ &= i\sqrt{\frac{\hbar m\omega}{2}}\left(\alpha^*(t) - \alpha(t)\right) = \\ &= \sqrt{2\hbar m\omega}\,\mathrm{Im}\,\alpha(t).\end{aligned} \tag{1.243}$$

For the variances, we calculate, for example,

$$\begin{aligned}\langle \hat{x}^2\rangle(t) &= \frac{\hbar}{2m\omega}\left(\langle a^2\rangle(t) + \langle a^{\dagger 2}\rangle(t) + \langle aa^{\dagger}\rangle(t) + \langle a^{\dagger}a\rangle(t)\right) = \\ &= \frac{\hbar}{2m\omega}\left(\alpha(t)^2 + \alpha^*(t)^2 + 1 + 2|\alpha(t)|^2\right) = \\ &= \frac{\hbar}{2m\omega}\left[\left(\alpha(t) + \alpha^*(t)\right)^2 + 1\right],\end{aligned} \tag{1.244}$$

then

$$\begin{aligned}(\Delta x(t))^2 &= \langle \hat{x}^2\rangle(t) - \langle \hat{x}\rangle^2(t) = \\ &= \frac{\hbar}{2m\omega}.\end{aligned} \tag{1.245}$$

Similarly, one finds

$$(\Delta p(t))^2 = \frac{\hbar m\omega}{2}, \tag{1.246}$$

which completes the proof. □

1.4 The Quantum Adiabatic Theorem

The quantum adiabatic theorem is a fundamental result that describes the behavior of quantum many-particle systems undergoing slow (adiabatic) evolution, i.e., when the Hamiltonian varies smoothly and sufficiently slowly in time. Suppose a quantum system is initially prepared in a given state, and then the system Hamiltonian is varied slowly in time. It is known that the system's state evolves according to the Schrödinger equation (see Eq. 1.86) with a time-dependent Hamiltonian. Remarkably, if the evolution is adiabatic and the spectrum of the Hamiltonian is discrete and nondegenerate, the system remains in the instantaneous eigenstate corresponding to its initial eigenstate throughout the evolution. Here we present a simple quantitative characterization of this principle.

Theorem 1.11 (The Quantum Adiabatic Theorem) *Let $\mathcal{H}(t)$ be a Hamiltonian that varies very slowly with time, and assume its spectrum is discrete and nondegenerate. Then, given two times t_1 and t_2 with $t_2 > t_1$, if at time t_1 the system is in the n-th eigenstate of $\mathcal{H}(t_1)$, it will be in the n-th eigenstate of $\mathcal{H}(t_2)$ at time t_2.*

Proof At each instant t, the eigenfunctions and eigenvalues satisfy the eigenvalue problem

$$\mathcal{H}(t)\varphi_m(t) = \mathcal{E}_m(t)\varphi_m(t). \tag{1.247}$$

The wave function can be expanded in the instantaneous eigenbasis as

$$\Psi(t) = \sum_m c_m(t)\varphi_m(t)e^{-\frac{i}{\hbar}\theta_m(t)}, \tag{1.248}$$

where each phase $\theta_m(t)$ is defined as follows

$$\theta_m(t) = \int_0^t \mathcal{E}_m(t')dt'. \tag{1.249}$$

Inserting this expansion into the time-dependent Schrödinger equation (see Eq. 1.86), we obtain

$$i\hbar \sum_m \left[\partial_t c_m(t)\varphi_m(t) + c_m(t)\partial_t\varphi_m(t) - \frac{i}{\hbar}c_m(t)\varphi_m(t)\partial_t\theta_m(t) \right] e^{-\frac{i}{\hbar}\theta_m(t)} =$$

$$= \sum_m c_m(t)\mathcal{H}(t)\varphi_m(t)e^{-\frac{i}{\hbar}\theta_m(t)}. \tag{1.250}$$

Since $\partial_t\theta_m(t) = \mathcal{E}_m(t)$, the terms containing $\mathcal{E}_m(t)$ cancel out on both sides, leaving

$$\sum_m \left[\partial_t c_m(t)\varphi_m(t) + c_m(t)\partial_t\varphi_m(t)\right] e^{-\frac{i}{\hbar}\theta_m(t)} = 0. \tag{1.251}$$

Multiplying from the left by $\langle\varphi_n(t)|$ and using orthonormality, we find

$$\partial_t c_n(t) = -\sum_m c_m(t)\langle\varphi_n(t)|\partial_t\varphi_m(t)\rangle e^{-\frac{i}{\hbar}(\theta_m(t)-\theta_n(t))}. \tag{1.252}$$

Separating the term with $m = n$, we get

$$\partial_t c_n(t) = -c_n(t)\langle\varphi_n(t)|\partial_t\varphi_n(t)\rangle - \sum_{m\neq n} c_m(t)\langle\varphi_n(t)|\partial_t\varphi_m(t)\rangle e^{-\frac{i}{\hbar}(\theta_m(t)-\theta_n(t))}. \tag{1.253}$$

Because the spectrum is nondegenerate, the phase factors $e^{-\frac{i}{\hbar}(\theta_m-\theta_n)}$ oscillate rapidly and average out, making the object

$$\sum_{m\neq n} c_m(t)\langle\varphi_n(t)|\partial_t\varphi_m(t)\rangle e^{-\frac{i}{\hbar}(\theta_m(t)-\theta_n(t))} \tag{1.254}$$

negligible. Furthermore, the matrix elements $\langle\varphi_n(t)|\partial_t\varphi_m(t)\rangle$ are themselves suppressed because they are proportional to $\langle\varphi_n(t)|\partial_t\mathcal{H}(t)|\varphi_m(t)\rangle$ relative to the energy gap $\mathcal{E}_n(t) - \mathcal{E}_m(t)$ (as we'll show in Eq. 1.258). Moreover, since $|c_n(t)|^2 \leq 1$, it remains to evaluate the terms

$$\langle \varphi_n(t)|\partial_t \varphi_m(t)\rangle. \tag{1.255}$$

Differentiating the eigenvalue equation with respect to time,

$$\partial_t \mathcal{H}(t)\varphi_m(t) + \mathcal{H}(t)\partial_t \varphi_m(t) = \partial_t \mathcal{E}_m(t)\varphi_m(t) + \mathcal{E}_m(t)\partial_t \varphi_m(t), \tag{1.256}$$

and projecting on $\varphi_n(t)$ with $n \neq m$, we get

$$\langle \varphi_n(t)|\partial_t \mathcal{H}(t)|\varphi_m(t)\rangle + \mathcal{E}_n(t)\langle \varphi_n(t)|\partial_t \varphi_m(t)\rangle = \mathcal{E}_m(t)\langle \varphi_n(t)|\partial_t \varphi_m(t)\rangle, \tag{1.257}$$

which yields

$$(\mathcal{E}_n(t) - \mathcal{E}_m(t))\,\langle \varphi_n(t)|\partial_t \varphi_m(t)\rangle = -\langle \varphi_n(t)|\partial_t \mathcal{H}(t)|\varphi_m(t)\rangle. \tag{1.258}$$

Since

$$0 < |\mathcal{E}_n(t) - \mathcal{E}_m(t)| < +\infty, \quad \forall\, n \neq m, \tag{1.259}$$

the quantity $\langle \varphi_n(t)|\partial_t \varphi_m(t)\rangle$ is negligible if the matrix elements of $\partial_t \mathcal{H}(t)$ are sufficiently small. Under this condition, it is legitimate to neglect the non-adiabatic term in the evolution of $c_n(t)$; this is known as the adiabatic approximation. The expression 1.258 is called the non-adiabatic term. In the adiabatic approximation, the time evolution of the coefficient c_n reduces to

$$\partial_t c_n(t) = -c_n(t)\langle \varphi_n(t)|\partial_t \varphi_n(t)\rangle. \tag{1.260}$$

From the orthonormality condition, it follows that

$$\langle \varphi_n(t)|\partial_t \varphi_n(t)\rangle + \langle \partial_t \varphi_n(t)|\varphi_n(t)\rangle = 0, \tag{1.261}$$

which implies

$$2\,\mathrm{Re}\left(\langle \varphi_n(t)|\partial_t \varphi_n(t)\rangle\right) = 0. \tag{1.262}$$

Therefore, $\langle \varphi_n(t)|\partial_t \varphi_n(t)\rangle$ is purely imaginary. Solving the first-order differential equation with separable variables, we have

$$\int \frac{dc_n(t)}{c_n(t)} = -\int \langle \varphi_n(t)|\partial_t \varphi_n(t)\rangle dt, \tag{1.263}$$

which gives

$$\ln c_n(t) = -\int_0^t \langle \varphi_n(t')|\partial_{t'} \varphi_n(t')\rangle dt' + \text{const.} \tag{1.264}$$

Exponentiating and fixing the integration constant by initial conditions yields

$$c_n(t) = c_n(0)\, e^{-\int_0^t \langle\varphi_n(t')|\partial_{t'}\varphi_n(t')\rangle\, dt'}. \tag{1.265}$$

The adiabatic phase, given by the exponent, is purely imaginary and is called the Berry phase acquired by the eigenstate during the adiabatic evolution. This confirms that, under adiabatic evolution, the system remains coherently aligned with the instantaneous n-th eigenstate, differing only by an accumulated geometric phase factor. □

1.5 One Particle Green's Function

In quantum mechanics, the one-particle Green's function offers a compact way to express the evolution of a wave function under the action of the Hamiltonian. Instead of solving the Schrödinger equation directly for each initial condition, one may use the Green's function as a propagator, which fully determines the dynamics once the spectrum of the Hamiltonian is known.

According to Eq. 1.88, in the Schrödinger representation the time evolution of a state is explicitly governed by the evolution operator $U(t, t')$. In Dirac notation, this leads to the following expression for the corresponding wave functions

$$\langle \mathbf{r}|\psi(t)\rangle = \langle \mathbf{r}|U(t, t')|\psi(t')\rangle, \tag{1.266}$$

where the time-evolution operator acts on the initial wave function to propagate the state to time t, and the position basis allows the dynamics to be expressed explicitly in terms of spatial and time variables. We insert 1.18 as follows

$$\psi(\mathbf{r}, t) = \int d\mathbf{r}'\, \langle \mathbf{r}|U(t, t')|\mathbf{r}'\rangle\, \langle \mathbf{r}'|\psi(t')\rangle, \tag{1.267}$$

and we have

$$\psi(\mathbf{r}, t) = \int d\mathbf{r}'\, G(\mathbf{r}, t, \mathbf{r}', t')\, \psi(\mathbf{r}', t'), \tag{1.268}$$

where

$$G(\mathbf{r}, t, \mathbf{r}', t') = \langle \mathbf{r}|U(t, t')|\mathbf{r}'\rangle \tag{1.269}$$

is called the Green's function. If $\mathcal{H}$ does not depend on time, then $U(t, t') = e^{-i\frac{\mathcal{H}}{\hbar}(t-t')}$ and the Green's function becomes

$$G(\mathbf{r}, t, \mathbf{r}', t') = \langle \mathbf{r}|e^{-i\frac{\mathcal{H}}{\hbar}(t-t')}|\mathbf{r}'\rangle. \tag{1.270}$$

Assume that the Hamiltonian has a discrete spectrum, i.e., $\mathcal{H}|\psi_\alpha\rangle = \mathcal{E}_\alpha|\varphi_\alpha\rangle$, then

$$
\begin{aligned}
G(\mathbf{r}, t, \mathbf{r}', t') &= \sum_{\alpha,\beta} \langle \mathbf{r}|\varphi_\alpha\rangle\langle\varphi_\alpha|e^{-i\frac{\mathcal{H}}{\hbar}(t-t')}|\varphi_\beta\rangle\langle\varphi_\beta|\mathbf{r}'\rangle = \\
&= \sum_{\alpha,\beta} \varphi_\alpha(\mathbf{r})\langle\varphi_\alpha|e^{-i\frac{\mathcal{E}_\beta}{\hbar}(t-t')}|\varphi_\beta\rangle\varphi^*_\beta(\mathbf{r}') = \\
&= \sum_{\alpha,\beta} \varphi_\alpha(\mathbf{r})e^{-i\frac{\mathcal{E}_\beta}{\hbar}(t-t')}\langle\varphi_\alpha|\varphi_\beta\rangle\varphi^*_\beta(\mathbf{r}') = \\
&= \sum_{\alpha,\beta} \varphi_\alpha(\mathbf{r})e^{-i\frac{\mathcal{E}_\beta}{\hbar}(t-t')}\delta_{\alpha,\beta}\varphi^*_\beta(\mathbf{r}') = \\
&= \sum_{\alpha} \varphi_\alpha(\mathbf{r})e^{-i\frac{\mathcal{E}_\alpha}{\hbar}(t-t')}\varphi^*_\alpha(\mathbf{r}'),
\end{aligned}
\tag{1.271}
$$

where we made use of Eq. 1.10.

1.5.1 Meaning of One-Particle Green's Function

Consider the one-particle Schrödinger equation (see Eq. 1.86) with an initial condition $\psi(t')$. If the Hamiltonian is time-independent, the wavefunction at a later time $t > t'$ is given by 1.87. We multiply Eq. 1.267 by the Heaviside function $\Theta(t-t')$ in order to enforce the principle of temporal causality, then

$$
\Theta(t-t')\psi(\mathbf{r}, t) = i\hbar \int \frac{d\mathbf{r}'}{i\hbar}\Theta(t-t')\,\langle\mathbf{r}|U(t, t')|\mathbf{r}'\rangle\psi(\mathbf{r}', t'),
\tag{1.272}
$$

$$
\Theta(t-t')\psi(\mathbf{r}, t) = i\hbar \int d\mathbf{r}' G^{(r)}(\mathbf{r}, t, \mathbf{r}', t')\psi(\mathbf{r}', t').
\tag{1.273}
$$

Definition 1.2 (Retarded One-Particle Green's Function) Let $U(t, t')$ be the unitary time-evolution operator generated by the Hamiltonian $\mathcal{H}$. The retarded one-particle Green's function is defined as

$$
G^{(r)}(\mathbf{r}, t, \mathbf{r}', t') = -\frac{i}{\hbar}\Theta(t-t')\langle\mathbf{r}|U(t, t')|\mathbf{r}'\rangle,
\tag{1.274}
$$

where $\Theta(t-t')$ denotes the Heaviside step function, which enforces the causality condition.

First, we observe that Eq. 1.274 does not depend on the initial condition, since the wavefunction $\psi(t')$ does not appear explicitly. Once the initial condition is specified, the retarded Green's function allows us to reconstruct the solution of the Schrödinger equation (see Eq. 1.86) at any later time $t > t'$. Furthermore, the retarded Green's function is itself a solution of the Schrödinger equation with the special initial condition

$$\psi(\mathbf{r}', t') = \delta(\mathbf{r}' - \mathbf{r}_0), \ \forall t > t', \ \forall \mathbf{r}_0. \tag{1.275}$$

Indeed, substituting Eq. 1.274 into Eq. 1.273, it follows

$$\Theta(t - t')\psi(\mathbf{r}, t) = i\hbar \int d\mathbf{r}' G^{(r)}(\mathbf{r}, t, \mathbf{r}', t')\delta(\mathbf{r}' - \mathbf{r}_0), \tag{1.276}$$

which reduces to

$$\Theta(t - t')\psi(\mathbf{r}, t) = i\hbar\, G^{(r)}(\mathbf{r}, t, \mathbf{r}_0, t'). \tag{1.277}$$

We now make explicit the statement that the one-particle Green's function is the distributional inverse of the operator $(i\hbar\partial_t - \mathcal{H})$ in the time domain. We act with $(i\hbar\partial_t - \mathcal{H})$ on the kernel $\Theta(t - t')\langle\mathbf{r}|U(t, t')|\mathbf{r}'\rangle$. Using the product rule and the Schrödinger equation for the propagator kernel,

$$i\hbar\, \partial_t \langle\mathbf{r}|U(t, t')|\mathbf{r}'\rangle = \langle\mathbf{r}|\mathcal{H}U(t, t')|\mathbf{r}'\rangle, \tag{1.278}$$

we obtain, in the sense of distributions,

$$\begin{aligned}(i\hbar\partial_t - \mathcal{H})&\big[\Theta(t - t')\langle\mathbf{r}|U(t, t')|\mathbf{r}'\rangle\big] = \\ &i\hbar\, \delta(t - t')\, \langle\mathbf{r}|U(t', t')|\mathbf{r}'\rangle + \Theta(t - t')\, (i\hbar\partial_t - \mathcal{H})\langle\mathbf{r}|U(t, t')|\mathbf{r}'\rangle.\end{aligned} \tag{1.279}$$

The second term vanishes for all $t \neq t'$, and since $U(t', t') = \mathbb{1}$, we have

$$(i\hbar\partial_t - \mathcal{H})\big[\Theta(t - t')\langle\mathbf{r}|U(t, t')|\mathbf{r}'\rangle\big] = i\hbar\, \delta(t - t')\, \delta(\mathbf{r} - \mathbf{r}'). \tag{1.280}$$

Multiplying both sides by $-\frac{i}{\hbar}$ (the prefactor in the definition of $G^{(r)}$), we obtain

$$(i\hbar\partial_t - \mathcal{H})\, G^{(r)}(\mathbf{r}, t, \mathbf{r}', t') = \delta(t - t')\, \delta(\mathbf{r} - \mathbf{r}'). \tag{1.281}$$

This equality holds in the sense of distributions and must be complemented by the causality condition $G^{(r)}(\mathbf{r}, t < t', \mathbf{r}', t') = 0$. This property, namely, that $G^{(r)}$ is the unique distribution which, upon the action of the differential operator $(i\hbar\partial_t - \mathcal{H})$, yields a delta source, is precisely what qualifies it as a Green's function in the mathematical sense. In the framework of linear differential operators, a Green's function is defined as the fundamental solution, or distributional inverse, of an operator L, satisfying

$$LG^{(r)}(x, x') = \delta(x - x'). \tag{1.282}$$

Accordingly, the object $G^{(r)}$ serves as the Green's function associated with the operator $(i\hbar\partial_t - \mathcal{H})$, describing the system's response to a pointlike excitation in space and time.

Ultimately, if the Hamiltonian $\mathcal{H}$ does not depend explicitly on time, from Eq. 1.271, $G^{(r)}(\mathbf{r}, t, \mathbf{r}', t')$ can be written as a function of the eigenstates $\{\varphi_\alpha\}$ as follows

$$G^{(r)}(\mathbf{r}, t, \mathbf{r}', t') = -\frac{i}{\hbar}\Theta(t - t')\sum_\alpha \varphi_\alpha(\mathbf{r})e^{-i\frac{\mathcal{E}_\alpha}{\hbar}(t-t')}\varphi_\alpha^*(\mathbf{r}'). \tag{1.283}$$

The function $G^{(r)}(\mathbf{r}, t, \mathbf{r}', t')$ is often referred to as the one-particle (or single-particle) propagator. The reason is that it determines how the wavefunction propagates in space and time: the kernel $\langle \mathbf{r}|U(t, t')|\mathbf{r}'\rangle$ gives the transition amplitude for a particle prepared at position $\mathbf{r}'$ at time t' to be found at position $\mathbf{r}$ at time t. However, the actual probability density at $\mathbf{r}$ is obtained only after integrating over all possible initial positions weighted by the initial wavefunction, namely through Eq. 1.267.

1.5.2 *One-Dimensional Free Particle Green's Function*

We now proceed to the explicit evaluation of the Green's function $G^{(0)}(x, t, x', t')$ corresponding to the Hamiltonian of a free particle confined to one spatial dimension. By definition (see Eq. 1.274), it is given by

$$G^{(0)}(x, t, x', t') = \langle x|e^{-\frac{i}{\hbar}\frac{\hat{p}^2}{2m}(t-t')}|x'\rangle, \tag{1.284}$$

where x denotes the position of the particle along the one-dimensional space. Since $\mathcal{H}_0$ depends only on the momentum, we insert the one-dimensional version of the completeness relation (see Eq. 1.20) into $G^{(0)}(x, t, x', t')$, obtaining

$$G^{(0)}(x, t, x', t') = \int dp \int dp' \langle x|p\rangle\langle p|e^{-\frac{i}{\hbar}\frac{\hat{p}^2}{2m}(t-t')}|p'\rangle\langle p'|x'\rangle. \tag{1.285}$$

Now, let us consider Eq. 1.77 in one dimension. By replacing $\mathbf{r}$ with x, it becomes

$$\langle x|k\rangle = \frac{1}{\sqrt{V}}e^{-ikx}. \tag{1.286}$$

Substituting $p = \hbar k$ then gives

$$\langle x|p\rangle = \frac{1}{\sqrt{V}}e^{-\frac{i}{\hbar}px}. \tag{1.287}$$

Finally, in the thermodynamic limit, in particular for $V \to \infty$, by adapting Eq. C.11 to one dimension and replacing k with p, a factor of $\hbar$ naturally arises, yielding

$$\langle x|p\rangle = \frac{1}{\sqrt{2\pi\hbar}} e^{-\frac{i}{\hbar}px}, \tag{1.288}$$

which can be used as follows

$$\begin{aligned} G^{(0)}(x,t,x',t') &= \int \frac{dp}{\sqrt{2\pi\hbar}} \int \frac{dp'}{\sqrt{2\pi\hbar}}\, e^{-\frac{i}{\hbar}(px-p'x')} \langle p| e^{-\frac{i}{\hbar}\frac{\hat{p}^2}{2m}(t-t')}|p'\rangle = \\ &= \int \frac{dp}{\sqrt{2\pi\hbar}} \int \frac{dp'}{\sqrt{2\pi\hbar}}\, e^{-\frac{i}{\hbar}(px-p'x')} e^{-\frac{i}{\hbar}\frac{p^2}{2m}(t-t')} \delta(p-p') = \\ &= \int \frac{dp}{\sqrt{2\pi\hbar}} \int \frac{dp'}{\sqrt{2\pi\hbar}}\, e^{-\frac{i}{\hbar}(px-p'x') - \frac{i}{\hbar}\frac{p^2}{2m}(t-t')} \delta(p-p') = \\ &= \int \frac{dp}{2\pi\hbar}\, e^{-\frac{i}{\hbar}p(x-x') - \frac{i}{\hbar}\frac{p^2}{2m}(t-t')} . \end{aligned} \tag{1.289}$$

Since the set of eigenvalues generally has no upper limit, we introduce an arbitrary phase factor δ to ensure the convergence of the series, i.e.,

$$G^{(0)}(x,t,x',t') = \int \frac{dp}{2\pi\hbar}\, e^{-\frac{i}{\hbar}p(x-x') - \frac{i}{\hbar}\frac{p^2}{2m}(t-t'-i\delta)} \tag{1.290}$$

The infinitesimal term $i\delta$ plays a crucial role: it ensures both the convergence of the momentum integral and the causal structure of the propagator, so that $G^{(0)}$ is properly defined as a distribution, as will be shown below. Note that in this form, if we set

$$A = \frac{i}{m\hbar}\left(t - t' - i\delta\right), \tag{1.291}$$

$$B = -\frac{i}{\hbar}(x - x'), \tag{1.292}$$

the propagator $G^{(0)}$ reduces to a Gaussian integral, that is,

$$\int_{-\infty}^{+\infty} dy\, e^{-\frac{A}{2}y^2 + By} = \sqrt{\frac{2\pi}{A}}\, e^{-\frac{B^2}{2}A}, \quad A, B \in \mathbb{C}, \ \mathrm{Re}\, A > 0 \tag{1.293}$$

then

$$G^{(0)}(x,t,x',t') = \frac{1}{2\pi\hbar}\sqrt{\frac{2\pi m\hbar}{i(t-t'-i\delta)}}\, e^{\frac{im(x-x')^2}{2\hbar(t-t'-i\delta)}} =$$

$$= \sqrt{\frac{m}{2\pi i\hbar(t - t' - i\delta)}}\, e^{\frac{im(x-x')^2}{2\hbar(t-t'-i\delta)}} \tag{1.294}$$

is a pseudo-Gaussian, because of the imaginary unit.

Theorem 1.12 (Short-Time Limit of the Free Green's Function) *In the limit $t \to t'^{+}$, the free particle propagator $G^{(0)}(x, t, x', t')$ given in Eq. 1.294 reduces to the Dirac delta distribution in the position variable, i.e.,*

$$\lim_{t\to t'^{+}} G^{(0)}(x, t, x', t') = \delta(x - x'). \tag{1.295}$$

Proof Since we have introduced the convergence factor $\delta > 0$, we set $t - t' = 0$. In this way Eq. 1.294 becomes

$$G^{(0)}(x, t, x', t') = \sqrt{\frac{m}{2\pi i\hbar(-i\delta)}}\, e^{\frac{im(x-x')^2}{2\hbar(-i\delta)}} = $$

$$= \sqrt{\frac{m}{2\pi\hbar\delta}}\, e^{-\frac{m(x-x')^2}{2\hbar\delta}}, \tag{1.296}$$

and subsequently we study the limit $\delta \to 0^{+}$. In particular, consider the limit $\delta \to 0^{+}$ of the integral along the real axis of $G^{(0)}$ multiplied by a rapidly decreasing function f, i.e.,

$$\lim_{\delta\to 0^{+}} \int_{-\infty}^{+\infty} dx'\, G^{(0)}(x, t, x', t') f(x') = \lim_{\delta\to 0^{+}} \int_{-\infty}^{+\infty} dx \sqrt{\frac{m}{2\pi\hbar\delta}}\, e^{-\frac{m(x-x')^2}{2\hbar\delta}} f(x). \tag{1.297}$$

We set

$$R = \frac{m}{\hbar}, \tag{1.298}$$

then

$$\lim_{\delta\to 0^{+}} \int_{-\infty}^{+\infty} dx' G^{(0)}(x, t, x', t') f(x') = \lim_{\delta\to 0^{+}} \int_{-\infty}^{+\infty} dx' \sqrt{\frac{R}{2\pi\delta}}\, e^{-\frac{R(x-x')^2}{2\delta}} f(x'), \tag{1.299}$$

and we set

$$z = \sqrt{\frac{R}{\delta}}\,(x - x'), \tag{1.300}$$

$$x' = x - \sqrt{\frac{\delta}{R}} z, \tag{1.301}$$

then

$$
\begin{aligned}
&\lim_{\delta\to 0^+}\int_{-\infty}^{+\infty} dx'\, G^{(0)}(x,t,x',t')f(x') \\
&= \lim_{\delta\to 0^+}\int_{-\infty}^{+\infty}\left(-\frac{dz}{\sqrt{2\pi}}\right)e^{-\frac{z^2}{2}}f\left(x-\sqrt{\frac{\delta}{R}}z\right) = \\
&= \lim_{\delta\to 0^+}\int_{+\infty}^{-\infty}\frac{dz}{\sqrt{2\pi}}e^{-\frac{z^2}{2}}f\left(x-\sqrt{\frac{\delta}{R}}z\right) = \\
&= f(x)\int_{+\infty}^{-\infty}\frac{dz}{\sqrt{2\pi}}\,e^{-\frac{z^2}{2}} = \\
&= f(x)\int_{-\infty}^{+\infty}\frac{dz}{\sqrt{2\pi}}\,e^{-\frac{z^2}{2}} = \\
&= f(x),
\end{aligned} \tag{1.302}
$$

and finally we get

$$
\lim_{t'\to t^+} G^{(0)}(x,t,x',t') = \delta(x-x'). \tag{1.303}
$$

□

Theorem 1.12 holds in any spatial dimension. In particular, for a free particle in three spatial dimensions the propagator satisfies

$$
\lim_{t\to t'^+} G^{(0)}(\mathbf{r},t,\mathbf{r}',t') = \delta(\mathbf{r}-\mathbf{r}'). \tag{1.304}
$$

Theorem 1.13 *Under the assumption $\delta < |t-t'| \ll 1$, we show that the support of $G^{(0)}$, (see Eq. 1.294) i.e., the spacetime region where $G^{(0)}$ is significantly different from zero, is characterized by*

$$
|x-x'| \simeq \sqrt{\frac{\hbar|t-t'|}{m}}. \tag{1.305}
$$

Proof Without loss of generality, we assume $t > t'$. Consider

$$
\int_{-\infty}^{+\infty} dx'\, G^{(0)}(x,t,x',t')f(x'), \tag{1.306}
$$

where f is a rapidly decreasing function. Neglecting δ, which by assumption is several orders of magnitude less than $|t-t'|$, we have

$$
\int_{-\infty}^{+\infty} dx' \sqrt{\frac{m}{2\pi i\hbar(t-t')}}\; e^{\frac{im(x-x')^2}{2\hbar(t-t')}}\, f(x'), \tag{1.307}
$$

we set

$$z^2 = \frac{m(x - x')^2}{\hbar(t - t')}, \tag{1.308}$$

then

$$|x - x'| = \sqrt{\frac{\hbar(t - t')}{m}} z, \tag{1.309}$$

and for $x > x'$,

$$x = x' + \sqrt{\frac{\hbar(t - t')}{m}} z, \tag{1.310}$$

we get

$$\frac{1}{2\pi i} \int_{-\infty}^{+\infty} dz\, e^{-\frac{iz^2}{2}} f\left(x' + \sqrt{\frac{\hbar(t - t')}{m}} z\right). \tag{1.311}$$

For $|z| \gg 1$, the positive and negative contributions of the integral largely cancel due to the oscillatory factor $e^{-\frac{iz^2}{2}}$. Therefore, only values of z near $|z| \sim 1$ contribute significantly, i.e.,

$$x \simeq x' \pm \sqrt{\frac{\hbar(t - t')}{m}}, \tag{1.312}$$

$$|x - x'| \simeq \sqrt{\frac{\hbar(t - t')}{m}}. \tag{1.313}$$

This identifies the characteristic spatial scale of the support of $G^{(0)}$ for short times. □

1.6 Density Matrix Formalism

A density matrix (or density operator) is a matrix that describes an ensemble of physical systems as quantum states, even if the ensemble contains only one system. The development of such a formalism is due to Landau, and its applications involve quantum mechanics to one particle as well as to many bodies. Density matrices are thus crucial tools in areas of quantum mechanics such as quantum statistical mechanics and open quantum systems.

1.6.1 Density Matrix

Recall that the probability that an observable G takes the value α_r at time t in the state $\psi(t)$, and the expectation value $\langle G\rangle$ of the observable, are given respectively by

$$P(G=\alpha_r|t)=\langle\psi(t)|P_r|\psi(t)\rangle, \tag{1.314}$$

$$\langle G\rangle(t)=\langle\psi(t)|G|\psi(t)\rangle, \tag{1.315}$$

where P_r is the projection operator onto the eigenspace corresponding to the eigenvalue α_r. We now wish to rewrite the equations above using the trace operator. To this end, we introduce the following definition.

Definition 1.3 (Pure State Density Operator) Given a normalized state vector $|\psi(t)\rangle \in H$, the density operator (or density matrix) associated with the pure state is defined as

$$\rho(t)=|\psi(t)\rangle\langle\psi(t)|. \tag{1.316}$$

This operator provides a convenient way to describe quantum states, allowing probabilities and expectation values to be written in terms of the trace. Indeed, from the definition 1.316, we have

Theorem 1.14 (Probability and Expectation Value in Terms of the Trace) *Let G be an observable with spectral decomposition $G=\sum_r \alpha_r P_r$, where P_r is the projection operator onto the eigenspace associated with eigenvalue α_r. Then, the probability of obtaining the outcome α_r at time t, and the expectation value of G at time t, are given by*

$$P(G=\alpha_r|t)=\mathrm{Tr}\left(\rho(t)P_r\right), \tag{1.317}$$

$$\langle G\rangle(t)=\mathrm{Tr}\left(\rho(t)G\right). \tag{1.318}$$

Proof Let $\{|\varphi_i\rangle\}_i$ be an orthonormal basis of the Hilbert space. We have

$$\begin{aligned}\mathrm{Tr}\left(\rho(t)P_r\right) &= \sum_i\langle\varphi_i|\rho(t)P_r|\varphi_i\rangle = \\ &= \sum_i\langle\varphi_i|\psi(t)\rangle\langle\psi(t)|P_r|\varphi_i\rangle =\end{aligned}$$

$$= \sum_i \langle\psi(t)|P_r|\varphi_i\rangle\langle\varphi_i|\psi(t)\rangle =$$

$$= \langle\psi(t)|P_r|\psi(t)\rangle, \tag{1.319}$$

which proves Eq. 1.317. Similarly,

$$\begin{aligned} \mathrm{Tr}\,(\rho(t)G) &= \sum_i \langle\varphi_i|\rho(t)G|\varphi_i\rangle = \\ &= \sum_i \langle\varphi_i|\psi(t)\rangle\langle\psi(t)|G|\varphi_i\rangle = \\ &= \sum_i \langle\psi(t)|G|\varphi_i\rangle\langle\varphi_i|\psi(t)\rangle = \\ &= \langle\psi(t)|G|\psi(t)\rangle, \end{aligned} \tag{1.320}$$

which proves Eq. 1.318. □

Now consider a system whose state is not precisely determined, but is instead described by a statistical ensemble of possible pure states. Such a situation is referred to as a mixed state. Concretely, suppose that the initial state is known to be one of $|\psi_1(0)\rangle$, $|\psi_2(0)\rangle$, ..., $|\psi_n(0)\rangle$, where each $|\psi_i(0)\rangle$ is normalized, though the set is not necessarily orthonormal. Associated with these states is a set of probabilities $\{p_1, \ldots, p_n\}$, satisfying $\sum_{i=1}^{n} p_i = 1$. Assume that a measurement of the observable G is performed on a system prepared in one of these states. Then, at time t, the corresponding evolved states are $|\psi_1(t)\rangle$, $|\psi_2(t)\rangle$, ..., $|\psi_n(t)\rangle$. The probability of obtaining a given eigenvalue α_r of G, averaged over the ensemble of possible initial states, is given by

$$P_i(G = \alpha_r|t) = \left\langle\psi_i(t)|P_i^r|\psi_i(t)\right\rangle. \tag{1.321}$$

When the state of the system is not precisely known, the probability of obtaining a given eigenvalue of an observable G is computed as an average over all possible states, each weighted by its corresponding probability. Formally, if the system can be in a set of states $\{|\psi_i\rangle\}$ with probabilities $\{p_i\}$, the probability of measuring the eigenvalue α_r is

$$P(G = \alpha_r|t) = \sum_{i=1}^{n} p_i \left\langle\psi_i(t)|P_i^r|\psi_i(t)\right\rangle, \tag{1.322}$$

and similarly, the expectation value will be provided by

$$\langle G\rangle(t) = \sum_{i=1}^{n} p_i \,\langle\psi_i(t)|G|\psi_i(t)\rangle\,. \tag{1.323}$$

From the above considerations, we give the following definition.

Definition 1.4 (Density Operator for a Mixed State) Let $\{|\psi_i(t)\rangle\}_{i=1}^n$ be normalized state vectors and $\{p_i\}_{i=1}^n$ a probability distribution, i.e., $p_i \geq 0$ and $\sum_{i=1}^n p_i = 1$. The density operator of the corresponding mixed state is defined as

$$\rho(t) = \sum_{i=1}^{n} p_i \, |\psi_i(t)\rangle\langle\psi_i(t)|. \tag{1.324}$$

If the system is in a well-defined quantum state, the expression 1.324 reduces to the pure state 1.316, which contains the maximal information allowed by quantum mechanics. By contrast, when the density operator has the general form 1.324, it reflects incomplete information about the system, arising from the lack of precise knowledge of its actual state.

Theorem 1.15 *The density operator for a mixed state, Eq. 1.324, also satisfies Eqs. 1.317 and 1.318.*

Proof This statement follows directly from the linearity of the trace operation (see Appendix J) and the definition of the density matrix. Since the trace is linear, both equations hold equally well for mixed states described by a statistical mixture operator. □

Theorem 1.16 *Let $\rho(t)$ be a density operator, describing the state of a quantum system. It may represent either a pure state (see Eq. 1.316), or a mixed state (see Eq. 1.324). In both cases, the density operator satisfies the following properties:*

(1) it is hermitian;
(2) $\mathrm{Tr}(\rho(t)) = 1, \ \forall\, t$;
(3) it is semipositive definite.

Proof

(1) The coefficients p_i are real and the matrix $|\psi_i(t)\rangle\langle\psi_i(t)|$ is hermitian, then the density matrix is hermitian. The same reasoning applies directly to a pure state.
(2) We have

$$\begin{aligned}
\mathrm{Tr}(\rho(t)) &= \sum_{j=1}^{+\infty}\sum_{i=1}^{n} p_i \langle\varphi_j|\psi_i(t)\rangle\langle\psi_i(t)|\varphi_j\rangle = \\
&= \sum_{j=1}^{+\infty}\sum_{i=1}^{n} p_i \langle\psi_i(t)|\varphi_j\rangle\langle\varphi_j|\psi_i(t)\rangle = \\
&= \sum_{i=1}^{n} p_i \langle\psi_i(t)|\psi_i(t)\rangle =
\end{aligned}$$

$$= \sum_{i=1}^{n} p_i =$$

$$= 1. \tag{1.325}$$

For a pure state, $\operatorname{Tr}\rho_{\text{pure}} = \langle\psi|\psi\rangle = 1$.

(3) We have

$$\langle\varphi|\rho(t)|\varphi\rangle = \left\langle \varphi \left| \sum_{i=1}^{n} p_i |\psi_i(t)\rangle\langle\psi_i(t)| \right| \varphi \right\rangle =$$

$$= \sum_{i=1}^{n} p_i |\langle\varphi|\psi_i(t)\rangle|^2 \geq 0, \quad \forall\, \varphi \in \mathcal{H}. \tag{1.326}$$

In the pure case, $\langle\varphi|\rho_{\text{pure}}|\varphi\rangle = |\langle\varphi|\psi\rangle|^2 \geq 0$, confirming the same property.

□

Theorem 1.17 *The density operator describing mixed states, i.e. Eq. 1.324, satisfies a convex sum of density matrices is a density matrix.*

Proof For the sake of simplicity, let us consider case of two density matrices, ρ_1 and ρ_2, and let be

$$\rho = \lambda\rho_1 + (1-\lambda)\rho_2, \;\; \lambda \in [0, 1] \tag{1.327}$$

the convex sum. Since λ is real and ρ_1, ρ_2 are hermitians, then ρ is hermitian. Since $\lambda \in [0, 1]$ and ρ_1, ρ_2 are positive definite, then ρ is positive definite. Finally

$$\operatorname{Tr}(\rho) = \lambda \operatorname{Tr}(\rho_1) + (1-\lambda) \operatorname{Tr}(\rho_2) =$$

$$= \lambda + 1 - \lambda =$$

$$= 1. \tag{1.328}$$

□

Ultimately, the time evolution of the density operator is governed by the von Neumann equation, which is the quantum-mechanical analogue of the classical Liouville equation. It provides the dynamical law describing how a density matrix $\rho(t)$ evolves under the action of the system Hamiltonian.

Theorem 1.18 (Von Neumann Equation for Pure and Mixed States) *Let $\rho(t)$ be a density operator, describing the state of a quantum system. It may represent either a pure state (see Eq. 1.316), or a mixed state (see Eq. 1.324). Let $\mathcal{H}(t)$ be a self-adjoint Hamiltonian operator generating the system's time evolution. Then the density operator $\rho(t)$ obeys the von Neumann equation*

$$i\hbar\,\frac{d\rho(t)}{dt} = \left[\mathcal{H}(t), \rho(t)\right]. \tag{1.329}$$

Proof For a pure state, from Eq. 1.316, one finds

$$\frac{d\rho(t)}{dt} = \frac{d|\psi(t)\rangle}{dt}\langle\psi(t)| + |\psi(t)\rangle\frac{d\langle\psi(t)|}{dt}. \tag{1.330}$$

Using the Schrödinger equation (see Eq. 1.86) for both the ket and its adjoint,

$$\frac{d|\psi(t)\rangle}{dt} = -\frac{i}{\hbar}\,\mathcal{H}(t)\,|\psi(t)\rangle, \tag{1.331}$$

$$\frac{d\langle\psi(t)|}{dt} = \frac{i}{\hbar}\,\langle\psi(t)|\,\mathcal{H}(t), \tag{1.332}$$

we obtain

$$\begin{aligned}\frac{d\rho(t)}{dt} &= -\frac{i}{\hbar}\mathcal{H}(t)|\psi(t)\rangle\langle\psi(t)| + \frac{i}{\hbar}|\psi(t)\rangle\langle\psi(t)|\mathcal{H}(t) = \\ &= -\frac{i}{\hbar}\left[\mathcal{H}(t), \rho(t)\right],\end{aligned} \tag{1.333}$$

and multiplying both sides by $i\hbar$ leads to the von Neumann equation.

If the density matrix describes a mixed state, then, since both differentiation and the commutator are linear operations, the previous result can be applied term by term to each component of the statistical mixture and from Eq. 1.324, one finds

$$\begin{aligned}\frac{d\rho(t)}{dt} &= \sum_j p_j \frac{d}{dt}\left(|\psi(t)\rangle\langle\psi(t)|\right) = \\ &= -\frac{i}{\hbar}\sum_j p_j\left[\mathcal{H}(t), |\psi(t)\rangle\langle\psi(t)|\right] = \\ &= -\frac{i}{\hbar}\left[\mathcal{H}(t), \rho(t)\right].\end{aligned} \tag{1.334}$$

Multiplying again by $i\hbar$ confirms the von Neumann equation for the mixed state.

□

A mixed quantum state generally exhibits a higher degree of uncertainty, or disorder, than a pure state. To quantitatively distinguish between pure and mixed states, we introduce the concept of purity.

Definition 1.5 (Purity) Let ρ be the density operator of a quantum system. It may represent either a pure state (see Eq. 1.316), or a mixed state (see Eq. 1.324). The *purity* of the state is defined as

$$\gamma(\rho) = \mathrm{Tr}(\rho^2). \tag{1.335}$$

Purity provides a quantitative measure of how close a quantum state is to being pure, as we demonstrate below.

Theorem 1.19 (Purity for Pure and Mixed States) *Let $\rho(t)$ be a density operator, describing the state of a quantum system. It may represent either a pure state (see Eq. 1.316), or a mixed state (see Eq. 1.324). Then the purity of the state is given by*

$$\gamma_{\text{pure}} = 1, \tag{1.336}$$

$$\gamma_{\text{mixed}} < 1. \tag{1.337}$$

Proof For a pure state, we have

$$\begin{aligned}\rho^2_{\text{pure}} &= (|\psi(t)\rangle\langle\psi(t)|)(|\psi(t)\rangle\langle\psi(t)|) = \\ &= |\psi(t)\rangle\langle\psi(t)| = \\ &= \rho_{\text{pure}},\end{aligned} \tag{1.338}$$

so that

$$\begin{aligned}\mathrm{Tr}(\rho^2_{\text{pure}}) &= \mathrm{Tr}(\rho_{\text{pure}}) = \\ &= 1.\end{aligned} \tag{1.339}$$

For a mixed state,

$$\rho^2_{\text{mixed}} = \sum_{i,j} p_i p_j |\psi(t)\rangle\langle\psi(t)|\psi(t)\rangle\langle\psi(t)|, \tag{1.340}$$

and taking the trace gives

$$\mathrm{Tr}(\rho^2_{\text{mixed}}) = \sum_{i,j} p_i p_j |\langle\psi(t)|\psi(t)\rangle|^2 \leq \sum_i p_i^2 < 1, \tag{1.341}$$

with equality if and only if the state is pure. □

1.6.2 *Entropy*

Here, we introduce a quantity that provides an alternative way, akin to purity (see Eq. 1.335), to distinguish a pure state from a statistical mixture. Unlike purity, however, this measure has the additional advantage of being directly connected to characteristic thermodynamic quantities.

Definition 1.6 (Entropy) The entropy of a quantum system described by a density operator ρ is defined as

$$S = -k_B \operatorname{Tr}(\rho \ln \rho), \tag{1.342}$$

where k_B denotes the Boltzmann constant and ρ is given in Eq. 1.316 or Eq. 1.324.

We have

Theorem 1.20 *The entropy given in Eq. 1.342 satisfies:*

(1) it is semipositive definite;
(2) it is minimum for a pure state Eq. 1.316 and maximum for a mixed state Eq. 1.324 with equal coefficients.

Proof

(1) Diagonalizing the density operator in its eigenbasis, we can write

$$\begin{aligned} S &= -k_B \operatorname{Tr}(\rho \ln \rho) = \\ &= -k_B \sum_{i=1}^{N} \langle \varphi_i | \rho \ln \rho | \varphi_i \rangle = \\ &= -k_B \sum_{i=1}^{N} p_i \ln p_i \,. \end{aligned} \tag{1.343}$$

Since $x \ln x \leq 0$ for $x \in (0, 1]$, it follows that $S \geq 0$.

(2) For a pure state we have

$$\begin{aligned} S_{\text{pure}} &= -k_B \operatorname{Tr}(\rho \ln \rho) = \\ &= -k_B \ln 1 = \\ &= 0. \end{aligned} \tag{1.344}$$

For mixed states, the entropy depends on the probabilities p_i: maximizing S therefore amounts to solving a constrained optimization problem, namely

$$\max \left\{ -k_B \sum_{i=1}^{n} p_i \ln p_i \right\}, \tag{1.345}$$

$$\sum_{i=1}^{n} p_i = 1. \tag{1.346}$$

We apply the method of Lagrange multipliers and consider the function

$$\begin{aligned} f(p_i) &= S - \lambda \sum_i p_i = \\ &= - \sum_i \left(k_B p_i \ln p_i + \lambda p_i \right), \end{aligned} \tag{1.347}$$

where λ is a Lagrange multiplier. It must be

$$\begin{aligned} \frac{\partial f}{\partial p_j} &= - \left[k_B \left(\ln p_j + 1 \right) + \lambda \right] = \\ &= 0, \end{aligned} \tag{1.348}$$

which implies

$$\begin{aligned} p_j &= \text{const.} = \\ &= \frac{1}{N}, \ \forall j, \end{aligned} \tag{1.349}$$

that is,

$$\rho = \frac{1}{N} \mathbb{1}, \tag{1.350}$$

and finally it has shown

$$\begin{aligned} S &= -k_B \operatorname{Tr} \{ \rho \ln \rho \} = \\ &= -k_B \sum_{i=1}^{N} \frac{1}{N} \ln \left(\frac{1}{N} \right) = \\ &= k_B \sum_{i=1}^{N} \frac{1}{N} \ln N = \\ &= k_B \ln N. \end{aligned} \tag{1.351}$$

□

Remark 1.3 (Maximum Entropy Principle) The result above shows that entropy reaches its maximum when all accessible states are equally probable. More generally, the physical state of a system subject to macroscopic constraints (such as a fixed average energy) is determined by the maximum entropy principle: among all density operators compatible with the given constraints, the actual state of the system is the one that maximizes the entropy 1.342.

1.6.3 Maxwell-Boltzmann Distribution

Consider a quantum system described within the framework of the canonical ensemble, i.e., in thermodynamic equilibrium at temperature T and characterized by an average energy $\langle \mathcal{E} \rangle$. Introducing the inverse temperature (also called thermodynamic beta)

$$\beta = \frac{1}{k_B T}, \tag{1.352}$$

we can state the following result.

Theorem 1.21 (Canonical Density Matrix: Maxwell-Boltzmann Distribution) *For a quantum system in thermodynamic equilibrium at temperature T, the density matrix that maximizes the entropy, according to the maximum entropy principle (see Remark 1.3), under the constraints of normalization and fixed average energy $\langle \mathcal{E} \rangle$, constitutes the equilibrium density matrix of the system in the canonical ensemble and is given by*

$$\rho = \frac{e^{-\beta \mathcal{H}}}{Z}, \tag{1.353}$$

where β is given in Eq. 1.352 and

$$Z = \mathrm{Tr}\, e^{-\beta \mathcal{H}} \tag{1.354}$$

is the partition function. Equivalently, in the energy eigenbasis $\{|\varphi_i\rangle\}$ with eigenvalues $\mathcal{E}_i$, the diagonal elements of ρ are

$$p_i = \frac{e^{-\beta \mathcal{E}_i}}{Z}, \tag{1.355}$$

so that the probability of occupying the state $|\varphi_i\rangle$ is exponentially suppressed with energy.

Proof Since the system is at thermodynamic equilibrium, we look for a time-independent density matrix operator, i.e.,

$$\frac{d\rho(t)}{dt} = 0, \tag{1.356}$$

from Von Neumann Eq. 1.329, it follows

$$\left[\mathcal{H}, \rho\right] = 0, \tag{1.357}$$

and thus Hamiltonian and density matrix share a common eigenbasis. Given

$$\mathcal{H} = \sum_{ru} \mathcal{E}_r |\varphi_{ru}\rangle\langle\varphi_{ru}|, \tag{1.358}$$

where $u = 1, ...g_r$ labels the degeneracy of the eigenvalue $\mathcal{E}_r$, and we define $g_r = \dim(H_r)$ as the degeneracy of the eigenvalue $\mathcal{E}_r$. Since the density matrix commutes with the Hamiltonian, then

$$\rho = \sum_{ru} p_r |\varphi_{ru}\rangle\langle\varphi_{ru}|. \tag{1.359}$$

We are interested in determining the coefficients p_r that make the entropy maximum: this is a constrained optimization problem with constraints

$$\begin{aligned}\sum_{ru} p_r &= \sum_r g_r p_r = \\ &= 1,\end{aligned} \tag{1.360}$$

$$\begin{aligned}\sum_{ru} p_r \mathcal{E}_r &= \sum_r g_r p_r \mathcal{E}_r = \\ &= \langle\mathcal{E}\rangle.\end{aligned} \tag{1.361}$$

Introducing Lagrange multipliers λ and μ, we define the function

$$h = S - \lambda \sum_r g_r p_r - \mu \sum_r g_r p_r \mathcal{E}_r. \tag{1.362}$$

Differentiating h with respect to p_j and setting the derivative to zero gives

$$\frac{\partial h}{\partial p_j} = -g_j \left[k_B(\ln p_j + 1) + \lambda + \mu\mathcal{E}_j\right] = 0, \tag{1.363}$$

$$\ln p_j = -\frac{\mu}{k_B}\mathcal{E}_j - \frac{\lambda}{k_B} - 1. \tag{1.364}$$

We now define

$$k_1 = \frac{\mu}{k_B}, \tag{1.365}$$

$$k_2 = -\frac{\lambda}{k_B} - 1, \tag{1.366}$$

so that

$$\ln p_j = -k_1 \mathcal{E}_j + k_2, \tag{1.367}$$

$$p_j = e^{k_2} e^{-k_1 \mathcal{E}_j}. \tag{1.368}$$

Imposing the normalization condition, we find

$$\begin{aligned}\sum_j g_j p_j &= \sum_j g_j e^{k_2} e^{-k_1 \mathcal{E}_j} = \\ &= e^{k_2} \sum_j g_j e^{-k_1 \mathcal{E}_j} = \\ &= 1,\end{aligned} \tag{1.369}$$

which yields

$$\begin{aligned}k_3 &= e^{k_2} = \\ &= \frac{1}{\sum_j g_j e^{-k_1 \mathcal{E}_j}} \equiv \\ &\equiv \frac{1}{Z},\end{aligned} \tag{1.370}$$

$$k_2 = \ln k_3 = -\ln Z. \tag{1.371}$$

Ultimately, for the canonical distribution, the average energy is given by

$$\begin{aligned}\langle \mathcal{E} \rangle &= \sum_j g_j p_j \mathcal{E}_j = \\ &= \frac{\sum_j g_j \mathcal{E}_j e^{-k_1 \mathcal{E}_j}}{\sum_j g_j e^{-k_1 \mathcal{E}_j}}.\end{aligned} \tag{1.372}$$

We can equivalently write

$$\langle \mathcal{E} \rangle = -\frac{\partial}{\partial k_1} \ln Z, \tag{1.373}$$

because

$$\begin{aligned}\frac{\partial Z}{\partial k_1} &= \frac{\partial}{\partial k_1} \sum_j g_j e^{-k_1 \mathcal{E}_j} = \\ &= -\sum_j g_j \mathcal{E}_j e^{-k_1 \mathcal{E}_j},\end{aligned} \tag{1.374}$$

and thus

$$
\begin{aligned}
-\frac{\partial}{\partial k_1}\ln Z &= -\frac{1}{Z}\frac{\partial Z}{\partial k_1} = \\
&= \frac{\sum_j g_j \mathcal{E}_j e^{-k_1 \mathcal{E}_j}}{Z} = \\
&= \langle \mathcal{E} \rangle .
\end{aligned}
\tag{1.375}
$$

This step is crucial, as it directly relates k_1 to the average energy. From thermodynamics, it is known that, in the canonical ensemble, the average energy is related to the partition function Z through the temperature T, i.e.,

$$
\langle \mathcal{E} \rangle = -\frac{\partial}{\partial \beta}\ln Z. \tag{1.376}
$$

Comparing this expression with the previous result, one immediately concludes that

$$
k_1 = \beta = \frac{1}{k_B T}. \tag{1.377}
$$

In other words, the Lagrange multiplier associated with the energy constraint coincides with the thermodynamic parameter β. □

1.6.4 Thermal Average

In many-body physics, systems whose interactions cannot be fully incorporated into a Hamiltonian or whose equations of motion do not admit exact solutions are said to be described by incomplete information. In such cases, the system is no longer characterized by a single pure state $|\psi(t)\rangle$, but rather by a statistical mixture of normalized states $\{|\psi_i\rangle\}_{i=1,\dots,N}$, each occurring with probability p_i. The corresponding density operator (or density matrix) for a pure state and for a statistical mixture are given in Eqs. 1.316 and 1.324, respectively. According to statistical mechanics, the density operator ρ describing a quantum system in thermal equilibrium at temperature T with a heat bath is defined as in Eq. 1.353. After diagonalizing the Hamiltonian, the eigenvalues $\{\mathcal{E}_r\}$ define the energy spectrum of the system, and the probability of occupying the state with energy $\mathcal{E}_r$ is

$$
p_r = \frac{e^{-\beta \mathcal{E}_r}}{Z}. \tag{1.378}
$$

The expectation value of a generic observable G at time t is given by Eq. 1.318. At thermal equilibrium, this expression takes the form

$$\langle G\rangle = \mathrm{Tr}\,(\rho G)\,, \qquad \rho = \frac{e^{-\beta\mathcal{H}}}{Z}, \tag{1.379}$$

where ρ is the density operator for the many-body system. Equation 1.379 defines the thermal average, which holds for systems at thermodynamic equilibrium.

Ultimately, note that, in the limit $T \to 0^+$, the density matrix given in Eq. 1.353 reduces to the projector onto the ground state (in the absence of degeneracy),

$$\rho|_{T\to 0^+} \to |\psi_{gs}\rangle\langle\psi_{gs}|, \tag{1.380}$$

and Eq. 1.355 becomes

$$\frac{e^{-\beta\mathcal{E}_i}}{Z} = \begin{cases} 1, & \text{ground state} \\ 0, & \text{otherwise} \end{cases}, \tag{1.381}$$

in other words, in the limit $T \to 0^+$, ensemble averages become quantum expectation values taken with respect to the ground state.

Chapter 2
Second Quantization for Non-relativistic Identical Particles

Abstract This chapter introduces the formalism of second quantization, a powerful framework for describing quantum many-body systems. Unlike the first quantization approach, which describes systems with a fixed number of particles and focuses on individual particle coordinates, second quantization works in Fock space, where the number of particles is not fixed but can vary. In this framework, states are characterized by their occupation numbers, and creation and annihilation operators allow one to move between Hilbert spaces with different particle numbers, adding or removing particles as needed. The formalism distinguishes between fermions and bosons through their respective (anti)commutation relations, enabling a compact description of systems with variable particle numbers. Observable operators, such as one-body, two-body, and general N-body operators, are systematically expressed in terms of these operators, providing a unified treatment of free and interacting systems. A key object is the particle density operator, which plays a central role in analysing correlations and many-body effects. As a concrete application, the jellium model is introduced: a system of interacting electrons in a uniform positive background. Despite its simplicity, it captures essential features of electron-electron interactions and serves as a testbed for methods like perturbation theory, Hartree-Fock, and linear response. Its translation-invariant structure makes it especially suitable for momentum-space analysis, where second quantization reveals its full power and conceptual clarity.

2.1 Introduction

In statistical mechanics, the two principal ensembles employed to describe systems in thermal equilibrium are the canonical and the grand canonical ensembles. In the canonical ensemble, the system is characterized by a fixed number of particles N, a fixed volume V, and a fixed temperature T. The probability of finding the system in a given microstate is determined by the Boltzmann factor associated with its energy, and all macroscopic observables are computed as averages over this fixed-N distribution. This ensemble is particularly suitable for describing isolated systems where the number of particles is strictly conserved, and it provides a clear

F. Tafuri et al., *Understanding Quantum Many-Body Theory*, UNITEXT for Physics, https://doi.org/10.1007/978-3-032-17849-7_2

conceptual framework for connecting microscopic dynamics with thermodynamic quantities. In contrast, the grand canonical ensemble allows for fluctuations in the number of particles, with the system exchanging both energy and particles with a reservoir. It is characterized by the parameters (μ, V, T), where μ denotes the chemical potential controlling the average particle number. This ensemble is especially convenient for open systems or for theoretical calculations where particle number fluctuations simplify the mathematical treatment.

In quantum many-body theory, however, the distinction between these ensembles is not as sharp or rigid as in classical statistical mechanics. In practical calculations, one often moves between the canonical and grand canonical descriptions depending on which formulation simplifies the problem. For instance, certain correlation functions or perturbative expansions may be more tractable in the grand canonical ensemble, while others are naturally expressed in the canonical framework. At the same time, the physical nature of the system may dictate the choice of ensemble: truly isolated systems are best described canonically, whereas systems in contact with a particle reservoir are more naturally treated grand canonically. Nevertheless, a central result of statistical mechanics assures us that, if the average particle number is N, then in the thermodynamic limit the statistical fluctuations of particle number relative to N vanish. This implies that canonical and grand canonical descriptions are effectively equivalent for large systems, justifying the flexibility in switching between ensembles.

Motivated by this equivalence and by the need for a formalism capable of describing systems with variable particle number, quantum many-body theory introduces the concept of a Hilbert space with variable particle number, called the Fock space. In this construction, the states of the system are labeled by the number of particles they contain, and creation and annihilation operators act as the fundamental building blocks of the theory. This approach allows a unified treatment of systems with arbitrary numbers of indistinguishable particles, simplifying the description of phenomena such as Bose-Einstein condensation or Fermi statistics. The formalism of second quantization thereby provides a natural generalization of the quantum-mechanical description to the grand canonical ensemble: in this framework, the flexibility of varying particle number is built directly into the structure of the Hilbert space and the operators, and many calculations that would be cumbersome in first quantization become straightforward.

2.2 System of Non-relativistic Identical Particles of Fixed Number

In order to describe a quantum system composed of several non-relativistic identical particles, it is necessary to construct the corresponding many-body Hilbert space. In the passage from one particle to N particles, the Hilbert space is constructed as the tensor product of the one-particle space H_1 (see Eq. 1.37), namely

$$
\begin{aligned}
H_N &= H_1^{\otimes N} = \\
&= (H_o \otimes H_s)^{\otimes N} = \\
&= H_o^{\otimes N} \otimes H_s^{\otimes N}.
\end{aligned} \tag{2.1}
$$

The orbital part is the tensor product of N copies of $\mathcal{L}^2(V)$, which is isomorphic to $\mathcal{L}^2(V^N)$ since the wavefunctions depend on the coordinates of all N particles, that is $\mathbf{r}_1, \ldots, \mathbf{r}_N$. The spin part is the tensor product of N copies of $\mathbb{C}^{2S+1}$, which represents the internal spin degrees of freedom of the system. Altogether, one obtains

$$
H_N = \mathcal{L}^2(V^N) \otimes (\mathbb{C}^{2S+1})^{\otimes N}. \tag{2.2}
$$

Accordingly, the scalar product for N particles is obtained by integrating over the N spatial coordinates and summing over all spin indices, i.e.,

$$
\begin{aligned}
&(\psi, \phi)_{H_N} = \\
&= \int dx_1 \int dx_2 \ldots \int dx_N \psi^*(x_1, \ldots, x_N)\phi(x_1, \ldots, x_N) \equiv \\
&\equiv \sum_{s_1=-S}^{S} \ldots \sum_{s_N=-S}^{S} \int_V d^3\mathbf{r}_1 \ldots \\
&\int_V d^3\mathbf{r}_N \psi^*(\mathbf{r}_1, s_1, \ldots, \mathbf{r}_N, s_N)\phi(\mathbf{r}_1, s_1, \ldots, \mathbf{r}_N, s_N),
\end{aligned} \tag{2.3}
$$

where we employed Eq. 1.43. A basis in H_N is given by the tensor product of one-particle bases, namely

$$
\{\varphi_{\alpha_1}(x_1)\varphi_{\alpha_2}(x_2) \ldots \varphi_{\alpha_N}(x_N)\}_{\alpha_1, \ldots, \alpha_N}, \tag{2.4}
$$

which consists of all possible permutations of the set of indices $\{\alpha_1, \ldots, \alpha_N\}$ (see Eqs. 1.44 and 1.45) for fixed $x_1, \ldots, x_N$. When dealing with systems of identical particles, the central issue is that exchanging two particles cannot lead to a physically distinct state. In other words, the labeling of the particles carries no physical meaning: what matters are the collective degrees of freedom of the system. To formalize this principle of indistinguishability, one introduces the notion of permutations of particle labels. For fixed spatial coordinates $x_1, \ldots, x_N$, a permutation acts on the indices $\{\alpha_1, \ldots, \alpha_N\}$ by reordering them, which corresponds to assigning the same set of individual degrees of freedom to the particles in a different way. Because of indistinguishability, wave functions that differ only by such a reordering represent the same physical state. Consequently, not all functions in the N—particle Hilbert space are admissible for describing identical particles: one must restrict to those that transform in a definite way under the action of permutations. Formally, a permutation σ is a bijective application between a finite subset of the natural numbers in itself such that σ takes the elements of the set $\{1, \ldots, N\}$ and maps them into itself with an ordering $\{\sigma(1), \ldots, \sigma(N)\}$. The action of a permutation is generally made explicit by a matrix notation as follows

$$\sigma(1, \ldots, N) = \begin{pmatrix} 1 & \ldots & N \\ \sigma(1) & \ldots & \sigma(N) \end{pmatrix}. \tag{2.5}$$

If the permutation involves only two elements, say of positions i and j, it is called transposition and is denoted by

$$\sigma_{ij} = \begin{pmatrix} 1 \ldots i \ldots j \ldots N \\ 1 \ldots j \ldots i \ldots N \end{pmatrix}. \tag{2.6}$$

The set of permutations, together with the composition operation defined by $\sigma \cdot \tau \equiv \sigma\tau$, forms a group (see Definition 14.2), which satisfies the following properties:

- Closure: for any two permutations σ, τ, the composition $\sigma \cdot \tau$ is also a permutation in the set.
- Associativity: for any three permutations σ, τ, ρ, the composition satisfies

$$(\sigma \cdot \tau) \cdot \rho = \sigma \cdot (\tau \cdot \rho). \tag{2.7}$$

- Identity element: there exists a permutation σ_{Id} that leaves all elements unchanged, given by

$$\sigma_{\text{Id}} = \begin{pmatrix} 1\ 2 \ldots i \ldots j \ldots N-1\ N \\ 1\ 2 \ldots i \ldots j \ldots N-1\ N \end{pmatrix}, \tag{2.8}$$

  such that $\sigma_{\text{Id}} \cdot \sigma = \sigma \cdot \sigma_{\text{Id}} = \sigma$ for all σ.
- Inverse element: for each permutation σ, there exists an inverse permutation σ^{-1}, obtained by applying the transpositions that compose σ in reverse order, such that $\sigma \cdot \sigma^{-1} = \sigma^{-1} \cdot \sigma = \sigma_{\text{Id}}$.

Every permutation can be expressed as a composition of transpositions. Although the specific number of transpositions used to represent a given permutation may vary, the parity of this number, that is, whether it is even or odd, remains invariant for each permutation. Accordingly, we distinguish between even and odd permutations: a permutation is said to be even if it can be written as a composition of an even number of transpositions, and odd if it requires an odd number of transpositions. Consequently, we define

$$\text{sgn}(\sigma) = \begin{cases} +1, & \sigma \text{ even} \\ -1, & \sigma \text{ odd} \end{cases}. \tag{2.9}$$

If σ is even (odd) then σ^{-1} is also even (odd), since the numbers of transpositions composing them are equal. Each permutation σ can be related to an operator P_σ in the space H_N and whose action is defined as

$$P_\sigma \Psi(x_1, \ldots, x_N) = \Psi(x_{\sigma(1)}, \ldots, x_{\sigma(N)}). \tag{2.10}$$

If the wave function does not describe a system of identical particles, in general the new wave function does not have a definite relation with respect to the old wave function. The set of all permutation operators forms a group under composition, called the permutation operator group, with identity element

$$P_{\mathrm{Id}} = \mathbb{1}, \tag{2.11}$$

and the inverse element

$$P_\sigma^{-1} = P_{\sigma^{-1}}. \tag{2.12}$$

The map

$$\sigma \longrightarrow P_\sigma \tag{2.13}$$

defines a group homomorphism (see Definition 14.5), i.e.,

$$\begin{aligned} P_{\sigma_1} P_{\sigma_2} \Psi(x_1, \ldots, x_N) &= P_{\sigma_1} \Psi(x_{\sigma_2(1)}, \ldots, x_{\sigma_2(N)}) = \\ &= \Psi\left(x_{\sigma_1(\sigma_2(1))}, \ldots, x_{\sigma_1(\sigma_2(N))}\right) \\ &= \Psi\left(x_{(\sigma_1\sigma_2)(1)}, \ldots, x_{(\sigma_1\sigma_2)(N)}\right), \end{aligned} \tag{2.14}$$

that is,

$$P_{\sigma_1} P_{\sigma_2} = P_{\sigma_1\sigma_2}. \tag{2.15}$$

The operator of a transposition of the individual degrees of freedom x_i, x_j will be denoted by P_{ij}.

Theorem 2.1 *Permutation operators satisfy the properties:*

1. $P_\sigma^\dagger = P_\sigma^{-1}$, *that is,* P_σ *is unitary (see Eqs. I.2 and I.3). In particular, all its eigenvalues have unit modulus.*
2. *Transposition operators are self-adjoint and have eigenvalues* $+1$ *and* -1.

Proof

1. Given

$$(\psi, P_\sigma \phi) = \int dx_1 \ldots \int dx_N \psi^*(x_1, \ldots, x_N) \phi(x_{\sigma(1)}, \ldots, x_{\sigma(N)}), \tag{2.16}$$

the map between two sequences induced by a permutation σ, i.e.,

$$y_i = x_{\sigma(i)} \quad \Leftrightarrow \quad x_i = y_{\sigma^{-1}(i)} \tag{2.17}$$

leaves the measurement element unchanged, that is,

$$dy_1 dy_2 \dots dy_N = dx_{\sigma(1)} dx_{\sigma(2)} \dots dx_{\sigma(N)}, \ \forall \, \sigma, \tag{2.18}$$

then

$$\begin{aligned}(\psi, P_\sigma \phi) &= \int dy_1 \dots \int dy_N \psi^*(y_{\sigma^{-1}(1)}, \dots, y_{\sigma^{-1}(N)}) \phi(y_1, \dots, y_N) = \\ &= \int dy_1 \dots \int dy_N \, P_{\sigma^{-1}} \psi^*(y_1, \dots, y_N) \phi(y_1, \dots, y_N) = \\ &= \left(P_{\sigma^{-1}} \psi, \phi\right) .\end{aligned} \tag{2.19}$$

2. Since P_{ij} is a permutation operator, it is unitary, i.e.

$$P_{ij} P_{ij}^\dagger = P_{ij}^\dagger P_{ij} = \mathbb{1} \tag{2.20}$$

Moreover, performing the same transposition twice gives back the identity, hence

$$P_{ij}^2 = \mathbb{1}, \tag{2.21}$$

$$P_{ij}^{-1} = P_{ij}. \tag{2.22}$$

Combining these two facts, we obtain

$$\begin{aligned}P_{ij}^\dagger &= P_{ij}^{-1} = \\ &= P_{ij},\end{aligned} \tag{2.23}$$

so P_{ij} is self-adjoint. Now let λ be an eigenvalue of P_{ij} with eigenvector $v \neq 0$, so $P_{ij} v = \lambda v$. Applying P_{ij} again and using Eq. 2.21 yields

$$\begin{aligned}v &= P_{ij}^2 v = \\ &= P_{ij}(\lambda v) = \\ &= \lambda^2 v.\end{aligned} \tag{2.24}$$

Hence $\lambda^2 = 1$, and therefore

$$\lambda \in \{-1, +1\}. \tag{2.25}$$

Thus every eigenvalue of P_{ij} is either $+1$ or -1.

□

According to the principle of indistinguishability, an operator can describe an observable of a system of identical particles only if it is invariant under the action of particle exchange, that is, such that

$$[A, P_\sigma] = 0, \ \forall \sigma. \tag{2.26}$$

In addition, the states $\psi(x_1, \ldots, x_N)$ describing the system of identical particles must have square moduli invariant under permutation

$$|\psi|^2 = |P_\sigma \psi|^2 . \tag{2.27}$$

In particular, under the action of a transposition P_{ij}, there may be at most a difference in a phase factor, i.e.,

$$\psi(x_1, \ldots, x_j, \ldots, x_i, \ldots, x_N) = e^{i\theta} \psi(x_1, \ldots, x_i, \ldots, x_j, \ldots, x_N). \tag{2.28}$$

Let us apply P_{ij} to both sides, so that

$$P_{ij}\psi(x_1, \ldots, x_j, \ldots, x_i, \ldots, x_N) = P_{ij}e^{i\theta} \psi(x_1, \ldots, x_i, \ldots, x_j, \ldots, x_N), \tag{2.29}$$

$$P_{ij}\psi(x_1, \ldots, x_j, \ldots, x_i, \ldots, x_N) = e^{i\theta} P_{ij}\psi(x_1, \ldots, x_i, \ldots, x_j, \ldots, x_N), \tag{2.30}$$

$$\psi(x_1, \ldots, x_i, \ldots, x_j, \ldots, x_N) = e^{i\theta} \psi(x_1, \ldots, x_j, \ldots, x_i, \ldots, x_N), \tag{2.31}$$

$$\psi(x_1, \ldots, x_i, \ldots, x_j, \ldots, x_N) = e^{2i\theta} \psi(x_1, \ldots, x_i, \ldots, x_j, \ldots, x_N), \tag{2.32}$$

then

$$\theta = 0, \tag{2.33}$$

$$\theta = \pi. \tag{2.34}$$

The wave functions describing systems of identical particles within the Hilbert space H_N exhibit a definite symmetry characteristic under the operation of transposition operator: they are either mapped onto themselves or onto their negative. Consequently, this leads to the fundamental distinction between:

- totally symmetric functions are such that $\theta = 0$ for any σ, they are therefore invariant for any permutation, i.e.

$$P_\sigma \psi(x_1, \ldots, x_N) = \psi(x_1, \ldots, x_N), \ \forall \sigma; \tag{2.35}$$

- totally antisymmetric functions are such that $\theta = 0$ with even σ and $\theta = \pi$ with odd σ, i.e.

$$P_\sigma \psi(x_1, \ldots, x_N) = \begin{cases} +\psi(x_1, \ldots, x_N), & \sigma \text{ even} \\ -\psi(x_1, \ldots, x_N), & \sigma \text{ odd} \end{cases} . \tag{2.36}$$

Totally symmetric and totally antisymmetric functions can be regarded as eigenfunctions of the particle-exchange (transposition) operators, corresponding to eigenvalues $+1$ and -1, respectively. As such, they span orthogonal subspaces of the total Hilbert space. This naturally leads to the definition of two subspaces of H_N: the antisymmetric subspace H_N^A and the symmetric subspace H_N^S, together with their associated projection operators, the antisymmetrization operator and the symmetrization operator.

Definition 2.1 (Antisymmetrization and Symmetrization Operators) Let H_N denote the Hilbert space of N-particle wavefunctions, and let P_σ be the permutation operator corresponding to a permutation σ of the N particles.

- The antisymmetrization operator $P^{(A,N)}$ is defined as

$$P^{(A,N)} = \frac{1}{N!} \sum_\sigma \operatorname{sgn}(\sigma) P_\sigma, \tag{2.37}$$

  where $\operatorname{sgn}(\sigma)$ denotes the sign of the permutation. $P^{(A,N)}$ projects any wavefunction onto the totally antisymmetric subspace $H_N^A \subset H_N$.
- The symmetrization operator $P^{(S,N)}$ is defined as

$$P^{(S,N)} = \frac{1}{N!} \sum_\sigma P_\sigma, \tag{2.38}$$

  which projects any wavefunction onto the totally symmetric subspace $H_N^S \subset H_N$.

The factor $(N!)^{-1}$ in Eqs. 2.37 and 2.38 will later be seen to contribute to the normalization of the projected states. However, it can already be noticed that its presence ensures the idempotency of the two operators. Indeed, we have

Theorem 2.2 *The antisymmetrization and symmetrization operators, i.e. 2.37 and 2.38 respectively, are projectors and in particular they project generic functions of Hilbert space H_N into two orthogonal eigenspaces, i.e., H_N^A and H_N^S, with eigenvalues -1 and $+1$, respectively.*

Proof First, we must verify that they are self-adjoint and idempotent. Consider the antisymmetrization operator and apply to it the adjoint operation, i.e.,

$$\begin{aligned}
\left[P^{(A,N)}\right]^{\dagger} &= \frac{1}{N!}\sum_{\sigma} \operatorname{sgn}(\sigma) P_{\sigma}^{\dagger} = \\
&= \frac{1}{N!}\sum_{\sigma} \operatorname{sgn}(\sigma) P_{\sigma^{-1}} = \\
&= \frac{1}{N!}\sum_{\sigma} \operatorname{sgn}(\sigma^{-1}) P_{\sigma^{-1}} = \\
&= \frac{1}{N!}\sum_{\sigma^{-1}} \operatorname{sgn}(\sigma^{-1}) P_{\sigma^{-1}} = \\
&= P^{(A,N)},
\end{aligned} \tag{2.39}$$

where we made use of $\operatorname{sgn}(\sigma^{-1}) = \operatorname{sgn}(\sigma)$ and the fact that we are summing over all possible permutations, so the summation index is a dummy index. It remains to verify the idempotency of $P^{(A,N)}$: let us apply to it a permutation operator

$$\begin{aligned}
P_{\sigma'} P^{(A,N)} &= \frac{1}{N!}\sum_{\sigma} \operatorname{sgn}(\sigma) P_{\sigma'} P_{\sigma} = \\
&= \frac{1}{N!}\sum_{\sigma} \operatorname{sgn}^2(\sigma') \operatorname{sgn}(\sigma) P_{\sigma'} P_{\sigma} = \\
&= \frac{1}{N!} \operatorname{sgn}(\sigma') \sum_{\sigma} \operatorname{sgn}(\sigma') \operatorname{sgn}(\sigma) P_{\sigma'} P_{\sigma} = \\
&= \frac{1}{N!} \operatorname{sgn}(\sigma') \sum_{\sigma} \operatorname{sgn}(\sigma'\sigma) P_{\sigma'\sigma} = \\
&= \operatorname{sgn}(\sigma') P^{(A,N)},
\end{aligned} \tag{2.40}$$

where we used the identity $1 = \operatorname{sgn}^2(\sigma')$ and exploited the dummy summation index, with σ' fixed and σ running over all permutations. It follows that

$$\begin{aligned}
P^{(A,N)} P^{(A,N)} &= \frac{1}{N!}\sum_{\sigma} \operatorname{sgn}(\sigma) P_{\sigma} P^{(A,N)} = \\
&= \frac{1}{N!}\sum_{\sigma} \operatorname{sgn}(\sigma) \operatorname{sgn}(\sigma) P^{(A,N)} = \\
&= \frac{1}{N!}\sum_{\sigma} \operatorname{sgn}^2(\sigma) P^{(A,N)} = \\
&= \frac{1}{N!}\left(\sum_{\sigma} 1\right) P^{(A,N)} =
\end{aligned}$$

$$= \frac{1}{N!} N! P^{(A,N)} =$$
$$= P^{(A,N)}, \tag{2.41}$$

so it is proved that $P^{(A,N)}$ is a projection operator.

It is similarly shown that $P^{(S,N)}$ is a projection operator, with the difference that it satisfies the property

$$P_{\sigma'} P^{(S,N)} = P^{(S,N)}, \ \forall\, \sigma'. \tag{2.42}$$

Let us verify that $P^{(S,N)}$ and $P^{(A,N)}$ project elements of H_N into symmetric and antisymmetric spaces, respectively. From

$$P^{(A,N)} \psi = \phi, \ \psi \in H_N, \tag{2.43}$$

it follows

$$P_\sigma \phi = \mathrm{sgn}(\sigma) \phi =$$
$$= \begin{cases} +\phi & \sigma \text{ even}, \\ -\phi & \sigma \text{ odd}, \end{cases} \tag{2.44}$$

$$P^{(A,N)} \psi \in H_N^A, \ \forall\, \psi \in H_N, \tag{2.45}$$

and from

$$P^{(S,N)} \psi' = \phi', \ \psi' \in H_N, \tag{2.46}$$

it follows

$$P_\sigma \phi' = \phi', \ \forall\, \sigma, \tag{2.47}$$

$$P^{(S,N)} \psi' \in H_N^S, \ \forall\, \psi' \in H_N. \tag{2.48}$$

The operators $P^{(S,N)}$ and $P^{(A,N)}$ project functions from H_N onto the physically relevant subspaces. We now show that the two subspaces H_N^A and H_N^S have an empty intersection. By applying a transposition operator, an odd permutation, we observe that

$$P_{ij} P^{(A,N)} \psi = -P^{(A,N)} \psi, \tag{2.49}$$

$$P_{ij} P^{(S,N)} \psi' = P^{(S,N)} \psi', \tag{2.50}$$

that is, the antisymmetric and symmetric wave functions are eigenvectors of the transposition operator, which is self-adjoint: the eigenvalues -1 and $+1$ correspond to orthogonal eigenspaces. □

From phenomenological evidence, the following classification is well established:

- particles whose quantum states lie in the antisymmetric subspace H_N^A exhibit half-integer spin and are identified as fermions;
- particles whose quantum states lie in the symmetric subspace H_N^S exhibit integer spin and are identified as bosons.

Remark 2.1 (Decomposition of the Two-Particle Space) A particularly simple case is that of two identical particles. In this situation any state can be decomposed into the sum of a completely symmetric and a completely antisymmetric part, so that the Hilbert space is given by

$$H_2 = H_2^S \oplus H_2^A. \tag{2.51}$$

For $N > 2$ the structure is richer: in addition to the symmetric and antisymmetric subspaces, one must also consider a third subspace consisting of functions with no definite symmetry under the permutation group. However, this additional component has no physical relevance, since identical particles are described exclusively by states that are either symmetric (bosons) or antisymmetric (fermions).

To conclude this section, we introduce a quantity that will be useful throughout the text.

Definition 2.2 (Statistical Index) Let a system consist of identical particles. The statistical index is defined by

$$\varepsilon = \begin{cases} +1, & \text{bosons} \\ -1, & \text{fermions} \end{cases}, \tag{2.52}$$

which accounts for the bosonic or fermionic nature of the associated operators.

2.3 Fermionic and Bosonic Bases for N Particles

In quantum many-body theory, the construction of appropriate bases for systems of identical particles relies on the symmetry properties of their wave functions. We now introduce some preliminary definitions that will be useful for the construction of bases in systems of identical particles. In particular, we recall the notions of determinant and permanent, which play a central role in the characterization of fermionic and bosonic states, respectively.

Definition 2.3 (Determinant) Given a $N \times N$ matrix A, the determinant is defined as

$$\det A = \sum_{\sigma} \text{sgn}(\sigma)\, a_{1,\sigma(1)} \cdots a_{N,\sigma(N)}, \tag{2.53}$$

where the sum runs over all permutations σ of $\{1, \ldots, N\}$. For each permutation we take one element from each row, from distinct columns, and multiply them together; the determinant is obtained by summing over all such products, weighted by the sign of the permutation. Equivalently, one can express the determinant by columns:

$$\det A = \sum_{\tau} \text{sgn}(\tau)\, a_{\tau(1),1} \cdots a_{\tau(N),N}. \tag{2.54}$$

This construction is directly related to the action of the antisymmetrization operator.

Definition 2.4 (Permanent) Given a $N \times N$ matrix A, the permanent is defined similarly to the determinant, but without the signs of the permutations:

$$\text{perm } A = \sum_{\sigma} a_{1,\sigma(1)} \cdots a_{N,\sigma(N)}. \tag{2.55}$$

This definition is the natural analogue of the symmetrization operator.

Starting from H_N with basis $\{\varphi_{\alpha_1}(x_1) \ldots \varphi_{\alpha_N}(x_N)\}_{\alpha_1,\ldots,\alpha_N}$, we determine a basis for the antisymmetric subspace H_N^A and a basis for the symmetric subspace H_N^S.

2.3.1 *Fermionic Basis for N Particles*

Let us consider the action of the antisymmetrization operator (see Eq. 2.37) to a basis of H_N, namely

$$\begin{aligned}
\varphi^{(A,N)}_{\alpha_1,\ldots,\alpha_N}(x_1, \ldots, x_N) &\equiv P^{(A,N)} \left\{\varphi_{\alpha_1}(x_1) \ldots \varphi_{\alpha_N}(x_N)\right\} = \\
&= \frac{1}{N!} \sum_{\sigma} \text{sgn}(\sigma) P_{\sigma} \left\{\varphi_{\alpha_1}(x_1) \ldots \varphi_{\alpha_N}(x_N)\right\} = \\
&= \frac{1}{N!} \sum_{\sigma} \text{sgn}(\sigma) \varphi_{\alpha_1}(x_{\sigma(1)}) \ldots \varphi_{\alpha_N}(x_{\sigma(N)}),
\end{aligned} \tag{2.56}$$

where the permutation acts on the index of the degree of freedom. If we consider the states $\{\alpha_i\}$ as row indices and the degrees of freedom $\{x_j\}$ as column indices, the function is obtained by multiplying $(N!)^{-1}$ by the determinant (see Eq. 2.53) of the matrix whose element is given by $\{\varphi_{\alpha_i}(x_j)\}$. However, if two indices in the determinant are exchanged, for instance assigning α_1 to the second particle and

α_2 to the first one, the determinant changes at most by an overall sign and therefore represents the same physical state (i.e. the same ray in Hilbert space). Consequently, configurations that differ only by a permutation of the one-particle labels should not be counted more than once when constructing a basis. We emphasize that the Slater determinant vanishes identically whenever two of the one-particle states coincide, due to the antisymmetry of the wavefunction. Hence, all one-particle states entering a nontrivial determinant must be distinct; in particular one must exclude choices that produce the null vector. Moreover, when the one-particle orbitals are chosen orthonormal, Slater determinants built from different sets of occupied orbitals are mutually orthogonal. If the orbitals are not orthonormal, the overlap between two determinants is given by the determinant of the corresponding overlap matrix. To avoid overcounting equivalent configurations and to ensure a convenient and unique labelling of nonzero many-body states, one customarily imposes an ordering condition on the one-particle quantum numbers, for example

$$\alpha_1 < \alpha_2 < \ldots < \alpha_N. \tag{2.57}$$

The object

$$\varphi^{(A,N)}_{\alpha_1,\ldots,\alpha_N}(x_1,\ldots,x_N) = \frac{1}{N!}\det\begin{pmatrix} \varphi_{\alpha_1}(x_1) & \ldots & \varphi_{\alpha_1}(x_N) \\ \vdots & \ddots & \vdots \\ \varphi_{\alpha_N}(x_1) & \ldots & \varphi_{\alpha_N}(x_N) \end{pmatrix} \tag{2.58}$$

turns out to be an element of H_N^A. Let us check whether $\varphi^{(A,N)}_{\alpha_1,\ldots,\alpha_N}$ can be a basis in H_N^A or whether we need to change its definition. Its completeness derives from the completeness of $\varphi_{\alpha_i}(x_1)$ in H_1: it is complete in H_N the product of bases and the projection of the product in H_N^A is also complete. Regarding orthonormality, we have

$$\begin{aligned}
&(\varphi^{(A,N)}_{(\alpha'_1,\ldots,\alpha'_N)}, \varphi^{(A,N)}_{(\alpha_1,\ldots,\alpha_N)}) = \\
&= \left(P^{(A,N)}\varphi_{\alpha'_1}(x_1)\ldots\varphi_{\alpha'_N}(x_N), P^{(A,N)}\varphi_{\alpha_1}(x_1)\ldots\varphi_{\alpha_N}(x_N)\right) = \\
&= \left(\varphi_{\alpha'_1}(x_1)\ldots\varphi_{\alpha'_N}(x_N), \left[P^{(A,N)}\right]^{\dagger} P^{(A,N)}\varphi_{\alpha_1}(x_1)\ldots\varphi_{\alpha_N}(x_N)\right) = \\
&= \left(\varphi_{\alpha'_1}(x_1)\ldots\varphi_{\alpha'_N}(x_N), P^{(A,N)}P^{(A,N)}\varphi_{\alpha_1}(x_1)\ldots\varphi_{\alpha_N}(x_N)\right) = \\
&= \left(\varphi_{\alpha'_1}(x_1)\ldots\varphi_{\alpha'_N}(x_N), P^{(A,N)}\varphi_{\alpha_1}(x_1)\ldots\varphi_{\alpha_N}(x_N)\right),
\end{aligned} \tag{2.59}$$

where we have employed Eqs. 2.39 and 2.41. We write

$$\frac{1}{N!}\sum_{\sigma}\mathrm{sgn}(\sigma)\int dx_1\ldots\int dx_N\varphi^*_{\alpha'_1}(x_1)\ldots\varphi^*_{\alpha'_N}(x_N)\varphi_{\alpha_{\sigma(1)}}(x_1)\ldots\varphi_{\alpha_{\sigma(N)}}(x_N), \tag{2.60}$$

where the determinant is written explicitly such that the permutation acts on the row indices (see Eq. 2.54), with the row and column indices identified with the subscripts of φ and x, respectively. By setting $\alpha_{\sigma(i)} = \alpha_i$, we observe that the integral is the product of N integrals of the form $\int dx \varphi^*_{\alpha'_i}(x)\varphi_{\alpha_j}(x) \equiv (\alpha'_i, \alpha_j)$, then the scalar product has been linked to

$$\left(\varphi^{(A,N)}_{(\alpha'_1,\ldots,\alpha'_N)}, \varphi^{(A,N)}_{(\alpha_1,\ldots,\alpha_N)}\right) = \frac{1}{N!}\det\begin{pmatrix} (\alpha'_1,\alpha_1) & \ldots & (\alpha'_1,\alpha_N) \\ \vdots & \ddots & \vdots \\ (\alpha'_N,\alpha_1) & \ldots & (\alpha'_N,\alpha_N) \end{pmatrix}. \tag{2.61}$$

We now show that if the two sets of indices differ by at least one particle state, the determinant vanishes. Suppose, for instance, that only $\alpha_1 \neq \alpha'_1$. In this case, the first row and the first column of the matrix 2.61 are entirely zero, which makes the determinant identically zero. Similarly, whenever any other pair of indices does not coincide, the corresponding rows and columns are both zero, so that the determinant vanishes. It follows that a nonzero determinant arises only when all one-particle states coincide pairwise, i.e., $\alpha'_i = \alpha_i$ for all i. Consequently, the Slater determinants constructed from distinct sets of single-particle states are mutually orthogonal. Now, let us examine the normalization and reconsider the matrix introduced in Eq. 2.61, whose generic elements (α'_i, α_j) represent the scalar products in the one-particle Hilbert space H_1. When the sets of indices coincide pairwise, the matrix reduces to the identity, its determinant equals 1, and the corresponding scalar product evaluates to $(N!)^{-1}$. Therefore, to ensure that the N-particle antisymmetric state $\varphi^{(A,N)}_{\alpha_1,\ldots,\alpha_N}(x_1,\ldots,x_N)$ defined in Eq. 2.58 is properly normalized, it is necessary to include a factor of $\sqrt{N!}$, i.e.,

$$\varphi^{(A,N)}_{\alpha_1,\ldots,\alpha_N}(x_1,\ldots,x_N) = \sqrt{N!}P^{(A,N)}\left\{\varphi_{\alpha_1}(x_1)\ldots\varphi_{\alpha_N}(x_N)\right\}. \tag{2.62}$$

From

$$\begin{aligned}\varphi^{(A,N)}_{\alpha_1,\ldots,\alpha_N}(x_1,\ldots,x_N) &= \sqrt{N!}\frac{1}{N!}\sum_\sigma \operatorname{sgn}(\sigma)P_\sigma\left\{\varphi_{\alpha_1}(x_1)\ldots\varphi_{\alpha_N}(x_N)\right\} = \\ &= \frac{1}{\sqrt{N!}}\sum_\sigma \operatorname{sgn}(\sigma)\varphi_{\alpha_1}(x_{\sigma(1)})\ldots\varphi_{\alpha_N}(x_{\sigma(N)}),\end{aligned} \tag{2.63}$$

the scalar product becomes

$$\begin{aligned}&\left(\varphi^{(A,N)}_{(\alpha'_1,\ldots,\alpha'_N)}, \varphi^{(A,N)}_{(\alpha_1,\ldots,\alpha_N)}\right) = \\ &= \left(\sqrt{N!}P^{(A,N)}\varphi_{\alpha'_1}(x_1)\ldots\varphi_{\alpha'_N}(x_N), \sqrt{N!}P^{(A,N)}\varphi_{\alpha_1}(x_1)\ldots\varphi_{\alpha_N}(x_N)\right) =\end{aligned}$$

$$= N! \left(\varphi_{\alpha'_1}(x_1) \dots \varphi_{\alpha'_N}(x_N),\, P^{(A,N)} \varphi_{\alpha_1}(x_1) \dots \varphi_{\alpha_N}(x_N) \right) =$$

$$= \det \begin{pmatrix} (\alpha'_1, \alpha_1) & \dots & (\alpha'_1, \alpha_N) \\ \vdots & \ddots & \vdots \\ (\alpha'_N, \alpha_1) & \dots & (\alpha'_N, \alpha_N) \end{pmatrix}. \tag{2.64}$$

Finally, we obtain

Definition 2.5 (Fermionic Basis) A fermionic basis of the antisymmetric subspace H_N^A is given by

$$\varphi^{(A,N)}_{\alpha_1,\dots,\alpha_N}(x_1, \dots, x_N) = \frac{1}{\sqrt{N!}} \det \begin{pmatrix} \varphi_{\alpha_1}(x_1) & \dots & \varphi_{\alpha_1}(x_N) \\ \vdots & \ddots & \vdots \\ \varphi_{\alpha_N}(x_1) & \dots & \varphi_{\alpha_N}(x_N) \end{pmatrix}, \tag{2.65}$$

with $\alpha_1 < \alpha_2 < \dots < \alpha_N$ and occupation numbers $N_{\alpha_k} \in \{0, 1\}$ satisfying $\sum_k N_{\alpha_k} = N$.

2.3.2 *Bosonic Basis for N Particles*

Let us consider the action of the symmetrization operator (see Eq. 2.38) on a basis of H_N, namely

$$\begin{aligned} \varphi^{(S,N)}_{\alpha_1,\dots,\alpha_N}(x_1, \dots, x_N) &\equiv P^{(S,N)} \left\{ \varphi_{\alpha_1}(x_1) \dots \varphi_{\alpha_N}(x_N) \right\} = \\ &= \frac{1}{N!} \sum_\sigma P_\sigma \left\{ \varphi_{\alpha_1}(x_1) \dots \varphi_{\alpha_N}(x_N) \right\} = \\ &= \frac{1}{N!} \sum_\sigma \varphi_{\alpha_1}(x_{\sigma(1)}) \dots \varphi_{\alpha_N}(x_{\sigma(N)}). \end{aligned} \tag{2.66}$$

We note that, in the case of bosons, the symmetrized wavefunction remains non-zero even when two or more indices are equal, i.e., the permanent does not vanish. The indices are arranged according to the convention

$$\alpha_1 \le \alpha_2 \le \dots \le \alpha_N, \tag{2.67}$$

which reproduces the experimental evidence that lower-energy states are occupied first. According to this convention, we define

$$\varphi^{(S,N)}_{\alpha_1,\ldots,\alpha_N}(x_1,\ldots,x_N) = \frac{1}{N!}\,\mathrm{perm}\begin{pmatrix} \varphi_{\alpha_1}(x_1) & \ldots & \varphi_{\alpha_1}(x_N) \\ \vdots & \ddots & \vdots \\ \varphi_{\alpha_N}(x_1) & \ldots & \varphi_{\alpha_N}(x_N) \end{pmatrix}, \tag{2.68}$$

which implies that the function lies in the symmetric subspace H_N^S. Let us now verify whether Eq. 2.68 can serve as a basis for H_N^S, or if its definition requires modification. Regarding orthogonality, the proof proceeds analogously to that presented for the fermionic basis, since the permanent also vanishes whenever a row and a column contain only zeros. Let us now turn to the normalization. Since the indices α_i can be repeated, consider the case where a particular index α_k appears N_{α_k} times, with $\sum_k N_{\alpha_k} = N$. In this case, instead of the matrix in Eq. 2.61, we obtain a block-diagonal matrix composed of blocks of size $N_{\alpha_k} \times N_{\alpha_k}$, each filled entirely with ones. Because the permanent of a block-diagonal matrix factorizes as the product of the permanents of its blocks, and since the permanent of each such block equals $N_{\alpha_k}!$, the total permanent is given by $N_{\alpha_1}!\ldots N_{\alpha_M}!$, where M denotes the number of distinct one-particle states. Therefore, each function $\varphi^{(S,N)}_{\alpha_1,\ldots,\alpha_N}(x_1,\ldots,x_N)$ must be multiplied by

$$\frac{1}{\sqrt{N_{\alpha_1}!}\cdots\sqrt{N_{\alpha_M}!}} \tag{2.69}$$

in order to compensate for the overcounting arising from repeated indices in the permanent. In addition, a factor of $\sqrt{N!}$ is needed to cancel the prefactor $(N!)^{-1}$ appearing in Eq. 2.68. Putting everything together, we obtain

Definition 2.6 (Bosonic Basis) A bosonic basis of the symmetric subspace H_N^S is given by

$$\varphi^{(S,N)}_{\alpha_1,\ldots,\alpha_N}(x_1,\ldots,x_N) = \frac{1}{\sqrt{N!}\sqrt{N_{\alpha_1}!}\ldots\sqrt{N_{\alpha_M}!}}\,\mathrm{perm}\begin{pmatrix} \varphi_{\alpha_1}(x_1) & \ldots & \varphi_{\alpha_1}(x_N) \\ \vdots & \ddots & \vdots \\ \varphi_{\alpha_N}(x_1) & \ldots & \varphi_{\alpha_N}(x_N) \end{pmatrix}, \tag{2.70}$$

where the states satisfy $\alpha_1 \leq \alpha_2 \leq \ldots \leq \alpha_M$, M denotes the number of distinct one-particle states, and occupation numbers N_{α_k} satisfy $\sum_{k=1}^{M} N_{\alpha_k} = N$.

2.4 Fock Space

After having considered the Hilbert space H_N for a fixed number of particles N, we now extend the description to systems where the number of identical particles is not fixed.

Definition 2.7 (Fock Space) The Fock space $\mathcal{F}$ is defined as the direct sum of the Hilbert spaces with a fixed number of particles,

$$\mathcal{F} = \bigoplus_{N=0}^{\infty} H_N, \tag{2.71}$$

where H_0 is the one-dimensional space associated with the vacuum state (no particles), H_1 corresponds to the one-particle Hilbert space introduced in Eq. 1.1, and for each $N \geq 2$, the spaces H_N are constructed as discussed in the previous sections.

The vacuum vector $|0\rangle$ is represented as

$$|0\rangle \equiv \psi^{(0)} = \begin{pmatrix} 1 \\ 0 \\ 0 \\ \vdots \end{pmatrix}, \tag{2.72}$$

and every Hilbert space H_N with $N \geq 1$ is ∞-dimensional. A vector $|\psi\rangle$ in Fock space is characterized by having components (or projections) in each Hilbert subspace corresponding to a fixed number of particles, that is,

$$|\psi\rangle = \begin{pmatrix} \psi^{(0)} \\ \psi^{(1)}(x_1) \\ \psi^{(2)}(x_1, x_2) \\ \vdots \end{pmatrix} = \begin{pmatrix} \psi^{(0)} \\ 0 \\ 0 \\ \vdots \end{pmatrix} + \begin{pmatrix} 0 \\ \psi^{(1)}(x_1) \\ 0 \\ \vdots \end{pmatrix} + \begin{pmatrix} 0 \\ 0 \\ \psi^{(2)}(x_1, x_2) \\ \vdots \end{pmatrix} + \dots . \tag{2.73}$$

Fock space becomes a Hilbert space when endowed with the scalar product

$$(\psi, \phi)_{\mathcal{F}} = \sum_{N=0}^{\infty} \left(\psi^{(N)}, \phi^{(N)}\right)_{H_N}, \tag{2.74}$$

where the general term of the sequence is the scalar product of the N-th components of the vector. In particular, if $\psi = \phi$ and the vector is normalized, the quantity

$$(\psi^{(N)}, \psi^{(N)})_{H_N} \tag{2.75}$$

can be interpreted as the probability that the system contains N particles. Obviously, the vector $\psi^{(N)}$ describing the state of a system of N particles reduces in Fock space to the vector with elements all null except the N-th element

$$\psi^{(N)} = \begin{pmatrix} 0 \\ \vdots \\ 0 \\ \psi^{(N)}(x_1, \ldots, x_N) \\ 0 \\ \vdots \\ \vdots \end{pmatrix}. \tag{2.76}$$

Equation 2.76 represents a state with a fixed particle number, i.e., a state in which the number of particles does not fluctuate but is exactly N. In this case, the scalar product in Fock space reduces to a single term, namely the scalar product in H_N, and due to the normalization of the N-particle state, this product equals one.

As in the case of systems with a fixed particle number, in Fock space we are likewise interested in the following relevant subspaces

$$\mathcal{F}^A = H_0 \bigoplus H_1 \bigoplus_{N=2}^{\infty} H_N^A, \tag{2.77}$$

$$\mathcal{F}^S = H_0 \bigoplus H_1 \bigoplus_{N=2}^{\infty} H_N^S, \tag{2.78}$$

since the vectors in such subspaces are the only ones describing systems of identical particles.

2.5 Many-Body Annihilation and Creation Operators

Having introduced the Fock space, we now have at our disposal a mathematical framework suitable for the description of systems with a variable number of particles. To fully exploit this structure, it is necessary to introduce operators that allow transitions between the different N-particle subspaces of the Fock space. Inspired by the formalism of the harmonic oscillator, where the creation and annihilation operators raise or lower the energy quanta of a single mode (see Sect. 1.3), we now generalize these concepts to the many-body setting. In quantum many-body context, the operators we construct will act not by creating or annihilating quanta of energy, but by creating and annihilating particles themselves.

Definition 2.8 (Annihilation Operator in Fock Space) Let $\varphi \in H_1$ be a one-particle state. The annihilation operator associated with φ is the linear operator $a_\varphi : \mathcal{F} \to \mathcal{F}$, acting on the Fock space $\mathcal{F}$, defined by

$$\left[a_\varphi|\psi\rangle\right]^{(N)}(x_1,\ldots,x_N) \equiv \sqrt{N+1}\int dx\,\varphi^*(x)\,\psi^{(N+1)}(x,x_1,\ldots,x_N), \tag{2.79}$$

where $\psi^{(N+1)}$ denotes the $(N+1)$-particle component of the Fock vector $|\psi\rangle \in \mathcal{F}$.

The explicit form of this operator will be justified in the following discussion. At this stage, it is sufficient to note that a_φ acts by removing one particle from the one-particle state described by the wave function φ.

Consider a Fock vector with a fixed number of particles N, whose only nonvanishing component is $\psi^{(N)}(x_1,\ldots,x_N)$ (see Eq. 2.76). Applying the annihilation operator a_φ defined in Eq. 2.79, the M-particle component of the resulting vector is

$$\begin{aligned}&[a_\varphi|\psi\rangle]^{(M)}(x_1,\ldots,x_M)\\ &\quad= \sqrt{M+1}\int dx\,\varphi^*(x)\,\psi^{(M+1)}(x,x_1,\ldots,x_M).\end{aligned} \tag{2.80}$$

Since $\psi^{(M+1)} = 0$ for every $M+1 \neq N$, with $N \geq 1$, the only nonvanishing component corresponds to $M+1 = N$, i.e., $M = N-1$. In this case, we have

$$\begin{aligned}&[a_\varphi|\psi\rangle]^{(N-1)}(x_1,\ldots,x_{N-1})\\ &\quad= \sqrt{N}\int dx\,\varphi^*(x)\,\psi^{(N)}(x,x_1,\ldots,x_{N-1}).\end{aligned} \tag{2.81}$$

All other components vanish identically:

$$[a_\varphi|\psi\rangle]^{(M)} = 0, \quad M \neq N-1. \tag{2.82}$$

Hence, the resulting Fock vector is a state with exactly $N-1$ particles, with its unique nonvanishing component given by the integral above, confirming that the annihilation operator removes one particle from the original Fock vector. For the special case $N = 0$, the Fock vector corresponds to the vacuum state. In this case, all components of $a_\varphi|\psi\rangle$ vanish identically, so that the resulting state is again the vacuum. This is precisely as expected from the definition of the annihilation operator in the harmonic oscillator, which annihilates the vacuum (see Eq. 1.174).

Definition 2.9 (Creation Operator in Fock Space) Let $\varphi \in H_1$ be a one-particle state. The creation operator associated with φ is the linear operator $a_\varphi^\dagger : \mathcal{F} \to \mathcal{F}$, acting on the Fock space $\mathcal{F}$, defined by

$$\begin{aligned}&a_\varphi^\dagger : \left[a_\varphi^\dagger|\psi\rangle\right]^{(N)}(x_1,\ldots,x_N) =\\ &= \frac{1}{\sqrt{N}}\sum_{j=1}^{N}\varepsilon^{j+1}\varphi(x_j)\psi^{(N-1)}(x_1,\ldots,x_{j-1},x_{j+1},\ldots,x_N),\ N \geq 1,\end{aligned} \tag{2.83}$$

and

$$\left[a_\varphi^\dagger|\psi\rangle\right]^{(0)} = 0. \tag{2.84}$$

Here $\psi^{(N-1)}$ denotes the $(N-1)$-particle component of the Fock vector $|\psi\rangle \in \mathcal{F}$, and ε is the statistical index given in Eq. 2.52.

The operator $a_\varphi^\dagger$ has no projection on H_0, since it must create particles.

Consider a Fock vector with a fixed number of particles N, whose only nonvanishing component is $\psi^{(N)}(x_1, \ldots, x_N)$. Applying the creation operator $a_\varphi^\dagger$ defined in Eq. 2.83, the M-particle component of the resulting vector is

$$\begin{aligned}&[a_\varphi^\dagger|\psi\rangle]^{(M)}(x_1, \ldots, x_M)\\ &= \frac{1}{\sqrt{M}} \sum_{j=1}^{M} \varepsilon^{j+1} \varphi(x_j)\, \psi^{(M-1)}(x_1, \ldots, x_{j-1}, x_{j+1}, \ldots, x_M),\ M \geq 1.\end{aligned} \tag{2.85}$$

Since $\psi^{(M-1)} = 0$ for every $M-1 \neq N$, with $N \geq 0$, the only nonvanishing component corresponds to $M-1 = N$, i.e., $M = N+1$. In this case, we have

$$\begin{aligned}&[a_\varphi^\dagger|\psi\rangle]^{(N+1)}(x_1, \ldots, x_{N+1})\\ &= \frac{1}{\sqrt{N+1}} \sum_{j=1}^{N+1} \varepsilon^{j+1} \varphi(x_j)\, \psi^{(N)}(x_1, \ldots, x_{j-1}, x_{j+1}, \ldots, x_{N+1}).\end{aligned} \tag{2.86}$$

All other components vanish identically:

$$\left[a_\varphi^\dagger|\psi\rangle\right]^{(M)} = 0, \quad M \neq N+1. \tag{2.87}$$

Hence, the resulting Fock vector is a state with exactly $N+1$ particles, with its unique nonvanishing component given by the sum above, confirming that the creation operator adds one particle to the original Fock vector.

Remark 2.2 (Notation for Many-Body Annihilation and Creation Operators) Given a basis $\{\varphi_\alpha\}$ of the one-particle Hilbert space H_1, we denote by $a_\alpha \equiv a_{\varphi_\alpha}$ the annihilation operator associated with the one-particle state φ_α, and similarly by $a_\alpha^\dagger \equiv a_{\varphi_\alpha}^\dagger$ the corresponding creation operator.

Theorem 2.3 *An annihilation operator a_φ can be expressed in terms of the set $\{a_{\varphi_\alpha}\}$ as*

$$a_\varphi = \sum_\alpha c_\alpha^*\, a_{\varphi_\alpha}, \tag{2.88}$$

where $\varphi = \sum_\alpha c_\alpha \varphi_\alpha$ *is a decomposition of* $\varphi \in \mathcal{H}_1$ *in an orthonormal one-particle basis* $\{\varphi_\alpha\}$.

Proof Let $|\psi\rangle \in \mathcal{F}$ be a Fock space vector. Since $\varphi = \sum_\alpha c_\alpha \varphi_\alpha$, we compute the action of a_φ on $|\psi\rangle$ at fixed particle number N:

$$
\begin{aligned}
\left[a_\varphi|\psi\rangle\right]^{(N)}(x_1,\ldots,x_N) &= \sqrt{N+1}\int dx\,\varphi^*(x)\,\psi^{(N+1)}(x,x_1,\ldots,x_N) = \\
&= \sqrt{N+1}\int dx\,\sum_\alpha c_\alpha^*\varphi_\alpha^*(x)\,\psi^{(N+1)}(x,x_1,\ldots,x_N) = \\
&= \sum_\alpha c_\alpha^*\sqrt{N+1}\int dx\,\varphi_\alpha^*(x)\,\psi^{(N+1)}(x,x_1,\ldots,x_N) = \\
&= \sum_\alpha c_\alpha^*\left[a_{\varphi_\alpha}|\psi\rangle\right]^{(N)}(x_1,\ldots,x_N),
\end{aligned}
\tag{2.89}
$$

which implies the thesis. □

We now present an important result concerning the relationship between the creation and annihilation operators introduced above.

Theorem 2.4 *The operator 2.79 is the adjoint of the operator 2.83, and vice versa. That is, for all* $\phi, \psi \in \mathcal{F}$ *and* $\varphi \in H_1$, *the following identity holds*

$$
\langle\phi|a_\varphi|\psi\rangle = \langle\psi|a_\varphi^\dagger|\phi\rangle^*. \tag{2.90}
$$

Proof Let us focus on the second term and compute the scalar product between the two wave functions in Fock space, namely

$$
\langle\psi|a_\varphi^\dagger|\phi\rangle^* = \sum_{N=1}^{\infty}\left(\left[a_\varphi^\dagger|\phi\rangle\right]^{(N)},\psi^{(N)}\right)_{H_N}, \tag{2.91}
$$

with the series starting at 1 instead of 0, since in the vacuum (particle-free) vector the creation operator has no components. Now, consider the generic term in the N-particle Hilbert space H_N, i.e.,

$$
\begin{aligned}
&\left(\left[a_\varphi^\dagger|\phi\rangle\right]^{(N)},\psi^{(N)}\right)_{H_N} = \\
&= \frac{1}{\sqrt{N}}\sum_{j=1}^{N}\varepsilon^{j+1}\int dx_1\ldots\int dx_j \\
&\times\ldots\int dx_N\varphi^*(x_j) \\
&\times\left(\phi^{(N-1)}(x_1,\ldots,x_{j-1},x_{j+1},\ldots,x_N)\right)^*\psi^{(N)}(x_1,\ldots,x_N) =
\end{aligned}
$$

$$
= \frac{1}{\sqrt{N}} \sum_{j=1}^{N} \varepsilon^{j+1} \int dx_1 \dots \int dx_{j-1} \int dx_{j+1} \dots \int dx_N
$$

$$
\left(\phi^{(N-1)}(x_1, \dots, x_{j-1}, x_{j+1}, \dots, x_N)\right)^* \int dx_j \varphi^*(x_j) \psi^{(N)}(x_1, \dots, x_N),
\tag{2.92}
$$

Let us define

$$
I_j = \int dx_j \varphi^*(x_j) \psi^{(N)}(x_1, \dots, x_N). \tag{2.93}
$$

Now we compute the integral I_j in the cases $j = 1$, $j = 2$ and $j = 3$ by rearranging the degrees of freedom of the function $\psi^{(N)}$. We want to bring out the action of the annihilation operator on the state vector. For $j = 1$

$$
I_1 = \int dx_1 \varphi^*(x_1) \psi^{(N)}(x_1, \dots, x_N), \tag{2.94}
$$

for $j = 2$ we have

$$
\begin{aligned}
I_2 &= \int dx_2 \varphi^*(x_2) \psi^{(N)}(x_1, x_2, \dots, x_N) = \\
&= \varepsilon \int dx_2 \varphi^*(x_2) \psi^{(N)}(x_2, x_1, \dots, x_N),
\end{aligned}
\tag{2.95}
$$

for $j = 3$ we have

$$
\begin{aligned}
I_3 &= \int dx_3\, \varphi^*(x_3) \psi^{(N)}(x_1, x_2, x_3, \dots, x_N) = \\
&= \varepsilon \int dx_3\, \varphi^*(x_3) \psi^{(N)}(x_1, x_3, x_2, \dots, x_N) = \\
&= \varepsilon^2 \int dx_3\, \varphi^*(x_3) \psi^{(N)}(x_3, x_1, x_2, \dots, x_N),
\end{aligned}
\tag{2.96}
$$

we want to derive a general pattern from it. Given the set $\{x_1, \dots, x_N\}$, we take the j-th term, x_j, replace it with x_1, so the transposition produces a factor ε. Now, x_1 is in position j, but we want to take x_1 to position 2, since we want to get the sequence $x_j, x_1, \dots, x_{j-1}, x_{j+1}, \dots, x_N$. $j - 2$ exchanges are needed to order the sub-sequence $x_2, \dots, x_{j-1}, x_{j+1}, \dots, x_N$. Together with the initial permutation, we have $j - 1$ exchanges, which originate the factor ε^{j-1}. The integral corresponding to a generic j-th term is therefore

$$
I_j = \varepsilon^{j-1} \int dx_j \varphi^*(x_j) \psi^{(N)}(x_j, x_1, \dots, x_{j-1}, x_{j+1}, \dots, x_N). \tag{2.97}
$$

Recall that in the definition of the creation operator there is a factor ε^{j+1}. Multiplying this by the factor ε^{j-1} arising from the permutation, we obtain $\varepsilon^{j+1}\varepsilon^{j-1} = \varepsilon^{2j} = 1$, which shows that the dependence on ε cancels out in the scalar product, yielding

$$\frac{1}{\sqrt{N}}\sum_{j=1}^{N}\int dx_j\varphi^*(x_j)\int dx_1\ldots\int dx_{j-1}\int dx_{j+1}\ldots\int dx_N \left(\phi^{(N-1)}(x_1,\ldots,x_{j-1},x_{j+1},\ldots,x_N)\right)^* \psi^{(N)}(x_j,x_1,\ldots,x_{j-1},x_{j+1},\ldots,x_N). \tag{2.98}$$

Now the choice of the degree of freedom over which to integrate does not imply any sign changes in the scalar product, since for every permutation in $\psi^{(N)}$ there will be the same one in the product $\varphi^*\phi^{(N-1)}$, so the integral with respect to x_j is on a dummy index. x_j is renamed and replaced with x, the particle index is lost, and the remaining variables are renamed as $x_1,\ldots,x_{N-1}$

$$\begin{aligned}
&\frac{1}{\sqrt{N}}\sum_{j=1}^{N}\int dx\varphi^*(x)\int dx_1\ldots\int dx_{j-1}\int dx_{j+1}\ldots\int dx_N \\
&\left(\phi^{(N-1)}(x_1,\ldots,x_{j-1},x_{j+1},\ldots,x_N)\right)^* \\
&\psi^{(N)}(x,x_1,\ldots,x_{j-1},x_{j+1},\ldots,x_N) = \\
&= \frac{1}{\sqrt{N}}\left(\sum_{j=1}^{N}1\right)\int dx\varphi^*(x)\int dx_1\ldots\int dx_{N-1}\left(\phi^{(N-1)}(x_1,\ldots,x_{N-1})\right)^* \\
&\psi^{(N)}(x,x_1,\ldots,x_{N-1}) = \\
&= \sqrt{N}\int dx\varphi^*(x)\int dx_1\ldots\int dx_{N-1}\left(\phi^{(N-1)}(x_1,\ldots,x_{N-1})\right)^* \\
&\psi^{(N)}(x,x_1,\ldots,x_{N-1}).
\end{aligned} \tag{2.99}$$

Finally, rearranging the terms and first performing the integral on x, it follows

$$\begin{aligned}
&\left(\left[a^\dagger_\varphi|\phi\rangle\right]^{(N)},\psi^{(N)}\right)_{H_N} \\
&= \int dx_1\ldots\int dx_{N-1}\left(\phi^{(N-1)}(x_1,\ldots,x_{N-1})\right)^* \\
&\sqrt{N}\int dx\varphi^*(x)\psi^{(N)}(x,x_1,\ldots,x_{N-1}) =
\end{aligned}$$

$$= \int dx_1 \ldots \int dx_{N-1} \left(\phi^{(N-1)}(x_1, \ldots, x_{N-1})\right)^* \left[a_\varphi|\phi\rangle\right]^{(N-1)} =$$
$$= \left(\phi^{(N-1)}, \left[a_\varphi|\phi\rangle\right]^{(N-1)}\right)_{H_{N-1}}, \tag{2.100}$$

then

$$\langle\psi|a_\varphi^\dagger|\phi\rangle^* = \sum_{N=1}^{\infty} \left(\phi^{(N-1)}, \left[a_\varphi|\phi\rangle\right]^{(N-1)}\right)_{H_{N-1}} =$$
$$= \sum_{N=0}^{\infty} \left(\phi^{(N)}, \left[a_\varphi|\phi\rangle\right]^{(N)}\right)_{H_N}, \tag{2.101}$$

which is the thesis. □

Remark 2.3 (Notation for Different Types of Many-Body Annihilation and Creation Operators) In the following, we will deal with creation and annihilation operators of different types, i.e., bosonic, fermionic, or of a more general nature, depending on the context. For this reason, we shall not always employ the same symbols a_α and $a_\alpha^\dagger$ to denote them. Different letters will be used to indicate their specific character, while the statistical properties associated with each operator will be either evident from the context or explicitly specified when necessary.

2.6 Algebra of Many-Body Annihilation and Creation Operators

In general, we will work with operators defined independently of any specific particle statistics. To this end, we introduce a formal operator that can accommodate either bosonic or fermionic behavior.

Definition 2.10 (Unified Notation for Commutators and Anticommutators) Given two generic operators A and B, both bosonic or both fermionic, we define the parentheses

$$[\cdot,\cdot]^{(\varepsilon)} \;:\; [A, B]^{(\varepsilon)} \equiv AB - \varepsilon BA, \tag{2.102}$$

which takes the form of a commutator or an anticommutator for bosons or fermions, respectively, in accordance with the statistical index defined in Eq. 2.52.

Here we show that the many-body creation and annihilation operators defined in Fock space satisfy the familiar (anti)commutation algebras (see Eqs. D.1–D.6).

Theorem 2.5 (Algebra of Many-Body Annihilation and Creation Operators) *The many-body annihilation and creation operators, 2.79 and 2.83, respectively, satisfy the following algebraic relations for all* $\varphi, \varphi' \in \mathcal{F}$:

$$\left[a_{\varphi}, a_{\varphi'}\right]^{(\varepsilon)} = 0, \tag{2.103}$$

$$\left[a^{\dagger}_{\varphi}, a^{\dagger}_{\varphi'}\right]^{(\varepsilon)} = 0, \tag{2.104}$$

$$\left[a_{\varphi}, a^{\dagger}_{\varphi'}\right]^{(\varepsilon)} = (\varphi, \varphi')_{H_1}, \tag{2.105}$$

where the brackets $[\cdot, \cdot]^{(\varepsilon)}$ *are defined in Eq. 2.102.*

Proof Let us prove Eq. 2.105. Given

$$\left[a_{\varphi}, a^{\dagger}_{\varphi'}\right]^{(\varepsilon)} |\psi^{(N)}\rangle = a_{\varphi} a^{\dagger}_{\varphi'} |\psi^{(N)}\rangle - \varepsilon\, a^{\dagger}_{\varphi'} a_{\varphi} |\psi^{(N)}\rangle, \tag{2.106}$$

we compute separately the action of each operator product on the N-th component of the wave function in Fock space. We have

$$\begin{aligned}
&\varepsilon \left[a^{\dagger}_{\varphi'} a_{\varphi} |\psi\rangle\right]^{(N)} (x_1, \ldots, x_N) = \\
&= \frac{\varepsilon}{\sqrt{N}} \sum_{j=1}^{N} \varepsilon^{j+1} \varphi'(x_j) \left[a_{\varphi} |\psi\rangle\right]^{(N-1)} (x_1, \ldots, x_{j-1}, x_{j+1}, \ldots, x_N) = \\
&= \frac{1}{\sqrt{N}} \sum_{j=1}^{N} \varepsilon^{j+2} \varphi'(x_j) \left[a_{\varphi} |\psi\rangle\right]^{(N-1)} (x_1, \ldots, x_{j-1}, x_{j+1}, \ldots, x_N) = \\
&= \frac{1}{\sqrt{N}} \sum_{j=1}^{N} \varepsilon^{j+2} \varphi'(x_j) \sqrt{N} \int dx \varphi^*(x) \psi^{(N)}(x, x_1, \ldots, x_{j-1}, x_{j+1}, \ldots, x_N) \equiv \\
&\equiv \sum_{j=1}^{N} \varepsilon^{j} \varphi'(x_j) \int dx \varphi^*(x) \psi^{(N)}(x, x_1, \ldots, x_{j-1}, x_{j+1}, \ldots, x_N),
\end{aligned} \tag{2.107}$$

and given that $\varphi'(x_j)$ does not depend on the integration variable, we have

$$\sum_{j=1}^{N} \int dx \varepsilon^{j} \varphi'(x_j) \varphi^*(x) \psi^{(N)}(x, x_1, \ldots, x_{j-1}, x_{j+1}, \ldots, x_N). \tag{2.108}$$

Regarding the second term, we have

$$\left[a_\varphi a^\dagger_{\varphi'}|\psi\rangle\right]^{(N)}(x_1,\ldots,x_N)$$
$$= \sqrt{N+1}\int dx\varphi^*(x)\left[a^\dagger_{\varphi'}|\psi\rangle\right]^{(N+1)}(x,x_1,\ldots,x_N), \tag{2.109}$$

where the order of the operators is now reversed: a particle is created before one is annihilated. We explicitly express the creation operator as a sum over permutations of the sequence $(x, x_1, \ldots, x_N)$. We proceed as follows

$$\left[a_\varphi a^\dagger_{\varphi'}|\psi\rangle\right]^{(N)}(x_1,\ldots,x_N) =$$
$$= \sqrt{N+1}\int dx\varphi^*(x)\left[a^\dagger_{\varphi'}|\psi\rangle\right]^{(N+1)}(x,x_1,\ldots,x_N) =$$
$$= \int dx\varphi^*(x)\Big\{\varepsilon^2\varphi'(x)\psi^{(N)}(x_1,\ldots,x_N)+$$
$$+\varepsilon^3\varphi'(x_1)\psi^{(N)}(x,x_2,\ldots,x_N) + \varepsilon^4\varphi'(x_2)\psi^{(N)}(x,x_1,x_3,\ldots,x_N)+\ldots\Big\}. \tag{2.110}$$

By summing the two terms, we obtain

$$\left[\left(a_\varphi a^\dagger_{\varphi'} - \varepsilon a^\dagger_{\varphi'}a_\varphi\right)|\psi\rangle\right]^{(N)} = \left(\int dx\varphi^*(x)\varphi'(x)\right)\psi^{(N)}(x_1,\ldots,x_N) =$$
$$= (\varphi,\varphi')_{H_1}|\psi^{(N)}\rangle,\ \forall\, N, \tag{2.111}$$

which is the thesis. □

Given $\varphi_\alpha \in H_1$, a_{φ_α}, $a^\dagger_{\varphi_\alpha}$, from 2.105 and from the orthonormality relation 1.50, one finds

$$\left[a_{\varphi_\alpha}, a^\dagger_{\varphi'_\alpha}\right]^{(\varepsilon)} = \delta_{\alpha,\alpha'}, \tag{2.112}$$

from which the fundamental properties of fermions and bosons are derived, as we shall see next.

To conclude this section, we present a result that enables the creation and annihilation operators to be expressed explicitly in terms of the transformation relating the two bases.

Theorem 2.6 *Given two one-particle bases* $\{\varphi_\alpha(x)\}$ *and* $\{\omega_\beta(x)\}$, *related by a linear transformation*

$$\omega_\beta(x) = \sum_\alpha m_{\alpha,\beta}\varphi_\alpha(x), \tag{2.113}$$

the many-body annihilation c_β and creation $c_\beta^\dagger$ operators (see Eqs. 2.79 and 2.83) associated with $\omega_\beta(x)$ are written as

$$c_\beta = \sum_\alpha m^*_{\alpha,\beta} b_\alpha, \tag{2.114}$$

$$c_\beta^\dagger = \sum_\alpha m_{\alpha,\beta} b_\alpha^\dagger, \tag{2.115}$$

being b_α and $b_\alpha^\dagger$ the annihilation and creation operators, respectively, associated with the basis $\varphi_\alpha(x)$.

Proof To prove the statement, we consider the action of the annihilation operator c_{ω_β} on an arbitrary state $|\psi\rangle \in H^{(N+1)}$, that is,

$$\begin{aligned}
&\left[c_{\omega_\beta}|\psi\rangle\right]^{(N)}(x_1,\ldots,x_N)\\
&= \sqrt{N+1}\int dx\omega^*_\beta(x)\psi^{(N+1)}(x,x_1,\ldots,x_N) =\\
&= \sqrt{N+1}\int dx\left(\sum_\alpha m_{\alpha,\beta}\varphi_\alpha(x)\right)^*\psi^{(N+1)}(x,x_1,\ldots,x_N) =\\
&= \sqrt{N+1}\int dx\left(\sum_\alpha m^*_{\alpha,\beta}\varphi^*_\alpha(x)\right)\psi^{(N+1)}(x,x_1,\ldots,x_N) =\\
&= \sum_\alpha m^*_{\alpha,\beta}\sqrt{N+1}\int dx\varphi^*_\alpha(x)\psi^{(N+1)}(x,x_1,\ldots,x_N) =\\
&= \sum_\alpha m^*_{\alpha,\beta}\left[b_{\varphi_\alpha}|\psi\rangle\right]^{(N)}(x_1,\ldots,x_N),\ \forall\,|\psi\rangle,
\end{aligned} \tag{2.116}$$

and the expression for the creation operator follows analogously. □

2.7 Fermionic and Bosonic Bases in Fock Space

Starting from a one-particle basis, we can build a basis for the Fock space. Creation and annihilation operators play a central role in this construction, as they provide the fundamental mechanism for generating fermionic and bosonic states. In particular, from Eq. 2.83, the action of the creation operator on the vacuum state is given by

$$\left[a_\alpha^\dagger|0\rangle\right]^{(1)}(x) = \varphi_\alpha(x), \tag{2.117}$$

meaning that the creation operator, when applied to the vacuum, creates a particle in the state $\varphi_\alpha(x) \in H_1$.

2.7.1 Fermionic Basis in Fock Space

We apply to the vacuum state the fermionic creation operators (see Eq. 2.83) associated with the subscripts α_1, α_2 , with $\alpha_1 \neq \alpha_2$, namely

$$\left[a^\dagger_{\alpha_1} a^\dagger_{\alpha_2} |0\rangle\right]^{(2)} (x_1, x_2) = \frac{1}{\sqrt{2}} \left(\varphi_{\alpha_1}(x_1)\varphi_{\alpha_2}(x_2) - \varphi_{\alpha_1}(x_2)\varphi_{\alpha_2}(x_1)\right) = \\ = \frac{1}{\sqrt{2}} \det \begin{pmatrix} \varphi_{\alpha_1}(x_1) & \varphi_{\alpha_1}(x_2) \\ \varphi_{\alpha_2}(x_1) & \varphi_{\alpha_2}(x_2) \end{pmatrix}, \tag{2.118}$$

which coincides with the basis of the antisymmetric space H_2^A (see Eq. 2.65, Remark 2.1), provided that $\alpha_1 < \alpha_2$. Conversely, if we compute

$$\left[a^\dagger_{\alpha_2} a^\dagger_{\alpha_1} |0\rangle\right]^{(2)}, \tag{2.119}$$

the rows of the corresponding matrix are reversed. Indeed, due to the fermionic anticommutation relation D.6, $a^\dagger_{\alpha_1} a^\dagger_{\alpha_2} = -a^\dagger_{\alpha_2} a^\dagger_{\alpha_1}$, the determinant changes sign, introducing only a global phase factor between the two states, which does not affect the physical state. Next, we consider the case of N fermions and construct a basis of the antisymmetric subspace H_{N+1}^A. This is done by applying the creation operator $a^\dagger_\alpha$ to the basis elements of H_N^A (see Eq. 2.65), i.e.,

$$\left[a^\dagger_\alpha \varphi^{(A,N)}_{\alpha_1,\ldots,\alpha_N}\right]^{(N+1)} (x_1, \ldots, x_{N+1}) \\ = \frac{1}{\sqrt{N+1}} \sum_{j=1}^{N+1} \varepsilon^{j+1} \varphi_\alpha(x_j) \varphi^{(A,N)}_{\alpha_1,\ldots,\alpha_N}(x_1, \ldots, x_{j-1}, x_{j+1}, \ldots, x_{N+1}), \tag{2.120}$$

$$\varphi^{(A,N)}_{\alpha_1,\ldots,\alpha_N}(x_1, \ldots, x_{j-1}, x_{j+1}, \ldots, x_{N+1}) = \\ = \frac{1}{\sqrt{N!}} \det \begin{pmatrix} \varphi_{\alpha_1}(x_1) & \ldots & \varphi_{\alpha_1}(x_{j-1}) & \varphi_{\alpha_1}(x_{j+1}) & \ldots & \varphi_{\alpha_1}(x_{N+1}) \\ \vdots & \ddots & \vdots & \vdots & \ddots & \vdots \\ \varphi_{\alpha_N}(x_1) & \ldots & \varphi_{\alpha_N}(x_{j-1}) & \varphi_{\alpha_N}(x_{j+1}) & \ldots & \varphi_{\alpha_N}(x_{N+1}) \end{pmatrix}, \tag{2.121}$$

where $\varepsilon = -1$ (see Eq. 2.52). So, it follows

$$\left[a_{\alpha}^{\dagger}\varphi_{\alpha_1,\ldots,\alpha_N}^{(A,N)}\right]^{(N+1)}(x_1,\ldots,x_{N+1})$$
$$= \frac{1}{\sqrt{(N+1)!}}\sum_{j=1}^{N+1}(-1)^{j+1}\varphi_{\alpha}(x_j)$$
$$\det\begin{pmatrix}\varphi_{\alpha_1}(x_1) & \ldots & \varphi_{\alpha_1}(x_{j-1}) & \varphi_{\alpha_1}(x_{j+1}) & \ldots & \varphi_{\alpha_1}(x_{N+1})\\ \vdots & \ddots & \vdots & \vdots & \ddots & \vdots\\ \varphi_{\alpha_N}(x_1) & \ldots & \varphi_{\alpha_N}(x_{j-1}) & \varphi_{\alpha_N}(x_{j+1}) & \ldots & \varphi_{\alpha_N}(x_{N+1})\end{pmatrix}, \tag{2.122}$$

which coincides with the Laplace expansion of the determinant. Recall that given a $(N+1)\times(N+1)$ matrix A, Laplace's rule provides a method of calculating the determinant given by

$$\det A = \sum_{j=1}^{N+1}(-1)^{i+j}A_{ij}\det M_{ij}, \tag{2.123}$$

where M_{ij} is the matrix obtained by deleting the i-th row and the j-th column of A and is called the minor of row i and column j. The object

$$\left[a_{\alpha}^{\dagger}\varphi_{\alpha_1,\ldots,\alpha_N}^{(A,N)}\right]^{(N+1)}(x_1,\ldots,x_{N+1}) \tag{2.124}$$

has the structure of a Laplace expansion, where the identifications are assumed to be that the index of α_i is the row index, the index of the degree of freedom is a column index, and the expansion is performed on the first row, at subscript α. It follows

$$\left[a_{\alpha}^{\dagger}\varphi_{\alpha_1,\ldots,\alpha_N}^{(A,N)}\right]^{(N+1)}(x_1,\ldots,x_{N+1}) = \frac{1}{\sqrt{(N+1)!}}\sum_{j=1}^{N+1}(-1)^{1+j}A_{1j}\det M_{1j}, \tag{2.125}$$

$$A = \begin{pmatrix}\varphi_{\alpha}(x_1) & \ldots & \varphi_{\alpha}(x_{j-1}) & \varphi_{\alpha}(x_j) & \varphi_{\alpha}(x_{j+1}) & \ldots & \varphi_{\alpha}(x_{N+1})\\ \varphi_{\alpha_1}(x_1) & \ldots & \varphi_{\alpha_1}(x_{j-1}) & \varphi_{\alpha_1}(x_j) & \varphi_{\alpha_1}(x_{j+1}) & \ldots & \varphi_{\alpha_1}(x_{N+1})\\ \vdots & \ddots & \vdots & \vdots & \vdots & \ddots & \vdots\\ \varphi_{\alpha_N}(x_1) & \ldots & \varphi_{\alpha_N}(x_{j-1}) & \varphi_{\alpha_N}(x_j) & \varphi_{\alpha_N}(x_{j+1}) & \ldots & \varphi_{\alpha_N}(x_{N+1})\end{pmatrix}, \tag{2.126}$$

and the minor with respect to row 1 and column j is the matrix

$$M_{1j} = \begin{pmatrix}\varphi_{\alpha_1}(x_1) & \ldots & \varphi_{\alpha_1}(x_{j-1}) & \varphi_{\alpha_1}(x_{j+1}) & \ldots & \varphi_{\alpha_1}(x_{N+1})\\ \vdots & \ddots & \vdots & \vdots & \ddots & \vdots\\ \varphi_{\alpha_N}(x_1) & \ldots & \varphi_{\alpha_N}(x_{j-1}) & \varphi_{\alpha_N}(x_{j+1}) & \ldots & \varphi_{\alpha_N}(x_{N+1})\end{pmatrix}. \tag{2.127}$$

Recall that it must be $\alpha_1 < \alpha_2 < \ldots < \alpha_N$, therefore, in order for the element

$$\left[a^\dagger_\alpha \varphi^{(A,N)}_{\alpha_1,\ldots,\alpha_N}\right]^{(N+1)}(x_1,\ldots,x_{N+1}) \tag{2.128}$$

to be a basis in H^A_{N+1}, we need to arrange the row of subscript α according to the above convention, assuming of course that it does not coincide with any of the subscripts already present, otherwise Pauli's principle for fermions would imply a null determinant. If, on the other hand, $\alpha \neq \alpha_i$ $\forall i$, let S_α be the number S_α of exchanges needed to move the row α to the right position. Depending on whether S_α is even or odd, we will have a sign change or not. Since the sign of the permutation can be written as $(-1)^{S_\alpha}$, we have

$$\left[a^\dagger_\alpha \varphi^{(A,N)}_{\alpha_1,\ldots,\alpha_N}\right]^{(N+1)}(x_1,\ldots,x_{N+1}) = (-1)^{S_\alpha}\varphi^{(A,N+1)}_{\alpha_1,\ldots,\alpha_{i-1},\alpha,\alpha_i,\ldots,\alpha_N}(x_1,\ldots,x_{N+1}). \tag{2.129}$$

In the occupation-number basis the action of the creation operator can be expressed as

$$\begin{aligned} &a^\dagger_\alpha |N_1,\ldots,N_\alpha,\ldots,N_\infty;N\rangle \\ &\quad = (1-N_\alpha)(-1)^{S_\alpha}|N_1,\ldots,1+N_\alpha,\ldots,N_\infty;N+1\rangle, \end{aligned} \tag{2.130}$$

where $(1 - N_\alpha)$ is justified as follows: we are adding the row associated with $\{\varphi_\alpha(x_j)\}$, if it was already present, $N_\alpha = 1$ and the resulting vector coincides with the null vector, conversely if α is not present, then $N_\alpha = 0$ and $a^\dagger_\alpha$ adds a fermion in a new state. On the other hand, the phase factor $(-1)^{S_\alpha}$ is due to the fact that, a priori, α is not added respecting the ordering convention. Generalizing, a basis in the space H^A_N is obtainable by means of

$$a^\dagger_{\alpha_1}\ldots a^\dagger_{\alpha_N}|0\rangle. \tag{2.131}$$

Let us now analyze how to go from H^A_N to H^A_{N-1}, studying the action of the annihilation operator a_α on a string of N creation operators, i.e., $a_\alpha a^\dagger_{\alpha_1}\cdots a^\dagger_{\alpha_N}|0\rangle$. Suppose that particles occupy the states $\alpha_1,\ldots,\alpha_N$ but not the state α. From the fermionic algebra D.4 with $\alpha \neq \alpha_i$, $a_\alpha a^\dagger_{\alpha_i} = -a^\dagger_{\alpha_i}a_\alpha$, the operator a_α anticommutes with all $a^\dagger_{\alpha_i}$ and therefore acts directly on the vacuum state $|0\rangle$, that is

$$\begin{aligned} a_\alpha a^\dagger_{\alpha_1}\ldots a^\dagger_{\alpha_N}|0\rangle &= (-1)^{S_\alpha}a^\dagger_{\alpha_1}\ldots a^\dagger_{\alpha_N}a_\alpha|0\rangle = \\ &= 0. \end{aligned} \tag{2.132}$$

On the other hand, if α is present in the string, a_α anticommutes with all $a^\dagger_{\alpha_i}$ with $\alpha \neq \alpha_i$, up to $a^\dagger_\alpha$. From fermionic algebra (see Eq. D.4) for $a_\alpha a^\dagger_\alpha$, we get

$$
\begin{aligned}
& a_\alpha a^\dagger_{\alpha_1} \dots a^\dagger_{\alpha_N} |0\rangle \\
& \quad = (-1)^{S_\alpha} a^\dagger_{\alpha_1} \dots a_\alpha a^\dagger_\alpha \dots a^\dagger_{\alpha_N} |0\rangle \; = \\
& \quad = (-1)^{S_\alpha} a^\dagger_{\alpha_1} \dots \left(1 - a^\dagger_\alpha a_\alpha\right) \dots a^\dagger_{\alpha_N} |0\rangle \; = \\
& \quad = (-1)^{S_\alpha} a^\dagger_{\alpha_1} \dots 1 \dots a^\dagger_{\alpha_N} |0\rangle - (-1)^{S_\alpha} (-1)^{S'_\alpha} a^\dagger_{\alpha_1} \dots a^\dagger_\alpha \dots a^\dagger_{\alpha_N} a_\alpha |0\rangle \; = \\
& \quad = (-1)^{S_\alpha} a^\dagger_{\alpha_1} \dots 1 \dots a^\dagger_{\alpha_N} |0\rangle \; = \\
& \quad = (-1)^{S_\alpha} |N_1, \dots, N_\alpha = 0, \dots, N_N, 0, 0, \dots, N_\infty; N\rangle, \qquad (2.133)
\end{aligned}
$$

where S'_α denotes the number of exchanges required to move a_α past $a^\dagger_\alpha$ until it acts on the vacuum $|0\rangle$, resulting in the null vector. This process leaves us with the vector $(-1)^{S_\alpha} a^\dagger_{\alpha_1} \dots 1 \dots a^\dagger_{\alpha_N} |0\rangle$. Note that the operator $a^\dagger_\alpha$ is replaced by the identity operator $\mathbb{1}$, which denotes the absence of a particle in the state α.

In the occupation-number basis, this can be succinctly expressed as

$$
a_\alpha |N_1, \dots, N_\infty; N\rangle = (-1)^{S_\alpha} N_\alpha \, |N_1, \dots, N_\alpha - 1, \dots, N_\infty; N - 1\rangle. \qquad (2.134)
$$

In the occupation-number basis, for fermions we have obtained

$$
\begin{cases}
a^\dagger_\alpha |N_1, \dots, N_\alpha, \dots, N_\infty; N\rangle \\
\quad = (-1)^{S_\alpha} (1 - N_\alpha) |N_1, \dots, 1 + N_\alpha, \dots; N + 1\rangle \\
a_\alpha |N_1, \dots, N_\alpha, \dots, N_\infty; N\rangle \\
\quad = (-1)^{S_\alpha} N_\alpha |N_1, \dots, 1 - N_\alpha, \dots; N - 1\rangle.
\end{cases} \qquad (2.135)
$$

2.7.2 Bosonic Basis in Fock Space

Consider the action of the bosonic creation operator (see Eq. 2.83) on a basis of H^S_N (see Eq. 2.70), namely

$$
\begin{aligned}
& \left[a^\dagger_\alpha \varphi^{(S,N)}_{\alpha_1,\dots,\alpha_N}\right]^{(N+1)} (x_1, \dots, x_{N+1}) = \\
& = \frac{1}{\sqrt{N+1}} \sum_{j=1}^{N+1} \varphi_\alpha(x_j) \frac{1}{\sqrt{N! N_{\alpha_1}! \dots N_{\alpha_M}!}} \\
& \quad \text{perm} \begin{pmatrix}
\varphi_{\alpha_1}(x_1) & \dots & \varphi_{\alpha_1}(x_{j-1}) & \varphi_{\alpha_1}(x_{j+1}) & \dots & \varphi_{\alpha_1}(x_{N+1}) \\
\vdots & \ddots & \vdots & \vdots & \ddots & \vdots \\
\varphi_{\alpha_N}(x_1) & \dots & \varphi_{\alpha_N}(x_{j-1}) & \varphi_{\alpha_N}(x_{j+1}) & \dots & \varphi_{\alpha_N}(x_{N+1})
\end{pmatrix}, \qquad (2.136)
\end{aligned}
$$

in order for it to be a basis, the indices of

$$\left[a_\alpha^\dagger \varphi_{\alpha_1,\ldots,\alpha_N}^{(S,N)}\right]^{(N+1)}(x_1,\ldots,x_{N+1}) \tag{2.137}$$

must be ordered in an ascending direction, and since α has been added, its row must be shifted until an ascending order of the indices is obtained again. Unlike the fermionic case, however, the inversion of the rows in the permanent does not alter the sign, then any phase factors between the matrices are not present. As for normalization, suppose that the index α is present N_α times, now it will be present $N_\alpha + 1$ times, so the normalization constant of the basis of H_{N+1}^S is $\left(\sqrt{(N_\alpha+1)!}\right)^{-1}$, since $N_\alpha!$ is in the denominator, we multiply and divide by $\sqrt{N_\alpha + 1}$, i.e.,

$$\begin{aligned}
&\left[a_\alpha^\dagger \varphi_{\alpha_1,\ldots,\alpha_N}^{(S,N)}\right]^{(N+1)}(x_1,\ldots,x_{N+1}) \\
&= \frac{1}{\sqrt{(N+1)!}}\frac{1}{\sqrt{N_{\alpha_1}!}\ldots\sqrt{N_{\alpha_M}!}}\frac{\sqrt{N_\alpha+1}}{\sqrt{N_\alpha+1}}\sum_{j=1}^{N+1} A_{1j}\,\mathrm{perm}\,M_{1j} = \\
&= \sqrt{N_\alpha+1}\,\varphi_{\{\alpha_1,\ldots,\alpha_N\}}^{(S,N+1)}(x_1,\ldots,x_{N+1}),
\end{aligned} \tag{2.138}$$

$$A = \begin{pmatrix}
\varphi_\alpha(x_1) & \ldots & \varphi_\alpha(x_{j-1}) & \varphi_\alpha(x_j) & \varphi_\alpha(x_{j+1}) & \ldots & \varphi_\alpha(x_{N+1}) \\
\varphi_{\alpha_1}(x_1) & \ldots & \varphi_{\alpha_1}(x_{j-1}) & \varphi_{\alpha_1}(x_j) & \varphi_{\alpha_1}(x_{j+1}) & \ldots & \varphi_{\alpha_1}(x_{N+1}) \\
\vdots & \ddots & \vdots & \vdots & \vdots & \ddots & \vdots \\
\varphi_{\alpha_N}(x_1) & \ldots & \varphi_{\alpha_N}(x_{j-1}) & \varphi_{\alpha_N}(x_j) & \varphi_{\alpha_N}(x_{j+1}) & \ldots & \varphi_{\alpha_N}(x_{N+1})
\end{pmatrix}, \tag{2.139}$$

and the minor with respect to row 1 and column j is given by the matrix

$$M_{1j} = \begin{pmatrix}
\varphi_{\alpha_1}(x_1) & \ldots & \varphi_{\alpha_1}(x_{j-1}) & \varphi_{\alpha_1}(x_{j+1}) & \ldots & \varphi_{\alpha_1}(x_{N+1}) \\
\vdots & \ddots & \vdots & \vdots & \ddots & \vdots \\
\varphi_{\alpha_N}(x_1) & \ldots & \varphi_{\alpha_N}(x_{j-1}) & \varphi_{\alpha_N}(x_{j+1}) & \ldots & \varphi_{\alpha_N}(x_{N+1})
\end{pmatrix}. \tag{2.140}$$

Let us make explicit the action of the creation operator on the occupation-number states

$$a_\alpha^\dagger |N_1,\ldots,N_\alpha,\ldots,N_\infty;N\rangle = \sqrt{N_\alpha+1}\,|N_1,\ldots,N_\alpha+1,\ldots,N_\infty;N+1\rangle, \tag{2.141}$$

note that the multiplicative factor is the same as that arising from the action of the creation operator on the eigenstates of the harmonic oscillator (see Sect. 1.3). To construct the bosonic basis, we start from the vacuum state and generate particles through successive applications of creation operators. Unlike the fermionic case, here it is possible to have terms of the form $\left(a_{\alpha_k}^\dagger\right)^{N_k}$, representing N_k particles occupying the state α_k. A basis in the space H_N^S is obtained as

$$\frac{(a^\dagger_{\alpha_1})^{N_1}}{\sqrt{N_{\alpha_1}!}} \cdots \frac{(a^\dagger_{\alpha_M})^{N_{\alpha M}}}{\sqrt{N_{\alpha_M}!}}|0\rangle. \tag{2.142}$$

To move from H_N^S to H_{N-1}^S, we need to study the action of the annihilation operator on the basis given in Eq. 2.142. Consider two possible cases: the state α is not in the string or the state is present and appears $N_\alpha \geq 1$ times. In the first case, a_α commutes with all other $a^\dagger_{\alpha_i}$, up to the state $|0\rangle$ and gives the empty vector; in the second case, the annihilation operator commutes with all creation operators with different indices up to the N_α creation operators $a^\dagger_\alpha$. Ultimately, we have to evaluate the commutator $[a, \left(a^\dagger\right)^N_\alpha]$, which has been derived previously for operators satisfying a bosonic algebra: from Eq. 1.159, that is,

$$a_\alpha \left(a^\dagger_\alpha\right)^{N_\alpha} = N_\alpha \left(a^\dagger_\alpha\right)^{N_\alpha - 1} + a^\dagger_\alpha a_\alpha, \tag{2.143}$$

the operator $a^\dagger_\alpha a_\alpha$ causes the annihilation operator to act on $|0\rangle$, yielding the vacuum vector, whereas the first term yields

$$\begin{aligned}
&a_\alpha \frac{(a^\dagger_{\alpha_1})^{N_1}}{\sqrt{N_{\alpha_1}!}} \cdots \frac{(a^\dagger_\alpha)^{N_\alpha}}{\sqrt{N_\alpha!}} \cdots \frac{(a^\dagger_{\alpha_M})^{N_{\alpha M}}}{\sqrt{N_{\alpha_M}!}}|0\rangle \\
&= \frac{(a^\dagger_{\alpha_1})^{N_1}}{\sqrt{N_{\alpha_1}!}} \cdots N_\alpha \frac{(a^\dagger_\alpha)^{N_\alpha - 1}}{\sqrt{N_\alpha!}} \cdots \frac{(a^\dagger_{\alpha_M})^{N_{\alpha M}}}{\sqrt{N_{\alpha_M}!}}|0\rangle = \\
&= \frac{(a^\dagger_{\alpha_1})^{N_1}}{\sqrt{N_{\alpha_1}!}} \cdots N_\alpha \frac{(a^\dagger_\alpha)^{N_\alpha - 1}}{\sqrt{N_\alpha}\sqrt{(N_\alpha - 1)!}} \cdots \frac{(a^\dagger_{\alpha_M})^{N_{\alpha M}}}{\sqrt{N_{\alpha_M}!}}|0\rangle = \\
&= \frac{(a^\dagger_{\alpha_1})^{N_1}}{\sqrt{N_{\alpha_1}!}} \cdots \sqrt{N_\alpha} \frac{(a^\dagger_\alpha)^{N_\alpha - 1}}{\sqrt{(N_\alpha - 1)!}} \cdots \frac{(a^\dagger_{\alpha_M})^{N_{\alpha M}}}{\sqrt{N_{\alpha_M}!}}|0\rangle
\end{aligned} \tag{2.144}$$

which in the occupation-number basis is written as

$$a_\alpha |N_1, \ldots, N_\infty; N\rangle = \sqrt{N_\alpha}|N_1, \ldots, N_\alpha - 1, \ldots, N_\infty; N - 1\rangle. \tag{2.145}$$

The operator a_α destroys a boson and has the same structure as the annihilation operator for the harmonic oscillator (see Sect. 1.3).

In the occupation-number basis, for bosons we have obtained

$$\begin{cases} a^\dagger_\alpha |N_1, \ldots, N_\alpha, \ldots, N_\infty; N\rangle = \sqrt{N_\alpha + 1}\,|N_1, \ldots, N_\alpha + 1, \ldots, N_\infty; N + 1\rangle \\ a_\alpha |N_1, \ldots, N_\alpha, \ldots, N_\infty; N\rangle = \sqrt{N_\alpha}\,|N_1, \ldots, N_\alpha - 1, \ldots, N_\infty; N - 1\rangle. \end{cases} \tag{2.146}$$

2.7.3 The Number Operator in Many-Body Systems

Let us analyze the action of the operator $a^\dagger_\alpha a_\alpha$ on a system of bosons or fermions, where the many-body annihilation and creation operators are defined in Eqs. 2.79 and 2.83. This operator annihilates a particle in the state α and subsequently recreates one in the same state, so that the initial and final Hilbert spaces coincide. The key point to be addressed is how the state is modified under this operation. The operator $a^\dagger_\alpha a_\alpha$ plays a fundamental role, as it counts the number of particles occupying the one-particle state α, and is therefore identified as the number operator associated with the mode α. A detailed understanding of its action is essential for describing the occupation properties of quantum many-body systems. In the fermionic case, its action is given by

$$
\begin{aligned}
& a^\dagger_\alpha a_\alpha |N_1, \ldots, N_\alpha, \ldots, N_\infty; N\rangle \\
& \quad = (-1)^{S_\alpha} N_\alpha a^\dagger_\alpha |N_1, \ldots, 1 - N_\alpha, \ldots, N_\infty; N-1\rangle = \\
& \quad = (-1)^{S_\alpha}(-1)^{S_\alpha} N_\alpha \left[1 - (1 - N_\alpha)\right] |N_1, \ldots, 1 + (1 - N_\alpha), \ldots, N_\infty; N\rangle = \\
& \quad = N_\alpha^2 |N_1, \ldots, 2 - N_\alpha, \ldots, N_\infty; N\rangle = \\
& \quad = N_\alpha |N_1, \ldots, N_\alpha, \ldots, N_\infty; N\rangle.
\end{aligned} \tag{2.147}
$$

The substitution $2 - N_\alpha \to N_\alpha$ in the last step is justified by considering the possible values of N_α. If the state α was initially unoccupied ($N_\alpha = 0$), the multiplicative factor $N_\alpha = 0$ annihilates the vector, yielding the null vector. In this case, whether we write $2 - N_\alpha$ or 0 at the α-th position makes no difference. Conversely, if the state was occupied ($N_\alpha = 1$), then $2 - N_\alpha = 1$ matches the occupation number correctly. Therefore, the final result is either the original state (if the particle was present) or the zero vector (if it was not).

For the bosonic case, we have instead

$$
\begin{aligned}
& a^\dagger_\alpha a_\alpha |N_1, \ldots, N_\alpha, \ldots, N_\infty; N\rangle \\
& \quad = \sqrt{N_\alpha}\, a^\dagger_\alpha |N_1, \ldots, N_\alpha - 1, \ldots, N_\infty; N-1\rangle = \\
& \quad = \sqrt{N_\alpha}\sqrt{N_\alpha - 1 + 1}\, |N_1, \ldots, N_\alpha - 1 + 1, \ldots, N_\infty; N\rangle = \\
& \quad = N_\alpha |N_1, \ldots, N_\alpha, \ldots, N_\infty; N\rangle.
\end{aligned} \tag{2.148}
$$

In both cases, the operator $a^\dagger_\alpha a_\alpha$ acts as a number operator, denoted as $\hat{N}_\alpha$ (see Eqs. 1.139 and D.12), counting the number of particles occupying the state φ_α.

2.8 Field Operators

In order to describe the creation or annihilation of particles at a specific point in space and with a given spin, it is convenient to introduce the field operators. While the operators a_α and $a_\alpha^\dagger$ create or annihilate particles in a given one-particle state φ_α, they do not act locally in space. By combining these operators with the corresponding one-particle wavefunctions, we define the annihilation and creation field operators $\hat{\psi}(x)$ and $\hat{\psi}^\dagger(x)$, which act directly at the spatial point and spin component specified by x. This construction allows us to represent physical observables, such as particle densities and interaction terms, in a local form suitable for quantum many-body systems.

Definition 2.11 (Field Operators) Let $\varphi_\alpha \subset H_1$ be a basis of the one-particle Hilbert space (see Sect. 1.1), and let $a_\alpha, a_\alpha^\dagger$ denote the many-body annihilation and creation operators associated with the one-particle state φ_α (see Eqs. 2.79 and 2.83). The corresponding field operators are defined as

$$\hat{\psi}(x) \equiv \sum_\alpha \varphi_\alpha(x) a_\alpha, \tag{2.149}$$

$$\hat{\psi}^\dagger(x) \equiv \sum_\alpha \varphi_\alpha^*(x) a_\alpha^\dagger, \tag{2.150}$$

Theorem 2.7 (Algebra of Field Operators) *The annihilation and creation field operators $\hat{\psi}(x)$ and $\hat{\psi}^\dagger(x)$, defined in Eqs. 2.149 and 2.150, satisfy the following algebraic relations*

$$\left[\hat{\psi}(x), \hat{\psi}(x')\right]^{(\varepsilon)} = 0, \tag{2.151}$$

$$\left[\hat{\psi}^\dagger(x), \hat{\psi}^\dagger(x')\right]^{(\varepsilon)} = 0, \tag{2.152}$$

$$\left[\hat{\psi}(x), \hat{\psi}^\dagger(x')\right]^{(\varepsilon)} = \delta(x - x'), \tag{2.153}$$

where the brackets $[\cdot, \cdot]^{(\varepsilon)}$ are defined in Eq. 2.102.

Proof Using the definitions of the field operators and the canonical (anti) commutation relations of the operators $a_\alpha, a_{\alpha'}^\dagger$ (see Theorem 2.5), one finds

$$\left[\hat{\psi}(x), \hat{\psi}(x')\right]^{(\varepsilon)} = \left[\sum_\alpha \varphi_\alpha(x) a_\alpha, \sum_\beta \varphi_\beta(x') a_\beta\right]^{(\varepsilon)} =$$

$$= \sum_{\alpha,\beta} \varphi_\alpha(x)\varphi_\beta(x')\left[a_\alpha, a_\beta\right]^{(\varepsilon)} =$$
$$= 0, \tag{2.154}$$

$$\left[\hat{\psi}^\dagger(x), \hat{\psi}^\dagger(x')\right]^{(\varepsilon)} = \left[\sum_\alpha \varphi_\alpha^*(x)a_\alpha^\dagger, \sum_\beta \varphi_\beta^*(x')a_\beta^\dagger\right]^{(\varepsilon)} =$$
$$= \sum_{\alpha,\beta} \varphi_\alpha^*(x)\varphi_\beta^*(x')\left[a_\alpha^\dagger, a_\beta^\dagger\right]^{(\varepsilon)} =$$
$$= 0. \tag{2.155}$$

In addition to the algebra of many-body operators, the completeness of the basis $\{\varphi_\alpha\}$ in the one-particle Hilbert space, that is, Eq. 1.51, yields

$$\left[\hat{\psi}(x), \hat{\psi}^\dagger(x')\right]^{(\varepsilon)} = \sum_\alpha \sum_{\alpha'} \varphi_\alpha(x)\, \varphi_{\alpha'}^*(x')\left[a_\alpha, a_{\alpha'}^\dagger\right]^{(\varepsilon)} =$$
$$= \sum_\alpha \varphi_\alpha(x)\, \varphi_\alpha^*(x') =$$
$$= \delta(x - x'). \tag{2.156}$$

□

Theorem 2.8 (Field Representation of Many-Body Annihilation and Creation Operators) *Let $\{\varphi_\alpha(x)\}$ be an orthonormal basis of the one-particle Hilbert space. Then the many-body annihilation and creation operators (see Eqs. 2.79 and 2.83) can be expressed in terms of the field operators (see Eqs. 2.149 and 2.150) as*

$$a_\alpha = \int dx\, \varphi_\alpha^*(x)\, \hat{\psi}(x), \tag{2.157}$$

$$a_\alpha^\dagger = \int dx\, \varphi_\alpha(x)\, \hat{\psi}^\dagger(x), \tag{2.158}$$

respectively.

Proof Starting from the definitions of the field operators, and exploiting the orthonormality of the basis functions, we obtain

$$\int dx \varphi_\alpha^*(x)\hat{\psi}(x) = \int dx \sum_{\alpha'} \varphi_\alpha^*(x)\varphi_{\alpha'}(x)a_{\alpha'} =$$

$$
\begin{aligned}
&= \sum_{\alpha'} \left(\int dx \varphi_\alpha^*(x) \varphi_{\alpha'}(x) \right) a_{\alpha'} = \\
&= \sum_{\alpha'} \delta_{\alpha,\alpha'} a_{\alpha'} = \\
&= a_\alpha ,
\end{aligned}
\tag{2.159}
$$

$$
\begin{aligned}
\int dx\, \varphi_\alpha(x)\, \hat{\psi}^\dagger(x) &= \int dx \sum_{\alpha'} \varphi_\alpha(x)\, \varphi_{\alpha'}^*(x)\, a_{\alpha'}^\dagger = \\
&= \sum_{\alpha'} \left(\int dx\, \varphi_\alpha(x)\, \varphi_{\alpha'}^*(x) \right) a_{\alpha'}^\dagger = \\
&= \sum_{\alpha'} \delta_{\alpha,\alpha'}\, a_{\alpha'}^\dagger = \\
&= a_\alpha^\dagger .
\end{aligned}
\tag{2.160}
$$

□

In order to clarify the physical meaning of the annihilation field operator, we evaluate its explicit action on a generic state $|\phi\rangle \in \mathcal{F}$. Specifically, we examine the N-particle component of the state, that is,

$$
\begin{aligned}
&\left[\hat{\psi}(x) |\phi\rangle \right]^{(N)} (x_1, \ldots, x_N) \\
&= \sum_\alpha \varphi_\alpha(x) \left[a_\alpha |\phi\rangle \right]^{(N)} (x_1, \ldots, x_N) = \\
&= \sum_\alpha \varphi_\alpha(x) \sqrt{N+1} \int dx'\, \varphi_\alpha^*(x')\, \phi^{(N+1)}(x', x_1, \ldots, x_N) = \\
&= \sqrt{N+1} \int dx' \left(\sum_\alpha \varphi_\alpha(x) \varphi_\alpha^*(x') \right) \phi^{(N+1)}(x', x_1, \ldots, x_N) = \\
&= \sqrt{N+1} \int dx'\, \delta(x - x')\, \phi^{(N+1)}(x', x_1, \ldots, x_N) = \\
&= \sqrt{N+1}\, \phi^{(N+1)}(x, x_1, \ldots, x_N).
\end{aligned}
\tag{2.161}
$$

Note that while the annihilation operator a_φ is defined with respect to the wavefunction $\varphi(x')$, the annihilation field operator $\hat{\psi}(x)$ is defined using the Dirac delta function $\delta(x - x')$. In this sense, $\hat{\psi}(x)$ can be viewed as an annihilation operator associated with the generalized function $\delta(x - x')$. Just as a_φ annihilates a boson or fermion in the state described by the wavefunction φ, the field operator $\hat{\psi}(x)$

annihilates a particle localized at the point $\mathbf{r}$ with spin s. In other words, $\hat{\psi}(x)$ annihilates bosons or fermions at a specific position in space and with a well-defined spin.

Analogously, let the creation field operator $\hat{\psi}^\dagger(x)$ act on a vector $|\phi\rangle \in \mathcal{F}$. We consider its N-particle component, that is,

$$\left[\hat{\psi}^\dagger(x)|\phi\rangle\right]^{(N)}(x_1,\ldots,x_N) =$$

$$= \sum_\alpha \varphi_\alpha^*(x)\left[a_\alpha^\dagger|\phi\rangle\right]^{(N)}(x_1,\ldots,x_N) =$$

$$= \sum_\alpha \varphi_\alpha^*(x)\frac{1}{\sqrt{N}}\sum_{j=1}^N \varepsilon^{j+1}\varphi_\alpha(x_j)\phi^{(N-1)}(x_1,\ldots,x_{j-1},x_{j+1},\ldots,x_N) =$$

$$= \frac{1}{\sqrt{N}}\sum_{j=1}^N\left(\sum_\alpha \varphi_\alpha^*(x)\varphi_\alpha(x_j)\right)\varepsilon^{j+1}\phi^{(N-1)}(x_1,\ldots,x_{j-1},x_{j+1},\ldots,x_N) =$$

$$= \frac{1}{\sqrt{N}}\sum_{j=1}^N \delta(x-x_j)\varepsilon^{j+1}\phi^{(N-1)}(x_1,\ldots,x_{j-1},x_{j+1},\ldots,x_N). \tag{2.162}$$

Note that, whereas the creation operator $a_\varphi^\dagger$ involves the function $\varphi(x_j)$ in its definition, the creation field operator $\hat{\psi}^\dagger(x)$ involves the Dirac delta function $\delta(x - x_j)$: in this sense, $\hat{\psi}^\dagger(x)$ acts like a creation operator associated with the generalized function $\delta(x - x_j)$. Just as $a_\varphi^\dagger$ creates a boson or fermion in the quantum state described by the wavefunction φ, the field operator $\hat{\psi}^\dagger(x)$ creates a particle at the spatial point $\mathbf{r}$ with spin s, that is, it creates bosons or fermions at a specific position in space and with a definite spin.

Moreover, thanks to the completeness of the one-particle basis, the field operators are defined independently of the particular choice of basis in the one-particle Hilbert space. In this text, however, field operators will often be expanded in terms of plane waves multiplied by standard spin eigenfunctions, namely according to Eq. 1.81. Therefore, we shall frequently employ this representation in the following

$$\hat{\psi}(\mathbf{r},s) = \sum_{\mathbf{k},\sigma}\frac{e^{i\mathbf{k}\cdot\mathbf{r}}}{\sqrt{V}}\chi_\sigma(s)a_{\mathbf{k},\sigma}, \tag{2.163}$$

$$\hat{\psi}^\dagger(\mathbf{r},s) = \sum_{\mathbf{k},\sigma}\frac{e^{-i\mathbf{k}\cdot\mathbf{r}}}{\sqrt{V}}\chi_\sigma^*(s)a_{\mathbf{k},\sigma}^\dagger. \tag{2.164}$$

Ultimately, the time evolution of field operators is often most conveniently expressed in the interaction picture (see Eq. 1.107). Using Eq. 2.187, the annihilation field operator evolves as

$$\hat{\psi}^{(0)}(x,t) = \sum_{\alpha} e^{-i\frac{\mathcal{E}_\alpha}{\hbar}t}\,\varphi_\alpha(x)\,a_\alpha, \tag{2.165}$$

while, by Eqs. 2.150 and 2.192, the creation field operator follows

$$\hat{\psi}^{\dagger(0)}(x,t) = \sum_{\alpha} e^{i\frac{\mathcal{E}_\alpha}{\hbar}t}\,\varphi^*_\alpha(x)\,a^\dagger_\alpha. \tag{2.166}$$

2.9 Representation in Fock Space of One-Body Operators

In this section, we consider one-body operators defined on the N-particle Hilbert space, i.e., operators that act independently on each particle. The goal is to show how such operators, initially defined in first quantization as sums over all particle indices, can be rewritten in Fock space using field operators. This approach naturally allows the treatment of systems with a variable number of particles and automatically incorporates the statistical properties of fermions and bosons. Recall that a one-body operator V_N defined on the N-particle Hilbert space H_N can be expressed in first quantization as a sum of one-particle operators, i.e.,

$$V_N \equiv \sum_{i=1}^{N} V(x_i), \tag{2.167}$$

where $V(x)$ acts on the one-particle Hilbert space.

Theorem 2.9 (One-Body Operators in Second Quantization) *Let V_N be a one-body operator acting on the N-particle Hilbert space H_N, of the form 2.167. In Fock space, its second-quantized form is*

$$\hat{V} = \int dx \hat{\psi}^\dagger(x) V(x) \hat{\psi}(x). \tag{2.168}$$

Proof In Fock space, given $|\phi\rangle, |\phi'\rangle \in \mathcal{F}^A$ (similarly for $\mathcal{F}^S$), the corresponding matrix element is defined as

$$\langle\phi|\hat{V}|\phi'\rangle_{\mathcal{F}} \equiv \sum_{N=1}^{\infty} \left(\phi^{(N)}, V_N \phi'^{(N)}\right)_{H_N}, \tag{2.169}$$

and we analyze the scalar product in the N-particle space H_N, i.e.

$$I \equiv \left(\phi^{(N)}, V_N \phi'^{(N)}\right)_{H_N} =$$

$$= \int dx_1 \ldots \int dx_N \left(\phi^{(N)}(x_1, \ldots, x_N)\right)^* \left(\sum_{i=1}^{N} V(x_i)\right) \phi'^{(N)}(x_1, \ldots, x_N) =$$
$$= \sum_{i=1}^{N} \int dx_1 \ldots \int dx_N \left(\phi^{(N)}(x_1, \ldots, x_N)\right)^* V(x_i)\phi'^{(N)}(x_1, \ldots, x_N). \tag{2.170}$$

Let us calculate the integral with respect to dx_i, i.e.,

$$I = \sum_{i=1}^{N} \int dx_1 \ldots \int dx_{i-1} \int dx_{i+1} \ldots$$
$$\int dx_N \int dx_i \left(\phi^{(N)}(x_1, \ldots, x_i, \ldots, x_N)\right)^* V(x_i)\phi'^{(N)}(x_1, \ldots, x_i, \ldots, x_N), \tag{2.171}$$

we move x_i to the first position in both ϕ and ϕ', the total number of permutations will be even, so the calculations to follow do not depend on the statistical nature of the particles. We set $x_i = x$ and rename the variables on which we do not integrate, i.e.,

$$\sum_{i=1}^{N} \int dx_1 \ldots \int dx_{N-1} \int dx \left(\phi^{(N)}(x, x_1, \ldots, x_{N-1})\right)^*$$
$$V(x)\phi'^{(N)}(x, x_1, \ldots, x_{N-1}) =$$
$$= \int dx_1 \ldots \int dx_{N-1} \int dx \left\{\sqrt{N} \left(\phi^{(N)}(x, x_1, \ldots, x_{N-1})\right)^*\right\}$$
$$V(x) \left\{\sqrt{N}\phi'^{(N)}(x, x_1, \ldots, x_{N-1})\right\}, \tag{2.172}$$

where we used the fact that the summand does not depend on the index i and the identity $\sum_{i=1}^{N} 1 = N = \sqrt{N}\sqrt{N}$. From Eq. 2.149, we write

$$I = \int dx_1 \ldots \int dx_{N-1} \int dx \left(\left[\hat{\psi}(x)|\phi\rangle\right]^{(N-1)} (x_1, \ldots, x_{N-1})\right)^*$$
$$V(x) \left(\left[\hat{\psi}(x)|\phi'\rangle\right]^{(N-1)} (x_1, \ldots, x_{N-1})\right) =$$
$$= \int dx \int dx_1 \ldots \int dx_{N-1} \left(\left[\hat{\psi}(x)|\phi\rangle\right]^{(N-1)} (x_1, \ldots, x_{N-1})\right)^*$$
$$V(x) \left(\left[\hat{\psi}(x)|\phi'\rangle\right]^{(N-1)} (x_1, \ldots, x_{N-1})\right) =$$

$$= \int dx \left(\left[\hat{\psi}(x)|\phi\rangle \right]^{(N-1)}, V(x) \left[\hat{\psi}(x)|\phi'\rangle \right]^{(N-1)} \right)_{H_{N-1}}, \tag{2.173}$$

and the series becomes

$$\sum_{N=1}^{\infty} \left(\phi^{(N)}, V_N \phi'^{(N)} \right)_{H_N}$$

$$= \sum_{N=1}^{\infty} \int dx \left(\left[\hat{\psi}(x)|\phi\rangle \right]^{(N-1)}, V(x) \left[\hat{\psi}(x)|\phi'\rangle \right]^{(N-1)} \right)_{H_{N-1}} =$$

$$= \sum_{N=0}^{\infty} \int dx \left(\left[\hat{\psi}(x)|\phi\rangle \right]^{(N)}, V(x) \left[\hat{\psi}(x)|\phi'\rangle \right]^{(N)} \right)_{H_N}. \tag{2.174}$$

Assume the order of summation and integration can be interchanged, that is,

$$\langle\phi|\hat{V}|\phi'\rangle_{\mathcal{F}} = \sum_{N=1}^{\infty} \left(\phi^{(N)}, V_N \phi'^{(N)} \right)_{H_N} =$$

$$= \int dx \sum_{N=0}^{\infty} \left(\left[\hat{\psi}(x)|\phi\rangle \right]^{(N)}, V(x) \left[\hat{\psi}(x)|\phi'\rangle \right]^{(N)} \right)_{H_N} =$$

$$= \int dx \left(\hat{\psi}(x)|\phi\rangle, V(x)\hat{\psi}(x)|\phi'\rangle \right)_{\mathcal{F}} =$$

$$= \int dx \langle\phi|\hat{\psi}^{\dagger}(x)V(x)\hat{\psi}(x)|\phi'\rangle_{\mathcal{F}}, \; \forall\, \phi, \phi' \in \mathcal{F}^A \; \left(\forall\, \phi, \phi' \in \mathcal{F}^S \right), \tag{2.175}$$

which is the thesis. □

The particle index is lost: two field operators create and annihilate particles at a given point in space with a specific spin component. The statistical nature of the particles is encoded in the creation and annihilation operators. Moreover, the quantity $\hat{V}$ plays a role analogous to an average, where the field operators take the place of the wave function and its complex conjugate in defining the expectation value of an observable. In first quantization, the wave function of a particle is a vector in a Hilbert space; in second quantization, it becomes an operator.

2.9.1 Kinetic Term and One-Body Potential

Let us consider a system composed of N non-interacting particles (either fermions or bosons), which may be subjected to an external one-body potential $V(x)$. The corresponding Hamiltonian, expressed in the first quantization formalism, is given by

$$\mathcal{H} = \sum_{i=1}^{N} \left(\frac{\hat{\mathbf{p}}_i^2}{2m} + V(x_i) \right). \tag{2.176}$$

Given the restriction of $\mathcal{H}$ to each subspace H_N of Fock space $\mathcal{F}$, our goal is to express $\mathcal{H}$ in second quantization. The second-quantized Hamiltonian is

$$\hat{\mathcal{H}} = \int dx\, \hat{\psi}^{\dagger}(x) \left(-\frac{\hbar^2 \nabla^2}{2m} + V(x) \right) \hat{\psi}(x). \tag{2.177}$$

Expanding the field operators in a one-particle basis $\{\varphi_\alpha(x)\}$

$$\hat{\mathcal{H}} = \sum_{\alpha,\alpha'} \mathcal{H}_{\alpha,\alpha'} a^{\dagger}_{\alpha'} a_\alpha, \tag{2.178}$$

$$\mathcal{H}_{\alpha,\alpha'} = \int dx\, \varphi^{*}_{\alpha'}(x) \left(-\frac{\hbar^2 \nabla^2}{2m} + V(x) \right) \varphi_\alpha(x). \tag{2.179}$$

By choosing the basis of eigenfunctions of the one-body Hamiltonian

$$\left(-\frac{\hbar^2 \nabla^2}{2m} + V(x) \right) \varphi_\alpha(x) = \mathcal{E}_\alpha \varphi_\alpha(x), \tag{2.180}$$

the Hamiltonian becomes diagonal, i.e.,

$$\hat{\mathcal{H}} = \sum_{\alpha} \mathcal{E}_\alpha a^{\dagger}_{\alpha} a_\alpha. \tag{2.181}$$

As a concrete example, if the external potential vanishes, $V(x) = 0$, the eigenfunctions reduce to plane waves, see Eq. 1.77, so that the quantum number α is given by Eq. 1.45, while the eigenvalues are those in Eq. 1.74, i.e. $\mathcal{E}_\alpha = \mathcal{E}_{\mathbf{k}}$. The eigenstates in the N-particle space H_N are (anti)symmetrized products of the one-particle eigenstates, i.e., Slater determinants or permanents. Since the number operator $a^{\dagger}_{\alpha} a_\alpha$ counts the number of particles, the Hamiltonian $\hat{\mathcal{H}}$ is written in terms of energy quanta. We have quantized the system; it exchanges energy only for multiples of $\mathcal{E}_\alpha$.

2.10 Time Evolution of Annihilation and Creation Operators in the Interaction Picture

Here, we evaluate the time evolution of the annihilation and creation operators in the interaction picture (see Eq. 1.107), where $\hat{\mathcal{H}}_0$ is of the form given in 2.181. The annihilation operator in the interaction picture is defined as

$$a_\alpha^{(0)}(t) = e^{i\frac{\hat{\mathcal{H}}_0}{\hbar}t} a_\alpha e^{-i\frac{\hat{\mathcal{H}}_0}{\hbar}t}. \tag{2.182}$$

Its time derivative is given by

$$\begin{aligned}
\frac{da_\alpha^{(0)}(t)}{dt} &= \frac{d\left(e^{i\frac{\hat{\mathcal{H}}_0}{\hbar}t} a_\alpha e^{-i\frac{\hat{\mathcal{H}}_0}{\hbar}t}\right)}{dt} = \\
&= \frac{de^{i\frac{\hat{\mathcal{H}}_0}{\hbar}t}}{dt} a_\alpha e^{-i\frac{\hat{\mathcal{H}}_0}{\hbar}t} + e^{i\frac{\hat{\mathcal{H}}_0}{\hbar}t} \frac{da_\alpha}{dt} e^{-i\frac{\hat{\mathcal{H}}_0}{\hbar}t} + e^{i\frac{\hat{\mathcal{H}}_0}{\hbar}t} a_\alpha \frac{de^{-i\frac{\hat{\mathcal{H}}_0}{\hbar}t}}{dt} = \\
&= \frac{de^{i\frac{\hat{\mathcal{H}}_0}{\hbar}t}}{dt} a_\alpha e^{-i\frac{\hat{\mathcal{H}}_0}{\hbar}t} + e^{i\frac{\hat{\mathcal{H}}_0}{\hbar}t} a_\alpha \frac{de^{-i\frac{\hat{\mathcal{H}}_0}{\hbar}t}}{dt} = \\
&= \frac{i}{\hbar}\hat{\mathcal{H}}_0 e^{i\frac{\hat{\mathcal{H}}_0}{\hbar}t} a_\alpha e^{-i\frac{\hat{\mathcal{H}}_0}{\hbar}t} - e^{i\frac{\hat{\mathcal{H}}_0}{\hbar}t} a_\alpha \frac{i}{\hbar}\hat{\mathcal{H}}_0 e^{-i\frac{\hat{\mathcal{H}}_0}{\hbar}t} = \\
&= \frac{i}{\hbar}\hat{\mathcal{H}}_0 e^{i\frac{\hat{\mathcal{H}}_0}{\hbar}t} a_\alpha e^{-i\frac{\hat{\mathcal{H}}_0}{\hbar}t} - e^{i\frac{\hat{\mathcal{H}}_0}{\hbar}t} a_\alpha e^{-i\frac{\hat{\mathcal{H}}_0}{\hbar}t} \frac{i}{\hbar}\hat{\mathcal{H}}_0,
\end{aligned} \tag{2.183}$$

where in the last step we used Eq. H.15. Equation 2.183 becomes

$$\frac{da_\alpha^{(0)}(t)}{dt} = \frac{i}{\hbar}\left[\hat{\mathcal{H}}_0, a_\alpha(t)\right]. \tag{2.184}$$

Note that 2.184 can be equivalently expressed in a form more convenient for calculations. Indeed, from Eq. H.15, it follows that

$$\begin{aligned}
\frac{da_\alpha^{(0)}(t)}{dt} &= \frac{1}{\hbar}\left[\hat{\mathcal{H}}_0, e^{i\frac{\hat{\mathcal{H}}_0}{\hbar}t} a_\alpha e^{-i\frac{\hat{\mathcal{H}}_0}{\hbar}t}\right] = \\
&= \frac{1}{\hbar}\left(\hat{\mathcal{H}}_0 e^{i\frac{\hat{\mathcal{H}}_0}{\hbar}t} a_\alpha e^{-i\frac{\hat{\mathcal{H}}_0}{\hbar}t} - e^{i\frac{\hat{\mathcal{H}}_0}{\hbar}t} a_\alpha e^{-i\frac{\hat{\mathcal{H}}_0}{\hbar}t}\hat{\mathcal{H}}_0\right) = \\
&= \frac{1}{\hbar}\left(e^{i\frac{\hat{\mathcal{H}}_0}{\hbar}t}\hat{\mathcal{H}}_0 a_\alpha e^{-i\frac{\hat{\mathcal{H}}_0}{\hbar}t} - e^{i\frac{\hat{\mathcal{H}}_0}{\hbar}t} a_\alpha \hat{\mathcal{H}}_0 e^{-i\frac{\hat{\mathcal{H}}_0}{\hbar}t}\right) = \\
&= \frac{1}{\hbar} e^{i\frac{\hat{\mathcal{H}}_0}{\hbar}t}\left[\hat{\mathcal{H}}_0, a_\alpha\right] e^{-i\frac{\hat{\mathcal{H}}_0}{\hbar}t}.
\end{aligned} \tag{2.185}$$

From Eq. D.13, it follows

$$\frac{da_\alpha^{(0)}(t)}{dt} = -\frac{i}{\hbar}\mathcal{E}_\alpha e^{i\frac{\hat{\mathcal{H}}_0}{\hbar}t} a_\alpha e^{-\frac{i\hat{\mathcal{H}}_0}{\hbar}t} = \\ = -\frac{i}{\hbar}\mathcal{E}_\alpha a_\alpha^{(0)}(t). \tag{2.186}$$

It is a linear operator differential equation with constant coefficients, whose solution is well known. Consequently, the time evolution of the annihilation operator in the thermal interaction picture is given by

$$a_\alpha^{(0)}(t) = e^{-i\frac{\mathcal{E}_\alpha}{\hbar}t} a_\alpha. \tag{2.187}$$

Similarly, given the interaction picture of the creation operator $a_\alpha^\dagger$, i.e.,

$$a_\alpha^{\dagger(0)}(t) = e^{i\frac{\hat{\mathcal{H}}_0}{\hbar}t} a_\alpha^\dagger e^{-i\frac{\hat{\mathcal{H}}_0}{\hbar}t}, \tag{2.188}$$

its time evolution can be derived in the same manner as for the annihilation operator, using Eq. D.14 instead of Eq. D.13), yielding

$$\frac{da_\alpha^{\dagger(0)}(t)}{dt} = \frac{i}{\hbar}\left[\hat{\mathcal{H}}_0, a_\alpha^{\dagger(0)}\right], \tag{2.189}$$

$$\frac{da_\alpha^{\dagger(0)}(t)}{dt} = \frac{i}{\hbar}e^{i\frac{\hat{\mathcal{H}}_0}{\hbar}t}\left[\hat{\mathcal{H}}_0, a_\alpha^\dagger\right]e^{-i\frac{\hat{\mathcal{H}}_0}{\hbar}t}, \tag{2.190}$$

$$\frac{da_\alpha^{\dagger(0)}(t)}{dt} = \frac{1}{\hbar}\mathcal{E}_\alpha a_\alpha^{\dagger(0)}(t). \tag{2.191}$$

$$a_\alpha^{\dagger(0)}(t) = e^{i\frac{\mathcal{E}_\alpha}{\hbar}t} a_\alpha^\dagger. \tag{2.192}$$

Note that, as in the Schrödinger picture, in the interaction picture the creation and annihilation operators remain Hermitian conjugates of each other.

2.11 Thermal Interaction Picture

Here, we adapt the interaction picture (see Eq. 1.107) formalism to imaginary times. Starting from the real-time evolution and performing a Wick rotation

$$t \to -i\tau, \tag{2.193}$$

the imaginary-time variable τ is taken to be real.

Definition 2.12 (Thermal Interaction Picture) Assuming the system is in thermal equilibrium at temperature T, we define the thermal interaction picture (also referred to as the Matsubara interaction picture) of an operator A, with respect to the Hamiltonian $\hat{\mathcal{H}}_0$ (see Eq. 1.85), within the framework of second quantization, as

$$A^{(0)}(\tau) = e^{\frac{\hat{\mathcal{H}}_0}{\hbar}\tau} A e^{-\frac{\hat{\mathcal{H}}_0}{\hbar}\tau}, \quad \tau \in [-\beta\hbar, \beta\hbar], \tag{2.194}$$

where τ is a real time variable related to imaginary times t through Eq. 2.193.

Imaginary-time evolution is thus obtained by analytic continuation of the real-time dynamics to $t = -i\tau$. This picture is termed thermal due to the presence of a finite, nonzero temperature in the formalism. Note that, from the definition of the thermal interaction picture, it follows that

$$\begin{aligned} A^{(0)}(\tau + \beta\hbar) &= e^{\frac{\hat{\mathcal{H}}_0}{\hbar}(\tau+\beta\hbar)} A e^{-\frac{\hat{\mathcal{H}}_0}{\hbar}(\tau+\beta\hbar)} = \\ &= e^{\beta\hat{\mathcal{H}}_0} e^{\frac{\hat{\mathcal{H}}_0}{\hbar}\tau} A e^{-\frac{\hat{\mathcal{H}}_0}{\hbar}\tau} e^{-\beta\hat{\mathcal{H}}_0} \equiv \\ &\equiv e^{\beta\hat{\mathcal{H}}_0} A^{(0)}(\tau) e^{-\beta\hat{\mathcal{H}}_0}, \end{aligned} \tag{2.195}$$

where we have used the commutativity of the exponential operators, since their exponents are proportional to the same operator $\hat{\mathcal{H}}_0$ (see Eq. H.20).

In the thermal interaction picture, the time evolution of states is governed by the corresponding evolution operator

$$U(\tau) = e^{\frac{\hat{\mathcal{H}}_0}{\hbar}\tau} e^{-\frac{\hat{\mathcal{H}}}{\hbar}\tau}. \tag{2.196}$$

From

$$U^{\dagger}(\tau) = e^{-\frac{\hat{\mathcal{H}}}{\hbar}\tau} e^{\frac{\hat{\mathcal{H}}_0}{\hbar}\tau}, \tag{2.197}$$

and

$$U^{-1}(\tau) = e^{\frac{\hat{\mathcal{H}}}{\hbar}\tau} e^{-\frac{\hat{\mathcal{H}}_0}{\hbar}\tau}, \tag{2.198}$$

note that the operator $U(\tau)$ is not a unitary transformation (see Eqs. I.2 and I.3), indeed

$$U^{\dagger}(\tau) \neq U^{-1}(\tau). \tag{2.199}$$

We now introduce two time-ordering operators that will be used in the subsequent discussion. Although the real-time variable is typically denoted by t, we use τ in order to remain consistent with the notation commonly adopted in the imaginary-time formalism.

Definition 2.13 (Wick and Dyson Time-Ordering Operators) Let $A(\tau')$ and $B(\tau'')$ be two arbitrary operators depending on the time variable τ, and let ε be the statistical index introduced in Eq. 2.52. The Wick time-ordering operator $\hat{T}_\tau$ is defined as

$$\hat{T}_\tau A(\tau')B(\tau'') = \begin{cases} A(\tau')B(\tau''), & \tau' > \tau'' \\ \varepsilon\, B(\tau'')A(\tau'), & \tau'' > \tau' \end{cases}, \tag{2.200}$$

that is, it arranges products of operators from left to right in decreasing order with respect to the time variable, taking into account the statistical nature of the operators.

The Dyson time-ordering operator, denoted by $\hat{T}_D$, is defined as

$$\hat{T}_D A(\tau')B(\tau'') = \begin{cases} A(\tau')B(\tau''), & \tau' > \tau'' \\ B(\tau'')A(\tau'), & \tau'' > \tau' \end{cases}, \tag{2.201}$$

where $\hat{T}_D$ orders operators from left to right in decreasing order of the time variable, regardless of their statistics.

We are now able to prove

Theorem 2.10 (The Dyson's Series) *Given an Hamiltonian of the form 1.85, within the framework of the second quantizazion, the time evolution of the operator $U(\tau)$ defined in Eq. 2.196 involves the thermal interaction pictures (see Eq. 2.194) of $\hat{\mathcal{H}}_I$, i.e.*

$$U(\tau) = \hat{T}_\tau \exp\left\{-\frac{1}{\hbar}\int_0^\tau d\tau_1 \hat{\mathcal{H}}_I^{(0)}(\tau_1)\right\}, \tag{2.202}$$

where $\hat{T}_\tau$ is given in Eq. 2.201. The expression 2.202 is known as Dyson's series.

Proof Let us compute the first time derivative of the operator $U(\tau)$, i.e.,

$$\frac{\partial U}{\partial \tau} = -\frac{1}{\hbar}\hat{\mathcal{H}}_I^{(0)}(\tau)U(\tau), \tag{2.203}$$

with the initial condition $U(0) = \mathbb{1}$, indeed

$$\frac{\partial U}{\partial \tau} = \frac{1}{\hbar}\hat{\mathcal{H}}_0 e^{\frac{\hat{\mathcal{H}}_0}{\hbar}\tau} e^{-\frac{\hat{\mathcal{H}}}{\hbar}\tau} - \frac{1}{\hbar} e^{\frac{\hat{\mathcal{H}}_0}{\hbar}\tau}\hat{\mathcal{H}} e^{-\frac{\hat{\mathcal{H}}}{\hbar}\tau} =$$

$$
\begin{aligned}
&= \frac{1}{\hbar} e^{\frac{\hat{\mathcal{H}}_0}{\hbar}\tau} \hat{\mathcal{H}}_0 e^{-\frac{\hat{\mathcal{H}}}{\hbar}\tau} - \frac{1}{\hbar} e^{\frac{\hat{\mathcal{H}}_0}{\hbar}\tau} \hat{\mathcal{H}} e^{-\frac{\hat{\mathcal{H}}}{\hbar}\tau} = \\
&= -\frac{1}{\hbar} e^{\frac{\hat{\mathcal{H}}_0}{\hbar}\tau} \hat{\mathcal{H}}_I e^{-\frac{\hat{\mathcal{H}}}{\hbar}\tau} = \\
&= -\frac{1}{\hbar} e^{\frac{\hat{\mathcal{H}}_0}{\hbar}\tau} \hat{\mathcal{H}}_I e^{-\frac{\hat{\mathcal{H}}_0}{\hbar}\tau} e^{\frac{\hat{\mathcal{H}}_0}{\hbar}\tau} e^{-\frac{\hat{\mathcal{H}}}{\hbar}\tau} = \\
&= -\frac{1}{\hbar} \hat{\mathcal{H}}_I^{(0)}(\tau) U(\tau),
\end{aligned}
\tag{2.204}
$$

where we used

$$
\left[\hat{\mathcal{H}}_0, e^{\frac{\hat{\mathcal{H}}_0}{\hbar}\tau} \right] = 0, \tag{2.205}
$$

which is justified by Eq. H.15, after which we used the identity $\mathbb{1} = e^{-\frac{\hat{\mathcal{H}}_0}{\hbar}\tau} e^{\frac{\hat{\mathcal{H}}_0}{\hbar}\tau}$ to bring out the thermal interaction picture of $\hat{\mathcal{H}}_I$. Let us integrate both members of Eq. 2.203 between 0 and τ, i.e.,

$$
U(\tau) - U(0) = -\frac{1}{\hbar} \int_0^{\tau} \hat{\mathcal{H}}_I^{(0)}(\tau') U(\tau') d\tau', \tag{2.206}
$$

$$
U(\tau) = U(0) - \frac{1}{\hbar} \int_0^{\tau} \hat{\mathcal{H}}_I^{(0)}(\tau') U(\tau') d\tau', \tag{2.207}
$$

$$
U(\tau) = \mathbb{1} - \frac{1}{\hbar} \int_0^{\tau} \hat{\mathcal{H}}_I^{(0)}(\tau') U(\tau') d\tau', \tag{2.208}
$$

and we insert the above integral relation of $U(\tau)$ into $U(\tau')$ as follows

$$
\begin{aligned}
U(\tau) &= \mathbb{1} - \frac{1}{\hbar} \int_0^{\tau} \hat{\mathcal{H}}_I^{(0)}(\tau') \left[\mathbb{1} - \frac{1}{\hbar} \int_0^{\tau'} \hat{\mathcal{H}}_I^{(0)}(\tau'') U(\tau'') d\tau'' \right] d\tau' = \\
&= \mathbb{1} - \frac{1}{\hbar} \int_0^{\tau} \hat{\mathcal{H}}_I^{(0)}(\tau') d\tau' + \left(-\frac{1}{\hbar} \right)^2 \int_0^{\tau} d\tau' \hat{\mathcal{H}}_I^{(0)}(\tau') \\
&\int_0^{\tau'} d\tau'' \hat{\mathcal{H}}_I^{(0)}(\tau'') U(\tau''),
\end{aligned}
\tag{2.209}
$$

and by iterating this procedure, we generate a series expansion. We now want to write the integral

$$
J_2 = \int_0^{\tau} d\tau' \int_0^{\tau'} d\tau'' \hat{\mathcal{H}}_I^{(0)}(\tau') \hat{\mathcal{H}}_I^{(0)}(\tau'') \tag{2.210}
$$

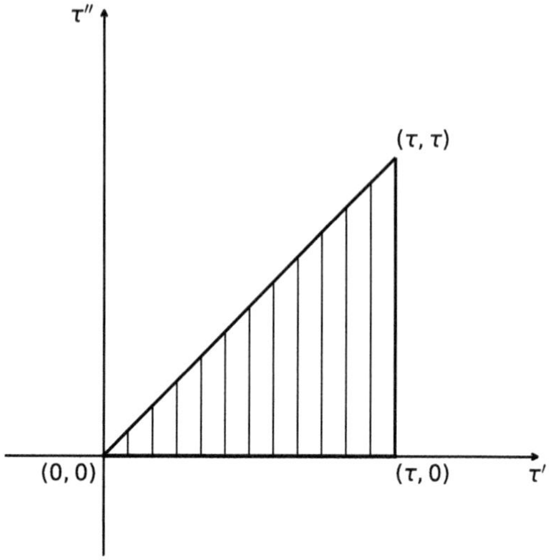

Fig. 2.1 Integration region corresponding to Eq. 2.211, where $0 \leq \tau'' \leq \tau' \leq \tau$. This domain reflects the time ordering $\tau'' \leq \tau'$ required by the Dyson series at second order. The integration is performed first over τ'' (from 0 to τ'), then over τ' (from 0 to τ), and matches the natural ordering of operators in $\hat{T}_D\{\hat{\mathcal{H}}_I^{(0)}(\tau')\hat{\mathcal{H}}_I^{(0)}(\tau'')\}$

in a physically more expressive form in which the integration limits are decoupled. By means of the Dyson time-ordering operator, we write

$$\begin{aligned} J_2 &= \int_0^{\tau} d\tau' \int_0^{\tau'} d\tau'' \hat{\mathcal{H}}_I^{(0)}(\tau')\hat{\mathcal{H}}_I^{(0)}(\tau'') = \\ &= \int_0^{\tau} d\tau' \int_0^{\tau'} d\tau'' \hat{T}_D \left\{ \hat{\mathcal{H}}_I^{(0)}(\tau')\hat{\mathcal{H}}_I^{(0)}(\tau'') \right\}. \end{aligned} \tag{2.211}$$

In Fig. 2.1 the region and direction of the above integration are shown.

This integral can also be calculated by reversing the order of integration, that is, according to the Fig. 2.2, i.e.,

$$\begin{aligned} J_2 &= \int_0^{\tau} d\tau' \int_0^{\tau'} d\tau'' \hat{\mathcal{H}}_I^{(0)}(\tau')\hat{\mathcal{H}}_I^{(0)}(\tau'') = \\ &= \int_0^{\tau} d\tau'' \int_{\tau''}^{\tau} d\tau' \hat{\mathcal{H}}_I^{(0)}(\tau')\hat{\mathcal{H}}_I^{(0)}(\tau''), \end{aligned} \tag{2.212}$$

and given that τ' and τ'' are dummy indices, they can be inverted in the last member of Eq. 2.212 as follows

$$\begin{aligned} J_2 &= \int_0^{\tau} d\tau' \int_{\tau'}^{\tau} d\tau'' \hat{\mathcal{H}}_I^{(0)}(\tau'')\hat{\mathcal{H}}_I^{(0)}(\tau') = \\ &= \int_0^{\tau} d\tau' \int_{\tau'}^{\tau} d\tau'' \hat{T}_D \left\{ \hat{\mathcal{H}}_I^{(0)}(\tau')\hat{\mathcal{H}}_I^{(0)}(\tau'') \right\}, \end{aligned} \tag{2.213}$$

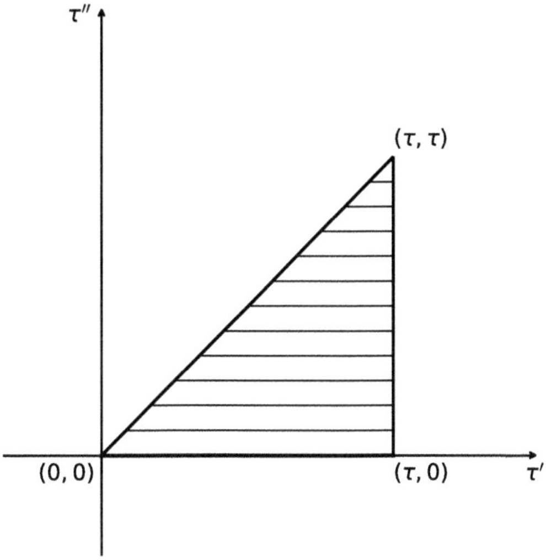

Fig. 2.2 Integration region corresponding to Eq. 2.212, where $0 \leq \tau'' \leq \tau$ and $\tau'' \leq \tau' \leq \tau$. This is an equivalent rewriting of the domain in Fig. 2.1, preserving the time ordering $\tau'' \leq \tau'$. The integration is performed first over τ'', and for each fixed τ'', over τ' from τ'' to τ

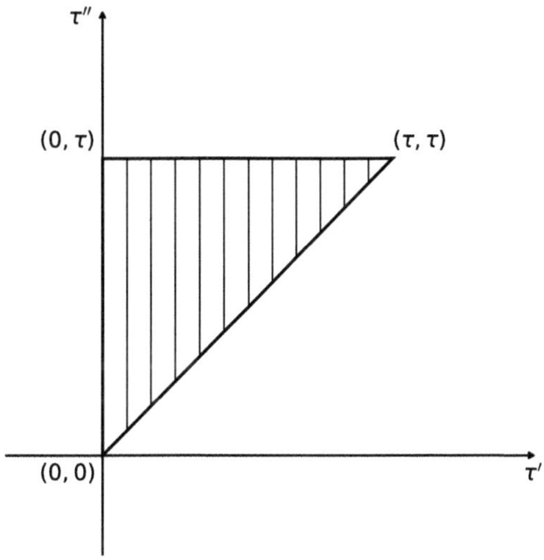

Fig. 2.3 Integration region corresponding to Eq. 2.213, where $0 \leq \tau' \leq \tau'' \leq \tau$. This region corresponds to the opposite time ordering, $\tau' \leq \tau''$, and appears in the second-order Dyson term when operators are explicitly reordered by $\hat{T}_D$. The integration is carried out first over τ', and then over τ'' from τ' to τ

where we have inserted the Dyson time-ordering operator $\hat{T}_D$ since $\tau'' \geq \tau'$. Now the integral is computed in the triangle above the diagonal $\tau' = \tau''$, as in the Fig. 2.3. Consequently, the integral J_2 can be calculated in the following two equivalent ways

$$J_2 = \int_0^{\tau} d\tau' \int_0^{\tau'} d\tau'' \hat{T}_D \left\{ \hat{\mathcal{H}}_I^{(0)}(\tau') \hat{\mathcal{H}}_I^{(0)}(\tau'') \right\}, \tag{2.214}$$

$$J_2 = \int_0^{\tau} d\tau' \int_{\tau'}^{\tau} d\tau'' \hat{T}_D \left\{ \hat{\mathcal{H}}_I^{(0)}(\tau') \hat{\mathcal{H}}_I^{(0)}(\tau'') \right\}, \tag{2.215}$$

and summing Eqs. 2.214 and 2.215, it follows

$$J_2 = \frac{1}{2}\int_0^\tau d\tau' \int_0^\tau d\tau'' \hat{T}_D \left\{ \hat{\mathcal{H}}_I^{(0)}(\tau')\hat{\mathcal{H}}_I^{(0)}(\tau'') \right\}. \tag{2.216}$$

Regarding $U(\tau)$, we perform iteration for n integrals and send n to infinity to compose the series

$$U(\tau) = \mathbb{1} + \sum_{n=1}^{\infty} \left(-\frac{1}{\hbar}\right)^n \frac{1}{n!} \int_0^\tau d\tau_1 \ldots \int_0^\tau d\tau_n \, \hat{T}_\tau \left\{ \hat{\mathcal{H}}_I^{(0)}(\tau_1) \ldots \hat{\mathcal{H}}_I^{(0)}(\tau_n) \right\}, \tag{2.217}$$

which is equivalent to Eq. 2.202. Note that the factor $\frac{1}{2}$ generalizes to $\frac{1}{n!}$ because the set of operators $\left\{ \hat{\mathcal{H}}_I^{(0)}(\tau_1), \ldots, \hat{\mathcal{H}}_I^{(0)}(\tau_n) \right\}$ can be reordered in $n!$ different ways. Ultimately, note the Dyson time-ordering operator $\hat{T}_D$ has been replaced by the Wick time-ordering operator $\hat{T}_\tau$ (see Remark 2.4). □

Remark 2.4 (Equivalence of Wick and Dyson Time-Ordering Operators for Two-Body Interactions) In many-body systems governed by Hamiltonians with two-body interactions, the Wick time-ordering operator $\hat{T}_\tau$ and the Dyson time-ordering operator $\hat{T}_D$ yield the same result. This equivalence arises because, in the second quantization formalism, each interaction term $\hat{\mathcal{H}}_I$ involves an even number of creation and annihilation operators. When exchanging two such interaction terms evaluated at different times, the total number of operator transpositions is even. Consequently, any sign changes due to fermionic statistics cancel out. Therefore, the time-ordering procedure becomes independent of particle statistics, and we can identify $\hat{T}_D \equiv \hat{T}_\tau$.

Together with the transformation $U(\tau)$, we define the matrix $S(\tau, \tau')$ such that

$$\psi(\tau) = S(\tau, \tau')\psi(\tau'). \tag{2.218}$$

From $\psi(\tau) = U(\tau)\psi(0)$ and from

$$\begin{aligned} \psi(\tau) &= S(\tau, \tau')\psi(\tau') = \\ &= S(\tau, \tau')U(\tau')\psi(0), \end{aligned} \tag{2.219}$$

we get

$$S(\tau, \tau')U(\tau') = U(\tau), \ \forall \ \psi(0), \tag{2.220}$$

which implies

$$S(\tau, \tau') = U(\tau)U^{-1}(\tau'). \tag{2.221}$$

By definition it follows

$$S(\tau, 0) = U(\tau), \tag{2.222}$$

$$S(0, \tau) = U^{-1}(\tau). \tag{2.223}$$

Theorem 2.11 *The $S(\tau, \tau')$ matrix (see Eq. 2.218) satisfies:*

1.

$$S(\tau, \tau) = \mathbb{1}, \tag{2.224}$$

which is the initial condition for the matrix $S(\tau, \tau')$;

2.

$$S(\tau_1, \tau_2)S(\tau_2, \tau_3) = S(\tau_1, \tau_3). \tag{2.225}$$

Proof Both statements follow directly from the definition of $S(\tau, \tau')$.

1. Observe that

$$\begin{aligned} S(\tau, \tau) &= U(\tau)U^{-1}(\tau) = \\ &= \mathbb{1}. \end{aligned} \tag{2.226}$$

2. A direct calculation yields

$$\begin{aligned} S(\tau_1, \tau_2)S(\tau_2, \tau_3) &= U(\tau_1)U^{-1}(\tau_2)U(\tau_2)U^{-1}(\tau_3) = \\ &= U(\tau_1)U^{-1}(\tau_3) \equiv \\ &\equiv S(\tau_1, \tau_3). \end{aligned} \tag{2.227}$$

□

As a function of the Hamiltonian, the matrix $S(\tau, \tau')$ is written as

$$\begin{aligned} S(\tau, \tau') &= U(\tau)U^{-1}(\tau') = \\ &= e^{\frac{\hat{\mathcal{H}}_0}{\hbar}\tau} e^{-\frac{\hat{\mathcal{H}}}{\hbar}\tau} e^{\frac{\hat{\mathcal{H}}}{\hbar}\tau'} e^{-\frac{\hat{\mathcal{H}}_0}{\hbar}\tau'} = \\ &= e^{\frac{\hat{\mathcal{H}}_0}{\hbar}\tau} e^{-\frac{\hat{\mathcal{H}}}{\hbar}(\tau-\tau')} e^{-\frac{\hat{\mathcal{H}}_0}{\hbar}\tau'}. \end{aligned} \tag{2.228}$$

Analogously to the evolution operator $U(\tau)$, the matrix $S(\tau, \tau')$ satisfies the differential equation

$$\frac{\partial S(\tau,\tau')}{\partial \tau} = -\frac{1}{\hbar}\hat{\mathcal{H}}_I^{(0)}(\tau)S(\tau,\tau'), \tag{2.229}$$

and, upon integration between τ' and τ, it admits the formal series expansion

$$S(\tau,\tau') = \mathbb{1} + \sum_{n=1}^{\infty}\left(-\frac{1}{\hbar}\right)^n \frac{1}{n!}\int_{\tau'}^{\tau} d\tau_1 \ldots \int_{\tau'}^{\tau} d\tau_n \, \hat{T}_\tau\left\{\hat{\mathcal{H}}_I^{(0)}(\tau_1)\ldots\hat{\mathcal{H}}_I^{(0)}(\tau_n)\right\}. \tag{2.230}$$

2.12 Time Evolution of Annihilation and Creation Operators in the Thermal Interaction Picture

In this section we follow steps analogous to those in Sect. 2.10, adapting the derivation to the thermal interaction picture. Within the framework of second quantization, the annihilation operator in imaginary time is defined as

$$a_\alpha^{(0)}(\tau) = e^{\frac{\hat{\mathcal{H}}_0}{\hbar}\tau} a_\alpha e^{-\frac{\hat{\mathcal{H}}_0}{\hbar}\tau}. \tag{2.231}$$

Consider now its derivative with respect to imaginary time, i.e.,

$$\begin{aligned}
\frac{da_\alpha^{(0)}(\tau)}{d\tau} &= \frac{d\left(e^{\frac{\hat{\mathcal{H}}_0}{\hbar}\tau} a_\alpha e^{-\frac{\hat{\mathcal{H}}_0}{\hbar}\tau}\right)}{d\tau} = \\
&= \frac{de^{\frac{\hat{\mathcal{H}}_0}{\hbar}\tau}}{d\tau} a_\alpha e^{-\frac{\hat{\mathcal{H}}_0}{\hbar}\tau} + e^{\frac{\hat{\mathcal{H}}_0}{\hbar}\tau}\frac{da_\alpha}{d\tau} e^{-\frac{\hat{\mathcal{H}}_0}{\hbar}\tau} + e^{\frac{\hat{\mathcal{H}}_0}{\hbar}\tau} a_\alpha \frac{de^{-\frac{\hat{\mathcal{H}}_0}{\hbar}\tau}}{d\tau} = \\
&= \frac{de^{\frac{\hat{\mathcal{H}}_0}{\hbar}\tau}}{d\tau} a_\alpha e^{-\frac{\hat{\mathcal{H}}_0}{\hbar}\tau} + e^{\frac{\hat{\mathcal{H}}_0}{\hbar}\tau} a_\alpha \frac{de^{-\frac{\hat{\mathcal{H}}_0}{\hbar}\tau}}{d\tau} = \\
&= \frac{\hat{\mathcal{H}}_0}{\hbar} e^{\frac{\hat{\mathcal{H}}_0}{\hbar}\tau} a_\alpha e^{-\frac{\hat{\mathcal{H}}_0}{\hbar}\tau} - e^{\frac{\hat{\mathcal{H}}_0}{\hbar}\tau} a_\alpha \frac{\hat{\mathcal{H}}_0}{\hbar} e^{-\frac{\hat{\mathcal{H}}_0}{\hbar}\tau} = \\
&= \frac{\hat{\mathcal{H}}_0}{\hbar} e^{\frac{\hat{\mathcal{H}}_0}{\hbar}\tau} a_\alpha e^{-\frac{\hat{\mathcal{H}}_0}{\hbar}\tau} - e^{\frac{\hat{\mathcal{H}}_0}{\hbar}\tau} a_\alpha e^{-\frac{\hat{\mathcal{H}}_0}{\hbar}\tau}\frac{\hat{\mathcal{H}}_0}{\hbar},
\end{aligned} \tag{2.232}$$

where in the last step we used Eq. H.15. Equation 2.232 becomes

$$\frac{da_\alpha^{(0)}(\tau)}{d\tau} = \frac{1}{\hbar}\left[\hat{\mathcal{H}}_0, a_\alpha^{(0)}(\tau)\right]. \tag{2.233}$$

Note that 2.233 can be equivalently expressed in a form more convenient for calculations. Indeed, from Eq. H.15, it follows that

$$\begin{aligned}\frac{da_\alpha^{(0)}(\tau)}{d\tau} &= \frac{1}{\hbar}\left[\hat{\mathcal{H}}_0, e^{\frac{\hat{\mathcal{H}}_0}{\hbar}\tau} a_\alpha e^{-\frac{\hat{\mathcal{H}}_0}{\hbar}\tau}\right] = \\ &= \frac{1}{\hbar}\left(\hat{\mathcal{H}}_0 e^{\frac{\hat{\mathcal{H}}_0}{\hbar}\tau} a_\alpha e^{-\frac{\hat{\mathcal{H}}_0}{\hbar}\tau} - e^{\frac{\hat{\mathcal{H}}_0}{\hbar}\tau} a_\alpha e^{-\frac{\hat{\mathcal{H}}_0}{\hbar}\tau}\hat{\mathcal{H}}_0\right) = \\ &= \frac{1}{\hbar}\left(e^{\frac{\hat{\mathcal{H}}_0}{\hbar}\tau}\hat{\mathcal{H}}_0 a_\alpha e^{-\frac{\hat{\mathcal{H}}_0}{\hbar}\tau} - e^{\frac{\hat{\mathcal{H}}_0}{\hbar}\tau} a_\alpha \hat{\mathcal{H}}_0 e^{-\frac{\hat{\mathcal{H}}_0}{\hbar}\tau}\right) = \\ &= \frac{1}{\hbar} e^{\frac{\hat{\mathcal{H}}_0}{\hbar}\tau}\left[\hat{\mathcal{H}}_0, a_\alpha\right] e^{-\frac{\hat{\mathcal{H}}_0}{\hbar}\tau}. \end{aligned} \tag{2.234}$$

From Eq. D.13, it follows

$$\begin{aligned}\frac{da_\alpha^{(0)}(\tau)}{d\tau} &= -\frac{1}{\hbar}\mathcal{E}_\alpha e^{\frac{\hat{\mathcal{H}}_0}{\hbar}\tau} a_\alpha e^{-\frac{\hat{\mathcal{H}}_0}{\hbar}\tau} = \\ &= -\frac{1}{\hbar}\mathcal{E}_\alpha a_\alpha^{(0)}(\tau). \end{aligned} \tag{2.235}$$

It is a linear operator differential equation with constant coefficients, whose solution is well known. Consequently, the time evolution of the annihilation operator in thermal interaction picture is given by

$$a_\alpha^{(0)}(\tau) = e^{-\frac{\mathcal{E}_\alpha}{\hbar}\tau} a_\alpha. \tag{2.236}$$

Similarly, given the thermal interaction picture of the creation operator $a_\alpha^\dagger$ within the framework of the second quantization, i.e.,

$$a_\alpha^{\dagger(0)}(\tau) = e^{\frac{\hat{\mathcal{H}}_0}{\hbar}\tau} a_\alpha^\dagger e^{-\frac{\hat{\mathcal{H}}_0}{\hbar}\tau}, \tag{2.237}$$

its time evolution can be derived in the same manner as for the annihilation operator, using Eq. D.14 instead of Eq. D.13), yielding

$$\frac{da_\alpha^{\dagger(0)}(\tau)}{d\tau} = \frac{1}{\hbar}\left[\hat{\mathcal{H}}_0, a_\alpha^{\dagger(0)}\right], \tag{2.238}$$

$$\frac{da_\alpha^{\dagger(0)}(\tau)}{d\tau} = \frac{1}{\hbar} e^{\frac{\hat{\mathcal{H}}_0}{\hbar}\tau}\left[\hat{\mathcal{H}}_0, a_\alpha^\dagger\right] e^{-\frac{\hat{\mathcal{H}}_0}{\hbar}\tau}, \tag{2.239}$$

$$\frac{da_\alpha^{\dagger(0)}(\tau)}{d\tau} = \frac{1}{\hbar}\mathcal{E}_\alpha a_\alpha^{\dagger(0)}(\tau). \tag{2.240}$$

$$a_\alpha^{\dagger(0)}(\tau) = e^{\frac{\mathcal{E}_\alpha}{\hbar}\tau} a_\alpha^\dagger. \tag{2.241}$$

Consequently, the operators $a_\alpha^{(0)}(\tau)$ and $a_\alpha^{\dagger(0)}(\tau)$ are no longer related by Hermitian conjugation in the thermal interaction picture, due to the real exponential factors.

Remark 2.5 (Finite Lifetimes) For interacting particles, the time evolution of the annihilation and creation operators a_α and $a_\alpha^\dagger$ in the thermal Heisenberg picture involves exponential factors containing generally complex energies of the form $\mathcal{E}_\alpha - i\delta$. The real part $\mathcal{E}_\alpha$ corresponds to the excitation energy, while the imaginary part $i\delta$ introduces a damping factor $e^{-\delta t}$. Physically, this damping represents the finite lifetime of quasiparticle excitations, arising from interaction-induced decay processes and scattering events within the system. Such a finite lifetime implies that excitations are not stable but decay over time, leading to a broadening of energy levels. Quantitatively, the decay rate δ and related lifetime can be computed systematically by means of Feynman's perturbation theory (see Chaps. 6 and 7), which accounts for the interactions through self-energy corrections in the Green's functions framework.

To conclude this section, we demonstrate an interesting property of creation and annihilation operators in the thermal interaction picture. We have

Theorem 2.12 *Let A_α denote either a creation or an annihilation operator associated with the one-particle state α. In the thermal interaction picture (see Eq. 2.194), this operator satisfies*

$$A_\alpha^{(0)}(\tau + \beta\hbar) = e^{\delta_\alpha \beta \mathcal{E}_\alpha} A_\alpha^{(0)}(\tau), \tag{2.242}$$

where δ_α is $+1$ or -1 depending on whether the A_α operator is a creation or annihilation operator, respectively, namely

$$\delta_\alpha = \begin{cases} +1, & A_\alpha = a_\alpha^\dagger \\ -1, & A_\alpha = a_\alpha \end{cases}. \tag{2.243}$$

Proof From Eq. 2.195, the expression can be rearranged as

$$A_\alpha^{(0)}(\tau + \beta\hbar) = e^{\frac{\hat{\mathcal{H}}_0}{\hbar}\tau} e^{\beta\hat{\mathcal{H}}_0} A_\alpha e^{-\beta\hat{\mathcal{H}}_0} e^{-\frac{\hat{\mathcal{H}}_0}{\hbar}\tau}, \tag{2.244}$$

where we have used the commutativity of the exponential operators (see Eq. H.20). We now focus on the operator $e^{\beta\hat{\mathcal{H}}_0} A_\alpha e^{-\beta\hat{\mathcal{H}}_0}$ and we make use of the conjugation formula (see Eq. H.5) with the identifications

$$X = \beta\hat{\mathcal{H}}_0, \tag{2.245}$$

$$Y = A_\alpha, \tag{2.246}$$

namely

$$e^{\beta\hat{\mathcal{H}}_0}A_\alpha e^{-\beta\hat{\mathcal{H}}_0} = A_\alpha + \left[\beta\hat{\mathcal{H}}_0, A_\alpha\right] + \frac{1}{2!}\left[\beta\hat{\mathcal{H}}_0, \left[\beta\hat{\mathcal{H}}_0, A_\alpha\right]\right] + \\ + \frac{1}{3!}\left[\beta\hat{\mathcal{H}}_0, \left[\beta\hat{\mathcal{H}}_0, \left[\beta\hat{\mathcal{H}}_0, A_\alpha\right]\right]\right] + \dots \tag{2.247}$$

Since A_α may represent either a creation or an annihilation operator, from Eqs. D.13 and D.14, we can write, in compact form as follows

$$\left[\beta\hat{\mathcal{H}}_0, A_\alpha\right] = \delta_\alpha \beta \mathcal{E}_\alpha A_\alpha, \tag{2.248}$$

where δ_α is defined in Eq. 2.243. Using the commutation relation derived above, one finds for the second nested commutator

$$\frac{1}{2!}\left[\beta\hat{\mathcal{H}}_0, \left[\beta\hat{\mathcal{H}}_0, A_\alpha\right]\right] = \frac{1}{2}\beta^2\left[\hat{\mathcal{H}}_0, \left[\hat{\mathcal{H}}_0, A_\alpha\right]\right] = \\ = \frac{1}{2}\delta_\alpha^2 \beta^2 \mathcal{E}_\alpha^2 A_\alpha. \tag{2.249}$$

All higher-order nested commutators are proportional to A_α, and the conjugation formula resums into an exponential

$$e^{\beta\hat{\mathcal{H}}_0}A_\alpha e^{-\beta\hat{\mathcal{H}}_0} = A_\alpha\left(1 + \delta_\alpha\beta\mathcal{E}_\alpha + \delta_\alpha^2\beta^2\mathcal{E}_\alpha^2 + \dots\right) = \\ = e^{\delta_\alpha\beta\mathcal{E}_\alpha}A_\alpha. \tag{2.250}$$

Upon substitution into Eq. 2.244 and by invoking the definition of the thermal interaction picture, the desired result follows. □

It is worth noting that the same result can be directly obtained from Eqs. 2.236 and 2.241 by setting $\tau = \beta\hbar$.

2.13 Average Occupation Number in a Non-interacting System

Here, we compute the thermal average (see Eq. 1.379) of the occupation number for non-interacting particles, providing a derivation within the framework of second quantization (see Theorem E.1 for a derivation based on statistical mechanics).

Theorem 2.13 (Average Occupation Number in a Non-interacting System: Second Quantization Formulation) *Consider a system of non-interacting particles with Hamiltonian $\hat{\mathcal{H}}_0$ as given in Eq. 2.181, and described within the grand* *canonical ensemble. Let a_α and $a_\alpha^\dagger$ be the fermionic or bosonic annihilation and creation operators defined in Eqs. 2.79 and 2.83. The full Hamiltonian is then of the form $\hat{\mathcal{H}}_0 - \mu\hat{N}$, where $\hat{N} = \sum_\alpha a_\alpha^\dagger a_\alpha$ denotes the total number operator. Then, Eq. E.2 can be written equivalently as follows*

$$\begin{aligned}\left\langle a_\alpha^\dagger a_\alpha \right\rangle &= \frac{1}{e^{\beta(\mathcal{E}_\alpha-\mu)} \pm 1} = \\ &= n_\varepsilon(\mathcal{E}_\alpha).\end{aligned} \tag{2.251}$$

Moreover, the thermal one-body correlator is diagonal in the one-particle basis, that is,

$$\left\langle a_{\alpha'}^\dagger a_\alpha \right\rangle = \delta_{\alpha,\alpha'} n_\varepsilon(\mathcal{E}_\alpha), \tag{2.252}$$

Proof From the definition of the thermal average (see Eq. 1.379), it follows

$$\begin{aligned}\left\langle a_\alpha a_\alpha^\dagger \right\rangle &= \frac{1}{Z_0}\,\mathrm{Tr}\left(e^{-\beta\left(\hat{\mathcal{H}}_0-\mu\hat{N}\right)} a_\alpha a_\alpha^\dagger\right) = \\ &= \frac{1}{Z_0}\,\mathrm{Tr}\left(a_\alpha^\dagger e^{-\beta\left(\hat{\mathcal{H}}_0-\mu\hat{N}\right)} a_\alpha\right) = \\ &= \frac{1}{Z_0}\,\mathrm{Tr}\left(a_\alpha a_\alpha^\dagger e^{-\beta\left(\hat{\mathcal{H}}_0-\mu\hat{N}\right)}\right) = \\ &= \frac{1}{Z_0}\,\mathrm{Tr}\left(a_\alpha e^{-\beta\left(\hat{\mathcal{H}}_0-\mu\hat{N}\right)} e^{\beta\left(\hat{\mathcal{H}}_0-\mu\hat{N}\right)} a_\alpha^\dagger e^{-\beta\left(\hat{\mathcal{H}}_0-\mu\hat{N}\right)}\right) = \\ &= \frac{1}{Z_0}\,\mathrm{Tr}\left(a_\alpha e^{-\beta\left(\hat{\mathcal{H}}_0-\mu\hat{N}\right)} e^{\beta(\mathcal{E}_\alpha-\mu)} a_\alpha^\dagger\right) = \\ &= \frac{e^{\beta(\mathcal{E}_\alpha-\mu)}}{Z_0}\,\mathrm{Tr}\left(a_\alpha e^{-\beta\left(\hat{\mathcal{H}}_0-\mu\hat{N}\right)} a_\alpha^\dagger\right) = \\ &= e^{\beta(\mathcal{E}_\alpha-\mu)}\left\langle a_\alpha^\dagger a_\alpha \right\rangle,\end{aligned} \tag{2.253}$$

where we have made Eq. 2.241 explicit with $\tau = \beta\hbar$. Now, taking the thermal average of Eq. 2.112, with $\alpha = \alpha'$, one finds

$$\left\langle a_\alpha a_\alpha^\dagger \right\rangle = \epsilon\left\langle a_\alpha^\dagger a_\alpha \right\rangle + 1. \tag{2.254}$$

Comparing this last equation with Eq. 2.253, i.e.,

$$e^{\beta(\mathcal{E}_\alpha-\mu)}\left\langle a_\alpha^\dagger a_\alpha \right\rangle = \epsilon\left\langle a_\alpha^\dagger a_\alpha \right\rangle + 1, \tag{2.255}$$

we solve for $\langle a_\alpha^\dagger a_\alpha \rangle$, that is,

$$\left(e^{\beta(\mathcal{E}_\alpha-\mu)} - \varepsilon\right)\left\langle a_\alpha^\dagger a_\alpha \right\rangle = 1, \tag{2.256}$$

and we obtain Eq. 2.251.

Now, if we take the thermal average of Eq. 2.112, with $\alpha \neq \alpha'$, one finds

$$\left\langle a_\alpha a_{\alpha'}^\dagger \right\rangle = \epsilon \left\langle a_{\alpha'}^\dagger a_\alpha \right\rangle, \tag{2.257}$$

and again, by comparing this last equation with the counterpart of Eq. 2.253 evaluated for different indices α and α', we obtain

$$e^{\beta(\mathcal{E}_\alpha-\mu)} \left\langle a_{\alpha'}^\dagger a_\alpha \right\rangle = \epsilon \left\langle a_{\alpha'}^\dagger a_\alpha \right\rangle, \tag{2.258}$$

it follows that $\left\langle a_{\alpha'}^\dagger a_\alpha \right\rangle$ vanishes for $\alpha \neq \alpha'$, that is, we obtain Eq. 2.252. □

Relation 2.253 expresses the Kubo-Martin-Schwinger (KMS) condition, which characterizes thermal equilibrium states. In particular, it demonstrates that thermal correlation functions exhibit KMS periodicity in imaginary time, resulting in Bose-Einstein or Fermi-Dirac statistics depending on the commutation relations of the annihilation and creation operators

2.14 Particle Density (Fluctuation) Operator

In this section, we introduce the particle density operator, which counts the number of particles at a given position in space. We then express it in the formalism of second quantization, allowing a natural treatment in terms of creation and annihilation operators. The Fourier components of the density operator, often referred to as density fluctuation operators, play a central role in describing collective excitations and correlations in many-body systems. We will also derive key commutation relations that are essential for applications in both fermionic and bosonic systems.

Definition 2.14 (Particle Density Operator in First Quantization) In first quantization, the particle density operator is defined as

$$\rho(\mathbf{r}) = \sum_{i=1}^{N} \delta\left(\mathbf{r} - \mathbf{r}_i\right). \tag{2.259}$$

Our next step is to express this operator within the framework of second quantization. From Theorem 2.9, one finds

$$\hat{\rho}(\mathbf{r}) = \sum_{s'=-S}^{S} \int d\mathbf{r}' \hat{\psi}^\dagger(\mathbf{r}', s') \delta\left(\mathbf{r}' - \mathbf{r}\right) \hat{\psi}(\mathbf{r}', s'), \tag{2.260}$$

that is,

Definition 2.15 (Particle Density Operator in Second Quantization) In second quantization, the particle density operator in Eq. 2.259 is given by

$$\hat{\rho}(\mathbf{r}) = \sum_{s=-S}^{S} \hat{\psi}^\dagger(\mathbf{r}, s) \hat{\psi}(\mathbf{r}, s). \tag{2.261}$$

The particle density fluctuation operator counts how many particles are in $\mathbf{r}$, regardless of the spin.

The following result establishes the explicit form of the Fourier transform of the particle density operator. This operator will recur frequently in the subsequent chapters, for instance in the discussion of the jellium model later in this chapter, and in the framework of linear response theory (see Chap. 10).

Theorem 2.14 (Fourier Transform of the Particle Density Operator) *The Fourier transform of the particle density operator (see Eq. 2.261) is given by*

$$\hat{\rho}_{\mathbf{q}} = \sum_{\mathbf{k},\sigma} a^\dagger_{\mathbf{k}+\mathbf{q},\sigma} a_{\mathbf{k},\sigma}. \tag{2.262}$$

Proof We want to compute

$$\hat{\rho}_{\mathbf{q}} = \int d^3\mathbf{r} e^{i\mathbf{q}\cdot\mathbf{r}} \hat{\rho}(\mathbf{r}), \tag{2.263}$$

which thanks to 2.261 becomes

$$\hat{\rho}_{\mathbf{q}} = \sum_{s=-S}^{S} \int d^3\mathbf{r} e^{i\mathbf{q}\cdot\mathbf{r}} \hat{\psi}^\dagger(\mathbf{r}, s) \hat{\psi}(\mathbf{r}, s). \tag{2.264}$$

Since the choice of basis is arbitrary, we expand the field operators in terms of plane waves combined with the standard spin eigenfunctions, namely Eqs. 2.163 and 2.164, obtaining

$$\hat{\rho}_{\mathbf{q}} = \frac{1}{V} \sum_{s=-S}^{S} \int d^3\mathbf{r} e^{i\mathbf{q}\cdot\mathbf{r}} \sum_{\mathbf{k}_1,\sigma_1} \sum_{\mathbf{k}_2,\sigma_2} e^{-i\mathbf{k}_1\cdot\mathbf{r}} e^{i\mathbf{k}_2\cdot\mathbf{r}} \chi^*_{\sigma_1}(s) \chi_{\sigma_2}(s) a^\dagger_{\mathbf{k}_1,\sigma_1} a_{\mathbf{k}_2,\sigma_2} =$$

$$= \frac{1}{V} \sum_{\mathbf{k}_1,\sigma_1} \sum_{\mathbf{k}_2,\sigma_2} \int d^3\mathbf{r} e^{-i(\mathbf{k}_1-\mathbf{k}_2-\mathbf{q})\cdot\mathbf{r}} \sum_{s=-S}^{S} \chi^*_{\sigma_1}(s)\chi_{\sigma_2}(s) a^\dagger_{\mathbf{k}_1,\sigma_1} a_{\mathbf{k}_2,\sigma_2} =$$

$$= \frac{1}{V} \sum_{\mathbf{k}_1,\sigma_1} \sum_{\mathbf{k}_2,\sigma_2} V \delta_{\mathbf{k}_1,\mathbf{k}_2+\mathbf{q}} \delta_{\sigma_1,\sigma_2} a^\dagger_{\mathbf{k}_1,\sigma_1} a_{\mathbf{k}_2,\sigma_2}. \tag{2.265}$$

Hence, by performing the sums over $\mathbf{k}_1$ and σ_1, the Kronecker deltas enforce the conditions $\sigma_1 = \sigma_2$ and $\mathbf{k}_1 = \mathbf{k}_2 + \mathbf{q}$. Renaming the dummy indices $\mathbf{k}_2 \to \mathbf{k}$ and $\sigma_2 \to \sigma$, the result immediately follows. □

Moreover, note that

$$\hat{\rho}^\dagger_{\mathbf{q}} = \sum_{\mathbf{k},\sigma} a^\dagger_{\mathbf{k},\sigma} a_{\mathbf{k}+\mathbf{q},\sigma}, \tag{2.266}$$

and if we set $\mathbf{k}' = \mathbf{k} + \mathbf{q}$, we finally have

$$\hat{\rho}^\dagger_{\mathbf{q}} = \sum_{\mathbf{k}',\sigma} a^\dagger_{\mathbf{k}'-\mathbf{q},\sigma} a_{\mathbf{k}',\sigma}, \tag{2.267}$$

which implies

$$\hat{\rho}^\dagger_{\mathbf{q}} = \hat{\rho}_{-\mathbf{q}}. \tag{2.268}$$

Remark 2.6 (On the Name "Particle Density Fluctuation Operator") We next explain why the operator defined in Eq. 2.261 is commonly referred to as the particle density fluctuation operator. In particular, given the Eq. 2.263, we define

$$\Delta\hat{\rho}(\mathbf{r}) = \hat{\rho}(\mathbf{r}) - \langle\hat{\rho}(\mathbf{r})\rangle. \tag{2.269}$$

For a homogeneous medium, the expectation value is independent of the position $\mathbf{r}$, so that

$$\langle\hat{\rho}(\mathbf{r})\rangle = \langle\hat{\rho}\rangle, \tag{2.270}$$

and, consequently, we have

$$\Delta\hat{\rho}_{\mathbf{q}} = \int d^3\mathbf{r} \Delta\hat{\rho}(\mathbf{r}) e^{i\mathbf{q}\cdot\mathbf{r}} =$$

$$= \hat{\rho}_{\mathbf{q}} - \langle\hat{\rho}\rangle \int d^3\mathbf{r} e^{i\mathbf{q}\cdot\mathbf{r}}, \tag{2.271}$$

but the integral $\int d^3\mathbf{r} e^{i\mathbf{q}\cdot\mathbf{r}}$ is nonzero only for $\mathbf{q} = \mathbf{0}$, then

$$\Delta\hat{\rho}_{\mathbf{q}} = \hat{\rho}_{\mathbf{q}},\ \mathbf{q} \neq \mathbf{0}. \tag{2.272}$$

This is the reason why the Fourier component $\hat{\rho}_{\mathbf{q}}$ (for $\mathbf{q} \neq 0$) is also referred to as the particle density fluctuation operator.

In what follows, we derive some useful relations for the fluctuation operator $\hat{\rho}_{\mathbf{q}}$, which will be employed in the subsequent chapters.

Theorem 2.15 (Commutators Involving the Fourier Transform of the Density Operator) *Let $\hat{\mathcal{H}}$ be a non-interacting many-body Hamiltonian of the form 2.181, in particular $\hat{\mathcal{H}} = \sum_{\mathbf{k},\sigma} \mathcal{E}_{\mathbf{k}} a^{\dagger}_{\mathbf{k},\sigma} a_{\mathbf{k},\sigma}$. Then, independently of the statistical nature of the operators $a_{\mathbf{k},\sigma}$, $a^{\dagger}_{\mathbf{k},\sigma}$, the Fourier transform of the particle density operator, that is, Eq. 2.262, satisfies the following commutators*

1.

$$\left[\hat{\rho}_{\mathbf{q}'}, \hat{\rho}_{\mathbf{q}''}\right] = 0,\ \forall\, \mathbf{q}', \mathbf{q}''; \tag{2.273}$$

2.

$$\left[\hat{\rho}^{\dagger}_{\mathbf{q}'}, \hat{\rho}^{\dagger}_{\mathbf{q}''}\right] = 0,\ \forall\, \mathbf{q}', \mathbf{q}''; \tag{2.274}$$

3.

$$\left[\hat{\rho}_{\mathbf{q}'}, \hat{\rho}^{\dagger}_{\mathbf{q}''}\right] = 0,\ \forall\, \mathbf{q}', \mathbf{q}''; \tag{2.275}$$

4.

$$\left[\hat{\rho}_{\mathbf{q}}, \sum_{\mathbf{k},\sigma} \mathcal{E}_{\mathbf{k}} a^{\dagger}_{\mathbf{k},\sigma} a_{\mathbf{k},\sigma}\right] = \sum_{\mathbf{k},\sigma} \left(\mathcal{E}_{\mathbf{k}} - \mathcal{E}_{\mathbf{k}+\mathbf{q}}\right) a^{\dagger}_{\mathbf{k}+\mathbf{q},\sigma} a_{\mathbf{k},\sigma}; \tag{2.276}$$

5.

$$\left[\left[\hat{\rho}_{\mathbf{q}}, \sum_{\mathbf{k},\sigma} \mathcal{E}_{\mathbf{k}} a^{\dagger}_{\mathbf{k},\sigma} a_{\mathbf{k},\sigma}\right], \hat{\rho}^{\dagger}_{\mathbf{q}}\right] = \sum_{\mathbf{k},\sigma} \left(\mathcal{E}_{\mathbf{k}+\mathbf{q}} - 2\mathcal{E}_{\mathbf{k}} + \mathcal{E}_{\mathbf{k}-\mathbf{q}}\right) a^{\dagger}_{\mathbf{k},\sigma} a_{\mathbf{k},\sigma}; \tag{2.277}$$

6.

$$\left[\hat{\rho}_{\mathbf{q}}, \sum_{\mathbf{k},\sigma} a^{\dagger}_{\mathbf{k},\sigma} a_{\mathbf{k},\sigma}\right] = 0, \tag{2.278}$$

which has a clear physical interpretation: the density operator $\hat{\rho}_{\mathbf{q}}$ conserves the total number of particles, it represents a fluctuation in the spatial distribution of a fixed number of particles, rather than a change in the total particle count;

7.
$$\left[\hat{\rho}_{\mathbf{q}}, \sum_{\mathbf{q}'} \hat{\rho}_{\mathbf{q}'} \hat{\rho}^{\dagger}_{\mathbf{q}'}\right] = 0. \tag{2.279}$$

Proof

1. We have to compute

$$\left[\hat{\rho}_{\mathbf{q}'}, \hat{\rho}_{\mathbf{q}''}\right] = \sum_{\mathbf{k}',\sigma'} \sum_{\mathbf{k}'',\sigma''} \left[a^{\dagger}_{\mathbf{k}'+\mathbf{q}',\sigma'} a_{\mathbf{k}',\sigma'}, a^{\dagger}_{\mathbf{k}''+\mathbf{q}'',\sigma''} a_{\mathbf{k}'',\sigma''}\right]. \tag{2.280}$$

From D.21, with the identifications

$$\alpha = \left(\mathbf{k}' + \mathbf{q}', \sigma'\right), \tag{2.281}$$

$$\beta = \left(\mathbf{k}'' + \mathbf{q}'', \sigma''\right), \tag{2.282}$$

$$\gamma = \left(\mathbf{k}', \sigma'\right), \tag{2.283}$$

$$\eta = \left(\mathbf{k}'', \sigma''\right), \tag{2.284}$$

one finds

$$\begin{aligned} &\left[a^{\dagger}_{\mathbf{k}'+\mathbf{q}',\sigma'} a_{\mathbf{k}',\sigma'}, a^{\dagger}_{\mathbf{k}''+\mathbf{q}'',\sigma''} a_{\mathbf{k}'',\sigma''}\right] \\ &\quad = \delta_{\mathbf{k}', \mathbf{k}''+\mathbf{q}''}\, \delta_{\sigma',\sigma''}\, a^{\dagger}_{\mathbf{k}'+\mathbf{q}',\sigma'} a_{\mathbf{k}'',\sigma''} - \delta_{\mathbf{k}'+\mathbf{q}', \mathbf{k}''}\, \delta_{\sigma',\sigma''}\, a^{\dagger}_{\mathbf{k}''+\mathbf{q}'',\sigma''} a_{\mathbf{k}',\sigma'}, \end{aligned} \tag{2.285}$$

$$\begin{aligned} \left[\hat{\rho}_{\mathbf{q}'}, \hat{\rho}_{\mathbf{q}''}\right] = \sum_{\mathbf{k}',\sigma'} \sum_{\mathbf{k}'',\sigma''} &\left(\delta_{\mathbf{k}', \mathbf{k}''+\mathbf{q}''}\, \delta_{\sigma',\sigma''}\, a^{\dagger}_{\mathbf{k}'+\mathbf{q}',\sigma'} a_{\mathbf{k}'',\sigma''}\right. \\ &\left.-\delta_{\mathbf{k}'+\mathbf{q}', \mathbf{k}''}\, \delta_{\sigma',\sigma''}\, a^{\dagger}_{\mathbf{k}''+\mathbf{q}'',\sigma''} a_{\mathbf{k}',\sigma'}\right). \end{aligned} \tag{2.286}$$

The sums over $\mathbf{k}''$ and σ'' enforce in the first term the conditions $\mathbf{k}'' = \mathbf{k}' - \mathbf{q}''$ and $\sigma'' = \sigma'$, while in the second term $\mathbf{k}'' = \mathbf{k}' + \mathbf{q}'$ and $\sigma'' = \sigma'$. Therefore, we obtain

$$\left[\hat{\rho}_{\mathbf{q}'}, \hat{\rho}_{\mathbf{q}''}\right] = \sum_{\mathbf{k}',\sigma'} a^\dagger_{\mathbf{k}'+\mathbf{q}',\sigma'} a_{\mathbf{k}'-\mathbf{q}'',\sigma'} - \sum_{\mathbf{k}',\sigma'} a^\dagger_{\mathbf{k}'+\mathbf{q}'+\mathbf{q}'',\sigma'} a_{\mathbf{k}',\sigma'} . \tag{2.287}$$

The equivalence of the two contributions, which ensures their cancellation, can be established by carrying out a change of variables in the first summation. Specifically, by introducing $\mathbf{k}''' = \mathbf{k}' - \mathbf{q}''$, the two sums are seen to coincide, indeed

$$\left[\hat{\rho}_{\mathbf{q}'}, \hat{\rho}_{\mathbf{q}''}\right] = \sum_{\mathbf{k}''',\sigma'} a^\dagger_{\mathbf{k}'''+\mathbf{q}'+\mathbf{q}'',\sigma'} a_{\mathbf{k}''',\sigma'} - \sum_{\mathbf{k}',\sigma'} a^\dagger_{\mathbf{k}'+\mathbf{q}'+\mathbf{q}'',\sigma'} a_{\mathbf{k}',\sigma'} , \tag{2.288}$$

which vanishes identically for all $\mathbf{q}'$ and $\mathbf{q}''$.

2. From Eq. 2.268 we obtain

$$\left[\hat{\rho}^\dagger_{\mathbf{q}'}, \hat{\rho}^\dagger_{\mathbf{q}''}\right] = \left[\hat{\rho}_{-\mathbf{q}'}, \hat{\rho}_{-\mathbf{q}''}\right] , \tag{2.289}$$

and by Eq. 2.273 the claim follows.

3. Similarly, from Eq. 2.268 we obtain

$$\left[\hat{\rho}_{\mathbf{q}'}, \hat{\rho}^\dagger_{\mathbf{q}''}\right] = \left[\hat{\rho}_{\mathbf{q}'}, \hat{\rho}_{-\mathbf{q}''}\right] , \tag{2.290}$$

and by Eq. 2.273 the claim follows.

4. We have to compute

$$\left[\hat{\rho}_{\mathbf{q}}, \sum_{\mathbf{k},\sigma} \mathcal{E}_{\mathbf{k}} a^\dagger_{\mathbf{k},\sigma} a_{\mathbf{k},\sigma}\right] = \sum_{\mathbf{k},\sigma} \sum_{\mathbf{k}',\sigma'} \mathcal{E}_{\mathbf{k}} \left[a^\dagger_{\mathbf{k}'+\mathbf{q},\sigma'} a_{\mathbf{k}',\sigma'}, a^\dagger_{\mathbf{k},\sigma} a_{\mathbf{k},\sigma}\right] . \tag{2.291}$$

From D.21, with the identifications

$$\alpha = \left(\mathbf{k}' + \mathbf{q}, \sigma'\right) , \tag{2.292}$$

$$\beta = \left(\mathbf{k}', \sigma'\right) , \tag{2.293}$$

$$\gamma = (\mathbf{k}, \sigma) , \tag{2.294}$$

$$\eta = (\mathbf{k}, \sigma) , \tag{2.295}$$

one finds

$$\left[a^\dagger_{\mathbf{k}'+\mathbf{q},\sigma'} a_{\mathbf{k}',\sigma'}, a^\dagger_{\mathbf{k},\sigma} a_{\mathbf{k},\sigma}\right] = \delta_{\mathbf{k}',\mathbf{k}} \delta_{\sigma',\sigma} a^\dagger_{\mathbf{k}'+\mathbf{q},\sigma'} a_{\mathbf{k},\sigma} - \delta_{\mathbf{k}'+\mathbf{q},\mathbf{k}} \delta_{\sigma',\sigma} a^\dagger_{\mathbf{k},\sigma} a_{\mathbf{k}',\sigma'} , \tag{2.296}$$

$$\left[\hat{\rho}_{\mathbf{q}}, \sum_{\mathbf{k},\sigma} \mathcal{E}_{\mathbf{k}} a^{\dagger}_{\mathbf{k},\sigma} a_{\mathbf{k},\sigma}\right]$$
$$= \sum_{\mathbf{k},\sigma} \sum_{\mathbf{k}',\sigma'} \mathcal{E}_{\mathbf{k}} \left(\delta_{\mathbf{k}',\mathbf{k}} \delta_{\sigma',\sigma} a^{\dagger}_{\mathbf{k}'+\mathbf{q},\sigma'} a_{\mathbf{k},\sigma} - \delta_{\mathbf{k}'+\mathbf{q},\mathbf{k}} \delta_{\sigma',\sigma} a^{\dagger}_{\mathbf{k},\sigma} a_{\mathbf{k}',\sigma'}\right). \tag{2.297}$$

The sums over $\mathbf{k}$ and σ enforce in the first term the conditions $\mathbf{k} = \mathbf{k}'$ and $\sigma = \sigma'$, while in the second term $\mathbf{k} = \mathbf{k}' + \mathbf{q}$ and $\sigma = \sigma'$. Therefore, we obtain

$$\left[\hat{\rho}_{\mathbf{q}}, \sum_{\mathbf{k},\sigma} \mathcal{E}_{\mathbf{k}} a^{\dagger}_{\mathbf{k},\sigma} a_{\mathbf{k},\sigma}\right] = \sum_{\mathbf{k}',\sigma'} \mathcal{E}_{\mathbf{k}'} a^{\dagger}_{\mathbf{k}'+\mathbf{q},\sigma'} a_{\mathbf{k}',\sigma'} - \sum_{\mathbf{k}',\sigma'} \mathcal{E}_{\mathbf{k}'+\mathbf{q}} a^{\dagger}_{\mathbf{k}'+\mathbf{q},\sigma'} a_{\mathbf{k}',\sigma'}, \tag{2.298}$$

which is the thesis.

5. From Eq. 2.268 we obtain

$$\left[\left[\hat{\rho}_{\mathbf{q}}, \sum_{\mathbf{k},\sigma} \mathcal{E}_{\mathbf{k}} a^{\dagger}_{\mathbf{k},\sigma} a_{\mathbf{k},\sigma}\right], \hat{\rho}^{\dagger}_{\mathbf{q}}\right] = \left[\left[\hat{\rho}_{\mathbf{q}}, \sum_{\mathbf{k},\sigma} \mathcal{E}_{\mathbf{k}} a^{\dagger}_{\mathbf{k},\sigma} a_{\mathbf{k},\sigma}\right], \hat{\rho}_{-\mathbf{q}}\right], \tag{2.299}$$

and from 2.276 we have to compute

$$\left[\left[\hat{\rho}_{\mathbf{q}}, \sum_{\mathbf{k},\sigma} \mathcal{E}_{\mathbf{k}} a^{\dagger}_{\mathbf{k},\sigma} a_{\mathbf{k},\sigma}\right], \hat{\rho}^{\dagger}_{\mathbf{q}}\right]$$
$$= \left[\sum_{\mathbf{k}',\sigma'} \left(\mathcal{E}_{\mathbf{k}'} - \mathcal{E}_{\mathbf{k}'+\mathbf{q}}\right) a^{\dagger}_{\mathbf{k}'+\mathbf{q},\sigma'} a_{\mathbf{k}',\sigma'}, \sum_{\mathbf{k}'',\sigma''} a^{\dagger}_{\mathbf{k}''-\mathbf{q},\sigma''} a_{\mathbf{k}'',\sigma''}\right] =$$
$$= \sum_{\mathbf{k}',\sigma'} \sum_{\mathbf{k}'',\sigma''} \left(\mathcal{E}_{\mathbf{k}'} - \mathcal{E}_{\mathbf{k}'+\mathbf{q}}\right) \left[a^{\dagger}_{\mathbf{k}'+\mathbf{q},\sigma'} a_{\mathbf{k}',\sigma'}, a^{\dagger}_{\mathbf{k}''-\mathbf{q},\sigma''} a_{\mathbf{k}'',\sigma''}\right]. \tag{2.300}$$

From D.21, with the identifications

$$\alpha = \left(\mathbf{k}' + \mathbf{q}, \sigma'\right), \tag{2.301}$$

$$\beta = \left(\mathbf{k}', \sigma'\right), \tag{2.302}$$

$$\gamma = \left(\mathbf{k}'' - \mathbf{q}, \sigma''\right), \tag{2.303}$$

$$\eta = \left(\mathbf{k}'', \sigma''\right), \tag{2.304}$$

we get

$$\left[a^{\dagger}_{\mathbf{k}'+\mathbf{q},\sigma'}a_{\mathbf{k}',\sigma'},\, a^{\dagger}_{\mathbf{k}''-\mathbf{q},\sigma''}a_{\mathbf{k}'',\sigma''}\right]$$
$$= \delta_{\mathbf{k}',\mathbf{k}''-\mathbf{q}}\delta_{\sigma',\sigma''}a^{\dagger}_{\mathbf{k}'+\mathbf{q},\sigma'}a_{\mathbf{k}'',\sigma''} - \delta_{\mathbf{k}'+\mathbf{q},\mathbf{k}''}\delta_{\sigma',\sigma''}a^{\dagger}_{\mathbf{k}''-\mathbf{q},\sigma''}a_{\mathbf{k}',\sigma'}, \tag{2.305}$$

$$\left[\left[\hat{\rho}_{\mathbf{q}}, \sum_{\mathbf{k},\sigma}\mathcal{E}_{\mathbf{k}}a^{\dagger}_{\mathbf{k},\sigma}a_{\mathbf{k},\sigma}\right], \hat{\rho}^{\dagger}_{\mathbf{q}}\right] =$$
$$= \sum_{\mathbf{k}',\sigma'}\sum_{\mathbf{k}'',\sigma''}(\mathcal{E}_{\mathbf{k}'}$$
$$-\mathcal{E}_{\mathbf{k}'+\mathbf{q}})\left(\delta_{\mathbf{k}',\mathbf{k}''-\mathbf{q}}\delta_{\sigma',\sigma''}a^{\dagger}_{\mathbf{k}'+\mathbf{q},\sigma'}a_{\mathbf{k}'',\sigma''} - \delta_{\mathbf{k}'+\mathbf{q},\mathbf{k}''}\delta_{\sigma',\sigma''}a^{\dagger}_{\mathbf{k}''-\mathbf{q},\sigma''}a_{\mathbf{k}',\sigma'}\right). \tag{2.306}$$

The sums over $\mathbf{k}'$ and σ' enforce the conditions $\mathbf{k}' = \mathbf{k}'' - \mathbf{q}$ and $\sigma' = \sigma''$ in both terms. Therefore, we obtain

$$\left[\left[\hat{\rho}_{\mathbf{q}}, \sum_{\mathbf{k},\sigma}\mathcal{E}_{\mathbf{k}}a^{\dagger}_{\mathbf{k},\sigma}a_{\mathbf{k},\sigma}\right], \hat{\rho}^{\dagger}_{\mathbf{q}}\right] =$$
$$= \sum_{\mathbf{k}'',\sigma''}\left(\mathcal{E}_{\mathbf{k}''-\mathbf{q}} - \mathcal{E}_{\mathbf{k}''}\right)a^{\dagger}_{\mathbf{k}'',\sigma''}a_{\mathbf{k}'',\sigma''} - \sum_{\mathbf{k}'',\sigma''}\left(\mathcal{E}_{\mathbf{k}''-\mathbf{q}} - \mathcal{E}_{\mathbf{k}''}\right)a^{\dagger}_{\mathbf{k}''-\mathbf{q},\sigma''}a_{\mathbf{k}''-\mathbf{q},\sigma''}. \tag{2.307}$$

By performing a change of variables $\mathbf{k}''' = \mathbf{k}'' - \mathbf{q}$ in the second term, we obtain

$$\left[\left[\hat{\rho}_{\mathbf{q}}, \sum_{\mathbf{k},\sigma}\mathcal{E}_{\mathbf{k}}a^{\dagger}_{\mathbf{k},\sigma}a_{\mathbf{k},\sigma}\right], \hat{\rho}^{\dagger}_{\mathbf{q}}\right] =$$
$$= \sum_{\mathbf{k}'',\sigma''}\left(\mathcal{E}_{\mathbf{k}''-\mathbf{q}} - \mathcal{E}_{\mathbf{k}''}\right)a^{\dagger}_{\mathbf{k}'',\sigma''}a_{\mathbf{k}'',\sigma''} - \sum_{\mathbf{k}''',\sigma''}\left(\mathcal{E}_{\mathbf{k}'''} - \mathcal{E}_{\mathbf{k}'''+\mathbf{q}}\right)a^{\dagger}_{\mathbf{k}''',\sigma''}a_{\mathbf{k}''',\sigma''}. \tag{2.308}$$

As $\mathbf{k}''$ and $\mathbf{k}'''$ are dummy summation indices, the stated result follows.

6. In this case, one may proceed, analogously to some of the previous derivations, by employing Eq. D.21. Alternatively, one can observe that

$$\left[\hat{\rho}_{\mathbf{q}}, \sum_{\mathbf{k},\sigma}a^{\dagger}_{\mathbf{k},\sigma}a_{\mathbf{k},\sigma}\right] \tag{2.309}$$

is a particular case of

$$\left[\hat{\rho}_{\mathbf{q}}, \sum_{\mathbf{k},\sigma} \mathcal{E}_{\mathbf{k}} a^{\dagger}_{\mathbf{k},\sigma} a_{\mathbf{k},\sigma}\right] \tag{2.310}$$

with $\mathcal{E}_{\mathbf{k}} = 1$ for all $\mathbf{k}$. Consequently, the assertion follows directly from Eq. 2.276.

7. From Jacobi identity F.1, we obtain

$$\left[\hat{\rho}_{\mathbf{q}}, \sum_{\mathbf{q}'} \hat{\rho}_{\mathbf{q}'} \hat{\rho}^{\dagger}_{\mathbf{q}'}\right] = \sum_{\mathbf{q}'} \left(\left[\hat{\rho}_{\mathbf{q}}, \hat{\rho}_{\mathbf{q}'}\right] \hat{\rho}^{\dagger}_{\mathbf{q}'} + \hat{\rho}_{\mathbf{q}'} \left[\hat{\rho}_{\mathbf{q}}, \hat{\rho}^{\dagger}_{\mathbf{q}'}\right]\right), \tag{2.311}$$

and from Eqs. 2.273 and 2.275 the stated result follows.

□

Remark 2.7 (Bosonic Behaviour of Density Fluctuation Operators Under Permutations) Regardless of whether the creation and annihilation operators in 2.262 and 2.268 are fermionic or bosonic, the density fluctuation operators $\hat{\rho}_{\mathbf{q}}$ and $\hat{\rho}^{\dagger}_{\mathbf{q}}$ effectively behave as bosonic objects. This is due to the fact that they are bilinear in the fundamental operators, and hence invariant under the exchange of their constituents: they always involve an even number of creation and annihilation operators, so the relative minus signs that would arise in the fermionic case cancel out. Moreover, they satisfy relations that mirror those of bosonic operators: in particular, Eqs. 2.273 and 2.274, play the same role as Eqs. D.2 and D.3, respectively. It should be noted, however, that the bosonic algebra between annihilation and creation operators is not strictly fulfilled: in particular, instead of Eq. D.1, one finds Eq. 2.275. For this reason, we shall refer to these operators as exhibiting bosonic behavior, even though they do not satisfy the full bosonic algebra.

2.15 Representation in Fock Space of Two-Body Operators

The representation of one-body operators in Fock space having been established, we now consider two-body operators. These operators describe interactions between pairs of particles and are initially defined in first quantization as sums over all distinct particle pairs. Our goal in this section is to rewrite such operators in the formalism of second quantization, where the pairwise interactions are naturally encoded using creation and annihilation field operators, while respecting the symmetry of the interaction potential. Consider the two-body operator

$$U_N = \frac{1}{2} \sum_{i \neq j}^{N} U(x_i, x_j), \tag{2.312}$$

where the interaction potential satisfies the symmetry condition

$$U(x_i, x_j) = U(x_j, x_i). \tag{2.313}$$

Theorem 2.16 (Two-Body Operators in Second Quantization) *Let U_N be a two-body operator acting on the N-particle Hilbert space H_N, of the form 2.312, where the interaction potential satisfies the symmetry condition 2.313. In Fock space, its second-quantized form is*

$$\hat{U} = \frac{1}{2}\int dx \int dx'\ \hat{\psi}^\dagger(x')\hat{\psi}^\dagger(x)U(x,x')\hat{\psi}(x)\hat{\psi}(x'). \tag{2.314}$$

Proof We want to express the operator 2.312 within the formalism of second quantization, i.e.,

$$\langle\phi|\hat{U}|\phi'\rangle_{\mathcal{F}} \equiv \sum_{N=2}^{\infty}\left(\phi^{(N)}, U_N\phi'^{(N)}\right)_{H_N}. \tag{2.315}$$

Let us study the scalar product in the space H_N, that is,

$$\begin{aligned}&\left(\phi^{(N)}, U_N\phi'^{(N)}\right)_{H_N}\\ &= \frac{1}{2}\sum_{i\neq j}^{N}\int dx_1\ldots\int dx_N\left(\phi^{(N)}(x_1,\ldots,x_N)\right)^* U(x_i,x_j)\phi'^{(N)}(x_1,\ldots,x_N),\end{aligned} \tag{2.316}$$

we fix an index i and divide the sum in the following way $\frac{1}{2}\sum_{i=1}^{N}\left\{\sum_{j<i}\cdots+\sum_{j>i}\cdots\right\}$. For example, for $j<i$

$$\begin{aligned}&\int dx_1\ldots\int dx_{j-1}\int dx_{j+1}\ldots\int dx_{i-1}\int dx_{i+1}\ldots\int dx_N\\ &\int dx_j\int dx_i\left(\phi^{(N)}(x_1,\ldots,x_j,\ldots,x_i,\ldots,x_N)\right)^*\\ &U(x_i,x_j)\left(\phi'^{(N)}(x_1,\ldots,x_j,\ldots,x_i,\ldots,x_N)\right).\end{aligned} \tag{2.317}$$

We set $x_j \equiv x$, $x_i \equiv x'$, rename the remaining $N-2$ coordinates as the sequence $x_1,\ldots,x_{N-2}$, and then reorder the entire set of variables so that x and x' appear in the first two positions, followed by $x_1,\ldots,x_{N-2}$. The number of permutations required to bring the coordinates into this order is even; hence, the sign of the total wavefunction remains unchanged, and the following expressions do not depend on the statistical nature of the particles. We get

$$\int dx_1 \ldots \int dx_{N-2} \int dx \int dx' \left(\phi^{(N)}(x, x', x_1, \ldots, x_{N-2})\right)^*$$
$$U(x', x)\phi'^{(N)}(x, x', x_1, \ldots, x_{N-2}). \tag{2.318}$$

We proceed in the same way for $j > i$, since the variables are dummy and $U(x, x') = U(x', x)$. The two sums do not depend on the indices and return

$$\sum_{j<i} 1 = \frac{N(N-1)}{2}, \tag{2.319}$$

then

$$\left(\phi^{(N)}, U_N \phi'^{(N)}\right)_{H_N}$$
$$= \frac{N(N-1)}{2} \int dx_1 \ldots \int dx_{N-2} \int dx \int dx'$$
$$\left(\phi^{(N)}(x, x', x_1, \ldots, x_{N-2})\right)^* U(x, x')\phi'^{(N)}(x, x', x_1, \ldots, x_{N-2}) =$$
$$= \frac{1}{2} \int dx_1 \ldots \int dx_{N-2} \int dx \int dx' \sqrt{N}\sqrt{N-1}$$
$$\left(\phi^{(N)}(x, x', x_1, \ldots, x_{N-2})\right)^* U(x, x')\sqrt{N}\sqrt{N-1}\phi'^{(N)}(x, x', x_1, \ldots, x_{N-2}), \tag{2.320}$$

where we have used

$$\frac{N(N-1)}{2} = \frac{\sqrt{N-1}\sqrt{N}\sqrt{N-1}\sqrt{N}}{2}. \tag{2.321}$$

We insert

$$\left[\hat{\psi}(x)\hat{\psi}(x')\phi\right]^{(N-2)}(x_1, \ldots, x_{N-2}) = \sqrt{N}\sqrt{N-1}\phi^{(N)}(x, x', x_1, \ldots, x_{N-2}), \tag{2.322}$$

into the scalar product in H_N, exchanging the order of integration between $\int dx_1 \ldots dx_{N-2}$ and $\int dx \int dx'$, thereby obtaining a scalar product in H_{N-2}, that is,

$$\langle \phi | \hat{U} | \phi' \rangle_{\mathcal{F}}$$
$$= \sum_{N=2}^{\infty} (\phi^{(N)}, U_N \phi'^{(N)})_N =$$

$$= \frac{1}{2}\sum_{N=2}^{\infty}\int dx \int dx' \Big(\Big[\hat{\psi}(x)\hat{\psi}(x')|\phi\rangle\Big]^{(N-2)},$$

$$U(x,x')\Big[\hat{\psi}(x)\hat{\psi}(x')|\phi'\rangle\Big]^{(N-2)}\Big)_{H_{N-2}} =$$

$$= \sum_{N=0}^{\infty}\frac{1}{2}\int dx \int dx' \Big(\Big[\hat{\psi}(x)\hat{\psi}(x')|\phi\rangle\Big]^{(N)},$$

$$U(x,x')\Big[\hat{\psi}(x)\hat{\psi}(x')||\phi'\rangle\Big]^{(N)}\Big)_{H_N}. \tag{2.323}$$

Assuming that it is possible to reverse the order between $\sum_{N=0}^{\infty}$ and $\int dx \int dx'$, we have

$$\langle\phi|\hat{U}|\phi'\rangle_{\mathcal{F}}$$

$$= \frac{1}{2}\int dx \int dx' \sum_{N=0}^{\infty}\Big(\Big[\hat{\psi}(x)\hat{\psi}(x')|\phi\rangle\Big]^{(N)},$$

$$U(x,x')\Big[\hat{\psi}(x)\hat{\psi}(x')|\phi'\rangle\Big]^{(N)}\Big)_{H_N} =$$

$$= \frac{1}{2}\int dx \int dx' \Big(\hat{\psi}(x)\hat{\psi}(x')|\phi\rangle, U(x,x')\hat{\psi}(x)\hat{\psi}(x')|\phi'\rangle\Big)_{\mathcal{F}} =$$

$$= \frac{1}{2}\int dx \int dx' \langle\phi|\hat{\psi}^{\dagger}(x')\hat{\psi}^{\dagger}(x)U(x,x')\hat{\psi}(x)\hat{\psi}(x')|\phi'\rangle_{\mathcal{F}}, \tag{2.324}$$

we formed a scalar product in $\mathcal{F}$ and the obtained relation holds $\forall\phi, \phi' \in \mathcal{F}$. □

Note that the order of action of the creation field operators is mirrored by that of the annihilation field operators, as shown by 2.314. Given a basis $\{\varphi_\alpha\} \in H_1$, according to Eq. 2.314, we have

$$\hat{U} = \frac{1}{2}\sum_{\substack{\alpha_1,\alpha_2,\\ \alpha_3,\alpha_4}}\Big[\int dx \int dx' \varphi^*_{\alpha_1}(x)\varphi^*_{\alpha_2}(x')U(x,x')\varphi_{\alpha_3}(x')\varphi_{\alpha_4}(x)\Big] a^{\dagger}_{\alpha_1}a^{\dagger}_{\alpha_2}a_{\alpha_3}a_{\alpha_4} =$$

$$= \frac{1}{2}\sum_{\substack{\alpha_1,\alpha_2,\\ \alpha_3,\alpha_4}}\langle\alpha_1\alpha_2|U|\alpha_3\alpha_4\rangle a^{\dagger}_{\alpha_1}a^{\dagger}_{\alpha_2}a_{\alpha_3}a_{\alpha_4}, \tag{2.325}$$

which is independent of both the basis and the statistical nature of the particles.

2.15.1 Two-Body Hamiltonian

In first quantization, a typical Hamiltonian of matter physics is given by

$$\begin{aligned}\mathcal{H} &= \sum_{i=1}^{N}\left[\frac{\hat{\mathbf{p}}_i^2}{2m} + V(x_i)\right] + \frac{1}{2}\sum_{i\neq j} U(x_i, x_j) \equiv \\ &\equiv \mathcal{H}_0 + \frac{1}{2}\sum_{i\neq j} U(x_i, x_j),\end{aligned} \tag{2.326}$$

where $\sum_{i=1}^{N} V(x_i)$ is an external field and $\frac{1}{2}\sum_{i\neq j}^{N} U(x_i, x_j)$ is a two-body operator. By applying Theorem 2.16 and expanding in the eigenbasis of $\mathcal{H}_0$, we obtain

$$\hat{\mathcal{H}} = \int dx \hat{\psi}^\dagger(x)\mathcal{H}_0(x)\hat{\psi}(x) + \frac{1}{2}\int dx \int dx' \; \hat{\psi}^\dagger(x)\hat{\psi}^\dagger(x')U(x,x')\hat{\psi}(x')\hat{\psi}(x), \tag{2.327}$$

$$\hat{\mathcal{H}} = \sum_{\alpha_1,\alpha_2} \langle \alpha_1|\mathcal{H}_0|\alpha_2\rangle a^\dagger_{\alpha_1} a_{\alpha_2} + \frac{1}{2}\sum_{\substack{\alpha_1,\alpha_2,\\ \alpha_3,\alpha_4}} \langle \alpha_1\alpha_2|U|\alpha_3\alpha_4\rangle a^\dagger_{\alpha_1} a^\dagger_{\alpha_2} a_{\alpha_3} a_{\alpha_4}. \tag{2.328}$$

The first term can be diagonalized; however, the eigenstates of $\hat{\mathcal{H}}_0$ are not eigenstates of the two-body operator. For this reason, one works in the basis of the eigenstates of $\hat{\mathcal{H}}_0$ and treats the two-body interaction as a perturbation. When particle-particle interactions are taken into account, the eigenstates of the Hamiltonian $\hat{\mathcal{H}}$ are expected to be linear combinations of Slater determinants (or permanents). The coefficients of these linear combinations cannot be determined exactly, since the eigenvalue problem is not analytically solvable. At this point, one typically abandons the description in terms of wave functions and introduces the many-body Green's function (see Chap. 6), which encodes correlations between observables, allowing one to capture the effects of interactions without explicitly tracking individual particle states (see Chap. 7).

2.16 Reduced Density Matrices

Consider a system of N particles, described by the pure state density operator ρ defined in Eq. 1.316. The matrix elements of the pure-state density operator in the coordinate representation is obtained from the N-particle wavefunction as

$$\begin{aligned}\rho(x_1,\ldots,x_N;x'_1,\ldots,x'_N) &= \langle x_1,\ldots,x_N|\rho|x'_1,\ldots,x'_N\rangle = \\ &= \psi(x_1,\ldots,x_N)\psi^*(x'_1,\ldots,x'_N),\end{aligned} \tag{2.329}$$

whose diagonal elements are given by

$$\rho(x_1, \ldots, x_N) = |\psi(x_1, \ldots, x_N)|^2 . \tag{2.330}$$

We define the first-order reduced density matrix, or one-particle reduced density matrix, as the object

$$\rho_1(x) = N \int dx_2 \int dx_3 \ldots \int dx_N \, |\psi(x, x_2, \ldots, x_N)|^2 , \tag{2.331}$$

which represents the probability of finding one of the N particles at x, that is, at position $\mathbf{r}$ with spin component s. We have

Theorem 2.17 (Representation of the One-Particle Reduced Density Matrix) *Let $\psi(x_1, \ldots, x_N)$ be a normalized wavefunction of a system of N indistinguishable particles, then*

$$\rho_1(x) = \frac{1}{N} \left\langle \sum_{i=1}^{N} \delta(x - x_i) \right\rangle , \tag{2.332}$$

where the expectation value $\langle \ldots \rangle$ is taken with respect to the state ψ.

Proof Given the quantity

$$\left\langle \sum_{i=1}^{N} \delta(x - x_i) \right\rangle$$
$$= \int dx_1 \cdots dx_N \left(\sum_{i=1}^{N} \delta(x - x_i) \right) |\psi(x_1, \ldots, x_N)|^2, \tag{2.333}$$

since the particles are indistinguishable, the wavefunction is symmetric (for bosons) or antisymmetric (for fermions), and in both cases the squared absolute value is symmetric with respect to any permutation of the indices, consequently, the object

$$\int dx_1 \cdots dx_N \delta(x - x_i) |\psi(x_1, \ldots, x_N)|^2 \tag{2.334}$$

does not depend on i, so we can fix $i = 1$ and write

$$\left\langle \sum_{i=1}^{N} \delta(x - x_i) \right\rangle = \sum_{i=1}^{N} \int dx_1 \cdots dx_N \delta(x - x_i) |\psi(x_1, \ldots, x_N)|^2 =$$
$$= N \int dx_1 \cdots dx_N \delta(x - x_1) |\psi(x_1, \ldots, x_N)|^2 . \tag{2.335}$$

Integrating over the variable x_1, we obtain

$$\left\langle \sum_{i=1}^{N} \delta(x - x_i) \right\rangle = N \int dx_2 \cdots dx_N |\psi(x, x_2, \ldots, x_N)|^2, \tag{2.336}$$

then

$$\begin{aligned}\rho_1(x) &= N \int dx_2 \cdots dx_N |\psi(x, x_2, \ldots, x_N)|^2 = \\ &= \left\langle \sum_{i=1}^{N} \delta(x - x_i) \right\rangle,\end{aligned} \tag{2.337}$$

from which the statement follows. □

Note that the particle density operator in first quantization (see Eq. 2.259) does not appear in Theorem 2.17, since it is defined differently, that is, it is independent of spin. Thanks to 2.332, we define the projection of the one-particle reduced density matrix onto the N-particle Fock subspace as

$$\begin{aligned}\tilde{\rho}_1(x) &= N\rho_1(x) = \\ &= \left\langle \sum_{i=1}^{N} \delta(x - x_i) \right\rangle.\end{aligned} \tag{2.338}$$

Similarly to the matrix 2.331, we define the two-body reduced density matrix, or pair reduced density matrix, as the object

$$\rho_2(x, x') = N(N-1) \int dx_3 \int dx_4 \ldots \int dx_N \left|\psi(x, x', x_3, \ldots, x_N)\right|^2, \tag{2.339}$$

which is the probability of finding a particle at x, that is, at position $\mathbf{r}$ with spin component s, and a particle at x', that is, at position $\mathbf{r}'$ with spin component s'. We have

Theorem 2.18 (Representation of the Two-Body Reduced Density Matrix) *Let $\psi(x_1, \ldots, x_N)$ be a normalized wavefunction of a system of N indistinguishable particles, then*

$$\rho_2(x, x') = \frac{1}{N(N-1)} \left\langle \sum_{i \neq j} \delta(x - x_i)\, \delta(x' - x_j) \right\rangle, \tag{2.340}$$

where the expectation value $\langle \ldots \rangle$ is taken with respect to the state ψ.

Proof Given the quantity

$$\left\langle \sum_{i \neq j} \delta(x - x_i)\,\delta(y - x_j) \right\rangle$$
$$= \int dx_1 \cdots dx_N \left(\sum_{i \neq j} \delta(x - x_i)\,\delta(x' - x_j) \right) |\psi(x_1, \ldots, x_N)|^2, \tag{2.341}$$

Since the particles are indistinguishable, the wavefunction is symmetric (for bosons) or antisymmetric (for fermions), and in both cases the squared absolute value is symmetric with respect to any permutation of the variables, it follows that the object

$$\int dx_1 \cdots dx_N\, \delta(x - x_i)\,\delta(x' - x_j)\, |\psi(x_1, \ldots, x_N)|^2 \tag{2.342}$$

does not depend on the pair of indices $i \neq j$, and given that the total number of ordered pairs (i, j) with $i \neq j$ is $N(N-1)$, we can choose an arbitrary pair, for example $i = 1$, $j = 2$, and we write

$$\left\langle \sum_{i \neq j} \delta(x - x_i)\,\delta(x' - x_j) \right\rangle$$
$$= \sum_{i \neq j} \int dx_1 \cdots dx_N \delta(x - x_i)\,\delta(x' - x_j) |\psi(x_1, \ldots, x_N)|^2 =$$
$$= N(N-1) \int dx_1 \cdots dx_N\, \delta(x - x_1)\,\delta(x' - x_2)\, |\psi(x_1, \ldots, x_N)|^2. \tag{2.343}$$

Integrating over the variables x_1, x_2, we obtain

$$\left\langle \sum_{i \neq j} \delta(x - x_i)\,\delta(x' - x_j) \right\rangle = N(N-1) \int dx_3 \cdots dx_N |\psi(x, x', x_3, \ldots, x_N)|^2, \tag{2.344}$$

then

$$\rho_2(x, x') = N(N-1) \int dx_3 \cdots dx_N |\psi(x, x', x_3, \ldots, x_N)|^2 =$$
$$= \left\langle \sum_{i \neq j} \delta(x - x_i)\,\delta(x' - x_j) \right\rangle, \tag{2.345}$$

from which the statement follows. □

Thanks to Eq. 2.340, we define the projection of the two-body reduced density matrix onto the N-particle Fock subspace as

$$\tilde{\rho}_2(x, x') = N(N-1)\rho_2(x, x') = \\ = \left\langle \sum_{i \neq j} \delta(x - x_i)\, \delta(x' - x_j) \right\rangle. \tag{2.346}$$

We now want to express the reduced matrices 2.331 and 2.339 in second quantization. From 2.332 and 2.340, one finds

$$\begin{aligned} \hat{\tilde{\rho}}_1(x) &= \int dx_1\, \hat{\psi}^\dagger(x_1)\, \langle \delta(x - x_1) \rangle\, \hat{\psi}(x_1) = \\ &= \left\langle \int dx_1\, \hat{\psi}^\dagger(x_1) \delta(x - x_1) \hat{\psi}(x_1) \right\rangle = \\ &= \left\langle \hat{\psi}^\dagger(x)\, \hat{\psi}(x) \right\rangle, \end{aligned} \tag{2.347}$$

$$\begin{aligned} \hat{\tilde{\rho}}_2(x, x') &= \frac{1}{2} \int dx_1 \int dx_2\, \hat{\psi}^\dagger(x_1)\, \hat{\psi}^\dagger(x_2)\, \left\langle \delta(x - x_1)\, \delta(x' - x_2) \right\rangle\, \hat{\psi}(x_2)\, \hat{\psi}(x_1) = \\ &= \frac{1}{2} \left\langle \int dx_1 \int dx_2\, \hat{\psi}^\dagger(x_1)\, \hat{\psi}^\dagger(x_2)\, \delta(x - x_1)\, \delta(x' - x_2)\, \hat{\psi}(x_2)\, \hat{\psi}(x_1) \right\rangle = \\ &= \frac{1}{2} \left\langle \hat{\psi}^\dagger(x)\, \hat{\psi}^\dagger(x')\, \hat{\psi}(x')\, \hat{\psi}(x) \right\rangle. \end{aligned} \tag{2.348}$$

The quantities introduced above allow us to define the pair correlation function

$$g(x, x') = \frac{\left\langle \psi^\dagger(x) \psi^\dagger(x') \psi(x') \psi(x) \right\rangle}{\left\langle \psi^\dagger(x) \psi(x) \right\rangle \left\langle \psi^\dagger(x') \psi(x') \right\rangle}, \tag{2.349}$$

which is directly proportional to the joint probability of finding one particle at position $\mathbf{r}$ and another at $\mathbf{r}'$. In particular, when the two particles are sufficiently far apart, the numerator factorizes as

$$\left\langle \hat{\psi}^\dagger(\mathbf{r}, s)\, \hat{\psi}^\dagger(\mathbf{r}', s')\, \hat{\psi}(\mathbf{r}', s')\, \hat{\psi}(\mathbf{r}, s) \right\rangle \\ \xrightarrow[|\mathbf{r}-\mathbf{r}'| \to +\infty]{} \left\langle \hat{\psi}^\dagger(\mathbf{r}, s)\, \hat{\psi}(\mathbf{r}, s) \right\rangle \left\langle \hat{\psi}^\dagger(\mathbf{r}', s')\, \hat{\psi}(\mathbf{r}', s') \right\rangle, \tag{2.350}$$

then

$$\lim_{|\mathbf{r}-\mathbf{r}'| \to +\infty} g(\mathbf{r}, s, \mathbf{r}', s') = 1. \tag{2.351}$$

2.17 Representation in Fock Space of n-Body Operators

The Fock space representations of one-body and two-body operators having been established, we now extend the analysis to n-body operators. These describe interactions among n particles and are initially defined in first quantization as sums over all distinct n-tuples of particles. Our goal in this section is to rewrite such operators in the formalism of second quantization, where the n-particle interactions are naturally encoded using creation and annihilation field operators, while respecting the symmetry of the interaction potential. Consider the n-body operator

$$M_N^{(n)} = \frac{1}{n!} \sum_{i_1 \neq i_2 \neq \ldots \neq i_n}^{N} W\left(x_{i_1}, \ldots, x_{i_n}\right), \tag{2.352}$$

where the interaction potential W satisfies the full symmetry condition

$$W\left(x_1, x_2, \ldots, x_n\right) = W\left(x_2, x_1, \ldots, x_n\right) = \ldots = W\left(x_n, \ldots, x_1\right), \tag{2.353}$$

Theorem 2.19 (n-Body Operators in Second Quantization) *Let $M_N^{(n)}$ be an n-body operator of the form 2.352, acting on the N-particle Hilbert space H_N, where the interaction potential satisfies the symmetry condition 2.353. In Fock space, its second-quantized form is*

$$\hat{M} = \frac{1}{n!} \int dx_1 \ldots \int dx_n \, \hat{\psi}^\dagger(x_1) \ldots \hat{\psi}^\dagger(x_n) W(x_1, \ldots, x_n) \hat{\psi}(x_n) \ldots \hat{\psi}(x_1). \tag{2.354}$$

2.18 Jellium Model

2.18.1 The Model in the First Quantization

Throughout this book, we will make extensive use of the jellium model, one of the simplest and most useful idealizations for describing electronic systems in solids. In this model, the ions are replaced by a uniformly distributed positive background that neutralizes the charge of the valence electrons. The absence of an explicit crystal structure allows for a clearer analysis of electron–electron interactions. Since the ionic mass is much greater than the electronic one, the positive background is assumed to be static. The jellium model thus provides a minimal yet effective theoretical framework, which will be used frequently in the analyses that follow. In first quantization, for a system of N electrons in a uniform background, the Hamiltonian takes the form

$$\mathcal{H} = \sum_{i=1}^{N} \frac{\hat{\mathbf{p}}_i^2}{2m_e} + \frac{1}{2} \sum_{i \neq j}^{N} \frac{q_e^2}{|\mathbf{r}_i - \mathbf{r}_j|} e^{-k_s |\mathbf{r}_i - \mathbf{r}_j|}$$

$$+ \frac{1}{2} \int d^3\mathbf{R} \int d^3\mathbf{R}' \frac{q_e^2}{|\mathbf{R} - \mathbf{R}'|} \rho(\mathbf{R}) \rho(\mathbf{R}') e^{-k_s |\mathbf{R} - \mathbf{R}'|} +$$

$$- q_e^2 \sum_{i=1}^{N} \int d^3\mathbf{R} \frac{\rho(\mathbf{R})}{|\mathbf{R} - \mathbf{r}_i|} e^{-k_s |\mathbf{R} - \mathbf{r}_i|}. \tag{2.355}$$

In this form, the Hamiltonian includes a screened Yukawa-type interaction instead of the bare Coulomb potential. The screening is controlled by the parameter $k_s > 0$, which represents the inverse screening length. Larger values of k_s correspond to stronger screening (i.e., shorter-range interactions), while in the limit

$$q_e^2 \lim_{k_s \to 0} \frac{e^{-k_s |\mathbf{r} - \mathbf{r}'|}}{|\mathbf{r} - \mathbf{r}'|} = \frac{q_e^2}{|\mathbf{r} - \mathbf{r}'|}, \tag{2.356}$$

one recovers the standard unscreened Coulomb interaction. The function $\rho(\mathbf{R})$ denotes the positive background charge density, the so-called jellium background, that neutralizes the electron charge. Assuming a uniform distribution, the background density is taken to be constant, that is

$$\rho(\mathbf{R}) = \rho = \frac{N}{V}, \tag{2.357}$$

where V is the volume of the system. This condition ensures overall charge neutrality, which is essential to avoid infrared divergences and to guarantee that the total energy per unit volume remains finite. We consider the thermodynamic limit, where both the number N of electrons and the volume V tend to infinity, while keeping the electronic density constant. Finally, we take the limit $k_s \to 0$ to recover the long-range Coulomb interaction. The order of these limits is crucial and cannot be reversed. Due to periodic boundary conditions, the system is invariant under spatial translations. Consequently, the integrand in

$$\frac{1}{2} \int d^3\mathbf{R} \int d^3\mathbf{R}' \frac{q_e^2}{|\mathbf{R} - \mathbf{R}'|} \rho(\mathbf{R}) \rho(\mathbf{R}') e^{-k_s |\mathbf{R} - \mathbf{R}'|} \tag{2.358}$$

depends only on the difference $\mathbf{R} - \mathbf{R}'$, and not on the individual positions. Using $\int d^3\mathbf{R}' = V$, one finds

$$\frac{1}{2} \int d^3\mathbf{R} \int d^3\mathbf{R}' \frac{q_e^2}{|\mathbf{R} - \mathbf{R}'|} \rho(\mathbf{R}) \rho(\mathbf{R}') e^{-k_s |\mathbf{R} - \mathbf{R}'|} = \frac{1}{2} q_e^2 \left(\frac{N}{V} \right)^2 V \int d^3\mathbf{R} \frac{e^{-k_s |\mathbf{R}|}}{|\mathbf{R}|}. \tag{2.359}$$

Switching to spherical coordinates and using

$$\int_0^{+\infty} x e^{-ax}\, dx = \frac{1}{a^2}, \quad a > 0, \tag{2.360}$$

we evaluate the integral as follows

$$\begin{aligned}\frac{1}{2}q_e^2\left(\frac{N}{V}\right)^2 V \int d^3\mathbf{R}\frac{e^{-k_s|\mathbf{R}|}}{|\mathbf{R}|} &= \frac{1}{2}q_e^2\left(\frac{N}{V}\right)^2 V \int_0^\infty 4\pi R^2 \frac{e^{-k_s R}}{R} dR = \\ &= \frac{q_e^2}{2}\left(\frac{N}{V}\right)^2 V 4\pi \frac{1}{k_s^2} = \\ &= \frac{q_e^2}{2}\frac{N^2}{V}\frac{4\pi}{k_s^2}.\end{aligned} \tag{2.361}$$

Now consider the interaction between electrons and the background

$$-q_e^2 \sum_{i=1}^N \int d^3\mathbf{R}\frac{\rho(\mathbf{R})}{|\mathbf{R}-\mathbf{r}_i|}e^{-k_s|\mathbf{R}-\mathbf{r}_i|} = -q_e^2\left(\frac{N}{V}\right)\sum_{i=1}^N \int d^3\mathbf{R}\frac{e^{-k_s|\mathbf{R}-\mathbf{r}_i|}}{|\mathbf{R}-\mathbf{r}_i|}. \tag{2.362}$$

The integral does not explicitly depend on the electronic position $\mathbf{r}_i$, due to translational invariance. Therefore, we may shift coordinates and write

$$\begin{aligned}-q_e^2 \sum_{i=1}^N \int d^3\mathbf{R}\frac{\rho(\mathbf{R})}{|\mathbf{R}-\mathbf{r}_i|}e^{-k_s|\mathbf{R}-\mathbf{r}_i|} &= -q_e^2\left(\frac{N}{V}\right)\sum_{i=1}^N \int d^3\mathbf{R}\frac{e^{-k_s|\mathbf{R}|}}{|\mathbf{R}|} = \\ &= -q_e^2\frac{N^2}{V}\frac{4\pi}{k_s^2}.\end{aligned} \tag{2.363}$$

Combining the two background-related terms, we obtain

$$\begin{aligned}&\frac{1}{2}\int d^3\mathbf{R}\int d^3\mathbf{R}'\frac{q_e^2}{|\mathbf{R}-\mathbf{R}'|}\rho(\mathbf{R})\rho(\mathbf{R}')e^{-k_s|\mathbf{r}-\mathbf{r}'|} \\ &-q_e^2\sum_{i=1}^N \int d^3\mathbf{R}\frac{\rho(\mathbf{R})}{|\mathbf{R}-\mathbf{r}_i|}e^{-k_s|\mathbf{R}-\mathbf{r}_i|} = -\frac{q_e^2}{2}\frac{N^2}{V}\frac{4\pi}{k_s^2}.\end{aligned} \tag{2.364}$$

Finally, the jellium Hamiltonian in first quantization is give by

$$\mathcal{H} = \sum_{i=1}^N \frac{\hat{\mathbf{p}}_i^2}{2m_e} + \frac{1}{2}\sum_{i\neq j}^N \frac{q_e^2}{|\mathbf{r}_i-\mathbf{r}_j|}e^{-k_s|\mathbf{r}_i-\mathbf{r}_j|} - \frac{q_e^2}{2}\frac{N^2}{V}\frac{4\pi}{k_s^2}. \tag{2.365}$$

2.18.2 The Model in the Second Quantization

After deriving the jellium Hamiltonian in first quantization (see Eq. 2.365), our goal is to reformulate it in second quantization. Note that, unlike the rest of the Chapter, here we use $C^\dagger$ and C to denote electron creation and annihilation operators instead of the more general $a^\dagger$ and a, since the treatment in this section is specialized to fermions (see Remark 2.3). We have

Theorem 2.20 (Jellium Hamiltonian in Second Quantization (Canonical Ensemble)) *The Hamiltonian of the jellium model, that is Eq. 2.365, in the canonical ensemble, in the limit $N, V \to \infty$ with constant density ρ, and for $k_s \to 0$, can be written in second quantization as*

$$\hat{\mathcal{H}} = \sum_{\mathbf{k},\sigma} \frac{\hbar^2 \mathbf{k}^2}{2m_e} C^\dagger_{\mathbf{k},\sigma} C_{\mathbf{k},\sigma} + \frac{1}{2V} \sum_{\substack{\mathbf{k}_1,\mathbf{k}_2,\\ \mathbf{q}\neq \mathbf{0}}} \sum_{\sigma_1,\sigma_2} \frac{4\pi q_e^2}{q^2} C^\dagger_{\mathbf{k}_1+\mathbf{q},\sigma_1} C^\dagger_{\mathbf{k}_2-\mathbf{q},\sigma_2} C_{\mathbf{k}_2,\sigma_2} C_{\mathbf{k}_1,\sigma_1}. \tag{2.366}$$

Proof From Theorem 2.16 and Eq. 2.365, we have to compute

$$\begin{aligned} \hat{\mathcal{H}} = &\int dx \hat{\psi}^\dagger(x) \frac{\hat{\mathbf{p}}^2}{2m_e} \hat{\psi}(x) \\ &+ \frac{1}{2} q_e^2 \int dx \int dx' \hat{\psi}^\dagger(x) \hat{\psi}^\dagger(x') \frac{e^{-k_s|\mathbf{r}-\mathbf{r}'|}}{|\mathbf{r}-\mathbf{r}'|} \hat{\psi}(x') \hat{\psi}(x) - \frac{q_e^2}{2} \frac{N^2}{V} \frac{4\pi}{k_s^2}. \end{aligned} \tag{2.367}$$

By means of the expansions 2.163 and 2.164, where the field operators are expressed in the plane-wave basis tensored with the standard eigenstates of $\hat{S}_z$, the kinetic operator takes a diagonal form. In particular, we obtain

$$\begin{aligned} \hat{\mathcal{H}} = &\sum_{\mathbf{k},\sigma} \frac{\hbar^2 \mathbf{k}^2}{2m_e} C^\dagger_{\mathbf{k},\sigma} C_{\mathbf{k},\sigma} + \frac{1}{2} \frac{q_e^2}{V^2} \sum_{\substack{\mathbf{k}_1,\mathbf{k}_2,\\ \mathbf{k}_3,\mathbf{k}_4}} \sum_{\substack{\sigma_1,\sigma_2,\\ \sigma_3,\sigma_4}} \sum_{s} \sum_{s'} \int d^3\mathbf{r} \\ &\int d^3\mathbf{r}' e^{-i\mathbf{k}_1\cdot\mathbf{r}} e^{-i\mathbf{k}_2\cdot\mathbf{r}'} e^{i\mathbf{k}_3\cdot\mathbf{r}'} e^{i\mathbf{k}_4\cdot\mathbf{r}} \\ &\chi^*_{\sigma_1}(s) \chi^*_{\sigma_2}(s') \chi_{\sigma_3}(s') \chi_{\sigma_4}(s) \frac{e^{-k_s|\mathbf{r}-\mathbf{r}'|}}{|\mathbf{r}-\mathbf{r}'|} C^\dagger_{\mathbf{k}_1,\sigma_1} C^\dagger_{\mathbf{k}_2,\sigma_2} C_{\mathbf{k}_3,\sigma_3} C_{\mathbf{k}_4,\sigma_4} \\ &- \frac{q_e^2}{2} \frac{N^2}{V} \frac{4\pi}{k_s^2}. \end{aligned} \tag{2.368}$$

Recall that we are working in a cubic box of volume $V = L^3$ with periodic boundary conditions, then the allowed wave vectors $\mathbf{k}$ are discrete (see Eq. 1.76). Now, given that the electron-electron interaction does not depend on the spin variables, we can make use of the spin orthonormality relation, namely Eq. 1.23 specialized to the standard spin basis. In the present case it implies

$$\sum_s \chi^*_{\sigma_1}(s)\chi_{\sigma_4}(s) = \delta_{\sigma_1,\sigma_4}, \tag{2.369}$$

$$\sum_{s'} \chi^*_{\sigma_2}(s')\chi_{\sigma_3}(s') = \delta_{\sigma_2,\sigma_3}, \tag{2.370}$$

and we sum over σ_3, σ_4, from which it follows that $\sigma_1 = \sigma_4$ e $\sigma_2 = \sigma_3$, that is,

$$\hat{\mathcal{H}} = \sum_{\mathbf{k},\sigma} \frac{\hbar^2\mathbf{k}^2}{2m_e} C^\dagger_{\mathbf{k},\sigma} C_{\mathbf{k},\sigma} + \frac{1}{2}\frac{q_e^2}{V^2} \sum_{\substack{\mathbf{k}_1,\mathbf{k}_2,\\ \mathbf{k}_3,\mathbf{k}_4}} \sum_{\sigma_1,\sigma_2} \int d^3\mathbf{r} \int d^3\mathbf{r}' e^{-i\mathbf{k}_1\cdot\mathbf{r}} e^{-i\mathbf{k}_2\cdot\mathbf{r}'} e^{i\mathbf{k}_3\cdot\mathbf{r}'} e^{i\mathbf{k}_4\cdot\mathbf{r}}$$

$$\frac{e^{-k_s|\mathbf{r}-\mathbf{r}'|}}{|\mathbf{r}-\mathbf{r}'|} C^\dagger_{\mathbf{k}_1,\sigma_1} C^\dagger_{\mathbf{k}_2,\sigma_2} C_{\mathbf{k}_3,\sigma_2} C_{\mathbf{k}_4,\sigma_1} - \frac{q_e^2}{2}\frac{N^2}{V}\frac{4\pi}{k_s^2}. \tag{2.371}$$

Thanks to translational invariance, we set $\mathbf{r} = \mathbf{r}' + \mathbf{r}''$, which implies $d\mathbf{r} = d\mathbf{r}''$, and we integrate with respect to $\mathbf{r}''$. Every function depending on $\mathbf{r}$ becomes a function of $\mathbf{r}' + \mathbf{r}''$, thus

$$\frac{1}{2}\frac{q_e^2}{V^2} \sum_{\substack{\mathbf{k}_1,\mathbf{k}_2\\ \mathbf{k}_3,\mathbf{k}_4}} \sum_{\sigma_1,\sigma_2} \int d^3\mathbf{r}' \int d^3\mathbf{r}'' e^{-i(\mathbf{k}_1+\mathbf{k}_2-\mathbf{k}_3-\mathbf{k}_4)\cdot\mathbf{r}'} e^{-i(\mathbf{k}_1-\mathbf{k}_4)\cdot\mathbf{r}''} \frac{e^{-k_s|\mathbf{r}''|}}{|\mathbf{r}''|} C^\dagger_{\mathbf{k}_1,\sigma_1}$$

$$C^\dagger_{\mathbf{k}_2,\sigma_2} C_{\mathbf{k}_3,\sigma_2} C_{\mathbf{k}_4,\sigma_1}. \tag{2.372}$$

The integral over $\mathbf{r}'$ yields a Kronecker delta (recall that $\mathbf{k}$ takes discrete values in the finite box) showing the conservation of total momentum, i.e.,

$$\int d^3\mathbf{r}'\, e^{-i(\mathbf{k}_1+\mathbf{k}_2-\mathbf{k}_3-\mathbf{k}_4)\cdot\mathbf{r}'} = V\delta_{\mathbf{k}_1+\mathbf{k}_2,\ \mathbf{k}_3+\mathbf{k}_4}. \tag{2.373}$$

Thus, the electron-electron interaction can be rewritten as

$$\frac{1}{2}\frac{q_e^2}{V} \sum_{\substack{\mathbf{k}_1,\mathbf{k}_2\\ \mathbf{k}_3,\mathbf{k}_4}} \sum_{\sigma_1,\sigma_2} \delta_{\mathbf{k}_1+\mathbf{k}_2,\ \mathbf{k}_3+\mathbf{k}_4} \left[\int d^3\mathbf{r}'' \frac{e^{-k_s|\mathbf{r}''|}}{|\mathbf{r}''|} e^{-i(\mathbf{k}_1-\mathbf{k}_4)\cdot\mathbf{r}''}\right]$$

$$C^\dagger_{\mathbf{k}_1,\sigma_1} C^\dagger_{\mathbf{k}_2,\sigma_2} C_{\mathbf{k}_3,\sigma_2} C_{\mathbf{k}_4,\sigma_1}. \tag{2.374}$$

Now we exploit the conservation of total momentum in the system. By imposing momentum conservation, we set

$$\mathbf{k}_1 = \mathbf{k}_4 + \mathbf{q}, \tag{2.375}$$

$$\mathbf{k}_2 = \mathbf{k}_3 - \mathbf{q}, \tag{2.376}$$

so that $\mathbf{k}_3$, $\mathbf{k}_4$, and the transfer momentum $\mathbf{q}$ remain independent variables, while $\mathbf{k}_1$ and $\mathbf{k}_2$ are determined functions of them. We then perform the sums over $\mathbf{k}_1$ and $\mathbf{k}_2$, using the Kronecker delta $\delta_{\mathbf{k}_1+\mathbf{k}_2,\mathbf{k}_3+\mathbf{k}_4}$, which enforces total momentum conservation. This change of variables reduces the number of independent sums. We therefore obtain

$$\frac{1}{2}\frac{q_e^2}{V}\sum_{\sigma_1\sigma_2}\sum_{\substack{\mathbf{k}_3,\mathbf{k}_4,\\ \mathbf{q}}} C^\dagger_{\mathbf{k}_4+\mathbf{q},\sigma_1} C^\dagger_{\mathbf{k}_3-\mathbf{q},\sigma_2} C_{\mathbf{k}_3,\sigma_2} C_{\mathbf{k}_4,\sigma_1} \int d^3\mathbf{r}\frac{e^{-k_s|\mathbf{r}|}}{|\mathbf{r}|}e^{-i\mathbf{q}\cdot\mathbf{r}}, \tag{2.377}$$

and we use the Fourier transform of the Yukawa potential, i.e.

$$\int d^3\mathbf{r} e^{-i\mathbf{q}\cdot\mathbf{r}}\frac{e^{-k_s|\mathbf{r}|}}{|\mathbf{r}|} = \frac{4\pi}{q^2+k_s^2}, \tag{2.378}$$

and one finds

$$\frac{1}{2}\frac{q_e^2}{V}\sum_{\sigma_1\sigma_2}\sum_{\substack{\mathbf{k}_3,\mathbf{k}_4,\\ \mathbf{q}}} \frac{4\pi}{q^2+k_s^2} C^\dagger_{\mathbf{k}_4+\mathbf{q},\sigma_1} C^\dagger_{\mathbf{k}_3-\mathbf{q},\sigma_2} C_{\mathbf{k}_3,\sigma_2} C_{\mathbf{k}_4,\sigma_1}. \tag{2.379}$$

Now, we decompose $\sum_{\mathbf{q}}$ into a term with $\mathbf{q} = \mathbf{0}$ and one with $\mathbf{q} \neq \mathbf{0}$, we rename $\mathbf{k}_3$ as $\mathbf{k}_1$ and $\mathbf{k}_4$ as $\mathbf{k}_2$,

$$\begin{aligned}\hat{\mathcal{H}} = &\sum_{\mathbf{k},\sigma}\frac{\hbar^2\mathbf{k}^2}{2m_e}C^\dagger_{\mathbf{k},\sigma}C_{\mathbf{k},\sigma} + \frac{q_e^2}{2V}\sum_{\substack{\mathbf{k}_1,\mathbf{k}_2,\\ \mathbf{q}\neq\mathbf{0}}}\sum_{\sigma_1\sigma_2}\frac{4\pi}{q^2+k_s^2}C^\dagger_{\mathbf{k}_1+\mathbf{q},\sigma_1}C^\dagger_{\mathbf{k}_2-\mathbf{q},\sigma_2}C_{\mathbf{k}_2,\sigma_2}C_{\mathbf{k}_1,\sigma_1} + \\ &+ \frac{q_e^2}{2V}\sum_{\mathbf{k}_1,\mathbf{k}_2}\sum_{\sigma_1,\sigma_2}\frac{4\pi}{k_s^2}C^\dagger_{\mathbf{k}_1,\sigma_1}C^\dagger_{\mathbf{k}_2,\sigma_2}C_{\mathbf{k}_2,\sigma_2}C_{\mathbf{k}_1,\sigma_1} - \frac{q_e^2}{2}\frac{N^2}{V}\frac{4\pi}{k_s^2}.\end{aligned} \tag{2.380}$$

We employ the fermionic algebra, specifically the relations D.6 and D.4, in the following manner

$$\begin{aligned}C^\dagger_{\mathbf{k}_1,\sigma_1}C^\dagger_{\mathbf{k}_2,\sigma_2}C_{\mathbf{k}_2,\sigma_2}C_{\mathbf{k}_1,\sigma_1} &= -C^\dagger_{\mathbf{k}_2,\sigma_2}C^\dagger_{\mathbf{k}_1,\sigma_1}C_{\mathbf{k}_2,\sigma_2}C_{\mathbf{k}_1,\sigma_1} = \\ &= -C^\dagger_{\mathbf{k}_2,\sigma_2}\left[\delta_{\mathbf{k}_1,\mathbf{k}_2}\delta_{\sigma_1,\sigma_2} - C_{\mathbf{k}_2,\sigma_2}C^\dagger_{\mathbf{k}_1,\sigma_1}\right]C_{\mathbf{k}_1,\sigma_1},\end{aligned} \tag{2.381}$$

and we have

$$
\begin{aligned}
&\frac{q_e^2}{2V}\sum_{\mathbf{k}_1,\sigma_1}\sum_{\mathbf{k}_2,\sigma_2}\frac{4\pi}{k_s^2}C^\dagger_{\mathbf{k}_1,\sigma_1}C^\dagger_{\mathbf{k}_2,\sigma_2}C_{\mathbf{k}_2,\sigma_2}C_{\mathbf{k}_1,\sigma_1} = \\
&= -\frac{q_e^2}{2V}\frac{4\pi}{k_s^2}\left(\sum_{\mathbf{k}_1,\sigma_1}\sum_{\mathbf{k}_2,\sigma_2}\delta_{\mathbf{k}_1,\mathbf{k}_2}\delta_{\sigma_1,\sigma_2}C^\dagger_{\mathbf{k}_2,\sigma_2}C_{\mathbf{k}_1,\sigma_1}\right.\\
&\qquad\left.-\sum_{\mathbf{k}_1,\sigma_1}\sum_{\mathbf{k}_2,\sigma_2}C^\dagger_{\mathbf{k}_2,\sigma_2}C_{\mathbf{k}_2,\sigma_2}C^\dagger_{\mathbf{k}_1,\sigma_1}C_{\mathbf{k}_1,\sigma_1}\right) = \\
&= -\frac{q_e^2}{2V}\frac{4\pi}{k_s^2}\left[\sum_{\mathbf{k}_1,\sigma_1}C^\dagger_{\mathbf{k}_1,\sigma_1}C_{\mathbf{k}_1,\sigma_1} - \left(\sum_{\mathbf{k}_1,\sigma_1}C^\dagger_{\mathbf{k}_1,\sigma_1}C_{\mathbf{k}_1,\sigma_1}\right)\left(\sum_{\mathbf{k}_2,\sigma_2}C^\dagger_{\mathbf{k}_2,\sigma_2}C_{\mathbf{k}_2,\sigma_2}\right)\right] = \\
&= -\frac{q_e^2}{2V}\frac{4\pi}{k_s^2}\left[\sum_{\mathbf{k},\sigma}C^\dagger_{\mathbf{k},\sigma}C_{\mathbf{k},\sigma} - \left(\sum_{\mathbf{k},\sigma}C^\dagger_{\mathbf{k},\sigma}C_{\mathbf{k},\sigma}\right)^2\right].
\end{aligned} \tag{2.382}
$$

Recall that we work within the Fock space under the constraint of fixed particle number. In this case, to ensure conservation of the average electron number, the total number operator $\hat{N} = \sum_{\mathbf{k},\sigma} C^\dagger_{\mathbf{k},\sigma}C_{\mathbf{k},\sigma}$ can be replaced by its eigenvalue N (i.e., $N\mathbb{1}$, with the identity omitted), and we have

$$
\frac{q_e^2}{2V}\sum_{\mathbf{k}_1,\sigma_1}\sum_{\mathbf{k}_2,\sigma_2}\frac{4\pi}{k_s^2}C^\dagger_{\mathbf{k}_1,\sigma_1}C^\dagger_{\mathbf{k}_2,\sigma_2}C_{\mathbf{k}_2,\sigma_2}C_{\mathbf{k}_1,\sigma_1} = -\frac{q_e^2}{2V}\frac{4\pi}{k_s^2}\left(N - N^2\right). \tag{2.383}
$$

The second term compensates the factor

$$
-\frac{q_e^2}{2}\frac{N^2}{V}\frac{4\pi}{k_s^2} \tag{2.384}
$$

in the Hamiltonian 2.380, while the first term

$$
-\frac{q_e^2}{2}\frac{4\pi}{k_s^2}\frac{N}{V} \equiv -\frac{q_e^2}{2}\frac{4\pi}{k_s^2}\rho \tag{2.385}
$$

is called *jellium background term* (or *constant background term*) and is typically neglected when studying dynamic properties of a system, as we show. Indeed, taking the limits for $N \to \infty$ and $V \to \infty$, so that ρ remains constant, we expect the energy to grow with the number N, but we are interested in the energy per particle, then the object

$$-\frac{\frac{q_e^2}{2}\frac{4\pi}{k_s^2}\rho}{N} \tag{2.386}$$

vanishes for $N \to \infty$. The remaining contributions, such as the kinetic energy and the two-body interaction terms, are extensive. Although the total interaction formally involves a number of particle pairs that grows as N^2, the long-range component is canceled by the neutralizing background, leading to a total potential energy that increases only linearly with the number of particles. This is the same behavior found in Hartree-Fock theory (see Chap. 8), where both the mean-field and exchange contributions scale proportionally to the system size. As a result, the energy per particle remains finite in the thermodynamic limit, since normalizing the total energy by N gives a well-defined, finite value as $N \to \infty$. In the subsequent limits $N, V \to \infty$, followed by $k_s \to 0$, one recovers the second-quantized Hamiltonian of the jellium model, namely Eq. 2.366. □

In the limit $k_s \to 0$, the Fourier transform of the Yukawa potential, multiplied by q_e^2 reduces to

$$\frac{4\pi q_e^2}{q^2}, \tag{2.387}$$

which is precisely the Fourier transform of the Coulomb potential. This shows that when the screening parameter k_s vanishes, the Yukawa potential naturally recovers the long-range Coulomb interaction.

Theorem 2.21 (Density-Operator Representation of the Jellium Hamiltonian) *The jellium Hamiltonian 2.366 can be rewritten in a more insightful form, namely in terms of the Fourier-transformed particle density operator, Eq. 2.262, its adjoint, and the total number operator* $\hat{N} = \sum_{\mathbf{k},\sigma} C^\dagger_{\mathbf{k},\sigma} C_{\mathbf{k},\sigma}$, *as follows*

$$\hat{\mathcal{H}} = \sum_{\mathbf{k},\sigma} \frac{\hbar^2 \mathbf{k}^2}{2m_e} C^\dagger_{\mathbf{k},\sigma} C_{\mathbf{k},\sigma} + \frac{1}{2V} \sum_{\mathbf{q}\neq\mathbf{0}} \frac{4\pi q_e^2}{q^2} \left(\hat{N} - \hat{\rho}_{\mathbf{q}} \hat{\rho}^\dagger_{\mathbf{q}} \right). \tag{2.388}$$

Proof Using fermionic algebra (see Eq. D.4) as follows

$$C^\dagger_{\mathbf{k}_2-\mathbf{q},\sigma_2} C_{\mathbf{k}_1,\sigma_1} = \delta_{\mathbf{k}_1,\mathbf{k}_2-\mathbf{q}} \delta_{\sigma_1,\sigma_2} - C_{\mathbf{k}_1,\sigma_1} C^\dagger_{\mathbf{k}_2-\mathbf{q},\sigma_2}, \tag{2.389}$$

we invert the order of the annihilation operators in the jellium Hamiltonian 2.366 and we obtain

$$\hat{\mathcal{H}} = \sum_{\mathbf{k},\sigma} \frac{\hbar^2 \mathbf{k}^2}{2m_e} C^\dagger_{\mathbf{k},\sigma} C_{\mathbf{k},\sigma} + \frac{1}{2V} \sum_{\substack{\mathbf{k}_1,\mathbf{k}_2,\\ \mathbf{q}\neq\mathbf{0}}} \sum_{\sigma_1,\sigma_2} \frac{4\pi q_e^2}{q^2} C^\dagger_{\mathbf{k}_1+\mathbf{q},\sigma_1} C^\dagger_{\mathbf{k}_2-\mathbf{q},\sigma_2} C_{\mathbf{k}_1,\sigma_1} C_{\mathbf{k}_2,\sigma_2} =$$

$$= \sum_{\mathbf{k},\sigma} \frac{\hbar^2 \mathbf{k}^2}{2m_e} C^\dagger_{\mathbf{k},\sigma} C_{\mathbf{k},\sigma} + \frac{1}{2V} \sum_{\substack{\mathbf{k}_1,\mathbf{k}_2, \\ \mathbf{q}\neq \mathbf{0}}} \sum_{\sigma_1,\sigma_2} \frac{4\pi q_e^2}{q^2} C^\dagger_{\mathbf{k}_1+\mathbf{q},\sigma_1}$$

$$\left(\delta_{\mathbf{k}_1,\mathbf{k}_2-\mathbf{q}}\delta_{\sigma_1,\sigma_2} - C_{\mathbf{k}_1,\sigma_1} C^\dagger_{\mathbf{k}_2-\mathbf{q},\sigma_2}\right) C_{\mathbf{k}_2,\sigma_2}. \tag{2.390}$$

We manipulate as follows

$$\sum_{\mathbf{k}_1,\mathbf{k}_2} \sum_{\sigma_1,\sigma_2} C^\dagger_{\mathbf{k}_1+\mathbf{q},\sigma_1} \left(\delta_{\mathbf{k}_1,\mathbf{k}_2-\mathbf{q}}\delta_{\sigma_1,\sigma_2} - C_{\mathbf{k}_1,\sigma_1} C^\dagger_{\mathbf{k}_2-\mathbf{q},\sigma_2}\right) C_{\mathbf{k}_2,\sigma_2} =$$

$$= \sum_{\mathbf{k}_1,\mathbf{k}_2} \sum_{\sigma_1,\sigma_2} C^\dagger_{\mathbf{k}_1+\mathbf{q},\sigma_1} \delta_{\mathbf{k}_1,\mathbf{k}_2-\mathbf{q}}\delta_{\sigma_1,\sigma_2} C_{\mathbf{k}_2,\sigma_2}$$

$$- \sum_{\mathbf{k}_1,\mathbf{k}_2} \sum_{\sigma_1,\sigma_2} C^\dagger_{\mathbf{k}_1+\mathbf{q},\sigma_1} C_{\mathbf{k}_1,\sigma_1} C^\dagger_{\mathbf{k}_2-\mathbf{q},\sigma_2} C_{\mathbf{k}_2,\sigma_2} =$$

$$= \sum_{\mathbf{k}_1,\mathbf{k}_2} \sum_{\sigma_1,\sigma_2} C^\dagger_{\mathbf{k}_1+\mathbf{q},\sigma_1} \delta_{\mathbf{k}_1,\mathbf{k}_2-\mathbf{q}}\delta_{\sigma_1,\sigma_2} C_{\mathbf{k}_2,\sigma_2} - \hat{\rho}_{\mathbf{q}}\hat{\rho}^\dagger_{\mathbf{q}}, \tag{2.391}$$

and summing over $\mathbf{k}_1$, σ_1, the deltas imply $\mathbf{k}_1 = \mathbf{k}_2 - \mathbf{q}$, $\sigma_1 = \sigma_2$, that is, the first term becomes

$$\sum_{\mathbf{k}_2,\sigma_2} C^\dagger_{\mathbf{k}_2,\sigma_2} C_{\mathbf{k}_2,\sigma_2} = \sum_{\mathbf{k}_2,\sigma_2} \hat{N}_{\mathbf{k}_2,\sigma_2} =$$

$$= \hat{N}, \tag{2.392}$$

hence, Eq. 2.388 is established. □

Remark 2.8 (Extension of Jellium the Model to the Grand Canonical Ensemble) In the jellium model, electrons immersed in a neutralizing background are traditionally described by a Hamiltonian that conserves the total particle number. To treat systems where the number of electrons can fluctuate, as often occurs in both experimental and theoretical contexts, it is convenient to adopt the grand canonical formalism. In this framework, the Hamiltonian is extended by the term $-\mu\hat{N}$, where μ is the chemical potential and $\hat{N}$ the total number operator, allowing for a consistent description of particle-number fluctuations in quantum many-body systems. In the grand canonical ensemble, the second-quantized Hamiltonian of the jellium model is straightforwardly adapted as follows

$$
\begin{aligned}
\hat{\mathcal{H}} = & \sum_{\mathbf{k},\sigma} \left(\frac{\hbar^2 \mathbf{k}^2}{2m_e} - \mu \right) C^\dagger_{\mathbf{k},\sigma} C_{\mathbf{k},\sigma} \\
& + \frac{1}{2V} \sum_{\substack{\mathbf{k}_1,\mathbf{k}_2, \\ \mathbf{q}\neq\mathbf{0}}} \sum_{\sigma_1,\sigma_2} \frac{4\pi q_e^2}{q^2} C^\dagger_{\mathbf{k}_1+\mathbf{q},\sigma_1} C^\dagger_{\mathbf{k}_2-\mathbf{q},\sigma_2} C_{\mathbf{k}_2,\sigma_2} C_{\mathbf{k}_1,\sigma_1}.
\end{aligned} \tag{2.393}
$$

Part II
A Quantum Many-Body Model of a Solid

Three Steps to Model a Solid: Preliminary Discussion

A paradigmatic example in quantum many-body theory is the study of the dynamical behavior of a solid. The simplest model assumes atoms arranged with a specific spatial periodicity. Each atom consists of a nucleus (ion) and valence electrons, treated collectively as point-like entities. Each atom contains Z_V valence electrons, which are responsible for forming chemical bonds between atoms, in addition to electrons in closed inner shells that remain tightly bound to the nucleus. When atoms are brought close together to form a crystal lattice, the orbitals associated with the inner closed shells remain largely unaltered, whereas the outermost orbitals are strongly affected by the presence of neighboring atoms. In other words, the atomic wave functions overlap. For each positive ion corresponding to the valence electrons, there are Z_V valence electrons present. The Hamiltonian describing such a solid within the formalism of first quantization is highly complex, namely,

$$\mathcal{H} = T^{(e)} + T^{(i)} + V^{(e-e)} + V^{(i-i)} + V^{(e-i)}, \tag{II.1}$$

where

$$T^{(e)} = \sum_{j=1}^{N_e} \frac{\hat{\mathbf{p}}_j^2}{2m_e}, \tag{II.2}$$

is the kinetic energy of the valence electrons, where N_e is the total number of valence electrons in the crystal,

$$T^{(i)} = \sum_{j=1}^{N_{\text{ion}}} \frac{\hat{\mathbf{P}}_j^2}{2M_j}, \tag{II.3}$$

is the kinetic energy of the ions, where N_{ion} is the total number of ions and M_j is the mass of the j-th ion,

$$V^{(e-e)} = \frac{1}{2} \sum_{\substack{i,j=1 \\ i \neq j}}^{N_e} \frac{q_e^2}{|\mathbf{r}_i - \mathbf{r}_j|}, \tag{II.4}$$

is the Coulomb repulsion energy between valence electrons, $\mathbf{r}_i$ being the electron position,

$$V^{(i-i)} = \frac{1}{2} \sum_{\substack{i,j=1 \\ i \neq j}}^{N_{\text{ion}}} \frac{Z_{V,i} Z_{V,j} q_e^2}{|\mathbf{R}_i - \mathbf{R}_j|}, \tag{II.5}$$

is the Coulomb repulsion energy between ions, where $\mathbf{R}_i$ are the ion positions, and $Z_{V,i}$, $Z_{V,j}$ are the valence electron numbers of atoms i and j, respectively,

$$V^{(e-i)} = -\sum_{i=1}^{N_e} \sum_{j=1}^{N_{\text{ion}}} \frac{Z_{V,j} q_e^2}{|\mathbf{r}_i - \mathbf{R}_j|}, \tag{II.6}$$

is the Coulomb attraction between ions and valence electrons. Charge neutrality requires that the total electronic charge compensates the ionic charge, namely

$$N_e = \sum_{I=1}^{N_{\text{ion}}} Z_{V,I}. \tag{II.7}$$

For a crystal composed of identical ions with valence Z_V, this reduces to

$$N_e = Z_V N_{\text{ion}}. \tag{II.8}$$

In general, with a particle density of $\rho \sim 10^{23}\,\text{cm}^{-3}$ ions and electrons coupled together, the eigenvalue problem is impossible to solve exactly, so approximations are necessary. First, the mass of ions is several orders of magnitude larger than that of electrons, so the dominant dynamics is governed by the electrons. Typical electron response times are on the order of femtoseconds ($\sim$fs), whereas ion response times are much slower, on the order of picoseconds or longer ($\gtrsim$ps). In the limit $M_j \to \infty$, ions can be treated as fixed classical points, and the problem reduces to describing approximately 10^{23} electrons interacting among themselves and with fixed nuclei according to the Hamiltonian

$$\mathcal{H}^{(e)} = T^{(e)} + V^{(e-e)} + V^{(e-i)}, \tag{II.9}$$

which represents the electronic energy in the fixed nuclei (Born-Oppenheimer) approximation. Although this simplification considerably reduces the complexity of the problem, obtaining exact solutions remains intractable. Nevertheless, let us assume that, despite the computational difficulty, we have succeeded in determining all the eigenfunctions of this fixed-nuclei Hamiltonian, denoted by $\varphi_n(\{\mathbf{r}\}; \{\mathbf{R}\})$, i.e.,

$$\mathcal{H}^{(e)} \varphi_n (\{\mathbf{r}\}; \{\mathbf{R}\}) = \mathcal{E}_n (\{\mathbf{R}\}) \, \varphi_n (\{\mathbf{r}\}; \{\mathbf{R}\}) \,, \tag{II.10}$$

we then investigate whether this electronic solution can be consistently related to an eigenstate of the full Hamiltonian

$$\mathcal{H} = \mathcal{H}^{(e)} + V^{(i-i)} + T^{(i)}. \tag{II.11}$$

Considering the total wavefunction expanded in the eigenbasis of $\mathcal{H}$, we investigate the possibility of expressing it as a superposition of the ionic eigenstates coupled with the electronic eigenfunctions as follows

$$\psi (\{\mathbf{r}\}, \{\mathbf{R}\}) = \sum_n c_n (\{\mathbf{R}\}) \, \varphi_n (\{\mathbf{r}\}; \{\mathbf{R}\}) \,. \tag{II.12}$$

The goal is to determine the coefficient functions $c_n(\{\mathbf{R}\})$ for each ionic configuration $\{\mathbf{R}\}$. We impose that $|\psi\rangle$ is an eigenstate of the full Hamiltonian $\mathcal{H}$. The electronic Hamiltonian $\mathcal{H}^{(e)}$ acts on $\varphi_n(\{\mathbf{r}\}; \{\mathbf{R}\})$ yielding the eigenvalues $\mathcal{E}_n(\{\mathbf{R}\})$. The ion-ion interaction term $V^{(i-i)}$ acts multiplicatively with respect to the ionic coordinates, whereas the ionic kinetic energy operator $T^{(i)}$, being a Laplacian in $\{\mathbf{R}\}$, acts on both the coefficient functions $c_n(\{\mathbf{R}\})$ and on the parametric dependence of $\varphi_n(\{\mathbf{r}\}; \{\mathbf{R}\})$ as follows

$$\nabla^2_{\mathbf{R}_j} (c_n \varphi_n) = \left(\nabla^2_{\mathbf{R}_j} c_n\right) \varphi_n + c_n \left(\nabla^2_{\mathbf{R}_j} \varphi_n\right) + 2 \nabla_{\mathbf{R}_j} c_n \cdot \nabla_{\mathbf{R}_j} \varphi_n . \tag{II.13}$$

The eigenvalue equation then becomes

$$A + \sum_n \left\{ \left[\mathcal{E}_n(\{\mathbf{R}\}) + V^{(i-i)} (\{\mathbf{R}\}) \right] c_n \varphi_n - \sum_j \frac{\hbar^2}{2M_j} (\nabla^2_{\mathbf{R}_j} c_n) \varphi_n \right\} = \mathcal{E} \sum_n c_n \varphi_n , \tag{II.14}$$

$$A = - \sum_n \sum_j \frac{\hbar^2}{2M_j} \left[c_n \left(\nabla^2_{\mathbf{R}_j} \varphi_n\right) + 2 \nabla_{\mathbf{R}_j} c_n \cdot \nabla_{\mathbf{R}_j} \varphi_n \right], \tag{II.15}$$

where we have omitted the explicit dependence of the eigenfunctions on both the electronic and ionic position variables to simplify the notation. We now multiply both sides from the left by φ^*_m and integrate over the electronic coordinates. Using the orthonormality of the set $\{\varphi_n\}$, we obtain

$$D + \left[\mathcal{E}_m(\{\mathbf{R}\}) + V^{(i-i)}(\{\mathbf{R}\}) - \sum_j \frac{\hbar^2}{2M_j} \nabla^2_{\mathbf{R}_j} \right] c_m = \mathcal{E} c_m, \tag{II.16}$$

where

$$D = -\sum_j \sum_n \frac{\hbar^2}{2M_j} \left\{ \langle \varphi_m | \nabla^2_{\mathbf{R}_j} | \varphi_n \rangle c_n + \langle \varphi_m | \nabla_{\mathbf{R}_j} | \varphi_n \rangle \cdot \nabla_{\mathbf{R}_j} c_n \right\} \tag{II.17}$$

is called the non-adiabatic term. Since the ions move, the electrons do not remain confined in the same electronic state φ_m, but undergo transitions from state m to other states n. Electrons and ions exchange energy and influence one another. Then, the term D represents an electron-ion interaction contribution, involving both first and second derivatives of the electronic wavefunctions $\{\varphi_n\}$ with respect to the ionic coordinates. In order to compute such derivatives, one must account for the variations of the electronic wavefunctions with respect to infinitesimal displacements of $\mathbf{R}_j$. However, because the electronic motion is much faster than the ionic motion, such variations are typically expected to be small.

To describe the dynamics of a solid, we proceed through three steps. Since the derivative couplings D are expected to be small owing to the much faster motion of the electrons compared to the ions, a first approximation, known as the Born-Oppenheimer approximation, consists in neglecting them, i.e., setting $D = 0$. Under this assumption, the total wave function factorizes into an electronic and an ionic part, and the electronic and ionic problems can be treated separately. The Born-Oppenheimer approximation is thus implemented through the following two steps:

1. The fixed-ion electronic problem

$$\mathcal{H}^{(e)} \varphi_n = \mathcal{E}_n \varphi_n \tag{II.18}$$

must, in principle, be solved for every nuclear configuration $\mathbf{R}$. In practice, this problem is intractable due to the enormous number of electrons involved, so we simplify the situation by fixing the ions at the lattice sites. The remaining challenge is then to describe the interactions among approximately $10^{23}\,\text{cm}^{-3}$ electrons, as well as their interactions with the static ionic lattice. We apply the Hartree-Fock method to $\mathcal{H}^{(e)}$, which transforms the original interacting many-body Hamiltonian into an effective Hamiltonian describing independent electrons moving in a mean-field potential. This many-body Hartree-Fock Hamiltonian, involving N valence electrons, can be written as a sum of one-particle contributions as follows

$$\mathcal{H}^{(e)} \xrightarrow[\text{Hartree Fock}]{} \mathcal{H}^{(e),N}_{\text{HF}} = \sum_{i=1}^{N} \left(\frac{\hat{\mathbf{p}}_i^{\,2}}{2m_e} + V(\mathbf{r}_i) \right), \tag{II.19}$$

where $V(\mathbf{r}_i)$ is a periodic effective potential. Consequently, since the electrons are treated as independent particles, we focus on the one-particle Hamiltonian

$$\mathcal{H}_{\text{HF}}^{(e)} = \frac{\hat{\mathbf{p}}^2}{2m_e} + V(\mathbf{r}), \tag{II.20}$$

reducing the problem to that of a particle moving in a periodic potential, known as the *Bloch eigenvalue problem* (or *Bloch problem*), i.e.,

$$\mathcal{H}_{\text{HF}}^{(e)}\varphi_n = \mathcal{E}_n\varphi_n, \tag{II.21}$$

and we will compute the so-called Bloch eigenfunctions.

Remark 2.9 Although the Bloch problem is formulated using a one-particle Hartree-Fock potential, residual electronic interactions remain and can be significant. In certain contexts, an additional local interaction term, such as the Hubbard Hamiltonian (see Chap. 3), is often incorporated into the full solid-state Hamiltonian.

2. Given the solutions of the fixed-ion electronic problem, we need to solve the ion problem. We expand the effective potential $V^{(i-i)}(\{\mathbf{R}\}) + \mathcal{E}_n(\{\mathbf{R}\})$ to second order with respect to its minimum, i.e., we consider its harmonic expansion. Now, there are on the order of $10^{23}\,\text{cm}^{-3}$ ionic variables coupled together, but owing to the harmonic expansion, this coupling is simple to treat. In fact, we consider the classical normal modes, find the normal coordinates and quantize them: in this way, we obtain the energy quantities of the lattice vibration field, which are called phonons.

In most cases this separation provides an excellent description of the system. However, in certain situations, such as in superconductivity or nonadiabatic transitions, the derivative couplings D cannot be neglected. To account for these effects, we proceed to the third step:

3. Our focus is on electron-phonon coupling.

At the end of the three steps, the Hamiltonian of a solid in second quantization has the form

$$\begin{aligned}\hat{\mathcal{H}} &= \sum_{n,\mathbf{k},\sigma} \mathcal{E}_{n,\mathbf{k}} C^{\dagger}_{n,\mathbf{k},\sigma} C_{n,\mathbf{k},\sigma} + \sum_{\mathbf{q},s} \hbar\omega_{\mathbf{q},s}\left(a^{\dagger}_{\mathbf{q},s} a_{\mathbf{q},s} + \frac{1}{2}\right) \\ &+ \sum_{\substack{n,n',\sigma,s,\\ \mathbf{k},\mathbf{G},\mathbf{q}}} T^{s}_{\mathbf{k},\mathbf{q},\mathbf{G}}\, C^{\dagger}_{n',\mathbf{k}+\mathbf{q}+\mathbf{G},\sigma} C_{n,\mathbf{k},\sigma}\left(a_{\mathbf{q},s} + a^{\dagger}_{-\mathbf{q},s}\right),\end{aligned} \tag{II.22}$$

where C, $C^{\dagger}$ are fermionic operators and a, $a^{\dagger}$ are bosonic operators. The first, second, and third terms of Eq. II.22 will be derived and analyzed in Chaps. 3, 4 and 5, respectively. This preliminary discussion thus provides a conceptual foundation that will guide the detailed development of each step in the following three chapters, offering the reader a comprehensive overview before delving into each individual phase.

Ultimately, in place of the Hamiltonian II.22, a simplified form of the solid Hamiltonian is given by

$$\hat{\mathcal{H}} = \sum_{\mathbf{k},\sigma} \mathcal{E}_{\mathbf{k}} C^{\dagger}_{\mathbf{k},\sigma} C_{\mathbf{k},\sigma} + \sum_{\mathbf{q}} \hbar\omega_{\mathbf{q}} a^{\dagger}_{\mathbf{q}} a_{\mathbf{q}} + \sum_{\mathbf{k},\mathbf{q},\sigma} M_{\mathbf{q}} C^{\dagger}_{\mathbf{k}+\mathbf{q},\sigma} C_{\mathbf{k},\sigma} \left(a_{\mathbf{q}} + a^{\dagger}_{-\mathbf{q}}\right), \tag{II.23}$$

where we have considered a single electronic band, the one crossing the chemical potential, and a single phonon branch, namely the one that couples most strongly to the electrons in the conduction band.

Chapter 3
Step I of the Model of a Solid: Bloch Electrons

Abstract This chapter presents the fundamental concept of the crystal lattice, the geometric framework underlying the structure of solids. A lattice represents the periodic arrangement of atoms in space, and its symmetry plays a key role in determining the electronic and structural properties of materials. The periodic nature of the lattice enables the application of powerful mathematical tools, most notably Bloch's theorem. This theorem states that the eigenfunctions of an electron in a periodic potential can be written as the product of a plane wave and a periodic function, yielding Bloch functions, which are central to the analysis of electronic band structures. While Bloch functions are naturally suited for momentum-space descriptions, an alternative basis, based on Wannier functions, provides a localized, real-space perspective. Wannier functions are particularly valuable for studying local phenomena and implementing numerical methods in condensed matter physics. This chapter offers a clear and rigorous introduction to these foundational ideas: starting from the geometric definition of lattices, proceeding to the formulation and implications of Bloch's theorem, and culminating in the construction and application of both Bloch and Wannier functions. Mastery of these concepts is essential for tackling advanced problems in solid-state physics and for modelling material properties at the microscopic level.

3.1 Bravais Lattices and Their Properties

In this section, we introduce geometric concepts from statistical physics that will be of fundamental importance in the subsequent analysis, and in particular for the proof of Bloch's theorem.

3.1.1 Direct Bravais Lattices

In geometry and crystallography, a Bravais lattice denotes an infinite set of points in space generated by all integer linear combinations of a basis of primitive

F. Tafuri et al., *Understanding Quantum Many-Body Theory*, UNITEXT for Physics, https://doi.org/10.1007/978-3-032-17849-7_3

translation vectors. It constitutes the fundamental translational structure underlying a crystalline solid. The Bravais lattice concept is used to formally define a crystalline arrangement and its frontiers: a crystal is made up of one or more atoms, called the basis or motif, at each lattice point. The basis may consist of atoms, molecules, or polymer strings of solid matter, and the lattice provides the locations of the basis. Placed the origin of the Cartesian axes on any point of the lattice, each point is identified by a vector. A Bravais lattice is generated by translation operations in the space of a set of vectors, called primitive vectors. The primitive vectors are linearly independent and their choice is not unique: the general definition of a Bravais lattice vector $\bar{R}$ in d dimension is

$$\bar{R} = \sum_{i=1}^{d} n_i \bar{a}_i, \tag{3.1}$$

where n_i are integers and $\bar{a}_i$ are the primitive vectors of the lattice. Group theory allows defining the number of possible Bravais lattices for each dimension of space. Two Bravais lattices are often considered equivalent if they have isomorphic symmetry groups. In this sense, there are 5 possible Bravais lattices in 2-dimensional space and 14 possible Bravais lattices in 3-dimensional space. The 14 3-dimensional Bravais lattices, grouped into seven crystal systems: cubic (see Fig. 3.1), tetragonal (see Fig. 3.2), orthorhombic (see Fig. 3.3), monoclinic (see Fig. 3.4), triclinic (see Fig. 3.5), hexagonal (see Fig. 3.6), and rhombohedral (see Fig. 3.7). Each lattice is defined by its symmetry, the lengths and angles of its primitive vectors, and can occur in primitive (P), body-centered (I), face-centered (F), or base-centered (C) forms. This classification encompasses all possible periodic arrangements of points in three-dimensional space consistent with crystallographic symmetry.

In particular, the lattice vectors in three dimensions have form

$$\begin{aligned} \mathbf{R} \equiv \mathbf{R_n} &= \\ &= n_1\mathbf{a}_1 + n_2\mathbf{a}_2 + n_3\mathbf{a}_3, \; n_i \in \mathbb{Z}, \end{aligned} \tag{3.2}$$

where $\{\mathbf{a}_i\}_{i=1,2,3}$ are the primitive lattice vectors of the direct lattice. Each triplet of integers $\mathbf{n} = (n_1, n_2, n_3) \in \mathbb{Z}^3$ uniquely and bijectively corresponds to a lattice vector $\mathbf{R_n}$, which identifies a point in the lattice. In what follows, we will use the notations $\mathbf{R}$ and $\mathbf{R_n}$ interchangeably, depending on whether or not it is necessary to make the dependence on $\mathbf{n}$ explicit.

Primitive unit cells are defined as unit cells with the smallest volume for a given lattice. There can be more than one way to choose a primitive cell for a given crystal and each choice will have a different primitive cell shape, but the primitive cell volume is the same for every choice and each choice will have the property that a one-to-one correspondence can be established between primitive unit cells and discrete lattice points over the associated lattice. Among all possible primitive cells for a given crystal, an obvious primitive cell may be the parallelepiped formed by a chosen set of primitive translation vectors, that is, the set of all

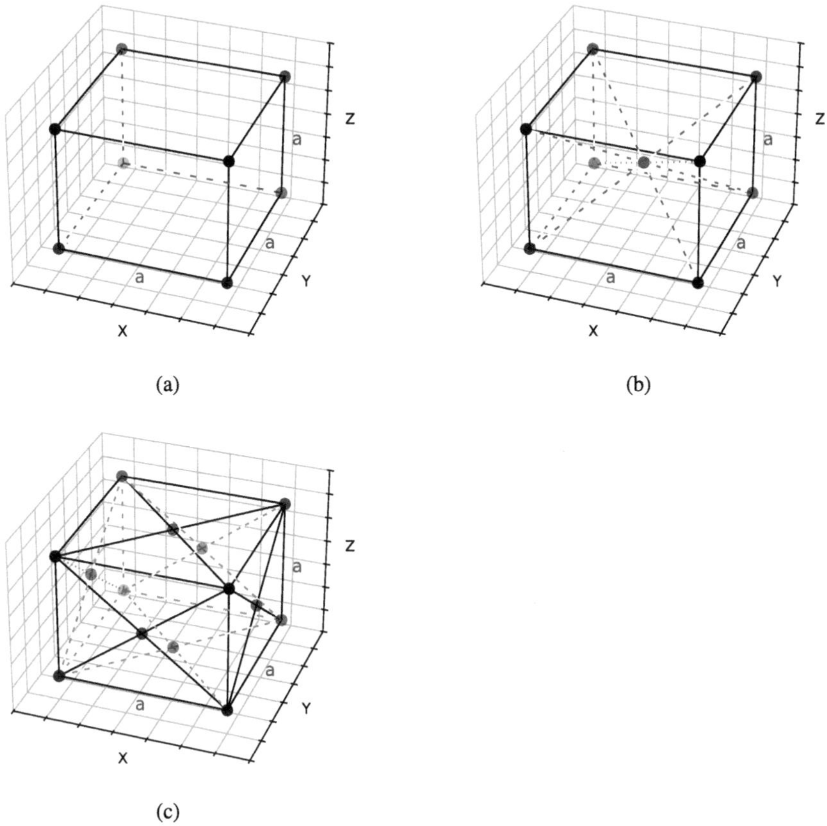

Fig. 3.1 (**a**) Simple cubic (SC): lattice points only at the corners of the cubic cell; coordination number 6. (**b**) Body-centered cubic (BCC): an additional lattice point at the center of the cubic cell; coordination number 8. (**c**) Face-centered cubic (FCC): lattice points at the corners and at the centers of all faces; coordination number 12 and the highest packing efficiency among cubic lattices

points $x_1\mathbf{a}_1 + x_2\mathbf{a}_2 + x_3\mathbf{a}_3$ where $0 \leq x_i < 1$ and $\{\mathbf{a}_i\}_{i=1,2,3}$ are the chosen primitive vector. This primitive cell does not always show the clear symmetry of a given crystal. The unique property of a crystal is that its atoms are arranged in a regular three-dimensional array called a lattice. All the properties attributed to crystalline materials stem from this highly ordered structure. Such a structure exhibits discrete translational symmetry. In order to model and study such a periodic system, one needs a mathematical framework to describe its symmetry and thereby draw conclusions about the material properties arising from it. The Wigner-Seitz cell is a means to achieve this. To construct a Wigner-Seitz cell, let us start by first picking a lattice point. After a point is chosen, lines are drawn to all nearby lattice points. At the midpoint of each line, another line is drawn normal to each of the first set of lines. The smallest area enclosed in this way is called the Wigner-Seitz

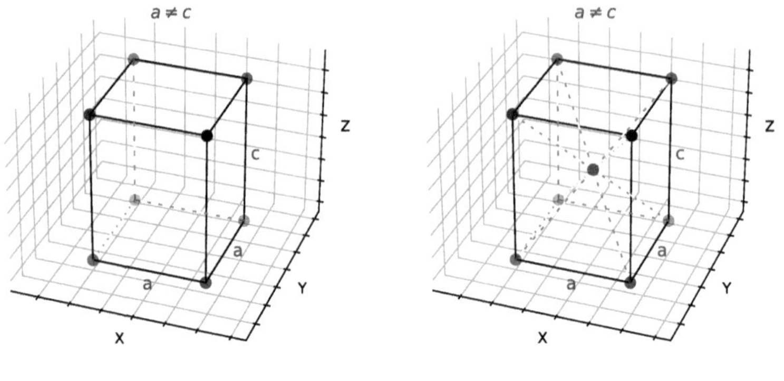

(a) (b)

Fig. 3.2 (**a**) Simple tetragonal (ST): lattice points only at the corners of the tetragonal cell; fourfold rotational symmetry along the c-axis. (**b**) Body-centered tetragonal (BCT): lattice points at the corners and one at the center of the tetragonal cell; fourfold symmetry preserved, with increased coordination along the c-axis

primitive cell. A Wigner-Seitz cell is an example of a primitive cell, which is a unit cell containing exactly one lattice point. For a given lattice, the Wigner-Seitz cell is uniquely defined as the region of space consisting of all points that are closer to a specific lattice point than to any other lattice point.

3.1.2 Reciprocal Bravais Lattices

Let $f(\mathbf{r})$ be a periodic function on a Bravais lattice, i.e., such that

$$f(\mathbf{r}+\mathbf{R}) = f(\mathbf{r}), \tag{3.3}$$

where $\mathbf{r}$ is defined within the unit cell. To write a Fourier expansion of such a function (see Theorem C.1), we must determine plane waves that satisfy

$$e^{i\mathbf{G}\cdot\mathbf{r}} = e^{i\mathbf{G}\cdot(\mathbf{r}+\mathbf{R})}, \tag{3.4}$$

i.e., we require a vector $\mathbf{G}$ such that

$$e^{i\mathbf{G}\cdot\mathbf{R}} = 1, \ \forall\mathbf{R}, \tag{3.5}$$

then

$$\mathbf{G}\cdot\mathbf{R} = 2\pi l, \ l \in \mathbb{Z}. \tag{3.6}$$

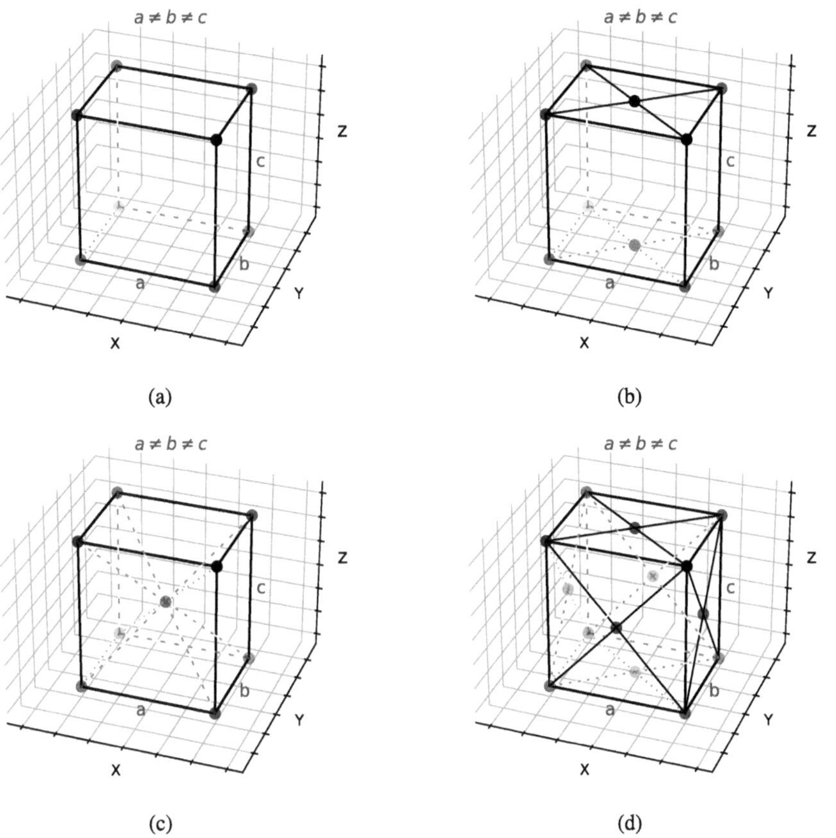

Fig. 3.3 Orthorhombic Bravais lattices: (**a**) Simple orthorhombic (SO): lattice points only at the corners of the orthorhombic cell. (**b**) Base-centered orthorhombic (BCO): lattice points at the corners and at the centers of one pair of opposing faces. (**c**) Body-centered orthorhombic (BO): lattice points at the corners and one at the center of the cell. (**d**) Face-centered orthorhombic (FCO): lattice points at the corners and at the centers of all faces; highest packing efficiency among orthorhombic lattices

It can be shown that Eq. 3.6 is satisfied if and only if we consider 3 vectors $\mathbf{b}_1$, $\mathbf{b}_2$, $\mathbf{b}_3$ such that

$$\mathbf{a}_i \cdot \mathbf{b}_j = 2\pi \delta_{ij}. \tag{3.7}$$

One solution is represented by

$$\mathbf{b}_1 = 2\pi \frac{\mathbf{a}_2 \times \mathbf{a}_3}{\mathbf{a}_1 \cdot [\mathbf{a}_2 \times \mathbf{a}_3]}, \tag{3.8}$$

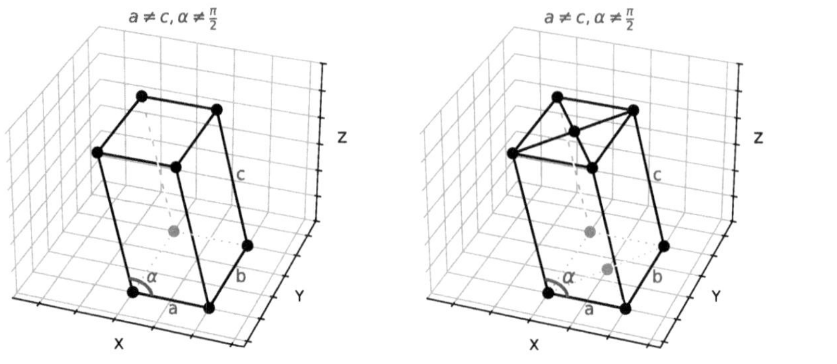

(a) (b)

Fig. 3.4 Monoclinic Bravais lattices: (**a**) Simple monoclinic (SM): lattice points only at the corners of the monoclinic cell; one oblique angle. (**b**) Base-centered monoclinic (BCM): lattice points at the corners and at the centers of one pair of opposing faces; monoclinic symmetry preserved along the unique axis

Fig. 3.5 Triclinic Bravais lattice: Simple triclinic (ST): lattice points only at the corners of the triclinic cell; three unequal edges, three oblique angles, and no rotational symmetry

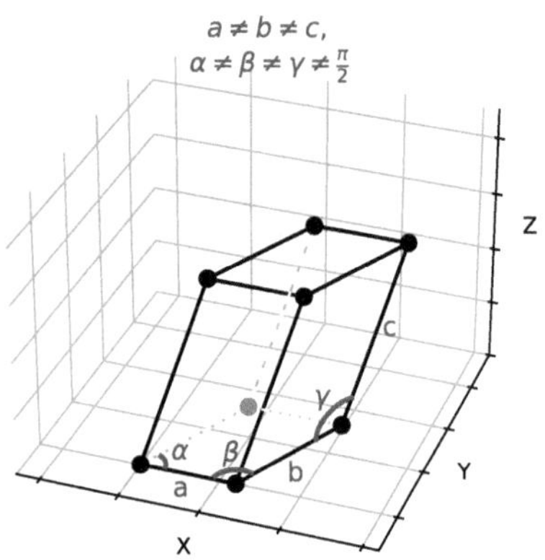

Fig. 3.6 Hexagonal Bravais lattice: Simple hexagonal (SH): lattice points at the corners of a hexagonal prism; two equal edges in the basal plane, a distinct height along the z-axis, and sixfold rotational symmetry around the vertical axis

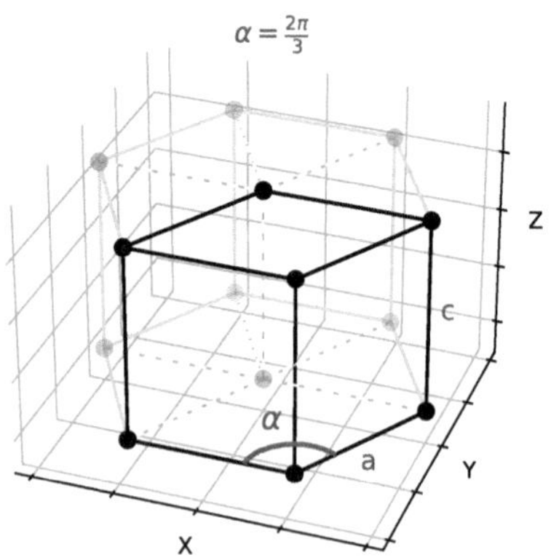

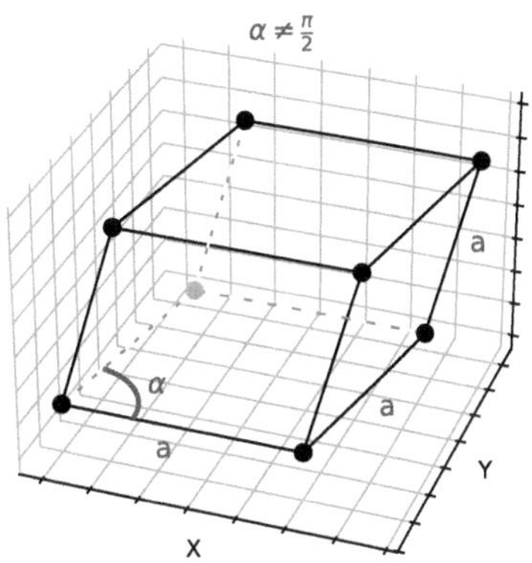

Fig. 3.7 Rhombohedral Bravais lattice: Simple rhombohedral (SR): lattice points at the corners of a rhombohedral (trigonal) cell; equal edge lengths and equal oblique angles; threefold rotational symmetry along the body diagonal

$$\mathbf{b}_2 = 2\pi \frac{\mathbf{a}_3 \times \mathbf{a}_1}{\mathbf{a}_1 \cdot [\mathbf{a}_2 \times \mathbf{a}_3]}, \tag{3.9}$$

$$\mathbf{b}_3 = 2\pi \frac{\mathbf{a}_1 \times \mathbf{a}_2}{\mathbf{a}_1 \cdot [\mathbf{a}_2 \times \mathbf{a}_3]}, \tag{3.10}$$

whose dimensions are given by

$$[b_i] = [\text{length}]^{-1}. \tag{3.11}$$

The set of the vectors **G** fulfilling condition 3.6 takes the form

$$\mathbf{G} = m_1\mathbf{b}_1 + m_2\mathbf{b}_2 + m_3\mathbf{b}_3, \; m_i \in \mathbb{Z}, \tag{3.12}$$

and given that the vectors $\{\mathbf{b}_i\}_{i=1,2,3}$ form a linearly independent set, they constitute a basis. The vectors $\{\mathbf{b}_i\}_{i=1,2,3}$ form a new lattice, called the reciprocal lattice. Note that both because of the presence of the numerical factor 2π in the product $\mathbf{a}_i \cdot \mathbf{b}_j$ and because of the physical dimensions of the $\{\mathbf{b}_i\}_{i=1,2,3}$, we are naturally induced to interpret these vectors as wave vectors. In addition, the quantity

$$V_d = \mathbf{a}_1 \cdot [\mathbf{a}_2 \times \mathbf{a}_3] \tag{3.13}$$

has a physical interpretation, i.e., it is the volume of the elementary cell of the direct lattice, i.e., $V_{\text{direct}} \equiv V_d$. The cell of the primitive vectors $\mathbf{a}_1$, $\mathbf{a}_2$, $\mathbf{a}_3$ on the Bravais lattice locates a reciprocal cell of vectors $\mathbf{b}_1$, $\mathbf{b}_2$, $\mathbf{b}_3$ that belong to a lattice that has the dimensions of the reciprocal of a length and is then called a reciprocal lattice. Importantly, the first Brillouin zone (1st BZ) of a crystal lattice is defined as the Wigner-Seitz cell in the reciprocal lattice. The volume of the elementary cell of the reciprocal lattice, i.e., $V_{\text{reciprocal}} \equiv V_r$, which dimensionally is the reciprocal of a volume ($[V_r] = [L]^{-3}$), is given by

$$V_r = \mathbf{b}_1 \cdot [\mathbf{b}_2 \times \mathbf{b}_3] . \tag{3.14}$$

It satisfies

$$V_r = \frac{(2\pi)^3}{V_d}, \tag{3.15}$$

and is equal to the volume of the first Brillouin zone. A periodic function on Bravais lattice, say $f(\mathbf{r})$, can be expanded by means of the reciprocal lattice vectors as follows

$$f(\mathbf{r}) = \sum_{\mathbf{G}} c_{\mathbf{G}} \frac{e^{i\mathbf{G}\cdot\mathbf{r}}}{\sqrt{V_d}}, \tag{3.16}$$

$$c_{\mathbf{G}} = \frac{1}{\sqrt{V_d}} \int_{V_d} d^3\mathbf{r} f(\mathbf{r}) e^{-i\mathbf{G}\cdot\mathbf{r}}, \tag{3.17}$$

where $\left\{ \frac{e^{i\mathbf{G}\cdot\mathbf{r}}}{\sqrt{V_d}} \right\}$ denotes a set of spatial plane waves that form an orthonormal and complete basis on the domain V_d, i.e., within each elementary cell, and are periodic over the entire direct lattice.

Let $f(\mathbf{k})$ be defined on the reciprocal lattice and such that

$$f(\mathbf{k}) = f(\mathbf{k} + \mathbf{G}), \ \forall \, \mathbf{G}, \tag{3.18}$$

where $\mathbf{k}$ takes values in the first Brillouin zone. Such a function can be expanded in Fourier series on the vectors of the direct lattice, i.e.,

$$f(\mathbf{k}) = \sum_{\mathbf{R}} c_{\mathbf{R}} \frac{e^{i\mathbf{k}\cdot\mathbf{R}}}{\sqrt{V_r}}, \tag{3.19}$$

with

$$c_{\mathbf{R}} = \frac{1}{\sqrt{V_r}} \int_{V_r} d^3\mathbf{k} f(\mathbf{k}) \, e^{-i\mathbf{k}\cdot\mathbf{R}}. \tag{3.20}$$

3.1.3 Simple Cubic Lattice

The simplest Bravais lattice is the simple cubic lattice (see Fig. 3.1a). Given a site, all nearest neighbors are at distance a from each other. A basis in the direct simple cubic lattice is given by the vectors

$$\mathbf{a}_1 = a(1, 0, 0), \tag{3.21}$$

$$\mathbf{a}_2 = a(0, 1, 0), \tag{3.22}$$

$$\mathbf{a}_3 = a(0, 0, 1), \tag{3.23}$$

with volume $V_d = a^3$. The reciprocal lattice of a simple cubic lattice is a simple cubic lattice, with primitive vectors

$$\mathbf{b}_1 = \frac{2\pi}{a}(1, 0, 0), \tag{3.24}$$

$$\mathbf{b}_2 = \frac{2\pi}{a}(0, 1, 0), \tag{3.25}$$

$$\mathbf{b}_3 = \frac{2\pi}{a}(0, 0, 1), \tag{3.26}$$

with volume $V_r = \frac{8\pi^3}{a^3}$. Finally, a basis for the first Brillouin zone of the simple cubic lattice consists of the wave vectors $\mathbf{k} = (k_z, k_y, k_z)$, whose components satisfy

$$k_i \in \left[-\frac{\pi}{a}, \frac{\pi}{a}\right], \; i = x, y, z. \tag{3.27}$$

This defines a cubic region in reciprocal space centered at the origin, which constitutes the first Brillouin zone for the simple cubic lattice.

3.2 Bloch's Theorem

In a crystalline solid, the lattice periodicity strongly constrains the electronic wave functions. Bloch's theorem formalizes this by showing that the eigenfunctions of a Hamiltonian describing an electron in a periodic potential can be written as a plane wave modulated by a lattice-periodic function. Within the fixed-nuclei approximation, applying the mean-field Hartree-Fock method yields an eigenvalue equation of the form given in Eq. II.21. Here, each electron experiences an average effective potential that accounts approximately for the interactions with all other electrons. Specifically, $\mathcal{H}_{\text{HF}}^{(e)}$ in Eq. II.20 describes an electron moving in a periodic potential $V(\mathbf{r})$ that reflects the translational symmetry of the crystal lattice, i.e.,

$$V(\mathbf{r} + \mathbf{R}) = V(\mathbf{r}), \tag{3.28}$$

or equivalently (see Eq. 3.2)

$$V(\mathbf{r} + \mathbf{R_n}) = V(\mathbf{r}). \tag{3.29}$$

In the context of a lattice system, we adapt the notion of spatial translation to reflect the discrete periodicity of the structure. Accordingly, we adapt the action of the translation operator on the lattice, as given in Eq. 15.162, as follows

$$\hat{T}_{\mathbf{n}} : \hat{T}_{\mathbf{n}} \psi(\mathbf{r}) = \psi\left(\mathbf{r} + \mathbf{R}_{\mathbf{n}}\right). \tag{3.30}$$

It is observed that

$$\left[\hat{T}_{\mathbf{n}}, \hat{T}_{\mathbf{n}'}\right] = 0 \tag{3.31}$$

since the order of the translations does not matter. In addition, the equations

$$\left[\hat{T}_{\mathbf{n}}, \frac{\hat{\mathbf{p}}^2}{2m}\right] = 0, \ \forall \mathbf{n}, \tag{3.32}$$

$$\left[\hat{T}_{\mathbf{n}}, V(\mathbf{r})\right] = 0, \ \forall \mathbf{n} \tag{3.33}$$

imply

$$\left[\hat{T}_{\mathbf{n}}, \mathcal{H}_{\mathrm{HF}}^{(e)}\right] = 0, \ \forall \mathbf{n}. \tag{3.34}$$

Consequently, the electronic Hartree-Fock Hamiltonian $\mathcal{H}_{\mathrm{HF}}^{(e)}$ and the translation operator $\hat{T}_{\mathbf{n}}$ have a common set of eigenfunctions. The aim of this section is to determine these eigenfunctions, known as Bloch eigenfunctions or Bloch waves. To this end, we first show

Theorem 3.1 (Bloch's Theorem) *Let $V(\mathbf{r})$ be a potential that is periodic with respect to a crystal lattice, i.e., of the form 3.28, then the solutions to the one-particle Schrödinger equation (see Eq. 1.86) with respect to the Hamiltonian given in Eq. II.20 can be chosen in the form*

$$\varphi_{n,\mathbf{k}}(\mathbf{r}) = e^{i\mathbf{k}\cdot\mathbf{r}}\, U_{n,\mathbf{k}}(\mathbf{r}), \tag{3.35}$$

where

- $\mathbf{k}$ *is a vector in the first Brillouin zone;*
- $n \in \mathbb{N}$ *is the band index;*
- $U_{n,\mathbf{k}}(\mathbf{r})$ *is a function that is periodic on the lattice, i.e.,*

$$U_{n,\mathbf{k}}\left(\mathbf{r} + \mathbf{R}\right) = U_{n,\mathbf{k}}(\mathbf{r}). \tag{3.36}$$

Proof Since $V(\mathbf{r})$ is periodic on the direct lattice with respect to the unit cell (see Eq. 3.13), it can be expanded as Fourier series (see Theorem C.1) as follows

$$V(\mathbf{r}) = \sum_{\mathbf{G}} V_{\mathbf{G}} \frac{e^{i\mathbf{G}\cdot\mathbf{r}}}{\sqrt{V_d}}, \tag{3.37}$$

$$V_{\mathbf{G}} = \frac{1}{\sqrt{V_d}} \int_{V_d} d^3\mathbf{r}\, V(\mathbf{r})\, e^{-i\mathbf{G}\cdot\mathbf{r}}, \tag{3.38}$$

where

$$\mathbf{G} = m_1\mathbf{b}_1 + m_2\mathbf{b}_2 + m_3\mathbf{b}_3,\ \ m_i \in \mathbb{Z}, \tag{3.39}$$

is a vector of the reciprocal lattice of the crystal lattice. Moreover, $V(\mathbf{r})$ is a real function, then

$$\begin{aligned} V^*(\mathbf{r}) &= \sum_{\mathbf{G}} V^*_{\mathbf{G}} \frac{e^{-i\mathbf{G}\cdot\mathbf{r}}}{\sqrt{V_d}} = \\ &= \sum_{-\mathbf{G}} V^*_{-\mathbf{G}} \frac{e^{i\mathbf{G}\cdot\mathbf{r}}}{\sqrt{V_d}}, \end{aligned} \tag{3.40}$$

$$V^*_{\mathbf{G}} = V_{-\mathbf{G}}. \tag{3.41}$$

If $\mathbf{G} = \mathbf{0}$, that is, the wave vector becomes a point on the reciprocal lattice, then $V_{\mathbf{G}=\mathbf{0}}$ becomes the mean value of $V(\mathbf{r})$ with respect to an elementary cell of the direct lattice, i.e., it is a constant. The energy $\mathcal{H}^{(e)}_{\text{HF}}$ is defined up to an arbitrary constant, then we can set

$$V_{\mathbf{G}=\mathbf{0}} = 0. \tag{3.42}$$

The lattice model is based on the assumption that, in each spatial direction, the system size tends to infinity, even though real crystals have finite dimensions. However, if the volume of the crystal is sufficiently large and the number of atoms at the surface is negligible compared to those in the *bulk*, the crystal can be approximated by an infinite lattice. Real boundary conditions are thus replaced by analytically more convenient ones. Periodic Born-von Karman boundary conditions are assumed. Consider a crystal with N_1 cells in the $\mathbf{a}_1$ direction, N_2 cells in the $\mathbf{a}_2$ direction and N_3 cells in the $\mathbf{a}_3$ direction, where $N_i \sim 10^{23}$ (recalling that the typical particle density is $\rho_i \sim 10^{23}\,\text{cm}^{-3}$) for $i = 1, 2, 3$. By repeating the sequence $N_i\mathbf{a}_i$ infinitely many times along the i-th direction, an infinite periodic structure—called the *supercrystal*—is generated as a replication of the crystal. As a result, the elementary cell of the supercrystal corresponds to the entire crystal. Since the elementary cell of the crystal is defined by the lattice vectors $\mathbf{a}_1$, $\mathbf{a}_2$, and $\mathbf{a}_3$, the elementary cell of the supercrystal is thus characterized by the vectors

$$\begin{cases} \tilde{\mathbf{a}}_1 = N_1 \mathbf{a}_1 \\ \tilde{\mathbf{a}}_2 = N_2 \mathbf{a}_2 \\ \tilde{\mathbf{a}}_3 = N_3 \mathbf{a}_3 \end{cases}, \tag{3.43}$$

where $\tilde{\mathbf{a}}_i$ are the primitive vectors of the supercrystal, and N_1, N_2, and N_3 are the number of unit cells along each lattice direction. The structure of the reciprocal lattice corresponding to this supercrystal can be determined, and it can be shown that the primitive vectors of the reciprocal lattice are given by

$$\tilde{\mathbf{b}}_i = \frac{\mathbf{b}_i}{N_i}, \; i = 1, 2, 3. \tag{3.44}$$

Let $\mathbf{q}$ be a vector of the supercrystal's reciprocal lattice. Then $\mathbf{q}$ can be written as a linear combination of the primitive reciprocal vectors with rational coefficients, i.e.,

$$\mathbf{q} = \frac{m_1}{N_1}\mathbf{b}_1 + \frac{m_2}{N_2}\mathbf{b}_2 + \frac{m_3}{N_3}\mathbf{b}_3, \; m_i \in \mathbb{Z}, \; N_i \in \mathbb{N}. \tag{3.45}$$

Given the periodic boundary conditions, the wave function is periodic on the supercrystal, i.e.,

$$\psi\left(\mathbf{r} + \tilde{\mathbf{R}}\right) = \psi(\mathbf{r}), \tag{3.46}$$

i.e., the wave function is replicated in each crystal translated by a vector of the supercrystal lattice of the form

$$\tilde{\mathbf{R}} = c_1\tilde{\mathbf{a}}_1 + c_2\tilde{\mathbf{a}}_2 + c_3\tilde{\mathbf{a}}_3, \; c_i \in \mathbb{Z}. \tag{3.47}$$

Consequently, $\psi(\mathbf{r})$ can be expanded as Fourier series by means of the vector $\mathbf{q}$ of the reciprocal lattice of the supercrystal, i.e.,

$$\psi(\mathbf{r}) = \sum_{\mathbf{q}} \varphi_{\mathbf{q}} \frac{e^{i\mathbf{q}\cdot\mathbf{r}}}{\sqrt{V}}, \tag{3.48}$$

with

$$\varphi_{\mathbf{q}} = \frac{1}{\sqrt{V}} \int_V d^3\mathbf{r}\, \psi(\mathbf{r})\, e^{-i\mathbf{q}\cdot\mathbf{r}}, \tag{3.49}$$

where we recall that V represents the volume of the crystal, which, under periodic boundary conditions, corresponds to the volume of the elementary cell of the supercrystal. Consequently, it is necessary to compute $\varphi_{\mathbf{q}}$, as this quantity leads to the solution of the Bloch eigenvalue problem and, ultimately, to the demonstration of Bloch's theorem. We therefore require that $\psi(\mathbf{r})$ satisfy the Bloch eigenvalue problem (see Eq. II.21), then

$$\left[\frac{\hat{\mathbf{p}}^2}{2m_e} + \sum_{\mathbf{G}} V_{\mathbf{G}} \frac{e^{i\mathbf{G}\cdot\mathbf{r}}}{\sqrt{V_d}}\right] \sum_{\mathbf{q}} \varphi_{\mathbf{q}} \frac{e^{i\mathbf{q}\cdot\mathbf{r}}}{\sqrt{V}} = \mathcal{E} \sum_{\mathbf{q}} \varphi_{\mathbf{q}} \frac{e^{i\mathbf{q}\cdot\mathbf{r}}}{\sqrt{V}}, \tag{3.50}$$

$$\frac{\hat{\mathbf{p}}^2}{2m_e} \sum_{\mathbf{q}} \varphi_{\mathbf{q}} \frac{e^{i\mathbf{q}\cdot\mathbf{r}}}{\sqrt{V}} = \sum_{\mathbf{q}} \varphi_{\mathbf{q}} \frac{\hbar^2\mathbf{q}^2}{2m_e} \frac{e^{i\mathbf{q}\cdot\mathbf{r}}}{\sqrt{V}}, \tag{3.51}$$

$$\sum_{\mathbf{G}} V_{\mathbf{G}} \frac{e^{i\mathbf{G}\cdot\mathbf{r}}}{\sqrt{V_d}} \sum_{\mathbf{q}} \varphi_{\mathbf{q}} \frac{e^{i\mathbf{q}\cdot\mathbf{r}}}{\sqrt{V}} = \sum_{\mathbf{G}} \sum_{\mathbf{q}} \frac{V_{\mathbf{G}}}{\sqrt{V_d}} \varphi_{\mathbf{q}} \frac{e^{i(\mathbf{q}+\mathbf{G})\cdot\mathbf{r}}}{\sqrt{V}}. \tag{3.52}$$

By definition, the components of the vectors $\mathbf{q}$ are rational, while those of the vectors $\mathbf{G}$ are integers. Therefore, the components of $\mathbf{q}+\mathbf{G}$ are also rational. Since the sum is taken over the $\mathbf{q}$ vectors, we perform the change of variables $\mathbf{q} \to \mathbf{q}-\mathbf{G}$ in $\varphi_{\mathbf{q}}$, and accordingly set $\mathbf{q}+\mathbf{G} \to \mathbf{q}$ in the plane-wave element $\frac{e^{i(\mathbf{q}+\mathbf{G})\cdot\mathbf{r}}}{\sqrt{V}}$. Given

$$\sum_{\mathbf{G}} \sum_{\mathbf{q}} \frac{V_{\mathbf{G}}}{\sqrt{V_d}} \varphi_{\mathbf{q}} \frac{e^{i(\mathbf{q}+\mathbf{G})\cdot\mathbf{r}}}{\sqrt{V}} = \sum_{\mathbf{G}} \sum_{\mathbf{q}} \frac{V_{\mathbf{G}}}{\sqrt{V_d}} \varphi_{\mathbf{q}-\mathbf{G}} \frac{e^{i\mathbf{q}\cdot\mathbf{r}}}{\sqrt{V}}, \tag{3.53}$$

we have

$$\sum_{\mathbf{q}} \left[\frac{\hbar^2\mathbf{q}^2}{2m_e} \varphi_{\mathbf{q}} + \sum_{\mathbf{G}} \frac{V_{\mathbf{G}}}{\sqrt{V_d}} \varphi_{\mathbf{q}-\mathbf{G}} - \mathcal{E}\varphi_{\mathbf{q}}\right] \frac{e^{i\mathbf{q}\cdot\mathbf{r}}}{\sqrt{V}} = 0, \tag{3.54}$$

which, since $\left\{\frac{e^{i\mathbf{q}\cdot\mathbf{r}}}{\sqrt{V}}\right\}$ forms a basis for functions defined on the crystal, implies that

$$\left(\frac{\hbar^2\mathbf{q}^2}{2m_e} - \mathcal{E}\right) \varphi_{\mathbf{q}} + \sum_{\mathbf{G}} \frac{V_{\mathbf{G}}}{\sqrt{V_d}} \varphi_{\mathbf{q}-\mathbf{G}} = 0, \ \forall\, \mathbf{q}. \tag{3.55}$$

Now, we write the division between an integer m_1 and a natural N_1 as the sum of an integer d_1 and the remainder natural S_1 divided by N_1, i.e.,

$$\frac{m_1}{N_1} = d_1 + \frac{S_1}{N_1}, \ m_1 \in \mathbb{Z}, \ N_1 \in \mathbb{N}, \ d_1 \in \mathbb{Z}, \ S_1 \in \mathbb{N}, \ 0 \leq S_1 \leq N_1 - 1. \tag{3.56}$$

From Eq. 3.45, one finds

$$\mathbf{q} = \left(d_1 + \frac{S_1}{N_1}\right) \mathbf{b}_1 + \left(d_2 + \frac{S_2}{N_2}\right) \mathbf{b}_2$$

$$+\left(d_3+\frac{S_3}{N_3}\right)\mathbf{b}_3,\ d_i\in\mathbb{Z},\ S_i\in\mathbb{N},\ \frac{S_i}{N_i}<1. \tag{3.57}$$

A vector $\mathbf{q}$ of the reciprocal lattice of the supercrystal can be written as

$$\mathbf{q}=\mathbf{G}'+\mathbf{k}, \tag{3.58}$$

that is, it is the sum of a reciprocal lattice vector $\mathbf{G}'$ and a vector $\mathbf{k}$ within the elementary cell of the reciprocal lattice. Since the choice of the elementary cell is arbitrary, $\mathbf{k}$ can equivalently be taken within the first Brillouin zone. The definitions of the vectors $\mathbf{k}$ depend on the rests $\frac{S_i}{N_i}$, with $0\le S_i\le N_i-1$, then there exist N_i such vectors. We write the vector $\mathbf{q}$ taking into account the previous considerations, i.e.,

$$\left[\frac{\hbar^2\left(\mathbf{k}+\mathbf{G}'\right)^2}{2m_e}-\mathcal{E}\right]\varphi_{\mathbf{k}+\mathbf{G}'}+\sum_{\mathbf{G}}\frac{V_{\mathbf{G}}}{\sqrt{V_d}}\varphi_{\mathbf{k}+\mathbf{G}'-\mathbf{G}}=0. \tag{3.59}$$

Now, all the wave vectors of the Bloch eigenvalue problem lie either in the reciprocal lattice of the crystal or in the first Brillouin zone. We set $\mathbf{G}'-\mathbf{G}=\mathbf{G}''$ in the sum with respect to $\mathbf{G}$, which becomes a sum over $\mathbf{G}''$, i.e.,

$$\left[\frac{\hbar^2\left(\mathbf{k}+\mathbf{G}'\right)^2}{2m_e}-\mathcal{E}\right]\varphi_{\mathbf{k}+\mathbf{G}'}+\sum_{\mathbf{G}''}\frac{V_{\mathbf{G}'-\mathbf{G}''}}{\sqrt{V_d}}\varphi_{\mathbf{k}+\mathbf{G}''}=0. \tag{3.60}$$

We rename the dummy index $\mathbf{G}''=\mathbf{G}$ in the second term as follows

$$\left[\frac{\hbar^2\left(\mathbf{k}+\mathbf{G}'\right)^2}{2m_e}-\mathcal{E}\right]\varphi_{\mathbf{k}+\mathbf{G}'}+\sum_{\mathbf{G}}\frac{V_{\mathbf{G}'-\mathbf{G}}}{\sqrt{V_d}}\varphi_{\mathbf{k}+\mathbf{G}}=0, \tag{3.61}$$

and we obtain the Bloch eigenvalue problem in explicit form, i.e.,

$$\sum_{\mathbf{G}}\left\{\left[\frac{\hbar^2\left(\mathbf{k}+\mathbf{G}'\right)^2}{2m_e}-\mathcal{E}\right]\delta_{\mathbf{G},\mathbf{G}'}+\frac{V_{\mathbf{G}'-\mathbf{G}}}{\sqrt{V_d}}\right\}\varphi_{\mathbf{k}+\mathbf{G}}=0. \tag{3.62}$$

The diagonal terms are kinetic, i.e., $\frac{\hbar^2(\mathbf{k}+\mathbf{G})^2}{2m_e}$, while the off-diagonal terms are $\frac{V_{\mathbf{G}'-\mathbf{G}}}{\sqrt{V_d}}$, being $V_{\mathbf{0}}=0$. The matrix in Eq. 3.62 is bloch-diagonal: for each $\mathbf{k}$ there is a block. All blocks have the same dimension, and the nonzero matrix elements are determined by the number of $\mathbf{G}$-vectors included in the expansion of the periodic effective potential $V(\mathbf{r})$. There exist as many blocks as there are $\mathbf{k}$ vectors, i.e., there are $N=N_1N_2N_3$ blocks. By fixing a particular $\mathbf{k}$, we diagonalize its corresponding block to obtain the eigenvectors $\varphi_{n,\mathbf{k}}$. Here, $\mathbf{k}$ labels the block (i.e., the crystal momentum), while n indexes the eigenstates within the block, naturally associated with the energy band index for each $\mathbf{k}$. Thus, $\mathbf{k}$ and n are known as Bloch indices. It follows that the corresponding eigenvalues of the Hamiltonian can be denoted by $\mathcal{E}_{n,\mathbf{k}}$, representing the energy of the n-th band at crystal momentum $\mathbf{k}$. Now, we have an explicit form of $\psi(\mathbf{r})$ of Eq. 3.48, in particular it is a superposition of Bloch waves, eigenfunctions of a block matrix, i.e.,

$$\varphi_{n,\mathbf{k}}(\mathbf{r}) = \sum_{\mathbf{G}} \frac{e^{i(\mathbf{k}+\mathbf{G})\cdot\mathbf{r}}}{\sqrt{V}} \varphi_{n,\mathbf{k}+\mathbf{G}}. \tag{3.63}$$

Now, if we rewrite eq. 3.63 in the form of 3.35, that is, if we define

$$U_{n,\mathbf{k}}(\mathbf{r}) = \sum_{\mathbf{G}} \frac{e^{i\mathbf{G}\cdot\mathbf{r}}}{\sqrt{V}} \varphi_{n,\mathbf{k}+\mathbf{G}}, \tag{3.64}$$

then

$$\begin{aligned} U_{n,\mathbf{k}}(\mathbf{r}+\mathbf{R}) &= \sum_{\mathbf{G}} \frac{e^{i\mathbf{G}\cdot(\mathbf{r}+\mathbf{R})}}{\sqrt{V}} \varphi_{n,\mathbf{k}+\mathbf{G}} = \\ &= \sum_{\mathbf{G}} \frac{e^{i\mathbf{G}\cdot\mathbf{r}}}{\sqrt{V}} \varphi_{n,\mathbf{k}+\mathbf{G}} = \\ &= U_{n,\mathbf{k}}(\mathbf{r}), \end{aligned} \tag{3.65}$$

that is, the function $U_{n,\mathbf{k}}(\mathbf{r})$ is periodic with the periodicity of the crystal lattice (i.e., with respect to unit cells), and the theorem is thus proved. □

The Bloch functions are chosen to be orthonormal over the entire crystal volume V, by imposing the normalization condition

$$\int_V d^3\mathbf{r}\varphi^*_{n,\mathbf{k}}(\mathbf{r})\varphi_{n',\mathbf{k}'}(\mathbf{r}) = \delta_{n,n'}\delta_{\mathbf{k},\mathbf{k}'}. \tag{3.66}$$

Since the Bloch functions can be written as Eq. 3.35, the normalization in Eq. 3.66 implies for the periodic functions $U_{n,\mathbf{k}}(\mathbf{r})$ the following relation

$$\int_{V_\mathrm{d}} d^3\mathbf{r}|U_{n,\mathbf{k}}(\mathbf{r})|^2 = \frac{1}{N}, \tag{3.67}$$

where (see Eq. 3.13)

$$N = \frac{V}{V_\mathrm{d}} \tag{3.68}$$

denotes the number of unit cells in the crystal. The periodic functions $U_{n,\mathbf{k}}(\mathbf{r})$ are, in general, not orthonormal within a single unit cell. The orthogonality of Bloch states with different wave vectors $\mathbf{k} \neq \mathbf{k}'$ arises only after summing over all lattice translations. Indeed, inserting the Bloch form 3.35 into the orthonormality condition 3.66 gives

$$\int_V d^3\mathbf{r} U^*_{n,\mathbf{k}}(\mathbf{r}) U_{n',\mathbf{k}'}(\mathbf{r})\, e^{i(\mathbf{k}'-\mathbf{k})\cdot\mathbf{r}}. \tag{3.69}$$

We decompose the integral over the total volume V as a sum of integrals over the volumes $V_\mathbf{n}$ of the individual unit cells labeled by $\mathbf{n}$. To do so, we express the position vector $\mathbf{r}$ as the sum of a position $\mathbf{r_0}$ within the reference unit cell $\mathbf{0}$ and a lattice translation vector $\mathbf{R_n}$, i.e., we set $\mathbf{r} \to \mathbf{r_0} + \mathbf{R_n}$. Since the functions $U_{n',\mathbf{k}'}(\mathbf{r_0} + \mathbf{R_n})$ are periodic with respect to lattice translations (see Eq. 3.36), this change of variables does not modify the form of the periodic functions, and the integral becomes

$$\sum_\mathbf{n} e^{i(\mathbf{k}'-\mathbf{k})\cdot\mathbf{R_n}} \int_{V_\mathrm{d}} d^3\mathbf{r}_0 U^*_{n,\mathbf{k}}(\mathbf{r}_0) U_{n',\mathbf{k}'}(\mathbf{r}_0) e^{i(\mathbf{k}'-\mathbf{k})\cdot\mathbf{r}_0}. \tag{3.70}$$

The lattice sum

$$\sum_\mathbf{n} e^{i(\mathbf{k}'-\mathbf{k})\cdot\mathbf{R_n}} = N\delta_{\mathbf{k},\mathbf{k}'} \tag{3.71}$$

enforces orthogonality between different crystal momenta. The remaining cell integral is not proportional to $\delta_{\mathbf{k},\mathbf{k}'}$; therefore,

$$\int_{V_\mathrm{d}} d^3\mathbf{r}_0 U^*_{n,\mathbf{k}}(\mathbf{r}_0) U_{n',\mathbf{k}'}(\mathbf{r}_0) \neq \delta_{n,n'}\delta_{\mathbf{k},\mathbf{k}'} \tag{3.72}$$

in general. Physically, this means that the orthogonality of Bloch states with different wave vectors is a global property of the crystal, enforced by translational symmetry, and not a local property of the periodic parts within a single unit cell.

Remark 3.1 (Notation for Wave Vectors in a Reciprocal Unit Cell) In what follows, we will adopt the convention of writing the wave vectors within the elementary cell of the reciprocal lattice as follows

$$\frac{S_1}{N_1}\mathbf{b}_1 + \frac{S_2}{N_2}\mathbf{b}_2 + \frac{S_3}{N_3}\mathbf{b}_3,\ S_i \in \mathbb{N},\ 0 \leq S_i \leq N_i - 1, \tag{3.73}$$

where $\mathbf{b}_i$ are the primitive vectors of the reciprocal lattice and N_i determines the discretization along each direction.

Now, consider the action of the momentum operator $\hat{\mathbf{p}} = -i\hbar\nabla$ on a Bloch wave $\varphi_{n,\mathbf{k}}(\mathbf{r})$, namely

$$\hat{\mathbf{p}}\varphi_{n,\mathbf{k}}(\mathbf{r}) = \hbar\mathbf{k}\varphi_{n,\mathbf{k}}(\mathbf{r}) - i\hbar\, e^{i\mathbf{k}\cdot\mathbf{r}}\,\nabla U_{n,\mathbf{k}}(\mathbf{r}). \tag{3.74}$$

In the case of many free particles, the Bloch eigenfunctions reduce to plane waves (see Eq. 1.77), which are eigenfunctions of the momentum operator.

In a crystal lattice, however, $\mathbf{k}$ is no longer the physical momentum but rather the crystal quasi-momentum, and Bloch waves are generally not eigenfunctions of the momentum operator with eigenvalue $\hbar\mathbf{k}$. It should be noted, however, that at specific high-symmetry points of the first Brillouin zone, such as the zone center Γ, certain components of $\mathbf{k}$ may correspond to approximate momentum eigenvalues.

Electrons in the lattice become quasiparticles whose properties are modified by the periodic potential. In particular, their inertial response is described by an effective mass m, different from the bare electron mass m_e. To describe these quasiparticles within Bloch's formalism, we refer to them as Bloch quasi-electrons. The effective mass of a Bloch quasi-electron depends on the wave vector $\mathbf{k}$ and is generally anisotropic, and can be characterized by a mass tensor defined as

$$\frac{1}{m_{ij}\,(n,\mathbf{k})} = \frac{1}{\hbar^2}\frac{\partial^2\mathcal{E}_{n,\mathbf{k}}}{\partial k_j \partial k_i}. \tag{3.75}$$

In the case of many free electrons, $\mathcal{E}_{n,\mathbf{k}} = \mathcal{E}_{\mathbf{k}} = \frac{\hbar^2\mathbf{k}^2}{2m_e}$, then the quantity $m_{ij}^{-1}\,(n,\mathbf{k})$ is a diagonal matrix and the diagonal terms are equal to m_e^{-1}. De Broglie's relation $|\mathbf{p}| = \hbar|\mathbf{k}|$ is no longer valid; it must be generalized, just as the definition of group velocity must be generalized: the group velocity of a Bloch wave is given by

$$v_{n,\mathbf{k}} = \frac{1}{\hbar}\nabla_{\mathbf{k}}\mathcal{E}_{n,\mathbf{k}}. \tag{3.76}$$

We now wish to emphasize more clearly the physical meaning of the crystal quasi-momentum. In the regime of weak and slowly varying electric and magnetic fields, one can show that its equation of motion takes the form

$$\dot{\mathbf{p}}_{n,\mathbf{k}} = q_e\left(\mathbf{E} + \frac{1}{c}\,\mathbf{v}_{n,\mathbf{k}} \times \mathbf{H}\right). \tag{3.77}$$

This equation, although formally identical to that of a free charged particle in vacuum, holds only on average, describing the effective dynamics of an electron moving in the periodic potential of the ions. It is important to note that, in the right-hand side of Eq. 3.77, only the external fields $\mathbf{E}$ and $\mathbf{H}$ appear. At first glance, it may seem that the electron does not experience the direct action of the ions or of the other electrons. However, these interactions are not absent: they are already incorporated into the definitions of 3.76 and 3.77. In other words, the electron is no longer a bare particle but a quasiparticle, an effective entity that includes the dressing due to its interactions with the periodic ionic potential, which gives rise to the band structure, and the other electrons, which contribute many-body corrections. Therefore, the semiclassical equation of motion for the momentum describes the dynamical behavior of a collective excitation of the electronic system, rather than that of an isolated electron.

We now aim to show that the quantity defined in Eq. 3.76 is related to the momentum operator $\hat{\mathbf{p}}$. To this end, we first recall a general result.

Theorem 3.2 (Hellmann-Feynman Theorem) *Let $A(\lambda)$ be a Hermitian operator depending smoothly on a real parameter λ, and let $|\psi(\lambda)\rangle$ be a normalized eigenstate of $A(\lambda)$ with eigenvalue $\mathcal{E}(\lambda)$, then the derivative of the eigenvalue with respect to λ is given by*

$$\frac{d\mathcal{E}(\lambda)}{d\lambda} = \left\langle \psi(\lambda) \middle| \frac{dA(\lambda)}{d\lambda} \middle| \psi(\lambda) \right\rangle. \tag{3.78}$$

Proof Differentiating the eigenvalue equation $A(\lambda)|\psi(\lambda)\rangle = \mathcal{E}(\lambda)|\psi(\lambda)\rangle$ with respect to λ gives

$$\frac{dA(\lambda)}{d\lambda}|\psi\rangle + A(\lambda)\frac{d|\psi\rangle}{d\lambda} = \frac{d\mathcal{E}(\lambda)}{d\lambda}|\psi\rangle + \mathcal{E}(\lambda)\frac{d|\psi\rangle}{d\lambda}. \tag{3.79}$$

Taking the inner product with $\langle\psi|$ from the left and using the Hermiticity of A, we obtain

$$\begin{aligned}\left\langle \psi(\lambda) \middle| \frac{dA(\lambda)}{d\lambda} \middle| \psi(\lambda) \right\rangle + \left\langle \psi(\lambda) \middle| A(\lambda) \middle| \frac{d\psi(\lambda)}{d\lambda} \right\rangle \\ = \frac{d\mathcal{E}(\lambda)}{d\lambda} \langle\psi(\lambda)|\psi(\lambda)\rangle + \mathcal{E}(\lambda) \left\langle \psi(\lambda) \middle| \frac{d\psi(\lambda)}{d\lambda} \right\rangle. \end{aligned} \tag{3.80}$$

Since $\langle\psi|\psi\rangle = 1$ and $\langle\psi|A(\lambda) = \mathcal{E}(\lambda)\langle\psi|$, the second terms on both sides cancel, leading to the thesis. □

We now return to Eq. 3.35 and consider the Bloch eigenstate $\varphi_{n,\mathbf{k}}(\mathbf{r})$, associated with the Hamiltonian $\mathcal{H}_{\text{HF}}^{(e)}$ given in Eq. II.20. We employ the periodic part $U_{n,\mathbf{k}}(\mathbf{r})$ through 3.35. Note that Eq. 3.74 can be written as

$$\begin{aligned}\hat{\mathbf{p}}\varphi_{n,\mathbf{k}}(\mathbf{r}) &= \hat{\mathbf{p}}\big(e^{i\mathbf{k}\cdot\mathbf{r}}U_{n,\mathbf{k}}(\mathbf{r})\big) = \\ &= -i\hbar\nabla\big(e^{i\mathbf{k}\cdot\mathbf{r}}U_{n,\mathbf{k}}(\mathbf{r})\big) = \\ &= e^{i\mathbf{k}\cdot\mathbf{r}}\big(\hat{\mathbf{p}} + \hbar\mathbf{k}\big)U_{n,\mathbf{k}}(\mathbf{r}), \end{aligned} \tag{3.81}$$

consequently

$$\hat{\mathbf{p}}^2\left(e^{i\mathbf{k}\cdot\mathbf{r}}U_{n,\mathbf{k}}(\mathbf{r})\right) = e^{i\mathbf{k}\cdot\mathbf{r}}\,(\hat{\mathbf{p}}+\hbar\mathbf{k})^2 U_{n,\mathbf{k}}(\mathbf{r}). \tag{3.82}$$

It follows that

$$\begin{aligned}\mathcal{H}^{(e)}_{\mathrm{HF}}\varphi_{n,\mathbf{k}} &= e^{i\mathbf{k}\cdot\mathbf{r}}\left[\frac{(\hat{\mathbf{p}}+\hbar\mathbf{k})^2}{2m_e} + V(\mathbf{r})\right]U_{n,\mathbf{k}} \equiv \\ &\equiv e^{i\mathbf{k}\cdot\mathbf{r}}\mathcal{H}^{(e)}_{\mathrm{HF},\mathbf{k}}U_{n,\mathbf{k}},\end{aligned} \tag{3.83}$$

where we have defined the **k**-dependent operator

$$\mathcal{H}^{(e)}_{\mathrm{HF},\mathbf{k}} \equiv \frac{(\hat{\mathbf{p}}+\hbar\mathbf{k})^2}{2m_e} + V(\mathbf{r}). \tag{3.84}$$

Hence the Bloch problem is equivalent to

$$\mathcal{H}^{(e)}_{\mathrm{HF},\mathbf{k}}U_{n,\mathbf{k}} = \mathcal{E}_{n,\mathbf{k}}\,U_{n,\mathbf{k}}. \tag{3.85}$$

Now, from

$$\begin{aligned}\nabla_{\mathbf{k}}\mathcal{H}^{(e)}_{\mathrm{HF},\mathbf{k}} &= \nabla_{\mathbf{k}}\frac{(\hat{\mathbf{p}}+\hbar\mathbf{k})^2}{2m_e} = \\ &= \frac{\hbar}{m_e}(\hat{\mathbf{p}}+\hbar\mathbf{k}).\end{aligned} \tag{3.86}$$

and from Eq. 3.78, it follows

$$\nabla_{\mathbf{k}}\mathcal{E}_{n,\mathbf{k}} = \frac{\hbar}{m_e}\big\langle U_{n,\mathbf{k}}\big|\,\hat{\mathbf{p}}+\hbar\mathbf{k}\,\big|U_{n,\mathbf{k}}\big\rangle. \tag{3.87}$$

From Eq. 3.81, one finds

$$\nabla_{\mathbf{k}}\mathcal{E}_{n,\mathbf{k}} = \frac{\hbar}{m_e}\langle\varphi_{n,\mathbf{k}}|\hat{\mathbf{p}}|\varphi_{n,\mathbf{k}}\rangle, \tag{3.88}$$

that is, from Eq. 3.76, we finally obtain

$$\mathbf{v}_{n,\mathbf{k}} = \frac{1}{m_e}\langle\varphi_{n,\mathbf{k}}|\hat{\mathbf{p}}|\varphi_{n,\mathbf{k}}\rangle. \tag{3.89}$$

This completes the detailed derivation, showing explicitly how the Bloch form of the wavefunction converts the $(\hat{\mathbf{p}}+\hbar\mathbf{k})$ expectation into the canonical momentum expectation on the full Bloch state, and hence yields the stated expression for the group velocity.

To conclude this section, we now prove that the Bloch eigenstates are also eigenfunctions of the lattice spatial translation operator $\hat{T}_{\mathbf{m}}$ (see Eq. 3.30). Here we deliberately use the index $\mathbf{m}$ — rather than $\mathbf{n}$ — to label lattice translations, in order to avoid any conflict with the Bloch band index n appearing in $\varphi_{n,\mathbf{k}}(\mathbf{r})$. One finds

$$\begin{aligned}\hat{T}_{\mathbf{m}}\varphi_{n,\mathbf{k}}(\mathbf{r}) &= \varphi_{n,\mathbf{k}}\left(\mathbf{r}+\mathbf{R}_{\mathbf{m}}\right) = \\ &= e^{i\mathbf{k}\cdot(\mathbf{r}+\mathbf{R}_{\mathbf{m}})}U_{n,\mathbf{k}}\left(\mathbf{r}+\mathbf{R}_{\mathbf{m}}\right) = \\ &= e^{i\mathbf{k}\cdot(\mathbf{r}+\mathbf{R}_{\mathbf{m}})}U_{n,\mathbf{k}}(\mathbf{r}) = \\ &= e^{i\mathbf{k}\cdot\mathbf{r}}e^{i\mathbf{k}\cdot\mathbf{R}_{\mathbf{m}}}U_{n,\mathbf{k}}(\mathbf{r}) = \\ &= e^{i\mathbf{k}\cdot\mathbf{R}_{\mathbf{m}}}\varphi_{n,\mathbf{k}}(\mathbf{r}),\end{aligned} \tag{3.90}$$

that is, $\varphi_{n,\mathbf{k}}(\mathbf{r})$ are eigenfunctions of the translation operator $\hat{T}_{\mathbf{m}}$. It is worth noting that the action of the translation operator on Bloch functions, given in Eq. 3.90, is perfectly consistent with the action of the spatial translation operator defined in Eq. 15.167, since the opposite signs in the exponential terms correctly reflect the sign convention associated with the translation vector.

3.3 Bloch Waves

Valence electrons determine the chemical bonding in a crystal, and thus it is of particular interest to study their corresponding eigenfunctions, denoted by $\varphi_{n,\mathbf{k}+\mathbf{G}}$. To this purpose, we first consider the electron problem in the absence of the ionic potential, treating it subsequently as a perturbation. This framework is known as the nearly-free electron model. In the case $V(\mathbf{r}) = 0$, the Hamiltonian matrix in Eq. 3.62 is diagonal, and the eigenvalues reduce to the kinetic energies of free electrons, that is

$$\begin{aligned}\mathcal{E}_{n,\mathbf{k}} &= \frac{\hbar^2}{2m_e}\left[\mathbf{k}+\mathbf{G}\right]^2 = \\ &= \frac{\hbar^2}{2m_e}\left[\mathbf{k}+m_1\mathbf{b}_1+m_2\mathbf{b}_2+m_3\mathbf{b}_3\right]^2.\end{aligned} \tag{3.91}$$

Let us consider a simple cubic lattice, $\mathbf{k}$ is defined in the first Brillouin zone, which is itself a simple cubic lattice of linear size $\frac{2\pi}{a}$ and components given by 3.27. Bloch's theorem 3.1 returns the following eigenfunctions

$$\varphi^{(0)}_{n,\mathbf{k}}(\mathbf{r}) = \frac{e^{i(\mathbf{k}+\mathbf{G})\cdot\mathbf{r}}}{\sqrt{V}}, \tag{3.92}$$

where the superscript (0) denotes the unperturbed (free-electron) eigenfunction, i.e., the eigenstate of the Hamiltonian in the absence of the ionic potential $V(\mathbf{r})$. Indeed, such functions are eigenvectors of a diagonal Hamiltonian; they are proportional to the N unit vectors of the standard basis. In other words, in the absence of an ionic potential, the Bloch eigenfunctions reduce to plane waves. It is important to note that the wavevectors within the first Brillouin zone are sufficient to fully describe the dynamics of many non-interacting free particles. For example, in one dimension, on a simple cubic lattice, $\mathbf{k}$ is replaced by the scalar k and $\mathbf{r}$ is replaced by x, and we have

$$\mathcal{E}_{0,k} = \frac{\hbar^2 k^2}{2m_e}, \tag{3.93}$$

$$\mathcal{E}_{-1,k} = \frac{\hbar^2}{2m_e}\left(k - \frac{2\pi}{a}\right)^2, \tag{3.94}$$

$$\mathcal{E}_{+1,k} = \frac{\hbar^2}{2m_e}\left(k + \frac{2\pi}{a}\right)^2, \tag{3.95}$$

which corresponds to three parabolic dispersion curves centered at 0, $-\frac{2\pi}{a}$, and $\frac{2\pi}{a}$, respectively. The free-electron parabola is originally defined over the entire wavevector domain. However, when k is restricted to the first Brillouin zone, the dispersion relation is correspondingly confined within that zone. The remaining portions of the free-particle spectrum can be recovered by translating the central parabola through reciprocal lattice vectors. Hence, it is sufficient to analyze the energy bands within the first Brillouin zone.

The effect of the periodic potential is that, in general, a perturbing potential shifts energy levels and may lift degeneracies. In this case, the parabolas centered at $-\frac{2\pi}{a}$ and $\frac{2\pi}{a}$ intersect the central one (centered at 0), resulting in two doubly degenerate energy levels. For example, at $k = \frac{\pi}{a}$, the degenerate eigenfunctions correspond to $n = 0$ and $n = -1$, i.e., $\frac{e^{i\frac{\pi}{a}x}}{\sqrt{L}}$, $\frac{e^{-i\frac{\pi}{a}x}}{\sqrt{L}}$. However, the presence of the periodic potential lifts this degeneracy, as we shall now see. According to perturbation theory, we must diagonalize a 2×2 matrix at $k = \frac{\pi}{a}$. For a system of free particles, the eigenvalues correspond to the diagonal terms. The off-diagonal elements, instead, are determined by the matrix elements of the potential between the two degenerate unperturbed states, i.e.,

$$\langle n = 0|V(x)|n = -1\rangle,\ k = \frac{\pi}{a}. \tag{3.96}$$

The resulting matrix takes the form

$$\begin{pmatrix} \frac{\hbar^2}{2m_e}\left(\frac{\pi}{a}\right)^2 & \tilde{V} \\ \tilde{V}^* & \frac{\hbar^2}{2m_e}\left(-\frac{\pi}{a}\right)^2 \end{pmatrix}, \tag{3.97}$$

where

$$\tilde{V} = \frac{1}{L}\int dx e^{-i\frac{\pi}{a}x} V(x) e^{i\frac{\pi}{a}x}. \tag{3.98}$$

From the eigenvalue equation

$$\left[\frac{\hbar^2}{2m_e}\left(\frac{\pi}{a}\right)^2 - \mathcal{E}\right]^2 - \left|\tilde{V}\right|^2 = 0, \tag{3.99}$$

we get

$$\mathcal{E}_{\pm} = \frac{\hbar^2}{2m_e}\left(\frac{\pi}{a}\right)^2 \pm \left|\tilde{V}\right|. \tag{3.100}$$

In other words, the periodic potential is responsible for the opening of energy gaps in solids.

To summarize: the free-electron spectrum, originally continuous, is folded into the first Brillouin zone due to the periodicity of the crystal lattice. This folding creates points of degeneracy where different energy branches intersect. The periodic potential lifts these degeneracies, resulting in the formation of energy gaps. Near the degenerate points, the energy bands split and slightly deviate from the free-electron parabolas, creating a gap. This behavior repeats itself at higher energies, giving rise to the characteristic band structure of solids. Bloch's theorem 3.1 provides the fundamental explanation for the existence and structure of these energy bands.

3.4 Wannier Functions

The tight-binding model is suited for describing valence electrons that are strongly localized around their respective nuclei, as is the case, for example, with the valence electrons of transition metals in the periodic table. The model begins with isolated atoms. Suppose we have N such atoms initially placed at infinite separation. These atoms are then gradually brought together to form a crystal, reducing the inter-atomic distance until it equals the lattice constant a. At infinite separation, each atomic energy level is N-fold degenerate. As atoms approach one another, electrons can tunnel to neighboring nuclei, a process known as hybridization, which lifts the degeneracy and causes the energy levels to split. This splitting gives rise to energy bands. Hence, in the tight-binding picture, the electronic band structure of the crystal emerges from the quantum-mechanical tunneling of electrons between the periodic potential wells created by the atoms. This phenomenon is called electron hopping. From Eq. 3.90, it follows

$$\left|\hat{T}_{\mathbf{n}}\varphi_{n,\mathbf{k}}\left(\mathbf{r}\right)\right|^{2} = \left|\varphi_{n,\mathbf{k}}(\mathbf{r})\right|^{2}, \tag{3.101}$$

where the lattice spatial translation operator $\hat{T}_{\mathbf{n}}$ is given in Eq. 3.30. Bloch eigenfunctions are extended waves: the probability of finding the electron at a point within an elementary cell is independent of the cell. As discussed earlier, any point in the reciprocal lattice space of a crystal can be obtained by a translation from the elementary cell, then

$$\varphi_{n,\mathbf{k}+\mathbf{G}}(\mathbf{r}) = \varphi_{n,\mathbf{k}}(\mathbf{r}), \tag{3.102}$$

in other words, the elementary cell where we compute $\varphi_{n,\mathbf{k}}(\mathbf{r})$ is arbitrary. Then $\varphi_{n,\mathbf{k}}(\mathbf{r})$ can be expanded as Fourier series on the direct lattice as

$$\varphi_{n,\mathbf{k}}(\mathbf{r}) = \sum_{\mathbf{R}} \frac{e^{i\mathbf{k}\cdot\mathbf{R}}}{\sqrt{N}} \omega_{n}\left(\mathbf{r},\mathbf{R}\right), \tag{3.103}$$

where $\omega_n\left(\mathbf{r},\mathbf{R}\right)$ are the Wannier functions.

Bloch waves and Wannier functions form equivalent bases for electronic states, as stated in the following theorem.

Theorem 3.3 *Let* $\{\mathbf{R}\}$ *be a finite set of vectors of the direct lattice, and let* $\{\mathbf{k}\}$ *be the corresponding discrete set of wavevectors in the first Brillouin zone (see Sect. 3.1). Then, the matrix*

$$\frac{e^{i\mathbf{k}\cdot\mathbf{R}}}{\sqrt{N}} \tag{3.104}$$

is a unitary transformation.

Proof From Eq. 3.73, the components of $\mathbf{k}$ are given by $\frac{S_i}{N_i}$, with integers satisfying $0 \le S_i \le N_i - 1$. Therefore, there are $N = N_1N_2N_3$ such vectors. On the other hand, the vectors $\mathbf{R}$ belong to the direct lattice, where the correspondence between lattice points and Wigner-Seitz cells is one-to-one, yielding the same number $N = N_1N_2N_3$ of $\mathbf{R}$ vectors. Hence, a one-to-one mapping exists between the sets $\{\mathbf{k}\}$ and $\{\mathbf{R}\}$, ensuring that the scalar product $\mathbf{k}\cdot\mathbf{R}$ is well defined. Consequently,

$$\mathbf{R}\cdot\mathbf{k} = \left(n_1\mathbf{a}_1 + n_2\mathbf{a}_2 + n_3\mathbf{a}_3\right)\cdot\left(\frac{S_1}{N_1}\mathbf{b}_1 + \frac{S_2}{N_2}\mathbf{b}_2 + \frac{S_3}{N_3}\mathbf{b}_3\right) =$$

$$= 2\pi \left[\frac{n_1 S_1}{N_1} + \frac{n_2 S_2}{N_2} + \frac{n_3 S_3}{N_3} \right],$$

$$n_i, S_i \in \mathbb{N},\ 0 \le n_i \le N_i - 1,\ 0 \le S_i \le N_i - 1, \tag{3.105}$$

where we employed Eq. 3.7. Consider

$$\sum_{\mathbf{R}} \frac{e^{i\mathbf{k}\cdot\mathbf{R}}}{\sqrt{N}} = \sum_{n_1,n_2,n_3} \frac{e^{i2\pi n_1 \frac{S_1}{N_1}}}{\sqrt{N_1}} \frac{e^{i2\pi n_2 \frac{S_2}{N_2}}}{\sqrt{N_2}} \frac{e^{i2\pi n_3 \frac{S_3}{N_3}}}{\sqrt{N_3}} =$$
$$= \left[\sum_{n_1=0}^{N_1-1} \frac{e^{i2\pi n_1 \frac{S_1}{N_1}}}{\sqrt{N_1}} \right] \left[\sum_{n_2=0}^{N_2-1} \frac{e^{i2\pi n_2 \frac{S_2}{N_2}}}{\sqrt{N_2}} \right] \left[\sum_{n_3=0}^{N_3-1} \frac{e^{i2\pi n_3 \frac{S_3}{N_3}}}{\sqrt{N_3}} \right]. \tag{3.106}$$

Each of these three terms is a function of $\mathbf{k}$, that is, of S_i, $i = 1, 2, 3$. Consider the first component, if $S_1 = 0$ (or equivalently if $k_1 = 0$), we have

$$\left. \sum_{n_1=0}^{N_1-1} \frac{e^{i2\pi n_1 \frac{S_1}{N_1}}}{\sqrt{N_1}} \right|_{S_1=0} = \frac{1}{\sqrt{N_1}} \sum_{n_1=0}^{N_1-1} 1 =$$
$$= \sqrt{N_1}, \tag{3.107}$$

and we want to prove that it is zero for $S_1 \neq 0$. If we assume that $S_1 \neq 0$ and that $e^{i2\pi n_1 \frac{S_1}{N_1}} \neq 1$, we have

$$\sum_{n_1=0}^{N_1-1} \frac{e^{i2\pi n_1 \frac{S_1}{N_1}}}{\sqrt{N_1}} = \sum_{n_1=0}^{N_1-1} \frac{e^{\left(i2\pi \frac{S_1}{N_1}\right) n_1}}{\sqrt{N_1}} =$$
$$= \frac{1}{\sqrt{N_1}} \frac{1 - \left(e^{i2\pi \frac{S_1}{N_1}}\right)^{N_1}}{1 - e^{i2\pi \frac{S_1}{N_1}}} =$$
$$= \frac{1}{\sqrt{N_1}} \frac{1 - e^{i2\pi S_1}}{1 - e^{i2\pi \frac{S_1}{N_1}}}. \tag{3.108}$$

Now, the quantity S_1 is a natural, so the above ratio is zero except for $S_1 \neq 0$, which we ruled out a priori. Consider the objects $\frac{e^{i\mathbf{k}_i\cdot\mathbf{R}}}{\sqrt{N}}$, $\frac{e^{i\mathbf{k}_j\cdot\mathbf{R}}}{\sqrt{N}}$, which correspond to two columns of the matrix under consideration. We now compute their scalar product by multiplying one of them by the adjoint of the other, namely $\frac{e^{-i\mathbf{k}_j\cdot\mathbf{R}}}{\sqrt{N}}$. Hence, we have

$$\sum_{\mathbf{R}} \frac{e^{i(\mathbf{k}_i-\mathbf{k}_j)\cdot\mathbf{R}}}{N} = \delta_{\mathbf{k}_i,\mathbf{k}_j}, \tag{3.109}$$

which shows that the matrix

$$\frac{e^{i\mathbf{k}\cdot\mathbf{R}}}{\sqrt{N}} \tag{3.110}$$

is unitary. □

Remark 3.2 (Discrete Fourier Transform) Given a set of N complex numbers $\{c_n\}$, with $n = 0, \ldots, N-1$, its discrete Fourier transform (DFT) is defined as the set

$$\tilde{c}_k = \sum_{n=0}^{N-1} \frac{e^{i\frac{2\pi kn}{N}}}{\sqrt{N}} c_n, \quad k = 0, \ldots, N-1. \tag{3.111}$$

According to Remark 3.2, Bloch eigenfunctions are the discrete Fourier transforms of Wannier functions. Since 3.104 is a unitary transformation, it is invertible and the inverse Fourier transform is indexed on the **k** vectors of the first Brillouin zone. Then, Wannier functions are defined by a discrete inverse Fourier transform of Bloch eigenfunctions (see Eq. 3.35), i.e.,

$$\omega_n(\mathbf{r}, \mathbf{R}) = \sum_{\mathbf{k}\in 1^{\text{st}}\,\text{BZ}} \frac{e^{-i\mathbf{k}\cdot\mathbf{R}}}{\sqrt{N}} \varphi_{n,\mathbf{k}}(\mathbf{r}). \tag{3.112}$$

To each lattice site **R** there corresponds a Wannier function, which is localized in real space and provides a complete orthonormal basis for the electronic states of the crystal. This stands in contrast to Bloch functions, which are delocalized and extend over the entire crystal. The unitary transformation allows one to go from the extended basis of Bloch functions to the localized basis of Wannier functions. A very useful property is the following

$$\omega_n\left(\mathbf{r}+\mathbf{R}', \mathbf{R}+\mathbf{R}'\right) = \omega_n(\mathbf{r}, \mathbf{R}), \tag{3.113}$$

where $\mathbf{R}'$ is any vector of the direct lattice. For $\mathbf{R}' = -\mathbf{R}$,

$$\begin{aligned} \omega_n\left(\mathbf{r}+\mathbf{R}', \mathbf{R}+\mathbf{R}'\right) &= \omega_n(\mathbf{r}-\mathbf{R}, \mathbf{0}) \equiv \\ &\equiv \omega_n(\mathbf{r}-\mathbf{R}), \end{aligned} \tag{3.114}$$

that is, it does not depend on the two indices separately but only on their difference vector. Wannier functions depend on n and $\mathbf{r} - \mathbf{R}$, hence the relation between the two functions can be written as

$$\omega_n(\mathbf{r}-\mathbf{R})=\sum_{\mathbf{k}\in 1^{\mathrm{st}}\,\mathrm{BZ}}\frac{e^{-i\mathbf{k}\cdot\mathbf{R}}}{\sqrt{N}}\varphi_{n,\mathbf{k}}(\mathbf{r}), \tag{3.115}$$

$$\varphi_{n,\mathbf{k}}(\mathbf{r})=\sum_{\mathbf{R}}\frac{e^{i\mathbf{k}\cdot\mathbf{R}}}{\sqrt{N}}\omega_n(\mathbf{r}-\mathbf{R}). \tag{3.116}$$

Ultimately, we show the following fundamental property of Wannier functions.

Theorem 3.4 (Orthonormality of Wannier Functions) *Wannier functions (see Eq. 3.112) form a complete orthonormal set with respect to both the band index and the lattice site, i.e.,*

$$\left\langle\omega_n(\mathbf{r},\mathbf{R})|\omega_m(\mathbf{r},\mathbf{R}')\right\rangle=\delta_{nm}\,\delta_{\mathbf{R},\mathbf{R}'}. \tag{3.117}$$

Proof Starting from the definition of the Wannier functions as the discrete inverse Fourier transform of the Bloch eigenfunctions (see Eq. 3.112), the scalar product between two Wannier functions reads

$$\langle\omega_n(\mathbf{r},\mathbf{R})|\omega_m(\mathbf{r},\mathbf{R}')\rangle=\int_V d^3\mathbf{r}\,\omega_n^*(\mathbf{r},\mathbf{R})\,\omega_m(\mathbf{r},\mathbf{R}'). \tag{3.118}$$

It follows

$$\left\langle\omega_n(\mathbf{r},\mathbf{R})|\omega_m(\mathbf{r},\mathbf{R}')\right\rangle=\sum_{\mathbf{k},\mathbf{k}'}\frac{e^{i(\mathbf{k}\cdot\mathbf{R}-\mathbf{k}'\cdot\mathbf{R}')}}{N}\int d^3\mathbf{r}\,\varphi_{n,\mathbf{k}}^*(\mathbf{r})\,\varphi_{m,\mathbf{k}'}(\mathbf{r}). \tag{3.119}$$

The Bloch functions form an orthonormal set (see Eq. 3.66), then

$$\left\langle\omega_n(\mathbf{r},\mathbf{R})|\omega_m(\mathbf{r},\mathbf{R}')\right\rangle=\delta_{n,m}\sum_{\mathbf{k}}\frac{e^{i\mathbf{k}\cdot(\mathbf{R}-\mathbf{R}')}}{N}. \tag{3.120}$$

Using the discrete orthogonality relation of the lattice Fourier transform,

$$\frac{1}{N}\sum_{\mathbf{k}}e^{i\mathbf{k}\cdot(\mathbf{R}-\mathbf{R}')}=\delta_{\mathbf{R},\mathbf{R}'}, \tag{3.121}$$

the thesis follows. □

3.5 Tight-Binding Model

In general, Wannier functions provide a complete representation of the electronic states in a crystal. However, their explicit computation is often challenging and computationally demanding. Therefore, in practical analyses, it is common to resort to approximations that allow expressing Wannier functions in terms of localized atomic orbitals. One such approximation is the tight-binding model, which assumes that electrons are primarily localized around atomic nuclei but can *hop* between neighboring sites via a hopping term determined by the overlap of adjacent orbitals. This simplification enables an analytical description of the electronic energy levels and the formation of energy bands as a function of the crystal momentum $\mathbf{k}$.

For an isolated atom, a valence electron—i.e., an electron residing in one of the outermost atomic orbitals—moves in the effective atomic potential $V^{\mathrm{at}}(\mathbf{r})$. Its stationary states are described by the orbital eigenfunctions $\varphi_n^{\mathrm{at}}(\mathbf{r})$, which satisfy the one-particle eigenvalue equation

$$\left[\frac{\hat{\mathbf{p}}^2}{2m_e} + V^{\mathrm{at}}(\mathbf{r})\right]\varphi_n^{\mathrm{at}}(\mathbf{r}) = \mathcal{E}_n^{\mathrm{at}}\varphi_n^{\mathrm{at}}(\mathbf{r}), \tag{3.122}$$

where $\mathcal{E}_n^{\mathrm{at}}$ denotes the corresponding orbital eigenvalue and the index n labels the atomic orbital under consideration.

In this context, we adopt an approximate but physically transparent scheme—known as the atomic-orbital approximation—in which the Wannier functions are replaced by localized atomic orbitals. Next, we express the Wannier functions in terms of these atomic—orbital eigenfunctions. Since the Bloch functions $\varphi_{n,\mathbf{k}}(\mathbf{r})$ (see Eq. 3.35), can be written in terms of Wannier functions (see Eq. 3.116), we may rewrite them by substituting the atomic-orbital expansion of the Wannier states, then

$$\varphi_{n,\mathbf{k}}(\mathbf{r}) = \sum_{\mathbf{R}} \frac{e^{i\mathbf{k}\cdot\mathbf{R}}}{\sqrt{N}}\varphi_n^{\mathrm{at}}(\mathbf{r}-\mathbf{R}), \tag{3.123}$$

the eigenvalue equation for $\mathcal{H}_{\mathrm{HF}}^{(e)}$ (see Eq. II.21) can be written as

$$\left[\frac{\hat{\mathbf{p}}^2}{2m_e} + V(\mathbf{r})\right]\left[\sum_{\mathbf{R}} \frac{e^{i\mathbf{k}\cdot\mathbf{R}}}{\sqrt{N}}\varphi_n^{\mathrm{at}}(\mathbf{r}-\mathbf{R})\right] = \mathcal{E}\sum_{\mathbf{R}} \frac{e^{i\mathbf{k}\cdot\mathbf{R}}}{\sqrt{N}}\varphi_n^{\mathrm{at}}(\mathbf{r}-\mathbf{R}), \tag{3.124}$$

$$\sum_{\mathbf{R}} e^{i\mathbf{k}\cdot\mathbf{R}}\left[\frac{\hat{\mathbf{p}}^2}{2m_e} + V(\mathbf{r})\right]\varphi_n^{\mathrm{at}}(\mathbf{r}-\mathbf{R}) = \mathcal{E}\sum_{\mathbf{R}} e^{i\mathbf{k}\cdot\mathbf{R}}\varphi_n^{\mathrm{at}}(\mathbf{r}-\mathbf{R}). \tag{3.125}$$

We multiply both members to the left by $\left(\varphi_n^{\mathrm{at}}(\mathbf{r})\right)^*$, integrate both members with respect to the measure $d^3\mathbf{r}$, i.e.,

$$\sum_{\mathbf{R}} e^{i\mathbf{k}\cdot\mathbf{R}} \int d^3\mathbf{r}\left(\varphi_n^{\mathrm{at}}(\mathbf{r})\right)^*\left[\frac{\hat{\mathbf{p}}^2}{2m_e} + V(\mathbf{r})\right]\varphi_n^{\mathrm{at}}(\mathbf{r}-\mathbf{R})$$

$$= \mathcal{E}\sum_{\mathbf{R}} e^{i\mathbf{k}\cdot\mathbf{R}} \int d^3\mathbf{r}\ \left(\varphi_n^{\mathrm{at}}(\mathbf{r})\right)^*\varphi_n^{\mathrm{at}}(\mathbf{r}-\mathbf{R}), \tag{3.126}$$

$$\sum_{\mathbf{R}} e^{i\mathbf{k}\cdot\mathbf{R}} \int d^3\mathbf{r} \left(\varphi_n^{\text{at}}(\mathbf{r})\right)^* \left[\frac{\hat{\mathbf{p}}^2}{2m_e} - \mathcal{E}\right] \varphi_n^{\text{at}}(\mathbf{r}-\mathbf{R})$$
$$= \sum_{\mathbf{R}} e^{i\mathbf{k}\cdot\mathbf{R}} \int d^3\mathbf{r} \left(\varphi_n^{\text{at}}(\mathbf{r})\right)^* [-V(\mathbf{r})]\, \varphi_n^{\text{at}}(\mathbf{r}-\mathbf{R}) . \tag{3.127}$$

We add $V^{\text{at}}(\mathbf{r}-\mathbf{R})$ to both members in the integrand function and recall the eigenvalue equation for the atomic potential, that is

$$\left[\frac{\hat{\mathbf{p}}^2}{2m_e} + V^{\text{at}}(\mathbf{r}-\mathbf{R})\right] \varphi_n^{\text{at}}(\mathbf{r}-\mathbf{R}) = \mathcal{E}_n^{\text{at}} \varphi_n^{\text{at}}(\mathbf{r}-\mathbf{R}) , \tag{3.128}$$

then

$$\sum_{\mathbf{R}} e^{i\mathbf{k}\cdot\mathbf{R}} \int d^3\mathbf{r} \left(\varphi_n^{\text{at}}(\mathbf{r})\right)^* \left[\mathcal{E}_n^{\text{at}} - \mathcal{E}\right] \varphi_n^{\text{at}}(\mathbf{r}-\mathbf{R})$$
$$= \sum_{\mathbf{R}} e^{i\mathbf{k}\cdot\mathbf{R}} \int d^3\mathbf{r} \left(\varphi_n^{\text{at}}(\mathbf{r})\right)^* \left[V^{\text{at}}(\mathbf{r}-\mathbf{R}) - V(\mathbf{r})\right] \varphi_n^{\text{at}}(\mathbf{r}-\mathbf{R}) . \tag{3.129}$$

On the left-hand side, the orbital functions are assumed to be orthonormal, an assumption justified by the fact that the Wannier functions from which they are derived already form an orthonormal set (see Theorem 3.4). The scalar products between the atomic orbital centered at $\mathbf{0}$ and those centered at $\mathbf{r}-\mathbf{R}$ vanish for all $\mathbf{R} \neq \mathbf{0}$, while for $\mathbf{R} = \mathbf{0}$ the scalar product equals unity, yielding the term $\mathcal{E}_n^{\text{at}} - \mathcal{E}$. On the right-hand side, when $\mathbf{R} = \mathbf{0}$, the integral reduces to a constant independent of $\mathbf{k}$, but dependent on the orbital n; let us denote this constant by $-\tilde{c}_n$. Conversely, for $\mathbf{R} \neq \mathbf{0}$, the quantity $\left[V^{\text{at}}(\mathbf{r}-\mathbf{R}) - V(\mathbf{r})\right]$ enables the interaction and overlap between $\left(\varphi_n^{\text{at}}(\mathbf{r})\right)^*$ and $\varphi_n^{\text{at}}(\mathbf{r}-\mathbf{R})$, thereby permitting electron transfer between an atom and its nearest neighbors. We then write

$$\mathcal{E}_n^{\text{at}} - \mathcal{E} = -\tilde{c}_n + \sum_{\boldsymbol{\delta}} e^{i\mathbf{k}\cdot\boldsymbol{\delta}}\, t_n , \tag{3.130}$$

$$\mathcal{E} = \mathcal{E}_n^{\text{at}} + \tilde{c}_n - \sum_{\boldsymbol{\delta}} e^{i\mathbf{k}\cdot\boldsymbol{\delta}}\, t_n , \tag{3.131}$$

where

$$t_n = \int d^3\mathbf{r} \left(\varphi_n^{\text{at}}(\mathbf{r})\right)^* \left[V^{\text{at}}(\mathbf{r}-\boldsymbol{\delta}) - V(\mathbf{r})\right] \varphi_n^{\text{at}}(\mathbf{r}-\boldsymbol{\delta}) \tag{3.132}$$

is the integral hopping and $\boldsymbol{\delta}$ is the vector connecting the atom in $\mathbf{r}$ with its nearest neighbors. From Eq. 3.132, if we add and subtract the kinetic term as follows

$$t_n = \int d^3\mathbf{r}\, \left(\varphi_n^{\text{at}}(\mathbf{r})\right)^* \left[V^{\text{at}}\left(\mathbf{r}-\boldsymbol{\delta}\right) + \frac{\hat{\mathbf{p}}^2}{2m_e} - \frac{\hat{\mathbf{p}}^2}{2m_e} - V(\mathbf{r}) \right] \varphi_n^{\text{at}}\left(\mathbf{r}-\boldsymbol{\delta}\right), \tag{3.133}$$

and we make use of the eigenvalue equation for the atomic orbital, then

$$\begin{aligned} t_n &= \int d^3\mathbf{r}\, \left(\varphi_n^{\text{at}}(\mathbf{r})\right)^* \mathcal{E}_n^{\text{at}} \varphi_n^{\text{at}}\left(\mathbf{r}-\boldsymbol{\delta}\right) \\ &\quad - \int d^3\mathbf{r}\, \left(\varphi_n^{\text{at}}(\mathbf{r})\right)^* \left[\frac{\hat{\mathbf{p}}^2}{2m_e} + V(\mathbf{r}) \right] \varphi_n^{\text{at}}\left(\mathbf{r}-\boldsymbol{\delta}\right) = \\ &= \mathcal{E}_n^{\text{at}} \int d^3\mathbf{r}\, \left(\varphi_n^{\text{at}}(\mathbf{r})\right)^* \varphi_n^{\text{at}}\left(\mathbf{r}-\boldsymbol{\delta}\right) \\ &\quad - \int d^3\mathbf{r}\, \left(\varphi_n^{\text{at}}(\mathbf{r})\right)^* \left[\frac{\hat{\mathbf{p}}^2}{2m_e} + V(\mathbf{r}) \right] \varphi_n^{\text{at}}\left(\mathbf{r}-\boldsymbol{\delta}\right). \end{aligned} \tag{3.134}$$

The first integral vanishes as a consequence of the completeness of the eigenfunctions $\varphi_n^{\text{at}}(\mathbf{r})$, whereas the second involves the electronic Hartree-Fock Hamiltonian $\mathcal{H}_{\text{HF}}^{(e)}$, then

$$-t_n = \int d^3\mathbf{r}\, \left(\varphi_n^{\text{at}}(\mathbf{r})\right)^* \mathcal{H}_{\text{HF}}^{(e)} \varphi_n^{\text{at}}\left(\mathbf{r}-\boldsymbol{\delta}\right). \tag{3.135}$$

The hopping term is responsible for the motion of an electron from one site to its nearest neighbors and depends on the n orbital. Of course, to describe the actual interactions, the contributions of the interactions of one atom with all atoms must be added. However, if the electrons are localized, such terms are negligible. Here, the energy depends on the quantum number n and the crystalline momentum $\mathbf{k}$. We write

$$\mathcal{E}_{n,\mathbf{k}} = \mathcal{E}_n^{\text{at}} + \tilde{c}_n - \sum_{\boldsymbol{\delta}} e^{i\mathbf{k}\cdot\boldsymbol{\delta}}\, t_n. \tag{3.136}$$

Now, if the sum $\sum_{\boldsymbol{\delta}} e^{i\mathbf{k}\cdot\boldsymbol{\delta}}\, t_n$ vanishes, the energy reduces to the atomic energy plus a constant $\tilde{c}_n$, which arises from the presence of all atoms in the lattice and represents the crystal field effect, the potential exerted on the electron by the surrounding atoms. Conversely, if the hopping integral is nonzero, the energy becomes $\mathbf{k}$-dependent, resulting in the formation of an energy band.

3.5.1 Application on a Simple Cubic Lattice

In the simple cubic lattice model (see Fig. 3.1a and Sect. 3.1.3), atoms are positioned at the vertices of a cube with side length a, where each site has six nearest neighbors, and the displacement vector $\boldsymbol{\delta}$ can take one of the following values: $(\pm a, 0, 0)$, $(0, \pm a, 0)$, or $(0, 0, \pm a)$. Our goal is to compute the energy band $\mathcal{E}_{n,\mathbf{k}}$ and to analyze its behavior in the limiting cases. In particular, we are interested in its expansion near the center of the Brillouin zone ($\mathbf{k} \to \mathbf{0}$). We have

$$\begin{aligned}
\mathcal{E}_{n,\mathbf{k}} &= \mathcal{E}_n^{\text{at}} + \tilde{c}_n - \sum_{\boldsymbol{\delta}} e^{i\mathbf{k}\cdot\boldsymbol{\delta}} \, (+t_n) = \\
&= \mathcal{E}_n^{\text{at}} + \tilde{c}_n - t_n \left[e^{ik_x a} + e^{-ik_x a} + e^{ik_y a} + e^{-ik_y a} + e^{ik_z a} + e^{-ik_z a} \right] = \\
&= \mathcal{E}_n^{\text{at}} + \tilde{c}_n - 2t_n \left[\cos(k_x a) + \cos(k_y a) + \cos(k_z a) \right], \\
& k_x, k_y, k_z \in \left[-\frac{\pi}{a}, \frac{\pi}{a} \right].
\end{aligned} \tag{3.137}$$

In the one-dimensional case, that is, for a linear lattice chain, the vector $\mathbf{k}$ is replaced by the scalar k, and the energy band becomes

$$\mathcal{E}_{n,k} = \mathcal{E}_n^{\text{at}} + \tilde{c}_n - 2t_n \cos(ka). \tag{3.138}$$

Since $k = \frac{2\pi}{\lambda}$, in the limit $k \to 0$, the wavelength λ of the Bloch function is strictly greater than the lattice step, that is

$$\cos(ka) \sim 1 - \frac{(ka)^2}{2}, \tag{3.139}$$

$$\mathcal{E}_n(k) = t_n (ka)^2 + \text{const.}, \tag{3.140}$$

in other words for small wave vectors, the energy band is approximated by a parabola. The previous equation can be rewritten as

$$\mathcal{E}_{n,k} \sim \frac{\hbar^2 k^2}{2m^*}, \tag{3.141}$$

up to a constant, where

$$m^*(k = 0) = \frac{\hbar^2}{2t_n a^2} \tag{3.142}$$

is the effective mass for $k = 0$. If $k \to 0$, then

$$\mathcal{E}_{n,k} \underset{k\to 0}{\sim} \frac{\hbar^2 k^2}{2m^*}, \tag{3.143}$$

that is, the energy spectrum approaches that of a free particle, with the electron mass effectively replaced by a renormalized effective mass. When the hopping amplitude vanishes, the energy band becomes completely flat, the effective mass diverges, and the electron is strictly localized at an atomic site. In contrast, increasing the hopping amplitude reduces the effective mass, allowing the electron to delocalize and propagate more easily through the lattice. The parameter t_n thus governs both the width of the energy band and the strength of the inter-site electron hopping. In three dimensions, the group velocity is given by (see Eq. 3.76)

$$\mathbf{v}_{n,\mathbf{k}} = \frac{1}{\hbar} 2t_n a \left(\sin(k_x a), \sin(k_y a), \sin(k_z a)\right), \tag{3.144}$$

which satisfies

$$\mathbf{v}_{n,\mathbf{k}} \underset{\mathbf{k}\to \mathbf{0}}{\sim} \frac{\hbar \mathbf{k}}{m^*}, \tag{3.145}$$

i.e., the system behaves as a free particle system, where the electron mass is replaced by an effective mass that accounts for the effects of the periodic potential.

3.6 Bloch Hamiltonian in Second Quantization

The Hamiltonian $\mathcal{H}_{\text{HF}}^{(e),N}$ given in Eq. II.19 is a one-body operator; according to Theorem 2.9, its representation in second quantization, is given by

$$\hat{\mathcal{H}}_{\text{HF}}^{(e)} = \int dx \hat{\psi}^\dagger(x) \mathcal{H}_{\text{HF}}^{(e)}(x) \hat{\psi}(x). \tag{3.146}$$

Note that, according to Theorem 2.9, the N-body Hamiltonian $\mathcal{H}_{\text{HF}}^{(e),N}$ introduced in Eq. II.19 is represented in second quantization by replacing the one-body operator $\mathcal{H}_{\text{HF}}^{(e)}$ of Eq. II.20 within the field-operator expression. Now, we analyze the equivalent representations in both the Bloch and Wannier bases.

3.6.1 Bloch's Description

We use as a basis the Bloch eigenfunctions (see Eq. 3.35) tensored with the standard eigenfunctions of the spin operator $\hat{S}_z$ (see Sect. 1.1). Accordingly, the annihilation and creation field operators are written as

$$\hat{\psi}(x) = \sum_{n,\mathbf{k},\sigma} \varphi_{n,\mathbf{k}}(\mathbf{r})\chi_\sigma(s)C_{n,\mathbf{k},\sigma}, \tag{3.147}$$

$$\hat{\psi}^\dagger(x) = \sum_{n,\mathbf{k},\sigma} \varphi^*_{n,\mathbf{k}}(\mathbf{r})\chi^*_\sigma(s)C^\dagger_{n,\mathbf{k},\sigma}, \tag{3.148}$$

then Bloch's description in second quantization is given by

$$\begin{aligned}
\hat{\mathcal{H}}^{(e)}_{\mathrm{HF}} &= \int dx \left(\sum_{n,\mathbf{k},\sigma} \varphi^*_{n,\mathbf{k}}(\mathbf{r})\, \chi^*_\sigma(s)\, C^\dagger_{n,\mathbf{k},\sigma} \right) \\
&\quad \mathcal{H}^{(e)}_{\mathrm{HF}} \left(\sum_{n',\mathbf{k}',\sigma'} \varphi_{n',\mathbf{k}'}(\mathbf{r})\, \chi_{\sigma'}(s)\, C_{n',\mathbf{k}',\sigma'} \right) = \\
&= \int dx \left(\sum_{n,\mathbf{k},\sigma} \varphi^*_{n,\mathbf{k}}(\mathbf{r})\, \chi^*_\sigma(s)\, C^\dagger_{n,\mathbf{k},\sigma} \right) \\
&\quad \mathcal{E}_{n',\mathbf{k}'} \left(\sum_{n',\mathbf{k}',\sigma'} \varphi_{n',\mathbf{k}'}(\mathbf{r})\, \chi_{\sigma'}(s)\, C_{n',\mathbf{k}',\sigma'} \right) = \\
&= \sum_{n,\mathbf{k},\sigma} \mathcal{E}_{n,\mathbf{k}}\, C^\dagger_{n,\mathbf{k},\sigma} C_{n,\mathbf{k},\sigma},
\end{aligned} \tag{3.149}$$

which represents the first term of the Hamiltonian II.22. Each eigenstate is a Slater determinant of Bloch wave functions, with eigenvalues given by the occupation numbers 0 or 1.

3.6.2 *Wannier's Description*

Let us expand the field operators in the basis of Wannier functions. The indices $\{n, \mathbf{R}, \sigma\}$ refer respectively to: n, the band or orbital index; $\mathbf{R}$, the position of the lattice site (i.e., the center of the corresponding unit cell); and σ, the spin projection. In the Wannier representation, the field operators are expressed as

$$\hat{\psi}(x) = \sum_{n,i,\sigma} \omega_n(\mathbf{r} - \mathbf{R}_i)\chi_\sigma(s)C_{n,i,\sigma}, \tag{3.150}$$

$$\hat{\psi}^\dagger(x) = \sum_{n,i,\sigma} \omega^*_n(\mathbf{r} - \mathbf{R}_i)\chi^*_\sigma(s)C^\dagger_{n,i,\sigma}, \tag{3.151}$$

and we have

$$\hat{\mathcal{H}}_{\rm HF}^{(e)} = \int dx\, \hat{\psi}^{\dagger}(x)\mathcal{H}_{\rm HF}^{(e)}\hat{\psi}(x) =$$
$$= \sum_{s} \sum_{\substack{n,i,\sigma \\ n',j,\sigma'}} \int d^3\mathbf{r}\, \omega_n^*(\mathbf{r}-\mathbf{R}_i)\chi_\sigma^*(s)\mathcal{H}_{\rm HF}^{(e)}\omega_{n'}(\mathbf{r}-\mathbf{R}_j)\chi_{\sigma'}(s)C_{n,i,\sigma}^{\dagger}C_{n',j,\sigma'}. \tag{3.152}$$

Equation 1.23 is applied with respect to s, yielding $\delta_{\sigma,\sigma'}$. Subsequently, we sum over σ', that is,

$$\hat{\mathcal{H}}_{\rm HF}^{(e)} = \sum_{\substack{n,i,\sigma \\ n',j}} \int d^3\mathbf{r}\omega_n^*(\mathbf{r}-\mathbf{R}_i)\mathcal{H}_{\rm HF}^{(e)}\omega_{n'}(\mathbf{r}-\mathbf{R}_j)C_{n,i,\sigma}^{\dagger}C_{n',j,\sigma}. \tag{3.153}$$

Since the one-body operator is diagonal in the basis of Bloch eigenfunctions rather than in the Wannier basis, we express the Wannier functions in terms of the Bloch functions and we replace the operator with its eigenvalue, as obtained from the solution of the corresponding eigenvalue problem as follows

$$\hat{\mathcal{H}}_{\rm HF}^{(e)} = \sum_{\substack{n,i,\sigma \\ n',j}} \sum_{\mathbf{k},\mathbf{k}'} \int d^3\mathbf{r}\varphi_{n,\mathbf{k}}^*(\mathbf{r})\mathcal{H}_{\rm HF}^{(e)}\varphi_{n',\mathbf{k}'}(\mathbf{r})\frac{e^{i\mathbf{k}\cdot\mathbf{R}_i}}{\sqrt{N}}\frac{e^{-i\mathbf{k}'\cdot\mathbf{R}_j}}{\sqrt{N}}C_{n,i,\sigma}^{\dagger}C_{n',j,\sigma} =$$
$$= \sum_{\substack{n,i,\sigma \\ n',j}} \sum_{\mathbf{k},\mathbf{k}'} \int d^3\mathbf{r}\, \mathcal{E}_{n',\mathbf{k}'}\varphi_{n,\mathbf{k}}^*(\mathbf{r})\varphi_{n',\mathbf{k}'}(\mathbf{r})\frac{e^{i\mathbf{k}\cdot(\mathbf{R}_i-\mathbf{R}_j)}}{N}C_{n,i,\sigma}^{\dagger}C_{n',j,\sigma}. \tag{3.154}$$

The Bloch eigenfunctions constitute a basis, then, upon integration over the spatial variable $\mathbf{r}$, all cross terms with $\mathbf{k} \neq \mathbf{k}'$ or $n \neq n'$ vanish, and only the diagonal contributions remain, that is

$$\hat{\mathcal{H}}_{\rm HF}^{(e)} = \sum_{n,\mathbf{k},i,j,\sigma} \mathcal{E}_{n,\mathbf{k}}\frac{e^{i\mathbf{k}\cdot(\mathbf{R}_i-\mathbf{R}_j)}}{N}C_{n,i,\sigma}^{\dagger}C_{n,j,\sigma} \equiv$$
$$\equiv - \sum_{n,i,j,\sigma} t_{ij}^n C_{n,i,\sigma}^{\dagger}C_{n,j,\sigma}, \tag{3.155}$$

with

$$t_{ij}^n = -\int d^3\mathbf{r}\, \omega_n^*(\mathbf{r}-\mathbf{R}_i)\mathcal{H}_{\rm HF}^{(e)}\omega_n(\mathbf{r}-\mathbf{R}_j) =$$
$$= -\sum_{\mathbf{k}} \mathcal{E}_{n,\mathbf{k}}\frac{e^{i\mathbf{k}\cdot(\mathbf{R}_i-\mathbf{R}_j)}}{N}. \tag{3.156}$$

A fully rigorous treatment based on Wannier functions is, however, rather cumbersome. To quantize the Hamiltonian $\mathcal{H}_{\text{HF}}^{(e),N}$ given in Eq. II.19, we once again employ the atomic-orbital approximation (see Eq. 3.123), replacing the Wannier functions with localized atomic orbitals φ_n^{at}. In this representation, the field operators take the form

$$\hat{\psi}(x) = \sum_{n,\mathbf{R},\sigma} \varphi_n^{\text{at}}(\mathbf{r}-\mathbf{R})\,\chi_\sigma(s) C_{n,\mathbf{R},\sigma}, \tag{3.157}$$

$$\hat{\psi}^\dagger(x) = \sum_{n,\mathbf{R},\sigma} \left(\varphi_n^{\text{at}}(\mathbf{r}-\mathbf{R})\right)^*\chi_\sigma^*(s) C_{n,\mathbf{R},\sigma}^\dagger, \tag{3.158}$$

where $C_{n,\mathbf{R},\sigma}$ ($C_{n,\mathbf{R},\sigma}^\dagger$) destroys (creates) a fermion on the atomic orbital n centered on the $\mathbf{R}$ site with spin component σ. Wannier operators do not act on delocalized waves, like Bloch's, but on a specific lattice atom. We consider only contributions from the same site as $\mathbf{R}$ or its nearest neighbors, so, substituting the Hamiltonian operator, we have

$$\hat{\mathcal{H}}_{\text{HF}}^{(e)} = \sum_{n,\mathbf{R},\sigma} \left(\mathcal{E}_n^{\text{at}} + \tilde{c}_n\right) C_{n,\mathbf{R},\sigma}^\dagger C_{n,\mathbf{R},\sigma} - \sum_{\substack{n,\mathbf{R},\\ \boldsymbol{\delta},\sigma}} t_n C_{n,\mathbf{R}-\boldsymbol{\delta},\sigma}^\dagger C_{n,\mathbf{R},\sigma}. \tag{3.159}$$

The first term is diagonal when $\mathbf{R} = \mathbf{R}'$, whereas the second term is off-diagonal, corresponding to $\mathbf{R}' = \mathbf{R} - \boldsymbol{\delta}$. This indicates that the Hamiltonian operator is not diagonal in the Wannier basis, which is consistent with the fact that it is the Bloch functions, rather than the Wannier functions, that constitute the eigenfunctions of the Bloch Hamiltonian. From Eqs. 2.114 and 2.115, with the following identifications

$$\begin{cases} b_{n,\mathbf{k},\sigma}, \quad b_{n,\mathbf{k},\sigma}^\dagger & \text{Bloch operators} \\ c_{n,\mathbf{R},\sigma}, \quad c_{n,\mathbf{R},\sigma}^\dagger & \text{Wannier operators} \\ m_{\mathbf{k}} = \dfrac{e^{-i\mathbf{k}\cdot\mathbf{R}}}{\sqrt{N}} & \text{coefficients of the unitary transformation} \end{cases}, \tag{3.160}$$

the annihilation operator in Wannier's basis is

$$c_{n,\mathbf{R},\sigma} = \sum_{\mathbf{k}\in 1^{\text{st}}\,\text{BZ}} \frac{e^{i\mathbf{k}\cdot\mathbf{R}}}{\sqrt{N}} b_{n,\mathbf{k},\sigma}, \tag{3.161}$$

and the creation operator in Wannier's basis is

$$c_{n,\mathbf{R},\sigma}^\dagger = \sum_{\mathbf{k}\in 1^{\text{st}}\,\text{BZ}} \frac{e^{-i\mathbf{k}\cdot\mathbf{R}}}{\sqrt{N}} b_{n,\mathbf{k},\sigma}^\dagger. \tag{3.162}$$

The first and second terms involve the following sums

$$\frac{1}{N}\sum_{\mathbf{R}} e^{i\mathbf{k}\cdot\mathbf{R}} e^{-i\mathbf{k}'\cdot\mathbf{R}} = \delta_{\mathbf{k},\mathbf{k}'}, \tag{3.163}$$

$$\frac{1}{N}\sum_{\mathbf{R}} e^{i\mathbf{k}\cdot(\mathbf{R}-\boldsymbol{\delta})} e^{-i\mathbf{k}'\cdot\mathbf{R}} = \delta_{\mathbf{k},\mathbf{k}'} e^{-i\mathbf{k}\cdot\boldsymbol{\delta}}, \tag{3.164}$$

respectively. In particular, the exponential $e^{-i\mathbf{k}\cdot\boldsymbol{\delta}}$ cancels out the corresponding positive exponential term present in the Hamiltonian operator, effectively removing the summation over $\boldsymbol{\delta}$. As a result, from

$$\hat{\mathcal{H}}_{\text{HF}}^{(e)} = \sum_{\substack{n,\mathbf{k},\\ \mathbf{R},\sigma}} \left(\mathcal{E}_n^{\text{at}} + \tilde{c}_n\right) b^{\dagger}_{n,\mathbf{k},\sigma} b_{n,\mathbf{k},\sigma} - \sum_{\substack{n,\mathbf{k},\\ \boldsymbol{\delta},\sigma}} t_n e^{i\mathbf{k}\cdot\boldsymbol{\delta}} b^{\dagger}_{n,\mathbf{k},\sigma} b_{n,\mathbf{k},\sigma}, \tag{3.165}$$

the Hamiltonian $\hat{\mathcal{H}}_{\text{HF}}^{(e),N}$ given in Eq. 3.146 becomes in second quantization

$$\hat{\mathcal{H}}_{\text{HF}}^{(e)} = \sum_{n,\mathbf{k},\sigma} \mathcal{E}_{n,\mathbf{k}} b^{\dagger}_{n,\mathbf{k},\sigma} b_{n,\mathbf{k},\sigma}, \tag{3.166}$$

that is, the Wannier and Bloch descriptions are equivalent within the framework of second quantization.

3.7 One-Dimensional Electronic Crystal Structure

In a one-dimensional diatomic chain, the interaction between two different atomic species leads to the formation of energy bands. Here, we construct a simple tight-binding model to illustrate the origin of band splitting and the resulting covalent bonding. Given a one-dimensional chain of lattice step a, at the center of each pair of sites is placed an atom of a species different from the first: if we add a repeated basis of atoms to each lattice point, we obtain what is called crystal structure. Assume that each atom has only one valence electron, we refer to the outermost orbital for each atom, say $\mathcal{E}_1$ for the first species and $\mathcal{E}_2$ for the second, and construct a model for the Hamiltonian. We assume that $\mathcal{E}_2 > \mathcal{E}_1$. From the two atomic levels of the valence electrons of the two types of isolated atoms, the two atoms will be coupled by a jump term. In second quantization,

$$\begin{aligned}\hat{\mathcal{H}} = &\sum_{\tau,\sigma} \mathcal{E}_1 c^{\dagger}_{\tau,\sigma} c_{\tau,\sigma} + \sum_{\tau,\sigma} \mathcal{E}_2 d^{\dagger}_{\tau+\frac{1}{2},\sigma} d_{\tau+\frac{1}{2},\sigma} + \\ &- t\sum_{\tau,\sigma}\left[c^{\dagger}_{\tau,\sigma} d_{\tau+\frac{1}{2},\sigma} + c^{\dagger}_{\tau,\sigma} d_{\tau-\frac{1}{2},\sigma} + d^{\dagger}_{\tau+\frac{1}{2},\sigma} c_{\tau,\sigma} + d^{\dagger}_{\tau-\frac{1}{2},\sigma} c_{\tau,\sigma}\right],\end{aligned} \tag{3.167}$$

which we write briefly as

$$\hat{\mathcal{H}} = \sum_{\tau,\sigma} \mathcal{E}_1 c^{\dagger}_{\tau,\sigma} c_{\tau,\sigma} + \sum_{\tau,\sigma} \mathcal{E}_2 d^{\dagger}_{\tau+\frac{1}{2},\sigma} d_{\tau+\frac{1}{2},\sigma} + \\ - t \sum_{\tau,\sigma} \left[c^{\dagger}_{\tau,\sigma} d_{\tau+\frac{1}{2},\sigma} + c^{\dagger}_{\tau,\sigma} d_{\tau-\frac{1}{2},\sigma} + \text{h.c.} \right], \tag{3.168}$$

where the abbreviation h.c. stands for the Hermitian conjugate. The operators $c^{\dagger}_{\tau,\sigma}$ and $c_{\tau,\sigma}$ correspond to the species $\mathcal{E}_1$, whereas $d^{\dagger}_{\tau+\frac{1}{2},\sigma}$ and $d_{\tau+\frac{1}{2},\sigma}$ correspond to the species $\mathcal{E}_2$. Here, τ denotes the site index, and σ labels the spin. The Hamiltonian $\hat{\mathcal{H}}$ is composed of three terms: the first represents the contribution of the first species, the second corresponds to the second species, and the third describes the hopping between the two species. Our objective is to diagonalize $\hat{\mathcal{H}}$ in order to find its eigenvalues and eigenvectors. We now turn to Bloch's formalism. Recall that the Bloch and Wannier annihilation operators are related by

$$c_{\tau,\sigma} = \sum_{k \in 1^{\text{st}}\,\text{BZ}} \frac{e^{ik\tau a}}{\sqrt{N}} c_{k,\sigma}, \tag{3.169}$$

$$d_{\tau+\frac{1}{2},\sigma} = \sum_{k \in 1^{\text{st}}\,\text{BZ}} \frac{e^{ik\left(\tau+\frac{1}{2}\right)a}}{\sqrt{N}} d_{k,\sigma}, \tag{3.170}$$

where, in the one-dimensional case, the vector $\mathbf{k}$ has been replaced by the scalar k. We know that the operators $c^{\dagger}_{\tau,\sigma} c_{\tau,\sigma}$ and $d^{\dagger}_{\tau+\frac{1}{2},\sigma} d_{\tau+\frac{1}{2},\sigma}$ are diagonal in both bases. Then,

$$\sum_{\tau,\sigma} c^{\dagger}_{\tau,\sigma} d_{\tau+\frac{1}{2},\sigma} = \frac{1}{N} \sum_{\sigma} \sum_{k} \sum_{k'} \sum_{\tau} e^{-i(k-k')\tau a} e^{i\frac{k'a}{2}} c^{\dagger}_{k,\sigma} d_{k',\sigma} = \\ = \frac{1}{N} \sum_{\sigma} \sum_{k} \sum_{k'} \delta_{k,k'} e^{i\frac{k'a}{2}} c^{\dagger}_{k,\sigma} d_{k',\sigma} = \\ = \sum_{k,\sigma} e^{ik\frac{a}{2}} c^{\dagger}_{k,\sigma} d_{k,\sigma}. \tag{3.171}$$

where we used the completeness property. Similarly,

$$\sum_{\tau,\sigma} c^{\dagger}_{\tau,\sigma} d_{\tau-\frac{1}{2},\sigma} = \sum_{\sigma} \sum_{k} e^{-ik\frac{a}{2}} c^{\dagger}_{k,\sigma} d_{k,\sigma}. \tag{3.172}$$

The Hamiltonian can be written as

$$\hat{\mathcal{H}} = \sum_{k,\sigma} \mathcal{E}_1 c^\dagger_{k,\sigma} c_{k,\sigma} + \sum_{k,\sigma} \mathcal{E}_2 d^\dagger_{k,\sigma} d_{k,\sigma} + \sum_{k,\sigma} \alpha_k \left[c^\dagger_{k,\sigma} d_{k,\sigma} + \text{h.c.} \right], \tag{3.173}$$

with

$$\begin{aligned} \alpha_k &= -t \left(e^{ik\frac{a}{2}} + e^{-ik\frac{a}{2}} \right) = \\ &= -2t \cos\left(\frac{ka}{2} \right). \end{aligned} \tag{3.174}$$

The hopping amplitude 3.174 is maximal at the Brillouin zone center ($k = 0$) and vanishes at the zone boundary ($k = \pm\frac{\pi}{a}$). Moreover, the hopping term between the two atomic species makes the matrix non-diagonal. We write the Hamiltonian in matrix form

$$\hat{\mathcal{H}} = \sum_{k,\sigma} \begin{pmatrix} c^\dagger_{k,\sigma} & d^\dagger_{k,\sigma} \end{pmatrix} \begin{pmatrix} \mathcal{E}_1 & \alpha_k \\ \alpha_k & \mathcal{E}_2 \end{pmatrix} \begin{pmatrix} c_{k,\sigma} \\ d_{k,\sigma} \end{pmatrix}, \tag{3.175}$$

we use the identity $\mathbb{1} = PP^{-1}$ as follows

$$\hat{\mathcal{H}} = \sum_{k,\sigma} \begin{pmatrix} c^\dagger_{k,\sigma} & d^\dagger_{k,\sigma} \end{pmatrix} PP^{-1} \begin{pmatrix} \mathcal{E}_1 & \alpha_k \\ \alpha_k & \mathcal{E}_2 \end{pmatrix} PP^{-1} \begin{pmatrix} c_{k,\sigma} \\ d_{k,\sigma} \end{pmatrix}, \tag{3.176}$$

where P is a 2×2 matrix whose columns are the eigenvectors of the matrix

$$M(k) = \begin{pmatrix} \mathcal{E}_1 & \alpha_k \\ \alpha_k & \mathcal{E}_2 \end{pmatrix}, \tag{3.177}$$

which, being symmetrical with respect to the real numbers, i.e., it is a self-adjoint matrix, can be diagonalized. The columns of P are real and orthogonal to each other, i.e., the object $P^{-1}MP$ is diagonal and the diagonal elements are the eigenvalues, say $\mathcal{E}^{(+)}(k)$ and $\mathcal{E}^{(-)}(k)$, i.e.,

$$P^{-1}MP = \begin{pmatrix} \mathcal{E}^{(+)}(k) & 0 \\ 0 & \mathcal{E}^{(-)}(k) \end{pmatrix}, \tag{3.178}$$

$$\hat{\mathcal{H}} = \sum_{k,\sigma} \begin{pmatrix} c^\dagger_{k,\sigma} & d^\dagger_{k,\sigma} \end{pmatrix} P \begin{pmatrix} \mathcal{E}^{(+)}(k) & 0 \\ 0 & \mathcal{E}^{(-)}(k) \end{pmatrix} P^{-1} \begin{pmatrix} c_{k,\sigma} \\ d_{k,\sigma} \end{pmatrix}, \tag{3.179}$$

$$\mathcal{E}^{(\pm)}(k) = \frac{\mathcal{E}_1 + \mathcal{E}_2}{2} \pm \frac{1}{2}\sqrt{(\mathcal{E}_1 - \mathcal{E}_2)^2 + 4\alpha_k^2}. \tag{3.180}$$

From 3.180, we infer that a gap of energy $\mathcal{E}_2 - \mathcal{E}_1 > 0$ is formed. We define two operators as

$$\begin{pmatrix} \tilde{c}_{k,\sigma} \\ \tilde{d}_{k,\sigma} \end{pmatrix} = P^{-1} \begin{pmatrix} c_{k,\sigma} \\ d_{k,\sigma} \end{pmatrix}, \tag{3.181}$$

that is, $\tilde{c}_{k,\sigma}$ and $\tilde{d}_{k,\sigma}$ are linear combination of the annihilation operators of both atomic species, then each operator destroys a quasi-momentum eigenstate that is a linear combination of the original orbitals from both atomic species. We apply the adjoint operator and obtain

$$\left(\tilde{c}^{\dagger}_{k,\sigma}\ \tilde{d}^{\dagger}_{k,\sigma}\right) = \left(c^{\dagger}_{k,\sigma}\ d^{\dagger}_{k,\sigma}\right) \left(P^{-1}\right)^{*}_{t}, \tag{3.182}$$

given that P is an orthogonal matrix, the transpose of the inverse of P is equal to P and the complex conjugate operation leaves the matrix unchanged, i.e.,

$$\begin{aligned} \left(P^{-1}\right)^{*}_{t} &= P^{*} = \\ &= P, \end{aligned} \tag{3.183}$$

then

$$\left(\tilde{c}^{\dagger}_{k,\sigma}\ \tilde{d}^{\dagger}_{k,\sigma}\right) = \left(c^{\dagger}_{k,\sigma}\ d^{\dagger}_{k,\sigma}\right) P, \tag{3.184}$$

which is just the term to the left of the matrix $P^{-1}MP$. Substituting in $\hat{\mathcal{H}}$ as follows

$$\begin{aligned} \hat{\mathcal{H}} &= \sum_{k,\sigma} \left(\tilde{c}^{\dagger}_{k,\sigma}\ \tilde{d}^{\dagger}_{k,\sigma}\right) \begin{pmatrix} \mathcal{E}^{(+)}(k) & 0 \\ 0 & \mathcal{E}^{(-)}(k) \end{pmatrix} \begin{pmatrix} \tilde{c}_{k,\sigma} \\ \tilde{d}_{k,\sigma} \end{pmatrix} = \\ &= \sum_{k,\sigma} \left\{ \mathcal{E}^{(+)}(k) \tilde{c}^{\dagger}_{k,\sigma} \tilde{c}_{k,\sigma} + \mathcal{E}^{(-)}(k) \tilde{d}^{\dagger}_{k,\sigma} \tilde{d}_{k,\sigma} \right\}, \end{aligned} \tag{3.185}$$

i.e., the Hamiltonian $\hat{\mathcal{H}}$ has been diagonalized. A hybrid level has formed between the two individual energy levels: the electrons of the two atoms transfer to the nearest neighbors and the $\mathcal{E}^{(\pm)}(k)$ bands originate. The coefficients of the linear combination depend on k, because the eigenvectors of the matrix $M(k)$ vary with the hopping amplitude α_k (see Eq. 3.174). The squared components of these eigenvectors determine the weight of each atomic orbital in the hybridized state. Explicitly, the probabilities of finding the electron on the original orbitals are

$$|\tilde{c}_k|^2 = \frac{1}{2}\left(1 \pm \frac{\mathcal{E}_2 - \mathcal{E}_1}{\sqrt{(\mathcal{E}_2 - \mathcal{E}_1)^2 + 4\alpha_k^2}}\right), \tag{3.186}$$

$$|\tilde{d}_k|^2 = 1 - |\tilde{c}_k|^2. \tag{3.187}$$

These expressions follow directly from the eigenvectors of $M(k)$. The new orbitals have equal weights on the original atomic orbitals only in special cases, such as $\mathcal{E}_1 = \mathcal{E}_2$ or at particular k points (e.g., $k = 0$). In general, the distribution of the electron between the two atoms depends on k.

3.8 One-Band Hubbard's Model

The Hubbard model constitutes one of the most fundamental theoretical frameworks for describing interacting electrons in a periodic potential. It provides a minimal microscopic description that balances two competing physical processes: the kinetic delocalization of electrons across a lattice and their local Coulomb repulsion. In essence, it represents the simplest nontrivial realization of electronic correlations within the tight-binding approximation.

The starting point of the model is the assumption that the relevant low-energy electronic states of a crystalline solid can be expressed in terms of localized Wannier orbitals, typically associated with a single partially filled band. Within this reduced Hilbert space, electron motion is described as hopping between adjacent lattice sites, while the electron–electron interaction is approximated by an effective local term acting when two electrons occupy the same site. This treatment naturally introduces two characteristic energy scales: the hopping amplitude, which governs the itinerant character of the electrons, and the on-site Coulomb repulsion, which penalizes double occupancy. The interplay between these two terms gives rise to a broad spectrum of collective phenomena, ranging from metallic and magnetic behavior to Mott insulating states.

The conceptual strength of the Hubbard model lies in its ability to capture the essential physics of strongly correlated systems while remaining analytically and computationally tractable. It serves as a unifying platform for exploring magnetism, superconductivity, and charge ordering in transition-metal oxides, high-temperature superconductors, and other complex materials. Furthermore, its versatility allows for systematic extensions, such as multi-band generalizations, long-range interactions, and inclusion of spin–orbit coupling, which are indispensable for realistic modeling of correlated electron materials.

In summary, the Hubbard model embodies the core idea that electronic correlations cannot be treated as a small perturbation in many materials of contemporary interest. Instead, they fundamentally reshape the electronic structure and give rise to emergent quantum phases that continue to challenge our theoretical understanding of condensed matter.

3.8.1 The Model

In the Hubbard model of solids, ions are fixed at the lattice sites, while electrons interact both with one another and with the ions via the Coulomb force. Under these assumptions, electron dynamics is described in first quantization by the Hamiltonian (see Eq. II.19)

$$\mathcal{H} = \mathcal{H}_{\mathrm{HF}}^{(e),N} + \mathcal{H}_2, \tag{3.188}$$

which has the same structure as Eq. 2.326. The two-body Hamiltonian

$$\begin{aligned}\mathcal{H}_2 &= \sum_{1\le i<j\le N} U(\hat{\mathbf{r}}_i, \hat{\mathbf{r}}_j) = \\ &= \sum_{1\le i<j\le N} U(|\mathbf{r}_i - \mathbf{r}_j|)\end{aligned} \tag{3.189}$$

describes the Coulomb interaction.

We use the basis of the eigenstates of the one-body Hamiltonian $\tilde{h}_1$, i.e., Bloch eigenfunctions, to write the Hubbard Hamiltonian 3.188 in second quantization. Instead of expanding in the basis of Bloch eigenfunctions, we employ the corresponding Wannier functions, which form a localized basis. In this representation, the field operators given by Eqs. 3.150 and 3.151. The one-body operator has already been expressed in the basis of Wannier functions (see Eqs. 3.155 and 3.156). As for the two-body operator, we have

$$\begin{aligned}\hat{\mathcal{H}}_2 &= \frac{1}{2}\int dx \int dx' \, \hat{\psi}^\dagger(x)\hat{\psi}^\dagger(x')U(|\mathbf{r}-\mathbf{r}'|)\hat{\psi}(x')\hat{\psi}(x) = \\ &= \frac{1}{2}\sum_{s,s'}\sum_{\substack{n_1,i,\sigma_1\\ n_2,j,\sigma_2}}\sum_{\substack{n_3,k,\sigma_3\\ n_4,l,\sigma_4}}\int d^3\mathbf{r} \\ &\quad \int d^3\mathbf{r}' \, \omega^*_{n_1}(\mathbf{r}-\mathbf{R}_i)\chi^*_{\sigma_1}(s)\omega^*_{n_2}(\mathbf{r}'-\mathbf{R}_j)\chi^*_{\sigma_2}(s')U(|\mathbf{r}-\mathbf{r}'|) \\ &\quad \omega_{n_3}(\mathbf{r}'-\mathbf{R}_l)\chi_{\sigma_3}(s')\omega_{n_4}(\mathbf{r}-\mathbf{R}_l)\chi_{\sigma_4}(s)C^\dagger_{n_1,i,\sigma_1}C^\dagger_{n_2,j,\sigma_2}C_{n_3,k,\sigma_3}C_{n_4,l,\sigma_4}.\end{aligned} \tag{3.190}$$

We make use of Eq. 1.23 as follows

$$\sum_s \chi^*_{\sigma_1}(s)\chi_{\sigma_4}(s) = \delta_{\sigma_1,\sigma_4}, \tag{3.191}$$

$$\sum_{s'} \chi^*_{\sigma_2}(s')\chi_{\sigma_3}(s') = \delta_{\sigma_2,\sigma_3}, \tag{3.192}$$

then we sum over σ_3 and σ_4, we define $\sigma_1 \equiv \sigma, \sigma_2 \equiv \sigma'$, then

$$\hat{\mathcal{H}}_2 = \frac{1}{2} \sum_{\substack{n_1,n_2 \\ n_3,n_4}} \sum_{i,j,k,l} \sum_{\sigma,\sigma'} \int d^3\mathbf{r} \int d^3\mathbf{r}'$$
$$\omega^*_{n_1}(\mathbf{r} - \mathbf{R}_i)\omega^*_{n_2}(\mathbf{r}' - \mathbf{R}_j)U(|\mathbf{r} - \mathbf{r}'|)\omega_{n_3}(\mathbf{r}' - \mathbf{R}_l)$$
$$\omega_{n_4}(\mathbf{r} - \mathbf{R}_l)C^\dagger_{n_1,i,\sigma} C^\dagger_{n_2,j,\sigma'} C_{n_3,k,\sigma'} C_{n_4,l,\sigma}, \tag{3.193}$$

and we set

$$h^{n_1n_2n_3n_4}_{ijkl} = \int d^3\mathbf{r} \int d^3\mathbf{r}'\; \omega^*_{n_1}(\mathbf{r} - \mathbf{R}_i)\omega^*_{n_2}(\mathbf{r}' - \mathbf{R}_j)U(|\mathbf{r} - \mathbf{r}'|)\omega_{n_3}$$
$$(\mathbf{r}' - \mathbf{R}_l)\omega_{n_4}(\mathbf{r} - \mathbf{R}_l). \tag{3.194}$$

The Hubbard Hamiltonian 3.188 in first quantization can be rewritten in second quantization as

$$\hat{\mathcal{H}} = - \sum_{n,i,j,\sigma} t^n_{ij} C^\dagger_{n,i,\sigma} C_{n,j,\sigma}$$
$$+ \frac{1}{2} \sum_{\substack{n_1,n_2 \\ n_3,n_4}} \sum_{i,j,k,l} \sum_{\sigma,\sigma'} h^{n_1n_2n_3n_4}_{ijkl} C^\dagger_{n_1,i,\sigma} C^\dagger_{n_2,j,\sigma'} C_{n_3,k,\sigma'} C_{n_4,l,\sigma}. \tag{3.195}$$

In the regime where the Fermi surface (see Appendix O) resides within a conduction band, interband coupling to energetically distant bands may be neglected. Consequently, in the single-band approximation, that is, we keep only one band of interest, say fixed n, and assuming that the Wannier states are orthonormal between different bands, we introduce the delta on the band index as follows

$$t^{nn'}_{ij} = t_{ij}\, \delta_{n,n'}, \tag{3.196}$$

$$t_{ij} \equiv - \int d^3\mathbf{r} \omega^*_n(\mathbf{r} - \mathbf{R}_i)\mathcal{H}^{(e)}_{\text{HF}} \omega_n(\mathbf{r} - \mathbf{R}_j). \tag{3.197}$$

In addition, in most materials, electrons can hop only between nearby sites, since the probability of an electron jumping over long distances is negligible. The hopping amplitude t_{ij} decreases rapidly with distance. For this reason, the summation over all sites i and j is typically replaced by a summation over nearest-neighbor pairs, denoted by $\langle i, j\rangle$, i.e.,

$$t_{ij} = \begin{cases} t, & \text{if } i, j \text{ are nearest neighbours } \langle i, j\rangle, \\ 0, & \text{otherwise.} \end{cases} \tag{3.198}$$

As for the two-body interaction potential, in order to move to the single-band approximation and retain only the dominant on-site interaction, one proceeds as follows. First, one restricts to equal bands:

$$h_{ijkl}^{n_1 n_2 n_3 n_4} = h_{ijkl} \delta_{n_1,n} \delta_{n_2,n} \delta_{n_3,n} \delta_{n_4,n}, \tag{3.199}$$

where n is the band under consideration. Assuming strongly localized Wannier functions, the dominant contribution arises when all site indices coincide ($i = j = k = l$), namely when both electrons occupy the same lattice site. The on-site Hubbard interaction U is defined as

$$\begin{aligned} U \equiv h_{iiii} &= \\ &= \int d^3\mathbf{r} \int d^3\mathbf{r}' \, |\omega_n(\mathbf{r} - \mathbf{R}_i)|^2 \, U(|\mathbf{r} - \mathbf{r}'|) \, |\omega_n(\mathbf{r}' - \mathbf{R}_i)|^2 . \end{aligned} \tag{3.200}$$

Therefore, the approximation that keeps only the on-site interaction can be expressed with spatial delta functions as

$$h_{ijkl}^{nnnn} = U \delta_{i,j} \, \delta_{j,k} \, \delta_{k,l}. \tag{3.201}$$

Beginning with the general second-quantized Hamiltonian in Eq. 3.195, and after restricting to a single band and retaining only the dominant on-site interaction, one arrives at the one-band Hubbard model

$$\hat{\mathcal{H}} = - \sum_{\langle i,j \rangle,\sigma} t_{ij} C_{i,\sigma}^{\dagger} C_{j,\sigma} + \frac{U}{2} \sum_{i,\sigma \neq \sigma'} C_{i,\sigma}^{\dagger} C_{i,\sigma'}^{\dagger} C_{i,\sigma'} C_{i,\sigma}. \tag{3.202}$$

The interaction term includes a sum over spin indices with the constraint $\sigma \neq \sigma'$, since the Pauli exclusion principle forbids two electrons with identical spin from occupying the same site i. The factor $\frac{1}{2}$ accounts for the fact that the sum includes both spin configurations (e.g., $\uparrow\downarrow$ and $\downarrow\uparrow$), thereby avoiding double counting of the interaction energy. Finally, we use fermionic algebra D.4 with $\sigma \neq \sigma'$ to write the two-body operator in a more physically suggestive form, indeed, from

$$\begin{aligned} C_{i,\sigma}^{\dagger} C_{i,\sigma'}^{\dagger} C_{i,\sigma'} C_{i,\sigma} &= -C_{i,\sigma}^{\dagger} C_{i,\sigma'}^{\dagger} C_{i,\sigma} C_{i,\sigma'} = \\ &= C_{i,\sigma}^{\dagger} C_{i,\sigma} C_{i,\sigma'}^{\dagger} C_{i,\sigma'} = \\ &= \hat{N}_{i,\sigma} \hat{N}_{i,\sigma'}, \end{aligned} \tag{3.203}$$

and finally

$$\hat{\mathcal{H}} = - \sum_{\langle i,j \rangle,\sigma} t_{ij} C_{i,\sigma}^{\dagger} C_{j,\sigma} + \frac{U}{2} \sum_{i,\sigma \neq \sigma'} \hat{N}_{i,\sigma} \hat{N}_{i,\sigma'}, \tag{3.204}$$

which is the one-band Hubbard Hamiltonian in second quantization. The first term describes the hopping of electrons between neighboring sites i and j with amplitude t_{ij}, conserving spin σ. The second term accounts for the on-site Coulomb repulsion, where U is the energy cost for having two electrons with opposite spins on the same site. The factor $\frac{1}{2}$ ensures that each pairwise interaction is counted only once, thereby avoiding double-counting of the interaction energy.

Chapter 4
Step II of the Model of a Solid: Phonons

Abstract This Chapter explores the vibrational dynamics of atoms in a crystalline solid, which oscillate around their equilibrium positions due to thermal and quantum fluctuations. Initially described as harmonic oscillations of nuclei linked by elastic forces, these vibrations give rise to collective modes—normal modes—that represent the classical solutions of coupled motion equations. To capture their full physical significance, a quantum description is required. By quantizing these normal modes, each mode is associated with a discrete energy quantum called a phonon, the fundamental excitation of lattice vibrations—analogous to photons in the electromagnetic field. This quantum formalism not only reveals the particle-like nature of vibrations but also provides an elegant framework for describing thermal and interaction phenomena in solids. Key applications include models of lattice heat capacity, such as the Einstein and Debye models, which offer pedagogically insightful approaches to understanding specific heat in different temperature regimes. The Chapter also introduces phonon creation and annihilation operators, which simplify the analysis of vibrational energy and scattering processes. Beginning with the classical theory and culminating in its quantum reformulation, this chapter demonstrates how the phonon naturally emerges as a central concept in the microscopic theory of crystalline solids.

4.1 Quantization of the Lattice Vibrations

4.1.1 The Harmonic Approximation

Ionic dynamics is described by an effective ion-ion potential of the form

$$V_{i-i}^{\mathrm{eff}} = V_{i-i} + \mathcal{E}_n\left(\{\mathbf{R}\}\right), \tag{4.1}$$

where V_{i-i} is the Coulomb repulsion and $\mathcal{E}_n$ ($\{\mathbf{R}$) denotes the electronic energy obtained by solving the fixed–ion electronic problem. We are in the low-energy physics scenario, so we replace $\mathcal{E}_n\left(\{\mathbf{R}\}\right)$ with the minimum electronic eigenvalue $\mathcal{E}_0\left(\{\mathbf{R}\}\right)$ as follows

F. Tafuri et al., *Understanding Quantum Many-Body Theory*, UNITEXT for Physics, https://doi.org/10.1007/978-3-032-17849-7_4

$$V_{i-i}^{\text{eff}} = V_{i-i} + \mathcal{E}_0 (\{\mathbf{R}\}) \,. \tag{4.2}$$

We aim to minimize the effective interionic potential V_{i-i}^{eff}, which is generally highly complex. To tackle this difficulty, it is necessary to resort to suitable approximations. In particular, we consider the second-order Taylor expansion of V_{i-i}^{eff}, as given in Eq. 4.2, around the stable equilibrium positions of the ions. This leads to the harmonic approximation, in which V_{i-i}^{eff} is replaced by a harmonic potential. Such an approximation is well justified by experimental evidence indicating that ions typically undergo only small displacements from their equilibrium positions at the lattice sites. We now consider a crystal lattice defined by the primitive vectors $\mathbf{a}_1$, $\mathbf{a}_2$, and $\mathbf{a}_3$, with ions positioned at the lattice vertices and τ distinct atomic species per unit cell. Recall that the entire lattice consists of $N = N_1 N_2 N_3$ unit cells, each labeled by a triplet of integers $\mathbf{n} = (n_1, n_2, n_3)$, with the elementary cell associated with $\mathbf{n}$ denoted by $\mathbf{R}_\mathbf{n}$ (see Eq. 3.2). Then, let $\mathbf{R}_{\mathbf{n},\mu}(t)$ denote the position at time t of an ion of species μ located in cell $\mathbf{n}$. This position can be expressed as

$$\mathbf{R}_{\mathbf{n},\mu}(t) = \mathbf{R}_\mathbf{n} + \mathbf{d}_\mu + \mathbf{S}_{\mathbf{n},\mu}(t), \ \left|\mathbf{S}_{\mathbf{n},\mu}(t)\right| \ll a_i, \ i = 1, 2, 3, \tag{4.3}$$

where $\mathbf{d}_\mu$ is the position vector of the atom of species μ ($\mu = 1, \dots, \tau$) within the unit cell indexed by $\mathbf{0}$ and $\mathbf{S}_{\mathbf{n},\mu}(t)$ represents the displacement from the equilibrium position $\mathbf{R}_\mathbf{n} + \mathbf{d}_\mu$ at time t. Accordingly, the second-order Taylor expansion of V_{i-i}^{eff} is performed with respect to the variables $\mathbf{R}_{\mathbf{n},\mu}$, around the equilibrium positions $\mathbf{R}_\mathbf{n} + \mathbf{d}_\mu$. In particular, one needs to compute: the Hessian matrix, the matrix of first derivatives evaluated at the minimum point, which is identically zero, and a constant corresponding to the value of the potential at the ions equilibrium positions. Since the potential is defined up to an additive constant, we assume, without loss of generality, that this constant is zero. This expansion of the potential is therefore expressed as

$$V_{i-i}^{\text{eff}} (\{\mathbf{R}\}) = \frac{1}{2} \sum_{\substack{\mathbf{n} \\ \mu,i}} \sum_{\substack{\mathbf{m} \\ \nu,k}} W_{\mathbf{m},\nu,k}^{\mathbf{n},\mu,i} S_{\mathbf{n},\mu,i} S_{\mathbf{m},\nu,k} + o\left(|S|^3\right), \tag{4.4}$$

where

$$W_{\mathbf{m},\nu,k}^{\mathbf{n},\mu,i} = \frac{\partial^2 V_{i-i}^{\text{eff}}}{\partial R_{\mathbf{m},\nu,k} \partial R_{\mathbf{n},\mu,i}} \tag{4.5}$$

is the multi-index object of the second derivatives of the potential with respect to the positions $R_{\mathbf{n},\mu,i}$, $R_{\mathbf{m},\nu,k}$ and is calculated for $\mathbf{S} = \mathbf{0}$. Here, i and k denote the Cartesian components of the corresponding position vectors. Since V_{i-i}^{eff} has a minimum at $\mathbf{S} = \mathbf{0}$, it is a positive definite form. Let us now study the classical Lagrangian for ion dynamics. The time-dependent quantity in $\mathbf{R}_{\mathbf{n},\mu}$ is $\mathbf{S}_{\mathbf{n},\mu}$, then the kinetic term is

$$T = \frac{1}{2} \sum_{\substack{\mathbf{n} \\ \mu,i}} M_\mu \dot{R}^2_{\mathbf{n},\mu,i} \equiv$$
$$\equiv \frac{1}{2} \sum_{\substack{\mathbf{n} \\ \mu,i}} M_\mu \dot{S}^2_{\mathbf{n},\mu,i}, \tag{4.6}$$

and the Lagrangian function is

$$\mathcal{L} = T - V =$$
$$= \frac{1}{2} \sum_{\substack{\mathbf{n} \\ \mu,i}} M_\mu \dot{S}^2_{\mathbf{n},\mu,i} - \frac{1}{2} \sum_{\substack{\mathbf{n} \\ \mu,i}} \sum_{\substack{\mathbf{m} \\ \nu,k}} W^{\mathbf{n},\mu,i}_{\mathbf{m},\nu,k} S_{\mathbf{n},\mu,i} S_{\mathbf{m},\nu,k}. \tag{4.7}$$

The equations of motion follow from the Euler-Lagrange equations

$$\frac{d}{dt}\frac{\partial \mathcal{L}}{\partial \dot{q}_h} - \frac{\partial \mathcal{L}}{\partial q_h} = 0, \tag{4.8}$$

with identifications $h \to (\mathbf{n}, \mu, i)$, $q \to S$, $q^h \to S_{\mathbf{n},\mu,i}$, the Euler-Lagrange equations for the ionic dynamics become

$$M_\mu \ddot{S}_{\mathbf{n},\mu,i} = -\sum_{\substack{\mathbf{m} \\ \nu,k}} W^{\mathbf{n},\mu,i}_{\mathbf{m},\nu,k} S_{\mathbf{m},\nu,k}, \tag{4.9}$$

which describe a system of many coupled harmonic oscillators (see Sect. 1.3). Specifically, the total number of vectors $\mathbf{n}$ corresponds to the number of elementary cells in the crystal lattice, namely $N_1 N_2 N_3$, where N_i is the number of cells along the i-th crystallographic direction (with $i = 1, 2, 3$). Since there are three spatial directions and τ distinct atomic species, the total number of coupled oscillators is

$$3\tau N_1 N_2 N_3. \tag{4.10}$$

Our goal is to diagonalize the Lagrangian. Because the Lagrangian is a quadratic form, this diagonalization can be carried out exactly. Through this process, we will determine the normal modes of the system. Let us first consider the quantities $W^{\mathbf{n},\mu,i}_{\mathbf{m},\nu,k}$. For a single harmonic oscillator, rather than a system of coupled oscillators, the quantity W reduces to the elastic constant of the oscillator. In the general case, these quantities are referred to as elastic *material constants* or *force constants*.

Theorem 4.1 (Fundamental Properties of the Force Constants) *The force constants, Eq. 4.5, satisfy the following properties.*

1. *Translational invariance, i.e.,*

$$W_{\mathbf{m},\nu,k}^{\mathbf{n},\mu,i} = W_{\mathbf{0},\nu,k}^{\mathbf{n}-\mathbf{m},\mu,i}. \tag{4.11}$$

2. *Index-exchange symmetry, i.e.,*

$$W_{\mathbf{m},\nu,k}^{\mathbf{n},\mu,i} = W_{\mathbf{n},\mu,i}^{\mathbf{m},\nu,k}. \tag{4.12}$$

3. *Reality condition, i.e.,*

$$\left(W_{\mathbf{m},\nu,k}^{\mathbf{n},\mu,i}\right)^* = W_{\mathbf{m},\nu,k}^{\mathbf{n},\mu,i}. \tag{4.13}$$

4. *Acoustic sum rule, i.e.,*

$$\sum_{\mathbf{m},\nu} W_{\mathbf{m},\nu,k}^{\mathbf{n},\mu,i}\Big|_{k\ \mathit{fixed}} = 0. \tag{4.14}$$

Proof

1. By definition, the force constants are obtained as second derivatives of the total energy with respect to atomic displacements, evaluated at equilibrium positions. If both cells labeled by **n** and **m** are translated by the same lattice vector $\mathbf{R_h}$, the relative positions of all atoms remain unchanged, and hence the derivatives of the energy are invariant, that is,

$$W_{\mathbf{m+h},\nu,k}^{\mathbf{n+h},\mu,i} = W_{\mathbf{m},\nu,k}^{\mathbf{n},\mu,i}, \ \forall\, \mathbf{R_h} \in \text{direct lattice}. \tag{4.15}$$

   In particular, choosing $\mathbf{R_h} = -\mathbf{R_m}$ yields Eq. 4.11. Therefore, the coefficients $W_{\mathbf{m},\nu,k}^{\mathbf{n},\mu,i}$ depend only on the difference $\mathbf{R_n} - \mathbf{R_m}$, and not on the individual lattice vectors $\mathbf{R_n}$ and $\mathbf{R_m}$.
2. Since the potential energy is an analytic function of the displacement variables, mixed partial derivatives commute. Explicitly,

$$\frac{\partial^2 V_{i-i}^{\text{eff}}}{\partial R_{\mathbf{m},\nu,k}\,\partial R_{\mathbf{n},\mu,i}} = \frac{\partial^2 V_{i-i}^{\text{eff}}}{\partial R_{\mathbf{n},\mu,i}\,\partial R_{\mathbf{m},\nu,k}}, \tag{4.16}$$

   and this commutativity directly implies the symmetry relation in Eq. 4.12.
3. Since the potential energy is a real-valued function of real variables, all such derivatives are real numbers.
4. The equation of motion for the displacement $S_{\mathbf{n},\mu,i}$ of atom $(\mathbf{n}, \mu)$ along direction i reads

$$M_\mu \ddot{S}_{\mathbf{n},\mu,i} = -\sum_{\mathbf{m},\nu,k} W_{\mathbf{m},\nu,k}^{\mathbf{n},\mu,i} S_{\mathbf{m},\nu,k}. \tag{4.17}$$

If an atom $(\mathbf{m}, \nu)$ is displaced along direction k, the term $W^{\mathbf{n},\mu,i}_{\mathbf{m},\nu,k} S_{\mathbf{m},\nu,k}$ represents (minus) the force exerted on atom $(\mathbf{n}, \mu)$ in direction i. Now consider a rigid translation of the entire lattice in direction k, i.e., all $S_{\mathbf{m},\nu,k}$ are equal and constant. In this case, all relative distances remain unchanged, and hence no restoring force can arise. Therefore, the total force on each atom must vanish, which is equivalent to Eq. 4.14.

□

4.1.2 Classical Formalism

From this point onward, in order to solve the Euler-Lagrange equations for ionic dynamics (see Eq. 4.9), we apply a series of transformations aimed at diagonalizing the Lagrangian and ultimately obtaining an equation that governs atomic oscillations. As a first step, we perform a change of variables from the original coordinates S to a new set of coordinates U, chosen to simplify the treatment of the atomic masses M_μ. Let $M = \sum_\mu M_\mu$ denote the total mass of the system, then we define

$$U_{\mathbf{n},\mu,i} = \sqrt{\frac{M_\mu}{M}} S_{\mathbf{n},\mu,i}, \tag{4.18}$$

then

$$\mathcal{L} = \frac{1}{2} M \sum_{\substack{\mathbf{n}\\ \mu,i}} \dot{U}^2_{\mathbf{n},\mu,i} - \frac{1}{2} \sum_{\substack{\mathbf{n}\\ \mu,i}} \sum_{\substack{\mathbf{m}\\ \nu,k}} D^{\mathbf{n},\mu,i}_{\mathbf{m},\nu,k} U_{\mathbf{n},\mu,i} U_{\mathbf{m},\nu,k}, \tag{4.19}$$

where

$$D^{\mathbf{n},\mu,i}_{\mathbf{m},\nu,k} = \frac{M}{\sqrt{M_\mu M_\nu}} W^{\mathbf{n},\mu,i}_{\mathbf{m},\nu,k} \tag{4.20}$$

are the *rescaled force constants*. The next goal is to decouple the system of coupled harmonic oscillators. To achieve this, we first aim to eliminate the double summation over the lattice indices $\mathbf{n}$ and $\mathbf{m}$ (see Eq. 4.11). This is conveniently done by transforming to reciprocal space, introducing wave vectors $\mathbf{q}$ defined on the elementary cell of the reciprocal lattice. Since the displacements $U_{\mathbf{n},\mu,i}$ are defined on the N sites of the direct lattice, we express $U_{\mathbf{n},\mu,i}$ as the action of a unitary transformation 3.104, with matrix elements $\frac{e^{i\mathbf{q}\cdot\mathbf{R_n}}}{\sqrt{N}}$, on the variables $C_{\mathbf{q},\mu,i}$, specifically

$$U_{\mathbf{n},\mu,i} = \sum_{\mathbf{q}} \frac{e^{i\mathbf{q}\cdot\mathbf{R_n}}}{\sqrt{N}} C_{\mathbf{q},\mu,i}, \tag{4.21}$$

where the wave vectors $\mathbf{q}$ are defined in Eq. 3.73. It is worth noting that the transformation between the real-space displacements $U_{\mathbf{n},\mu,i}$ and the reciprocal-space variables $C_{\mathbf{q},\mu,i}$ is formally analogous to the relation between Wannier functions and Bloch waves. The Lagrangian is modified as

$$\mathcal{L} = \frac{1}{2}M \sum_{\mathbf{n},\mu,i} \sum_{\mathbf{q}} \sum_{\mathbf{q}'} \frac{1}{N} e^{i\mathbf{q}\cdot\mathbf{R_n}} e^{i\mathbf{q}'\cdot\mathbf{R_n}} \dot{C}_{\mathbf{q},\mu,i} \dot{C}_{\mathbf{q}',\mu,i}$$
$$- \frac{1}{2} \sum_{\substack{\mathbf{n}\\ \mu,i}} \sum_{\substack{\mathbf{m}\\ \nu,k}} D^{\mathbf{n},\mu,i}_{\mathbf{m},\nu,k} \sum_{\mathbf{q}} \sum_{\mathbf{q}'} \frac{1}{N} e^{i\mathbf{q}\cdot\mathbf{R_n}} e^{i\mathbf{q}'\cdot\mathbf{R_m}} C_{\mathbf{q},\mu,i} C_{\mathbf{q}',\nu,k}. \tag{4.22}$$

In the first term, the quantities $\dot{C}_{\mathbf{q},\mu,i}\dot{C}_{\mathbf{q},\mu,i}$ do not depend on the lattice index $\mathbf{n}$, therefore

$$\frac{1}{N} \sum_{\mathbf{n}} e^{i\mathbf{q}\cdot\mathbf{R_n}} e^{i\mathbf{q}'\cdot\mathbf{R_n}} = \delta_{\mathbf{q},-\mathbf{q}'}, \tag{4.23}$$

$$\mathcal{L} = \frac{1}{2}M \sum_{\mu,i} \sum_{\mathbf{q}} \sum_{\mathbf{q}'} \delta_{\mathbf{q},-\mathbf{q}'} \dot{C}_{\mathbf{q},\mu,i} \dot{C}_{\mathbf{q}',\mu,i}$$
$$- \frac{1}{2} \sum_{\substack{\mathbf{n}\\ \mu,i}} \sum_{\substack{\mathbf{m}\\ \nu,k}} D^{\mathbf{n},\mu,i}_{\mathbf{m},\nu,k} \sum_{\mathbf{q}} \sum_{\mathbf{q}'} \frac{1}{N} e^{i\mathbf{q}\cdot\mathbf{R_n}} e^{i\mathbf{q}'\cdot\mathbf{R_m}} C_{\mathbf{q},\mu,i} C_{\mathbf{q}',\nu,k} =$$
$$= \frac{1}{2}M \sum_{\substack{\mathbf{q}\\ \mu,i}} \dot{C}_{\mathbf{q},\mu,i} \dot{C}_{-\mathbf{q},\mu,i} - \frac{1}{2} \sum_{\substack{\mathbf{n}\\ \mu,i}} \sum_{\substack{\mathbf{m}\\ \nu,k}} D^{\mathbf{n},\mu,i}_{\mathbf{m},\nu,k} \sum_{\mathbf{q}} \sum_{\mathbf{q}'} \frac{1}{N} e^{i\mathbf{q}\cdot\mathbf{R_n}} e^{i\mathbf{q}'\cdot\mathbf{R_m}} C_{\mathbf{q},\mu,i} C_{\mathbf{q}',\nu,k}. \tag{4.24}$$

We cannot apply the relation $\delta_{\mathbf{q},-\mathbf{q}'}$ to the second term because the force constants depend explicitly on the lattice index $\mathbf{n}$. Instead, we utilize the symmetry properties of D (and equivalently W) and insert the identity operator in the form $\mathbb{1} = e^{-i\mathbf{q}\cdot\mathbf{R_m}} e^{i\mathbf{q}\cdot\mathbf{R_m}}$ to rewrite the Lagrangian as follows

$$\mathcal{L} = \frac{1}{2}M \sum_{\substack{\mathbf{q}\\ \mu,i}} \dot{C}_{\mathbf{q},\mu,i} \dot{C}_{-\mathbf{q},\mu,i}$$
$$- \frac{1}{2} \sum_{\substack{\mathbf{n}\\ \mu,i}} \sum_{\substack{\mathbf{m}\\ \nu,k}} D^{\mathbf{n},\mu,i}_{\mathbf{m},\nu,k} \sum_{\mathbf{q}} \sum_{\mathbf{q}'} \frac{1}{N} e^{i\mathbf{q}\cdot\mathbf{R_n}} e^{-i\mathbf{q}\cdot\mathbf{R_m}} e^{i\mathbf{q}\cdot\mathbf{R_m}} e^{i\mathbf{q}'\cdot\mathbf{R_m}} C_{\mathbf{q},\mu,i} C_{\mathbf{q}',\nu,k} =$$
$$= \frac{1}{2}M \sum_{\substack{\mathbf{q}\\ \mu,i}} \dot{C}_{\mathbf{q},\mu,i} \dot{C}_{-\mathbf{q},\mu,i}$$

$$-\frac{1}{2}\sum_{\substack{\mathbf{n}\\ \mu,i}}\sum_{\substack{\mathbf{m}\\ \nu,k}} D_{\mathbf{0},\nu,k}^{\mathbf{n}-\mathbf{m},\mu,i}\sum_{\mathbf{q}}\sum_{\mathbf{q}'}\frac{1}{N}e^{i\mathbf{q}\cdot(\mathbf{R_n}-\mathbf{R_m})}e^{i(\mathbf{q}+\mathbf{q}')\cdot\mathbf{R_m}}C_{\mathbf{q},\mu,i}C_{\mathbf{q}',\nu,k}, \tag{4.25}$$

where, in the last step, we have made explicit that the force constants depend only on the difference between the lattice vectors. We set $\mathbf{R_n} \to \mathbf{R_m} + \mathbf{R_h},\ \forall\,\mathbf{R_m}$, then

$$\mathcal{L} = \frac{1}{2}M\sum_{\substack{\mathbf{q}\\ \mu,i}}\dot{C}_{\mathbf{q},\mu,i}\dot{C}_{-\mathbf{q},\mu,i}$$
$$-\frac{1}{2}\sum_{\substack{\mathbf{h}\\ \mu,i}}\sum_{\nu,k}\sum_{\mathbf{q}} D_{\mathbf{0},\nu,k}^{\mathbf{h},\mu,i}e^{i\mathbf{q}\cdot\mathbf{R_h}}\sum_{\mathbf{m}}\sum_{\mathbf{q}'}\frac{1}{N}e^{i(\mathbf{q}+\mathbf{q}')\cdot\mathbf{R_m}}C_{\mathbf{q},\mu,i}C_{\mathbf{q}',\nu,k}, \tag{4.26}$$

and we define

$$\left[\tilde{D}(\mathbf{q})\right]_{\substack{\mu,\nu\\ i,k}} = \sum_{\mathbf{h}} D_{\mathbf{0},\nu,k}^{\mathbf{h},\mu,i}\, e^{i\mathbf{q}\cdot\mathbf{R_h}}, \tag{4.27}$$

so that the dependence of the force constants on the positional index of the cell disappears, allowing us to write

$$\frac{1}{N}\sum_{\mathbf{m}} e^{i(\mathbf{q}+\mathbf{q}')\cdot\mathbf{R_m}} = \delta_{\mathbf{q},-\mathbf{q}'}, \tag{4.28}$$

$$\mathcal{L} = \frac{1}{2}M\sum_{\substack{\mathbf{q}\\ \mu,i}}\dot{C}_{\mathbf{q},\mu,i}\dot{C}_{-\mathbf{q},\mu,i} - \frac{1}{2}\sum_{\mathbf{q}}\sum_{\substack{\mu,i\\ \nu,k}}\left[\tilde{D}(\mathbf{q})\right]_{\substack{\mu,\nu\\ i,k}} C_{\mathbf{q},\mu,i}C_{-\mathbf{q},\nu,k}. \tag{4.29}$$

Now, each term indexed by $(\mathbf{q}, \mu, i)$ couples only with terms indexed by $\{(-\mathbf{q}, \mu, i)\}_{\mu,i}$, so the Lagrangian becomes block diagonal. For each wave vector $\mathbf{q}$, there is a block of dimension $3\tau \times 3\tau$, and the total number of such blocks equals the number of $\mathbf{q}$-vectors, i.e., N blocks. Thus, the Lagrangian decomposes into N independent blocks. The matrices defined in 4.27 are Hermitian, indeed,

$$\begin{aligned}\left[\left[\tilde{D}(\mathbf{q})\right]_{\substack{\mu,\nu\\ i,k}}\right]^{\dagger} &= \sum_{\mathbf{h}} D_{\mathbf{0},\mu,i}^{\mathbf{h},\nu,k}\, e^{-i\mathbf{q}\cdot\mathbf{R_h}} = \\ &= \sum_{\mathbf{h}} D_{\mathbf{0},\mu,i}^{-\mathbf{h},\nu,k}\, e^{i\mathbf{q}\cdot\mathbf{R_h}} = \\ &= \sum_{\mathbf{h}} D_{\mathbf{h},\mu,i}^{\mathbf{0},\nu,k}\, e^{i\mathbf{q}\cdot\mathbf{R_h}} =\end{aligned}$$

$$= \sum_{\mathbf{h}} D^{\mathbf{h},\mu,i}_{\mathbf{0},\nu,k} \, e^{i\mathbf{q}\cdot\mathbf{R_h}} \equiv$$

$$\equiv \left[\tilde{D}(\mathbf{q})\right]_{\substack{\mu,\nu \\ i,k}} . \tag{4.30}$$

The matrices D are real, so the adjoint operation coincides with the transpose, which in our notation corresponds to exchanging the order of the index pairs (μ, ν) and (i, k) in the definition of $\tilde{D}(\mathbf{q})$. We substituted $\mathbf{R_h} \to -\mathbf{R_h}$ since $\mathbf{h}$ is a dummy summation index, then applied property 4.11, and finally reordered the indices using property 4.11. Therefore, for each $\mathbf{q}$, the matrix $\tilde{D}(\mathbf{q})$ has 3τ real eigenvalues and 3τ corresponding eigenvectors.

Let us now define $\mathbf{C_q}$ as the column vector composed of the 3τ components $C_{\mathbf{q},\mu,i}$ for each fixed $\mathbf{q}$. Under this notation, the Lagrangian becomes

$$\mathcal{L} = \frac{1}{2} M \sum_{\mathbf{q}} \dot{\mathbf{C}}^T_{-\mathbf{q}} \cdot \dot{\mathbf{C}}_{\mathbf{q}} - \frac{1}{2} \sum_{\mathbf{q}} \mathbf{C}^T_{-\mathbf{q}} \tilde{D}(\mathbf{q}) \, \mathbf{C}_{\mathbf{q}}, \tag{4.31}$$

where we have rearranged the order of $\dot{\mathbf{C}}_{\mathbf{q}}$, $\dot{\mathbf{C}}_{-\mathbf{q}}$, $\mathbf{C}_{-\mathbf{q}}$, and $\mathbf{C_q}$ to write the first term as a row-by-column scalar product and the second term as a matrix product. The matrix $\tilde{D}(\mathbf{q})$ is Hermitian by construction. Fixing $\mathbf{q}$, we denote by $P(\mathbf{q})$ the $3\tau \times 3\tau$ matrix whose columns are the eigenvectors of $\tilde{D}(\mathbf{q})$. The object

$$P^{-1}(\mathbf{q})\tilde{D}(\mathbf{q})P(\mathbf{q}) = K(\mathbf{q}) \tag{4.32}$$

is a diagonal matrix whose diagonal elements are the 3τ real eigenvalues of $\tilde{D}(\mathbf{q})$, and we have

$$\begin{aligned}
\mathcal{L} &= \frac{1}{2} M \sum_{\mathbf{q}} \dot{\mathbf{C}}^T_{-\mathbf{q}} \cdot \dot{\mathbf{C}}_{\mathbf{q}} - \frac{1}{2} \sum_{\mathbf{q}} \mathbf{C}^T_{-\mathbf{q}} P(\mathbf{q})P^{-1}(\mathbf{q})\tilde{D}(\mathbf{q})P(\mathbf{q})P^{-1}(\mathbf{q})\mathbf{C}_{\mathbf{q}} = \\
&= \frac{1}{2} M \sum_{\mathbf{q}} \dot{\mathbf{C}}^T_{-\mathbf{q}} \cdot \dot{\mathbf{C}}_{\mathbf{q}} - \frac{1}{2} \sum_{\mathbf{q}} \mathbf{C}^T_{-\mathbf{q}} P(\mathbf{q})K(\mathbf{q})P^{-1}(\mathbf{q})\mathbf{C}_{\mathbf{q}}.
\end{aligned} \tag{4.33}$$

Due to the similarity transformation $P^{-1}(\mathbf{q})\tilde{D}(\mathbf{q})P(\mathbf{q})$, the matrix $\tilde{D}(\mathbf{q})$ becomes diagonal in the basis formed by the vectors

$$\mathbf{C}^T_{-\mathbf{q}} P(\mathbf{q}), \tag{4.34}$$

$$P^{-1}(\mathbf{q})\mathbf{C}_{\mathbf{q}}, \tag{4.35}$$

which are respectively row and column vectors. We then define the column vector

$$\mathbf{Q_q} = P^{-1}(\mathbf{q})\mathbf{C}_{\mathbf{q}}, \tag{4.36}$$

and we aim to compute

$$\mathbf{Q}^T_{-\mathbf{q}} = \left(P^{-1}(-\mathbf{q})\mathbf{C}_{-\mathbf{q}}\right)^T = \\ = \mathbf{C}^T_{-\mathbf{q}}\left(P^{-1}(-\mathbf{q})\right)^T. \tag{4.37}$$

Since $P(\mathbf{q})$ is a change-of-basis matrix, i.e., a unitary transformation, we have

$$P^\dagger(-\mathbf{q}) = P^{-1}(-\mathbf{q}), \tag{4.38}$$

$$\left(P^*(-\mathbf{q})\right)^T = P^{-1}(-\mathbf{q}), \tag{4.39}$$

$$P^*(-\mathbf{q}) = \left(P^{-1}(-\mathbf{q})\right)^T, \tag{4.40}$$

$$\mathbf{Q}^T_{-\mathbf{q}} = \mathbf{C}^T_{-\mathbf{q}}P^*(-\mathbf{q}). \tag{4.41}$$

The quantity $\mathbf{Q}^T_{-\mathbf{q}}$ can be manipulated by exploiting the properties of $\tilde{D}(\mathbf{q})$. From Eq. 4.27, the matrices $\tilde{D}(\mathbf{q})$ satisfy

$$\left[\tilde{D}(\mathbf{q})\right]^* = \tilde{D}(-\mathbf{q}). \tag{4.42}$$

Then, we manipulate the eigenvalue equation as follows

$$\tilde{D}(\mathbf{q})V = \lambda V, \tag{4.43}$$

$$\left[\tilde{D}(\mathbf{q})\right]^* V^* = \lambda V^*, \tag{4.44}$$

$$\tilde{D}(-\mathbf{q})V^* = \lambda V^*, \tag{4.45}$$

$$\left[\tilde{D}(-\mathbf{q})\right]^* V = \lambda V, \tag{4.46}$$

therefore, $\tilde{D}(\mathbf{q})$ and $\left[\tilde{D}(-\mathbf{q})\right]^*$ share the same eigenvalues, which are real, as well as the eigenvectors, consequently we have

$$P^*(-\mathbf{q}) = P(\mathbf{q}). \tag{4.47}$$

The object

$$\mathbf{Q}^T_{-\mathbf{q}} = \mathbf{C}^T_{-\mathbf{q}}P(\mathbf{q}) \tag{4.48}$$

is the row vector appearing in the Lagrangian

$$\mathcal{L} = \frac{1}{2}M\sum_{\mathbf{q}}\dot{\mathbf{C}}^T_{-\mathbf{q}}P(\mathbf{q})P^{-1}(\mathbf{q})\dot{\mathbf{C}}_{\mathbf{q}} - \frac{1}{2}\sum_{\mathbf{q}}\mathbf{C}^T_{-\mathbf{q}}P(\mathbf{q})K(\mathbf{q})P^{-1}(\mathbf{q})\mathbf{C}_{\mathbf{q}} =$$
$$= \frac{1}{2}M\sum_{\mathbf{q}}\dot{\mathbf{Q}}^T_{-\mathbf{q}}\cdot\dot{\mathbf{Q}}_{\mathbf{q}} - \frac{1}{2}\sum_{\mathbf{q}}\mathbf{Q}^T_{-\mathbf{q}}K(\mathbf{q})\,\mathbf{Q}_{\mathbf{q}}. \tag{4.49}$$

Since $K(\mathbf{q})$ is diagonal, we can express the scalar products in the Lagrangian in terms of the components of $\dot{\mathbf{Q}}^T_{-\mathbf{q}}$, $\dot{\mathbf{Q}}^T_{-\mathbf{q}}$, $\mathbf{Q}^T_{-\mathbf{q}}$ and $\mathbf{Q}^T_{-\mathbf{q}}$, i.e.,

$$\mathcal{L} = \frac{1}{2}M\sum_{\mathbf{q}}\sum_{s=1}^{3\tau}\dot{Q}_{-\mathbf{q},s}\dot{Q}_{\mathbf{q},s} - \frac{1}{2}\sum_{\mathbf{q}}\sum_{s=1}^{3\tau}Q_{-\mathbf{q},s}\,[K(\mathbf{q})]_{s,s}\,Q_{\mathbf{q},s}. \tag{4.50}$$

The vectors $\mathbf{Q}_{\mathbf{q}}$ and $\mathbf{Q}_{-\mathbf{q}}$ are indexed by a single index s in both the kinetic and potential terms.

The displacements $S_{\mathbf{n},\mu,i}$ can now be rewritten in terms of the variables $Q_{\mathbf{q},s}$. For convenience, we group the indices (μ, i) into a single index l. The vector $\mathbf{C}_{\mathbf{q}}$ can then be written as a function of $\mathbf{Q}_{\mathbf{q}}$ as

$$\mathbf{C}_{\mathbf{q}} = P(\mathbf{q})\mathbf{Q}_{\mathbf{q}}. \tag{4.51}$$

Now, recalling the transformations between $S_{\mathbf{n},l}$ and $U_{\mathbf{n},l}$ (see Eq. 4.18), and between $U_{\mathbf{n},l}$ and $C_{\mathbf{q},l}$ (see Eq. 4.21), we have

$$S_{\mathbf{n},l} = \sqrt{\frac{M}{M_\mu}}\sum_{\mathbf{q}}\frac{e^{i\mathbf{q}\cdot\mathbf{R}_\mathbf{n}}}{\sqrt{N}}\sum_{s=1}^{3\tau}[P(\mathbf{q})]_{l,s}\,Q_{\mathbf{q},s}. \tag{4.52}$$

Equivalently, the displacement $S_{\mathbf{n},\mu,i}$ in cell $\mathbf{n}$ is represented as a linear combination of the variables $Q_{\mathbf{q},s}$. Moreover, the objects $S_{\mathbf{n},\mu,i}$ are real quantities, while the $Q_{\mathbf{q},s}$ are generally complex; thus, the number of $Q_{\mathbf{q},s}$ variables is twice that of the $S_{\mathbf{n},\mu,i}$. Consequently, there must be a constraint on the $Q_{\mathbf{q},s}$ to ensure they form a linearly independent system. Since $S_{\mathbf{n},l}$ is real, it follows that

$$S^*_{\mathbf{n},l} = \sqrt{\frac{M}{M_\mu}}\sum_{\mathbf{q}}\frac{e^{-i\mathbf{q}\cdot\mathbf{R}_\mathbf{n}}}{\sqrt{N}}\sum_{s=1}^{3\tau}\left[P^*(\mathbf{q})\right]_{l,s}Q^*_{\mathbf{q},s}, \tag{4.53}$$

we rename the dummy index $\mathbf{q} \to -\mathbf{q}$ as follows

$$S^*_{\mathbf{n},l} = \sqrt{\frac{M}{M_\mu}}\sum_{\mathbf{q}}\frac{e^{i\mathbf{q}\cdot\mathbf{R}_\mathbf{n}}}{\sqrt{N}}\sum_{s=1}^{3\tau}\left[P^*(-\mathbf{q})\right]_{l,s}Q^*_{-\mathbf{q},s}, \tag{4.54}$$

and this must be equal to $S_{\mathbf{n},l}$. Recall that

$$\left[P^*(-\mathbf{q})\right]_{l,s} = [P(\mathbf{q})]_{l,s}, \tag{4.55}$$

then

$$Q^*_{-\mathbf{q},s} = Q_{\mathbf{q},s}. \tag{4.56}$$

Among the $6N\tau$ variables $Q_{\mathbf{q},s}$, only $3N\tau$ are linearly independent. The Lagrangian becomes

$$\mathcal{L} = \frac{1}{2}M\sum_{\mathbf{q}}\sum_{s=1}^{3\tau}|\dot{Q}_{\mathbf{q},s}|^2 - \frac{1}{2}\sum_{\mathbf{q}}\sum_{s=1}^{3\tau}[K(\mathbf{q})]_{s,s}|Q_{\mathbf{q},s}|^2, \tag{4.57}$$

where $Q_{\mathbf{q},s}$ denote the classical normal modes. The Euler-Lagrange equations (see Eq. 4.8) are given by

$$\frac{d}{dt}\frac{\partial\mathcal{L}}{\partial\dot{Q}^*_{\mathbf{q},s}} - \frac{\partial\mathcal{L}}{\partial Q^*_{\mathbf{q},s}} = 0,\ \forall\,\mathbf{q},s, \tag{4.58}$$

where we used the complex conjugates of $Q_{\mathbf{q},s}$ and $\dot{Q}_{\mathbf{q},s}$. To properly handle terms like $Q_{\mathbf{q},s}Q^*_{\mathbf{q},s}$ and $\dot{Q}_{\mathbf{q},s}\dot{Q}^*_{\mathbf{q},s}$, we must consider both the components at $\mathbf{q}$ and $-\mathbf{q}$. These components are related and can be expressed in terms of each other, which removes the factor $\frac{1}{2}$. Consequently, the Euler-Lagrange equations yield

$$M\ddot{Q}_{\mathbf{q},s} = -[K(\mathbf{q})]_{s,s}Q_{\mathbf{q},s}. \tag{4.59}$$

The Lagrangian we have derived is positive definite; hence, the eigenvalues of $\tilde{D}(\mathbf{q})$ and $K(\mathbf{q})$ are all positive. We define

$$[K(\mathbf{q})]_{s,s} = M\omega^2_{\mathbf{q},s}, \tag{4.60}$$

Note that, from Eq. 4.42, such eigenvalues satisfy

$$\omega_{\mathbf{q},s} = \omega_{-\mathbf{q},s},\ \forall\mathbf{q},s. \tag{4.61}$$

Equation 4.59 becomes

$$\ddot{Q}_{\mathbf{q},s} = -\omega^2_{\mathbf{q},s}Q_{\mathbf{q},s}, \tag{4.62}$$

which corresponds to a system of $3\tau N$ independent second-order differential equations, each describing a simple harmonic motion with solutions given by

$$Q_{\mathbf{q},s}(t) = A_{\mathbf{q},s}e^{i\omega_{\mathbf{q},s}t} + B_{\mathbf{q},s}e^{-i\omega_{\mathbf{q},s}t}. \tag{4.63}$$

The displacement field

$$S_{\mathbf{n},l} = \sqrt{\frac{M}{M_\mu}} \sum_{\mathbf{q}} \frac{e^{i\mathbf{q}\cdot\mathbf{R_n}}}{\sqrt{N}} \sum_{s=1}^{3\tau} [P(\mathbf{q})]_{l,s}\, Q_{\mathbf{q},s} \tag{4.64}$$

is a superposition of $3\tau N$ normal modes, each of which satisfies the equation of a simple harmonic motion. Recall that the index l groups the indices (μ, i), and therefore takes on 3τ values. For each atomic species μ, we now collect the three components $S_{\mathbf{n},\mu,i}$, corresponding to the Cartesian directions $i = 1, 2, 3$, into a vector $\mathbf{S}_{\mathbf{n},\mu}$, which represents the displacement of the atom of species μ in the cell labeled by $\mathbf{n}$. Thus, on the left-hand side we have the substitution $S_{\mathbf{n},\mu,i} \to \mathbf{S}_{\mathbf{n},\mu}$. On the right-hand side, only the quantity $[P(\mathbf{q})]_{l,s}$ depends on the index l. Here, $[P(\mathbf{q})]_{l,s}$ denotes the entry in the l-th row and s-th column of the matrix $P(\mathbf{q})$, which is a $3\tau \times 3\tau$ unitary matrix whose columns are the eigenvectors of the dynamical matrix $\tilde{D}(\mathbf{q})$, i.e.,

$$P(\mathbf{q}) = \begin{pmatrix} \varepsilon^{(1)}_{1,1}(\mathbf{q}) & \varepsilon^{(2)}_{1,1}(\mathbf{q}) & \dots & \varepsilon^{(3\tau)}_{1,1}(\mathbf{q}) \\ \varepsilon^{(1)}_{1,2}(\mathbf{q}) & \varepsilon^{(2)}_{1,2}(\mathbf{q}) & \dots & \varepsilon^{(3\tau)}_{1,2}(\mathbf{q}) \\ \varepsilon^{(1)}_{1,3}(\mathbf{q}) & \varepsilon^{(2)}_{1,3}(\mathbf{q}) & \dots & \varepsilon^{(3\tau)}_{1,3}(\mathbf{q}) \\ \vdots & \vdots & \ddots & \vdots \\ \varepsilon^{(1)}_{\tau,1}(\mathbf{q}) & \varepsilon^{(2)}_{\tau,1}(\mathbf{q}) & \dots & \varepsilon^{(3\tau)}_{\tau,1}(\mathbf{q}) \\ \varepsilon^{(1)}_{\tau,2}(\mathbf{q}) & \varepsilon^{(2)}_{\tau,2}(\mathbf{q}) & \dots & \varepsilon^{(3\tau)}_{\tau,2}(\mathbf{q}) \\ \varepsilon^{(1)}_{\tau,3}(\mathbf{q}) & \varepsilon^{(2)}_{\tau,3}(\mathbf{q}) & \dots & \varepsilon^{(3\tau)}_{\tau,3}(\mathbf{q}) \end{pmatrix}. \tag{4.65}$$

Each column of $P(\mathbf{q})$ consists of 3τ components, where the index s spans both the atomic species μ and the Cartesian direction i. We group the three scalar components $\varepsilon^{(s)}_{\mu,1}(\mathbf{q})$, $\varepsilon^{(s)}_{\mu,2}(\mathbf{q})$, and $\varepsilon^{(s)}_{\mu,3}(\mathbf{q})$, corresponding to the three Cartesian directions associated with atom μ, into a single three–component vector $\boldsymbol{\varepsilon}^{(s)}_{\mu}(\mathbf{q})$. Then, the displacement of the atom of species μ in the unit cell located at $\mathbf{n}$ is given by

$$\begin{aligned} \mathbf{S}_{\mathbf{n},\mu} &= \sqrt{\frac{M}{M_\mu}} \sum_{\mathbf{q}} \sum_{s=1}^{3\tau} \frac{e^{i\mathbf{q}\cdot\mathbf{R_n}}}{\sqrt{N}}\, \boldsymbol{\varepsilon}^{(s)}_{\mu}(\mathbf{q})\, Q_{\mathbf{q},s} = \\ &= \sqrt{\frac{M}{M_\mu}} \sum_{\mathbf{q}} \sum_{s=1}^{3\tau} \frac{e^{i\mathbf{q}\cdot\mathbf{R_n}}}{\sqrt{N}}\, \boldsymbol{\varepsilon}^{(s)}_{\mu}(\mathbf{q}) \left(A_{\mathbf{q},s} e^{i\omega_{\mathbf{q},s} t} + B_{\mathbf{q},s} e^{-i\omega_{\mathbf{q},s} t} \right). \end{aligned} \tag{4.66}$$

Due to this structure, where the factor $e^{i\mathbf{q}\cdot\mathbf{R_n}}$ encodes time oscillations and $Q_{\mathbf{q},s}$ describes temporal oscillations, the displacement $\mathbf{S}_{\mathbf{n},\mu}(t)$ results as a linear combination of forward and backward propagating plane waves. Each term in the sum corresponds to a mode characterized by a wavevector $\mathbf{q}$, a polarization vector

$\boldsymbol{\varepsilon}_{\mu}^{(s)}(\mathbf{q})$, and a time-dependent amplitude $Q_{\mathbf{q},s}$. The polarization vector $\boldsymbol{\varepsilon}_{\mu}^{(s)}(\mathbf{q})$ defines the direction and relative amplitude of the displacement of atom μ in the unit cell for the s-th normal mode with wavevector $\mathbf{q}$. The atomic displacements can thus be expressed as superpositions of these normal modes, each characterized by a well-defined wavevector and polarization, forming a complete basis for the lattice vibrational motion. From

$$\begin{aligned}\mathcal{L} &\equiv \mathcal{L}\left(Q_{\mathbf{q},s}, \dot{Q}_{\mathbf{q},s}\right) = \\ &= \frac{1}{2} M \sum_{\mathbf{q}} \sum_{s=1}^{3\tau} |\dot{Q}_{\mathbf{q},s}|^2 - \frac{1}{2} \sum_{\mathbf{q}} \sum_{s=1}^{3\tau} [K(\mathbf{q})]_{s,s} \, |Q_{\mathbf{q},s}|^2, \end{aligned} \tag{4.67}$$

we compute the related Hamiltonian, that is,

$$\mathcal{H} = \sum_h P_h \dot{Q}_h - \mathcal{L}\left(Q_h, \dot{Q}_h(P_h)\right), \tag{4.68}$$

where $h \equiv (\mathbf{q}, s)$ and P_h is the momentum and is defined as

$$P_h = \frac{\partial \mathcal{L}}{\partial \dot{Q}_h^*}. \tag{4.69}$$

Here, P_h is the canonical momentum conjugate to Q_h^*. We have

$$\begin{aligned} P_{\mathbf{q},s} &= \frac{\partial \mathcal{L}}{\partial \dot{Q}_{\mathbf{q},s}^*} = \\ &= M \dot{Q}_{\mathbf{q},s}, \end{aligned} \tag{4.70}$$

then

$$\dot{Q}_{\mathbf{q},s} = \frac{P_{\mathbf{q},s}}{M}, \tag{4.71}$$

consequently $P_{\mathbf{q},s}$ fulfill Eq. 4.56, in the same way as the $\dot{Q}_{\mathbf{q},s}$. The variables $P_{\mathbf{q},s}$ form a set of $3\tau N$ independent momenta, and the object

$$\mathcal{H} = \frac{1}{2M} \sum_{\mathbf{q}} \sum_{s=1}^{3\tau} |P_{\mathbf{q},s}|^2 + \sum_{\mathbf{q}} \sum_{s=1}^{3\tau} \frac{1}{2} [K(\mathbf{q})]_{s,s} \, |Q_{\mathbf{q},s}|^2 \tag{4.72}$$

is the Hamiltonian of $3\tau N$ independent simple harmonic motions.

4.1.3 Quantum Formalism

In quantum mechanics, the objects $Q_{\mathbf{q},s}$ and $P_{\mathbf{q},s}$ are promoted to operators, so the constraints 4.56 imply

$$\hat{Q}^{\dagger}_{\mathbf{q},s} = \hat{Q}_{-\mathbf{q},s}, \tag{4.73}$$

$$\hat{P}^{\dagger}_{\mathbf{q},s} = \hat{P}_{-\mathbf{q},s}, \tag{4.74}$$

Since in the classical formulation the fundamental canonical variables are $Q^{*}_{\mathbf{q},s}$ and $P_{\mathbf{q},s}$ (see Eq. 4.70), upon quantization we postulate the canonical commutation relations

$$\left[\hat{Q}^{\dagger}_{\mathbf{q},s}, \hat{P}_{\mathbf{q}',s'}\right] = i\hbar\, \delta_{\mathbf{q},\mathbf{q}'}\, \delta_{s,s'}, \ \forall\, \mathbf{q}, \mathbf{q}', s, s'. \tag{4.75}$$

From Eq. 4.75 and using the hermiticity conditions above, one immediately obtains

$$\left[\hat{Q}_{\mathbf{q},s}, \hat{P}_{\mathbf{q}',s'}\right] = i\hbar\, \delta_{\mathbf{q},-\mathbf{q}'}\, \delta_{s,s'}, \ \forall\, \mathbf{q}, \mathbf{q}', s, s', \tag{4.76}$$

which coincides with the standard commutation relation for the phonon normal-mode operators. In addition, the following commutation relations hold:

$$\left[\hat{Q}_{\mathbf{q},s}, \hat{Q}_{\mathbf{q}',s'}\right] = 0, \ \forall\, \mathbf{q}, \mathbf{q}', s, s', \tag{4.77}$$

$$\left[\hat{P}_{\mathbf{q},s}, \hat{P}_{\mathbf{q}',s'}\right] = 0, \ \forall\, \mathbf{q}, \mathbf{q}', s, s'. \tag{4.78}$$

Starting from the operators $\hat{Q}_{\mathbf{q},s}$ and $\hat{P}_{\mathbf{q},s}$, we define

$$\hat{Q}_{\mathbf{q},s} = \sqrt{\frac{\hbar}{2M\omega_{\mathbf{q},s}}}\left(a_{\mathbf{q},s} + a^{\dagger}_{-\mathbf{q},s}\right), \tag{4.79}$$

$$\hat{P}_{\mathbf{q},s} = i\sqrt{\frac{M\hbar\omega_{\mathbf{q},s}}{2}}\left(a^{\dagger}_{\mathbf{q},s} - a_{-\mathbf{q},s}\right), \tag{4.80}$$

whose adjoint are given by

$$\begin{aligned}\hat{Q}^{\dagger}_{\mathbf{q},s} &= \hat{Q}_{-\mathbf{q},s} = \\ &= \sqrt{\frac{\hbar}{2M\omega_{\mathbf{q},s}}}\left(a^{\dagger}_{\mathbf{q},s} + a_{-\mathbf{q},s}\right),\end{aligned} \tag{4.81}$$

$$\hat{P}^{\dagger}_{\mathbf{q},s} = \hat{P}_{-\mathbf{q},s} =$$
$$= -i\sqrt{\frac{M\hbar\omega_{\mathbf{q},s}}{2}}\left(a_{\mathbf{q},s} - a^{\dagger}_{-\mathbf{q},s}\right). \tag{4.82}$$

We write

$$a_{\mathbf{q},s} + a^{\dagger}_{-\mathbf{q},s} = \sqrt{\frac{2M\omega_{\mathbf{q},s}}{\hbar}}\hat{Q}_{\mathbf{q},s}, \tag{4.83}$$

$$a^{\dagger}_{\mathbf{q},s} - a_{-\mathbf{q},s} = -i\sqrt{\frac{2}{M\hbar\omega_{\mathbf{q},s}}}\hat{P}_{\mathbf{q},s}, \tag{4.84}$$

$$a^{\dagger}_{\mathbf{q},s} + a_{-\mathbf{q},s} = \sqrt{\frac{2M\omega_{\mathbf{q},s}}{\hbar}}\hat{Q}_{-\mathbf{q},s}, \tag{4.85}$$

$$a_{\mathbf{q},s} - a^{\dagger}_{-\mathbf{q},s} = i\sqrt{\frac{2}{M\hbar\omega_{\mathbf{q},s}}}\hat{P}_{-\mathbf{q},s}. \tag{4.86}$$

We sum Eqs. 4.83 and 4.86,

$$a_{\mathbf{q},s} = \frac{1}{2}\sqrt{\frac{2M\omega_{\mathbf{q},s}}{\hbar}}\hat{Q}_{\mathbf{q},s} + \frac{i}{2}\sqrt{\frac{2}{M\hbar\omega_{\mathbf{q},s}}}\hat{P}_{-\mathbf{q},s}, \tag{4.87}$$

and finally

$$a_{\mathbf{q},s} = \sqrt{\frac{M\omega_{\mathbf{q},s}}{2\hbar}}\hat{Q}_{\mathbf{q},s} + i\sqrt{\frac{1}{2M\hbar\omega_{\mathbf{q},s}}}\hat{P}_{-\mathbf{q},s}. \tag{4.88}$$

Then, we sum Eqs. 4.85 and 4.84,

$$a^{\dagger}_{\mathbf{q},s} = \frac{1}{2}\sqrt{\frac{2M\omega_{\mathbf{q},s}}{\hbar}}\hat{Q}_{-\mathbf{q},s} - \frac{i}{2}\sqrt{\frac{2}{M\hbar\omega_{\mathbf{q},s}}}\hat{P}_{\mathbf{q},s} \tag{4.89}$$

and finally

$$a^{\dagger}_{\mathbf{q},s} = \sqrt{\frac{M\omega_{\mathbf{q},s}}{2\hbar}}\hat{Q}_{-\mathbf{q},s} - i\sqrt{\frac{1}{2M\hbar\omega_{\mathbf{q},s}}}\hat{P}_{\mathbf{q},s}. \tag{4.90}$$

The operators $a_{\mathbf{q},s}$ and $a^{\dagger}_{\mathbf{q},s}$ are not adjoint operators of each other. In fact,

$$a^{\dagger}_{\mathbf{q},s} = a_{-\mathbf{q},s}, \ \forall s, \tag{4.91}$$

then, the Hermitian conjugation always involves a change of sign of the wave vector. Moreover, in analogy with Theorem 1.4, we prove

Theorem 4.2 *The operators $a_{\mathbf{q},s}$ and $a^{\dagger}_{\mathbf{q},s}$, that is, Eqs. 4.88 and 4.90, satisfy the bosonic commutation relations (see Eqs. D.1–D.3):*

$$\left[a_{\mathbf{q},s}, a^{\dagger}_{\mathbf{q}',s'}\right] = \delta_{\mathbf{q},-\mathbf{q}'}\delta_{s,s'}, \tag{4.92}$$

$$\left[a_{\mathbf{q},s}, a_{\mathbf{q}',s'}\right] = 0, \ \forall\, \mathbf{q}, \mathbf{q}', s, s', \tag{4.93}$$

$$\left[a^{\dagger}_{\mathbf{q},s}, a^{\dagger}_{\mathbf{q}',s'}\right] = 0, \ \forall\, \mathbf{q}, \mathbf{q}', s, s'. \tag{4.94}$$

Proof Consider the commutator

$$\begin{aligned}
\left[a_{\mathbf{q},s}, a^{\dagger}_{\mathbf{q}',s'}\right] = & \sqrt{\frac{M\omega_{\mathbf{q},s}}{2\hbar}}\sqrt{\frac{M\omega_{\mathbf{q}',s'}}{2\hbar}}\left[\hat{Q}_{\mathbf{q},s}, \hat{Q}_{-\mathbf{q}',s'}\right] + \\
& + i\sqrt{\frac{1}{2M\hbar\omega_{\mathbf{q},s}}}\sqrt{\frac{M\omega_{\mathbf{q}',s'}}{2\hbar}}\left[\hat{P}_{-\mathbf{q},s}, \hat{Q}_{-\mathbf{q}',s'}\right] + \\
& - i\sqrt{\frac{M\omega_{\mathbf{q},s}}{2\hbar}}\sqrt{\frac{1}{2M\hbar\omega_{\mathbf{q}',s'}}}\left[\hat{Q}_{\mathbf{q},s}, \hat{P}_{\mathbf{q}',s'}\right] + \\
& + i(-i)\sqrt{\frac{1}{2M\hbar\omega_{\mathbf{q},s}}}\sqrt{\frac{1}{2M\hbar\omega_{\mathbf{q}',s'}}}\left[\hat{P}_{-\mathbf{q},s}, \hat{P}_{\mathbf{q}',s'}\right].
\end{aligned} \tag{4.95}$$

From 4.77, 4.78 and 4.76, one finds

$$\begin{aligned}
\left[a_{\mathbf{q},s}, a^{\dagger}_{\mathbf{q}',s'}\right] = & i\sqrt{\frac{M\omega_{\mathbf{q},s}}{2\hbar}}\sqrt{\frac{1}{2M\hbar\omega_{\mathbf{q}',s'}}}(-i)\hbar\,\delta_{\mathbf{q},-\mathbf{q}'}\delta_{s,s'} + \\
& - i\sqrt{\frac{M\omega_{\mathbf{q},s}}{2\hbar}}\sqrt{\frac{1}{2M\hbar\omega_{\mathbf{q}',s'}}}i\hbar\,\delta_{\mathbf{q},-\mathbf{q}'}\delta_{s,s'} = \\
= & \sqrt{\frac{\omega_{\mathbf{q},s}}{\omega_{\mathbf{q}',s'}}}\delta_{\mathbf{q},-\mathbf{q}'}\delta_{s,s'}.
\end{aligned} \tag{4.96}$$

Since the Kronecker delta $\delta_{\mathbf{q},-\mathbf{q}'}$ enforces $\mathbf{q} = -\mathbf{q}'$, and using 4.61, we obtain $\omega_{\mathbf{q},s} = \omega_{\mathbf{q}',s'}$. Therefore the prefactor

$$\sqrt{\frac{\omega_{\mathbf{q},s}}{\omega_{\mathbf{q}',s'}}} \tag{4.97}$$

equals 1 whenever the delta is nonzero, then Eq. 4.92 follows.

Now, consider the commutator

$$\begin{aligned}\left[a_{\mathbf{q},s}, a_{\mathbf{q}',s'}\right] = &\sqrt{\frac{M\omega_{\mathbf{q},s}}{2\hbar}}\sqrt{\frac{M\omega_{\mathbf{q}',s'}}{2\hbar}}[\hat{Q}_{\mathbf{q},s}, \hat{Q}_{\mathbf{q}',s'}]+\\ &+i\sqrt{\frac{M\omega_{\mathbf{q},s}}{2\hbar}}\sqrt{\frac{1}{2M\hbar\omega_{\mathbf{q}',s'}}}[\hat{Q}_{\mathbf{q},s}, \hat{P}_{-\mathbf{q}',s'}]+\\ &+i\sqrt{\frac{1}{2M\hbar\omega_{\mathbf{q},s}}}\sqrt{\frac{M\omega_{\mathbf{q}',s'}}{2\hbar}}[\hat{P}_{-\mathbf{q},s}, \hat{Q}_{\mathbf{q}',s'}]+\\ &+ii\sqrt{\frac{1}{2M\hbar\omega_{\mathbf{q},s}}}\sqrt{\frac{1}{2M\hbar\omega_{\mathbf{q}',s'}}}[\hat{P}_{-\mathbf{q},s}, \hat{P}_{-\mathbf{q}',s'}].\end{aligned} \tag{4.98}$$

From 4.77, 4.78 and 4.76, one finds

$$\begin{aligned}\left[a_{\mathbf{q},s}, a_{\mathbf{q}',s'}\right] = &i(i)\sqrt{\frac{M\omega_{\mathbf{q},s}}{2\hbar}}\sqrt{\frac{1}{2M\hbar\omega_{\mathbf{q}',s'}}}\hbar\delta_{\mathbf{q},-\mathbf{q}'}\delta_{s,s'}+\\ &+i(-i)\sqrt{\frac{1}{2M\hbar\omega_{\mathbf{q},s}}}\sqrt{\frac{M\omega_{\mathbf{q}',s'}}{2\hbar}}\hbar\delta_{\mathbf{q}',-\mathbf{q}}\delta_{s,s'} =\\ = &-\sqrt{\frac{\omega_{\mathbf{q},s}}{\omega_{\mathbf{q}',s'}}}\delta_{\mathbf{q},-\mathbf{q}'}\delta_{s,s'} + \sqrt{\frac{\omega_{\mathbf{q},s}}{\omega_{\mathbf{q}',s'}}}\delta_{\mathbf{q},-\mathbf{q}'}\delta_{s,s'},\end{aligned} \tag{4.99}$$

hence Eq. 4.93 is verified.

Ultimately, consider the commutator

$$\begin{aligned}\left[a^{\dagger}_{\mathbf{q},s}, a^{\dagger}_{\mathbf{q}',s'}\right] = &\sqrt{\frac{M\omega_{\mathbf{q},s}}{2\hbar}}\sqrt{\frac{M\omega_{\mathbf{q}',s'}}{2\hbar}}[\hat{Q}_{-\mathbf{q},s}, \hat{Q}_{-\mathbf{q}',s'}]+\\ &-i\sqrt{\frac{M\omega_{\mathbf{q},s}}{2\hbar}}\sqrt{\frac{1}{2M\hbar\omega_{\mathbf{q}',s'}}}[\hat{Q}_{-\mathbf{q},s}, \hat{P}_{\mathbf{q}',s'}]+\\ &-i\sqrt{\frac{1}{2M\hbar\omega_{\mathbf{q},s}}}\sqrt{\frac{M\omega_{\mathbf{q}',s'}}{2\hbar}}[\hat{P}_{\mathbf{q},s}, \hat{Q}_{-\mathbf{q}',s'}]+\end{aligned}$$

$$+(-i)(-i)\sqrt{\frac{1}{2M\hbar\omega_{\mathbf{q},s}}}\sqrt{\frac{1}{2M\hbar\omega_{\mathbf{q}',s'}}}[\hat{P}_{\mathbf{q},s},\hat{P}_{\mathbf{q}',s'}]. \tag{4.100}$$

From 4.77, 4.78 and 4.76, one finds

$$\begin{aligned}\left[a^{\dagger}_{\mathbf{q},s},a^{\dagger}_{\mathbf{q}',s'}\right] &= -i(i)\sqrt{\frac{M\omega_{\mathbf{q},s}}{2\hbar}}\sqrt{\frac{1}{2M\hbar\omega_{\mathbf{q}',s'}}}\hbar\delta_{\mathbf{q}',-\mathbf{q}}\delta_{s,s'}+\\ &\quad -i(-i)\sqrt{\frac{1}{2M\hbar\omega_{\mathbf{q},s}}}\sqrt{\frac{M\omega_{\mathbf{q}',s'}}{2\hbar}}\hbar\delta_{\mathbf{q},-\mathbf{q}'}\delta_{s,s'} = \\ &= \sqrt{\frac{\omega_{\mathbf{q},s}}{\omega_{\mathbf{q}',s'}}}\delta_{\mathbf{q},-\mathbf{q}'}\delta_{s,s'} - \sqrt{\frac{\omega_{\mathbf{q},s}}{\omega_{\mathbf{q}',s'}}}\delta_{\mathbf{q},-\mathbf{q}'}\delta_{s,s'},\end{aligned} \tag{4.101}$$

hence, we have the 4.94. □

Thanks to Theorem 4.2 and to the definitions of the operators $a_{\mathbf{q},s}$ and $a^{\dagger}_{\mathbf{q},s}$ (Eqs. 4.88 and 4.90), from their algebra we can rigorously derive that $a_{\mathbf{q},s}$ and $a^{\dagger}_{\mathbf{q},s}$ act as annihilation and creation operators, respectively. By means of these operators, we can now express the Hamiltonian of the lattice vibrations in a fully quantum-mechanical form, i.e., in the second quantization formalism.

Theorem 4.3 (Quantized Hamiltonian of Lattice Vibrations) *The Hamiltonian 4.72 is written as a function of the operators $a_{\mathbf{q},s}$ and $a^{\dagger}_{\mathbf{q},s}$ (see Eqs. 4.88 and 4.90) as*

$$\hat{\mathcal{H}} = \sum_{\mathbf{q}}\sum_{s=1}^{3\tau}\hbar\omega_{\mathbf{q},s}\left(a^{\dagger}_{\mathbf{q},s}a_{\mathbf{q},s}+\frac{1}{2}\right), \tag{4.102}$$

where the summation over $\mathbf{q}$ *is restricted to the first Brillouin zone.*

Proof We start from the Hamiltonian given in Eq. 4.72 and we replace the classical variables with their operator counterparts, obtaining

$$\hat{\mathcal{H}} = \frac{1}{2M}\sum_{\mathbf{q}}\sum_{s=1}^{3\tau}\hat{P}^{\dagger}_{\mathbf{q},s}\hat{P}_{\mathbf{q},s} + \frac{1}{2}\sum_{\mathbf{q}}\sum_{s=1}^{3\tau}M\omega^2_{\mathbf{q},s}\,\hat{Q}^{\dagger}_{\mathbf{q},s}\hat{Q}_{\mathbf{q},s}. \tag{4.103}$$

Using Eqs. 4.79 and 4.80, we compute separately the two contributions, that is,

$$\begin{aligned}\frac{1}{2M}\hat{P}^{\dagger}_{\mathbf{q},s}\hat{P}_{\mathbf{q},s} &= \frac{1}{2M}\frac{M\hbar\omega_{\mathbf{q},s}}{2}\left(a_{\mathbf{q},s}-a^{\dagger}_{-\mathbf{q},s}\right)\left(a^{\dagger}_{\mathbf{q},s}-a_{-\mathbf{q},s}\right) = \\ &= \frac{\hbar\omega_{\mathbf{q},s}}{4}\left(a_{\mathbf{q},s}a^{\dagger}_{\mathbf{q},s}-a_{\mathbf{q},s}a_{-\mathbf{q},s}-a^{\dagger}_{-\mathbf{q},s}a^{\dagger}_{\mathbf{q},s}+a^{\dagger}_{-\mathbf{q},s}a_{-\mathbf{q},s}\right),\end{aligned} \tag{4.104}$$

$$\frac{M}{2}\omega_{\mathbf{q},s}^2\,\hat{Q}_{\mathbf{q},s}^\dagger\,\hat{Q}_{\mathbf{q},s}=\frac{1}{2}M\omega_{\mathbf{q},s}^2\frac{\hbar}{2M\omega_{\mathbf{q},s}}\left(a_{\mathbf{q},s}^\dagger+a_{-\mathbf{q},s}\right)\left(a_{\mathbf{q},s}+a_{-\mathbf{q},s}^\dagger\right)=$$
$$=\frac{\hbar\omega_{\mathbf{q},s}}{4}\left(a_{\mathbf{q},s}^\dagger a_{-\mathbf{q},s}+a_{\mathbf{q},s}^\dagger a_{-\mathbf{q},s}^\dagger+a_{-\mathbf{q},s}a_{\mathbf{q},s}+a_{-\mathbf{q},s}a_{-\mathbf{q},s}^\dagger\right). \tag{4.105}$$

By adding the two contributions, and recalling Eqs. 4.93 and 4.94, we get

$$\frac{1}{2M}\hat{P}_{\mathbf{q},s}^\dagger\,\hat{P}_{\mathbf{q},s}+\frac{M}{2}\omega_{\mathbf{q},s}^2\,\hat{Q}_{\mathbf{q},s}^\dagger\,\hat{Q}_{\mathbf{q},s}$$
$$=\frac{\hbar\omega_{\mathbf{q},s}}{4}\left(a_{\mathbf{q},s}a_{\mathbf{q},s}^\dagger+a_{-\mathbf{q},s}^\dagger a_{-\mathbf{q},s}+a_{\mathbf{q},s}^\dagger a_{-\mathbf{q},s}+a_{-\mathbf{q},s}a_{-\mathbf{q},s}^\dagger\right). \tag{4.106}$$

Now, consider the generic contribution to the Hamiltonian associated with the pair of wavevector $\pm\,\mathbf{q}$, with $\mathbf{q}\neq\mathbf{0}$. To simplify the notation, we omit the index s (the following steps are understood to hold at fixed s). Under these assumptions, we have

$$\hat{\mathcal{H}}_{\mathbf{q}}=\frac{\hbar\omega_{\mathbf{q}}}{4}\left(a_{\mathbf{q}}a_{\mathbf{q}}^\dagger+a_{-\mathbf{q}}^\dagger a_{-\mathbf{q}}+a_{\mathbf{q}}^\dagger a_{-\mathbf{q}}+a_{-\mathbf{q}}a_{-\mathbf{q}}^\dagger\right). \tag{4.107}$$

From Eq. 4.92 with $\mathbf{q}=\mathbf{q}'$, we get

$$a_{\mathbf{q}}a_{\mathbf{q}}^\dagger=a_{\mathbf{q}}^\dagger a_{\mathbf{q}}, \tag{4.108}$$

$$a_{-\mathbf{q}}a_{\mathbf{q}}^\dagger=a_{\mathbf{q}}^\dagger a_{-\mathbf{q}}+1, \tag{4.109}$$

then

$$\hat{\mathcal{H}}_{\mathbf{q}}=\frac{\hbar\omega_{\mathbf{q}}}{4}\left(a_{\mathbf{q}}^\dagger a_{\mathbf{q}}+a_{-\mathbf{q}}^\dagger a_{-\mathbf{q}}+2\,a_{\mathbf{q}}^\dagger a_{-\mathbf{q}}+1\right). \tag{4.110}$$

By symmetry, upon exchanging $\mathbf{q}\leftarrow-\mathbf{q}$,

$$\hat{\mathcal{H}}_{-\mathbf{q}}=\frac{\hbar\omega_{\mathbf{q}}}{4}\left(a_{\mathbf{q}}^\dagger a_{\mathbf{q}}+a_{-\mathbf{q}}^\dagger a_{-\mathbf{q}}+2\,a_{-\mathbf{q}}^\dagger a_{\mathbf{q}}+1\right). \tag{4.111}$$

The sum of the pair of wavevector $\pm\,\mathbf{q}$ is therefore

$$\hat{\mathcal{H}}_{\mathbf{q}}+\hat{\mathcal{H}}_{-\mathbf{q}}=\frac{\hbar\omega_{\mathbf{q}}}{2}\left(a_{\mathbf{q}}^\dagger a_{\mathbf{q}}+a_{-\mathbf{q}}^\dagger a_{-\mathbf{q}}+a_{\mathbf{q}}^\dagger a_{-\mathbf{q}}+a_{-\mathbf{q}}^\dagger a_{\mathbf{q}}+1\right). \tag{4.112}$$

We now introduce a new set of operators as symmetric and antisymmetric linear combinations of $a_{\mathbf{q}}$ and $a_{-\mathbf{q}}$. This is simply a change of basis, i.e., a unitary linear combination, which makes explicit the symmetry between $\mathbf{q}$ and $-\,\mathbf{q}$ and allows us to express the Hamiltonian in terms of independent operators. In particular, we define

$$b_{\mathbf{q}}^{(+)} = \frac{a_{\mathbf{q}} + a_{-\mathbf{q}}}{\sqrt{2}}, \;\; \mathbf{q} \neq \mathbf{0}, \tag{4.113}$$

$$b_{\mathbf{q}}^{(-)} = -i\frac{a_{\mathbf{q}} - a_{-\mathbf{q}}}{\sqrt{2}}, \;\; \mathbf{q} \neq \mathbf{0}, \tag{4.114}$$

which satisfy

$$\left[b_{\mathbf{q}}^{(\pm)}, b_{\mathbf{q}'}^{(\pm)\dagger}\right] = \delta_{\mathbf{q},\mathbf{q}'}, \tag{4.115}$$

$$\left[b_{\mathbf{q}}^{(+)}, b_{\mathbf{q}'}^{(-)\dagger}\right] = 0. \tag{4.116}$$

Indeed, from Theorem 4.2, one finds

$$\begin{aligned}
\left[b_{\mathbf{q}}^{(+)}, b_{\mathbf{q}'}^{(+)\dagger}\right] &= \frac{1}{2}\left[a_{\mathbf{q}} + a_{-\mathbf{q}}, a_{\mathbf{q}'}^{\dagger} + a_{-\mathbf{q}'}^{\dagger}\right] = \\
&= \frac{1}{2}\left(\left[a_{\mathbf{q}}, a_{\mathbf{q}'}^{\dagger}\right] + \left[a_{\mathbf{q}}, a_{-\mathbf{q}'}^{\dagger}\right] + \left[a_{-\mathbf{q}}, a_{\mathbf{q}'}^{\dagger}\right] + \left[a_{-\mathbf{q}}, a_{-\mathbf{q}'}^{\dagger}\right]\right) = \\
&= \frac{1}{2}\left(\delta_{\mathbf{q},-\mathbf{q}'} + \delta_{\mathbf{q},\mathbf{q}'} + \delta_{\mathbf{q},\mathbf{q}'} + \delta_{\mathbf{q},-\mathbf{q}'}\right) = \\
&= \delta_{\mathbf{q},\mathbf{q}'},
\end{aligned} \tag{4.117}$$

$$\begin{aligned}
\left[b_{\mathbf{q}}^{(-)}, b_{\mathbf{q}'}^{(-)\dagger}\right] &= -\frac{1}{2}\left[a_{\mathbf{q}} - a_{-\mathbf{q}}, a_{\mathbf{q}'}^{\dagger} - a_{-\mathbf{q}'}^{\dagger}\right] = \\
&= -\frac{1}{2}\left(\left[a_{\mathbf{q}}, a_{\mathbf{q}'}^{\dagger}\right] - \left[a_{\mathbf{q}}, a_{-\mathbf{q}'}^{\dagger}\right] - \left[a_{-\mathbf{q}}, a_{\mathbf{q}'}^{\dagger}\right] + \left[a_{-\mathbf{q}}, a_{-\mathbf{q}'}^{\dagger}\right]\right) = \\
&= -\frac{1}{2}\left(\delta_{\mathbf{q},-\mathbf{q}'} - \delta_{\mathbf{q},\mathbf{q}'} - \delta_{\mathbf{q},\mathbf{q}'} + \delta_{\mathbf{q},\mathbf{q}'}\right) = \\
&= \delta_{\mathbf{q},\mathbf{q}'}.
\end{aligned} \tag{4.118}$$

In particular, one finds

$$\begin{aligned}
b_{\mathbf{q}}^{(+)\dagger} b_{\mathbf{q}}^{(+)} + b_{\mathbf{q}}^{(-)\dagger} b_{\mathbf{q}}^{(-)} &= \frac{1}{2}(a_{\mathbf{q}}^{\dagger} + a_{-\mathbf{q}}^{\dagger})(a_{\mathbf{q}} + a_{-\mathbf{q}}) + \frac{1}{2}(a_{\mathbf{q}}^{\dagger} - a_{-\mathbf{q}}^{\dagger})(a_{\mathbf{q}} - a_{-\mathbf{q}}) = \\
&= \frac{1}{2}\big[a_{\mathbf{q}}^{\dagger}a_{\mathbf{q}} + a_{\mathbf{q}}^{\dagger}a_{-\mathbf{q}} + a_{-\mathbf{q}}^{\dagger}a_{\mathbf{q}} + a_{-\mathbf{q}}^{\dagger}a_{-\mathbf{q}} + \\
&\quad + a_{\mathbf{q}}^{\dagger}a_{\mathbf{q}} - a_{\mathbf{q}}^{\dagger}a_{-\mathbf{q}} - a_{-\mathbf{q}}^{\dagger}a_{\mathbf{q}} + a_{-\mathbf{q}}^{\dagger}a_{-\mathbf{q}}\big] = \\
&= a_{\mathbf{q}}^{\dagger}a_{\mathbf{q}} + a_{-\mathbf{q}}^{\dagger}a_{-\mathbf{q}},
\end{aligned} \tag{4.119}$$

$$\begin{aligned} b_{\mathbf{q}}^{(+)\dagger} b_{\mathbf{q}}^{(+)} - b_{\mathbf{q}}^{(-)\dagger} b_{\mathbf{q}}^{(-)} &= \frac{1}{2}(a_{\mathbf{q}}^{\dagger} + a_{-\mathbf{q}}^{\dagger})(a_{\mathbf{q}} + a_{-\mathbf{q}}) - \frac{1}{2}(a_{\mathbf{q}}^{\dagger} - a_{-\mathbf{q}}^{\dagger})(a_{\mathbf{q}} - a_{-\mathbf{q}}) = \\ &= \frac{1}{2}\big[a_{\mathbf{q}}^{\dagger} a_{\mathbf{q}} + a_{\mathbf{q}}^{\dagger} a_{-\mathbf{q}} + a_{-\mathbf{q}}^{\dagger} a_{\mathbf{q}} + a_{-\mathbf{q}}^{\dagger} a_{-\mathbf{q}} + \\ &\quad - a_{\mathbf{q}}^{\dagger} a_{\mathbf{q}} + a_{\mathbf{q}}^{\dagger} a_{-\mathbf{q}} + a_{-\mathbf{q}}^{\dagger} a_{\mathbf{q}} - a_{-\mathbf{q}}^{\dagger} a_{-\mathbf{q}}\big] = \\ &= a_{\mathbf{q}}^{\dagger} a_{-\mathbf{q}} + a_{-\mathbf{q}}^{\dagger} a_{\mathbf{q}}, \end{aligned} \tag{4.120}$$

which shows that the total number operator is preserved under the transformation. By summing 4.119 and 4.120, one finds

$$\hat{\mathcal{H}}_{\mathbf{q}} + \hat{\mathcal{H}}_{-\mathbf{q}} = \hbar\omega_{\mathbf{q}} \left(b_{\mathbf{q}}^{(+)\dagger} b_{\mathbf{q}}^{(+)} + b_{\mathbf{q}}^{(-)\dagger} b_{\mathbf{q}}^{(-)} + 1 \right). \tag{4.121}$$

The operators $b_{\mathbf{q}}^{(\pm)}$ describe independent oscillators with the same frequency $\omega_{\mathbf{q}}$, and the total zero-point energy associated with the pair is $\hbar\omega_{\mathbf{q}}$. Let us now return to the original creation and annihilation operators. From Eqs. 4.119 and 4.120, we obtain

$$b_{\mathbf{q}}^{(+)\dagger} b_{\mathbf{q}}^{(+)} = \frac{1}{2} \left(a_{\mathbf{q}}^{\dagger} a_{\mathbf{q}} + a_{\mathbf{q}}^{\dagger} a_{-\mathbf{q}} + a_{-\mathbf{q}}^{\dagger} a_{\mathbf{q}} + a_{-\mathbf{q}}^{\dagger} a_{-\mathbf{q}} \right), \tag{4.122}$$

$$b_{\mathbf{q}}^{(-)\dagger} b_{\mathbf{q}}^{(-)} = \frac{1}{2} \left(a_{\mathbf{q}}^{\dagger} a_{\mathbf{q}} - a_{\mathbf{q}}^{\dagger} a_{-\mathbf{q}} - a_{-\mathbf{q}}^{\dagger} a_{\mathbf{q}} + a_{-\mathbf{q}}^{\dagger} a_{-\mathbf{q}} \right). \tag{4.123}$$

By summing them, we obtain

$$b_{\mathbf{q}}^{(+)\dagger} b_{\mathbf{q}}^{(+)} + b_{\mathbf{q}}^{(-)\dagger} b_{\mathbf{q}}^{(-)} = a_{\mathbf{q}}^{\dagger} a_{\mathbf{q}} + a_{-\mathbf{q}}^{\dagger} a_{-\mathbf{q}}, \tag{4.124}$$

then

$$\hat{\mathcal{H}}_{\mathbf{q}} + \hat{\mathcal{H}}_{-\mathbf{q}} = \hbar\omega_{\mathbf{q}} \left(a_{\mathbf{q}}^{\dagger} a_{\mathbf{q}} + a_{-\mathbf{q}}^{\dagger} a_{-\mathbf{q}} + 1 \right), \quad \mathbf{q} \neq \mathbf{0}. \tag{4.125}$$

If we divide the first Brillouin zone into two complementary halves, we denote by SBZ^+ the semispace containing a single representative of each pair $\{\mathbf{q}, -\mathbf{q}\}$ with $\mathbf{q} \neq \mathbf{0}$. Analogously, SBZ^- denotes the opposite semispace, which contains the corresponding negative momenta. With this convention, one has the identity

$$\sum_{\mathbf{q}\in\mathrm{SBZ}^+} f(\mathbf{q}) + \sum_{\mathbf{q}\in\mathrm{SBZ}^-} f(\mathbf{q}) = \sum_{\mathbf{q}\in 1^{\mathrm{st}}\mathrm{BZ}} f(\mathbf{q}), \tag{4.126}$$

which will be used in the following. Therefore, the $\mathbf{q} \neq \mathbf{0}$ part of the Hamiltonian reads

$$\sum_{\mathbf{q}\in \mathrm{SBZ}^+} \left(\hat{\mathcal{H}}_{\mathbf{q}} + \hat{\mathcal{H}}_{-\mathbf{q}}\right) = \sum_{\mathbf{q}\in \mathrm{SBZ}^+} \hbar\omega_{\mathbf{q}} \left(a_{\mathbf{q}}^{\dagger} a_{\mathbf{q}} + a_{-\mathbf{q}}^{\dagger} a_{-\mathbf{q}} + 1\right). \tag{4.127}$$

We rewrite

$$a_{\mathbf{q}}^{\dagger} a_{\mathbf{q}} + \frac{1}{2} + a_{-\mathbf{q}}^{\dagger} a_{-\mathbf{q}} + \frac{1}{2}, \tag{4.128}$$

such that

$$\sum_{\mathbf{q}\in \mathrm{SBZ}^+} \left(\hat{\mathcal{H}}_{\mathbf{q}} + \hat{\mathcal{H}}_{-\mathbf{q}}\right) = \sum_{\mathbf{q}\in \mathrm{SBZ}^+} \hbar\omega_{\mathbf{q}} \left(a_{\mathbf{q}}^{\dagger} a_{\mathbf{q}} + \frac{1}{2}\right) + \sum_{\mathbf{q}\in \mathrm{SBZ}^+} \hbar\omega_{\mathbf{q}} \left(a_{-\mathbf{q}}^{\dagger} a_{-\mathbf{q}} + \frac{1}{2}\right). \tag{4.129}$$

In the second summation, we perform the change of variable $\mathbf{q} \to -\mathbf{q}$, making use of Eq. 4.61, thus

$$\sum_{\mathbf{q}\in \mathrm{SBZ}^+} \left(\hat{\mathcal{H}}_{\mathbf{q}} + \hat{\mathcal{H}}_{-\mathbf{q}}\right) = \sum_{\mathbf{q}\in \mathrm{SBZ}^+} \hbar\omega_{\mathbf{q}} \left(a_{\mathbf{q}}^{\dagger} a_{\mathbf{q}} + \frac{1}{2}\right) + \sum_{\mathbf{q}\in \mathrm{SBZ}^-} \hbar\omega_{\mathbf{q}} \left(a_{\mathbf{q}}^{\dagger} a_{\mathbf{q}} + \frac{1}{2}\right). \tag{4.130}$$

Since $\mathrm{SBZ}^+ \cup \mathrm{SBZ}^-$ covers all $\mathbf{q} \neq \mathbf{0}$, we then obtain

$$\sum_{\mathbf{q}\neq\mathbf{0}} \hat{\mathcal{H}}_{\mathbf{q}} = \sum_{\mathbf{q}\neq\mathbf{0}} \hbar\omega_{\mathbf{q}} \left(a_{\mathbf{q}}^{\dagger} a_{\mathbf{q}} + \tfrac{1}{2}\right). \tag{4.131}$$

The wavevector $\mathbf{0}$ has no opposite partner $-\mathbf{0}$ and contributes as a single harmonic oscillator

$$\hat{\mathcal{H}}_{\mathbf{0}} = \frac{\hbar\omega_{\mathbf{0}}}{2} \left(a_{\mathbf{0}}^{\dagger} a_{\mathbf{0}} + \tfrac{1}{2}\right). \tag{4.132}$$

Combining Eqs. 4.131 and 4.132, and restoring the index s, we arrive at the standard form commonly found in the literature, that is,

$$\hat{\mathcal{H}} = \sum_{\mathbf{q}\neq\mathbf{0}} \hbar\omega_{\mathbf{q}} \left(a_{\mathbf{q}}^{\dagger} a_{\mathbf{q}} + \frac{1}{2}\right) + \frac{\hbar\omega_{\mathbf{0}}}{2} \left(a_{\mathbf{0}}^{\dagger} a_{\mathbf{0}} + \frac{1}{2}\right). \tag{4.133}$$

If the wavevector $\mathbf{0}$ is absent or treated separately, the Hamiltonian is often written in the more compact form of Eq. 4.102, with the understanding that the summation runs only over the truly independent degrees of freedom (e.g., a Brillouin semispace). □

The Hamiltonian 4.102 corresponds to the second term in the solid-state Hamiltonian II.22, and it describes $3\tau N$ independent harmonic oscillators, representing the ionic vibrations within the effective potential generated by the electrons. The field $\mathbf{S}_{\mathbf{n},\mu}$ of lattice vibrations has been quantized, and lattice energy exchanges occur for integer multiples of $\hbar\omega_{\mathbf{q},s}$. The field energy quanta of lattice vibrations are named *phonons*, and the atomic displacement field is given by the operator

$$\begin{aligned}\mathbf{S}_{\mathbf{n},\mu} &= \sqrt{\frac{M}{M_\mu}} \sum_{\mathbf{q}} \sum_{s=1}^{3\tau} \frac{e^{i\mathbf{q}\cdot\mathbf{R}_\mathbf{n}}}{\sqrt{N}} \boldsymbol{\varepsilon}_\mu^{(s)}(\mathbf{q}) \hat{Q}_{\mathbf{q},s} = \\ &= \sqrt{\frac{M}{M_\mu}} \sum_{\mathbf{q}} \sum_{s=1}^{3\tau} \frac{e^{i\mathbf{q}\cdot\mathbf{R}_\mathbf{n}}}{\sqrt{N}} \boldsymbol{\varepsilon}_\mu^{(s)}(\mathbf{q}) \sqrt{\frac{\hbar}{2M\omega_{\mathbf{q},s}}} \left(a_{\mathbf{q},s} e^{-i\omega_{\mathbf{q},s}t} + a^\dagger_{-\mathbf{q},s} e^{i\omega_{\mathbf{q},s}t}\right) = \\ &= \sum_{\mathbf{q}} \sum_{s=1}^{3\tau} \frac{e^{i\mathbf{q}\cdot\mathbf{R}_\mathbf{n}}}{\sqrt{N}} \boldsymbol{\varepsilon}_\mu^{(s)}(\mathbf{q}) \sqrt{\frac{\hbar}{2M_\mu\omega_{\mathbf{q},s}}} \left(a_{\mathbf{q},s} e^{-i\omega_{\mathbf{q},s}t} + a^\dagger_{-\mathbf{q},s} e^{i\omega_{\mathbf{q},s}t}\right).\end{aligned} \tag{4.134}$$

For simplicity, we will use the same symbol to denote both the classical field and the corresponding operator.

The amplitudes of the classical oscillations, $A_{\mathbf{q},s}$ and $B_{\mathbf{q},s}$, have become operators. If we fix s, we obtain a so-called branch. For a given branch, varying the wavevector $\mathbf{q}$ produces the ionic oscillation bands of the solid. The index s labels the phonon bands. These bands are classified as acoustic or optical depending on whether the phonon frequency $\omega_{\mathbf{q},s}$ tends to zero or remains finite, respectively, as $\mathbf{q} \to \mathbf{0}$. If the polarization vector of the phonon wave, that is $\boldsymbol{\varepsilon}_\mu(\mathbf{q})$, is parallel to $\mathbf{q}$, the phonon mode is longitudinal; if perpendicular, it is transverse. Unlike photons, which are exclusively transverse electromagnetic waves, phonons can be either longitudinal or transverse elastic waves.

4.2 1D Lattice with One Atom Per Cell: Acoustic Branch

Consider a linear chain with lattice constant a and a single atom per unit cell, i.e., $\tau = 1$. We impose Born-von Karman boundary conditions, meaning $S(N) = S(0)$, so that the ends of the chain are joined, and the N atoms are arranged along a circumference. The eigenvalues $M\omega_\mathbf{q}^2$ of the matrix $\tilde{D}(\mathbf{q})$, which appear on the diagonal of $K(\mathbf{q})$, are determined by calculating the force constants $W_{\mathbf{m},\nu,k}^{\mathbf{n},\mu,i}$ (see Eq. 4.5). Since there is only one atom per cell and the lattice is one-dimensional, the indices simplify: the lattice vectors $\mathbf{n}$, $\mathbf{m}$, and wavevector $\mathbf{q}$ reduce to scalars n, m, and q, and the atomic (μ, ν) and directional (i, k) indices disappear. The force constants reduce to the form W_m^n. In the simplest model, each atom interacts only with its nearest neighbors via an elastic potential with elastic constant k_{el}, namely

$$V = \frac{k_{\text{el}}}{2} \sum_{n=0}^{N-1} (S_n - S_{n+1})^2 . \tag{4.135}$$

Remark 4.1 (Elastic Potential in a Linear Chain) The potential V given in Eq. 4.135 can be written as

$$V = \frac{k_{\mathrm{el}}}{2} \sum_{n=0}^{N-1} \left(2S_n^2 - S_n S_{n+1} - S_n S_{n-1} \right). \tag{4.136}$$

Indeed, from

$$\begin{aligned}(S_n - S_{n+1})^2 &= (S_n - S_{n+1})(S_n - S_{n+1}) = \\ &= S_n^2 - S_n S_{n+1} - S_{n+1} S_n + S_{n+1}^2, \end{aligned} \tag{4.137}$$

we now consider the last term. Since the atoms in the chain are identical simple harmonic oscillators and periodic boundary conditions are assumed, the system is translationally invariant. As a consequence, quantities associated with a given lattice site do not depend on the specific choice of the site index. In particular, nearest-neighbour products are invariant under a relabelling of the lattice sites. By performing a shift of the summation index (made valid by the periodic boundary conditions), one obtains

$$\sum_{n=0}^{N-1} S_{n+1} S_n = \sum_{n=0}^{N-1} S_n S_{n-1}. \tag{4.138}$$

From Eq. 4.136, we write

$$\begin{aligned} V &= \frac{k_{\mathrm{el}}}{2} \sum_{n=0}^{N-1} \left(2S_n^2 - S_n S_{n+1} - S_n S_{n-1} \right) = \\ &= \frac{k_{\mathrm{el}}}{2} \sum_{n=0}^{N-1} \sum_{m=0}^{N-1} \left(2S_n S_m \delta_{n,m} - S_n S_m \delta_{m,n+1} - S_n S_m \delta_{m,n-1} \right). \end{aligned} \tag{4.139}$$

Indeed,

$$V = \frac{1}{2} W_m^n S_n S_m, \tag{4.140}$$

with

$$W_m^n = k_{\mathrm{el}} \left(2\delta_{n,m} - \delta_{n,m+1} - \delta_{n,n-1} \right). \tag{4.141}$$

Let us calculate $\tilde{D}(q)$, we have

$$
\begin{aligned}
\tilde{D}(q) &= \sum_h W_0^h e^{iqha} = \\
&= 2k_{\text{el}} - k_{\text{el}}\left(e^{iqa} + e^{-iqa}\right) = \\
&= 2k_{\text{el}} - 2k_{\text{el}}\cos(qa) = \\
&= 2k_{\text{el}}\left(1 - \cos(qa)\right) \equiv \\
&\equiv k_q,
\end{aligned}
\tag{4.142}
$$

where k_q is the eigenvalue of $K(q)$, since $\tilde{D}(q)$ is a 1×1 matrix. Note that $k_q \geq 0$, as it must be. From $k_q = M\omega_q^2$, we get

$$
\begin{aligned}
\omega_q^2 &= \frac{2k_{\text{el}}}{M}\left(1 - \cos(qa)\right) = \\
&= \frac{4k_{\text{el}}}{M}\sin^2\left(\frac{qa}{2}\right),
\end{aligned}
\tag{4.143}
$$

$$
\omega_q = \sqrt{\frac{4k_{\text{el}}}{M}}\left|\sin\left(\frac{qa}{2}\right)\right|, \quad -\frac{\pi}{a} \leq q \leq \frac{\pi}{a}. \tag{4.144}
$$

In the limit $qa \to 0$, we get

$$
\omega_q \sim \sqrt{\frac{2k_{\text{el}}}{M}}|q|\frac{a}{2}. \tag{4.145}
$$

Since $|q| = \frac{2\pi}{\lambda}$, in the limit $qa \to 0$ we have $\lambda \gg a$: the lattice is seen as a continuous structure and ω_q becomes linear with respect to $|q|$. We write

$$
\begin{aligned}
\omega_q &\sim \sqrt{\frac{2k_{\text{el}}}{M}}|q|\frac{a}{2} = \\
&= V_s|q|,
\end{aligned}
\tag{4.146}
$$

where

$$
V_s = \sqrt{\frac{2k_{\text{el}}}{M}}\frac{a}{2} \tag{4.147}
$$

is a velocity, signaling the existence of a dispersion relation.

Let us consider the displacement vector $\mathbf{S}_{n,\mu}$. The vector $\boldsymbol{\varepsilon}_\mu(q)$ characterizes the relative displacement of an atom with respect to its nearest neighbors, representing longitudinal compression and rarefaction waves. This interpretation is consistent with modeling the solid as a one-dimensional ring. Sound waves propagate through the lattice with velocity V_s, corresponding to the acoustic modes. For $|q| = \frac{\pi}{a}$, the associated wavelength $\lambda = \frac{2\pi}{|q|} = 2a$ becomes comparable to the lattice constant a. In this regime, the continuum approximation fails, and the discrete nature of the lattice must be taken into account. Here, atoms oscillate out of phase with their nearest neighbors, giving rise to standing wave patterns. Conversely, at $q = 0$, all atoms oscillate in phase, corresponding to a uniform translation of the lattice. In one dimension, the spatial modulation of the displacement is captured by the phase factor e^{iqna}, which satisfies periodic boundary conditions and reflects the translational symmetry of the lattice, indeed it satisfies

$$\lim_{q \to 0} e^{iqna} = 1, \tag{4.148}$$

corresponding to in-phase waves, and

$$\lim_{|q| \to \frac{\pi}{a}} e^{iqna} = e^{in\pi}, \tag{4.149}$$

which implies an alternating displacement pattern, where nearest-neighbor atoms oscillate in opposite phases.

4.3 1D Lattice with Two Atoms Per Cell: Acoustic Branches and Optical Branches

Given a linear chain of lattice step a with two atomic species per cell, of mass M_1 and M_2, we use a model analogous to Sect. 4.2. The potential is

$$V = \frac{k_{\mathrm{el}}}{2} \sum_{n=0}^{N-1} \left(S_n^{(1)} - S_n^{(2)} \right)^2 + \frac{k_{\mathrm{el}}}{2} \sum_{n=0}^{N-1} \left(S_n^{(2)} - S_{n+1}^{(1)} \right)^2, \tag{4.150}$$

where the first and second terms refer to pairs of nearest-neighbor atoms in the same cell (intra-cell) or in different cells (inter-cell), respectively. From

$$\begin{aligned} V = \frac{k_{\mathrm{el}}}{2} \sum_{n=0}^{N-1} \Big[\left(S_n^{(1)}\right)^2 + \left(S_n^{(2)}\right)^2 - 2S_n^{(1)} S_n^{(2)} + \left(S_n^{(2)}\right)^2 + \left(S_{n+1}^{(1)}\right)^2 - 2S_n^{(2)} S_{n+1}^{(1)} \Big] = \\ = \frac{k_{\mathrm{el}}}{2} \sum_{n=0}^{N-1} \Big[\left(S_n^{(1)}\right)^2 + 2\left(S_n^{(2)}\right)^2 - 2S_n^{(1)} S_n^{(2)} + \left(S_{n+1}^{(1)}\right)^2 - 2S_n^{(2)} S_{n+1}^{(1)} \Big], \end{aligned} \tag{4.151}$$

we use

$$\sum_{n=0}^{N-1}\left(S_{n+1}^{(1)}\right)^2=\sum_{n=0}^{N-1}\left(S_n^{(1)}\right)^2, \tag{4.152}$$

$$\sum_{n=0}^{N-1}2S_n^{(2)}S_{n+1}^{(1)}=\sum_{n=0}^{N-1}\left(S_n^{(2)}S_{n+1}^{(1)}+S_n^{(1)}S_{n-1}^{(2)}\right), \tag{4.153}$$

then

$$\begin{aligned}V=&\frac{k_{\text{el}}}{2}\sum_{n=0}^{N-1}\left[\left(S_n^{(1)}\right)^2+2\left(S_n^{(2)}\right)^2-2S_n^{(1)}S_n^{(2)}+\left(S_n^{(1)}\right)^2-S_n^{(2)}S_{n+1}^{(1)}-S_n^{(1)}S_{n-1}^{(2)}\right]=\\=&\frac{k_{\text{el}}}{2}\sum_{n=0}^{N-1}\left[2\left(S_n^{(1)}\right)^2+2\left(S_n^{(2)}\right)^2-2S_n^{(1)}S_n^{(2)}-S_n^{(2)}S_{n+1}^{(1)}-S_n^{(1)}S_{n-1}^{(2)}\right].\end{aligned} \tag{4.154}$$

Here, **n** and **m** are replaced by scalars n and m, $\tau=2$, $i=k=1$ and $\mu,\nu=1,2$, then the force constants are

$$\begin{aligned}W_{0,\nu}^{n-m,\mu}\equiv W_{0,\nu}^{h,\mu}=&\\=&k_{\text{el}}\left(2\delta_{h,0}\delta_{\mu,\nu}-\delta_{h,0}\delta_{\mu,1}\delta_{\nu,2}-\delta_{h,0}\delta_{\mu,2}\delta_{\nu,1}-\delta_{h,-a}\delta_{\mu,2}\delta_{\nu,1}\right.\\&\left.-\delta_{h,a}\delta_{\mu,1}\delta_{\nu,2}\right).\end{aligned} \tag{4.155}$$

We have

$$D_{m,\nu}^{n,\mu}=\sqrt{\frac{M}{M_\mu M_\nu}}W_{m,\nu}^{n,\mu}, \tag{4.156}$$

$$\left[\tilde{D}(q)\right]_\nu^\mu=\sum_h D_{0,\nu}^{h,\mu}e^{iqR_h} \tag{4.157}$$

and $\mu,\ \nu=1,2$, then $\tilde{D}(q)$ is a 2×2 matrix given by

$$\tilde{D}(q)=\begin{pmatrix}k_{\text{el}}\dfrac{2M}{M_1} & -k_{\text{el}}\dfrac{M}{\sqrt{M_1M_2}}(1+e^{iqa})\\ -k_{\text{el}}\dfrac{M}{\sqrt{M_1M_2}}(1+e^{-iqa}) & k_{\text{el}}\dfrac{2M}{M_2}\end{pmatrix}, \tag{4.158}$$

and it is hermitian. Let us compute the eigenvalues $M\omega_q^2$ of $\tilde{D}(q)$. We divide the matrix $\tilde{D}(q)$ by M and we have

$$\det \begin{pmatrix} k_{\mathrm{el}}\dfrac{2}{M_1} - \omega^2 & -k_{\mathrm{el}}\dfrac{1}{\sqrt{M_1 M_2}}(1+e^{iqa}) \\ -k_{\mathrm{el}}\dfrac{1}{\sqrt{M_1 M_2}}(1+e^{-iqa}) & k_{\mathrm{el}}\dfrac{2}{M_2} - \omega^2 \end{pmatrix} = 0, \tag{4.159}$$

$$\left(\frac{2k_{\mathrm{el}}}{M_1} - \omega^2\right)\left(\frac{2k_{\mathrm{el}}}{M_2} - \omega^2\right) - \frac{k_{\mathrm{el}}^2}{M_1 M_2}(1+e^{iqa})(1+e^{-iqa}) = 0, \tag{4.160}$$

$$\omega^4 - \left(\frac{2k_{\mathrm{el}}}{M_1} + \frac{2k_{\mathrm{el}}}{M_2}\right)\omega^2 + \frac{k_{\mathrm{el}}^2}{M_1 M_2} - \frac{k_{\mathrm{el}}^2}{M_1 M_2}(2 + 2\cos(qa)) = 0, \tag{4.161}$$

$$\omega^4 - 2k_{\mathrm{el}}\left(\frac{1}{M_1} + \frac{1}{M_2}\right)\omega^2 + \frac{2k_{\mathrm{el}}^2}{M_1 M_2}(1 - \cos(qa)) = 0, \tag{4.162}$$

then

$$\omega_\pm^2(q) = k_{\mathrm{el}}\left[\left(\frac{1}{M_1} + \frac{1}{M_2}\right) \pm \sqrt{\left(\frac{1}{M_1} + \frac{1}{M_2}\right)^2 - \frac{2}{M_1 M_2}(1 - \cos(qa))}\,\right]. \tag{4.163}$$

From Eq. 4.163, it follows

$$\omega_-^2(q = 0) = 0, \tag{4.164}$$

$$\omega_+^2(q = 0) = 2k_{\mathrm{el}}\frac{(M_1 + M_2)}{M_1 M_2}, \tag{4.165}$$

$$\omega_\pm^2\left(q = \frac{\pi}{a}\right) = k_{\mathrm{el}}\left[\left(\frac{1}{M_1} + \frac{1}{M_2}\right) \pm \sqrt{\left(\frac{1}{M_1} + \frac{1}{M_2}\right)^2 - \frac{2}{M_1 M_2}(1 - \cos(\pi))}\,\right] = \\ = k_{\mathrm{el}}\left[\left(\frac{1}{M_1} + \frac{1}{M_2}\right) \pm \frac{|M_2 - M_1|}{M_1 M_2}\right]. \tag{4.166}$$

For instance, assuming $M_2 > M_1$, then $\omega_+^2 = \frac{2k_{\mathrm{el}}}{M_1}$, $\omega_-^2 = \frac{2k_{\mathrm{el}}}{M_2}$.

Now, we rewrite Eq. 4.163 in an equivalent form that is more convenient for analyzing the small-momentum limit. We set

$$x = qa, \tag{4.167}$$

$$A = \frac{1}{M_1} + \frac{1}{M_2}, \tag{4.168}$$

$$B = \frac{1}{M_1 M_2}, \tag{4.169}$$

then

$$\omega_{\pm}^2(x) = k_{\rm el}\left(A \pm \sqrt{A^2 - 2B(1 - \cos(x))}\right). \tag{4.170}$$

In the limit $x \to 0$, one has

$$A^2 - 2B(1 - \cos x) \sim A^2 - Bx^2, \tag{4.171}$$

therefore

$$\sqrt{A^2 - Bx^2} = A\sqrt{1 - \frac{B}{A^2}x^2}. \tag{4.172}$$

From

$$\sqrt{1+u} \sim 1 + \frac{u}{2}, \qquad u \to 0, \tag{4.173}$$

with $u = -\frac{B}{A^2}x^2$, one finds

$$\sqrt{A^2 - Bx^2} \sim A\left(1 - \frac{B}{2A^2}x^2\right). \tag{4.174}$$

Then

$$\sqrt{A^2 - Bx^2} \sim A - \frac{B}{2A}x^2. \tag{4.175}$$

and, in the small-momentum limit, Eq. 4.170 reduces to

$$\omega_{\pm}^2(x)\Big|_{x\to 0} = k_{\rm el}\left[A \pm \left(A - \frac{B}{2A}x^2\right)\right]. \tag{4.176}$$

Restoring the definitions of x, A, B, one finds

$$\omega_{-}^2(qa)\Big|_{q\to 0} = \frac{k_{\rm el}}{2\,(M_1 + M_2)}(qa)^2, \tag{4.177}$$

$$\omega_{+}^2(qa)\Big|_{q\to 0} = 2k_{\rm el}\left(\frac{1}{M_1} + \frac{1}{M_2}\right) - \frac{k_{\rm el}}{2}\frac{1}{M_1 + M_2}(qa)^2, \tag{4.178}$$

which correspond to the acoustic and optical branches, respectively, in the small-momentum limit.

4.4 3D Lattice with Many Atoms Per Cell

Here, we study the general case of $\tau \geq 2$ atomic species per cell in three dimensions, specifically in a simple cubic lattice, and show that three acoustic frequencies exist. From

$$
\begin{aligned}
\left[\tilde{D}(\mathbf{q})\right]_{\substack{\mu,\nu \\ i,k}} &= \sum_{\mathbf{h}} D^{\mathbf{h},\mu,i}_{\mathbf{0},\nu,k} \, e^{i\mathbf{q}\cdot\mathbf{R}_h} = \\
&= \sum_{\mathbf{h}} \sqrt{\frac{M^2}{M_\mu M_\nu}} W^{\mathbf{h},\mu,i}_{\mathbf{0},\nu,k} \, e^{i\mathbf{q}\cdot\mathbf{R}_h},
\end{aligned} \tag{4.179}
$$

we have

$$
\begin{aligned}
\left[\tilde{D}(\mathbf{0})\right]_{\substack{\mu,\nu \\ i,k}} &= \sum_{\mathbf{h}} \sqrt{\frac{M^2}{M_\mu M_\nu}} W^{\mathbf{h},\mu,i}_{\mathbf{0},\nu,k} = \\
&= \sum_{\mathbf{h}} \sqrt{\frac{M^2}{M_\mu M_\nu}} W^{\mathbf{0},\mu,i}_{-\mathbf{h},\nu,k} = \\
&= \sum_{\mathbf{h}} \sqrt{\frac{M^2}{M_\mu M_\nu}} W^{\mathbf{0},\mu,i}_{\mathbf{h},\nu,k} .
\end{aligned} \tag{4.180}
$$

In the last step we used that $\mathbf{h}$ is a dummy index.

Theorem 4.4 *The matrix 4.180 has three eigenvalues equal to zero.*

Proof We seek vectors $\mathbf{V}$ that satisfy the eigenvalue equation, i.e.,

$$
\sum_{\nu,k} \left[\tilde{D}(\mathbf{0})\right]_{\substack{\mu,\nu \\ i,k}} V_{\nu,k} = 0, \tag{4.181}
$$

$$
\sum_{\mathbf{h},\nu,k} \sqrt{\frac{M^2}{M_\mu M_\nu}} W^{\mathbf{0},\mu,i}_{\mathbf{h},\nu,k} V_{\nu,k} = 0, \tag{4.182}
$$

$$
\sum_{\mathbf{h},\nu,k} \sqrt{\frac{1}{M_\nu}} W^{\mathbf{0},\mu,i}_{\mathbf{h},\nu,k} V_{\nu,k} = 0. \tag{4.183}
$$

We now insert the following three vectors

$$V_{\nu,k}^{(1)} = \sqrt{M_\nu}\delta_{k,1}, \tag{4.184}$$

$$V_{\nu,k}^{(2)} = \sqrt{M_\nu}\delta_{k,2}, \tag{4.185}$$

$$V_{\nu,k}^{(3)} = \sqrt{M_\nu}\delta_{k,3}, \tag{4.186}$$

into the eigenvalue equation

$$\sum_{\mathbf{h},\nu,k} \sqrt{\frac{1}{M_\nu}} W_{\mathbf{h},\nu,k}^{\mathbf{0},\mu,i} V_{\nu,k}^{(j)} = 0, \;\; j = 1, 2, 3. \tag{4.187}$$

Indeed, from 4.14, it follows that

$$\sum_{\mathbf{h},\nu,k} \sqrt{\frac{1}{M_\nu}} W_{\mathbf{h},\nu,k}^{\mathbf{0},\mu,i} \sqrt{M_\nu}\delta_{k,j} = \sum_{\mathbf{h},\nu,k} W_{\mathbf{h},\nu,k}^{\mathbf{0},\mu,i} \delta_{k,j} =$$

$$= \sum_{\mathbf{h},\nu} W_{\mathbf{h},\nu,j}^{\mathbf{0},\mu,i}\Big|_{j \text{ fixed}} = 0, \;\; j = 1, 2, 3. \tag{4.188}$$

We have shown that among the 3τ branches, at least three correspond to acoustic modes. □

It can be shown that only three branches are acoustic, one corresponding to longitudinal motion and the other two to transverse motion. Let us calculate the displacements $S_{\mathbf{n},\mu,i}^{(\mathrm{ac})}$ of these acoustic branches. By definition, the eigenvectors $V_{\nu,k}^{(j)}$ are equal to the displacements $U_{\mathbf{n},\mu,i}$, hence

$$\begin{aligned} S_{\mathbf{n},\mu,i}^{(\mathrm{ac})} &= \sqrt{\frac{M}{M_\mu}} U_{\mathbf{n},\mu,i} = \\ &= \sqrt{\frac{M}{M_\mu}} V_{\nu,k}^{(i)} = \\ &= \sqrt{\frac{M}{M_\mu}} \sqrt{M_\mu}\delta_{k,i} = \\ &= \sqrt{M}\delta_{k,i}, \;\; i = 1, 2, 3. \end{aligned} \tag{4.189}$$

In other words, the displacements of the acoustic branches depend neither on the atom nor on the cell: all atoms move in phase, and these longitudinal motions are responsible for the propagation of sound waves in the solid. Let us turn to the optical branches. The matrix $\tilde{D}(\mathbf{q})$ is Hermitian, so the eigenvectors $V_{\nu,k}^{(s)}$ corresponding to different eigenvalues are orthogonal to each other and therefore satisfy

$$\sum_{\nu,k} V_{\nu,k}^{(s)*}(\mathbf{q}) V_{\nu,k}^{(s')}(\mathbf{q}) = \delta_{s,s'}. \tag{4.190}$$

Consider an optical branch and any of the three acoustic branches, i.e.,

$$\sum_{\nu,k} V_{\nu,k}^{(o)*}(\mathbf{q}) V_{\nu,k}^{(a)}(\mathbf{q}) = 0. \tag{4.191}$$

If $\mathbf{q} = \mathbf{0}$, the eigenvectors of the acoustic branches are known. Moreover, the matrix $\tilde{D}(\mathbf{0})$ is real and symmetric, so its eigenvectors can be chosen to be real.

$$\begin{aligned} 0 = \sum_{\nu,k} V_{\nu,k}^{(o)}(\mathbf{0}) V_{\nu,k}^{(a)}(\mathbf{0}) = \\ = \sum_{\nu,k} V_{\nu,k}^{(o)}(\mathbf{0}) \sqrt{M_\mu} \delta_{k,j}, \end{aligned} \tag{4.192}$$

$$\sum_{\nu} V_{\nu,j}^{(o)}(\mathbf{0}) \sqrt{M_\mu} = 0, \tag{4.193}$$

$$\sum_{\nu} V_{\nu,j}^{(o)}(\mathbf{0}) \sqrt{\frac{M}{M_\mu}} \sqrt{M_\mu} \sqrt{\frac{M_\mu}{M}} = 0, \tag{4.194}$$

$$\sum_{\nu} S_{\mathbf{n}=\mathbf{0},\mu,j}(\mathbf{q} = \mathbf{0}) \sqrt{M_\mu} \sqrt{\frac{M_\mu}{M}} = 0, \tag{4.195}$$

$$\frac{1}{\sqrt{M}} \sum_{\nu} S_{\mathbf{n}=\mathbf{0},\mu,j}(\mathbf{q} = \mathbf{0}) M_\nu = 0, \tag{4.196}$$

$$\frac{1}{M} \sum_{\nu} S_{\mathbf{n}=\mathbf{0},\mu,j}(\mathbf{q} = \mathbf{0}) M_\nu = 0. \tag{4.197}$$

Equation 4.197 states that the center of mass of the optical branches remains stationary. This implies that atoms in the optical modes move out of phase with respect to their neighbors, effectively forming oscillating electric dipoles within the unit cell. The optical branch is so named because the frequency it attains at $\mathbf{q} = \mathbf{0}$ typically lies in the infrared region, and can interact with light.

4.5 Lattice Specific Heat

The phonon Hamiltonian 4.102 provides an adequate description of the lattice contribution to the specific heat in metals. Within the Born-Oppenheimer approximation, the total specific heat arises from both electronic and ionic degrees of freedom; the lattice specific heat specifically accounts for the ionic part. We have

$$\left\langle\hat{\mathcal{H}}_{\text{ret}}\right\rangle = \sum_{s} \int D_s(\omega) \frac{\hbar\omega}{e^{\beta\hbar\omega_{\mathbf{q},s}} - 1} d\omega, \tag{4.198}$$

where

$$D_s(\omega) = \sum_{\mathbf{q}} \delta\left(\omega - \omega_{\mathbf{q},s}\right) \tag{4.199}$$

is the density of modes or density of states of the s branch and it takes account for the number of vibrations for each energy.

4.5.1 Einstein's Approximation of the Density of Modes

Einstein's approximation of the density of modes consists of setting

$$\omega_{\mathbf{q},s} = \omega_0 > 0, \ \forall \mathbf{q}, s, \tag{4.200}$$

that is, all branches become optical. We have

$$D_s(\omega) = N\delta(\omega - \omega_0), \tag{4.201}$$

$$\begin{aligned} \left\langle\hat{\mathcal{H}}_{\text{ret}}\right\rangle &= \sum_{s} \int D_s(\omega) \frac{\hbar\omega}{e^{\beta\hbar\omega_{\mathbf{q},s}} - 1} d\omega = \\ &= \sum_{s} N \frac{\hbar\omega_0}{e^{\beta\hbar\omega_0} - 1} = \\ &= 3\tau N \frac{\hbar\omega_0}{e^{\frac{\hbar\omega_0}{k_B T}} - 1}. \end{aligned} \tag{4.202}$$

Let us compute the energy of a mole: for each species we consider an Avogadro number of atoms (recall $N = N_A$, $N_A k_B = R$), then

$$\mathcal{E}_{\text{mol}} = 3\tau N_A \frac{\hbar\omega_0}{e^{\frac{\hbar\omega_0}{k_B T}} - 1} \equiv$$
$$\equiv 3\tau N_A \frac{\hbar\omega_0}{e^{\frac{T_E}{T}} - 1}, \tag{4.203}$$

where we have introduced the Einstein temperature

$$T_E \equiv \frac{\hbar\omega_0}{k_B}. \tag{4.204}$$

The molar specific heat at constant volume is therefore

$$\begin{aligned} C_V &= \frac{\partial \mathcal{E}_{\text{mol}}}{\partial T} = \\ &= \frac{3\tau N_A \hbar\omega_0}{\left(e^{\frac{T_E}{T}} - 1\right)^2} e^{\frac{T_E}{T}} \frac{T_E}{T} = \\ &= 3\tau N_A \hbar\omega_0 \frac{1}{\left(e^{\frac{T_E}{T}} - 1\right)^2} e^{\frac{T_E}{T}} \frac{T_E}{T} \frac{k_B}{k_B} = \\ &= \frac{3\tau R}{\left(e^{\frac{T_E}{T}} - 1\right)^2} \left(\frac{T_E}{T}\right)^2 e^{\frac{T_E}{T}}. \end{aligned} \tag{4.205}$$

Let us analyze the behavior of the specific heat in the limits of high and low temperature. At high temperature, i.e., $T \gg T_E$, the specific heat approaches $C_V \to 3\tau R$, which is consistent with the equipartition theorem: classically, each quadratic degree of freedom contributes an average energy of $\frac{k_B T}{2}$. Therefore, the total classical energy is given by $3\tau \frac{k_B T}{2} 2N$, where the factor 2 accounts for the two degrees of freedom of the harmonic oscillator. Here, the potential and kinetic energies are considered as independent contributions. Differentiating this energy with respect to temperature yields $C_V = 3\tau N k_B$, matching the Dulong-Petit law. On the other hand, Einstein's approximation fails to correctly describe the low-temperature behavior. For $T \ll T_E$, it predicts an exponential decay of the specific heat, $C_V \sim e^{-\frac{T_E}{T}}$, which contradicts experimental observations. This discrepancy arises because Einstein's model neglects the low-energy excitations associated with the acoustic branches near $\mathbf{q} \to \mathbf{0}$. Only by including these acoustic modes can the isochoric specific heat be accurately reproduced. Indeed, as temperature decreases below a certain threshold, the excitation of these modes becomes energetically inaccessible, causing them to cease contributing to the specific heat. Thus, to correctly capture the low-temperature behavior, we must go beyond Einstein's approximation and explicitly consider the contributions from the acoustic branches.

4.5.2 Debye's Approximation of the Density of Modes

In order to simplify the analysis of lattice vibrations, Debye approximates the crystal as a continuous elastic medium. This is justified for long-wavelength vibrations, where the wavelength λ is much larger than the lattice constant a ($\lambda \gg a$). In this limit, the discrete spectrum of normal modes becomes effectively continuous, and the dispersion relation can be approximated as linear, that is,

$$\omega_{\mathbf{q},s} = |\mathbf{q}|\, V_s, \tag{4.206}$$

where the velocity V_s can have only two values: V_t for the two transverse modes and V_l for the longitudinal mode. From spatial periodicity (see Appendix C)

$$\begin{aligned} q_x(x + L_x) &= q_x x + q_x L_x = \\ &= qx + q_x n_x \lambda = \\ &= q_x x + \frac{2\pi}{\lambda} n_x \lambda = \\ &= q_x x + 2\pi n_x, \end{aligned} \tag{4.207}$$

where n_x is the number of waves with wave vector $\mathbf{q}$ and in the direction x of the lattice, of length L_x. Given the wave vector (q_x, q_y, q_z), the density of the number of waves with wave vector between $\mathbf{q}$ and $\mathbf{q} + d\mathbf{q}$ is

$$dn_x dn_y dn_z = \frac{L_x L_y L_z}{(2\pi)^3} dq_x dq_y dq_z. \tag{4.208}$$

From Eq. 4.199, in the thermodynamic limit C.11, one finds

$$\begin{aligned} D_s(\omega) &= \frac{L_x L_y L_z}{(2\pi)^3} \int d^3\mathbf{q}\, \delta\left(\omega - |\mathbf{q}| V_s\right) \equiv \\ &\equiv \frac{V}{8\pi^3} \int d^3\mathbf{q}\, \delta\left(\omega - |\mathbf{q}| V_s\right), \end{aligned} \tag{4.209}$$

where V is the volume of the crystal. Using polar coordinates,

$$D_s(\omega) = 4\pi \frac{V}{8\pi^3} \int \mathbf{q}^2 dq\, \delta\left(\omega - |\mathbf{q}| V_s\right), \tag{4.210}$$

$$D_s(\omega) = 4\pi \frac{V}{8\pi^3} \left(\frac{\omega_{\mathbf{q},s}}{V_s}\right)^2 \int dq\, \delta\left(\omega - |\mathbf{q}| V_s\right). \tag{4.211}$$

Now, we use the following property of Dirac's delta

$$\int \delta\left[f(x)\right]dx = \int \sum_i \frac{\delta(x - x_i)}{|f'(x_i)|}dx, \;\; f(x_i) \neq 0, \; \forall\, i, \tag{4.212}$$

x_i being the zeros of f. In our case, the function is $\omega(\mathbf{q})$ and the first derivative is equal to V_s. The density of modes is written as

$$D_s(\omega) = 4\pi \frac{V}{8\pi^3}\left(\frac{\omega}{V_s}\right)^2 \frac{1}{V_s}, \tag{4.213}$$

then the energy becomes

$$\begin{aligned}
\mathcal{E} &= \sum_s \int D_s(\omega)\frac{\hbar\omega}{e^{\frac{\hbar\omega}{k_BT}} - 1}d\omega = \\
&= \sum_s \int 4\pi \frac{V}{8\pi^3}\left(\frac{\omega}{V_s}\right)^2 \frac{1}{V_s}\frac{\hbar\omega}{e^{\frac{\hbar\omega}{k_BT}} - 1}d\omega = \\
&= 4\pi \frac{V}{8\pi^3}\sum_s \left(\frac{1}{V_s}\right)^3 \int \frac{\hbar\omega^3}{e^{\frac{\hbar\omega}{k_BT}} - 1}d\omega = \\
&= 4\pi \frac{V}{8\pi^3}\sum_s \left(\frac{1}{V_s}\right)^3 \int \frac{\hbar\omega^3}{e^{\frac{\hbar\omega}{k_BT}} - 1}d\omega \\
&= 4\pi \frac{V}{8\pi^3}\left[\frac{2}{V_t^3} + \frac{1}{V_l^3}\right] \int \frac{\hbar\omega^3}{e^{\frac{\hbar\omega}{k_BT}} - 1}d\omega.
\end{aligned} \tag{4.214}$$

The definition of $D_s(\omega)$ implies a constraint, indeed

$$\begin{aligned}
\sum_s \int d\omega D_s(\omega) &= \sum_{\mathbf{q},s} \int d\omega \delta\left(\omega - \omega_{\mathbf{q},s}\right) = \\
&= \sum_{\mathbf{q},s} 1 = \\
&= 3N,
\end{aligned} \tag{4.215}$$

i.e., the total number of acoustic phonons must be equal to the number of elementary cells in the lattice, so

$$\begin{aligned}
\sum_s \int d\omega D_s(\omega) &= \int_0^{\omega_D} d\omega \, 4\pi \frac{V}{8\pi^3}\left[\frac{2}{V_t^3} + \frac{1}{V_l^3}\right]\omega^2 = \\
&= 3N.
\end{aligned} \tag{4.216}$$

Debye's approximation implies that a maximum frequency ω_D, the Debye frequency, exists. The integral with respect to the measure $d\omega$ gives

$$4\pi \frac{V}{8\pi^3}\left[\frac{2}{V_t^3}+\frac{1}{V_l^3}\right]\frac{\omega_D^3}{3}=3N, \tag{4.217}$$

from which we compute ω_D. The Debye frequency is related to a temperature

$$T_D=\frac{\hbar\omega_D}{k_B}, \tag{4.218}$$

which is the Debye temperature. We insert into the energy $\mathcal{E}$ the quantity

$$4\pi \frac{V}{8\pi^3}\left[\frac{2}{V_t^3}+\frac{1}{V_l^3}\right]=\frac{9N}{\omega_D^3}, \tag{4.219}$$

then

$$\mathcal{E}=\frac{9N\hbar}{\omega_D^3}\int_0^{\omega_D}d\omega\frac{\omega^3}{e^{\frac{\hbar\omega}{k_BT}}-1}. \tag{4.220}$$

We multiply and divide by $\hbar^3$

$$\mathcal{E}=\frac{9N}{(\hbar\omega_D)^3}\hbar\int_0^{\omega_D}d\omega\frac{(\hbar\omega)^3}{e^{\frac{\hbar\omega}{k_BT}}-1}, \tag{4.221}$$

and we set $x=\dfrac{\hbar\omega}{k_BT}$, then

$$\begin{aligned}\mathcal{E}&=\frac{9N}{(\hbar\omega_D)^3}\hbar\int_0^{\frac{\hbar\omega_D}{k_BT}}dx\frac{k_BT}{\hbar}(k_BT)^3\frac{x^3}{e^x-1}=\\&=\frac{9N(k_BT)^4}{(\hbar\omega_D)^3}\int_0^{\frac{\hbar\omega_D}{k_BT}}dx\frac{x^3}{e^x-1}=\\&=\frac{9N(k_BT)^4}{(\hbar\omega_D)^3}\int_0^{\frac{T_D}{T}}dx\frac{x^3}{e^x-1}=\\&=9Nk_BT\left(\frac{T}{T_D}\right)^3\int_0^{\frac{T_D}{T}}dx\frac{x^3}{e^x-1}.\end{aligned} \tag{4.222}$$

For one mole, we have

$$\mathcal{E}_{\text{mol}} = 9RT\left(\frac{T}{T_D}\right)^3 \int_0^{\frac{T_D}{T}} dx \frac{x^3}{e^x - 1}, \tag{4.223}$$

and the molar specific heat is

$$C_{V,\text{mol}} = \frac{\partial \mathcal{E}_{\text{mol}}}{\partial T}. \tag{4.224}$$

We now examine the behavior of the molar internal energy and specific heat in the high- and low-temperature limits, relative to the Debye temperature T_D. In the high-temperature regime, where $T \gg T_D$, we have

$$\frac{T}{T_D} \gg 1. \tag{4.225}$$

In this limit, the upper limit of integration in the Debye model becomes small, so the integrand is significant only for $x \ll 1$, and we can use the approximation $e^x \approx 1 + x$. The molar internal energy then becomes

$$\begin{aligned} \mathcal{E}_{\text{mol}}(T \gg T_D) &\sim 9RT\left(\frac{T}{T_D}\right)^3 \int_0^{\frac{T_D}{T}} dx\, x^2 = \\ &= 9RT\left(\frac{T}{T_D}\right)^3 \left(\frac{T_D}{T}\right)^3 \frac{1}{3} = \\ &= 3RT. \end{aligned} \tag{4.226}$$

As a result, the molar specific heat at constant volume is

$$C_{V,\text{mol}} = 3R, \tag{4.227}$$

which coincides with the classical Dulong-Petit law. In the low-temperature regime, where $T \ll T_D$, the upper limit of integration becomes large, and the full form of the Bose-Einstein distribution must be retained. In this case, the molar internal energy behaves as $\mathcal{E}_{\text{mol}}(T \ll T_D) \propto T^4$, so the molar specific heat scales as

$$C_{V,\text{mol}} \propto T^3. \tag{4.228}$$

This T^3 dependence agrees with experimental observations and confirms the success of Debye's model at low temperatures.

Chapter 5
Step III of the Model of a Solid: Electron-Phonon Interaction

Abstract This chapter derives the electron-phonon interaction Hamiltonian, a fundamental component in understanding the electronic properties of solids. Starting from microscopic principles, we develop a model that captures how electrons interact with lattice vibrations. The derivation leads to a compact, second-quantized expression for the interaction Hamiltonian, providing a practical and powerful tool for analysing many-body effects and transport phenomena in crystalline solids.

5.1 Electron-Phonon Interaction

The electron-ion interaction is given by

$$V^{(e-i)} = \sum_j \sum_{\mathbf{n},\mu} V_\mu \left[\mathbf{r}_j - \mathbf{R_n} - \mathbf{d}_\mu - \mathbf{S}_{\mathbf{n},\mu}\right], \tag{5.1}$$

where we recall that: $\mathbf{r}_j$ is the position of the j-th electron, $\mathbf{R_n}$ (see Eq. 3.2) is the position of the n-th Bravais lattice site, $\mathbf{d}_\mu$ is the position of atom μ within the unit cell, relative to $\mathbf{R_n}$, $\mathbf{S}_{\mathbf{n},\mu}$ (see Eq. 4.134) is the displacement of atom μ in cell $\mathbf{n}$ from its equilibrium position, accounting for lattice vibrations. We add and subtract the potential energy evaluated at $\mathbf{S}_{\mathbf{n},\mu} = \mathbf{0}$, i.e., at the ionic equilibrium positions, that is

$$V^{(e-i)} = \sum_j \sum_{\mathbf{n},\mu} V_\mu \left[\mathbf{r}_j - \mathbf{R_n} - \mathbf{d}_\mu\right] + \sum_j \sum_{\mathbf{n},\mu} \left\{ V_\mu \left[\mathbf{r}_j - \mathbf{R_n} - \mathbf{d}_\mu - \mathbf{S}_{\mathbf{n},\mu}\right] \right.$$
$$\left. - V_\mu \left[\mathbf{r}_j - \mathbf{R_n} - \mathbf{d}_\mu\right] \right\}. \tag{5.2}$$

The first term originates from the fixed ions at the equilibrium positions, and thus it is included in the electronic problem studied in the previous chapter. The second term corresponds to the interaction between electrons and ions when the latter are displaced from their equilibrium positions. Therefore, the quantity that characterizes the electron-phonon interaction, which we will now analyze, is

F. Tafuri et al., *Understanding Quantum Many-Body Theory*, UNITEXT for Physics, https://doi.org/10.1007/978-3-032-17849-7_5

$$\sum_j \sum_{\mathbf{n},\mu} \left\{ V_\mu \left[\mathbf{r}_j - \mathbf{R_n} - \mathbf{d}_\mu - \mathbf{S}_{\mathbf{n},\mu} \right] - V_\mu \left[\mathbf{r}_j - \mathbf{R_n} - \mathbf{d}_\mu \right] \right\}. \tag{5.3}$$

In a crystalline solid, the ions oscillate with small amplitudes around their equilibrium positions, consequently, we expand the potential $V_\mu \left(\mathbf{r}_j - \mathbf{R_n} - \mathbf{d}_\mu - \mathbf{S}_{\mathbf{n},\mu} \right)$ as a Taylor series with respect to $\mathbf{S}_{\mathbf{n},\mu}$. We have

$$\begin{aligned} &V_\mu \left[\mathbf{r}_j - \mathbf{R_n} - \mathbf{d}_\mu - \mathbf{S}_{\mathbf{n},\mu} \right] - V_\mu \left[\mathbf{r}_j - \mathbf{R_n} - \mathbf{d}_\mu \right] \\ &\quad \simeq - \sum_j \sum_{\mathbf{n},\mu} \mathbf{S}_{\mathbf{n},\mu} \cdot \nabla_{\mathbf{S}_{\mathbf{n},\mu}} V_\mu \left[\mathbf{r}_j - \mathbf{R_n} - \mathbf{d}_\mu - \mathbf{S}_{\mathbf{n},\mu} \right] \big|_{\mathbf{S}_{\mathbf{n},\mu} = \mathbf{0}}, \end{aligned} \tag{5.4}$$

where the minus sign in the second term arises from the functional dependence of $V^{(e-i)}$ on $-\mathbf{S}_{\mathbf{n},\mu}$. We aim to express this quantity in second quantization, where the electron-ion potential $V^{(e-i)}$ is transformed into the electron-phonon Hamiltonian $\hat{\mathcal{H}}_{e-ph}$. Since V_μ (and consequently $\nabla_{\mathbf{S}_{\mathbf{n},\mu}} V_\mu$) is a one-body potential, we use Theorem 2.9, i.e.,

$$\hat{\mathcal{H}}_{e-ph} = - \sum_{\mathbf{n},\mu} \mathbf{S}_{\mathbf{n},\mu} \cdot \int dx \, \hat{\psi}^\dagger(x) \, \nabla_{\mathbf{S}_{\mathbf{n},\mu}} V_\mu \left[\mathbf{r} - \mathbf{R_n} - \mathbf{d}_\mu - \mathbf{S}_{\mathbf{n},\mu} \right] \big|_{\mathbf{S}_{\mathbf{n},\mu} = \mathbf{0}} \, \hat{\psi}(x), \tag{5.5}$$

and we employ the basis of Bloch waves multiplied by the standard spin eigenfunctions, that is Eqs. 3.147 and 3.148, then

$$\begin{aligned} \hat{\mathcal{H}}_{e-ph} &= - \sum_{\mathbf{n},\mu} \mathbf{S}_{\mathbf{n},\mu} \cdot \sum_{\sigma,\sigma'} \sum_{n,n'} \sum_{\mathbf{k},\mathbf{k}'} C^\dagger_{n',\mathbf{k}',\sigma'} C_{n,\mathbf{k},\sigma} \\ &\quad \sum_s \int_V d^3\mathbf{r} \, \varphi^*_{n',\mathbf{k}'}(\mathbf{r}) \chi^*_{\sigma'}(s) \, \nabla_{\mathbf{S}_{\mathbf{n},\mu}} V_\mu \left[\mathbf{r} - \mathbf{R_n} - \mathbf{d}_\mu - \mathbf{S}_{\mathbf{n},\mu} \right] \big|_{\mathbf{S}_{\mathbf{n},\mu} = \mathbf{0}} \\ &\quad \times \varphi_{n,\mathbf{k}}(\mathbf{r}) \chi_\sigma(s) = \\ &= - \sum_{\mathbf{n},\mu} \mathbf{S}_{\mathbf{n},\mu} \cdot \sum_{\sigma,\sigma'} \sum_{n,n'} \sum_{\mathbf{k},\mathbf{k}'} C^\dagger_{n',\mathbf{k}',\sigma} C_{n,\mathbf{k},\sigma} \delta_{\sigma,\sigma'} \\ &\quad \int_V d^3\mathbf{r} \, \varphi^*_{n',\mathbf{k}'}(\mathbf{r}) \, \nabla_{\mathbf{S}_{\mathbf{n},\mu}} V_\mu \left[\mathbf{r} - \mathbf{R_n} - \mathbf{d}_\mu - \mathbf{S}_{\mathbf{n},\mu} \right] \big|_{\mathbf{S}_{\mathbf{n},\mu} = \mathbf{0}} \varphi_{n,\mathbf{k}}(\mathbf{r}) = \\ &= - \sum_{\mathbf{n},\mu} \mathbf{S}_{\mathbf{n},\mu} \cdot \sum_{\sigma} \sum_{n,n'} \sum_{\mathbf{k},\mathbf{k}'} C^\dagger_{n',\mathbf{k}',\sigma} C_{n,\mathbf{k},\sigma} \int_V d^3\mathbf{r} \, \varphi^*_{n',\mathbf{k}'}(\mathbf{r}) \, \nabla_{\mathbf{S}_{\mathbf{n},\mu}} V_\mu \\ &\quad \times \left[\mathbf{r} - \mathbf{R_n} - \mathbf{d}_\mu - \mathbf{S}_{\mathbf{n},\mu} \right] \big|_{\mathbf{S}_{\mathbf{n},\mu} = \mathbf{0}} \varphi_{n,\mathbf{k}}(\mathbf{r}), \end{aligned} \tag{5.6}$$

where V is the crystal volume, and we have used the orthonormality of the spin functions (see Eq. 1.23). From the structure of $\hat{\mathcal{H}}_{e-ph}$, it follows that an electron is annihilated in band n with quasi-momentum $\mathbf{k}$ and spin component σ, and simultaneously created in band n' with quasi-momentum $\mathbf{k}'$ and the same spin component σ. The band index n' may coincide with n or differ from it: thus, the electron-phonon interaction can induce both intra-band and inter-band transitions. The Hamiltonian $\hat{\mathcal{H}}_{e-ph}$ involves the quantized displacement $\mathbf{S}_{\mathbf{n},\mu}$ defined in Eq. 4.134, which is expressed in terms of the phonon annihilation and creation operators (see Eqs. 4.88 and 4.90). This corresponds respectively to the annihilation of a phonon with momentum $\mathbf{q}$ or the creation of one with momentum $-\mathbf{q}$. Through interaction with a phonon of momentum $\pm\,\mathbf{q}$, the electron undergoes a transition from state $\mathbf{k}$ to $\mathbf{k}'$, with conservation of total momentum requiring that

$$\mathbf{k}' = \mathbf{k} + \mathbf{q}. \tag{5.7}$$

Given the expression for $\mathbf{S}_{\mathbf{n},\mu}$, the next step is to evaluate the matrix element of the electron–phonon interaction Hamiltonian. Recall that periodic boundary conditions have been imposed: the crystal of volume V is replicated infinitely many times along each crystallographic direction i. Consequently, the potential V_μ is periodic throughout the entire supercrystal, i.e.,

$$V_\mu\left[\mathbf{r} - \mathbf{R_n} - \mathbf{d}_\mu - \mathbf{S}_{\mathbf{n},\mu}\right] = \sum_{\mathbf{q}_1} \frac{e^{i\mathbf{q}_1\cdot(\mathbf{r}-\mathbf{R_n}-\mathbf{d}_\mu-\mathbf{S}_{\mathbf{n},\mu})}}{\sqrt{N}} V_\mu(\mathbf{q}_1), \tag{5.8}$$

where (see Eq. 3.45)

$$\mathbf{q}_1 = \frac{\tilde{n}_1}{N_1}\mathbf{b}_1 + \frac{\tilde{n}_2}{N_2}\mathbf{b}_2 + \frac{\tilde{n}_3}{N_3}\mathbf{b}_3, \; \tilde{n}_i \in \mathbb{Z} \tag{5.9}$$

denote the wave vectors for each cell in the reciprocal lattice of the supercrystal. We define

$$\tilde{V}_\mu(\mathbf{q}_1) = V_\mu(\mathbf{q}_1)e^{-i\mathbf{q}_1\cdot\mathbf{d}_\mu}, \tag{5.10}$$

then

$$V_\mu\left[\mathbf{r} - \mathbf{R_n} - \mathbf{d}_\mu - \mathbf{S}_{\mathbf{n},\mu}\right] = \sum_{\mathbf{q}_1} \frac{e^{i\mathbf{q}_1\cdot(\mathbf{r}-\mathbf{R_n}-\mathbf{S}_{\mathbf{n},\mu})}}{\sqrt{N}} \tilde{V}_\mu(\mathbf{q}_1), \tag{5.11}$$

$$\nabla_{\mathbf{S}_{\mathbf{n},\mu}} V_\mu\left[\mathbf{r} - \mathbf{R_n} - \mathbf{d}_\mu - \mathbf{S}_{\mathbf{n},\mu}\right]\Big|_{\mathbf{S}_{\mathbf{n},\mu}=\mathbf{0}} = \sum_{\mathbf{q}_1}(-i\mathbf{q}_1)\frac{e^{i\mathbf{q}_1\cdot(\mathbf{r}-\mathbf{R_n})}}{\sqrt{N}} \tilde{V}_\mu(\mathbf{q}_1), \tag{5.12}$$

where $(-i\mathbf{q}_1)$ in the last equation is derived from the gradient operation. From the previous results, we have

$$\hat{\mathcal{H}}_{e-ph} = -\sum_{\mathbf{n},\mu} \mathbf{S}_{\mathbf{n},\mu} \cdot \sum_{\sigma} \sum_{n,n'} \sum_{\mathbf{k},\mathbf{k}'} C^{\dagger}_{n',\mathbf{k}',\sigma} C_{n,\mathbf{k},\sigma} \int_V d^3\mathbf{r}\, \varphi^*_{n',\mathbf{k}'}(\mathbf{r})$$

$$\times \sum_{\mathbf{q}_1} (-i\mathbf{q}_1) \frac{e^{i\mathbf{q}_1\cdot(\mathbf{r}-\mathbf{R}_\mathbf{n})}}{\sqrt{N}} \tilde{V}_\mu(\mathbf{q}_1)\varphi_{n,\mathbf{k}}(\mathbf{r}) =$$

$$= \sum_{\mathbf{n},\mu} \mathbf{S}_{\mathbf{n},\mu} \cdot \sum_{\sigma} \sum_{n,n'} \sum_{\mathbf{k},\mathbf{k}'} C^{\dagger}_{n',\mathbf{k}',\sigma} C_{n,\mathbf{k},\sigma} \int_V d^3\mathbf{r}\, \varphi^*_{n',\mathbf{k}'}(\mathbf{r})$$

$$\times \sum_{\mathbf{q}_1} i\mathbf{q}_1 \frac{e^{i\mathbf{q}_1\cdot(\mathbf{r}-\mathbf{R}_\mathbf{n})}}{\sqrt{N}} \tilde{V}_\mu(\mathbf{q}_1)\varphi_{n,\mathbf{k}}(\mathbf{r}). \tag{5.13}$$

Now, $\frac{e^{i\mathbf{q}\cdot\mathbf{R}_\mathbf{n}}}{\sqrt{N}}$ (appearing in $\mathbf{S}_{\mathbf{n},\mu}$) and $\frac{e^{i\mathbf{q}_1\cdot(\mathbf{r}-\mathbf{R}_\mathbf{n})}}{\sqrt{N}}$ (appearing in $\nabla_{\mathbf{S}_{\mathbf{n},\mu}} V_\mu \mid_{\mathbf{S}_{\mathbf{n},\mu}=\mathbf{0}}$) both depend on $\mathbf{n}$. Note that the wave vectors $\mathbf{q}$ and $\mathbf{q}_1$ belong to different domains: specifically, $\mathbf{q}$ belongs to the first Brillouin zone (see Eq. 3.73), while $\mathbf{q}_1$ takes values in the reciprocal lattice vectors of the superlattice. Consequently, the sum over $\mathbf{n}$ does not produce a Kronecker delta. Therefore, no immediate wave vector selection occurs, and more elaborate algebraic manipulations are required to evaluate the interaction terms. From

$$\sum_{\mathbf{n}} \frac{e^{i\mathbf{q}\cdot\mathbf{R}_\mathbf{n}}}{\sqrt{N}} \frac{e^{-i\mathbf{q}_1\cdot\mathbf{R}_\mathbf{n}}}{\sqrt{N}} = \sum_{n_1,n_2,n_3} \frac{e^{i(\mathbf{q}-\mathbf{q}_1)\cdot\mathbf{R}_\mathbf{n}}}{N}, \tag{5.14}$$

we have

$$\frac{1}{N}\sum_{n_1,n_2,n_3} e^{i(\mathbf{q}-\mathbf{q}_1)\cdot\mathbf{R}_\mathbf{n}} = \frac{1}{N}\left(\sum_{n_1=0}^{N_1-1} e^{i\frac{n_1}{N_1}(\tilde{\tilde{n}}_1-\tilde{n}_1)\mathbf{a}_1\cdot\mathbf{b}_1}\right)\left(\sum_{n_2=0}^{N_2-1} e^{i\frac{n_2}{N_2}(\tilde{\tilde{n}}_2-\tilde{n}_2)\mathbf{a}_2\cdot\mathbf{b}_2}\right)$$

$$\times \left(\sum_{n_3=0}^{N_3-1} e^{i\frac{n_3}{N_3}(\tilde{\tilde{n}}_3-\tilde{n}_3)\mathbf{a}_3\cdot\mathbf{b}_3}\right) =$$

$$= \frac{1}{N_1}\left(\sum_{n_1=0}^{N_1-1} e^{i2\pi\frac{n_1}{N_1}(\tilde{\tilde{n}}_1-\tilde{n}_1)}\right)\frac{1}{N_2}\left(\sum_{n_2=0}^{N_2-1} e^{i2\pi\frac{n_2}{N_2}(\tilde{\tilde{n}}_2-\tilde{n}_2)}\right)\frac{1}{N_3}$$

$$\times \left(\sum_{n_3=0}^{N_3-1} e^{i2\pi\frac{n_3}{N_3}(\tilde{\tilde{n}}_3-\tilde{n}_3)}\right), \tag{5.15}$$

which is the product of three truncated geometric progressions of complex common ratio of unit modulus. Let us consider the first one

$$\sum_{n_1=0}^{N_1-1}\left(e^{i\frac{2\pi}{N_1}(\tilde{\tilde{n}}_1-\tilde{n}_1)}\right)^{n_1}=\begin{cases}N_1,\ e^{i\frac{2\pi}{N_1}(\tilde{\tilde{n}}_1-\tilde{n}_1)}=1\\ \dfrac{1-\left(e^{i\frac{2\pi}{N_1}(\tilde{\tilde{n}}_1-\tilde{n}_1)}\right)^{N_1}}{1-e^{i\frac{2\pi}{N_1}(\tilde{\tilde{n}}_1-\tilde{n}_1)}},\ e^{i\frac{2\pi}{N_1}(\tilde{\tilde{n}}_1-\tilde{n}_1)}\neq 1\end{cases}. \tag{5.16}$$

From

$$\begin{aligned}\left(e^{i\frac{2\pi}{N_1}(\tilde{\tilde{n}}_1-\tilde{n}_1)}\right)^{N_1}&=e^{i2\pi(\tilde{\tilde{n}}_1-\tilde{n}_1)}=\\&=e^{i2\pi m}=\\&=1,\ m\in\mathbb{Z},\end{aligned}\tag{5.17}$$

the quantity above vanishes and the truncated geometric progression is nonnull only if

$$e^{i\frac{2\pi}{N_1}(\tilde{\tilde{n}}_1-\tilde{n}_1)}=1, \tag{5.18}$$

$$\frac{\tilde{\tilde{n}}_1-\tilde{n}_1}{N_1}\in\mathbb{Z}, \tag{5.19}$$

but $\tilde{\tilde{n}}_1\in[0, N_1-1]$, while $\tilde{n}_1\in\mathbb{Z}$, so

$$\tilde{n}_1=\tilde{\tilde{n}}_1+m_1N_1,\ m_1\in\mathbb{Z}. \tag{5.20}$$

We insert the previous results into

$$\begin{aligned}\mathbf{q}_1&=\left(\frac{\tilde{\tilde{n}}_1}{N_1}+m_1\right)\mathbf{b}_1+\left(\frac{\tilde{\tilde{n}}_2}{N_2}+m_2\right)\mathbf{b}_2+\left(\frac{\tilde{\tilde{n}}_3}{N_3}+m_3\right)\mathbf{b}_3\equiv\\&\equiv\mathbf{q}+\mathbf{G},\end{aligned}\tag{5.21}$$

where we recall that $\mathbf{q}$ belongs to the first Brillouin zone of the reciprocal lattice of the crystal, and $\mathbf{G}$ is an arbitrary vector of the reciprocal lattice. Therefore, combining the results of the three truncated geometric series, the sum over $\mathbf{n}$ can be expressed as

$$\sum_{\mathbf{n}}\frac{e^{i(\mathbf{q}-\mathbf{q}_1)\cdot\mathbf{R}_\mathbf{n}}}{N}=\delta_{\mathbf{q}_1,\mathbf{q}+\mathbf{G}}. \tag{5.22}$$

We get

$$\hat{\mathcal{H}}_{e-ph} = \sum_{\mathbf{n},\mu}\sum_{\mathbf{q}}\sum_{s=1}^{3\tau} \frac{e^{i\mathbf{q}\cdot\mathbf{R_n}}}{\sqrt{N}} \boldsymbol{\varepsilon}_{\mu}^{(s)}(\mathbf{q})\sqrt{\frac{\hbar}{2M_\mu\omega_{\mathbf{q},s}}}\left(a_{\mathbf{q},s}+a^\dagger_{-\mathbf{q},s}\right)$$
$$\times\sum_{\sigma}\sum_{n,n'}\sum_{\mathbf{k},\mathbf{k}'} C^\dagger_{n',\mathbf{k}',\sigma}C_{n,\mathbf{k},\sigma}$$
$$\int_V d^3\mathbf{r}\,\varphi^*_{n',\mathbf{k}'}(\mathbf{r})\sum_{\mathbf{q}_1} i\mathbf{q}_1\frac{e^{i\mathbf{q}_1\cdot(\mathbf{r}-\mathbf{R_n})}}{\sqrt{N}}\tilde{V}_\mu(\mathbf{q}_1)\varphi_{n,\mathbf{k}}(\mathbf{r}) =$$
$$=\sum_{\mu}\sum_{\mathbf{q}}\sum_{s=1}^{3\tau}\boldsymbol{\varepsilon}_{\mu}^{(s)}(\mathbf{q})\sqrt{\frac{\hbar}{2M_\mu\omega_{\mathbf{q},s}}}\left(a_{\mathbf{q},s}+a^\dagger_{-\mathbf{q},s}\right)\sum_{\sigma}\sum_{n,n'}\sum_{\mathbf{k},\mathbf{k}'} C^\dagger_{n',\mathbf{k}',\sigma}C_{n,\mathbf{k},\sigma}$$
$$\int_V d^3\mathbf{r}\,\varphi^*_{n',\mathbf{k}'}(\mathbf{r})\sum_{\mathbf{q}_1} i\mathbf{q}_1\delta_{\mathbf{q}_1,\mathbf{q}+\mathbf{G}}e^{i\mathbf{q}_1\cdot\mathbf{r}}\tilde{V}_\mu(\mathbf{q}_1)\varphi_{n,\mathbf{k}}(\mathbf{r}) =$$
$$=i\sum_{\mu}\sum_{\mathbf{q}}\sum_{s=1}^{3\tau}\boldsymbol{\varepsilon}_{\mu}^{(s)}(\mathbf{q})\sqrt{\frac{\hbar}{2M_\mu\omega_{\mathbf{q},s}}}\left(a_{\mathbf{q},s}+a^\dagger_{-\mathbf{q},s}\right)\sum_{\sigma}\sum_{n,n'}\sum_{\mathbf{k},\mathbf{k}'} C^\dagger_{n',\mathbf{k}',\sigma}C_{n,\mathbf{k},\sigma}$$
$$\int_V d^3\mathbf{r}\,\varphi^*_{n',\mathbf{k}'}(\mathbf{r})\sum_{\mathbf{q}_1} \mathbf{q}_1\delta_{\mathbf{q}_1,\mathbf{q}+\mathbf{G}}e^{i\mathbf{q}_1\cdot\mathbf{r}}\tilde{V}_\mu(\mathbf{q}_1)\varphi_{n,\mathbf{k}}(\mathbf{r}), \tag{5.23}$$

and we sum over $\mathbf{q}_1$, then

$$\hat{\mathcal{H}}_{e-ph} = i\sum_{\mu}\sum_{\mathbf{q}}\sum_{s=1}^{3\tau}\boldsymbol{\varepsilon}_{\mu}^{(s)}(\mathbf{q})\sqrt{\frac{\hbar}{2M_\mu\omega_{\mathbf{q},s}}}\left(a_{\mathbf{q},s}+a^\dagger_{-\mathbf{q},s}\right)\sum_{\sigma}\sum_{n,n'}\sum_{\mathbf{k},\mathbf{k}'} C^\dagger_{n',\mathbf{k}',\sigma}C_{n,\mathbf{k},\sigma}$$
$$\int_V d^3\mathbf{r}\,\varphi^*_{n',\mathbf{k}'}(\mathbf{r})\,(\mathbf{q}+\mathbf{G})\,e^{i(\mathbf{q}+\mathbf{G})\cdot\mathbf{r}}\tilde{V}_\mu\,(\mathbf{q}+\mathbf{G})\,\varphi_{n,\mathbf{k}}(\mathbf{r}). \tag{5.24}$$

We now rewrite the integral

$$\int_V d^3\mathbf{r}\,\varphi^*_{n',\mathbf{k}'}(\mathbf{r})e^{i(\mathbf{q}+\mathbf{G})\cdot\mathbf{r}}\varphi_{n,\mathbf{k}}(\mathbf{r}) = \int_V d^3\mathbf{r}\,U^*_{n',\mathbf{k}'}(\mathbf{r})e^{-i\mathbf{k}'\cdot\mathbf{r}}e^{i(\mathbf{q}+\mathbf{G})\cdot\mathbf{r}}U_{n,\mathbf{k}}(\mathbf{r})e^{i\mathbf{k}\cdot\mathbf{r}}, \tag{5.25}$$

as a sum of integrals over the individual unit cells, following Eq. 3.70. Then (see Eq. 3.13)

$$\int_V d^3\mathbf{r}\,U^*_{n',\mathbf{k}'}(\mathbf{r})e^{-i\mathbf{k}'\cdot\mathbf{r}}e^{i(\mathbf{q}+\mathbf{G})\cdot\mathbf{r}}U_{n,\mathbf{k}}(\mathbf{r})e^{i\mathbf{k}\cdot\mathbf{r}} =$$
$$=\sum_{\mathbf{n}}\int_{V_\mathrm{d}} d^3\mathbf{r_0}\,U^*_{n',\mathbf{k}'}(\mathbf{r_0}+\mathbf{R_n})e^{-i\mathbf{k}'\cdot(\mathbf{r_0}+\mathbf{R_n})}e^{i(\mathbf{q}+\mathbf{G})\cdot(\mathbf{r_0}+\mathbf{R_n})}$$
$$\times\,U_{n,\mathbf{k}}(\mathbf{r_0}+\mathbf{R_n})e^{i\mathbf{k}\cdot(\mathbf{r_0}+\mathbf{R_n})} =$$

$$
\begin{aligned}
&= \sum_{\mathbf{n}} \int_{V_d} d^3\mathbf{r_0}\, U^*_{n',\mathbf{k}'}(\mathbf{r_0}) e^{-i\mathbf{k}'\cdot(\mathbf{r_0}+\mathbf{R_n})} e^{i(\mathbf{q}+\mathbf{G})\cdot(\mathbf{r_0}+\mathbf{R_n})} U_{n,\mathbf{k}}(\mathbf{r_0}) e^{i\mathbf{k}\cdot(\mathbf{r_0}+\mathbf{R_n})} = \\
&= \sum_{\mathbf{n}} \int_{V_d} d^3\mathbf{r_0}\, U^*_{n',\mathbf{k}'}(\mathbf{r_0}) e^{-i(\mathbf{k}'-\mathbf{k}-\mathbf{q}-\mathbf{G})\cdot\mathbf{r_0}} e^{-i(\mathbf{k}'-\mathbf{k}-\mathbf{q}-\mathbf{G})\cdot\mathbf{R_n}} U_{n,\mathbf{k}}(\mathbf{r_0}) \equiv \\
&\equiv N \sum_{\mathbf{n}} \int_{V_d} d^3\mathbf{r_0}\, U^*_{n',\mathbf{k}'}(\mathbf{r_0}) e^{-i(\mathbf{k}'-\mathbf{k}-\mathbf{q}-\mathbf{G})\cdot\mathbf{r_0}} \frac{e^{-i(\mathbf{k}'-\mathbf{k}-\mathbf{q}-\mathbf{G})\cdot\mathbf{R_n}}}{N} U_{n,\mathbf{k}}(\mathbf{r_0}),
\end{aligned}
\tag{5.26}
$$

where at the last step we multiplied and divided by N to get

$$
\sum_{\mathbf{n}} \frac{e^{-i(\mathbf{k}'-\mathbf{k}-\mathbf{q}-\mathbf{G})\cdot\mathbf{R_n}}}{N} = \delta_{\mathbf{k}',\mathbf{k}+\mathbf{q}+\mathbf{G}},
\tag{5.27}
$$

and recall that $\mathbf{k}'$, $\mathbf{k}$ and $\mathbf{q}$ are wave vectors of the first Brillouin zone, while $\mathbf{G}$ is any vector of the reciprocal crystal lattice. We have

$$
\begin{aligned}
&\int_V d^3\mathbf{r} \varphi^*_{n',\mathbf{k}'}(\mathbf{r}) e^{i(\mathbf{q}+\mathbf{G})\cdot\mathbf{r}} \varphi_{n,\mathbf{k}}(\mathbf{r}) \\
&= N \int_{V_d} d^3\mathbf{r_0}\, U^*_{n',\mathbf{k}'}(\mathbf{r_0}) e^{-i(\mathbf{k}'-\mathbf{k}-\mathbf{q}-\mathbf{G})\cdot\mathbf{r_0}} \delta_{\mathbf{k}',\mathbf{k}+\mathbf{q}+\mathbf{G}} U_{n,\mathbf{k}}(\mathbf{r_0}),
\end{aligned}
\tag{5.28}
$$

where we emphasize that the integral over the unit cell cannot be simplified further, because the periodic functions $U_{n,\mathbf{k}}$ do not satisfy orthonormality relations over a single cell (see Eq. 3.72). The Hamiltonian becomes

$$
\begin{aligned}
\hat{\mathcal{H}}_{e-ph} &= i \sum_{\mu} \sum_{\mathbf{q}} \sum_{s=1}^{3\tau} \boldsymbol{\varepsilon}^{(s)}_{\mu}(\mathbf{q}) \sqrt{\frac{\hbar}{2M_\mu \omega_{\mathbf{q},s}}} \left(a_{\mathbf{q},s} + a^\dagger_{-\mathbf{q},s} \right) \cdot \sum_{\sigma} \sum_{n,n'} \sum_{\mathbf{k},\mathbf{k}'} C^\dagger_{n',\mathbf{k}',\sigma} C_{n,\mathbf{k},\sigma} \\
&(\mathbf{q}+\mathbf{G}) \left[N \int_{V_d} d^3\mathbf{r_0}\, U^*_{\mathbf{k}',n'}(\mathbf{r_0}) e^{-i(\mathbf{k}'-\mathbf{k}-\mathbf{q}-\mathbf{G})\cdot\mathbf{r_0}} \delta_{\mathbf{k}',\mathbf{k}+\mathbf{q}+\mathbf{G}} U_{\mathbf{k},n}(\mathbf{r_0}) \right].
\end{aligned}
\tag{5.29}
$$

We perform the sum over $\mathbf{k}'$, which activates the delta function and enforces momentum conservation. As a result, the electron-phonon interaction Hamiltonian takes the form

$$
\hat{\mathcal{H}}_{e-ph} = i \sum_{\substack{n,n',\sigma,\\ s,\mu,\\ \mathbf{k},\mathbf{G},\mathbf{q}}} C^\dagger_{n',\mathbf{k}+\mathbf{q}+\mathbf{G},\sigma} C_{n,\mathbf{k},\sigma} \left(a_{\mathbf{q},s} + a^\dagger_{-\mathbf{q},s} \right) \left(\sqrt{\frac{\hbar}{2M_\mu \omega_{\mathbf{q},s}}} \right)
$$

$$\left\{\boldsymbol{\varepsilon}^{(s)}_{\mu}(\mathbf{q})\cdot(\mathbf{q}+\mathbf{G})\right\}\tilde{V}_{\mu}\left(\mathbf{q}+\mathbf{G}\right)\left[N\int_{V_{\mathrm{d}}}d^3\mathbf{r}\,U^*_{n',\mathbf{k}+\mathbf{q}+\mathbf{G}}(\mathbf{r})U_{n,\mathbf{k}}(\mathbf{r})\right], \tag{5.30}$$

which we rewrite as

$$\hat{\mathcal{H}}_{e-ph}=\sum_{\substack{n,n',\sigma,s,\\ \mathbf{k},\mathbf{G},\mathbf{q}}} T^s_{\mathbf{k},\mathbf{q},\mathbf{G}}\,C^{\dagger}_{n',\mathbf{k}+\mathbf{q}+\mathbf{G},\sigma}C_{n,\mathbf{k},\sigma}\left(a_{\mathbf{q},s}+a^{\dagger}_{-\mathbf{q},s}\right), \tag{5.31}$$

which represents the third term of the Hamiltonian II.22, and we have defined

$$T^s_{\mathbf{k},\mathbf{q},\mathbf{G}}=i\sum_{\mu}\left(\sqrt{\frac{\hbar}{2M_{\mu}\omega_{\mathbf{q},s}}}\right)\left\{\boldsymbol{\varepsilon}^{(s)}_{\mu}(\mathbf{q})\cdot(\mathbf{q}+\mathbf{G}))\right\}\tilde{V}_{\mu}\left(\mathbf{q}+\mathbf{G}\right)$$
$$\times\left[N\int_{V_{\mathrm{d}}}d^3\mathbf{r}\,U^*_{n',\mathbf{k}+\mathbf{q}+\mathbf{G}}(\mathbf{r})U_{n,\mathbf{k}}(\mathbf{r})\right]. \tag{5.32}$$

These matrix elements $T^s_{\mathbf{k},\mathbf{q},\mathbf{G}}$ can be understood as describing interaction vertices where an electron absorbs or emits a phonon, with the conservation of crystal momentum explicitly enforced. The distinction between normal and Umklapp processes is automatically taken into account by whether a reciprocal lattice vector $\mathbf{G}$ is required to bring the resulting electron momentum back into the first Brillouin zone. An electron is annihilated in the state with momentum $\mathbf{k}$ and created in the state with momentum $\mathbf{k}+\mathbf{q}+\mathbf{G}$, where $\mathbf{q}$ is the phonon momentum. Let us now consider the constraint $\mathbf{k}'=\mathbf{k}+\mathbf{q}+\mathbf{G}$. The resulting vector $\mathbf{k}'$ must lie within the first Brillouin zone, since $\mathbf{k}$ and $\mathbf{q}$ are both defined within the first Brillouin zone of the reciprocal lattice. However, the sum $\mathbf{k}+\mathbf{q}$ may fall outside the first Brillouin zone. In such cases, a reciprocal lattice vector $\mathbf{G}$ is introduced to map the momentum back into the first Brillouin zone. This gives rise to what are known as Umklapp processes. If instead $\mathbf{k}+\mathbf{q}$ already lies within the first Brillouin zone, then $\mathbf{G}=\mathbf{0}$, and we refer to these as normal processes. In the case of normal processes, the vibrational modes contributing to the electron-phonon interaction and satisfying

$$\boldsymbol{\varepsilon}_{\mu}(\mathbf{q})\cdot\mathbf{q}\neq 0 \tag{5.33}$$

are longitudinal. Transverse phonons contribute to the dynamics only for Umklapp processes, with $\mathbf{G}\neq\mathbf{0}$.

The possible processes are two. First one is given by $C^{\dagger}_{n',\mathbf{k}',\sigma}C_{n,\mathbf{k},\sigma}a_{\mathbf{q},s}$, that is, a phonon with momentum $\mathbf{q}$ and an electron with momentum $\mathbf{k}$ are destroyed in the n band; an electron is created with momentum $\mathbf{k}'$, in the n' band, with the unchanged spin component. Second process is given by $C^{\dagger}_{n',\mathbf{k}',\sigma}C_{n,\mathbf{k},\sigma}a^{\dagger}_{-\mathbf{q},s}$, that is, an electron with momentum $\mathbf{k}$ in the n band is destroyed; a phonon with momentum $-\mathbf{q}$ and an electron with momentum $\mathbf{k}$ in the n' band with the unchanged spin component are created.

5.2 Simplification of the Hamiltonian of Electron-Phonon Interaction

In practice, we restrict the electron-phonon interaction to the conduction band (i.e., the highest electronic band) and consider only the phonon branch that couples most strongly to electrons and is experimentally relevant. Moreover, only intra-band processes are retained, and Umklapp processes are neglected, which corresponds to setting the reciprocal lattice vector $\mathbf{G} = \mathbf{0}$. Under these assumptions, the electron-phonon Hamiltonian simplifies to

$$\hat{\mathcal{H}}_{e-ph} = \sum_{\mathbf{q},\mathbf{k},\sigma} T_{\mathbf{k},\mathbf{q}} C^{\dagger}_{\mathbf{k}+\mathbf{q},\sigma} C_{\mathbf{k},\sigma} \left(a_{\mathbf{q}} + a^{\dagger}_{-\mathbf{q}} \right), \tag{5.34}$$

with a single electronic band ($n = n' = 1$) and a single phonon branch ($s = 1$). A further simplification arises in models such as the Fröhlich or Holstein Hamiltonians, where the matrix element $T_{\mathbf{k},\mathbf{q}}$ depends only on the phonon momentum $\mathbf{q}$. In this case, the electron-phonon Hamiltonian can be rewritten as

$$\hat{\mathcal{H}}_{e-ph} = \sum_{\mathbf{q},\mathbf{k},\sigma} M_{\mathbf{q}} C^{\dagger}_{\mathbf{k}+\mathbf{q},\sigma} C_{\mathbf{k},\sigma} \left(a_{\mathbf{q}} + a^{\dagger}_{-\mathbf{q}} \right). \tag{5.35}$$

Since the relation $\hat{\mathcal{H}}^{\dagger} = \hat{\mathcal{H}}$ holds, it follows that

$$M^{*}_{\mathbf{q}} = M_{-\mathbf{q}}. \tag{5.36}$$

Ultimately, note that from Eq. 5.5, the Hamiltonian in Eq. 5.35 can be expressed in terms of the (bosonic or fermionic) field operators $\hat{\psi}$ and $\hat{\psi}^{\dagger}$ (see Eqs. 2.149 and 2.150). In particular, Eq. 5.35 can be written as

$$\hat{\mathcal{H}}_{e-ph} = \sum_{\mathbf{q}} M_{\mathbf{q}} \left(a_{\mathbf{q}} + a^{\dagger}_{-\mathbf{q}} \right) \int dx \hat{\psi}^{\dagger}(x) e^{i\mathbf{q}\cdot\mathbf{r}} \hat{\psi}(x). \tag{5.37}$$

Part III
Developments of Quantum Many-Body Theory

Chapter 6
Many-Body Green's Functions

Abstract This chapter introduces many-body Green's functions, essential tools for analysing interacting quantum systems. Defined for both fermions and bosons, Green's functions provide a unified framework to describe time evolution, correlations, and system responses under interactions. Their analytical structure encodes key physical observables, including thermodynamic quantities like particle number in interacting systems, thus connecting quantum field theory with experimental measurements. A central focus is on spectral functions, which capture the excitation spectrum and are directly linked to the density of states. Through sum rules, these functions obey constraints that reflect fundamental conservation laws, serving as valuable checks on theoretical and numerical approaches. The chapter also distinguishes between imaginary-time and real-time Green's functions: the former are pivotal for studying equilibrium at finite temperature via tools like Matsubara summation, while the latter are essential for describing real-time dynamics and experimental observables. Analytic continuation connects the two, allowing the formalism to bridge theoretical models and measurable quantities. We illustrate these ideas with the Independent Boson Model, an exactly solvable system where a particle interacts with a bosonic environment, highlighting the impact of such couplings on dynamics and excitation properties. Another key application includes impurity problems such as the Anderson single-impurity model.

6.1 The Retarded One Particle Green's Function in Quantum Many-Body Theory

Here we aim to generalize the definition of the one-particle retarded Green's function (see Eq. 1.274) to a quantum many-body system. Let $\mathcal{H}$ be a one-body Hamiltonian with eigenstates $\{\varphi_\alpha(x)\}$ and eigenvalues $\{\mathcal{E}_\alpha\}$. In the one-particle picture, the general solution of the Schrödinger equation (see Eq. 1.86) reads

$$\psi(x,t) = \sum_\alpha c_\alpha \, e^{-i\frac{\mathcal{E}_\alpha}{\hbar}t} \varphi_\alpha(x), \tag{6.1}$$

F. Tafuri et al., *Understanding Quantum Many-Body Theory*, UNITEXT for Physics, https://doi.org/10.1007/978-3-032-17849-7_6

where we used Eq. 2.187. In the second-quantized framework, the role of the one-particle wave function is then taken over by the field operators (see Eqs. 2.149 and 2.150): the annihilation operator $\hat{\psi}(x, t)$ removes a particle at space-time point $x = (\mathbf{r}, s)$, while its adjoint $\hat{\psi}^\dagger(x, t)$ creates one.

The transition from the one-particle wave function with its unitary time evolution to the many-body formalism is achieved: the time dependence is now carried by the field operators in the interaction picture (see Eq. 1.107), which create or annihilate particles at given space-time points. As consequence, we might expect the many-body Green's function to be

$$-\frac{i}{\hbar}\Theta(t - t')\langle\hat{\psi}(x, t)\hat{\psi}^\dagger(x', t')\rangle, \tag{6.2}$$

where we employed the Heisenberg picture (see Eq. 1.102) and $\langle \ldots \rangle$ denotes the thermal average (see Eq. 1.379). However, the expression in Eq. 6.2 is not consistent with the expected physical properties of the Green's function. In particular, it fails to yield the correct short-time singularity (see Eqs. 1.295 and 1.304), that is,

$$\lim_{t \to t'^+} -\frac{i}{\hbar}\Theta(t - t')\langle\hat{\psi}(x, t)\hat{\psi}^\dagger(x', t')\rangle \neq -\frac{i}{\hbar}\delta(x - x'). \tag{6.3}$$

From Eq. 2.153, we infer

Definition 6.1 (Retarded Many-Body Green's Function of Field Operators) The retarded many-body Green's function associated with the field operators (see Eqs. 2.149 and 2.150), expressed in the Heisenberg picture (see Eq. 1.102), is defined as

$$\begin{aligned} G^{(r)}(x, t, x', t') &\equiv G^{(r)}_{\hat{\psi}(x), \hat{\psi}^\dagger(x')}(t, t') = \\ &= -\frac{i}{\hbar}\Theta(t - t')\langle[\hat{\psi}(x, t), \hat{\psi}^\dagger(x', t')]^{(\varepsilon)}\rangle, \; t > t', \end{aligned} \tag{6.4}$$

where $x = (\mathbf{r}, s)$ and $x' = (\mathbf{r}', s')$, as defined in Eq. 1.43, and $[\cdot, \cdot]^{(\varepsilon)}$ denotes the general commutator defined in Eq. 2.102.

Remark 6.1 (Notation for the Many-Body Green's Function of Field Operators) Note that the same symbol $G^{(r)}(x, t, x', t')$ has been used to denote both the one-particle propagator (see Eqs. 1.274 and 1.284) and the many-body Green's function associated with the field operators (see Eq. 6.4). It should be recalled that, in the one-particle case, the variable x denotes a one-dimensional spatial coordinate (a scalar position variable), whereas in the many-body case it denotes the pair $x = (\mathbf{r}, s)$ (see Eq. 1.43), where $\mathbf{r}$ is the position vector and s the spin variable. Unless otherwise specified, in the remainder of this book the symbol $G^{(r)}(x, t, x', t')$ will always refer to the many-body Green's function of the field operators.

For notational compactness, we avoid writing explicit operator labels such as $G^{(r)}_{\hat{\psi}(x),\hat{\psi}^\dagger(x')}(t,t')$, and we simply use $G^{(r)}(x,t,x',t')$ to emphasize the dependence on the continuous variables x and x'.
In the remainder of this book the symbol $G^{(r)}(x,t,x',t')$ will always refer to the many-body Green's function of the field operators.

The quantity appearing in the thermal average in Eq. 6.4 can be written explicitly as

$$\left[\hat{\psi}(x,t),\hat{\psi}^\dagger(x',t')\right]^{(\varepsilon)} = \hat{\psi}(x,t)\hat{\psi}^\dagger(x',t') - \varepsilon\hat{\psi}^\dagger(x',t')\hat{\psi}(x,t),\ t > t'. \tag{6.5}$$

These are respectively referred to as the particle term (where a particle is first created and then destroyed) and the hole term (where a particle is first destroyed and then created). Both particle and hole propagation are present, which represents the correct conceptual generalization of the propagator to a quantum many-body system. We now demonstrate that Eq. 6.4 indeed satisfies the defining properties of a propagator. As a first check, we consider the case of non-interacting particles and verify that the free many-body propagator exhibits the same structure as the one-particle propagator, namely Eq. 1.283. To this end, given the Hamiltonian of free non-interacting particles, that is, Eq. 2.181, we have

$$\begin{aligned}
G^{(r)(0)}(x,t,x',t') &= -\frac{i}{\hbar}\Theta(t-t')\sum_{\alpha,\alpha'}\varphi_\alpha(x)e^{-i\frac{\mathcal{E}_\alpha}{\hbar}t}e^{i\frac{\mathcal{E}_{\alpha'}}{\hbar}t'}\varphi^*_{\alpha'}(x')\left\langle\left[a_\alpha,a^\dagger_{\alpha'}\right]^{(\varepsilon)}\right\rangle = \\
&= -\frac{i}{\hbar}\Theta(t-t')\sum_{\alpha,\alpha'}\varphi_\alpha(x)e^{-i\frac{\mathcal{E}_\alpha}{\hbar}t}e^{i\frac{\mathcal{E}_{\alpha'}}{\hbar}t'}\varphi^*_{\alpha'}(x')\delta_{\alpha,\alpha'} = \\
&= -\frac{i}{\hbar}\Theta(t-t')\sum_{\alpha}\varphi_\alpha(x)e^{-i\frac{\mathcal{E}_\alpha}{\hbar}(t-t')}\varphi^*_{\alpha}(x'),
\end{aligned} \tag{6.6}$$

which, in fact, has the same structure as Eq. 1.283. Then, for both non-interacting and interacting particles, Eq. 6.4 satisfies

$$\lim_{t\to t'^+} -\frac{i}{\hbar}\Theta(t-t')\left\langle\left[\psi(x,t),\psi^\dagger(x',t')\right]^{(\varepsilon)}\right\rangle = -\frac{i}{\hbar}\delta(x-x'). \tag{6.7}$$

Finally, we can introduce

Definition 6.2 (Retarded Many-Body Green's Function) Given the operators A and B, the retarded many-body Green's function associated with A and B at times t and t' is defined as

$$G^{(r)}_{A,B}(t,t') = -\frac{i}{\hbar}\Theta(t-t')\left\langle\left[A(t),B(t')\right]^{(\varepsilon)}\right\rangle \equiv$$

$$\equiv -\frac{i}{\hbar}\Theta(t-t')\left[\langle A(t)B(t')\rangle - \varepsilon\langle B(t')A(t)\rangle\right], \tag{6.8}$$

where $[\cdot,\cdot]^{(\varepsilon)}$ denotes the general commutator defined in Eq. 2.102 and the operators $A(t)$ and $B(t)$ are operators in Heisenberg picture (see Eq. 1.102). Equation 6.8 characterizes the causal correlation between the operators A and B at times t and t', vanishing when $t < t'$.

Many functions of physical interest are written in the form given in Eq. 6.8, as we shall see in the following.

Remark 6.2 (Note on Terminology) In one-particle quantum mechanics, the term propagator refers to the Green's function that describes the amplitude for a particle to propagate from one space-time point to another. It directly solves the linear equation of motion for a non-interacting particle. In contrast, within quantum many-body theory, the analogous quantities are known as many-body Green's functions. These functions describe the effective propagation of particles that are influenced by interactions with other particles in the system. Because of these interactions, many-body Green's functions incorporate complex correlation effects, such as particle creation and annihilation processes, and cannot be simply interpreted as one-particle propagators. Therefore, although the term propagator is sometimes used informally in quantum many-body theory, the more precise and rigorous terminology is Green's function or many-body Green's function. However, for stylistic convenience, throughout this text we will often use the two terms interchangeably, keeping in mind the subtle conceptual differences.

At the conclusion of this section, we present an important property of Green's functions that will be extensively used in the following. In many-body physics and quantum statistical mechanics, the concept of thermal equilibrium plays a fundamental role in determining the properties of correlation functions. When a system is coupled to a thermal reservoir at temperature T, it reaches a stationary state characterized by invariance under time translations. This time-translation invariance implies that dynamical quantities such as the retarded Green's function depend only on the elapsed time between events, rather than on the absolute times themselves. The following theorem formalizes this property.

Theorem 6.1 (Time-Translation Invariance of Green's Functions in Thermal Equilibrium) *At thermal equilibrium, the retarded many-body Green's function given in Eq. 6.8 exhibits time-translation invariance, i.e.,*

$$G^{(r)}_{A,B}(t,t') = G^{(r)}_{A,B}(t-t'). \tag{6.9}$$

Proof Given the retarded Green's function 6.8, regarding the first thermal average (see Eq. 1.379), we have

$$
\begin{aligned}
\left\langle A(t)B(t')\right\rangle &= \frac{1}{Z}\,\mathrm{Tr}\left[e^{-\beta\hat{\mathcal{H}}}e^{i\frac{\hat{\mathcal{H}}}{\hbar}t}Ae^{-i\frac{\hat{\mathcal{H}}}{\hbar}t}e^{i\frac{\hat{\mathcal{H}}}{\hbar}t'}Be^{-i\frac{\hat{\mathcal{H}}}{\hbar}t'}\right] = \\
&= \frac{1}{Z}\,\mathrm{Tr}\left[e^{-\beta\hat{\mathcal{H}}}e^{i\frac{\hat{\mathcal{H}}}{\hbar}t}Ae^{-i\frac{\hat{\mathcal{H}}}{\hbar}(t-t')}Be^{-i\frac{\hat{\mathcal{H}}}{\hbar}t'}\right] = \\
&= \frac{1}{Z}\,\mathrm{Tr}\left[e^{-i\frac{\hat{\mathcal{H}}}{\hbar}t'}e^{-\beta\hat{\mathcal{H}}}e^{i\frac{\hat{\mathcal{H}}}{\hbar}t}Ae^{-i\frac{\hat{\mathcal{H}}}{\hbar}(t-t')}B\right] = \\
&= \frac{1}{Z}\,\mathrm{Tr}\left[e^{-\beta\hat{\mathcal{H}}}e^{-i\frac{\hat{\mathcal{H}}}{\hbar}t'}e^{i\frac{\hat{\mathcal{H}}}{\hbar}t}Ae^{-i\frac{\hat{\mathcal{H}}}{\hbar}(t-t')}B\right] = \\
&= \frac{1}{Z}\,\mathrm{Tr}\left[e^{-\beta\hat{\mathcal{H}}}e^{i\frac{\hat{\mathcal{H}}}{\hbar}(t-t')}Ae^{-i\frac{\hat{\mathcal{H}}}{\hbar}(t-t')}B\right] = \\
&= \left\langle A(t-t')B(0)\right\rangle ,
\end{aligned}
\tag{6.10}
$$

where we have used the cyclic property of the trace (see Eq. J.3), and then reversed the order of the product $e^{-i\frac{\hat{\mathcal{H}}}{\hbar}t'}e^{-\beta\hat{\mathcal{H}}}$, which is justified by Eq. H.20. Similarly, we obtain

$$
\begin{aligned}
\left\langle B(t')A(t)\right\rangle &= \frac{1}{Z}\,\mathrm{Tr}\left[e^{-\beta\hat{\mathcal{H}}}e^{i\frac{\hat{\mathcal{H}}}{\hbar}t'}Be^{-i\frac{\hat{\mathcal{H}}}{\hbar}t'}e^{i\frac{\hat{\mathcal{H}}}{\hbar}t}Ae^{-i\frac{\hat{\mathcal{H}}}{\hbar}t}\right] = \\
&= \frac{1}{Z}\,\mathrm{Tr}\left[e^{-\beta\hat{\mathcal{H}}}e^{i\frac{\hat{\mathcal{H}}}{\hbar}t'}Be^{i\frac{\hat{\mathcal{H}}}{\hbar}(t-t')}Ae^{-i\frac{\hat{\mathcal{H}}}{\hbar}t}\right] = \\
&= \frac{1}{Z}\,\mathrm{Tr}\left[e^{i\frac{\hat{\mathcal{H}}}{\hbar}t'}e^{-\beta\hat{\mathcal{H}}}Be^{i\frac{\hat{\mathcal{H}}}{\hbar}(t-t')}Ae^{-i\frac{\hat{\mathcal{H}}}{\hbar}t}\right] = \\
&= \frac{1}{Z}\,\mathrm{Tr}\left[e^{-\beta\hat{\mathcal{H}}}Be^{i\frac{\hat{\mathcal{H}}}{\hbar}(t-t')}Ae^{-i\frac{\hat{\mathcal{H}}}{\hbar}t}e^{i\frac{\hat{\mathcal{H}}}{\hbar}t'}\right] = \\
&= \frac{1}{Z}\,\mathrm{Tr}\left[e^{-\beta\hat{\mathcal{H}}}Be^{i\frac{\hat{\mathcal{H}}}{\hbar}(t-t')}Ae^{-i\frac{\hat{\mathcal{H}}}{\hbar}(t-t')}\right] = \\
&= \left\langle B(0)A(t-t')\right\rangle .
\end{aligned}
\tag{6.11}
$$

By combining both results, we observe that the thermal averages depend only on the time difference $t - t'$, confirming time-translation invariance of the thermal expectation values in the Heisenberg picture with a time-independent Hamiltonian, that is, establishing the thesis. □

6.2 Matsubara Green's Function

In order to formulate many-body quantum theory at finite temperature, it is convenient to extend the operator formalism to imaginary times (see Eq. 2.193).

Definition 6.3 (Thermal Heisenberg Picture) The thermal Heisenberg picture is obtained by adapting the Heisenberg picture (see Eq. 1.102) to imaginary times. For a generic Hamiltonian $\hat{\mathcal{H}}$ given in 1.85, within the second-quantization formalism the operator A in the thermal Heisenberg picture is defined as

$$A(\tau) = e^{\frac{\hat{\mathcal{H}}}{\hbar}\tau} A\, e^{-\frac{\hat{\mathcal{H}}}{\hbar}\tau}, \quad \tau \in [-\beta\hbar, \beta\hbar], \tag{6.12}$$

where τ is a real time variable related to imaginary times t through Eq. 2.193.

Note that from the identity $\mathbb{1} = e^{-\frac{\hat{\mathcal{H}}_0}{\hbar}\tau} e^{\frac{\hat{\mathcal{H}}_0}{\hbar}\tau}$, one finds

$$\begin{aligned} A(\tau) &= e^{\frac{\hat{\mathcal{H}}}{\hbar}\tau} A e^{-\frac{\hat{\mathcal{H}}}{\hbar}\tau} = \\ &= e^{\frac{\hat{\mathcal{H}}}{\hbar}\tau} e^{-\frac{\hat{\mathcal{H}}_0}{\hbar}\tau} e^{\frac{\hat{\mathcal{H}}_0}{\hbar}\tau} A e^{-\frac{\hat{\mathcal{H}}_0}{\hbar}\tau} e^{\frac{\hat{\mathcal{H}}_0}{\hbar}\tau} e^{-\frac{\hat{\mathcal{H}}}{\hbar}\tau} = \\ &= U^{-1}(\tau) A^{(0)}(\tau) U(\tau) \equiv \\ &\equiv S(0,\tau) A^{(0)}(\tau) S(\tau, 0), \end{aligned} \tag{6.13}$$

i.e., the thermal Heisenberg representation of an operator can be expressed in terms of its thermal interaction representation and the matrix S.

Remark 6.3 (Thermal Interaction Picture and Matsubara Formalism) The property 6.13 implies that any thermal equilibrium propagator (i.e., any imaginary-time ordered correlation function) can be expressed in terms of operators in the thermal interaction picture and the operator S defined on the imaginary-time. This reformulation is the formal basis of the Matsubara approach and of diagrammatic perturbation theory at finite temperature (see Chap. 7).

For reasons that will become clear shortly, we introduce the following

Definition 6.4 (Matsubara (Thermal) Green's Function) Let A and B be two operators in the thermal Heisenberg picture (see Eq. 6.12), with respect to a Hamiltonian of the form 1.85, within the second-quantization formalism. The Matsubara (thermal) Green's function associated with A and B at imaginary times τ' and τ'' is defined as

$$G^{(m)}_{A,B}(\tau', \tau'') = -\frac{1}{\hbar} \left\langle \hat{T}_\tau A(\tau') B(\tau'') \right\rangle, \tag{6.14}$$

where $\hat{T}_\tau$ is the Wick time-ordering operator defined in Eq. 2.200.

Theorem 6.1 also applies to the Matsubara Green's function. Consequently, from now on, whenever the system is in thermodynamic equilibrium, the Matsubara Green's function will be written concisely as

$$
\begin{aligned}
G^{(m)}_{A,B}(\tau,0) &\equiv G^{(m)}_{A,B}(\tau-0) \equiv \\
&\equiv G^{(m)}_{A,B}(\tau) = \\
&= -\frac{1}{\hbar}\left\langle \hat{T}_\tau A(\tau)B(0)\right\rangle = \\
&= -\frac{1}{\hbar}\left\langle \Theta(\tau)A(\tau)B(0) + \varepsilon\Theta(-\tau)B(0)A(\tau)\right\rangle = \\
&= -\frac{1}{\hbar}\left[\Theta(\tau)\left\langle A(\tau)B(0)\right\rangle + \varepsilon\Theta(-\tau)\left\langle B(0)A(\tau)\right\rangle\right].
\end{aligned}
\tag{6.15}
$$

where the action of the Wick time-ordering operator $\hat{T}_\tau$ has been made explicit.

6.2.1 Properties of Bosonic and Fermionic Matsubara Green's Functions

Consider the Matsubara Green's function for generic operators A and B (see Eq. 6.14). For convenience, we assume the system to be in thermal equilibrium, so that the Green's function depends only on a single time variable (see Theorem 6.1). Given the instant $\tau - \beta\hbar$, with $\tau \in (0, \beta\hbar]$, we evaluate the thermal average (see Eq. 1.379) of the second term in Eq. 6.15, that is,

$$
\begin{aligned}
G^{(m)}_{A,B}(\tau-\beta\hbar) &= -\frac{\varepsilon}{\hbar Z}\,\mathrm{Tr}\left(e^{-\beta\hat{\mathcal{H}}}Be^{\frac{\hat{\mathcal{H}}}{\hbar}(\tau-\beta\hbar)}Ae^{-\frac{\hat{\mathcal{H}}}{\hbar}(\tau-\beta\hbar)}\right) = \\
&= -\frac{\varepsilon}{\hbar Z}\,\mathrm{Tr}\left(e^{-\beta\hat{\mathcal{H}}}e^{\frac{\hat{\mathcal{H}}}{\hbar}\tau}Ae^{-\frac{\hat{\mathcal{H}}}{\hbar}\tau}e^{\beta\hat{\mathcal{H}}}e^{-\beta\hat{\mathcal{H}}}B\right) = \\
&= -\frac{\varepsilon}{\hbar Z}\,\mathrm{Tr}\left(e^{-\beta\hat{\mathcal{H}}}e^{\frac{\hat{\mathcal{H}}}{\hbar}\tau}Ae^{-\frac{\hat{\mathcal{H}}}{\hbar}\tau}B\right) = \\
&= \varepsilon G^{(m)}_{A,B}(\tau).
\end{aligned}
\tag{6.16}
$$

Then, if $\tau - \beta\hbar \in (-\beta\hbar, 0]$, i.e., if $\tau \in (0, \beta\hbar]$, the thermal Green's function computed at time $\tau - \beta\hbar$ is equal to the thermal Green's function evaluated at $\tau \in (0, \beta\hbar]$, multiplied by ε. It follows

$$
G^{(m)}_{A,B}(\tau-\beta\hbar) = G^{(m)}_{A,B}(\tau),\;\; \tau \in (0, \beta\hbar], \quad \text{(bosons)} \tag{6.17}
$$

that is, the bosonic thermal Green's function is a periodic function of period $\beta\hbar$, and

$$
G^{(m)}_{A,B}(\tau-\beta\hbar) = -G^{(m)}_{A,B}(\tau),\;\; \tau \in (0, \beta\hbar], \quad \text{(fermions)} \tag{6.18}
$$

that is, the fermionic thermal Green's function is an antisymmetric, periodic function of period $\beta\hbar$. The calculations involving Matsubara Green's functions can be consistently carried out by restricting the variable τ to the interval $(0, \beta\hbar]$, where the formalism is well-defined and periodic (or antiperiodic) boundary conditions apply.

Starting from the Fourier expansion, that is, Theorem C.1, adapted to the interval $[-L, L]$ instead of $[-\frac{L}{2}, \frac{L}{2}]$, and taking into account that the Matsubara Green's function $G^{(m)}_{A,B}(\tau)$ is periodic on the interval $[-\beta\hbar, \beta\hbar]$, we can expand it in a Fourier series over any interval of width $2\beta\hbar$ as

$$G^{(m)}_{A,B}(\tau) = \frac{1}{\beta\hbar}\sum_{n\in\mathbb{Z}} G^{(m)}_{A,B}(i\omega_n)e^{-i\omega_n\tau}, \tag{6.19}$$

$$\omega_n = \frac{n\pi}{\beta\hbar}, \quad n \in \mathbb{Z}, \tag{6.20}$$

$$G^{(m)}_{A,B}(i\omega_n) = \frac{1}{2}\int_{-\beta\hbar}^{\beta\hbar} d\tau G^{(m)}_{A,B}(\tau)e^{i\omega_n\tau}. \tag{6.21}$$

Note that, in addition to adapting the interval from $[-\frac{L}{2}, \frac{L}{2}]$ to $[-L, L]$, we are also using a different normalization convention for the Fourier series: here, the prefactor $(\beta\hbar)^{-1}$ appears only in the series, while the Fourier coefficients $G^{(m)}_{A,B}(i\omega_n)$ are defined without an additional square-root factor, unlike the convention used in Theorem C.1. As discussed in Remark C.1, these choices are purely conventional and do not affect the underlying mathematics.

From

$$G^{(m)}_{A,B}(i\omega_n) = \frac{1}{2}\left(\int_{-\beta\hbar}^{0} d\tau G^{(m)}_{A,B}(\tau)e^{i\omega_n\tau} + \int_{0}^{\beta\hbar} d\tau G^{(m)}_{A,B}(\tau)e^{i\omega_n\tau}\right), \tag{6.22}$$

we compute the first integral. We set $\tau' = \tau + \beta\hbar$ ($\tau' \in (0, \beta\hbar)$) and we employ Eq. 6.16, then

$$\begin{aligned} G^{(m)}_{A,B}(i\omega_n) &= \frac{1}{2}\left(\int_{0}^{\beta\hbar} d\tau' G^{(m)}_{A,B}(\tau' - \beta\hbar)e^{i\omega_n(\tau'-\beta\hbar)}\right. \\ &\quad \left. + \int_{0}^{\beta\hbar} d\tau G^{(m)}_{A,B}(\tau)e^{i\omega_n\tau}\right) = \\ &= \frac{1}{2}\left(\int_{0}^{\beta\hbar} d\tau' \varepsilon G^{(m)}_{A,B}(\tau')e^{i\omega_n(\tau'-\beta\hbar)} + \int_{0}^{\beta\hbar} d\tau G^{(m)}_{A,B}(\tau)e^{i\omega_n\tau}\right), \end{aligned} \tag{6.23}$$

and we rename the dummy index τ', so

$$G^{(m)}_{A,B}(i\omega_n) = \frac{1}{2}\int_0^{\beta\hbar} d\tau G^{(m)}_{A,B}(\tau)e^{i\omega_n\tau}\left(1+\varepsilon e^{-i\omega_n\beta\hbar}\right) \equiv$$
$$\equiv \frac{1}{2}\int_0^{\beta\hbar} d\tau G^{(m)}_{A,B}(\tau)e^{i\omega_n\tau}\left(1+\varepsilon e^{-in\pi}\right). \tag{6.24}$$

There are two cases, depending on the statistical nature of the particles. For bosons, if n is even, the factor $\frac{1}{2}$ is simplified by 2, if n is odd the Green's function $G^{(m)}_{A,B}(i\omega_n)$ is null: only the even frequencies are non-null and the factor $\frac{1}{2}$ disappears. For fermions, if n is odd, the factor $\frac{1}{2}$ simplifies to 2, if n is even, the Green's function $G^{(m)}_{A,B}(i\omega_n)$ is null: only the odd frequencies are non-null and the factor $\frac{1}{2}$ disappears. Any Matsubara Green's function can be written as

$$G^{(m)}_{A,B}(\tau) = \frac{1}{\beta\hbar}\sum_{n\in\mathbb{Z}} G^{(m)}_{A,B}(i\omega_n)e^{-i\omega_n\tau}, \tag{6.25}$$

with

$$G^{(m)}_{A,B}(i\omega_n) = \int_0^{\beta\hbar} d\tau G^{(m)}_{A,B}(\tau)e^{i\omega_n\tau}, \tag{6.26}$$

where the frequencies

$$i\omega_n = \begin{cases} \dfrac{i2n\pi}{\beta\hbar}, & \text{bosons} \\ \dfrac{i(2n+1)\pi}{\beta\hbar}, & \text{fermions} \end{cases} \tag{6.27}$$

belong to the imaginary axis.

Ultimately, note that the statistical index given in Eq. 2.52 satisfies

$$e^{i\omega_n\beta\hbar} \equiv \varepsilon. \tag{6.28}$$

6.3 Spectral Function and Analytical Extension of Green's Function

The thermal Green function $G^{(m)}_{A,B}(i\omega_n)$ given in Eq. 6.21 is defined on the imaginary frequency axis, but our interest lies in its behavior on the real frequency axis. Let $\hat{\mathcal{H}}$ be a Hamiltonian that includes interactions among particles, and suppose that it has been diagonalized, i.e., we assume its eigenvalues $\{\mathcal{E}_n\}$ and corresponding eigenstates $\{|n\rangle\}$ are known. We express the retarded many-body Green's function 6.8 in terms of this eigenbasis as follows

$$G^{(r)}_{A,B}(t) = -\frac{i}{\hbar}\,\Theta(t)\,\frac{1}{Z}\,\mathrm{Tr}\left\{e^{-\beta\hat{\mathcal{H}}}\left[e^{i\frac{\hat{\mathcal{H}}}{\hbar}t}\,A\,e^{-i\frac{\hat{\mathcal{H}}}{\hbar}t}\,B - \varepsilon\,B\,e^{i\frac{\hat{\mathcal{H}}}{\hbar}t}\,A\,e^{-i\frac{\hat{\mathcal{H}}}{\hbar}t}\right]\right\} =$$

$$= -\frac{i}{\hbar}\,\Theta(t)\,\frac{1}{Z}\sum_{n,m}\left[\langle n|e^{-\beta\hat{\mathcal{H}}}e^{i\frac{\hat{\mathcal{H}}}{\hbar}t}Ae^{-i\frac{\hat{\mathcal{H}}}{\hbar}t}|m\rangle\langle m|B|n\rangle\right.$$

$$\left. - \varepsilon\,\langle n|B|m\rangle\langle m|e^{-\beta\hat{\mathcal{H}}}e^{i\frac{\hat{\mathcal{H}}}{\hbar}t}Ae^{-i\frac{\hat{\mathcal{H}}}{\hbar}t}|n\rangle\right] =$$

$$= -\frac{i}{\hbar}\,\Theta(t)\,\frac{1}{Z}\sum_{n,m}\left[\langle n|A|m\rangle\langle m|B|n\rangle\,e^{-\beta\mathcal{E}_n}\,e^{i\frac{\mathcal{E}_n-\mathcal{E}_m}{\hbar}t}\right.$$

$$\left. -\varepsilon\,\langle n|B|m\rangle\langle m|A|n\rangle\,e^{-\beta\mathcal{E}_m}\,e^{i\frac{\mathcal{E}_m-\mathcal{E}_n}{\hbar}t}\right]. \tag{6.29}$$

we exchange the dummy indices n, m in the second term and we get

$$G^{(r)}_{A,B}(t) = -\frac{i}{\hbar}\,\Theta(t)\,\frac{1}{Z}\sum_{n,m}\langle n|A|m\rangle\langle m|B|n\rangle\,e^{i\frac{\mathcal{E}_n-\mathcal{E}_m}{\hbar}t}\left[e^{-\beta\mathcal{E}_n} - \varepsilon\,e^{-\beta\mathcal{E}_m}\right] \equiv$$

$$\equiv -\frac{i}{\hbar}\,\Theta(t)\,\frac{1}{Z}\sum_{n,m}A_{nm}\,B_{mn}\,e^{i\frac{\mathcal{E}_n-\mathcal{E}_m}{\hbar}t}\left[e^{-\beta\mathcal{E}_n} - \varepsilon\,e^{-\beta\mathcal{E}_m}\right], \tag{6.30}$$

At the thermodynamic equilibrium, the retarded Green's function can be written in the form

$$G^{(r)}_{A,B}(t) = \Theta(t)C(t), \tag{6.31}$$

where

$$C(t) = -\frac{i}{\hbar}\left\langle [A(t), B(0)]^{(\varepsilon)}\right\rangle \tag{6.32}$$

is the correlation function of A and B at thermodynamic equilibrium. For $t \to \infty$, $C(t)$ is zero or constant. We set $z = \omega + i\delta$, with $\omega, \delta \in \mathbb{R}$, and we apply a Fourier transform, i.e.,

$$G^{(r)}_{A,B}(z) = \int_{-\infty}^{+\infty} dt\; G^{(r)}_{A,B}(t)e^{izt} \;\equiv$$

$$\equiv \int_{0}^{+\infty} dt\; G^{(r)}_{A,B}(t)e^{i\omega t}\,e^{-\delta t}, \tag{6.33}$$

where at the last step we used Heaviside's function $\Theta(t)$. Conversely, if we know $G^{(\mathrm{r})}_{A,B}(z)$, to calculate $G^{(\mathrm{r})}_{A,B}(t)$ it is sufficient to apply the Fourier antitransform, i.e.,

$$G^{(r)}_{A,B}(t) = \int_{-\infty}^{+\infty} \frac{d\omega}{2\pi} G^{(r)}_{A,B}(\omega + i\delta) e^{-i(\omega+i\delta)t}, \ \forall\, \delta > 0, \tag{6.34}$$

where the prescription $\omega + i\delta$ ensures the correct analytic structure in the upper half of the complex frequency plane. Upon determining the eigenvalue spectrum and the associated eigenstates of $\hat{\mathcal{H}}$, the Green's function can be cast into an explicit form within the energy eigenbasis representation as

$$\begin{aligned} G^{(r)}_{A,B}(z) &= \int_0^{\infty} dt\, G^{(r)}_{A,B}(t) e^{izt} = \\ &= -\frac{i}{\hbar}\frac{1}{Z} \sum_{n,m} A_{nm} B_{mn} \left(e^{-\beta\mathcal{E}_n} - \varepsilon e^{-\beta\mathcal{E}_m}\right) \int_0^{\infty} dt\, e^{izt} e^{i\frac{\mathcal{E}_n - \mathcal{E}_m}{\hbar}t}. \end{aligned} \tag{6.35}$$

For every $\delta > 0$, the exponential function $e^{-\delta t}$ guarantees that $G^{(r)}_{A,B}(z)$ is analytic in the upper complex half-plane, indeed

$$\begin{aligned} G^{(r)}_{A,B}(z) &= -\frac{i}{\hbar Z} \sum_{n,m} A_{nm} B_{mn} \left(e^{-\beta\mathcal{E}_n} - \varepsilon e^{-\beta\mathcal{E}_m}\right) \int_0^{\infty} dt\, e^{i\left(z+\frac{\mathcal{E}_n-\mathcal{E}_m}{\hbar}\right)t} = \\ &= -\frac{i}{\hbar Z} \sum_{n,m} A_{nm} B_{mn} \left(e^{-\beta\mathcal{E}_n} - \varepsilon e^{-\beta\mathcal{E}_m}\right) \\ &\quad \int_0^{\infty} dt \frac{i\left(z+\frac{\mathcal{E}_n-\mathcal{E}_m}{\hbar}\right)}{i\left(z+\frac{\mathcal{E}_n-\mathcal{E}_m}{\hbar}\right)} e^{i\left(z+\frac{\mathcal{E}_n-\mathcal{E}_m}{\hbar}\right)t} = \\ &= -\frac{i}{\hbar Z} \sum_{n,m} A_{nm} B_{mn} \left(e^{-\beta\mathcal{E}_n} - \varepsilon e^{-\beta\mathcal{E}_m}\right) \frac{1}{i\left(z+\frac{\mathcal{E}_n-\mathcal{E}_m}{\hbar}\right)} \\ &\quad \left[e^{i\left(z+\frac{\mathcal{E}_n-\mathcal{E}_m}{\hbar}\right)t}\right]_0^{\infty} = \\ &= -\frac{i}{\hbar Z} \sum_{n,m} A_{nm} B_{mn} \left(e^{-\beta\mathcal{E}_n} - \varepsilon e^{-\beta\mathcal{E}_m}\right) \frac{1}{i\left(z+\frac{\mathcal{E}_n-\mathcal{E}_m}{\hbar}\right)} \\ &\quad \left[e^{i\left(\omega+\frac{\mathcal{E}_n-\mathcal{E}_m}{\hbar}\right)t} e^{-\delta t}\right]_0^{\infty} = \\ &= -\frac{1}{\hbar Z} \sum_{n,m} A_{nm} B_{mn} \left(e^{-\beta\mathcal{E}_n} - \varepsilon e^{-\beta\mathcal{E}_m}\right) \frac{1}{\left(z+\frac{\mathcal{E}_n-\mathcal{E}_m}{\hbar}\right)} \\ &\quad \left[e^{i\left(\omega+\frac{\mathcal{E}_n-\mathcal{E}_m}{\hbar}\right)t} e^{-\delta t}\right]_0^{\infty}. \end{aligned} \tag{6.36}$$

The last term is the product of an oscillating factor for a decreasing monotone function, in the limit for $t \to \infty$ tends to zero, so

$$
\begin{aligned}
G^{(r)}_{A,B}(z) &= -\frac{1}{\hbar Z}\sum_{n,m} A_{nm}B_{mn}\left(e^{-\beta \mathcal{E}_n} - \varepsilon e^{-\beta \mathcal{E}_m}\right)\frac{1}{\left(z + \frac{\mathcal{E}_n - \mathcal{E}_m}{\hbar}\right)}(0-1) = \\
&= \frac{1}{\hbar Z}\sum_{n,m} A_{nm}B_{mn}\left(e^{-\beta \mathcal{E}_n} - \varepsilon e^{-\beta \mathcal{E}_m}\right)\frac{1}{\left(z + \frac{\mathcal{E}_n - \mathcal{E}_m}{\hbar}\right)} = \\
&= \frac{1}{Z}\sum_{n,m} A_{nm}B_{mn}\frac{e^{-\beta \mathcal{E}_n} - \varepsilon e^{-\beta \mathcal{E}_m}}{\hbar\left(z + \frac{\mathcal{E}_n - \mathcal{E}_m}{\hbar}\right)},
\end{aligned}
\tag{6.37}
$$

and the Green's function is complex or not depending on the denominator $\left(z + \frac{\mathcal{E}_n - \mathcal{E}_m}{\hbar}\right)$. Before proceeding, it is worth noting that from the structure of Eq. 6.37 one can infer a very useful property. We have

Theorem 6.2 (High-Frequency Limit of Many-Body Retarded Green's Function) *For large $|z|$, the retarded many-body Green's function $G^{(r)}_{A,B}(z)$ given in Eq. 6.37 decreases as $(\hbar z)^{-1}$. More precisely, it satisfies*

$$
\lim_{|z|\to\infty} G^{(r)}_{A,B}(z) = \frac{1}{\hbar z}\left\langle [A, B]^{(\varepsilon)} \right\rangle. \tag{6.38}
$$

Proof Starting from Eq. 6.37, for $|z| \to \infty$, we can expand the denominator in powers of z^{-1} as follows

$$
\frac{1}{z + \frac{\mathcal{E}_n - \mathcal{E}_m}{\hbar}} = \frac{1}{z}\left(1 - \frac{\mathcal{E}_n - \mathcal{E}_m}{\hbar z} + O\left(\frac{1}{z^2}\right)\right). \tag{6.39}
$$

This expansion is valid in the regime where $|z|$ is large compared with $|\mathcal{E}_n - \mathcal{E}_m|/\hbar$ for the relevant energy eigenvalues, or more generally, in the complex limit $|z| \to \infty$. By inserting this expansion into Eq. 6.37, it follows

$$
\begin{aligned}
&G^{(r)}_{A,B}(z)\Big|_{|z|\to\infty} \\
&= \frac{1}{Z}\sum_{n,m} A_{nm}B_{mn}\frac{e^{-\beta \mathcal{E}_n} - \varepsilon e^{-\beta \mathcal{E}_m}}{\hbar}\left[\frac{1}{z} - \frac{\mathcal{E}_n - \mathcal{E}_m}{\hbar z^2} + O\left(\frac{1}{z^3}\right)\right] = \\
&= \frac{1}{\hbar z Z}\sum_{n,m} A_{nm}B_{mn}\left(e^{-\beta \mathcal{E}_n} - \varepsilon e^{-\beta \mathcal{E}_m}\right) + O\left(\frac{1}{z^2}\right).
\end{aligned}
\tag{6.40}
$$

We now observe that the double sums appearing in the expression for $G^{(r)}_{A,B}(z)$ can be rewritten in terms of thermal expectation values (see Eq. 1.379). Hence, the leading term in Eq. 6.40 becomes

$$G^{(r)}_{A,B}(z)\Big|_{|z|\to\infty} = \frac{1}{\hbar z}\left(\langle AB\rangle - \varepsilon\langle BA\rangle\right) + O\left(\frac{1}{z^2}\right) = \\ = \frac{1}{\hbar z}\left\langle [A,B]^{(\varepsilon)}\right\rangle + O\left(\frac{1}{z^2}\right), \tag{6.41}$$

which completes the proof. □

Now, we write Eq. 6.37 in integral form as follows

$$G^{(r)}_{A,B}(z) = \int_{-\infty}^{\infty} \frac{d\omega'}{2\pi} \frac{A_{AB}(\omega')}{\hbar(z-\omega')}, \tag{6.42}$$

where we define

Definition 6.5 (Spectral Function of the Operators A **and** B**)** Given two operators A and B and their associated retarded many-body Green's function given in Eq. 6.8, the spectral function associated with these operators is defined as

$$A_{A,B}(\omega') = \frac{2\pi}{Z}\sum_{n,m} A_{nm}B_{mn}\left(e^{-\beta\mathcal{E}_n} - \varepsilon e^{-\beta\mathcal{E}_m}\right)\delta\left[\omega' - \frac{\mathcal{E}_m - \mathcal{E}_n}{\hbar}\right]. \tag{6.43}$$

In the most physically relevant case, one considers $B = A^\dagger$, yielding

$$A_{A,A^\dagger}(\omega') = \frac{2\pi}{Z}\sum_{n,m} |A_{nm}|^2\left(e^{-\beta\mathcal{E}_n} - \varepsilon e^{-\beta\mathcal{E}_m}\right)\delta\left[\omega' - \frac{\mathcal{E}_m - \mathcal{E}_n}{\hbar}\right]. \tag{6.44}$$

Remark 6.4 (Spectral Function and its Connection to ARPES) Consider a system that is translationally invariant in space, so that all correlation functions depend only on relative coordinates. Under this assumption, the one-particle excitations can be described in terms of their momentum $\mathbf{k}$ and energy ω. When the two operators are taken to be the annihilation and creation operators, respectively, that is, $C_{\mathbf{k}}$ and $C^\dagger_{\mathbf{k}}$, the resulting function is the one-particle spectral function $A(\mathbf{k},\omega)$, which describes the probability of removing or adding a particle with momentum $\mathbf{k}$ and energy ω. The spectral function $A(\mathbf{k},\omega)$ plays a central role in quantum many-body theory, as it encodes the probability of adding or removing a particle with momentum $\mathbf{k}$ and energy ω. It reflects the intrinsic electronic structure of the system, including renormalizations due to interactions and the finite lifetime of quasiparticles. Importantly, the spectral function is directly connected to experimental observables. In particular, it is accessible via Angle-Resolved Photoemission Spectroscopy (ARPES), a powerful technique for probing the electronic structure of materials. ARPES involves illuminating a material with ultraviolet or X-ray photons to eject electrons from its surface. By measuring the kinetic energy and emission angle of the photoemitted electrons, one reconstructs their in-plane momentum and binding energy. The resulting photoemission intensity is, under suitable approximations, proportional to the product $A(\mathbf{k}, \omega)n_{-1}(\hbar\omega)$, where $n_{-1}(\hbar\omega)$ is the Fermi-Dirac distribution (see Eq. E.4). This enables direct access to the occupied part of the spectral function, making ARPES an invaluable tool for studying dispersions, energy gaps, correlation effects, and emergent many-body phenomena in quantum materials.

We now consider the thermal Green's function in equilibrium, which depends only on τ (see Eq. 6.14). Moreover, suppose that the eigenvalues and eigenfunctions of the interacting Hamiltonian are known. For $\tau > 0$, the time-ordering operator in Eq. 6.14 can be omitted and we get

$$\begin{aligned} G^{(m)}_{A,B}(\tau) &= -\frac{1}{\hbar}\langle A(\tau)B(0)\rangle = \\ &= -\frac{1}{\hbar Z}\mathrm{Tr}\left(e^{-\beta\hat{\mathcal{H}}}e^{\frac{\hat{\mathcal{H}}}{\hbar}\tau}Ae^{-\frac{\hat{\mathcal{H}}}{\hbar}\tau}B\right) = \\ &= -\frac{1}{\hbar Z}\sum_{n,m}A_{nm}B_{mn}e^{-\beta\mathcal{E}_n}e^{\frac{\mathcal{E}_n-\mathcal{E}_m}{\hbar}\tau}. \end{aligned} \tag{6.45}$$

From $G^{(r)}_{A,B}(t)$, we obtained $G^{(r)}_{A,B}(z)$ in the upper half of the complex plane. We now apply the Fourier transform to the Matsubara Green's function $G^{(m)}_{A,B}(\tau)$ (see Eq. 6.21) as follows

$$\begin{aligned} G^{(m)}_{A,B}(i\omega_n) &= \int_0^{\beta\hbar} G^{(m)}_{A,B}(\tau)e^{i\omega_n\tau}d\tau = \\ &= -\frac{1}{\hbar Z}\sum_{n,m}A_{nm}B_{mn}e^{-\beta\mathcal{E}_n}\int_0^{\beta\hbar}d\tau\, e^{i\omega_n\tau}e^{\frac{\mathcal{E}_n-\mathcal{E}_m}{\hbar}\tau} = \\ &= -\frac{1}{\hbar Z}\sum_{n,m}A_{nm}B_{mn}e^{-\beta\mathcal{E}_n}\frac{e^{i\omega_n\beta\hbar}e^{\beta(\mathcal{E}_n-\mathcal{E}_m)}-1}{i\omega_n+\frac{\mathcal{E}_n-\mathcal{E}_m}{\hbar}} = \\ &= -\frac{1}{\hbar Z}\sum_{n,m}A_{nm}B_{mn}e^{-\beta\mathcal{E}_n}\frac{\varepsilon e^{\beta(\mathcal{E}_n-\mathcal{E}_m)}-1}{i\omega_n+\frac{\mathcal{E}_n-\mathcal{E}_m}{\hbar}} = \\ &= -\frac{1}{\hbar Z}\sum_{n,m}A_{nm}B_{mn}\frac{\varepsilon e^{-\beta\mathcal{E}_n}e^{\beta(\mathcal{E}_n-\mathcal{E}_m)}-e^{-\beta\mathcal{E}_n}}{i\omega_n+\frac{\mathcal{E}_n-\mathcal{E}_m}{\hbar}} = \\ &= \frac{1}{\hbar Z}\sum_{n,m}A_{nm}B_{mn}\frac{e^{-\beta\mathcal{E}_n}-\varepsilon e^{-\beta\mathcal{E}_n}e^{\beta(\mathcal{E}_n-\mathcal{E}_m)}}{i\omega_n+\frac{\mathcal{E}_n-\mathcal{E}_m}{\hbar}} \equiv \\ &\equiv \frac{1}{Z}\sum_{n,m}A_{nm}B_{mn}\frac{e^{-\beta\mathcal{E}_n}-\varepsilon e^{-\beta\mathcal{E}_m}}{\hbar\left(i\omega_n+\frac{\mathcal{E}_n-\mathcal{E}_m}{\hbar}\right)}, \end{aligned} \tag{6.46}$$

where we employed Eq. 6.28. We now express $G^{(m)}_{A,B}(i\omega_n)$ in integral form:

$$G^{(m)}_{A,B}(i\omega_n) = \int_{-\infty}^{\infty} \frac{d\omega'}{2\pi} \frac{\tilde{A}_{A,B}(\omega')}{\hbar(i\omega_n - \omega')}, \tag{6.47}$$

from which by comparison, $\tilde{A}_{A,B}(\omega')$ is the spectral function found above, then

$$G^{(m)}_{A,B}(i\omega_n) = \int_{-\infty}^{\infty} \frac{d\omega'}{2\pi} \frac{A_{A,B}(\omega')}{\hbar(i\omega_n - \omega')}. \tag{6.48}$$

On the other hand, the Fourier transform of the retarded many-body Green's function is given by

$$G^{(r)}_{A,B}(z) = \int_{-\infty}^{+\infty} \frac{d\omega'}{2\pi} \frac{A_{A,B}(\omega')}{\hbar(z - \omega')}, \quad z = \omega + i\delta, \ \delta > 0, \tag{6.49}$$

in other words, the spectral function obtained from the Fourier transform of the retarded Green's function $G^{(r)}_{A,B}(z)$ coincides with that of the Fourier transform of the Matsubara Green's function $G^{(m)}_{A,B}(i\omega_n)$, as defined in Eq. 6.43. According to the Abrikosov-Gor'kov theorem, if the Matsubara Green's function $G^{(m)}_{A,B}(i\omega_n)$ is known along the imaginary frequency axis $i\omega_n$, it admits infinitely many analytic continuations into the upper half of the complex plane. However, if one further imposes the condition that, asymptotically, the function scales as z^{-1} for large $|z|$, the analytic continuation becomes unique. This is precisely the case here (see Eq. 6.38). We do not present a proof of this theorem here. Instead, our attention is directed toward the Fourier transform of the Matsubara Green's function evaluated along the real frequency axis. To relate the values on the imaginary axis to those on the real axis, we consider complex frequencies of the form $z = \omega + i\delta$, with $\omega, \delta \in \mathbb{R}$ and $\delta > 0$, and then take the limit

$$G^{(r)}_{A,B}(\omega) = \lim_{\delta \to 0^+} G^{(m)}_{A,B}(\omega + i\delta). \tag{6.50}$$

Theorem 6.3 *Let A and B be two generic operators. The associated the spectral function $A_{A,B}(\omega)$ (see Eq. 6.43) satisfies the following identity*

$$\int_{-\infty}^{+\infty} \frac{d\omega}{2\pi} \frac{A_{A,B}(\omega)}{1 - \varepsilon e^{-\beta\hbar\omega}} = \langle AB \rangle, \tag{6.51}$$

where $\langle \ldots \rangle$ denotes the thermal average (see Eq. 1.379).

Proof We compute the integral on the left-hand side of Eq. 6.51. Using the explicit expression of the spectral function, that is, Eq. 6.43, we obtain,

$$
\begin{aligned}
&\int_{-\infty}^{+\infty} \frac{d\omega}{2\pi} \frac{A_{A,B}(\omega)}{1-\varepsilon e^{-\beta\hbar\omega}} \\
&= \frac{2\pi}{Z} \sum_{n,m} A_{nm} B_{mn} \left(e^{-\beta\mathcal{E}_n} - \varepsilon e^{-\beta\mathcal{E}_m}\right) \\
&\int_{-\infty}^{+\infty} \frac{d\omega}{2\pi} \frac{1}{1-\varepsilon e^{-\beta\hbar\omega}} \delta\left[\omega - \frac{\mathcal{E}_m - \mathcal{E}_n}{\hbar}\right] = \\
&= \frac{2\pi}{Z} \sum_{n,m} A_{nm} B_{mn} \left(e^{-\beta\mathcal{E}_n} - \varepsilon e^{-\beta\mathcal{E}_m}\right) \frac{1}{2\pi} \frac{1}{1-\varepsilon e^{-\beta\hbar\left(\frac{\mathcal{E}_m-\mathcal{E}_n}{\hbar}\right)}} = \\
&= \frac{1}{Z} \sum_{n,m} A_{nm} B_{mn} e^{-\beta\mathcal{E}_n} \left(1 - \varepsilon e^{-\beta(\mathcal{E}_m - \mathcal{E}_n)}\right) \frac{1}{1-\varepsilon e^{-\beta(\mathcal{E}_m-\mathcal{E}_n)}} \equiv \\
&\equiv \sum_{n,m} \frac{e^{-\beta\mathcal{E}_n}}{Z} \langle n|A|m\rangle\langle m|B|n\rangle = \\
&= \sum_{n} \frac{e^{-\beta\mathcal{E}_n}}{Z} \langle n|AB|n\rangle,
\end{aligned}
\tag{6.52}
$$

which is the thesis. □

6.4 Matsubara Green's Function for Non-interacting Particles

In this section, we explicitly compute the Matsubara Green's function in the non-interacting case, corresponding to the creation and annihilation field operators of a system of free particles in thermal equilibrium (see Theorem 6.1), also referred to as the free Matsubara Green's function. Specifically, we adapt Eq. 6.14 to field operators, setting $A = \hat{\psi}$ and $B = \hat{\psi}^\dagger$, and replacing the full Hamiltonian $\hat{\mathcal{H}}$, i.e., the second-quantized form of Eq. 1.85, with the non-interacting Hamiltonian $\hat{\mathcal{H}}_0$ given in 2.181. That is, we consider the Matsubara Green's function for a system of free particles given by

$$
G^{(m)(0)}(x, x', \tau) = -\frac{1}{\hbar} \left\langle \hat{T}_\tau \hat{\psi}^{(0)}(x, \tau) \hat{\psi}^{\dagger(0)}(x', 0) \right\rangle_0 . \tag{6.53}
$$

Remark 6.5 (Thermal Interaction Picture in the Free Matsubara Green's Function) Since the free Matsubara Green's function involves operators evolving under the non-interacting Hamiltonian $\hat{\mathcal{H}}_0$ of the form 2.181, the imaginary-time evolution in its definition is taken in the thermal interaction picture (see Eq. 2.194) rather than in the thermal Heisenberg picture (see Eq. 6.12). This reflects the fact that, for the free Green's function, the operators evolve only with respect to the non-interacting part of the Hamiltonian, while the effects of interactions are not included.

We expand the field operators in the basis of the eigenfunctions of $\mathcal{H}_0$ and assume for simplicity $\tau > 0$, then

$$
\begin{aligned}
G^{(m)(0)}(x, x', \tau) &= -\frac{1}{\hbar} \sum_{\alpha,\alpha'} \varphi_\alpha(x)\varphi^*_{\alpha'}(x') \left\langle \hat{T}_\tau a^{(0)}_\alpha(\tau) a^{\dagger(0)}_{\alpha'}(0) \right\rangle_0 = \\
&= -\frac{1}{\hbar} \sum_{\alpha,\alpha'} \varphi_\alpha(x)\varphi^*_{\alpha'}(x') e^{-\frac{\mathcal{E}_\alpha \tau}{\hbar}} \left\langle a_\alpha a^\dagger_{\alpha'} \right\rangle_0 ,
\end{aligned} \tag{6.54}
$$

where we used Eqs. 2.236 and 2.241. The terms obtained by summing over α with the condition $\alpha' \neq \alpha$ are given by

$$
\sum_{\substack{\alpha' \\ \alpha' \neq \alpha}} \varphi_\alpha(x)\varphi^*_{\alpha'}(x') e^{-\frac{\mathcal{E}_\alpha \tau}{\hbar}} \left\langle a_\alpha a^\dagger_{\alpha'} \right\rangle_0 = \sum_{\substack{\alpha' \\ \alpha' \neq \alpha}} \varphi_\alpha(x)\varphi^*_{\alpha'}(x') e^{-\frac{\mathcal{E}_\alpha \tau}{\hbar}} \left\langle \varepsilon a^\dagger_{\alpha'} a_\alpha \right\rangle_0 , \tag{6.55}
$$

where we used D.9, given that $\alpha \neq \alpha'$. From 2.252, we have

$$
\begin{aligned}
\sum_{\substack{\alpha' \\ \alpha' \neq \alpha}} \varphi_\alpha(x)\varphi^*_{\alpha'}(x') e^{-\frac{\mathcal{E}_\alpha \tau}{\hbar}} \left\langle \varepsilon a^\dagger_{\alpha'} a_\alpha \right\rangle_0 &= \sum_{\substack{\alpha' \\ \alpha' \neq \alpha}} \varphi_\alpha(x)\varphi^*_{\alpha'}(x') e^{-\frac{\mathcal{E}_\alpha \tau}{\hbar}} \varepsilon \delta_{\alpha,\alpha'} n_\varepsilon(\mathcal{E}_\alpha) = \\
&= 0.
\end{aligned} \tag{6.56}
$$

Consequently, the free Matsubara Green's function at thermodynamic equilibrium becomes

$$
G^{(m)(0)}(x, x', \tau) = -\frac{1}{\hbar} \sum_\alpha \varphi_\alpha(x)\varphi^*_\alpha(x') e^{-\frac{\mathcal{E}_\alpha \tau}{\hbar}} \left(1 + \frac{\varepsilon}{e^{\beta \mathcal{E}_\alpha} - \varepsilon} \right). \tag{6.57}
$$

We apply a time Fourier transform, that is

$$
G^{(m)(0)}(x, x', i\omega_n) = \int_0^{\beta\hbar} d\tau e^{i\omega_n \tau} G^{(m)(0)}(x, x', \tau), \tag{6.58}
$$

and since only $e^{-\frac{\mathcal{E}_\alpha \tau}{\hbar}}$ depends explicitly on τ, we have

$$
\begin{aligned}
&G^{(m)(0)}(x, x', i\omega_n) \\
&= \left(-\frac{1}{\hbar} \sum_\alpha \varphi_\alpha(x)\varphi^*_\alpha(x') \left[1 + \frac{\varepsilon}{e^{\beta \mathcal{E}_\alpha} - \varepsilon} \right] \right) \int_0^{\beta\hbar} d\tau e^{i\omega_n \tau} e^{-\frac{\mathcal{E}_\alpha}{\hbar}\tau} =
\end{aligned}
$$

$$= \left(-\frac{1}{\hbar}\sum_{\alpha} \varphi_{\alpha}(x)\varphi_{\alpha}^{*}(x')\left[\frac{e^{\beta\mathcal{E}_{\alpha}} - \varepsilon + \varepsilon}{e^{\beta\mathcal{E}_{\alpha}} - \varepsilon}\right]\right)\left(\frac{1}{i\omega_n - \frac{\mathcal{E}_{\alpha}}{\hbar}}\left[e^{i\omega_n - \frac{\mathcal{E}_{\alpha}}{\hbar}}\right]_0^{\beta\hbar}\right) =$$
$$= -\frac{1}{\hbar}\sum_{\alpha} \varphi_{\alpha}(x)\varphi_{\alpha}^{*}(x')\frac{e^{\beta\mathcal{E}_{\alpha}}}{e^{\beta\mathcal{E}_{\alpha}} - \varepsilon}\frac{1}{i\omega_n - \frac{\mathcal{E}_{\alpha}}{\hbar}}\left[e^{i\omega_n\beta\hbar}e^{-\beta\hbar} - 1\right] =$$
$$= -\frac{1}{\hbar}\sum_{\alpha} \varphi_{\alpha}(x)\varphi_{\alpha}^{*}(x')\frac{e^{\beta\mathcal{E}_{\alpha}}}{e^{\beta\mathcal{E}_{\alpha}} - \varepsilon}\frac{1}{i\omega_n - \frac{\mathcal{E}_{\alpha}}{\hbar}}\left(\varepsilon e^{-\beta\mathcal{E}_{\alpha}} - 1\right) =$$
$$= -\frac{1}{\hbar}\sum_{\alpha} \varphi_{\alpha}(x)\varphi_{\alpha}^{*}(x')\frac{1}{i\omega_n - \frac{\mathcal{E}_{\alpha}}{\hbar}}\frac{\varepsilon - e^{\beta\mathcal{E}_{\alpha}}}{e^{\beta\mathcal{E}_{\alpha}} - \varepsilon} =$$
$$= \sum_{\alpha} \frac{\varphi_{\alpha}(x)\varphi_{\alpha}^{*}(x')}{\hbar\left[i\omega_n - \frac{\mathcal{E}_{\alpha}}{\hbar}\right]}. \tag{6.59}$$

This final expression highlights that the Matsubara Green's function in frequency space depends solely on the spectrum $\{\mathcal{E}_{\alpha}\}$ and the eigenfunctions $\{\varphi_{\alpha}(x)\}$, and it is valid for both bosons and fermions through the appropriate choice of the Matsubara frequencies ω_n.

6.5 Equations-of-Motion Method

Given the retarded many-body Green's function $G^{(r)}_{A,B}(t)$ at the thermodynamic equilibrium of two generic operators A and B, i.e., the Eqs. 6.31 and 6.32, we define $z = \omega + i\delta$, with $\omega, \delta \in \mathbb{R}$, $\delta > 0$, and integrate by parts the quantity

$$\int_0^{\infty} dt\ \dot{C}(t)e^{izt} = C(t)e^{izt}\Big|_0^{\infty} - iz\int_0^{\infty} dt\ C(t)e^{izt}, \tag{6.60}$$

with

$$C(t)e^{izt}\Big|_0^{\infty} = -C(0). \tag{6.61}$$

We have

$$\int_0^{\infty} dt\ \dot{C}(t)e^{izt} = -C(0) - iz\int_0^{\infty} dt\ C(t)e^{izt}, \tag{6.62}$$

$$\int_0^{\infty} dt\ \dot{C}(t)e^{izt} = -C(0) - iz\int_0^{\infty} dt\ \Theta(t)C(t)e^{izt}, \tag{6.63}$$

$$\int_0^\infty dt\, \dot{C}(t)e^{izt} = -C(0) - iz\int_0^\infty dt\, G^{(r)}_{A,B}(t)e^{izt}, \tag{6.64}$$

$$\int_0^\infty dt\, \dot{C}(t)e^{izt} = -C(0) - izG^{(r)}_{A,B}(z), \tag{6.65}$$

$$G^{(r)}_{A,B}(z) = -\frac{1}{iz}C(0) - \frac{1}{iz}\int_0^\infty dt\, \dot{C}(t)e^{izt}, \tag{6.66}$$

$$G^{(r)}_{A,B}(z) = \left(-\frac{1}{iz}\right)\left(-\frac{i}{\hbar}\left\langle [A, B]^{(\varepsilon)}\right\rangle\right) - \frac{1}{iz}\int_0^\infty dt\, \dot{C}(t)e^{izt}, \tag{6.67}$$

$$G^{(r)}_{A,B}(z) = \frac{1}{\hbar z}\left\langle [A, B]^{(\varepsilon)}\right\rangle - \frac{1}{iz}\int_0^\infty dt\, \dot{C}(t)e^{izt}, \tag{6.68}$$

$$G^{(r)}_{A,B}(z) = \frac{1}{\hbar z}\left\langle [A, B]^{(\varepsilon)}\right\rangle - \frac{1}{iz}\int_0^\infty dt\left(\left(-\frac{i}{\hbar}\right)\frac{1}{i\hbar}\left\langle\left[\left[A(t), \hat{\mathcal{H}}\right], B\right]^{(\varepsilon)}\right\rangle\right)e^{izt}, \tag{6.69}$$

$$G^{(r)}_{A,B}(z) = \frac{1}{\hbar z}\left\langle [A, B]^{(\varepsilon)}\right\rangle + \frac{1}{iz}\int_0^\infty dt \frac{1}{\hbar^2}\left\langle\left[\left[A(t), \hat{\mathcal{H}}\right], B\right]^{(\varepsilon)}\right\rangle e^{izt}, \tag{6.70}$$

$$G^{(r)}_{A,B}(z) = \frac{1}{\hbar z}\left\langle [A, B]^{(\varepsilon)}\right\rangle + \frac{1}{i\hbar^2 z}\int_0^\infty dt\left\langle\left[\left[A(t), \hat{\mathcal{H}}\right], B\right]^{(\varepsilon)}\right\rangle e^{izt}, \tag{6.71}$$

$$G^{(r)}_{A,B}(z) = \frac{1}{\hbar z}\left\langle [A, B]^{(\varepsilon)}\right\rangle - \frac{i}{\hbar^2 z}\int_0^\infty dt\left\langle\left[\left[A(t), \hat{\mathcal{H}}\right], B\right]^{(\varepsilon)}\right\rangle e^{izt}, \tag{6.72}$$

$$G^{(r)}_{A,B}(z) = \frac{1}{\hbar z}\left\langle [A, B]^{(\varepsilon)}\right\rangle + \frac{1}{\hbar z}\int_0^\infty dt\left(-\frac{i}{\hbar}\left\langle\left[\left[A(t), \hat{\mathcal{H}}\right], B\right]^{(\varepsilon)}\right\rangle\right)e^{izt}, \tag{6.73}$$

$$G^{(r)}_{A,B}(z) = \frac{1}{\hbar z}\left\langle [A, B]^{(\varepsilon)}\right\rangle + \frac{1}{\hbar z}G^{(r)}_{\left[A,\hat{\mathcal{H}}\right],B}(z). \tag{6.74}$$

In other words, we obtain a new retarded Green's function: apart from the term $\frac{1}{\hbar z}\left\langle [A, B]^{(\varepsilon)}\right\rangle$, the function $G^{(r)}_{A,B}(z)$ is related to a new Green's function, namely $G^{(r)}_{[A,\hat{\mathcal{H}}],B}(z)$. From the relation connecting $G^{(r)}_{A,B}(z)$ with $G^{(r)}_{[A,\hat{\mathcal{H}}],B}(z)$, successive integrations by parts show that $G^{(r)}_{A,B}(z)$ is not only connected to $G^{(r)}_{[A,\hat{\mathcal{H}}],B}(z)$, but also to higher-order Green's functions such as $G^{(r)}_{[[A,\hat{\mathcal{H}}],\hat{\mathcal{H}}],B}(z)$, and so forth for each subsequent integration.

Now, consider a system described by a free Hamiltonian $\hat{\mathcal{H}}_0$ of the form 2.181. The corresponding free many-body Green's function at thermodynamic equilibrium, defined with respect to the annihilation and creation operators $A = a_\alpha$ and $B = a^\dagger_\alpha$ (see Eqs. 1.44, 1.45, 2.79 and 2.83), can then be expressed as follows

$$
\begin{aligned}
G^{(r)(0)}_{a_\alpha, a^\dagger_\alpha}(t) \equiv G^{(r)(0)}_\alpha(t) = \\
= -\frac{i}{\hbar}\Theta(t)\left\langle\left[a^{(0)}_\alpha(t), a^\dagger_\alpha(0)\right]^{(\varepsilon)}\right\rangle,
\end{aligned} \tag{6.75}
$$

where the creation and annihilation operators are expressed in the interaction picture (see Eq. 1.107). The many-body Green's function $G^{(r)(0)}_\alpha(t)$ describes the time evolution of the α state. From Eq. 2.102, it follows

$$
\left[a^{(0)}_\alpha(t), a^\dagger_\alpha(0)\right]^{(\varepsilon)} = a^{(0)}_\alpha(t)a^\dagger_\alpha(0) - \varepsilon a^\dagger_\alpha(0)a^{(0)}_\alpha(t). \tag{6.76}
$$

Note that the term $a^{(0)}_\alpha(t)a^\dagger_\alpha(0)$ represents a process in which a particle is created in state α at time 0 and annihilated at time t: this is a particle term, and it propagates forward in time. The term $-\varepsilon a^\dagger_\alpha(0)a^{(0)}_\alpha(t)$ represents a process in which a particle in state α is annihilated at time t and then created at time 0. This corresponds to a hole term that propagates backward in time. We apply the equations-of-motion method (see Eq. 6.74). From Eq. D.7, one finds

$$
\begin{aligned}
\left\langle\left[a_\alpha, a^\dagger_\alpha\right]^{(\varepsilon)}\right\rangle = \left\langle\delta_{\alpha,\alpha}\,\mathbb{1}\right\rangle = \\
= 1,
\end{aligned} \tag{6.77}
$$

and from Eq. D.13, we get $\left[a_\alpha, \hat{\mathcal{H}}_0\right] = \mathcal{E}_\alpha a_\alpha$. Then, we can determine the propagator using the equations-of-motion method as follows

$$
\begin{aligned}
G^{(r)(0)}_\alpha(z) &= \frac{1}{\hbar z}\left\langle\left[a_\alpha, a^\dagger_\alpha\right]^{(\varepsilon)}\right\rangle + \frac{1}{\hbar z}G^{(r)}_{\left[a_\alpha,\hat{\mathcal{H}}_0\right],a^\dagger}(z) = \\
&= \frac{1}{\hbar z} + \frac{1}{\hbar z}G^{(r)}_{\mathcal{E}_\alpha a_\alpha, a^\dagger_\alpha}(z) = \\
&= \frac{1}{\hbar z} + \frac{1}{\hbar z}\mathcal{E}_\alpha G^{(r)(0)}_\alpha(z),
\end{aligned} \tag{6.78}
$$

which can be rearranged as an algebraic equation in $G^{(r)(0)}_\alpha(z)$, that is,

$$
\begin{aligned}
G^{(r)(0)}_\alpha(z) &= \frac{1}{\hbar z}\left[\frac{1}{1-\frac{\mathcal{E}_\alpha}{\hbar z}}\right] = \\
&= \frac{1}{\hbar}\left[\frac{1}{z-\frac{\mathcal{E}_\alpha}{\hbar}}\right],
\end{aligned} \tag{6.79}
$$

and we directly derive the rational expression for the Green's function, i.e.,

$$G_\alpha^{(r)(0)}(\omega+i\delta) = \frac{1}{\hbar}\left[\frac{1}{\omega+i\delta-\frac{\mathcal{E}_\alpha}{\hbar}}\right] \equiv$$

$$\equiv \frac{1}{\hbar}\left[\frac{1}{\left(\omega-\frac{\mathcal{E}_\alpha}{\hbar}\right)+i\delta}\right], \tag{6.80}$$

and multiply and divide by the complex conjugate of the denominator, i.e.,

$$G_\alpha^{(r)(0)}(\omega+i\delta) = \frac{1}{\hbar}\left[\frac{\omega-\frac{\mathcal{E}_\alpha}{\hbar}-i\delta}{\left(\omega-\frac{\mathcal{E}_\alpha}{\hbar}\right)^2+\delta^2}\right]. \tag{6.81}$$

The imaginary part of the Fourier transform of the retarded Green's function

$$\operatorname{Im} G_\alpha^{(r)(0)}(\omega+i\delta) = \frac{1}{\hbar}\operatorname{Im}\left[\frac{\omega-\frac{\mathcal{E}_\alpha}{\hbar}-i\delta}{\left(\omega-\frac{\mathcal{E}_\alpha}{\hbar}\right)^2+\delta^2}\right] =$$

$$= -\frac{1}{\hbar}\frac{\delta}{\left(\omega-\frac{\mathcal{E}_\alpha}{\hbar}\right)^2+\delta^2} \tag{6.82}$$

is a measurable quantity. The imaginary part of the retarded Green's function for free particles is a Lorentzian function and in the limit $\delta \to 0^+$ tends to a Dirac's delta. Indeed, from Breit Wigner formula P.1, one finds

$$\lim_{\delta\to 0^+} \operatorname{Im} G_\alpha^{(r)(0)}(\omega+i\delta) = -\lim_{\delta\to 0^+}\frac{1}{\hbar}\frac{\pi}{\pi}\frac{\delta}{\left(\omega-\frac{\mathcal{E}_\alpha}{\hbar}\right)^2+\delta^2}, \tag{6.83}$$

$$\operatorname{Im} G_\alpha^{(r)(0)}(\omega) = -\frac{\pi}{\hbar}\delta\left[\omega-\frac{\mathcal{E}_\alpha}{\hbar}\right]. \tag{6.84}$$

We now proceed to determine the spectral function for free particles. To this end, let us first consider the general case: starting from Eq. 6.37 with $B = A^\dagger$ and $z = \omega + i\delta$, one finds

$$G^{(r)}_{A,A^\dagger}(\omega+i\delta)$$

$$= \frac{1}{Z}\sum_{n,m}|A_{nm}|^2\frac{e^{-\beta\mathcal{E}_n}-\varepsilon e^{-\beta\mathcal{E}_m}}{\hbar\left(\omega+i\delta+\frac{\mathcal{E}_n-\mathcal{E}_m}{\hbar}\right)} =$$

$$
= \frac{1}{Z}\sum_{n,m} |A_{nm}|^2 \frac{e^{-\beta\mathcal{E}_n} - \varepsilon e^{-\beta\mathcal{E}_m}}{\hbar\left(\omega + i\delta + \frac{\mathcal{E}_n - \mathcal{E}_m}{\hbar}\right)\left(\omega + \frac{\mathcal{E}_n - \mathcal{E}_m}{\hbar} - i\delta\right)}
$$

$$
\left(\omega + \frac{\mathcal{E}_n - \mathcal{E}_m}{\hbar} - i\delta\right) =
$$

$$
= \frac{1}{Z}\sum_{n,m} |A_{nm}|^2 \frac{e^{-\beta\mathcal{E}_n} - \varepsilon e^{-\beta\mathcal{E}_m}}{\hbar\left(\left(\omega + \frac{\mathcal{E}_n - \mathcal{E}_m}{\hbar}\right)^2 + \delta^2\right)}\left(\omega + \frac{\mathcal{E}_n - \mathcal{E}_m}{\hbar} - i\delta\right). \tag{6.85}
$$

From this expression, we extract the imaginary part as follows

$$
\begin{aligned}
&-2\hbar\,\mathrm{Im}\,G^{(r)}_{A,A^\dagger}(\omega) \\
&= \lim_{\delta\to 0^+}\left(-2\hbar\,\mathrm{Im}\,G^{(r)}_{A,A^\dagger}(\omega + i\delta)\right) = \\
&= \lim_{\delta\to 0^+}\left(-2\hbar\frac{1}{Z}\sum_{n,m}|A_{nm}|^2\frac{(-\delta)}{\hbar\left(\left(\omega + \frac{\mathcal{E}_n - \mathcal{E}_m}{\hbar}\right)^2 + \delta^2\right)}\left(e^{-\beta\mathcal{E}_n} - \varepsilon e^{-\beta\mathcal{E}_m}\right)\right) = \\
&= \lim_{\delta\to 0^+}\left(2\frac{1}{Z}\sum_{n,m}|A_{nm}|^2\frac{\delta}{\left(\left(\omega + \frac{\mathcal{E}_n - \mathcal{E}_m}{\hbar}\right)^2 + \delta^2\right)}\left(e^{-\beta\mathcal{E}_n} - \varepsilon e^{-\beta\mathcal{E}_m}\right)\right) = \\
&= \frac{1}{Z}\sum_{n,m}|A_{nm}|^2 2\left(e^{-\beta\mathcal{E}_n} - \varepsilon e^{-\beta\mathcal{E}_m}\right)\pi\delta\left[\omega + \frac{\mathcal{E}_n - \mathcal{E}_m}{\hbar}\right] \equiv \\
&\equiv \frac{2\pi}{Z}\sum_{n,m}|A_{nm}|^2\left(e^{-\beta\mathcal{E}_n} - \varepsilon e^{-\beta\mathcal{E}_m}\right)\delta\left[\omega + \frac{\mathcal{E}_n - \mathcal{E}_m}{\hbar}\right],
\end{aligned} \tag{6.86}
$$

that is equal to the spectral function 6.43 computed for $B = A^\dagger$. Finally, we get

$$
A_{A,A^\dagger}(\omega) = -2\hbar\,\mathrm{Im}\,G^{(r)}_{A,A^\dagger}(\omega), \tag{6.87}
$$

that is, the spectral function of the operators A and $A^\dagger$, calculated as a function of ω, is related to the imaginary part of the retarded Green's function of the same operators evaluated at ω. In the case of free particles, the spectral function is represented by a Dirac delta distribution centered at the excitation energies of the system. Indeed, by comparison with the imaginary part of the free many Green's function in Eq. 6.84, we find

$$A_\alpha(\omega) = 2\pi\delta\left[\omega - \frac{\mathcal{E}_\alpha}{\hbar}\right]. \tag{6.88}$$

6.5.1 Green's Function of Free Particles

For a system of non-interacting particles, the equations-of-motion method (see Eq. 6.74) yields the retarded many-body Green's function associated with the one-particle state α in the form

$$G_\alpha^{(r)(0)}(z) = \frac{1}{\hbar}\frac{1}{z - \frac{\mathcal{E}_\alpha}{\hbar}}, \tag{6.89}$$

where the pole $z = \frac{\mathcal{E}_\alpha}{\hbar}$ lies on the real axis. The corresponding spectral function is then

$$-2\hbar\,\mathrm{Im}\,G_\alpha^{(r)(0)}(\omega) = 2\pi\delta\left[\omega - \frac{\mathcal{E}_\alpha}{\hbar}\right]. \tag{6.90}$$

In the general case, the spectral function $A_{A,B}$ is given by a sum of Dirac delta distributions, reflecting the full excitation spectrum of the system. For free particles, however, the spectral function corresponding to the state α reduces to a single delta peak centered at $\frac{\mathcal{E}_\alpha}{\hbar}$, i.e., at the frequency associated with that particular state. To evaluate the Green's function in the time domain, one may perform the inverse Fourier transform. In the case of free particles, the time evolution of the annihilation operator a_α is determined by its interaction-picture representation (see Eq. 2.187), which leads to

$$\begin{aligned} G_\alpha^{(r)(0)}(t) &= -\frac{1}{\hbar}\Theta(t)\left\langle\left[a_\alpha^{(0)}(t), a_\alpha^\dagger(0)\right]^{(\varepsilon)}\right\rangle = \\ &= -\frac{1}{\hbar}\Theta(t)e^{-i\frac{\mathcal{E}_\alpha}{\hbar}t}. \end{aligned} \tag{6.91}$$

Hence, the retarded many-body Green's function for the free state α exhibits a purely oscillatory time dependence, characterized by a phase factor.

6.5.2 Application of the Equations-of-Motion Method to Anderson's Single-Impurity Model

Anderson's single-impurity model describes fermions interacting with an impurity, for example electrons in a metal coupled to a single discrete state of fixed energy.

In this treatment, the spin of the particles is neglected. The system is assumed to be in thermodynamic equilibrium; therefore, all physical quantities are computed as thermal averages (see Eq. 1.379). The equations-of-motion method (see Eq. 6.74) will be employed to demonstrate that the Anderson's single-impurity model is exactly solvable and that the hierarchy of equations closes at the second order. The Anderson single-impurity model treats the solid as a continuous medium hosting a band of non-interacting conduction electrons, represented by the fermionic annihilation and creation operators $C_{\mathbf{k}}$ and $C_{\mathbf{k}}^{\dagger}$, associated with the dispersion relation $\mathcal{E}_{\mathbf{k}}$. The conduction band spans the finite energy range

$$\mathcal{E}_{\mathbf{k}} \in [\mathcal{E}_{\min}, \mathcal{E}_{\max}], \ \forall \mathbf{k}. \tag{6.92}$$

In addition, a localized fermionic level of fixed energy $\mathcal{E}$, described by the operators C and $C^{\dagger}$, accounts for the impurity embedded within this continuum. Consequently, the Hamiltonian of Anderson's single-impurity model is given by

$$\hat{\mathcal{H}} = \sum_{\mathbf{k}} \mathcal{E}_{\mathbf{k}} C_{\mathbf{k}}^{\dagger} C_{\mathbf{k}} + \mathcal{E} C^{\dagger} C + \sum_{\mathbf{k}} A_{\mathbf{k}} \left(C_{\mathbf{k}}^{\dagger} C + C^{\dagger} C_{\mathbf{k}} \right), \tag{6.93}$$

where the first term is the continuous level of free electrons; $\mathcal{E} C^{\dagger} C$ is the impurity state; $A_{\mathbf{k}}$ quantifies the interaction between electrons and impurity for each momentum $\mathbf{k}$. Since both the conduction and impurity states correspond to fermionic modes, the operators in $\hat{\mathcal{H}}$ obey the canonical anticommutation relations (see Eq. D.4):

$$\{C_{\mathbf{k}}, C\} = 0, \tag{6.94}$$

$$\left\{C_{\mathbf{k}}, C^{\dagger}\right\} = 0, \tag{6.95}$$

$$\left\{C_{\mathbf{k}}^{\dagger}, C\right\} = 0, \tag{6.96}$$

$$\left\{C_{\mathbf{k}}^{\dagger}, C^{\dagger}\right\} = 0. \tag{6.97}$$

In terms of the interaction, the operator $C^{\dagger} C_{\mathbf{k}}$ describes a process in which an electron transitions from the continuum to the impurity, while $C_{\mathbf{k}}^{\dagger} C$ corresponds to an electron moving from the impurity to the continuum. From Eq. 6.8, the Green's function of the impurity is given by

$$\begin{aligned} G_{C,C^{\dagger}}^{(r)}(t) &= -\frac{i}{\hbar} \Theta(t) \left\langle \left[C(t), C^{\dagger}(0) \right]^{(\varepsilon)} \right\rangle = \\ &= -\frac{i}{\hbar} \Theta(t) \left\langle \left\{ C(t), C^{\dagger}(0) \right\} \right\rangle. \end{aligned} \tag{6.98}$$

From the equations-of-motion method (see Eq. 6.74), it follows

$$
\begin{aligned}
zG^{(r)}_{C,C^\dagger}(z) &= \frac{1}{\hbar}\left\langle\left\{C, C^\dagger\right\}\right\rangle + \frac{1}{\hbar}G^{(r)}_{\left[C,\hat{\mathcal{H}}\right],C^\dagger}(z) \\
&= \frac{1}{\hbar} + \frac{1}{\hbar}G^{(r)}_{\left[C,\hat{\mathcal{H}}\right],C^\dagger}(z),
\end{aligned}
\tag{6.99}
$$

then we are led to examine the commutator

$$
\begin{aligned}
\left[C,\hat{\mathcal{H}}\right] &= \left[C, \sum_{\mathbf{k}} \mathcal{E}_{\mathbf{k}} C^\dagger_{\mathbf{k}} C_{\mathbf{k}} + \mathcal{E}C^\dagger C + \sum_{\mathbf{k}} A_{\mathbf{k}}\left(C^\dagger_{\mathbf{k}}C + C^\dagger C_{\mathbf{k}}\right)\right] = \\
&= \sum_{\mathbf{k}} \mathcal{E}_{\mathbf{k}}\left[C, C^\dagger_{\mathbf{k}}C_{\mathbf{k}}\right] + \mathcal{E}\left[C, C^\dagger C\right] + \sum_{\mathbf{k}} A_{\mathbf{k}}\left\{\left[C, C^\dagger_{\mathbf{k}}C\right] + \left[C, C^\dagger C_{\mathbf{k}}\right]\right\}.
\end{aligned}
\tag{6.100}
$$

Regarding the first two commutators, we employ Eq. D.13 adapted to the case of fermionic operators. In particular, the commutator $\left[C, C^\dagger C\right]$ yields C, whereas $\left[C, C^\dagger_{\mathbf{k}}C_{\mathbf{k}}\right]$ vanishes since the fermionic levels are distinct. We therefore obtain

$$
\left[C,\hat{\mathcal{H}}\right] = \mathcal{E}C + \sum_{\mathbf{k}} A_{\mathbf{k}}\left\{\left[C, C^\dagger_{\mathbf{k}}C\right] + \left[C, C^\dagger C_{\mathbf{k}}\right]\right\},
\tag{6.101}
$$

and from Jacobi identity F.1, it follows

$$
\begin{aligned}
\left[C,\hat{\mathcal{H}}\right] &= \mathcal{E}C + \sum_{\mathbf{k}} A_{\mathbf{k}}\left\{\left[C, C^\dagger_{\mathbf{k}}\right]C + C^\dagger_{\mathbf{k}}\left[C, C\right] + \left[C, C^\dagger\right]C_{\mathbf{k}} + C^\dagger\left[C, C_{\mathbf{k}}\right]\right\} = \\
&= \mathcal{E}C + \sum_{\mathbf{k}} A_{\mathbf{k}}\left\{\left[C, C^\dagger_{\mathbf{k}}\right]C + \left[C, C^\dagger\right]C_{\mathbf{k}} + C^\dagger\left[C, C_{\mathbf{k}}\right]\right\} = \\
&= \mathcal{E}C + \sum_{\mathbf{k}} A_{\mathbf{k}}\left(CC^\dagger_{\mathbf{k}}C - C^\dagger_{\mathbf{k}}CC + CC^\dagger C_{\mathbf{k}}\right. \\
&\qquad \left. - C^\dagger CC_{\mathbf{k}} + C^\dagger CC_{\mathbf{k}} - C^\dagger C_{\mathbf{k}}C\right) = \\
&= \mathcal{E}C + \sum_{\mathbf{k}} A_{\mathbf{k}}\left(CC^\dagger_{\mathbf{k}}C - C^\dagger_{\mathbf{k}}CC + CC^\dagger C_{\mathbf{k}} - C^\dagger C_{\mathbf{k}}C\right).
\end{aligned}
\tag{6.102}
$$

From Eq. 6.96, it follows that $CC^\dagger_{\mathbf{k}}C = -C^\dagger_{\mathbf{k}}CC$. Consequently, we obtain

$$
\left[C,\hat{\mathcal{H}}\right] = \mathcal{E}C + \sum_{\mathbf{k}} A_{\mathbf{k}}\left(-2C^\dagger_{\mathbf{k}}CC + CC^\dagger C_{\mathbf{k}} - C^\dagger C_{\mathbf{k}}C\right),
\tag{6.103}
$$

and since the successive action of two annihilation operators yields the vacuum state, we can rewrite

$$\left[C, \hat{\mathcal{H}}\right] = \mathcal{E}C + \sum_{\mathbf{k}} A_{\mathbf{k}} \left(CC^{\dagger}C_{\mathbf{k}} - C^{\dagger}C_{\mathbf{k}}C \right). \tag{6.104}$$

From Eq. D.4, it follows

$$\begin{aligned}\left[C, \hat{\mathcal{H}}\right] &= \mathcal{E}C + \sum_{\mathbf{k}} A_{\mathbf{k}} \left\{ \left(\mathbb{1} - C^{\dagger}C \right) C_{\mathbf{k}} + C^{\dagger}CC_{\mathbf{k}} \right\} = \\ &= \mathcal{E}C + \sum_{\mathbf{k}} A_{\mathbf{k}} C_{\mathbf{k}},\end{aligned} \tag{6.105}$$

and Eq. 6.99 becomes

$$zG^{(r)}_{C,C^{\dagger}}(z) = \frac{1}{\hbar} + \frac{\mathcal{E}}{\hbar} G^{(r)}_{C,C^{\dagger}}(z) + \frac{1}{\hbar} \sum_{\mathbf{k}} A_{\mathbf{k}} G^{(r)}_{C_{\mathbf{k}},C^{\dagger}}(z). \tag{6.106}$$

We apply equations-of-motion method (see Eq. 6.74) for $G^{(r)}_{C_{\mathbf{k}},C^{\dagger}}(z)$, i.e.,

$$\begin{aligned}zG^{(r)}_{C_{\mathbf{k}},C^{\dagger}}(z) &= \frac{1}{\hbar} \left\langle \left\{ C_{\mathbf{k}}, C^{\dagger} \right\} \right\rangle + \frac{1}{\hbar} G^{(r)}_{\left[C_{\mathbf{k}},\hat{\mathcal{H}}\right],C^{\dagger}}(z) = \\ &= \frac{1}{\hbar} G^{(r)}_{\left[C_{\mathbf{k}},\hat{\mathcal{H}}\right],C^{\dagger}}(z).\end{aligned} \tag{6.107}$$

where Eq. 6.95 has been employed, and we must now analyze the commutator

$$\begin{aligned}\left[C_{\mathbf{k}}, \hat{\mathcal{H}}\right] &= \left[C_{\mathbf{k}}, \sum_{\mathbf{k}'} \mathcal{E}_{\mathbf{k}'} C^{\dagger}_{\mathbf{k}'} C_{\mathbf{k}'} + \mathcal{E}C^{\dagger}C + \sum_{\mathbf{k}'} A_{\mathbf{k}'} \left(C^{\dagger}_{\mathbf{k}'} C + C^{\dagger} C_{\mathbf{k}'} \right) \right] = \\ &= \sum_{\mathbf{k}'} \mathcal{E}_{\mathbf{k}'} \left[C_{\mathbf{k}}, C^{\dagger}_{\mathbf{k}'} C_{\mathbf{k}'} \right] + \mathcal{E} \left[C_{\mathbf{k}}, C^{\dagger}C \right] \\ &\quad + \sum_{\mathbf{k}'} A_{\mathbf{k}'} \left(\left[C_{\mathbf{k}}, C^{\dagger}_{\mathbf{k}'} C \right] + \left[C_{\mathbf{k}}, C^{\dagger} C_{\mathbf{k}'} \right] \right).\end{aligned} \tag{6.108}$$

Once again, we employ Eq. D.13 adapted to fermionic operators. In particular, the commutator $\left[C_{\mathbf{k}}, C^{\dagger}_{\mathbf{k}'} C_{\mathbf{k}'}\right]$ yields $\delta_{\mathbf{k},\mathbf{k}'} C_{\mathbf{k}}$, whereas $\left[C_{\mathbf{k}}, C^{\dagger}C\right]$ vanishes, since the fermionic levels are distinct. Accordingly, we obtain

$$\left[C_{\mathbf{k}}, \hat{\mathcal{H}}\right] = \mathcal{E}_{\mathbf{k}} C_{\mathbf{k}} + \sum_{\mathbf{k}'} A_{\mathbf{k}'} \left\{ \left[C_{\mathbf{k}}, C^{\dagger}_{\mathbf{k}'} C \right] + \left[C_{\mathbf{k}}, C^{\dagger} C_{\mathbf{k}'} \right] \right\}. \tag{6.109}$$

At this stage, we choose to follow an alternative route to the one based on the Jacobi identity employed earlier. Instead of expanding nested commutators, we take advantage of the fermionic anticommutation relations directly. In particular, we apply the commutator–anticommutator identity G.1, then

$$\begin{aligned}\left[C_{\mathbf{k}}, \hat{\mathcal{H}}\right] &= \mathcal{E}_{\mathbf{k}} C_{\mathbf{k}} + \sum_{\mathbf{k}'} A_{\mathbf{k}'} \left(\left\{C_{\mathbf{k}}, C^{\dagger}_{\mathbf{k}'}\right\} C - C^{\dagger}_{\mathbf{k}'} \left\{C_{\mathbf{k}}, C\right\} \right. \\ &\quad \left. + \left\{C_{\mathbf{k}}, C^{\dagger}\right\} C_{\mathbf{k}'} - C^{\dagger} \left\{C_{\mathbf{k}}, C_{\mathbf{k}'}\right\} \right) = \\ &= \mathcal{E}_{\mathbf{k}} C_{\mathbf{k}} + \sum_{\mathbf{k}'} A_{\mathbf{k}'} \delta_{\mathbf{k},\mathbf{k}'} C = \\ &= \mathcal{E}_{\mathbf{k}} C_{\mathbf{k}} + A_{\mathbf{k}} C, \end{aligned} \tag{6.110}$$

where Eqs. D.4, 6.94, 6.95 and D.5 have been employed. Eq. 6.107 becomes

$$zG^{(r)}_{C_{\mathbf{k}},C^{\dagger}}(z) = \frac{1}{\hbar} \mathcal{E}_{\mathbf{k}} G^{(r)}_{C_{\mathbf{k}},C^{\dagger}}(z) + \frac{A_{\mathbf{k}}}{\hbar} G^{(r)}_{C,C^{\dagger}}(z), \tag{6.111}$$

then

$$G^{(r)}_{C_{\mathbf{k}},C^{\dagger}}(z) = \frac{A_{\mathbf{k}}}{\hbar} G^{(r)}_{C,C^{\dagger}}(z) \frac{1}{z - \frac{\mathcal{E}_{\mathbf{k}}}{\hbar}}. \tag{6.112}$$

We insert 6.112 into Eq. 6.106 as follows

$$\begin{aligned} zG^{(r)}_{C,C^{\dagger}}(z) &= \frac{1}{\hbar} + \frac{\mathcal{E}}{\hbar} G^{(r)}_{C,C^{\dagger}}(z) + \frac{1}{\hbar} \sum_{\mathbf{k}} A_{\mathbf{k}} G^{(r)}_{C_{\mathbf{k}},C^{\dagger}}(z) = \\ &= \frac{1}{\hbar} + \frac{\mathcal{E}}{\hbar} G^{(r)}_{C,C^{\dagger}}(z) + \sum_{\mathbf{k}} \frac{A^2_{\mathbf{k}}}{\hbar^2} G^{(r)}_{C,C^{\dagger}}(z) \frac{1}{z - \frac{\mathcal{E}_{\mathbf{k}}}{\hbar}}, \end{aligned} \tag{6.113}$$

and we get

$$G^{(r)}_{C,C^{\dagger}}(z) = \frac{1}{\hbar} \frac{1}{z - \frac{\mathcal{E}}{\hbar} - \sum_{\mathbf{k}} \frac{A^2_{\mathbf{k}}}{\hbar^2} \frac{1}{z - \frac{\mathcal{E}_{\mathbf{k}}}{\hbar}}}, \tag{6.114}$$

where we defined the improper self-energy (see Chap. 7) as the complex function

$$\Sigma^*(z) \equiv \sum_{\mathbf{k}} \frac{A^2_{\mathbf{k}}}{\hbar} \frac{1}{z - \frac{\mathcal{E}_{\mathbf{k}}}{\hbar}}. \tag{6.115}$$

The improper self-energy will be further analyzed in Chap. 7. In the non-interacting limit, that is, when $A_{\mathbf{k}} = 0$ for all $\mathbf{k}$, the many-body Green's function reduces to its free-particle form (see Eq. 6.89), characterized by a simple pole on the real axis. The corresponding spectral function is then given by $2\pi\delta\left(\omega - \frac{\mathcal{E}}{\hbar}\right)$. The Green's function of an Anderson's single-impurity is given by

$$G^{(r)}_{C,C^{\dagger}}(z) = \frac{1}{\hbar}\frac{1}{z - \frac{\mathcal{E}}{\hbar} - \frac{\Sigma^*(z)}{\hbar}}. \tag{6.116}$$

By means of Feynman's perturbative theory (see Chap. 7, Eq. 7.89) we will show that, for any system in thermodynamic equilibrium, the corresponding many-body Green's function necessarily takes the form given in Eq. 6.116. In this expression, the energy $\mathcal{E}$ of the unperturbed level explicitly appears, while the function $\Sigma^*(z)$ contains the infinite sum of all Feynman diagrams representing interaction effects. How does $\Sigma^*(z)$ modify the free Green's function? The Green's function of the impurity, $G^{(r)}_{C,C^{\dagger}}(z)$, may exhibit a pole z_0 either on the real axis or in the complex lower half-plane. Suppose that z_0 lies in the complex lower half-plane: if $G^{(r)}_{C,C^{\dagger}}(z)$ has a pole at z_0, then necessarily $\Sigma^*(z_0) \neq 0$. Indeed, if $\Sigma^*(z_0) = 0$, the pole of the Green's function would be located at $z = \frac{\mathcal{E}}{\hbar}$, as in the non-interacting case, which contradicts the assumption that the pole lies in the complex plane. Then

$$\begin{aligned} G^{(r)}_{C,C^{\dagger}}(z) &= \frac{1}{\hbar}\frac{1}{z - \frac{\mathcal{E}}{\hbar} - \frac{\Sigma^*(z)}{\hbar}} \approx \\ &\approx \frac{1}{\hbar}\frac{1}{z - \frac{\mathcal{E}}{\hbar} - \frac{\Sigma^*(z_0)}{\hbar}}, \end{aligned} \tag{6.117}$$

where $\Sigma^*(z_0)$ has real and imaginary parts. We set $\Sigma^*(z_0) = \widetilde{\mathcal{E}} + i\gamma$. If it were worth $\gamma > 0$

$$G^{(r)}_{C,C^{\dagger}}(z) \approx \frac{1}{\hbar}\frac{1}{z - \frac{\mathcal{E}}{\hbar} - \frac{\widetilde{\mathcal{E}}}{\hbar} - i\frac{\gamma}{\hbar}}, \tag{6.118}$$

that is, the Green's function would have the pole $z_0 = \frac{\mathcal{E}}{\hbar} + \frac{\widetilde{\mathcal{E}}}{\hbar} + i\frac{\gamma}{\hbar}$ in the complex upper half-plane, but this is not possible, so it must be $\gamma < 0$. We set $\Sigma^*(z_0) = \widetilde{\mathcal{E}} - i\gamma$, $\gamma > 0$, then

$$\begin{aligned} G^{(r)}_{C,C^{\dagger}}(z) &\approx \frac{1}{\hbar}\frac{1}{z - \frac{\mathcal{E}}{\hbar} - \frac{\widetilde{\mathcal{E}}}{\hbar} + i\frac{\gamma}{\hbar}} = \\ &= \frac{1}{\hbar}\frac{z - \frac{\mathcal{E}}{\hbar} - \frac{\widetilde{\mathcal{E}}}{\hbar} - i\frac{\gamma}{\hbar}}{\left(z - \frac{\mathcal{E}}{\hbar} - \frac{\widetilde{\mathcal{E}}}{\hbar}\right)^2 + \frac{\gamma^2}{\hbar^2}}. \end{aligned} \tag{6.119}$$

Thus, the role of the improper self-energy becomes clear: in the case of free particles, the Green's function exhibits a pole on the real axis, while in the presence of interactions, this pole is shifted into the complex lower half-plane. This behavior is observed in the frequency domain. Naturally, one may wonder how this modification manifests itself in the time domain. We will employ the residue theorem K.1 to compute the inverse Fourier transform of the Green's function, thereby revealing its behavior in time. We obtain

$$\begin{aligned} G^{(r)}_{C,C^\dagger}(t) &= \int_\Gamma \frac{dz}{2\pi} e^{-izt} G^{(r)}_{C,C^\dagger}(z) = \\ &= \frac{1}{\hbar}\int_\Gamma \frac{dz}{2\pi} \frac{e^{-izt}}{z - \frac{\mathcal{E}}{\hbar} - \frac{\widetilde{\mathcal{E}}}{\hbar} + i\frac{\gamma}{\hbar}}, \end{aligned} \tag{6.120}$$

where Γ is a line parallel to the real axis and belongs to the upper half-plane. Assume $t > 0$. The pole is $z_0 = \frac{\mathcal{E}}{\hbar} + \frac{\widetilde{\mathcal{E}}}{\hbar} - i\frac{\gamma}{\hbar}$, then to obtain a circuit containing z_0, we consider the lower half-plane. The line integral of the function around the circuit is equal to $2\pi i$ times the sum of residues of the function at the points. Since for $R \to +\infty$, we apply Jordan's lemma, it remains to calculate the integral along the line. For the simple pole of the first order we have

$$\begin{aligned} G^{(r)}_{C,C^\dagger}(t) &= \frac{1}{\hbar}\frac{1}{2\pi}(-2\pi i)\,\Theta(t) e^{-i\left[\frac{\mathcal{E}+\widetilde{\mathcal{E}}}{\hbar} - i\frac{\gamma}{\hbar}\right]t} = \\ &= -\frac{i}{\hbar}\Theta(t) e^{-i\left[\frac{\mathcal{E}+\widetilde{\mathcal{E}}}{\hbar} - i\frac{\gamma}{\hbar}\right]t}, \end{aligned} \tag{6.121}$$

where we used Heaviside's $\Theta(t)$ so that it is $t > 0$. If $t < 0$, the circuit must belong to the upper half-plane, where there are no poles, and the integral is null, consequently the propagator is also null. The improper self-energy has two effects on the Green's function: the real part produces renormalization of energy levels and the imaginary part generates decays in time. We are interested in $z = \omega + i\delta$ in the limit $\delta \to 0^+$ and we analyze how this choice modifies the spectral function, that is,

$$\begin{aligned} A_{C,C^\dagger}(\omega) &= \lim_{\delta\to 0^+}\left(-2\hbar \operatorname{Im} G^{(r)}_{C,C^\dagger}(\omega + i\delta)\right) = \\ &= -2\left[\frac{\frac{\gamma}{\hbar}}{\left(\omega - \frac{\mathcal{E}+\widetilde{\mathcal{E}}}{\hbar}\right)^2 + \frac{\gamma^2}{\hbar^2}}\right], \end{aligned} \tag{6.122}$$

which is a Lorentzian function. In conclusion, if the approximation $\Sigma^*(z) \simeq \Sigma^*(z_0)$ holds, the improper self-energy leads to several important physical consequences. First, it induces a renormalization of the energy levels, shifting them from their unperturbed positions on the real axis. Secondly, it introduces a finite decay rate

γ, associated with the imaginary part of the self-energy. As a result, the spectral function acquires a Lorentzian profile with a width given by

$$\frac{\gamma}{\hbar} = \frac{1}{\tau}, \tag{6.123}$$

where τ is the lifetime of the excitation. In the limit $\gamma \to 0^+$, the Lorentzian function tends toward a Dirac delta. However, this delta is no longer centered at the unperturbed energy $\mathcal{E}$, but rather at $\mathcal{E} + \widetilde{\mathcal{E}}$, due to the real part of the self-energy remaining finite. The smaller the value of γ, the larger the corresponding lifetime τ: as $\gamma \to 0$, we have $\tau \to \infty$, indicating that the excitation becomes stable, with an infinite lifetime. Conversely, for finite γ, the lifetime τ is also finite, and the excitation with energy $\mathcal{E} + \widetilde{\mathcal{E}}$ decays into other states after a finite time.

The physical properties of the the Anderson model depend crucially on the position of the impurity level $\mathcal{E}$ relative to the conduction band (see Eq. 6.92).

- When the impurity energy lies within the conduction band,

$$\mathcal{E}_{\min} < \mathcal{E} < \mathcal{E}_{\max}, \tag{6.124}$$

the denominator of $\Sigma^*(z)$ can vanish for real z, so that the self-energy acquires a finite imaginary part: from Eq. 6.115, it follows

$$\mathrm{Im}\ \Sigma^*(\omega) = -\pi \sum_{\mathbf{k}} A_{\mathbf{k}}^2 \delta\left(\omega - \frac{\mathcal{E}_{\mathbf{k}}}{\hbar}\right) \neq 0. \tag{6.125}$$

The imaginary part $\gamma = -\,\mathrm{Im}\ \Sigma^*(\omega)$ represents the decay rate of the impurity excitation into the continuum, giving rise to a finite lifetime (see Eq. 6.123)

$$\tau = \frac{\hbar}{\gamma}. \tag{6.126}$$

Consequently, the spectral function takes the Lorentzian form

$$A_{C,C^\dagger}(\omega) \propto \frac{\frac{\gamma}{\hbar}}{(\omega - \omega_0)^2 + \left(\frac{\gamma}{\hbar}\right)^2}, \tag{6.127}$$

signaling a broadened resonance. The impurity state is therefore unstable and decays over time due to hybridization with the conduction band.

- When the impurity energy lies outside the conduction band,

$$\mathcal{E} < \mathcal{E}_{\min} \quad \text{or} \quad \mathcal{E} > \mathcal{E}_{\max}, \tag{6.128}$$

the denominator in Eq. 6.115 never vanishes for real z, and the self-energy $\Sigma^*(z)$ is purely real, i.e.,

$$\mathrm{Im}\,\Sigma^*(\omega) = 0. \tag{6.129}$$

In this case, the impurity cannot decay into the continuum, since no available conduction states exist at that energy. The Green's function thus has a real, isolated pole

$$z_0 = \frac{\mathcal{E} + \widetilde{\mathcal{E}}}{\hbar}, \qquad \mathrm{Im}\,\Sigma^*(z_0) = 0, \tag{6.130}$$

which corresponds to a bound state localized around the impurity with infinite lifetime. The corresponding spectral function reduces to a Dirac delta, i.e.,

$$A_{C,C^\dagger}(\omega) = 2\pi\hbar\delta\left(\omega - \frac{\mathcal{E} + \widetilde{\mathcal{E}}}{\hbar}\right), \tag{6.131}$$

indicating a stationary, stable excitation.

6.6 Independent Boson Model

We introduce a Hamiltonian describing a single fermionic impurity linearly coupled to a bosonic environment, that is,

$$\hat{\mathcal{H}} = \mathcal{E} b^\dagger b + \sum_{\mathbf{q}} \hbar\omega_{\mathbf{q}}\, a^\dagger_{\mathbf{q}} a_{\mathbf{q}} + \sum_{\mathbf{q}} M_{\mathbf{q}} \left(a_{\mathbf{q}} + a^\dagger_{\mathbf{q}}\right) b^\dagger b, \tag{6.132}$$

where the first term represents the energy of the impurity, the second term corresponds to a phonon bath (see Chap. 4), and the third term accounts for the interaction between the impurity and the phonons. Note that the phonon operators considered here are associated with the same momentum vector $\mathbf{q}$. The precise sign convention is not essential here, since the physical results remain unaffected. The operators b and $b^\dagger$ describe the impurity, while $a_{\mathbf{q}}$ and $a^\dagger_{\mathbf{q}}$ describe the phonons. Consequently, they commute, i.e.,

$$\left[b, a_{\mathbf{q}}\right] = 0, \tag{6.133}$$

$$\left[b^\dagger, a_{\mathbf{q}}\right] = 0, \tag{6.134}$$

$$\left[b, a^\dagger_{\mathbf{q}}\right] = 0, \tag{6.135}$$

$$\left[b^\dagger, a_{\mathbf{q}}^\dagger\right] = 0, \tag{6.136}$$

and the coupling constants $M_{\mathbf{q}}$ are real. This Hamiltonian, commonly known in the literature as the independent boson model (IBM), provides the simplest framework for studying a single impurity interacting with a bosonic environment. Despite its simplicity, it captures essential features such as the dressing of the impurity by the phonon field, giving rise to a fermionic quasiparticle known as the single-level polaron. The term $\mathcal{E} b^\dagger b$ represents a single nondispersive energy level, e.g., an impurity in a metal, and it interacts with phonons (see Chap. 4) with energy $\hbar\omega_{\mathbf{q}}$. The impurity, through the interaction term, can either create or destroy a phonon. Eq. 6.132 is the simplest Hamiltonian model containing an interaction term, since it is linear and exactly diagonalizable, as we shall now prove. To diagonalize the Hamiltonian, we employ a suitable unitary operator U. Given a generic operator S, the condition for $U = e^{-S}$ to be unitary is that $S^\dagger = -S$, i.e., S must be anti-Hermitian. We diagonalize the initial Hamiltonian according to the transformation

$$e^S \hat{\mathcal{H}} e^{-S} = \hat{\tilde{\mathcal{H}}}, \tag{6.137}$$

where, for the Hamiltonian 6.132, the generator S is given by

$$S = \sum_{\mathbf{q}} r_{\mathbf{q}}\, b^\dagger b \left(a_{\mathbf{q}}^\dagger - a_{\mathbf{q}}\right), \tag{6.138}$$

where

$$r_{\mathbf{q}} = \frac{M_{\mathbf{q}}}{\hbar\omega_{\mathbf{q}}} \tag{6.139}$$

is a dimensionless constant. Note that S scales with the ratio between the interaction energy and the oscillator energy; consequently, the amplitude of the operator grows as the coupling strength increases. To apply the exponential transformation of S, we make use of the BCH formula, that is, Eq. H.4. Note that the phononic term transforms as

$$e^S b^\dagger b\, e^{-S} = \left(e^S b^\dagger e^{-S}\right)\left(e^S b\, e^{-S}\right), \tag{6.140}$$

that is, the operator $b^\dagger b$ is mapped into the product of the individually transformed operators. Now consider the transformation of the bosonic annihilation operator $a_{\mathbf{q}}$. We first compute the commutator

$$[S, a_{\mathbf{q}}] = \left[\sum_{\mathbf{q}'} r_{\mathbf{q}'} b^\dagger b \left(a_{\mathbf{q}'}^\dagger - a_{\mathbf{q}'}\right),\, a_{\mathbf{q}}\right] =$$

$$= b^\dagger b \left(\sum_{\mathbf{q}'} r_{\mathbf{q}'} \left[a^\dagger_{\mathbf{q}'}, a_{\mathbf{q}} \right] - \sum_{\mathbf{q}'} r_{\mathbf{q}'} \left[a_{\mathbf{q}'}, a_{\mathbf{q}} \right] \right) =$$

$$= b^\dagger b \sum_{\mathbf{q}'} r_{\mathbf{q}'} \left[a^\dagger_{\mathbf{q}'}, a_{\mathbf{q}} \right] =$$

$$= b^\dagger b \sum_{\mathbf{q}'} r_{\mathbf{q}'} (-\delta_{\mathbf{q}',\mathbf{q}}) =$$

$$= -r_{\mathbf{q}}\, b^\dagger b, \tag{6.141}$$

$$\left[S, a^\dagger_{\mathbf{q}} \right] = \left[\sum_{\mathbf{q}'} r_{\mathbf{q}'} b^\dagger b \left(a^\dagger_{\mathbf{q}'} - a_{\mathbf{q}'} \right),\, a^\dagger_{\mathbf{q}} \right] =$$

$$= b^\dagger b \left(\sum_{\mathbf{q}'} r_{\mathbf{q}'} \left[a^\dagger_{\mathbf{q}'}, a^\dagger_{\mathbf{q}} \right] - \sum_{\mathbf{q}'} r_{\mathbf{q}'} \left[a_{\mathbf{q}'}, a^\dagger_{\mathbf{q}} \right] \right) =$$

$$= b^\dagger b \left(0 - \sum_{\mathbf{q}'} r_{\mathbf{q}'} \delta_{\mathbf{q}',\mathbf{q}} \right) =$$

$$= -r_{\mathbf{q}}\, b^\dagger b, \tag{6.142}$$

then we find that the bosonic operators transform as

$$e^S a_{\mathbf{q}} e^{-S} = a_{\mathbf{q}} - r_{\mathbf{q}} b^\dagger b, \tag{6.143}$$

$$e^S a^\dagger_{\mathbf{q}} e^{-S} = a^\dagger_{\mathbf{q}} - r_{\mathbf{q}} b^\dagger b, \tag{6.144}$$

since all higher-order commutators vanish identically. For the impurity operators, we compute the commutators

$$[S, b] = \sum_{\mathbf{q}'} r_{\mathbf{q}'} \left(a^\dagger_{\mathbf{q}'} - a_{\mathbf{q}'} \right) \left[b^\dagger b, b \right] =$$

$$= -b \sum_{\mathbf{q}'} r_{\mathbf{q}'} \left(a^\dagger_{\mathbf{q}'} - a_{\mathbf{q}'} \right), \tag{6.145}$$

$$\left[S, b^\dagger \right] = \sum_{\mathbf{q}'} r_{\mathbf{q}'} \left(a^\dagger_{\mathbf{q}'} - a_{\mathbf{q}'} \right) \left[b^\dagger b, b^\dagger \right] =$$

$$= b^\dagger \sum_{\mathbf{q}'} r_{\mathbf{q}'} \left(a^\dagger_{\mathbf{q}'} - a_{\mathbf{q}'} \right). \tag{6.146}$$

As a result, the impurity operators transform as

$$e^S b\, e^{-S} = b \left(\mathbb{1} - S' + \frac{1}{2!} S'^2 - \cdots \right), \tag{6.147}$$

$$e^S b^\dagger\, e^{-S} = b^\dagger \left(\mathbb{1} + S' + \frac{1}{2!} S'^2 + \cdots \right), \tag{6.148}$$

where the exponential series follows from the conjugation formula H.5, and the operator S' is defined as (see Eq. 6.138)

$$S' = \sum_{\mathbf{q}'} r_{\mathbf{q}'} \left(a^\dagger_{\mathbf{q}'} - a_{\mathbf{q}'} \right). \tag{6.149}$$

The operators b and $b^\dagger$ satisfy

$$e^S b\, e^{-S} = b\, e^{-\sum_{\mathbf{q}} r_{\mathbf{q}} \left(a^\dagger_{\mathbf{q}} - a_{\mathbf{q}} \right)}, \tag{6.150}$$

$$e^S b^\dagger\, e^{-S} = b^\dagger\, e^{\sum_{\mathbf{q}} r_{\mathbf{q}} \left(a^\dagger_{\mathbf{q}} - a_{\mathbf{q}} \right)}. \tag{6.151}$$

$$\begin{aligned} e^S b^\dagger b\, e^{-S} &= \left(e^S b^\dagger e^{-S} \right) \left(e^S b\, e^{-S} \right) \\ &= b^\dagger b. \end{aligned} \tag{6.152}$$

Consequently the Hamiltonian 6.132 transforms as

$$\begin{aligned} \hat{\tilde{\mathcal{H}}} &= \mathcal{E} e^S b^\dagger b\, e^{-S} + \sum_{\mathbf{q}} \hbar\omega_{\mathbf{q}}\, e^S a^\dagger_{\mathbf{q}} a_{\mathbf{q}}\, e^{-S} + \sum_{\mathbf{q}} M_{\mathbf{q}}\, e^S (a_{\mathbf{q}} + a^\dagger_{\mathbf{q}})\, b^\dagger b\, e^{-S} = \\ &= \mathcal{E} b^\dagger b + \sum_{\mathbf{q}} \hbar\omega_{\mathbf{q}}\, e^S a^\dagger_{\mathbf{q}}\, e^{-S} e^S a_{\mathbf{q}}\, e^{-S} + \sum_{\mathbf{q}} M_{\mathbf{q}}\, e^S (a_{\mathbf{q}} + a^\dagger_{\mathbf{q}})\, e^{-S} \cdot e^S b^\dagger b\, e^{-S} = \\ &= \mathcal{E} b^\dagger b + \sum_{\mathbf{q}} \hbar\omega_{\mathbf{q}} \left(a^\dagger_{\mathbf{q}} - \frac{M_{\mathbf{q}}}{\hbar\omega_{\mathbf{q}}} b^\dagger b \right) \left(a_{\mathbf{q}} - \frac{M_{\mathbf{q}}}{\hbar\omega_{\mathbf{q}}} b^\dagger b \right) \\ &\quad + \sum_{\mathbf{q}} M_{\mathbf{q}}\, e^S (a_{\mathbf{q}} + a^\dagger_{\mathbf{q}})\, e^{-S} b^\dagger b = \\ &= \mathcal{E} b^\dagger b + \sum_{\mathbf{q}} \hbar\omega_{\mathbf{q}} \left(a^\dagger_{\mathbf{q}} - \frac{M_{\mathbf{q}}}{\hbar\omega_{\mathbf{q}}} b^\dagger b \right) \left(a_{\mathbf{q}} - \frac{M_{\mathbf{q}}}{\hbar\omega_{\mathbf{q}}} b^\dagger b \right) \end{aligned}$$

$$
\begin{aligned}
&+ b^\dagger b \sum_{\mathbf{q}} M_{\mathbf{q}}\, e^S (a_{\mathbf{q}} + a_{\mathbf{q}}^\dagger)\, e^{-S} = \\
&= \mathcal{E} b^\dagger b + \sum_{\mathbf{q}} \hbar\omega_{\mathbf{q}} \left[a_{\mathbf{q}}^\dagger a_{\mathbf{q}} + \frac{M_{\mathbf{q}}^2}{(\hbar\omega_{\mathbf{q}})^2} b^\dagger b - \frac{M_{\mathbf{q}}}{\hbar\omega_{\mathbf{q}}} b^\dagger b\, a_{\mathbf{q}}^\dagger - \frac{M_{\mathbf{q}}}{\hbar\omega_{\mathbf{q}}} b^\dagger b\, a_{\mathbf{q}} \right] \\
&+ b^\dagger b \sum_{\mathbf{q}} M_{\mathbf{q}} \left(a_{\mathbf{q}}^\dagger + a_{\mathbf{q}} - \frac{2M_{\mathbf{q}}}{\hbar\omega_{\mathbf{q}}} b^\dagger b \right) = \\
&= \mathcal{E} b^\dagger b + \sum_{\mathbf{q}} \left[\hbar\omega_{\mathbf{q}}\, a_{\mathbf{q}}^\dagger a_{\mathbf{q}} + \frac{M_{\mathbf{q}}^2}{\hbar\omega_{\mathbf{q}}} b^\dagger b - M_{\mathbf{q}} b^\dagger b\, a_{\mathbf{q}}^\dagger - M_{\mathbf{q}} b^\dagger b\, a_{\mathbf{q}} \right] \\
&+ \sum_{\mathbf{q}} \left[M_{\mathbf{q}} b^\dagger b\, a_{\mathbf{q}}^\dagger + M_{\mathbf{q}} b^\dagger b\, a_{\mathbf{q}} - \frac{2M_{\mathbf{q}}^2}{\hbar\omega_{\mathbf{q}}} (b^\dagger b)^2 \right] = \\
&= \mathcal{E} b^\dagger b + \sum_{\mathbf{q}} \left[\hbar\omega_{\mathbf{q}}\, a_{\mathbf{q}}^\dagger a_{\mathbf{q}} + \frac{M_{\mathbf{q}}^2}{\hbar\omega_{\mathbf{q}}} b^\dagger b - \frac{2M_{\mathbf{q}}^2}{\hbar\omega_{\mathbf{q}}} (b^\dagger b)^2 \right].
\end{aligned}
\tag{6.153}
$$

Since the operator $b^\dagger b$ is fermionic, we have

$$
(b^\dagger b)^2 = b^\dagger b,
\tag{6.154}
$$

then

$$
\begin{aligned}
\hat{\tilde{\mathcal{H}}} &= \mathcal{E}\, b^\dagger b + \sum_{\mathbf{q}} \left(\hbar\omega_{\mathbf{q}}\, a_{\mathbf{q}}^\dagger a_{\mathbf{q}} + \frac{M_{\mathbf{q}}^2}{\hbar\omega_{\mathbf{q}}} b^\dagger b - \frac{2M_{\mathbf{q}}^2}{\hbar\omega_{\mathbf{q}}} b^\dagger b \right) = \\
&= \mathcal{E}\, b^\dagger b + \sum_{\mathbf{q}} \left(\hbar\omega_{\mathbf{q}}\, a_{\mathbf{q}}^\dagger a_{\mathbf{q}} - \frac{M_{\mathbf{q}}^2}{\hbar\omega_{\mathbf{q}}} b^\dagger b \right).
\end{aligned}
\tag{6.155}
$$

We can thus finally express the Hamiltonian 6.132 in diagonal form as follows

$$
\begin{aligned}
\hat{\tilde{\mathcal{H}}} &= \sum_{\mathbf{q}} \hbar\omega_{\mathbf{q}}\, a_{\mathbf{q}}^\dagger a_{\mathbf{q}} + \left(\mathcal{E} - \sum_{\mathbf{q}} \frac{M_{\mathbf{q}}^2}{\hbar\omega_{\mathbf{q}}} \right) b^\dagger b \equiv \\
&\equiv \sum_{\mathbf{q}} \hbar\omega_{\mathbf{q}}\, a_{\mathbf{q}}^\dagger a_{\mathbf{q}} + \tilde{\mathcal{E}}\, b^\dagger b,
\end{aligned}
\tag{6.156}
$$

with

$$
\tilde{\mathcal{E}} = \mathcal{E} - \sum_{\mathbf{q}} \frac{M_{\mathbf{q}}^2}{\hbar\omega_{\mathbf{q}}}.
\tag{6.157}
$$

By means of the new operators, the Hamiltonian is given by the sum of a bosonic and a fermionic term, so the eigenstates are given by the tensor product of the bosonic and fermionic eigenstates. In the occupation-number basis, the bosonic basis is given by (see Eq. 2.142)

$$|N_{\mathbf{q}_1}, \ldots, N_{\mathbf{q}_N}\rangle = \frac{(a^\dagger_{\mathbf{q}_1})^{N_{\mathbf{q}_1}}}{\sqrt{N_{\mathbf{q}_1}!}} \cdots \frac{(a^\dagger_{\mathbf{q}_N})^{N_{\mathbf{q}_N}}}{\sqrt{N_{\mathbf{q}_N}!}}|0\rangle_b, \tag{6.158}$$

with eigenvalues

$$N_{\mathbf{q}_1}\hbar\omega_{\mathbf{q}_1} + \cdots + N_{\mathbf{q}_N}\hbar\omega_{\mathbf{q}_N}. \tag{6.159}$$

Due to the impurity, the fermionic state in the occupation-number basis is

$$|N_b\rangle = b^\dagger|0\rangle_f, \tag{6.160}$$

and the corresponding eigenvalue is given in Eq. 6.157. Within the diagonalized Hamiltonian $\hat{\tilde{\mathcal{H}}}$, the impurity couples to the phonon field and becomes dressed, giving rise to a quasiparticle, i.e., the single-level polaron, with an effective energy $\tilde{\mathcal{E}}$. The Green's function of this quasiparticle reduces to a delta function, reflecting the exact solvability of the model and the fact that the quasiparticle is stable, with no intrinsic decay. In practice, however, experiments do not directly probe the Green's function of the dressed impurity. Instead, the measurable quantity corresponds to the Green's function of the bare impurity, i.e., the original fermionic operator prior to the unitary transformation. Our objective is therefore to determine the form of the Green's function associated with the bare impurity operator, which is physically relevant and experimentally accessible. In summary, the eigenstates and eigenvalues of $\hat{\tilde{\mathcal{H}}}$ are given by

$$|N_{\mathbf{q}_1}, \ldots, N_{\mathbf{q}_N}\rangle b^\dagger|0\rangle_f, \tag{6.161}$$

$$N_{\mathbf{q}_1}\hbar\omega_{\mathbf{q}_1} + \cdots + N_{\mathbf{q}_N}\hbar\omega_{\mathbf{q}_N} + \mathcal{E} - \sum_{\mathbf{q}} \frac{M^2_{\mathbf{q}}}{\hbar\omega_{\mathbf{q}}}. \tag{6.162}$$

respectively. It is then natural to determine the eigenstates and eigenvalues of the Hamiltonian $\hat{\mathcal{H}}$ in the old basis. To this end, we multiply to the left by e^{-S} the eigenvalue equation for the diagonalized Hamiltonian, obtaining

$$e^{-S}\hat{\tilde{\mathcal{H}}}|\psi_n\rangle = \mathcal{E}_n e^{-S}|\psi_n\rangle, \tag{6.163}$$

and from

$$e^{-S}\hat{\tilde{\mathcal{H}}} = e^{-S}e^{S}\hat{\mathcal{H}}e^{-S} = \\ = \hat{\mathcal{H}}e^{-S}, \tag{6.164}$$

we find

$$\hat{\mathcal{H}}e^{-S}|\psi_n\rangle = \mathcal{E}_n e^{-S}|\psi_n\rangle, \tag{6.165}$$

that is, $e^{-S}|\psi_n\rangle$ is an eigenstate of the Hamiltonian $\hat{\mathcal{H}}$ with the same eigenvalue $\mathcal{E}_n$ as $|\psi_n\rangle$. This is expected, since e^{-S} represents a unitary transformation and thus preserves the spectrum of the Hamiltonian. The eigenstates of $\hat{\mathcal{H}}$ are given by

$$e^{-S}|N_{\mathbf{q}_1}, \ldots, N_{\mathbf{q}_N}\rangle b^\dagger|0\rangle_f, \tag{6.166}$$

which, in the original basis, are written as

$$e^{-S}|N_{\mathbf{q}_1}, \ldots, N_{\mathbf{q}_N}\rangle b^\dagger|0\rangle_f = e^{-\sum_{\mathbf{q}} r_{\mathbf{q}}(a_{\mathbf{q}}^\dagger - a_{\mathbf{q}})b^\dagger b}|N_{\mathbf{q}_1}, \ldots, N_{\mathbf{q}_N}\rangle b^\dagger|0\rangle_f. \tag{6.167}$$

Since $b^\dagger b$ acts as a projector onto the occupied impurity state, on the subspace spanned by $b^\dagger|0\rangle_f$ it behaves like the identity operator. Therefore, when acting on this state, we can replace $b^\dagger b$ by the identity operator $\mathbb{1}$, and we write

$$e^{-S}|N_{\mathbf{q}_1}, \ldots, N_{\mathbf{q}_N}\rangle b^\dagger|0\rangle_f = e^{-\sum_{\mathbf{q}} r_{\mathbf{q}}(a_{\mathbf{q}}^\dagger - a_{\mathbf{q}})}|N_{\mathbf{q}_1}, \ldots, N_{\mathbf{q}_N}\rangle b^\dagger|0\rangle_f. \tag{6.168}$$

The exponential operator on the right-hand side is of the form e^{A+B}, so we can use Glauber's formula, that is Eq. H.18. We set

$$A = -\sum_{\mathbf{q}} r_{\mathbf{q}}\, a_{\mathbf{q}}^\dagger, \tag{6.169}$$

$$B = \sum_{\mathbf{q}} r_{\mathbf{q}}\, a_{\mathbf{q}}, \tag{6.170}$$

and we find that the eigenfunctions in the old basis are

$$e^{-S}|N_{\mathbf{q}_1}, \ldots, N_{\mathbf{q}_N}\rangle b^\dagger|0\rangle_f = e^{-\frac{1}{2}\sum_{\mathbf{q}} r_{\mathbf{q}}^2} e^{-\sum_{\mathbf{q}} r_{\mathbf{q}} a_{\mathbf{q}}^\dagger} e^{\sum_{\mathbf{q}} r_{\mathbf{q}} a_{\mathbf{q}}}\, |N_{\mathbf{q}_1}, \ldots, N_{\mathbf{q}_N}\rangle b^\dagger|0\rangle_f. \tag{6.171}$$

In the transformed basis, all phonons occupy their ground state, so the dressed ground state reads

$$|\tilde{G}_s\rangle = |0, 0, 0, \ldots, 0\rangle\, b^\dagger|0\rangle_f, \tag{6.172}$$

whereas in the old basis, this state is expressed as

$$|G_s\rangle = e^{-\frac{1}{2}\sum_{\mathbf{q}} r_{\mathbf{q}}^2} e^{-\sum_{\mathbf{q}} r_{\mathbf{q}} a_{\mathbf{q}}^\dagger} e^{\sum_{\mathbf{q}} r_{\mathbf{q}} a_{\mathbf{q}}} |0, \ldots, 0\rangle\, b^\dagger |0\rangle. \tag{6.173}$$

Since

$$e^{\sum_{\mathbf{q}} r_{\mathbf{q}} a_{\mathbf{q}}} |0, \ldots, 0\rangle = |0, \ldots, 0\rangle, \tag{6.174}$$

the exponential acts as the identity operator on the vacuum, so the ground state becomes

$$\begin{aligned} |G_s\rangle &= e^{-\frac{1}{2}\sum_{\mathbf{q}} r_{\mathbf{q}}^2} e^{-\sum_{\mathbf{q}} r_{\mathbf{q}} a_{\mathbf{q}}^\dagger} |0, \ldots, 0\rangle\, b^\dagger |0\rangle_f = \\ &= e^{-\frac{1}{2}\sum_{\mathbf{q}} r_{\mathbf{q}}^2} \left(\prod_{\mathbf{q}} e^{-r_{\mathbf{q}} a_{\mathbf{q}}^\dagger} |0\rangle_{b,\mathbf{q}} \right) b^\dagger |0\rangle_f . \end{aligned} \tag{6.175}$$

Note that the fundamental state of Hamiltonian 6.132 is a coherent state (see Sect. 1.3), being the product of coherent states. Now, referring to the old basis, we calculate the number of phonons with momentum $\mathbf{q}$. If the interaction is null, the average number of phonons $\left\langle a_{\mathbf{q}}^\dagger a_{\mathbf{q}} \right\rangle$ in the ground state vanishes. Otherwise, the average number of phonons can be calculated as

$$\begin{aligned} \sum_{\mathbf{q}} \left\langle G_s \left| a_{\mathbf{q}}^\dagger a_{\mathbf{q}} \right| G_s \right\rangle &= \sum_{\mathbf{q}} \left\langle \tilde{G}_s \left| e^S a_{\mathbf{q}}^\dagger e^{-S}\, e^S a_{\mathbf{q}} e^{-S} \right| \tilde{G}_s \right\rangle = \\ &= \sum_{\mathbf{q}} r_{\mathbf{q}}^2 \equiv \\ &\equiv \sum_{\mathbf{q}} \frac{M_{\mathbf{q}}^2}{(\hbar \omega_{\mathbf{q}})^2}, \end{aligned} \tag{6.176}$$

where in the last step we employed Eq. 6.139. The greater the interaction strength $M_{\mathbf{q}}$, the greater the number of phonons excited.

6.6.1 Green's Function in the Independent Boson Model

Here, we aim to compute the retarded Green's function of the single-level polaron. From Eq. 6.8, it follows

$$\begin{aligned} G^{(r)}_{b,b^\dagger}(t) &= -\frac{i}{\hbar}\Theta(t) \left\langle \left\{ b(t), b^\dagger(0) \right\} \right\rangle = \\ &= -\frac{i}{\hbar}\Theta(t) \left\langle b(t) b^\dagger(0) + b^\dagger(0) b(t) \right\rangle = \end{aligned}$$

$$= -\frac{i}{\hbar}\Theta(t)\frac{1}{Z}\mathrm{Tr}\left[e^{-\beta\hat{\mathcal{H}}}\left(b(t)b^\dagger(0) + b^\dagger(0)b(t)\right)\right] =$$
$$= -\frac{i}{\hbar}\Theta(t)\frac{1}{Z}\mathrm{Tr}\left[e^{-\beta\hat{\mathcal{H}}}\left(e^{\frac{i}{\hbar}\hat{\mathcal{H}}t}b\,e^{-\frac{i}{\hbar}\hat{\mathcal{H}}t}b^\dagger + b^\dagger\,e^{\frac{i}{\hbar}\hat{\mathcal{H}}t}b\,e^{-\frac{i}{\hbar}\hat{\mathcal{H}}t}\right)\right], \tag{6.177}$$

where, in the final step, we have explicitly expressed the Heisenberg picture (see Eq. 1.102). Since we can diagonalize $\hat{\hat{\mathcal{H}}} = e^S\hat{\mathcal{H}}e^{-S}$, but not $\hat{\mathcal{H}}$ directly, we apply the similarity transformation to each operator: we multiply to the left by e^S, and to the right by e^{-S}, and use the cyclic property of the trace to move the last e^S to the front

$$G^{(r)}_{b,b^\dagger}(t) = -\frac{i}{\hbar}\Theta(t)\frac{1}{Z}\mathrm{Tr}\Big[e^S e^{-\beta\hat{\mathcal{H}}}e^{-S}\Big(e^S e^{\frac{i}{\hbar}\hat{\mathcal{H}}t}e^{-S}e^S b\,e^{-S}e^S e^{-\frac{i}{\hbar}\hat{\mathcal{H}}t}e^{-S}e^S b^\dagger e^{-S} +$$
$$+ e^S b^\dagger e^{-S}e^S e^{\frac{i}{\hbar}\hat{\mathcal{H}}t}e^{-S}e^S b\,e^{-S}e^S e^{-\frac{i}{\hbar}\hat{\mathcal{H}}t}e^{-S}\Big)\Big]. \tag{6.178}$$

From the relations

$$e^S e^{-\beta\hat{\mathcal{H}}}e^{-S} = e^{-\beta\hat{\hat{\mathcal{H}}}}, \tag{6.179}$$

$$e^S e^{i\frac{\hat{\mathcal{H}}}{\hbar}t}e^{-S} = e^{i\frac{\hat{\hat{\mathcal{H}}}}{\hbar}t}, \tag{6.180}$$

$$e^S b e^{-S} = b\,e^{-\sum_{\mathbf{q}} r_{\mathbf{q}}\left(a^\dagger_{\mathbf{q}} - a_{\mathbf{q}}\right)}, \tag{6.181}$$

$$e^S b^\dagger e^{-S} = b^\dagger\,e^{\sum_{\mathbf{q}} r_{\mathbf{q}}\left(a^\dagger_{\mathbf{q}} - a_{\mathbf{q}}\right)}, \tag{6.182}$$

Equation 6.178 becomes

$$G^{(r)}_{b,b^\dagger}(t) = -\frac{i}{\hbar}\,\Theta(t)\,\frac{1}{Z}\,\mathrm{Tr}\left[e^{-\beta\hat{\hat{\mathcal{H}}}}\left(e^{i\frac{\hat{\hat{\mathcal{H}}}}{\hbar}t}\,b\,e^{-\sum_{\mathbf{q}} r_{\mathbf{q}}(a^\dagger_{\mathbf{q}} - a_{\mathbf{q}})}\,e^{-i\frac{\hat{\hat{\mathcal{H}}}}{\hbar}t}\,b^\dagger\,e^{\sum_{\mathbf{q}} r_{\mathbf{q}}(a^\dagger_{\mathbf{q}} - a_{\mathbf{q}})} +\right.\right.$$
$$\left.\left.+ b^\dagger\,e^{\sum_{\mathbf{q}} r_{\mathbf{q}}(a^\dagger_{\mathbf{q}} - a_{\mathbf{q}})}\,e^{i\frac{\hat{\hat{\mathcal{H}}}}{\hbar}t}\,b\,e^{-\sum_{\mathbf{q}} r_{\mathbf{q}}(a^\dagger_{\mathbf{q}} - a_{\mathbf{q}})}\,e^{-i\frac{\hat{\hat{\mathcal{H}}}}{\hbar}t}\right)\right]. \tag{6.183}$$

Since the Hamiltonian $\hat{\hat{\mathcal{H}}}$ is diagonal, in the Heisenberg picture (see Eq. 1.102), the operators $a_{\mathbf{q}}$, $a^\dagger_{\mathbf{q}}$, b, $b^\dagger$ evolve multiplicatively by a phase factor determined by their respective frequencies, in a manner analogous to their evolution in the interaction picture (see Eqs. 2.187 and 2.192). Therefore, Eq. 6.183 becomes

$$G^{(r)}_{b,b^\dagger}(t) = -\frac{i}{\hbar}\,\Theta(t)\,\frac{1}{Z}\,\mathrm{Tr}\left[e^{-\beta\hat{\hat{\mathcal{H}}}}\left(e^{-i\frac{\tilde{\mathcal{E}}}{\hbar}t}\,b\,e^{-\sum_{\mathbf{q}} r_{\mathbf{q}}(a^\dagger_{\mathbf{q}}e^{i\omega_{\mathbf{q}}t} - a_{\mathbf{q}}e^{-i\omega_{\mathbf{q}}t})}\right.\right.$$

$$b^\dagger\, e^{\sum_{\mathbf{q}} r_{\mathbf{q}}(a_{\mathbf{q}}^\dagger - a_{\mathbf{q}})} + b^\dagger\, e^{\sum_{\mathbf{q}} r_{\mathbf{q}}(a_{\mathbf{q}}^\dagger - a_{\mathbf{q}})} e^{-i\frac{\tilde{\mathcal{E}}}{\hbar}t}\, b\, e^{-\sum_{\mathbf{q}} r_{\mathbf{q}}(a_{\mathbf{q}}^\dagger e^{i\omega_{\mathbf{q}}t} - a_{\mathbf{q}} e^{-i\omega_{\mathbf{q}}t})}\Big)\Big]. \tag{6.184}$$

If we define the operator

$$d(t) \equiv e^{\sum_{\mathbf{q}} r_{\mathbf{q}}\left(a_{\mathbf{q}}^\dagger e^{i\omega_{\mathbf{q}}t} - a_{\mathbf{q}} e^{-i\omega_{\mathbf{q}}t}\right)}, \tag{6.185}$$

we can express the time-evolved dressing operator in a compact form as follows

$$G^{(r)}_{b,b^\dagger}(t) = -\frac{i}{\hbar}\,\Theta(t)\,\frac{1}{Z}\,\mathrm{Tr}\left\{e^{-i\frac{\tilde{\mathcal{E}}}{\hbar}t} e^{-\beta\hat{\tilde{\mathcal{H}}}}\left[(bd)(t)\,\left(b^\dagger d^\dagger\right)(0) + \left(b^\dagger d^\dagger\right)(0)\,(bd)(t)\right]\right\}. \tag{6.186}$$

The second term vanishes: indeed, since the operator combination $(b^\dagger d^\dagger)(0)(bd)(t)$ acting on the impurity vacuum state yields the null state, it can be identified as the null operator over the entire interval $[0, t]$. Consequently, we have

$$G^{(r)}_{b,b^\dagger}(t) = -\frac{i}{\hbar}\,\Theta(t)\,\frac{1}{Z}\,\mathrm{Tr}\left[e^{-i\frac{\tilde{\mathcal{E}}}{\hbar}t} e^{-\beta\hat{\tilde{\mathcal{H}}}}(bd)(t)\,\left(b^\dagger d^\dagger\right)(0)\right]. \tag{6.187}$$

We now explicitly insert the time-evolved operators and write

$$G^{(r)}_{b,b^\dagger}(t) = -\frac{i}{\hbar}\,\Theta(t)\,\frac{1}{Z}\,\mathrm{Tr}\left[e^{-i\frac{\tilde{\mathcal{E}}}{\hbar}t} e^{-\beta\hat{\tilde{\mathcal{H}}}} b\, e^{-\sum_{\mathbf{q}} r_{\mathbf{q}}(a_{\mathbf{q}}^\dagger e^{i\omega_{\mathbf{q}}t} - a_{\mathbf{q}} e^{-i\omega_{\mathbf{q}}t})}\, b^\dagger\, e^{\sum_{\mathbf{q}} r_{\mathbf{q}}(a_{\mathbf{q}}^\dagger - a_{\mathbf{q}})}\right]. \tag{6.188}$$

We evaluate the above trace with respect to the basis given in Eq. 6.161, in which the Hamiltonian 6.156 is diagonal: the bosonic and fermionic operators act independently on the eigenstates of $\hat{\tilde{\mathcal{H}}}$, and Eq. 6.188 becomes

$$\begin{aligned} G^{(r)}_{b,b^\dagger}(t) &= -\frac{i}{\hbar}\,\Theta(t)\,\frac{1}{Z_f}\,\mathrm{Tr}\left[e^{-i\frac{\tilde{\mathcal{E}}}{\hbar}t} e^{-\beta\tilde{\mathcal{E}}b^\dagger b} b b^\dagger\right] \\ &\quad \frac{1}{Z_b}\,\mathrm{Tr}\left[e^{-\beta\sum_{\mathbf{q}}\hbar\omega_{\mathbf{q}} a_{\mathbf{q}}^\dagger a_{\mathbf{q}}} e^{-\sum_{\mathbf{q}} r_{\mathbf{q}}(a_{\mathbf{q}}^\dagger e^{i\omega_{\mathbf{q}}t} - a_{\mathbf{q}} e^{-i\omega_{\mathbf{q}}t})} e^{\sum_{\mathbf{q}} r_{\mathbf{q}}(a_{\mathbf{q}}^\dagger - a_{\mathbf{q}})}\right] = \\ &= -\frac{i}{\hbar}\,\Theta(t)\, e^{-i\frac{\tilde{\mathcal{E}}}{\hbar}t}\,\frac{1}{Z_f}\,\mathrm{Tr}\left[e^{-\beta\tilde{\mathcal{E}}b^\dagger b} b b^\dagger\right] \end{aligned}$$

$$\frac{1}{Z_b} \mathrm{Tr}\left[e^{-\beta\sum_{\mathbf{q}}\hbar\omega_{\mathbf{q}}a_{\mathbf{q}}^{\dagger}a_{\mathbf{q}}}e^{-\sum_{\mathbf{q}}r_{\mathbf{q}}(a_{\mathbf{q}}^{\dagger}e^{i\omega_{\mathbf{q}}t}-a_{\mathbf{q}}e^{-i\omega_{\mathbf{q}}t})}e^{\sum_{\mathbf{q}}r_{\mathbf{q}}(a_{\mathbf{q}}^{\dagger}-a_{\mathbf{q}})}\right], \tag{6.189}$$

where the partition function factorizes as a product of the bosonic and fermionic partition functions, i.e.,

$$Z = Z_f Z_b, \tag{6.190}$$

with

$$Z_f = \mathrm{Tr}\left[e^{-\beta\tilde{\mathcal{E}}b^{\dagger}b}\right], \tag{6.191}$$

$$Z_b = \mathrm{Tr}\left[e^{-\beta\sum_{\mathbf{q}}\hbar\omega_{\mathbf{q}}a_{\mathbf{q}}^{\dagger}a_{\mathbf{q}}}\right]. \tag{6.192}$$

Since we are working in the canonical ensemble and the fermionic Hilbert space is restricted to occupations 0 and 1, the fermionic trace can be evaluated explicitly. In going from Eq. 6.186 to Eq. 6.187, we have assumed that at time $t = 0$ the impurity level is unoccupied; as a consequence, the fermionic trace yields unity. Consequently, only the bosonic contribution remains.

$$\begin{aligned}
&\frac{1}{Z_b} \mathrm{Tr}\left[e^{-\beta\sum_{\mathbf{q}}\hbar\omega_{\mathbf{q}}a_{\mathbf{q}}^{\dagger}a_{\mathbf{q}}}\, e^{-\sum_{\mathbf{q}}r_{\mathbf{q}}(a_{\mathbf{q}}^{\dagger}e^{i\omega_{\mathbf{q}}t}-a_{\mathbf{q}}e^{-i\omega_{\mathbf{q}}t})}e^{\sum_{\mathbf{q}}r_{\mathbf{q}}(a_{\mathbf{q}}^{\dagger}-a_{\mathbf{q}})}\right] \\
&\quad = \frac{1}{Z_b} \sum_{N_{\mathbf{q}_1},\ldots,N_{\mathbf{q}_N},\ldots,N_{\infty}} \\
&\left\langle N_{\mathbf{q}_1},\ldots,N_{\mathbf{q}_N},\ldots,N_{\infty}\right| e^{-\beta\sum_{\mathbf{q}}\hbar\omega_{\mathbf{q}}a_{\mathbf{q}}^{\dagger}a_{\mathbf{q}}}e^{-\sum_{\mathbf{q}}r_{\mathbf{q}}(a_{\mathbf{q}}^{\dagger}e^{i\omega_{\mathbf{q}}t}-a_{\mathbf{q}}e^{-i\omega_{\mathbf{q}}t})} \\
&\qquad e^{\sum_{\mathbf{q}}r_{\mathbf{q}}(a_{\mathbf{q}}^{\dagger}-a_{\mathbf{q}})}\left| N_{\mathbf{q}_1},\ldots,N_{\mathbf{q}_N},\ldots N_{\infty}\right\rangle.
\end{aligned} \tag{6.193}$$

Each exponential operator acts individually on the state labeled by $N_{\mathbf{q}}$, so the trace over all phonon modes can be factorized into a product over $\mathbf{q}$, with each factor being a series over the infinite admissible values of $N_{\mathbf{q}}$. That is,

$$\frac{1}{Z_b} \mathrm{Tr}\left[e^{-\beta\sum_{\mathbf{q}}\hbar\omega_{\mathbf{q}}a_{\mathbf{q}}^{\dagger}a_{\mathbf{q}}}\, e^{-\sum_{\mathbf{q}}r_{\mathbf{q}}(a_{\mathbf{q}}^{\dagger}e^{i\omega_{\mathbf{q}}t}-a_{\mathbf{q}}e^{-i\omega_{\mathbf{q}}t})}\, e^{\sum_{\mathbf{q}}r_{\mathbf{q}}(a_{\mathbf{q}}^{\dagger}-a_{\mathbf{q}})}\right] =$$

$$= \prod_{\mathbf{q}} \frac{1}{Z_{b,\mathbf{q}}} \sum_{N_{\mathbf{q}}=0}^{\infty} \left\langle N_{\mathbf{q}}\right| e^{-\beta\hbar\omega_{\mathbf{q}}a_{\mathbf{q}}^{\dagger}a_{\mathbf{q}}}\, e^{-r_{\mathbf{q}}(a_{\mathbf{q}}^{\dagger}e^{i\omega_{\mathbf{q}}t}-a_{\mathbf{q}}e^{-i\omega_{\mathbf{q}}t})}\, e^{r_{\mathbf{q}}(a_{\mathbf{q}}^{\dagger}-a_{\mathbf{q}})}\left| N_{\mathbf{q}}\right\rangle =$$

$$= \prod_{\mathbf{q}} \frac{1}{Z_{b,\mathbf{q}}} \sum_{N_{\mathbf{q}}=0}^{\infty} e^{-\beta\hbar\omega_{\mathbf{q}} N_{\mathbf{q}}} \left\langle N_{\mathbf{q}} \left| e^{-r_{\mathbf{q}}(a_{\mathbf{q}}^{\dagger} e^{i\omega_{\mathbf{q}} t} - a_{\mathbf{q}} e^{-i\omega_{\mathbf{q}} t})} \, e^{r_{\mathbf{q}}(a_{\mathbf{q}}^{\dagger} - a_{\mathbf{q}})} \right| N_{\mathbf{q}} \right\rangle, \tag{6.194}$$

where we have made the density operator explicit in terms of the number operator $a_{\mathbf{q}}^{\dagger} a_{\mathbf{q}}$ and replaced it with its eigenvalue $N_{\mathbf{q}}$. The bosonic partition function for each mode $\mathbf{q}$ factorizes as follows

$$\begin{aligned} Z_{b,\mathbf{q}} &= \mathrm{Tr}\left[e^{-\beta\hbar\omega_{\mathbf{q}} a_{\mathbf{q}}^{\dagger} a_{\mathbf{q}}}\right] = \\ &= \sum_{N_{\mathbf{q}}=0}^{\infty} e^{-\beta\hbar\omega_{\mathbf{q}} N_{\mathbf{q}}} = \\ &= \frac{1}{1 - e^{-\beta\hbar\omega_{\mathbf{q}}}}. \end{aligned} \tag{6.195}$$

Therefore, the bosonic trace becomes

$$\prod_{\mathbf{q}} \left(1 - e^{-\beta\hbar\omega_{\mathbf{q}}}\right) \sum_{N_{\mathbf{q}}=0}^{\infty} e^{-\beta\hbar\omega_{\mathbf{q}} N_{\mathbf{q}}} \left\langle N_{\mathbf{q}} \left| e^{-r_{\mathbf{q}}(a_{\mathbf{q}}^{\dagger} e^{i\omega_{\mathbf{q}} t} - a_{\mathbf{q}} e^{-i\omega_{\mathbf{q}} t})} \, e^{r_{\mathbf{q}}(a_{\mathbf{q}}^{\dagger} - a_{\mathbf{q}})} \right| N_{\mathbf{q}} \right\rangle. \tag{6.196}$$

Now, from the Glauber formula H.18 in the form

$$e^{-A+B} = e^{-A} e^{B} e^{\frac{1}{2}[A,B]}, \tag{6.197}$$

we set

$$A = r_{\mathbf{q}} a_{\mathbf{q}}^{\dagger} e^{i\omega_{\mathbf{q}} t}, \tag{6.198}$$

$$B = r_{\mathbf{q}} a_{\mathbf{q}} e^{-i\omega_{\mathbf{q}} t}, \tag{6.199}$$

and we get

$$\begin{aligned} e^{-r_{\mathbf{q}}\left(a_{\mathbf{q}}^{\dagger} e^{i\omega_{\mathbf{q}} t} - a_{\mathbf{q}} e^{-i\omega_{\mathbf{q}} t}\right)} &= e^{-r_{\mathbf{q}} a_{\mathbf{q}}^{\dagger} e^{i\omega_{\mathbf{q}} t}} \, e^{r_{\mathbf{q}} a_{\mathbf{q}} e^{-i\omega_{\mathbf{q}} t}} \, e^{-\frac{1}{2} r_{\mathbf{q}}^2} = \\ &= e^{-\frac{1}{2} r_{\mathbf{q}}^2} \, e^{-r_{\mathbf{q}} a_{\mathbf{q}}^{\dagger} e^{i\omega_{\mathbf{q}} t}} \, e^{r_{\mathbf{q}} a_{\mathbf{q}} e^{-i\omega_{\mathbf{q}} t}}. \end{aligned} \tag{6.200}$$

Similarly, from the Glauber formula H.18 in the form

$$e^{A-B} = e^{A} e^{-B} e^{\frac{1}{2}[A,B]}, \tag{6.201}$$

we set

$$A = r_{\mathbf{q}} a_{\mathbf{q}}^{\dagger}, \tag{6.202}$$

$$B = r_{\mathbf{q}} a_{\mathbf{q}}, \tag{6.203}$$

and we get

$$e^{r_{\mathbf{q}}(a_{\mathbf{q}}^{\dagger} - a_{\mathbf{q}})} = e^{r_{\mathbf{q}} a_{\mathbf{q}}^{\dagger}} e^{-r_{\mathbf{q}} a_{\mathbf{q}}} e^{-\frac{1}{2} r_{\mathbf{q}}^2} = \\ = e^{-\frac{1}{2} r_{\mathbf{q}}^2} e^{r_{\mathbf{q}} a_{\mathbf{q}}^{\dagger}} e^{-r_{\mathbf{q}} a_{\mathbf{q}}}, \tag{6.204}$$

consequently the bosonic trace becomes

$$\prod_{\mathbf{q}} \left(1 - e^{-\beta\hbar\omega_{\mathbf{q}}}\right) \sum_{N_{\mathbf{q}}=0}^{\infty} e^{-\beta\hbar\omega_{\mathbf{q}} N_{\mathbf{q}}} \\ \left\langle N_{\mathbf{q}} \left| e^{-r_{\mathbf{q}}^2} e^{-r_{\mathbf{q}} a_{\mathbf{q}}^{\dagger} e^{i\omega_{\mathbf{q}} t}} e^{r_{\mathbf{q}} a_{\mathbf{q}} e^{-i\omega_{\mathbf{q}} t}} e^{r_{\mathbf{q}} a_{\mathbf{q}}^{\dagger}} e^{-r_{\mathbf{q}} a_{\mathbf{q}}} \right| N_{\mathbf{q}} \right\rangle. \tag{6.205}$$

Now, regarding the product $e^{r_{\mathbf{q}} a_{\mathbf{q}} e^{-i\omega_{\mathbf{q}} t}} e^{r_{\mathbf{q}} a_{\mathbf{q}}^{\dagger}}$, we want to rewrite it so that the product $e^{r_{\mathbf{q}} a_{\mathbf{q}} e^{-i\omega_{\mathbf{q}} t}}$ appears to the right of the operator $e^{r_{\mathbf{q}} a_{\mathbf{q}}^{\dagger}}$. We write

$$e^{r_{\mathbf{q}} a_{\mathbf{q}} e^{-i\omega_{\mathbf{q}} t}} e^{r_{\mathbf{q}} a_{\mathbf{q}}^{\dagger}} \equiv e^{r_{\mathbf{q}} a_{\mathbf{q}}^{\dagger}} \left(e^{-r_{\mathbf{q}} a_{\mathbf{q}}^{\dagger}} e^{r_{\mathbf{q}} a_{\mathbf{q}} e^{-i\omega_{\mathbf{q}} t}} e^{r_{\mathbf{q}} a_{\mathbf{q}}^{\dagger}} \right), \tag{6.206}$$

then consider the product $e^{r_{\mathbf{q}} a_{\mathbf{q}} e^{-i\omega_{\mathbf{q}} t}} e^{r_{\mathbf{q}} a_{\mathbf{q}}^{\dagger}}$. Using the Glauber formula H.18, we set

$$A = r_{\mathbf{q}} a_{\mathbf{q}} e^{-i\omega_{\mathbf{q}} t}, \tag{6.207}$$

$$B = r_{\mathbf{q}} a_{\mathbf{q}}^{\dagger}, \tag{6.208}$$

and we get

$$e^{r_{\mathbf{q}} a_{\mathbf{q}} e^{-i\omega_{\mathbf{q}} t}} e^{r_{\mathbf{q}} a_{\mathbf{q}}^{\dagger}} = e^{\frac{1}{2} r_{\mathbf{q}}^2 \left[a_{\mathbf{q}} e^{-i\omega_{\mathbf{q}} t}, a_{\mathbf{q}}^{\dagger}\right]} e^{r_{\mathbf{q}} a_{\mathbf{q}} e^{-i\omega_{\mathbf{q}} t} + r_{\mathbf{q}} a_{\mathbf{q}}^{\dagger}} = \\ = e^{\frac{1}{2} r_{\mathbf{q}}^2 e^{-i\omega_{\mathbf{q}} t} \left[a_{\mathbf{q}}, a_{\mathbf{q}}^{\dagger}\right]} e^{r_{\mathbf{q}} a_{\mathbf{q}} e^{-i\omega_{\mathbf{q}} t} + r_{\mathbf{q}} a_{\mathbf{q}}^{\dagger}} = \\ = e^{\frac{1}{2} r_{\mathbf{q}}^2 e^{-i\omega_{\mathbf{q}} t}} e^{r_{\mathbf{q}} a_{\mathbf{q}} e^{-i\omega_{\mathbf{q}} t} + r_{\mathbf{q}} a_{\mathbf{q}}^{\dagger}}, \tag{6.209}$$

then

$$e^{-r_{\mathbf{q}} a_{\mathbf{q}}^{\dagger}} e^{r_{\mathbf{q}} a_{\mathbf{q}} e^{-i\omega_{\mathbf{q}} t}} e^{r_{\mathbf{q}} a_{\mathbf{q}}^{\dagger}} = e^{\frac{1}{2} r_{\mathbf{q}}^2 e^{-i\omega_{\mathbf{q}} t}} e^{-r_{\mathbf{q}} a_{\mathbf{q}}^{\dagger}} e^{r_{\mathbf{q}} a_{\mathbf{q}} e^{-i\omega_{\mathbf{q}} t} + r_{\mathbf{q}} a_{\mathbf{q}}^{\dagger}}. \tag{6.210}$$

Now we repeat the same procedure for the product $e^{-r_{\mathbf{q}}a^\dagger_{\mathbf{q}}}e^{r_{\mathbf{q}}a_{\mathbf{q}}e^{-i\omega_{\mathbf{q}}t}+r_{\mathbf{q}}a^\dagger_{\mathbf{q}}}$. Applying the Glauber formula H.18, we set

$$A = -r_{\mathbf{q}}a^\dagger_{\mathbf{q}}, \tag{6.211}$$

$$B = r_{\mathbf{q}}a_{\mathbf{q}}e^{-i\omega_{\mathbf{q}}t} + r_{\mathbf{q}}a^\dagger_{\mathbf{q}}, \tag{6.212}$$

and we get

$$\begin{aligned} e^{-r_{\mathbf{q}}a^\dagger_{\mathbf{q}}}\, e^{r_{\mathbf{q}}a_{\mathbf{q}}e^{-i\omega_{\mathbf{q}}t}+r_{\mathbf{q}}a^\dagger_{\mathbf{q}}} &= e^{\frac{1}{2}r^2_{\mathbf{q}}\left[-a^\dagger_{\mathbf{q}},a_{\mathbf{q}}e^{-i\omega_{\mathbf{q}}t}+a^\dagger_{\mathbf{q}}\right]}\, e^{-r_{\mathbf{q}}a^\dagger_{\mathbf{q}}+r_{\mathbf{q}}a_{\mathbf{q}}e^{-i\omega_{\mathbf{q}}t}+r_{\mathbf{q}}a^\dagger_{\mathbf{q}}} = \\ &= e^{\frac{1}{2}r^2_{\mathbf{q}}\left[-a^\dagger_{\mathbf{q}},a_{\mathbf{q}}e^{-i\omega_{\mathbf{q}}t}\right]}\, e^{r_{\mathbf{q}}a_{\mathbf{q}}e^{-i\omega_{\mathbf{q}}t}} = \\ &= e^{\frac{1}{2}r^2_{\mathbf{q}}e^{-i\omega_{\mathbf{q}}t}\left[-a^\dagger_{\mathbf{q}},a_{\mathbf{q}}\right]}\, e^{r_{\mathbf{q}}a_{\mathbf{q}}e^{-i\omega_{\mathbf{q}}t}} = \\ &= e^{\frac{1}{2}r^2_{\mathbf{q}}e^{-i\omega_{\mathbf{q}}t}}\, e^{r_{\mathbf{q}}a_{\mathbf{q}}e^{-i\omega_{\mathbf{q}}t}}. \end{aligned} \tag{6.213}$$

We group the results as follows

$$\begin{aligned} e^{r_{\mathbf{q}}a_{\mathbf{q}}e^{-i\omega_{\mathbf{q}}t}}\, e^{r_{\mathbf{q}}a^\dagger_{\mathbf{q}}} &\equiv e^{r_{\mathbf{q}}a^\dagger_{\mathbf{q}}}\left(e^{-r_{\mathbf{q}}a^\dagger_{\mathbf{q}}}\, e^{r_{\mathbf{q}}a_{\mathbf{q}}e^{-i\omega_{\mathbf{q}}t}}\, e^{r_{\mathbf{q}}a^\dagger_{\mathbf{q}}}\right) = \\ &= e^{r_{\mathbf{q}}a^\dagger_{\mathbf{q}}}\left(e^{\frac{1}{2}r^2_{\mathbf{q}}e^{-i\omega_{\mathbf{q}}t}}\, e^{-r_{\mathbf{q}}a^\dagger_{\mathbf{q}}}\, e^{r_{\mathbf{q}}a_{\mathbf{q}}e^{-i\omega_{\mathbf{q}}t}+r_{\mathbf{q}}a^\dagger_{\mathbf{q}}}\right) = \\ &= e^{r_{\mathbf{q}}a^\dagger_{\mathbf{q}}}\left(e^{\frac{1}{2}r^2_{\mathbf{q}}e^{-i\omega_{\mathbf{q}}t}}\, e^{\frac{1}{2}r^2_{\mathbf{q}}e^{-i\omega_{\mathbf{q}}t}}\, e^{r_{\mathbf{q}}a_{\mathbf{q}}e^{-i\omega_{\mathbf{q}}t}}\right) = \\ &= e^{r^2_{\mathbf{q}}e^{-i\omega_{\mathbf{q}}t}}\, e^{r_{\mathbf{q}}a^\dagger_{\mathbf{q}}}\, e^{r_{\mathbf{q}}a_{\mathbf{q}}e^{-i\omega_{\mathbf{q}}t}}, \end{aligned} \tag{6.214}$$

that is, we have thus obtained the desired ordering, and we can reorder the operators in the bosonic trace as follows

$$\begin{aligned} &\prod_{\mathbf{q}}\left(1-e^{-\beta\hbar\omega_{\mathbf{q}}}\right)\sum_{N_{\mathbf{q}}=0}^{\infty} e^{-\beta\hbar\omega_{\mathbf{q}}N_{\mathbf{q}}} \\ &\quad\left\langle N_{\mathbf{q}}\left| e^{-r^2_{\mathbf{q}}}\, e^{-r_{\mathbf{q}}a^\dagger_{\mathbf{q}}e^{i\omega_{\mathbf{q}}t}}\, e^{r_{\mathbf{q}}a_{\mathbf{q}}e^{-i\omega_{\mathbf{q}}t}}\, e^{r_{\mathbf{q}}a^\dagger_{\mathbf{q}}}\, e^{-r_{\mathbf{q}}a_{\mathbf{q}}}\right| N_{\mathbf{q}}\right\rangle = \\ &= \prod_{\mathbf{q}}\left(1-e^{-\beta\hbar\omega_{\mathbf{q}}}\right)\sum_{N_{\mathbf{q}}=0}^{\infty} e^{-\beta\hbar\omega_{\mathbf{q}}N_{\mathbf{q}}} \\ &\quad\left\langle N_{\mathbf{q}}\left| e^{-r^2_{\mathbf{q}}}\, e^{-r_{\mathbf{q}}a^\dagger_{\mathbf{q}}e^{i\omega_{\mathbf{q}}t}}\, e^{r^2_{\mathbf{q}}e^{-i\omega_{\mathbf{q}}t}}\, e^{r_{\mathbf{q}}a^\dagger_{\mathbf{q}}e^{i\omega_{\mathbf{q}}t}}\, e^{r_{\mathbf{q}}a_{\mathbf{q}}e^{-i\omega_{\mathbf{q}}t}}\, e^{-r_{\mathbf{q}}a_{\mathbf{q}}}\right| N_{\mathbf{q}}\right\rangle = \end{aligned}$$

$$= \prod_{\mathbf{q}} \left(1 - e^{-\beta\hbar\omega_{\mathbf{q}}}\right) \sum_{N_{\mathbf{q}}=0}^{\infty} e^{-\beta\hbar\omega_{\mathbf{q}} N_{\mathbf{q}}}$$

$$\left\langle N_{\mathbf{q}} \left| e^{-r_{\mathbf{q}}^2} e^{r_{\mathbf{q}}^2 e^{-i\omega_{\mathbf{q}} t}} e^{-r_{\mathbf{q}} a_{\mathbf{q}}^\dagger e^{i\omega_{\mathbf{q}} t}} e^{r_{\mathbf{q}} a_{\mathbf{q}}^\dagger e^{i\omega_{\mathbf{q}} t}} e^{r_{\mathbf{q}} a_{\mathbf{q}} e^{-i\omega_{\mathbf{q}} t}} e^{-r_{\mathbf{q}} a_{\mathbf{q}}} \right| N_{\mathbf{q}} \right\rangle =$$

$$= \prod_{\mathbf{q}} \left(1 - e^{-\beta\hbar\omega_{\mathbf{q}}}\right) \sum_{N_{\mathbf{q}}=0}^{\infty} e^{-\beta\hbar\omega_{\mathbf{q}} N_{\mathbf{q}}}$$

$$\left\langle N_{\mathbf{q}} \left| e^{-r_{\mathbf{q}}^2\left(1-e^{-i\omega_{\mathbf{q}} t}\right)} e^{-r_{\mathbf{q}} a_{\mathbf{q}}^\dagger e^{i\omega_{\mathbf{q}} t}} e^{r_{\mathbf{q}} a_{\mathbf{q}}^\dagger e^{i\omega_{\mathbf{q}} t}} e^{r_{\mathbf{q}} a_{\mathbf{q}} e^{-i\omega_{\mathbf{q}} t}} e^{-r_{\mathbf{q}} a_{\mathbf{q}}} \right| N_{\mathbf{q}} \right\rangle =$$

$$= \prod_{\mathbf{q}} e^{-r_{\mathbf{q}}^2\left(1-e^{-i\omega_{\mathbf{q}} t}\right)} \left(1 - e^{-\beta\hbar\omega_{\mathbf{q}}}\right) \sum_{N_{\mathbf{q}}=0}^{\infty} e^{-\beta\hbar\omega_{\mathbf{q}} N_{\mathbf{q}}}$$

$$\left\langle N_{\mathbf{q}} \left| e^{r_{\mathbf{q}}\left(1-e^{i\omega_{\mathbf{q}} t}\right) a_{\mathbf{q}}^\dagger} e^{-r_{\mathbf{q}}\left(1-e^{-i\omega_{\mathbf{q}} t}\right) a_{\mathbf{q}}} \right| N_{\mathbf{q}} \right\rangle. \tag{6.215}$$

By temporarily ignoring the dependence on $\mathbf{q}$, consider the expression

$$e^{-r\left(1-e^{-i\omega t}\right) a} |N\rangle, \tag{6.216}$$

We now expand the exponential as a power series

$$e^{-r\left(1-e^{-i\omega t}\right) a} |N\rangle = \sum_{m=0}^{N} \frac{(-1)^m}{m!} \left[r\left(1 - e^{-i\omega t}\right)\right]^m a^m |N\rangle. \tag{6.217}$$

Here the sum terminates at $m = N$, since $a^m|N\rangle = 0$ for $m > N$, because the annihilation operator cannot remove more quanta than present in the state (for example, see Eq. 1.169). The action of a^m on the number state $|N\rangle$ gives

$$a^m |N\rangle = \sqrt{\frac{N!}{(N-m)!}} |N - m\rangle, \tag{6.218}$$

thus, we obtain

$$e^{-ra\left(1-e^{-i\omega t}\right)} |N\rangle = \sum_{m=0}^{N} \frac{(-1)^m}{m!} \left[r\left(1 - e^{-i\omega t}\right)\right]^m \sqrt{\frac{N!}{(N-m)!}} |N - m\rangle. \tag{6.219}$$

Similarly, consider

$$
\begin{aligned}
\langle N|e^{ra^\dagger(1-e^{i\omega t})} &= \sum_{m'=0}^{N} \frac{1}{m'!}\left[r\left(1-e^{i\omega t}\right)\right]^{m'} \langle N|(a^\dagger)^{m'} = \\
&= \sum_{m'=0}^{N} \frac{1}{m'!}\left[r\left(1-e^{i\omega t}\right)\right]^{m'} \sqrt{\frac{N!}{(N-m')!}}\,\langle N-m'|.
\end{aligned} \tag{6.220}
$$

The scalar products between terms of the two series are proportional to $\langle N-m'|N-m\rangle = \delta_{m,m'}$. Therefore, the full matrix element becomes

$$
\begin{aligned}
&\left\langle N\left|e^{r\left(1-e^{i\omega t}\right)a^\dagger}\, e^{-r\left(1-e^{-i\omega t}\right)a}\right|N\right\rangle \\
&= \sum_{m=0}^{N} \frac{(-1)^m}{(m!)^2}\left[r^2(1-e^{i\omega t})(1-e^{-i\omega t})\right]^m \frac{N!}{(N-m)!} = \\
&= \sum_{m=0}^{N} \frac{(-1)^m}{(m!)^2}\left[r^2|1-e^{-i\omega t}|^2\right]^m \frac{N!}{(N-m)!} = \\
&= \sum_{m=0}^{N} \frac{(-1)^m}{m!}\left(r^2|1-e^{-i\omega t}|^2\right)^m \frac{N!}{m!(N-m)!} = \\
&= L_N\left(|\lambda|^2\right),
\end{aligned} \tag{6.221}
$$

where L_N is the Laguerre polynomial of order N, and we define the complex variable

$$
\lambda \equiv r(1-e^{-i\omega t}), \tag{6.222}
$$

so that $|\lambda|^2 = r^2|1-e^{-i\omega t}|^2$. We set

$$
z = e^{-\beta\hbar\omega}, \tag{6.223}
$$

then, from the generating function for the Laguerre polynomials

$$
\sum_{n=0}^{\infty} z^n L_N\left(|\lambda|^2\right) = \frac{1}{1-z}\exp\left(-\frac{z}{1-z}|\lambda|^2\right), \tag{6.224}
$$

we can compute the weighted sum

$$
\left(1-e^{-\beta\hbar\omega}\right)\sum_{N=0}^{\infty} e^{-\beta\hbar\omega N} L_N\left(|\lambda|^2\right) = (1-z)\sum_{N=0}^{\infty} z^n L_N\left(|\lambda|^2\right) =
$$

$$
\begin{aligned}
&= (1-z)\frac{1}{1-z}\exp\left(-\frac{z}{1-z}|\lambda|^2\right) = \\
&= \exp\left(-\frac{z}{1-z}|\lambda|^2\right). \qquad (6.225)
\end{aligned}
$$

We also note the identity

$$
\begin{aligned}
\frac{z}{1-z} &= \frac{e^{-\beta\hbar\omega}}{1-e^{-\beta\hbar\omega}} = \\
&= \frac{1}{e^{\beta\hbar\omega}-1} \equiv \\
&\equiv n_1(\hbar\omega), \qquad (6.226)
\end{aligned}
$$

where $n_1(\hbar\omega)$ is the Bose-Einstein statistics given in Eq. E.3, then we get

$$
\left(1-e^{-\beta\hbar\omega}\right)\sum_{N=0}^{\infty} e^{-\beta\hbar\omega N} L_N\left(|\lambda|^2\right) = \exp\left(-n_1(\hbar\omega)|\lambda|^2\right). \qquad (6.227)
$$

Restoring the $\mathbf{q}$ subscript and including the multiplicative factor $e^{-r_{\mathbf{q}}^2\left(1-e^{-i\omega_{\mathbf{q}}t}\right)}$, the bosonic trace finally becomes

$$
\prod_{\mathbf{q}} \exp\left\{-r_{\mathbf{q}}^2\left[\left(1-e^{-i\omega_{\mathbf{q}}t}\right) + \left|1-e^{-i\omega_{\mathbf{q}}t}\right|^2 n_1(\hbar\omega_{\mathbf{q}})\right]\right\}. \qquad (6.228)
$$

Note that from

$$
\begin{aligned}
\left|1-e^{-i\omega_{\mathbf{q}}t}\right|^2 &= \left(1-e^{-i\omega_{\mathbf{q}}t}\right)\left(1-e^{i\omega_{\mathbf{q}}t}\right) = \\
&= 2 - e^{i\omega_{\mathbf{q}}t} - e^{-i\omega_{\mathbf{q}}t}, \qquad (6.229)
\end{aligned}
$$

$$
\begin{aligned}
&\left(1-e^{-i\omega_{\mathbf{q}}t}\right) + \left|1-e^{-i\omega_{\mathbf{q}}t}\right|^2 n_1\left(\hbar\omega_{\mathbf{q}}\right) \\
&= \left(1-e^{-i\omega_{\mathbf{q}}t}\right) + \left(2 - e^{i\omega_{\mathbf{q}}t} - e^{-i\omega_{\mathbf{q}}t}\right) n_1\left(\hbar\omega_{\mathbf{q}}\right) = \\
&= 1 - e^{-i\omega_{\mathbf{q}}t} + 2n_1\left(\hbar\omega_{\mathbf{q}}\right) - n_1\left(\hbar\omega_{\mathbf{q}}\right)e^{i\omega_{\mathbf{q}}t} - n_1\left(\hbar\omega_{\mathbf{q}}\right)e^{-i\omega_{\mathbf{q}}t} = \\
&= 1 + 2n_1\left(\hbar\omega_{\mathbf{q}}\right) - \left(1+n_1\left(\hbar\omega_{\mathbf{q}}\right)\right)e^{-i\omega_{\mathbf{q}}t} - n_1\left(\hbar\omega_{\mathbf{q}}\right)e^{i\omega_{\mathbf{q}}t} = \\
&= \left(n_1\left(\hbar\omega_{\mathbf{q}}\right)+1\right)\left(1-e^{-i\omega_{\mathbf{q}}t}\right) + n_1\left(\hbar\omega_{\mathbf{q}}\right)\left(1-e^{i\omega_{\mathbf{q}}t}\right), \qquad (6.230)
\end{aligned}
$$

the bosonic trace finally becomes

$$\prod_{\mathbf{q}} \exp\left\{-r_{\mathbf{q}}^2\left[(n_1(\hbar\omega_{\mathbf{q}})+1)(1-e^{-i\omega_{\mathbf{q}}t})+n_1(\hbar\omega_{\mathbf{q}})(1-e^{i\omega_{\mathbf{q}}t})\right]\right\}. \tag{6.231}$$

By grouping the bosonic and fermionic traces, the retarded Green's function in the independent boson model at thermodynamic equilibrium in 6.189 becomes

$$G^{(r)}_{b,b^\dagger}(t) = -\frac{i}{\hbar}\Theta(t)e^{-i\frac{\tilde{\mathcal{E}}}{\hbar}t} \prod_{\mathbf{q}} \exp\left\{-r_{\mathbf{q}}^2\left[(n_1(\hbar\omega_{\mathbf{q}})+1)(1-e^{-i\omega_{\mathbf{q}}t})+n_1(\hbar\omega_{\mathbf{q}})(1-e^{i\omega_{\mathbf{q}}t})\right]\right\}. \tag{6.232}$$

We now define the phonon correlation function $\phi(t)$ as

$$\phi(t) \equiv \sum_{\mathbf{q}} r_{\mathbf{q}}^2\left[\left(1+n_1(\hbar\omega_{\mathbf{q}})\right)\left(1-e^{-i\omega_{\mathbf{q}}t}\right)+n_1(\hbar\omega_{\mathbf{q}})\left(1-e^{i\omega_{\mathbf{q}}t}\right)\right], \tag{6.233}$$

so that the Green's function can be written in the compact and well-known form

$$G^{(r)}_{b,b^\dagger}(t) = -\frac{i}{\hbar}\Theta(t)e^{-i\frac{\tilde{\mathcal{E}}}{\hbar}t}e^{-\phi(t)}. \tag{6.234}$$

6.7 Sum Rules for the Moments of the Spectral Function

We now show a set of exact sum rules for the spectral function, which follow from fundamental properties of the Green's function.

Theorem 6.4 (Zeroth Moment Sum Rule of the Spectral Function) *For two arbitrary operators A and B, the zeroth-order moment of the corresponding spectral function 6.43 satisfies*

$$\int_{-\infty}^{+\infty} \frac{d\omega}{2\pi} A_{A,B}(\omega) = \left\langle [A, B]^{(\varepsilon)} \right\rangle, \tag{6.235}$$

$[\cdot,\cdot]^{(\varepsilon)}$ *denotes the general commutator defined in Eq. 2.102.*

Proof

$$\int_{-\infty}^{+\infty} \frac{d\omega}{2\pi} A_{A,B}(\omega)$$

$$= \int_{-\infty}^{+\infty} \frac{d\omega}{2\pi} \frac{2\pi}{Z} \sum_{n,m} A_{nm} B_{mn} \left(e^{-\beta \mathcal{E}_n} - \varepsilon e^{-\beta \mathcal{E}_m}\right) \delta \left[\omega - \frac{\mathcal{E}_m - \mathcal{E}_n}{\hbar}\right] =$$

$$= \frac{2\pi}{Z} \sum_{n,m} A_{nm} B_{mn} \left(e^{-\beta \mathcal{E}_n} - \varepsilon e^{-\beta \mathcal{E}_m}\right) \int_{-\infty}^{+\infty} \frac{d\omega}{2\pi} \delta \left[\omega - \frac{\mathcal{E}_m - \mathcal{E}_n}{\hbar}\right] =$$

$$= \frac{2\pi}{Z} \sum_{n,m} A_{nm} B_{mn} \left(e^{-\beta \mathcal{E}_n} - \varepsilon e^{-\beta \mathcal{E}_m}\right) \frac{1}{2\pi} =$$

$$= \frac{1}{Z} \sum_{n,m} A_{nm} B_{mn} \left(e^{-\beta \mathcal{E}_n} - \varepsilon e^{-\beta \mathcal{E}_m}\right). \tag{6.236}$$

We compute the product

$$\frac{1}{Z} \sum_{n,m} A_{nm} B_{mn} e^{-\beta \mathcal{E}_n} - \frac{1}{Z} \sum_{n,m} A_{nm} B_{mn} \varepsilon e^{-\beta \mathcal{E}_m}, \tag{6.237}$$

we write the two traces as

$$\frac{1}{Z} \sum_{n,m} \langle n|A|m\rangle \langle m|B|n\rangle \, e^{-\beta \mathcal{E}_n} = \frac{1}{Z} \sum_{n} \langle n|AB|n\rangle \, e^{-\beta \mathcal{E}_n}, \tag{6.238}$$

$$-\frac{1}{Z} \sum_{n,m} \langle n|A|m\rangle \langle m|B|n\rangle \, \varepsilon \, e^{-\beta \mathcal{E}_m} = -\frac{1}{Z} \sum_{n,m} \langle m|B|n\rangle \langle n|A|m\rangle \, \varepsilon \, e^{-\beta \mathcal{E}_m} =$$

$$= -\frac{1}{Z} \sum_{m} \langle m|BA|m\rangle \, \varepsilon \, e^{-\beta \mathcal{E}_m}. \tag{6.239}$$

We sum with respect to n and m, respectively, and we obtain the thermal average (see Eq. 1.379) of the operator $[A, B]^{(\varepsilon)}$, that is, the thesis. □

Theorem 6.5 (First Moment Sum Rule of the Spectral Function) *For two arbitrary operators A, B, the first-order moment of the spectral function 6.43 satisfies*

$$\int_{-\infty}^{+\infty} \frac{d\omega}{2\pi} \omega A_{A,B}(\omega) = \frac{1}{\hbar} \left\langle \left[\left[A, \hat{\mathcal{H}}\right], B\right]^{(\varepsilon)} \right\rangle, \tag{6.240}$$

Proof

$$\int_{-\infty}^{+\infty} \frac{d\omega}{2\pi} \omega A_{A,B}(\omega)$$

$$= \int_{-\infty}^{+\infty} \frac{d\omega}{2\pi} \frac{2\pi}{Z} \omega \sum_{n,m} A_{nm} B_{mn} \left(e^{-\beta \mathcal{E}_n} - \varepsilon e^{-\beta \mathcal{E}_m} \right) \delta \left[\omega - \frac{\mathcal{E}_m - \mathcal{E}_n}{\hbar} \right] =$$

$$= \frac{2\pi}{Z} \sum_{n,m} A_{nm} B_{mn} \left(e^{-\beta \mathcal{E}_n} - \varepsilon e^{-\beta \mathcal{E}_m} \right) \int_{-\infty}^{+\infty} \frac{d\omega}{2\pi} \omega \delta \left[\omega - \frac{\mathcal{E}_m - \mathcal{E}_n}{\hbar} \right] =$$

$$= \frac{2\pi}{Z} \sum_{n,m} A_{nm} B_{mn} \left(e^{-\beta \mathcal{E}_n} - \varepsilon e^{-\beta \mathcal{E}_m} \right) \frac{1}{2\pi} \left(\frac{\mathcal{E}_m - \mathcal{E}_n}{\hbar} \right) =$$

$$= \frac{1}{Z} \sum_{n,m} A_{nm} B_{mn} \left(e^{-\beta \mathcal{E}_n} - \varepsilon e^{-\beta \mathcal{E}_m} \right) \left(\frac{\mathcal{E}_m - \mathcal{E}_n}{\hbar} \right), \tag{6.241}$$

note that this time we cannot directly use the trace property used earlier because of the product for $(\mathcal{E}_m - \mathcal{E}_n)$. We write

$$\frac{1}{Z} \sum_{n,m} A_{nm} B_{mn} e^{-\beta \mathcal{E}_n} \frac{\mathcal{E}_m - \mathcal{E}_n}{\hbar} - \frac{\varepsilon}{Z} \sum_{n,m} A_{nm} B_{mn} e^{-\beta \mathcal{E}_m} \frac{\mathcal{E}_m - \mathcal{E}_n}{\hbar}, \tag{6.242}$$

then we consider the first term and we get

$$\frac{1}{Z} \sum_{n,m} A_{nm} B_{mn} e^{-\beta \mathcal{E}_n} \frac{\mathcal{E}_m - \mathcal{E}_n}{\hbar}$$

$$= \frac{1}{\hbar Z} \sum_{n,m} A_{nm} B_{mn} e^{-\beta \mathcal{E}_n} \mathcal{E}_m - \frac{1}{\hbar Z} \sum_{n,m} A_{nm} B_{mn} e^{-\beta \mathcal{E}_n} \mathcal{E}_n \equiv$$

$$\equiv \frac{1}{\hbar Z} \sum_{n,m} \langle n|A|m\rangle \langle m|B|n\rangle\, e^{-\beta \mathcal{E}_n}\, \mathcal{E}_m - \frac{1}{\hbar Z} \sum_{n,m} \langle n|A|m\rangle \langle m|B|n\rangle\, e^{-\beta \mathcal{E}_n}\, \mathcal{E}_n =$$

$$= \frac{1}{\hbar Z} \sum_{n,m} \langle n|A\mathcal{E}_m|m\rangle \langle m|B|n\rangle\, e^{-\beta \mathcal{E}_n} - \frac{1}{\hbar Z} \sum_{n,m} \langle n|\mathcal{E}_n A|m\rangle \langle m|B|n\rangle\, e^{-\beta \mathcal{E}_n} =$$

$$= \frac{1}{\hbar Z} \sum_{n,m} \langle n|A\hat{\mathcal{H}}|m\rangle \langle m|B|n\rangle\, e^{-\beta \mathcal{E}_n} - \frac{1}{\hbar Z} \sum_{n,m} \langle n|\hat{\mathcal{H}}A|m\rangle \langle m|B|n\rangle\, e^{-\beta \mathcal{E}_n} =$$

$$= \frac{1}{\hbar Z} \sum_{n,m} \langle n|[A, \hat{\mathcal{H}}]|m\rangle \langle m|B|n\rangle\, e^{-\beta \mathcal{E}_n}. \tag{6.243}$$

Similarly for the second term, we have

$$- \frac{\varepsilon}{\hbar Z} \sum_{n,m} \langle n|[A, \hat{\mathcal{H}}]|m\rangle \langle m|B|n\rangle\, e^{-\beta \mathcal{E}_m} = - \frac{\varepsilon}{\hbar Z} \sum_{n,m} \langle m|B|n\rangle \langle n|[A, \hat{\mathcal{H}}]|m\rangle\, e^{-\beta \mathcal{E}_m}. \tag{6.244}$$

We sum with respect to m in the first term and with respect to n in the second term, combine the sums under one index, and obtain the thesis. □

In principle, it is possible to compute moments of all orders: each increase in the power of ω corresponds to an additional commutator with the Hamiltonian $\hat{\mathcal{H}}$. This approach is known as the moments method.

6.7.1 Application to the Independent Boson Model in the Einstein Approximation

In this section, we illustrate an application of the spectral-function sum rules to a particular case of the independent boson model (see Eqs. 6.233 and 6.234), which can be solved analytically in a fully exact manner. We adopt the Einstein approximation, in which the phonon frequencies $\omega_{\mathbf{q}}$ are assumed to be dispersionless, i.e.,

$$\omega_{\mathbf{q}} = \omega_0 > 0, \ \forall \mathbf{q}, \tag{6.245}$$

and we consider the limit $T \to 0^+$. It is worth noting that an identical approximation has already been introduced (see Eq. 4.200). In the zero-temperature limit, the Bose-Einstein distribution $n_1(\hbar\omega_{\mathbf{q}})$ vanish, so that Eq. 6.233 becomes

$$\begin{aligned} \phi(t)|_{T\to 0^+,\omega_{\mathbf{q}}=\omega_0} &= \left(1 - e^{-i\omega_0 t}\right)\sum_{\mathbf{q}} r_{\mathbf{q}}^2 \equiv \\ &\equiv g^2\left(1 - e^{-i\omega_0 t}\right), \end{aligned} \tag{6.246}$$

where we have defined (see Eq. 6.139)

$$\begin{aligned} g^2 &= \sum_{\mathbf{q}} r_{\mathbf{q}}^2 = \\ &= \frac{1}{(\hbar\omega_0)^2}\sum_{\mathbf{q}} M_{\mathbf{q}}^2. \end{aligned} \tag{6.247}$$

Consequently, Eq. 6.234 becomes

$$G_{b,b^\dagger}^{(r)}(t)\Big|_{T\to 0^+,\omega_{\mathbf{q}}=\omega_0} = -\frac{i}{\hbar}\Theta(t)e^{-i\frac{\tilde{\mathcal{E}}}{\hbar}t}e^{-g^2\left(1-e^{-i\omega_0 t}\right)}, \tag{6.248}$$

where the eigenvalue of the dressed impurity in the Einstein approximation is given by (see Eq. 6.157)

$$\tilde{\mathcal{E}} = \mathcal{E} - \frac{1}{\hbar\omega_0}\sum_{\mathbf{q}} M_{\mathbf{q}}^2 \equiv$$

$$\equiv \mathcal{E} - \hbar\omega_0 g^2. \tag{6.249}$$

Now, we write Eq. 6.248 as

$$\begin{aligned} G^{(r)}_{b,b^\dagger}(t)\Big|_{T\to 0^+,\omega_{\mathbf{q}}=\omega_0} &= -\frac{i}{\hbar}\Theta(t)e^{-i\frac{\tilde{\mathcal{E}}}{\hbar}t}e^{-g^2}e^{g^2e^{-i\omega_0 t}} = \\ &= -\frac{i}{\hbar}\Theta(t)e^{-i\frac{\tilde{\mathcal{E}}}{\hbar}t}e^{-g^2}\sum_{n=0}\frac{g^{2n}}{n!}e^{-i\omega_0 nt}, \end{aligned} \tag{6.250}$$

where we have explicitly expanded the exponential $e^{g^2e^{-i\omega_0 t}}$ in its Taylor series. We now apply the Fourier transform and obtain

$$G^{(r)}_{b,b^\dagger}(\omega+i\delta)\Big|_{T\to 0^+,\omega_{\mathbf{q}}=\omega_0} = -\frac{i}{\hbar}e^{-g^2}\sum_{n=0}\frac{g^{2n}}{n!}\int_0^{+\infty} dt e^{i\omega t}e^{-\delta t}e^{-i\frac{\tilde{\mathcal{E}}}{\hbar}t}e^{-i\omega_0 nt}, \tag{6.251}$$

where $\delta > 0$ guarantees the convergence of the integral, leading to

$$G^{(r)}_{b,b^\dagger}(\omega+i\delta)\Big|_{T\to 0^+,\omega_{\mathbf{q}}=\omega_0} = \frac{1}{\hbar}e^{-g^2}\sum_{n=0}\frac{g^{2n}}{n!}\frac{1}{\omega+i\delta-\left(\frac{\tilde{\mathcal{E}}}{\hbar}+n\omega_0\right)}. \tag{6.252}$$

From Eq. 6.87, we derive the spectral function for the independent boson model in the Einstein approximation, i.e.,

$$A_{b,b^\dagger}(\omega)\big|_{T\to 0^+,\omega_{\mathbf{q}}=\omega_0} = 2\pi e^{-g^2}\sum_{n=0}\frac{g^{2n}}{n!}\delta\left[\omega-\left(\frac{\tilde{\mathcal{E}}}{\hbar}+n\omega_0\right)\right]. \tag{6.253}$$

Equation 6.253 shows that, within the Einstein approximation at zero temperature, the spectral function consists of a discrete set of infinitely sharp δ-peaks. In realistic situations, however, these lines acquire a finite width due to impurity scattering, phonon lifetime effects, or finite temperature, which introduce spectral broadening and slightly shift the peak positions.

We now apply the sum rules to the spectral function in Eq. 6.253. Concerning the zeroth moment, since b and $b^\dagger$ denote fermionic impurity operators (see Eq. D.4), then the right-hand side of Eq. 6.235 equals unity, and we obtain the identity

$$\begin{aligned} &\int_{-\infty}^{+\infty}\frac{d\omega}{2\pi}A_{b,b^\dagger}(\omega)\big|_{T\to 0^+,\omega_{\mathbf{q}}=\omega_0} \\ &= e^{-g^2}\sum_{n=0}\frac{g^{2n}}{n!}\int_{-\infty}^{+\infty}d\omega\,\delta\left[\omega-\left(\frac{\tilde{\mathcal{E}}}{\hbar}+n\omega_0\right)\right] = \end{aligned}$$

$$
\begin{aligned}
&= e^{-g^2} \sum_{n=0} \frac{g^{2n}}{n!} = \\
&= 1. \qquad (6.254)
\end{aligned}
$$

We now turn to the first moment. Instead of explicitly writing the right-hand side of Eq. 6.240, we directly evaluate the corresponding integral, which can be computed straightforwardly. Indeed,

$$
\begin{aligned}
&\int_{-\infty}^{+\infty} \frac{d\omega}{2\pi} \omega \, A_{b,b^\dagger}(\omega)\big|_{T\to 0^+,\omega_{\mathbf{q}}=\omega_0} \\
&= e^{-g^2} \sum_{n=0} \frac{g^{2n}}{n!} \int_{-\infty}^{+\infty} d\omega \, \omega \delta\left[\omega - \left(\frac{\tilde{\mathcal{E}}}{\hbar} + n\omega_0\right)\right] = \\
&= e^{-g^2} \sum_{n=0} \frac{g^{2n}}{n!} \left(\frac{\tilde{\mathcal{E}}}{\hbar} + n\omega_0\right) = \\
&= \frac{\tilde{\mathcal{E}}}{\hbar} e^{-g^2} \sum_{n=0} \frac{g^{2n}}{n!} + \omega_0 e^{-g^2} \sum_{n=0} \frac{g^{2n}}{n!} n = \\
&= \frac{\tilde{\mathcal{E}}}{\hbar} + \omega_0 e^{-g^2} \sum_{n=1} \frac{g^{2n}}{n!} n, \qquad (6.255)
\end{aligned}
$$

where in the last step the summation starts from $n = 1$, since the term with $n = 0$ vanishes. From

$$
\begin{aligned}
e^{-g^2} \sum_{n=1} \frac{g^{2n}}{n!} n &= e^{-g^2} \sum_{n=1} \frac{g^{2n}}{(n-1)!} = \\
&= e^{-g^2} g^2 \sum_{n=1} \frac{g^{2(n-1)}}{(n-1)!} = \\
&= e^{-g^2} g^2 \sum_{n=0} \frac{g^{2n}}{n!} = \\
&= g^2, \qquad (6.256)
\end{aligned}
$$

one finds

$$
\int_{-\infty}^{+\infty} \frac{d\omega}{2\pi} \omega \, A_{b,b^\dagger}(\omega)\big|_{T\to 0^+,\omega_{\mathbf{q}}=\omega_0} = \frac{\tilde{\mathcal{E}}}{\hbar} + \omega_0 g^2. \qquad (6.257)
$$

Note that the quantities obtained in Eqs. 6.254 and 6.257 follow a Poisson probability distribution. This result is indeed expected, since the spectral function in Eq. 6.253 is directly related to a Poisson distribution with parameter g^2, whose n-th coefficient is given by

$$P_n = e^{-g^2} \frac{g^{2n}}{n!}. \tag{6.258}$$

The ground-state contribution to the spectral function arises from the term with $n = 0$, and it is given by

$$P_0 = e^{-g^2}, \tag{6.259}$$

which is directly connected to the coupling strength (see Eq. 6.247). In what follows, we consider two distinct limiting regimes.

- Weak coupling limit ($g^2 \ll 1$).
  In the limit of weak electron–phonon interaction, the Poisson distribution is strongly peaked at $n = 0$, so that higher phonon contributions are exponentially suppressed. The spectral function is then dominated by the $n = 0$ term:

$$A_{b,b^\dagger}(\omega)\big|_{T\to 0^+,\omega_{\mathbf{q}}=\omega_0} \approx e^{-g^2} \delta\left[\omega - \frac{\tilde{\mathcal{E}}}{\hbar}\right], \tag{6.260}$$

  which represents a single coherent quasiparticle peak corresponding to the dressed impurity (polaron ground state). The system behaves nearly as a free impurity with a small energy renormalization $\tilde{\mathcal{E}}$.
- Strong coupling limit ($g^2 \gg 1$).
  When the coupling strength is large, the Poisson distribution becomes broad and can be approximated by a Gaussian distribution in n:

$$P(n) \simeq \frac{1}{\sqrt{2\pi g^2}} \exp\left[-\frac{(n-g^2)^2}{2g^2}\right]. \tag{6.261}$$

  The spectral weight is then spread over many phonon sidebands, and the discrete structure of the spectrum becomes quasi-continuous. The spectral function can be viewed as a Gaussian envelope centered at

$$\omega_c = \frac{\tilde{\mathcal{E}}}{\hbar} + \omega_0 g^2. \tag{6.262}$$

  Using the definition of the renormalized energy (see Eq. 6.249), one finds that the Gaussian is centered around the bare impurity frequency,

$$\omega_c = \frac{\mathcal{E}}{\hbar}. \tag{6.263}$$

Therefore, in the strong-coupling regime the spectral weight redistributes symmetrically around the original impurity level.

For weak coupling, the impurity preserves its quasiparticle nature: the phonon dressing is weak, and the associated phonon cloud remains small and spatially localized. Consequently, the spectral function is dominated by a single, sharp coherent peak corresponding to the polaron ground state, with only faint phonon replicas carrying exponentially suppressed spectral weight. In this regime, the impurity behaves as a well-defined quasiparticle characterized by a renormalized energy $\tilde{\mathcal{E}}$ and a long lifetime.

In the strong-coupling regime, the situation changes qualitatively. The impurity becomes strongly dressed by a large number of phonons, forming a heavy composite object. The discrete phonon sidebands, whose intensities follow a Poisson distribution, overlap and coalesce into a quasi-continuous Gaussian envelope centered at the bare impurity frequency $\frac{\mathcal{E}}{\hbar}$ (see Eq. 6.263). Physically, this reflects the fact that, although the impurity is strongly entangled with the phonon field, the polaronic energy renormalization $\tilde{\mathcal{E}} = \mathcal{E} - \hbar\omega_0 g^2$ (see Eq. 6.249) is compensated by the average phonon energy $\hbar\omega_0 g^2$ carried by the surrounding cloud. This compensation can be understood from the first-moment sum rule: the weighted average of all spectral components, consistent with the same relation, constrains the spectral centroid to coincide with the bare impurity frequency. Consequently, in the frequency domain the impurity appears almost free again, yet embedded in an incoherent and strongly broadened phononic background where its quasiparticle identity is largely lost: the quasiparticles of the system are polarons rather than the original fermions.

6.8 Application of Kramers-Krönig Relations to Retarded Green's Function

The first Kramers-Krönig relation (see Appendix L) can be used to calculate the real part of many-body Green's function, after the imaginary part has been measured. From Eq. 6.49, one finds

$$\begin{aligned} G^{(r)}_{A,B}(z) &= \int_{-\infty}^{+\infty} \frac{d\omega'}{2\pi} \frac{A_{A,B}(\omega')}{\hbar(z-\omega')} = \\ &= \int_{-\infty}^{+\infty} \frac{d\omega'}{2\pi} \frac{A_{AB}(\omega')}{\hbar(\omega-\omega'+i\delta)} = \\ &= \frac{1}{\hbar}\int_{-\infty}^{+\infty} \frac{d\omega'}{2\pi} \frac{A_{A,B}(\omega')}{(\omega-\omega')^2+\delta^2}(\omega-\omega'-i\delta), \end{aligned} \tag{6.264}$$

we apply the limit $\delta \to 0^+$. In the case of physical interest $B = A^\dagger$ the spectral function is a real function, then the contributions to the integral from the real and imaginary parts are

$$
\begin{aligned}
G^{(r)}_{A,B}(\omega) &= -\frac{i}{\hbar}\lim_{\delta\to 0^+}\int_{-\infty}^{+\infty}\frac{d\omega'}{2\pi}\frac{\delta}{(\omega-\omega')^2+\delta^2}A_{A,B}(\omega') \\
&+\frac{1}{\hbar}\lim_{\delta\to 0^+}\int_{-\infty}^{+\infty}\frac{d\omega'}{2\pi}A_{A,B}(\omega')\frac{\omega-\omega'}{(\omega-\omega')^2+\delta^2} = \\
&= -\frac{i}{\hbar}\int_{-\infty}^{+\infty}\frac{d\omega'}{2}\left(\lim_{\delta\to 0^+}\frac{1}{\pi}\frac{\delta}{(\omega-\omega')^2+\delta^2}\right)A_{A,B}(\omega') \\
&+\frac{1}{\hbar}\int_{-\infty}^{+\infty}\frac{d\omega'}{2\pi}\frac{A_{A,B}(\omega')}{\omega-\omega'}\left(\lim_{\delta\to 0^+}\frac{(\omega-\omega')^2}{(\omega-\omega')^2+\delta^2}\right) = \\
&= -\frac{i}{\hbar}\int_{-\infty}^{+\infty}\frac{d\omega'}{2}\delta(\omega-\omega')A_{A,B}(\omega') \\
&+\frac{1}{\hbar}\int_{-\infty}^{+\infty}\frac{d\omega'}{2\pi}\frac{A_{A,B}(\omega')}{\omega-\omega'}\left(\lim_{\delta\to 0^+}\frac{(\omega-\omega')^2}{(\omega-\omega')^2+\delta^2}\right) = \\
&= -\frac{i}{2\hbar}A_{A,B}(\omega) \\
&+\frac{1}{\hbar}\int_{-\infty}^{+\infty}\frac{d\omega'}{2\pi}\frac{A_{A,B}(\omega')}{\omega-\omega'}\left(\lim_{\delta\to 0^+}\frac{(\omega-\omega')^2}{(\omega-\omega')^2+\delta^2}\right),
\end{aligned}
\tag{6.265}
$$

where in the first term we triggered Dirac's delta and in the second term we multiplied and divided by $\omega - \omega'$, finally in both we have reversed the limit and integral operations. The quantity

$$
\lim_{\delta\to 0^+}\frac{(\omega-\omega')^2}{(\omega-\omega')^2+\delta^2}
\tag{6.266}
$$

approaches 1 for $|\omega - \omega'| \gg \delta$, and tends to 0 for $|\omega - \omega'| \ll \delta$. Then, in the limit $\delta \to 0^+$ the singularity is for $\omega = \omega'$, in which case the function

$$
\frac{A_{A,B}(\omega')}{\omega-\omega'}
\tag{6.267}
$$

has a first-order pole. Accordingly, we can rewrite the second term in $G^{(r)}_{A,B}(\omega)$ as a principal value integral over frequency. This yields the representation:

$$
G^{(r)}_{A,B}(\omega) = -\frac{i}{2\hbar}A_{A,B}(\omega) + \frac{1}{\hbar}\,\mathrm{P.V.}\int_{-\infty}^{+\infty}\frac{d\omega'}{2\pi}\frac{A_{A,B}(\omega')}{\omega-\omega'}.
\tag{6.268}
$$

In the physically relevant case where $B = A^{\dagger}$, the spectral function $A_{A,A^{\dagger}}(\omega)$ is real-valued for real arguments. Consequently, we have

$$G^{(r)}_{A,A^{\dagger}}(\omega) = -\frac{i}{2\hbar}A_{A,A^{\dagger}}(\omega) + \frac{1}{\hbar}\ \text{P. V.} \int_{-\infty}^{+\infty} \frac{d\omega'}{2\pi} \frac{A_{A,A^{\dagger}}(\omega')}{\omega - \omega'}. \tag{6.269}$$

Taking the imaginary part of both sides gives

$$\begin{aligned} \text{Im}\, G^{(r)}_{A,A^{\dagger}}(\omega) &= \text{Im}\left(-\frac{i}{2\hbar}A_{A,A^{\dagger}}(\omega)\right) + \text{Im}\left(\frac{1}{\hbar}\ \text{P. V.} \int_{-\infty}^{+\infty} \frac{d\omega'}{2\pi} \frac{A_{A,A^{\dagger}}(\omega')}{\omega - \omega'}\right) = \\ &= \frac{1}{2\hbar}A_{A,A^{\dagger}}(\omega), \end{aligned} \tag{6.270}$$

which can be inverted to express the spectral function in terms of the imaginary part of the Green's function, i.e.,

$$A_{A,A^{\dagger}}(\omega) = -2\hbar\, \text{Im}\, G^{(r)}_{A,A^{\dagger}}(\omega). \tag{6.271}$$

By substituting this result into the principal value integral, we explicitly recover the first Kramers-Krönig relation, that is, Eq. L.4. This fundamental relation expresses the real part of the retarded Green's function as a transform of its imaginary part. This identity reflects the analyticity of $G^{(r)}(\omega)$ in the upper half of the complex plane and ultimately encodes the causality of the physical system. The relation shows that the dispersive (real) part of the response is completely determined by its absorptive (imaginary) part, in accordance with the general principles of linear response theory.

6.9 Average Occupation Number for Interacting Particles

Here we want to show an important application of many-body Green's function. In quantum many-body theory, understanding how particles occupy available quantum states is fundamental. The number operator plays a crucial role in this context, as it counts the occupation of a given many particles state. The thermal average (see Eq. 1.379) of the occupation number of fermions or bosons in the state $\alpha = (\mathbf{k}, \sigma)$ (see Eq. 1.45) with respect to a Hamiltonian of free particles, (see Eq. 2.181), is given by 2.251 in absence of interaction. Let us now consider a Hamiltonian of the form 1.85 within the framework of the second quantization, and let us compute how the average occupancy number changes in the state $(\mathbf{k}, \sigma)$ in an interacting system. We set $A = C_{\mathbf{k},\sigma}$, $B = C^{\dagger}_{\mathbf{k},\sigma}$ in the property 6.51 of the spectral function, that is,

$$\left\langle C_{\mathbf{k},\sigma} C^{\dagger}_{\mathbf{k},\sigma} \right\rangle = \int_{-\infty}^{+\infty} \frac{d\omega}{2\pi} \frac{A_{\mathbf{k},\sigma}(\omega)}{1 - \varepsilon e^{-\beta\hbar\omega}}, \tag{6.272}$$

and we make use of Eq. D.7, as follows

$$\langle \mathbb{1} \rangle + \varepsilon \left\langle C^{\dagger}_{\mathbf{k},\sigma} C_{\mathbf{k},\sigma} \right\rangle = \int_{-\infty}^{+\infty} \frac{d\omega}{2\pi} \frac{A_{\mathbf{k},\sigma}(\omega)}{1 - \varepsilon e^{-\beta\hbar\omega}}. \tag{6.273}$$

From 6.235, we write

$$\begin{aligned} 1 &= \left\langle \left[C_{\mathbf{k},\sigma}, C^{\dagger}{}_{\mathbf{k},\sigma} \right]^{(\varepsilon)} \right\rangle = \\ &= \int_{-\infty}^{+\infty} \frac{d\omega}{2\pi} A_{\mathbf{k},\sigma}(\omega), \end{aligned} \tag{6.274}$$

and we have

$$\begin{aligned} \varepsilon \left\langle C^{\dagger}_{\mathbf{k},\sigma} C_{\mathbf{k},\sigma} \right\rangle &= \int_{-\infty}^{+\infty} \frac{d\omega}{2\pi} \frac{A_{\mathbf{k},\sigma}(\omega)}{1 - \varepsilon e^{-\beta\hbar\omega}} - \int_{-\infty}^{+\infty} \frac{d\omega}{2\pi} A_{\mathbf{k},\sigma}(\omega) = \\ &= \int_{-\infty}^{+\infty} \frac{d\omega}{2\pi} A_{\mathbf{k},\sigma}(\omega) \left[\frac{\varepsilon e^{-\beta\hbar\omega}}{1 - \varepsilon e^{-\beta\hbar\omega}} \right], \end{aligned} \tag{6.275}$$

that is,

$$\begin{aligned} \left\langle \hat{N}_{\mathbf{k},\sigma} \right\rangle \equiv \left\langle C^{\dagger}_{\mathbf{k},\sigma} C_{\mathbf{k},\sigma} \right\rangle &= \\ &= \int_{-\infty}^{+\infty} \frac{d\omega}{2\pi} \frac{A_{\mathbf{k},\sigma}(\omega)}{e^{\beta\omega} - \varepsilon} = \\ &= \int_{-\infty}^{+\infty} \frac{d\omega}{2\pi} A_{\mathbf{k},\sigma}(\omega)\, n_{\varepsilon}(\omega), \end{aligned} \tag{6.276}$$

where $n_{\varepsilon}(\omega)$ is given by Eq. E.2. Consequently, the many-body Green's function, by means of the spectral function, allows calculation of the average occupation number and such object, due to interaction, is written as an integral of the Bose-Einstein or Fermi-Dirac statistics, that is, Eqs. E.3 and E.4, respectively.

Equation 6.276 holds for both discrete and continuous **k**.

6.10 Pole Close to the Real Axis: Quasiparticle Picture

As previously discussed, in Chap. 7 we will demonstrate that any retarded many-body Green's function can be expressed in the form of Eq. 6.116, where $\Sigma^*(z)$ is a complex function, known as the improper self-energy, that shifts the poles of the free Green's function from the real axis (in the non-interacting case) into the lower half of the complex plane (in the presence of interactions). The imaginary part of the self-energy determines the damping factor γ: the larger the shift of the pole into the complex plane, the stronger the damping. Let us denote by z_0 a pole of the retarded Green's function $G^{(r)}(z)$ lying in the lower half-plane. We define $z_0 \equiv \Omega - i\gamma$, $\gamma > 0$, and impose the condition for z_0 to be a pole, that is,

$$\Omega - i\gamma - \frac{\mathcal{E}}{\hbar} - \frac{1}{\hbar}\Sigma^*(\Omega - i\gamma) = 0. \tag{6.277}$$

To analyze the behavior of the Green's function near the pole, we assume that the pole is close to the real axis, i.e., $\gamma \ll |\Omega|$. Under this assumption, we perform a first-order Taylor expansion of $\Sigma^*(z)$ around the real point Ω, i.e.,

$$\begin{aligned}\Sigma^*(\Omega - i\gamma) &\simeq \Sigma^*(\Omega) - i\gamma \left.\frac{\partial \Sigma^*}{\partial z}\right|_{z=\Omega} = \\ &= \mathrm{Re}\,\Sigma^*(\Omega) + i\,\mathrm{Im}\,\Sigma^*(\Omega) - i\gamma\left[\left.\frac{\partial\,\mathrm{Re}\,\Sigma^*(\omega)}{\partial\omega}\right|_{\omega=\Omega} + i\left.\frac{\partial\,\mathrm{Im}\,\Sigma^*(\omega)}{\partial\omega}\right|_{\omega=\Omega}\right],\end{aligned} \tag{6.278}$$

where we used that $\Sigma^*(z)$ is analytic and so its derivative does not depend on direction and we can calculate it on the real axis.

$$\begin{aligned}&\Omega - i\gamma - \frac{\mathcal{E}}{\hbar} - \frac{1}{\hbar}\Bigg(\mathrm{Re}\,\Sigma^*(\Omega) + i\,\mathrm{Im}\,\Sigma^*(\Omega) - i\gamma\left.\frac{\partial\,\mathrm{Re}\,\Sigma^*(\omega)}{\partial\omega}\right|_{\omega=\Omega} \\ &+\gamma\left.\frac{\partial\,\mathrm{Im}\,\Sigma^*(\omega)}{\partial\omega}\right|_{\omega=\Omega}\Bigg) = 0,\end{aligned} \tag{6.279}$$

$$\Omega - i\gamma - \frac{\mathcal{E}}{\hbar} - \frac{1}{\hbar}\left(\mathrm{Re}\,\Sigma^*(\Omega) + i\,\mathrm{Im}\,\Sigma^*(\Omega) - i\gamma\left.\frac{\partial\,\mathrm{Re}\,\Sigma^*(\omega)}{\partial\omega}\right|_{\omega=\Omega}\right) = 0, \tag{6.280}$$

we have neglected $\gamma \left.\frac{\partial\,\mathrm{Im}\,\Sigma^*(\omega)}{\partial\omega}\right|_{\omega=\Omega}$ because the partial derivative of the imaginary part is already proportional to γ, so the product for γ is a second-order term and can be neglected. By setting equal 0 imaginary part and real part, we get respectively

$$\gamma = -\frac{\frac{1}{\hbar}\,\mathrm{Im}\,\Sigma^*(\Omega)}{1 - \frac{1}{\hbar}\left.\frac{\partial\,\mathrm{Re}\,\Sigma^*(\omega)}{\partial\omega}\right|_{\omega=\Omega}}, \tag{6.281}$$

$$\Omega = \frac{\mathcal{E}}{\hbar} + \frac{\mathrm{Re}\,\Sigma^*(\Omega)}{\hbar}. \tag{6.282}$$

From the first order expansion in γ it is deduced that the imaginary part of the improper self-energy determines the decay times, while the real part implies the shift of energy levels, which is to be determined self-consistently as zero of Eq. 6.282.

The energy is no longer $\frac{\mathcal{E}}{\hbar}$. Given $G^{(r)}(\omega)$, we take the limit for $\delta \to 0^{+}$ and obtain $G^{(r)}(\omega)$. Here, the goal is to study Green's function $G^{(r)}(\omega)$ for $\omega \approx \Omega$ and to do this, $\Sigma^{*}(z)$ must be expanded respect to Ω. We get

$$G^{(r)}(\omega) = \frac{1}{\hbar}\frac{1}{\omega - \frac{\mathcal{E}}{\hbar} - \frac{\mathrm{Re}\,\Sigma^{*}(\Omega)}{\hbar} - i\frac{\mathrm{Im}\,\Sigma^{*}(\Omega)}{\hbar} - \frac{1}{\hbar}\left.\frac{\partial\,\mathrm{Re}\,\Sigma^{*}(\omega)}{\partial\omega}\right|_{\omega=\Omega}(\omega-\Omega)} =$$

$$= \frac{1}{\hbar}\frac{1}{\omega - \Omega - i\frac{\mathrm{Im}\,\Sigma^{*}(\Omega)}{\hbar} - \frac{1}{\hbar}\left.\frac{\partial\,\mathrm{Re}\,\Sigma^{*}(\omega)}{\partial\omega}\right|_{\omega=\Omega}(\omega-\Omega)}, \tag{6.283}$$

where in the denominator we used $z = \omega + i\delta$ and omitted the second-order term containing the partial derivative of the imaginary part. In the limit $\delta \to 0^{+}$, $i\delta$ does not imply singularity since $\Sigma^{*}(z)$ includes a finite imaginary part. The form of $G^{(r)}(\omega)$ derived here is generally applicable to Green's functions, provided that the imaginary part of $\Sigma^{*}(\Omega)$ is sufficiently small. Since $\Sigma^{*}(\Omega)$ reflects the strength of the interactions, this expansion is restricted to the weak-coupling regime. In essence, the interacting propagator is developed as a perturbation around the propagator of the noninteracting system. It follows

$$G^{(r)}(\omega) = \frac{1}{\hbar}\frac{1}{(\omega-\Omega)\left(1 - \frac{1}{\hbar}\left.\frac{\partial\,\mathrm{Re}\,\Sigma^{*}(\omega)}{\partial\omega}\right|_{\omega=\Omega}\right) - i\frac{\mathrm{Im}\,\Sigma^{*}(\Omega)}{\hbar}}, \tag{6.284}$$

we employ Eq. 6.281 as follows

$$\frac{1}{\hbar}\,\mathrm{Im}\,\Sigma^{*}(\Omega) = -\gamma\left(1 - \frac{1}{\hbar}\left.\frac{\partial\,\mathrm{Re}\,\Sigma^{*}(\omega)}{\partial\omega}\right|_{\omega=\Omega}\right), \tag{6.285}$$

and we get

$$\mathrm{Im}\,G^{(r)}(\omega) \approx \frac{1}{\hbar}\left[\frac{1}{1 - \frac{1}{\hbar}\left.\frac{\partial\,\mathrm{Re}\,\Sigma^{*}(\omega)}{\partial\omega}\right|_{\omega=\Omega}}\right]\left[-\frac{\gamma}{(\omega-\Omega)^{2}+\gamma^{2}}\right]. \tag{6.286}$$

Multiplying by $-2\hbar\pi$, we have

$$A(\omega) \simeq 2\pi\left(\frac{1}{1 - \frac{1}{\hbar}\left.\frac{\partial\,\mathrm{Re}\,\Sigma^{*}(\omega)}{\partial\omega}\right|_{\omega=\Omega}}\right)\left[-\frac{\gamma}{(\omega-\Omega)^{2}+\gamma^{2}}\right], \tag{6.287}$$

$$A(\omega) \simeq 2\pi r\left[-\frac{\gamma}{(\omega-\Omega)^{2}+\gamma^{2}}\right], \tag{6.288}$$

with

$$r = \left(\frac{1}{1 - \frac{1}{\hbar} \frac{\partial \operatorname{Re} \Sigma^*(\omega)}{\partial \omega} \Big|_{\omega = \Omega}} \right). \tag{6.289}$$

Equation 6.287 is valid for small γ and describes a Lorentzian function normalized to unity, with a peak centered at Ω and a width γ. In the vicinity of Ω, the spectral function assumes a Lorentzian shape, with its peak at the real part Ω and its width determined by the imaginary part γ. The factor r is strictly less than 1. To see this, consider a Hamiltonian of the form 1.85 in the framework of second quantization, with $\hat{\mathcal{H}}_0$ given by 2.181. In the weak-interaction limit, the Green's function at real frequencies, to lowest order, takes the form of 6.283. Using Eq. 6.235 adapted for the spectral function of noninteracting fermions, denoted by $A_\alpha(\omega)$, we obtain

$$\int_{-\infty}^{+\infty} d\omega \frac{A_\alpha(\omega)}{2\pi} = 1. \tag{6.290}$$

When interactions are no longer weak, the spectral function generally becomes highly non-trivial and, in most cases, its detailed profile is essentially unpredictable in closed form. Nevertheless, we have encountered an important example in which strong interactions lead to a qualitatively different and analytically tractable behavior: within the Einstein approximation to the Anderson model (see Sect. 6.7.1), the spectral function approaches a Gaussian shape in the strong-coupling regime.

For fermions, $A_\alpha(\omega)$ is a real and positive-defined function. Its integral with respect to the measure ω should equal 1, and around Ω, the function $A_\alpha(\omega)$ is a Lorentzian normalized to 1. Substituting Eq. 6.287 into the integral, the result is r. Then, if the interactions are weak, we have

$$A(\omega) = 2\pi r \left[-\frac{\gamma}{(\omega - \Omega)^2 + \gamma^2} \right] + f(\omega), \tag{6.291}$$

where $f(\omega)$ is a function whose integral must give $1 - r$. We expect the parameter r to be close to 1, since in the free (non-interacting) case we have exactly $r = 1$, and we are currently considering a perturbative regime. In summary, when the interactions are weak, the Dirac delta peak characterizing the spectral function of a free particle is broadened into a Lorentzian profile, and the residue r of the Green's function at the pole becomes strictly less than 1.

Chapter 7
Feynman's Perturbative Theory of the Thermal Green's Function

Abstract This chapter develops the diagrammatic framework for interacting quantum systems, beginning with the Dyson equation, which relates the full Green's function to the noninteracting one via the self-energy, thus organizing interaction effects systematically for perturbation theory. The thermal Wick's theorem is introduced to generalize operator contractions to finite-temperature averages, forming the basis for expressing perturbative expansions in terms of Feynman diagrams. These diagrams provide a graphical language to encode and compute interaction contributions, making complex quantum processes more tractable and interpretable. The chapter also treats the summation over Matsubara frequencies, a vital technique in finite-temperature field theory that moves time-dependent quantities into frequency space to evaluate thermal Green's functions. Finally, the subsequent sections show how Green's functions and their diagrammatic expansions can be used to compute the expectation value of the particle density operator in interacting systems, thus linking the formalism to experimentally accessible observables. Overall, the chapter provides both the theoretical infrastructure and practical tools central to many-body and condensed matter physics.

7.1 Dyson Expansion of the Thermal Green's Function

In the present section, we adopt Feynman's formulation to develop the finite-temperature theory of many-body Green's functions. The starting point is the evaluation of the partition function within the framework of second quantization, which allows us to reformulate thermodynamic and dynamical quantities in terms of the interaction picture. This leads to a representation of the Matsubara Green's function as an expansion in powers of the interaction.

A fundamental step in the computation of the thermal Green's function is the evaluation of the partition function Z, associated with a generic Hamiltonian of the form 1.85, within the framework of second quantization, with $\hat{\mathcal{H}}_0$ given in Eq. 2.181. Specifically, we proceed with the following sequence of manipulations

F. Tafuri et al., *Understanding Quantum Many-Body Theory*, UNITEXT for Physics, https://doi.org/10.1007/978-3-032-17849-7_7

$$
\begin{aligned}
Z \equiv \operatorname{Tr} e^{-\beta\hat{\mathcal{H}}} &= \\
&= \operatorname{Tr} e^{-\beta\hat{\mathcal{H}}_0} e^{\beta\hat{\mathcal{H}}_0} e^{-\beta\hat{\mathcal{H}}} = \\
&= \operatorname{Tr} e^{-\beta\hat{\mathcal{H}}_0}\, U(\beta\hbar) = \\
&= \frac{\operatorname{Tr} e^{-\beta\hat{\mathcal{H}}_0}}{\operatorname{Tr} e^{-\beta\hat{\mathcal{H}}_0}} \operatorname{Tr} e^{-\beta\hat{\mathcal{H}}_0}\, U(\beta\hbar) = \\
&= \operatorname{Tr} e^{-\beta\hat{\mathcal{H}}_0} \frac{\operatorname{Tr} e^{-\beta\hat{\mathcal{H}}_0}\, U(\beta\hbar)}{\operatorname{Tr} e^{-\beta\hat{\mathcal{H}}_0}} = \\
&= \operatorname{Tr} e^{-\beta\hat{\mathcal{H}}_0} \langle U(\beta\hbar)\rangle_0 \equiv \\
&\equiv Z_0\, \langle U(\beta\hbar)\rangle_0,
\end{aligned} \tag{7.1}
$$

where Z_0 and $\langle\ldots\rangle_0$ denote the partition function and the thermal average (see Eq. 1.379) with respect to the Hamiltonian $\hat{\mathcal{H}}_0$, respectively, and $U(\beta\hbar)$ is the evolution operator in the thermal interaction picture (see Eq. 2.196). The Matsubara Green's function for the operators A and B (see Eq. 6.14) at thermal equilibrium (see Theorem 6.1), for $\tau > 0$, takes the form

$$
\begin{aligned}
G^{(m)}_{A,B}(\tau) &= -\frac{1}{\hbar}\left\langle \hat{T}_\tau A(\tau)B(0)\right\rangle = \\
&= -\frac{1}{\hbar}\langle A(\tau)B(0)\rangle = \\
&= -\frac{1}{\hbar}\frac{\operatorname{Tr} e^{-\beta\hat{\mathcal{H}}}A(\tau)B(0)}{\operatorname{Tr} e^{-\beta\hat{\mathcal{H}}}} = \\
&= -\frac{1}{\hbar}\frac{\operatorname{Tr} e^{-\beta\hat{\mathcal{H}}}A(\tau)B(0)}{\operatorname{Tr} e^{-\beta\hat{\mathcal{H}}_0}\langle U(\beta\hbar)\rangle_0} = \\
&= -\frac{1}{\hbar}\frac{1}{\langle U(\beta\hbar)\rangle_0}\frac{\operatorname{Tr} e^{-\beta\hat{\mathcal{H}}}A(\tau)B(0)}{\operatorname{Tr} e^{-\beta\hat{\mathcal{H}}_0}} = \\
&= -\frac{1}{\hbar}\frac{1}{\langle U(\beta\hbar)\rangle_0}\frac{\operatorname{Tr} e^{-\beta\hat{\mathcal{H}}_0}e^{\beta\hat{\mathcal{H}}_0}e^{-\beta\hat{\mathcal{H}}}A(\tau)B(0)}{\operatorname{Tr} e^{-\beta\hat{\mathcal{H}}_0}} = \\
&= -\frac{1}{\hbar}\frac{1}{\langle U(\beta\hbar)\rangle_0}\frac{\operatorname{Tr} e^{-\beta\hat{\mathcal{H}}_0}U(\beta\hbar)A(\tau)B(0)}{\operatorname{Tr} e^{-\beta\hat{\mathcal{H}}_0}}, \quad \tau > 0,
\end{aligned} \tag{7.2}
$$

where recall the operator A is expressed in the thermal Heisenberg picture (see Eq. 6.12), and we employed Eq. 2.196 with $\tau = \beta\hbar$. Recalling the imaginary-time evolution operator $S(\tau, \tau')$ (see Eq. 2.218), and using Eq. 2.222, with $\tau = \beta\hbar$, and the property 6.13 for operator A, one finds

$$
\begin{aligned}
G^{(m)}_{A,B}(\tau) &= -\frac{1}{\hbar}\frac{1}{\langle U(\beta\hbar)\rangle_0}\frac{\mathrm{Tr}\left(e^{-\beta\hat{\mathcal{H}}_0}U(\beta\hbar)A(\tau)B(0)\right)}{\mathrm{Tr}\left(e^{-\beta\hat{\mathcal{H}}_0}\right)} = \\
&= -\frac{1}{\hbar}\frac{1}{\langle U(\beta\hbar)\rangle_0}\frac{\mathrm{Tr}\left(e^{-\beta\hat{\mathcal{H}}_0}S(\beta\hbar,0)S(0,\tau)A^{(0)}(\tau)S(\tau,0)B(0)\right)}{\mathrm{Tr}\left(e^{-\beta\hat{\mathcal{H}}_0}\right)} = \\
&= -\frac{1}{\hbar}\frac{1}{\langle U(\beta\hbar)\rangle_0}\frac{\mathrm{Tr}\left(e^{-\beta\hat{\mathcal{H}}_0}S(\beta\hbar,\tau)A^{(0)}(\tau)S(\tau,0)B(0)\right)}{\mathrm{Tr}\left(e^{-\beta\hat{\mathcal{H}}_0}\right)} = \\
&\equiv -\frac{1}{\hbar}\frac{\langle S(\beta\hbar,\tau)A^{(0)}(\tau)S(\tau,0)B(0)\rangle_0}{\langle U(\beta\hbar)\rangle_0}, \quad \tau > 0.
\end{aligned}
\tag{7.3}
$$

If the order of $A^{(0)}(\tau)$ and $S(\tau,0)$ can be reversed, the thermal Green's function is given by

$$
\begin{aligned}
G^{(m)}_{A,B}(\tau) &= -\frac{1}{\hbar}\frac{\langle S(\beta\hbar,\tau)S(\tau,0)A^{(0)}(\tau)B(0)\rangle_0}{\langle U(\beta\hbar)\rangle_0} = \\
&= -\frac{1}{\hbar}\frac{\langle S(\beta\hbar,0)A^{(0)}(\tau)B(0)\rangle_0}{\langle U(\beta\hbar)\rangle_0} = \\
&= -\frac{1}{\hbar}\frac{\langle U(\beta\hbar)A^{(0)}(\tau)B(0)\rangle_0}{\langle U(\beta\hbar)\rangle_0}, \quad \tau > 0,
\end{aligned}
\tag{7.4}
$$

which, upon expanding $U(\beta\hbar)$ in a Dyson series (see Eq. 2.202), becomes

$$
\begin{aligned}
G^{(m)}_{A,B}(\tau) &= \frac{-\frac{1}{\hbar}\langle A^{(0)}(\tau)B(0)\rangle_0}{1+\sum_{n=1}^{\infty}\left(-\frac{1}{\hbar}\right)^n\frac{1}{n!}\int_0^{\beta\hbar}d\tau_1\cdots\int_0^{\beta\hbar}d\tau_n\left\langle\hat{T}_\tau\left\{\hat{\mathcal{H}}_I^{(0)}(\tau_1)\cdots\hat{\mathcal{H}}_I^{(0)}(\tau_n)\right\}\right\rangle_0} + \\
&+ \frac{-\frac{1}{\hbar}\sum_{n=1}^{\infty}\left(-\frac{1}{\hbar}\right)^n\frac{1}{n!}\int_0^{\beta\hbar}d\tau_1\cdots\int_0^{\beta\hbar}d\tau_n\left\langle\hat{T}_\tau\left\{\hat{\mathcal{H}}_I^{(0)}(\tau_1)\cdots\hat{\mathcal{H}}_I^{(0)}(\tau_n)A^{(0)}(\tau)B(0)\right\}\right\rangle_0}{1+\sum_{n=1}^{\infty}\left(-\frac{1}{\hbar}\right)^n\frac{1}{n!}\int_0^{\beta\hbar}d\tau_1\cdots\int_0^{\beta\hbar}d\tau_n\left\langle\hat{T}_\tau\left\{\hat{\mathcal{H}}_I^{(0)}(\tau_1)\cdots\hat{\mathcal{H}}_I^{(0)}(\tau_n)\right\}\right\rangle_0},
\end{aligned}
\tag{7.5}
$$

where recall $\tau > 0$ and $\hat{T}_\tau$ is the Wick time-ordering operator (see Eq. 2.200). We have

Theorem 7.1 (Dyson Expansion of Matsubara Green's Functions in the Thermal Interaction Picture with Free Hamiltonian Evolution) *Let $\langle \ldots \rangle_0$ denote the thermal average (see Eq. 1.379) with respect to the free Hamiltonian given in Eq. 2.181, within the framework of second quantization, and let $S(\tau, \tau')$ be the evolution operator given in Eq. 2.218. Then Eqs. 7.3 and 7.5 are equivalent. Consequently, the Matsubara Green's function defined in Eq. 6.15 can be expressed equivalently as in Eq. 7.5, in which the interaction Hamiltonians evolve according to free Hamiltonian, and the operators originally defined in the thermal Heisenberg picture (see Eq. 6.12) are replaced by their counterparts in the thermal interaction picture (see Eq. 2.194).*

Proof Consider the n-th term of the series in the numerator of Eq. 7.5, i.e.,

$$\int_0^{\beta\hbar} d\tau_1 \ldots \int_0^{\beta\hbar} d\tau_n \left\langle \hat{T}_\tau \left\{ \hat{\mathcal{H}}_I^{(0)}(\tau_1) \ldots \hat{\mathcal{H}}_I^{(0)}(\tau_n) \right\} A^{(0)}(\tau) B(0) \right\rangle_0 . \tag{7.6}$$

Recall that here the imaginary-time variable τ takes values in the interval $[0, \beta\hbar)$. We assume $0 < \tau < \beta\hbar$; hence, τ partitions the integration domain into the two disjoint subintervals $[0, \tau]$ and $[\tau, \beta\hbar]$. Accordingly, we write

$$\left(\int_0^{\tau} d\tau_1 + \int_\tau^{\beta\hbar} d\tau_1 \right) \left(\int_0^{\tau} d\tau_2 + \int_\tau^{\beta\hbar} d\tau_2 \right) \ldots \left(\int_0^{\tau} d\tau_n + \int_\tau^{\beta\hbar} d\tau_n \right), \tag{7.7}$$

then consider n_1 integrals of the form $\int_\tau^{\beta\hbar} d\tau_i$ and $m_1 = n - n_1$ integrals of the form $\int_0^\tau d\tau_i$. The explicit ordering of such a product corresponds to the correct time ordering, which proceeds in decreasing order from left to right, and is given by

$$\begin{aligned} &\int_\tau^{\beta\hbar} d\tau_1 \ldots \int_\tau^{\beta\hbar} d\tau_{n_1} \\ &\left\langle \hat{T}_\tau \left\{ \hat{\mathcal{H}}_I^{(0)}(\tau_1) \ldots \hat{\mathcal{H}}_I^{(0)}(\tau_{n_1}) \right\} A^{(0)}(\tau) \int_0^{\tau} d\tau_{m_1+1} \ldots \right. \\ &\qquad \left. \int_0^{\tau} d\tau_n \, \hat{T}_\tau \left\{ \hat{\mathcal{H}}_I^{(0)}(\tau_{m_1+1}) \ldots \hat{\mathcal{H}}_I^{(0)}(\tau_n) \right\} B(0) \right\rangle_0 , \end{aligned} \tag{7.8}$$

which corresponds to one of the $\frac{n!}{n_1! m_1!}$ possible combinations obtained by selecting n_1 integrals out of a total of n. However, all these products are equivalent since the Wick time-ordering operator $\hat{T}_\tau$ (see Eq. 2.200) reorders them identically within each integral. Therefore, we can represent the entire sum by the above expression multiplied by the combinatorial factor $\frac{n!}{n_1! m_1!}$, i.e.,

$$\frac{n!}{n_1! m_1!} \int_\tau^{\beta\hbar} d\tau_1 \ldots \int_\tau^{\beta\hbar} d\tau_{n_1}$$

$$
\left\langle \hat{T}_\tau \left\{ \hat{\mathcal{H}}_I^{(0)}(\tau_1) \ldots \hat{\mathcal{H}}_I^{(0)}(\tau_{n_1}) \right\} A^{(0)}(\tau) \int_0^\tau d\tau_{m_1+1} \ldots \right.
$$

$$
\left. \int_0^\tau d\tau_n \, \hat{T}_\tau \left\{ \hat{\mathcal{H}}_I^{(0)}(\tau_{m_1+1}) \ldots \hat{\mathcal{H}}_I^{(0)}(\tau_n) \right\} B(0) \right\rangle_0 . \tag{7.9}
$$

To generate all terms of the integral in Eq. 7.7, both n_1 and m_1 must vary from 0 to n, subject to the constraint $n_1 + m_1 = n$. Hence, we have

$$
\sum_{n_1=0}^{\infty} \sum_{m_1=0}^{\infty} \delta_{n_1+m_1,n} \frac{n!}{n_1! m_1!} \int_\tau^{\beta\hbar} d\tau_1 \ldots \int_\tau^{\beta\hbar} d\tau_{n_1}
$$

$$
\left\langle \hat{T}_\tau \left\{ \hat{\mathcal{H}}_I^{(0)}(\tau_1) \ldots \hat{\mathcal{H}}_I^{(0)}(\tau_{n_1}) \right\} A^{(0)}(\tau) \int_0^\tau d\tau_{m_1+1} \ldots \right.
$$

$$
\left. \int_0^\tau d\tau_n \, \hat{T}_\tau \left\{ \hat{\mathcal{H}}_I^{(0)}(\tau_{m_1+1}) \ldots \hat{\mathcal{H}}_I^{(0)}(\tau_n) \right\} B(0) \right\rangle_0 , \tag{7.10}
$$

where $\delta_{n_1+m_1,n}$ enforces the constraint on n and allows the summations to be formally extended to infinity. Consequently, the series in the numerator of Eq. 7.5 becomes

$$
\sum_{n=0}^{\infty} \sum_{n_1=0}^{\infty} \sum_{m_1=0}^{\infty} \left(-\frac{1}{\hbar}\right)^n \frac{1}{n!} \delta_{n_1+m_1,n} \frac{n!}{n_1! m_1!} \int_\tau^{\beta\hbar} d\tau_1 \ldots \int_\tau^{\beta\hbar} d\tau_{n_1}
$$

$$
\left\langle \hat{T}_\tau \left\{ \hat{\mathcal{H}}_I^{(0)}(\tau_1) \ldots \hat{\mathcal{H}}_I^{(0)}(\tau_{n_1}) \right\} A^{(0)}(\tau) \int_0^\tau d\tau_{m_1+1} \ldots \right.
$$

$$
\left. \int_0^\tau d\tau_n \hat{T}_\tau \left\{ \hat{\mathcal{H}}_I^{(0)}(\tau_{m_1+1}) \ldots \hat{\mathcal{H}}_I^{(0)}(\tau_n) \right\} B(0) \right\rangle_0 =
$$

$$
= \sum_{n=0}^{\infty} \sum_{n_1=0}^{\infty} \sum_{m_1=0}^{\infty} \left(-\frac{1}{\hbar}\right)^{n_1} \left(-\frac{1}{\hbar}\right)^{m_1} \delta_{n_1+m_1,n} \frac{1}{n_1! m_1!} \int_\tau^{\beta\hbar} d\tau_1 \ldots \int_\tau^{\beta\hbar} d\tau_{n_1}
$$

$$
\left\langle \hat{T}_\tau \left\{ \hat{\mathcal{H}}_I^{(0)}(\tau_1) \ldots \hat{\mathcal{H}}_I^{(0)}(\tau_{n_1}) \right\} A^{(0)}(\tau) \int_0^\tau d\tau_{m_1+1} \ldots \right.
$$

$$
\left. \int_0^\tau d\tau_n \hat{T}_\tau \left\{ \hat{\mathcal{H}}_I^{(0)}(\tau_{m_1+1}) \ldots \hat{\mathcal{H}}_I^{(0)}(\tau_n) \right\} B(0) \right\rangle_0 , \tag{7.11}
$$

where $n!$ has been simplified, and the power $(-\hbar)^{-n}$ is expressed as a function of n_1 and m_1. Summing over n, the $\delta_{n_1+m_1,n}$ is triggered and removed, and the two sums factorize as

$$\sum_{n_1=0}^{\infty}\left(-\frac{1}{\hbar}\right)^{n_1}\frac{1}{n_1!}\int_{\tau}^{\beta\hbar}d\tau_1\ldots\int_{\tau}^{\beta\hbar}d\tau_{n_1}\hat{T}_\tau\left\{\hat{\mathcal{H}}_I^{(0)}(\tau_1)\ldots\hat{\mathcal{H}}_I^{(0)}(\tau_{n_1})\right\}=S(\beta\hbar,\tau),\tag{7.12}$$

$$\sum_{m_1=0}^{\infty}\left(-\frac{1}{\hbar}\right)^{m_1}\frac{1}{m_1!}\int_0^{\tau}d\tau_{m_1+1}\ldots$$
$$\int_0^{\tau}d\tau_n\hat{T}_\tau\left\{\hat{\mathcal{H}}_I^{(0)}(\tau_{m_1+1})\ldots\hat{\mathcal{H}}_I^{(0)}(\tau_n)\right\}=S(\tau,0),\tag{7.13}$$

that is, the operator

$$S(\beta\hbar,\tau)A^{(0)}(\tau)S(\tau,0)B(0)\tag{7.14}$$

has been reconstructed. Therefore, Eqs. 7.3 and 7.5 are equal. □

Remark 7.1 (Thermal Interaction and Thermal Heisenberg Pictures in Dyson Expansion Formulation) Due to Theorem 7.1, any Matsubara Green's function 6.15 can be expressed in terms of thermal averages (see Eq. 1.379), where the average over the full interacting Hamiltonian $\hat{\mathcal{H}}$ is replaced by an average taken with respect to the non-interacting Hamiltonian $\hat{\mathcal{H}}_0$. Moreover, operators originally defined in the thermal Heisenberg picture (see Eq. 6.12) are replaced by their counterparts in the thermal interaction picture (see Eq. 2.194), so that their time evolution is governed solely by $\hat{\mathcal{H}}_0$. Moreover, according to Theorem 1, the operator B remains in the thermal Heisenberg picture; however, since it is evaluated at $\tau = 0$, it coincides with the operator itself. Therefore, since B, being always taken at time zero, does not depend on the chosen picture and is identical to its own Schrödinger representation, we will omit specifying the picture for B when applying Theorem 7.1, while still explicitly indicating its time dependence in parentheses.

7.2 Thermal Wick's Theorem

Consider thermal averages (see Eq. 1.379) of the form

$$\left\langle A_{\alpha_1}^{(1)}\ldots A_{\alpha_{2N}}^{(2N)}\right\rangle_0,\tag{7.15}$$

where each A is proportional to a creation or annihilation operator and does not depend explicitly on time. We emphasize that the superscript (i) on $A^{(i)}$ specifies solely the position of the operator within the ordered product, whereas the subscript α_i denotes the one-particle quantum state on which the operator acts. Moreover, the subscript 0 emphasizes that the averages must be evaluated on the non-interacting

Hamiltonian $\hat{\mathcal{H}}_0$ of the form 2.181. The calculation of these thermal averages leads to thermal Wick's theorem, whose proof we present in this section.

At first, we introduce a definition preparatory to thermal Wick's theorem.

Definition 7.1 (Contraction and Thermal Contraction) Given two operators $A^{(i)}_{\alpha_i}$ and $A^{(j)}_{\alpha_j}$, and the brackets introduced in Eq. 2.102, from

$$\begin{aligned} C^{ij} &= \left[A^{(i)}_{\alpha_i}, A^{(j)}_{\alpha_j}\right]^{(\varepsilon)} \equiv \\ &\equiv A^{(i)}_{\alpha_i} A^{(j)}_{\alpha_j} - \varepsilon A^{(j)}_{\alpha_j} A^{(i)}_{\alpha_i}, \end{aligned} \tag{7.16}$$

a contraction is defined as one of the quantities $A^{(i)}_{\alpha_i} A^{(j)}_{\alpha_j}$ or $\varepsilon A^{(j)}_{\alpha_j} A^{(i)}_{\alpha_i}$. Furthermore, a thermal contraction is defined as the corresponding thermal expectation value, $\left\langle A^{(i)}_{\alpha_i} A^{(j)}_{\alpha_j}\right\rangle$ or $\left\langle \varepsilon A^{(j)}_{\alpha_j} A^{(i)}_{\alpha_i}\right\rangle$.

Remark 7.2 (On the Definition of Contraction) In the following, for brevity, we will refer to both $A^{(i)}_{\alpha_i} A^{(j)}_{\alpha_j}$ and $\left\langle A^{(i)}_{\alpha_i} A^{(j)}_{\alpha_j}\right\rangle$ simply as the contraction. The precise meaning, operator or thermal average, will always be clear from the context.

Since, by assumption, each operator A is either a creation or an annihilation operator, the operator C^{ij} takes the following form:

1. $\alpha_i = \alpha_j \equiv \alpha$, $A^{(i)}_{\alpha} = a_\alpha$, $A^{(j)}_{\alpha} = a^\dagger_\alpha$. Then

$$\begin{aligned} C^{ij} &= a_\alpha a^\dagger_\alpha - \varepsilon a^\dagger_\alpha a_\alpha = \\ &= \left[a_\alpha, a^\dagger_\alpha\right]^{(\varepsilon)} = \\ &= \mathbb{1}, \end{aligned} \tag{7.17}$$

where we use the standard commutation (anticommutation) relations (see Eq. D.7 with $\alpha = \beta$).

2. $\alpha_i = \alpha_j \equiv \alpha$, $A^{(i)}_{\alpha} = a^\dagger_\alpha$, $A^{(j)}_{\alpha} = a_\alpha$. Then

$$\begin{aligned} C^{ij} &= a^\dagger_\alpha a_\alpha - \varepsilon a_\alpha a^\dagger_\alpha = \\ &= a^\dagger_\alpha a_\alpha - \varepsilon\left(\mathbb{1} + \varepsilon a^\dagger_\alpha a_\alpha\right) = \\ &= a^\dagger_\alpha a_\alpha - \varepsilon \mathbb{1} - \varepsilon^2 a^\dagger_\alpha a_\alpha = \\ &= a^\dagger_\alpha a_\alpha - \varepsilon \mathbb{1} - a^\dagger_\alpha a_\alpha = \\ &= -\varepsilon \mathbb{1}, \end{aligned} \tag{7.18}$$

where we employed Eq. D.8.

3. $\alpha_i \neq \alpha_j$. In this case, from Eq. D.7 with $\alpha \neq \beta$, one finds

$$C^{ij} = 0. \tag{7.19}$$

Consequently, the object $\langle C^{ij} \rangle_0$ takes the form

$$\left\langle C^{ij} \right\rangle_0 = \begin{cases} +1 \ , & \alpha_i = \alpha_j \equiv \alpha, \ A_\alpha^{(i)} = a_\alpha, \ A_\alpha^{(j)} = a_\alpha^\dagger \\ -\varepsilon \ , & \alpha_i = \alpha_j \equiv \alpha, \ A_\alpha^{(i)} = a_\alpha^\dagger, \ A_\alpha^{(j)} = a_\alpha \\ 0 \ , & \text{otherwise} \end{cases} . \tag{7.20}$$

With this definition and preliminary properties at hand, we are in a position to formulate and prove thermal Wick's theorem.

Theorem 7.2 (Thermal Wick's Theorem for Time-Independent Operators) *Let $A_{\alpha_1}^{(1)}, \ldots, A_{\alpha_{2N}}^{(2N)}$ denote time-independent annihilation or creation operators (see Eqs. 2.79 and 2.83), which may be either bosonic or fermionic. The thermal average $\langle \ldots \rangle_0$ (see Eq. 1.379), taken with respect to the non-interacting Hamiltonian $\hat{\mathcal{H}}_0$ given in Eq. 2.181, is given by the sum over all possible pairwise thermal contractions:*

$$\left\langle A_{\alpha_1}^{(1)} \ldots A_{\alpha_{2N}}^{(2N)} \right\rangle_0 =$$
$$= \sum_\sigma \varepsilon^{\mathrm{sign}(\sigma)} \left\langle A_{\alpha_{\sigma(1)}}^{(\sigma(1))} A_{\alpha_{\sigma(2)}}^{(\sigma(2))} \right\rangle_0 \left\langle A_{\alpha_{\sigma(3)}}^{(\sigma(3))} A_{\alpha_{\sigma(4)}}^{(\sigma(4))} \right\rangle_0 \cdots \left\langle A_{\alpha_{\sigma(2N-1)}}^{(\sigma(2N-1))} A_{\alpha_{\sigma(2N)}}^{(\sigma(2N))} \right\rangle_0 , \tag{7.21}$$

where ε is the statistical index defined in Eq. 2.52, and the sum runs over all permutations σ of the indices $1, \ldots, 2N$ that generate complete pairings, i.e., permutations such that

$$\begin{aligned} \sigma : \sigma(1) < \sigma(2), \ \sigma(3) < \sigma(4), \ \ldots, \ \sigma(2N-1) < \sigma(2N), \\ \sigma(1) < \sigma(3) < \ldots < \sigma(2N-1). \end{aligned} \tag{7.22}$$

A thermal contraction $\left\langle A_{\alpha_i}^{(i)} A_{\alpha_j}^{(j)} \right\rangle_0$ can be non-vanishing only if $A_{\alpha_i}^{(i)}$ and $A_{\alpha_j}^{(j)}$ are a creation and an annihilation operator associated with the same quantum state (in either order).

Proof For simplicity, we prove the theorem in the case of four operators, namely

$$\left\langle A_{\alpha_1}^{(1)} A_{\alpha_2}^{(2)} A_{\alpha_3}^{(3)} A_{\alpha_4}^{(4)} \right\rangle_0 . \tag{7.23}$$

We employ the operator C^{ij} introduced in Eq. 7.16. Now, we use several times the relation (see Eq. 2.102)

$$A^{(i)}_{\alpha_i} A^{(j)}_{\alpha_j} = C^{ij} + \varepsilon A^{(j)}_{\alpha_j} A^{(i)}_{\alpha_i} \tag{7.24}$$

in the thermal average as follows

$$\begin{aligned}
&\left\langle A^{(1)}_{\alpha_1} A^{(2)}_{\alpha_2} A^{(3)}_{\alpha_3} A^{(4)}_{\alpha_4} \right\rangle_0 = \\
&= \left\langle \left(C^{12} + \varepsilon A^{(2)}_{\alpha_2} A^{(1)}_{\alpha_1} \right) A^{(3)}_{\alpha_3} A^{(4)}_{\alpha_4} \right\rangle_0 = \\
&= \left\langle C^{12} \right\rangle_0 \left\langle A^{(3)}_{\alpha_3} A^{(4)}_{\alpha_4} \right\rangle_0 + \varepsilon \left\langle A^{(2)}_{\alpha_2} A^{(1)}_{\alpha_1} A^{(3)}_{\alpha_3} A^{(4)}_{\alpha_4} \right\rangle_0 = \\
&= \left\langle C^{12} \right\rangle_0 \left\langle A^{(3)}_{\alpha_3} A^{(4)}_{\alpha_4} \right\rangle_0 + \varepsilon \left\langle A^{(2)}_{\alpha_2} \left(C^{13} + \varepsilon A^{(3)}_{\alpha_3} A^{(1)}_{\alpha_1} \right) A^{(4)}_{\alpha_4} \right\rangle_0 = \\
&= \left\langle C^{12} \right\rangle_0 \left\langle A^{(3)}_{\alpha_3} A^{(4)}_{\alpha_4} \right\rangle_0 + \varepsilon \left\langle C^{13} \right\rangle_0 \left\langle A^{(2)}_{\alpha_2} A^{(4)}_{\alpha_4} \right\rangle_0 + \varepsilon^2 \left\langle A^{(2)}_{\alpha_2} A^{(3)}_{\alpha_3} A^{(1)}_{\alpha_1} A^{(4)}_{\alpha_4} \right\rangle_0 = \\
&\equiv \left\langle C^{12} \right\rangle_0 \left\langle A^{(3)}_{\alpha_3} A^{(4)}_{\alpha_4} \right\rangle_0 + \varepsilon \left\langle C^{13} \right\rangle_0 \left\langle A^{(2)}_{\alpha_2} A^{(4)}_{\alpha_4} \right\rangle_0 + \left\langle A^{(2)}_{\alpha_2} A^{(3)}_{\alpha_3} A^{(1)}_{\alpha_1} A^{(4)}_{\alpha_4} \right\rangle_0 = \\
&= \left\langle C^{12} \right\rangle_0 \left\langle A^{(3)}_{\alpha_3} A^{(4)}_{\alpha_4} \right\rangle_0 + \varepsilon \left\langle C^{13} \right\rangle_0 \left\langle A^{(2)}_{\alpha_2} A^{(4)}_{\alpha_4} \right\rangle_0 \\
&\quad + \left\langle A^{(2)}_{\alpha_2} A^{(3)}_{\alpha_3} \left(C^{14} + \varepsilon A^{(4)}_{\alpha_4} A^{(1)}_{\alpha_1} \right) \right\rangle_0 = \\
&= \left\langle C^{12} \right\rangle_0 \left\langle A^{(3)}_{\alpha_3} A^{(4)}_{\alpha_4} \right\rangle_0 + \varepsilon \left\langle C^{13} \right\rangle_0 \left\langle A^{(2)}_{\alpha_2} A^{(4)}_{\alpha_4} \right\rangle_0 \\
&\quad + \left\langle C^{14} \right\rangle_0 \left\langle A^{(2)}_{\alpha_2} A^{(3)}_{\alpha_3} \right\rangle_0 + \varepsilon \left\langle A^{(2)}_{\alpha_2} A^{(3)}_{\alpha_3} A^{(4)}_{\alpha_4} A^{(1)}_{\alpha_1} \right\rangle_0 .
\end{aligned} \tag{7.25}$$

Note that, since C^{ij} is proportional to the identity operator, every thermal average in which C^{ij} appears together with other operators factorizes into the product of $\langle C^{ij} \rangle_0$ and the thermal average of the remaining operators. This factorization has been used throughout the derivation above. In order to clarify whether

$$\varepsilon \left\langle A^{(2)}_{\alpha_2} A^{(3)}_{\alpha_3} A^{(4)}_{\alpha_4} A^{(1)}_{\alpha_1} \right\rangle_0 \tag{7.26}$$

can be cast into a form proportional to the first term, we write it as

$$\begin{aligned}
\left\langle A^{(2)}_{\alpha_2} A^{(3)}_{\alpha_3} A^{(4)}_{\alpha_4} A^{(1)}_{\alpha_1} \right\rangle_0 &\equiv \frac{1}{Z_0} \operatorname{Tr} \left(e^{-\beta \hat{\mathcal{H}}_0} A_2 A_3 A_4 A_1 \right) = \\
&= \frac{1}{Z_0} \operatorname{Tr} \left(A^{(2)}_{\alpha_2} A^{(3)}_{\alpha_3} A^{(4)}_{\alpha_4} A^{(1)}_{\alpha_1} e^{-\beta \hat{\mathcal{H}}_0} \right) = \\
&= \frac{1}{Z_0} \operatorname{Tr} \left(A^{(2)}_{\alpha_2} A^{(3)}_{\alpha_3} A^{(4)}_{\alpha_4} e^{-\beta \hat{\mathcal{H}}_0} e^{\beta \hat{\mathcal{H}}_0} A^{(1)}_{\alpha_1} e^{-\beta \hat{\mathcal{H}}_0} \right) =
\end{aligned}$$

$$
\begin{aligned}
&= \frac{1}{Z_0} \operatorname{Tr}\left(A^{(2)}_{\alpha_2} A^{(3)}_{\alpha_3} A^{(4)}_{\alpha_4} e^{-\beta \hat{\mathcal{H}}_0} e^{\delta_1 \beta \mathcal{E}_{\alpha_1}} A^{(1)}_{\alpha_1}\right) \equiv \\
&\equiv \frac{e^{\delta_1 \beta \mathcal{E}_{\alpha_1}}}{Z_0} \operatorname{Tr}\left(A^{(2)}_{\alpha_2} A^{(3)}_{\alpha_3} A^{(4)}_{\alpha_4} e^{-\beta \hat{\mathcal{H}}_0} A^{(1)}_{\alpha_1}\right) = \\
&= \frac{e^{\delta_1 \beta \mathcal{E}_{\alpha_1}}}{Z_0} \operatorname{Tr}\left(e^{-\beta \hat{\mathcal{H}}_0} A^{(1)}_{\alpha_1} A^{(2)}_{\alpha_2} A^{(3)}_{\alpha_3} A^{(4)}_{\alpha_4}\right) \equiv \\
&\equiv e^{\delta_1 \beta \mathcal{E}_{\alpha_1}} \left\langle A^{(1)}_{\alpha_1} A^{(2)}_{\alpha_2} A^{(3)}_{\alpha_3} A^{(4)}_{\alpha_4} \right\rangle_0 . \qquad (7.27)
\end{aligned}
$$

We have used the cyclic property of the trace (see Eq. J.3) by shifting $e^{-\beta \hat{\mathcal{H}}_0}$ to the right; we inserted the identity $\mathbb{1} = e^{-\beta \hat{\mathcal{H}}_0} e^{\beta \hat{\mathcal{H}}_0}$ between $A^{(4)}_{\alpha_4}$ and $A^{(1)}_{\alpha_1}$; by employing the thermal interaction picture (see Eq. 2.194), the time evolution of $A^{(1)}_{\alpha_1}$ at $\tau = \beta\hbar$ can be expressed in a form that incorporates Eqs. 2.241 and 2.236, that is (equivalently, the same result can be obtained from Eq. 2.242 in the particular case $\tau = 0$)

$$
\begin{aligned}
e^{\beta \hat{\mathcal{H}}_0} A^{(1)}_{\alpha_1} e^{-\beta \hat{\mathcal{H}}_0} &= A^{(1)(0)}_{\alpha_1}(\beta\hbar) = \\
&= e^{\delta_1 \beta \mathcal{E}_{\alpha_1}} A^{(1)(0)}_{\alpha_1}(0) \equiv \\
&\equiv e^{\delta_1 \beta \mathcal{E}_{\alpha_1}} A^{(1)}_{\alpha_1} , \qquad (7.28)
\end{aligned}
$$

where $\mathcal{E}_{\alpha_1}$ denotes the one-particle energy of the quantum state associated with $A^{(1)}_{\alpha_1}$, and δ_1 is given in Eq. 2.243; ultimately, we have used the cyclic property of the trace first to bring $A^{(1)}_{\alpha_1}$ to the left and then to move $e^{-\beta \hat{\mathcal{H}}_0}$ back to the right. As a result, one finds

$$
\begin{aligned}
&\left\langle A^{(1)}_{\alpha_1} A^{(2)}_{\alpha_2} A^{(3)}_{\alpha_3} A^{(4)}_{\alpha_4} \right\rangle_0 = \\
&= \left\langle C^{12} \right\rangle_0 \left\langle A^{(3)}_{\alpha_3} A^{(4)}_{\alpha_4} \right\rangle_0 + \varepsilon \left\langle C^{13} \right\rangle_0 \left\langle A^{(2)}_{\alpha_2} A^{(4)}_{\alpha_4} \right\rangle_0 + \left\langle C^{14} \right\rangle_0 \left\langle A^{(2)}_{\alpha_2} A^{(3)}_{\alpha_3} \right\rangle_0 \\
&+ \varepsilon e^{\delta_1 \beta \mathcal{E}_{\alpha_1}} \left\langle A^{(1)}_{\alpha_1} A^{(2)}_{\alpha_2} A^{(3)}_{\alpha_3} A^{(4)}_{\alpha_4} \right\rangle_0 , \qquad (7.29)
\end{aligned}
$$

$$
\begin{aligned}
&\left(1 - \varepsilon e^{\delta_1 \beta \mathcal{E}_{\alpha_1}}\right) \left\langle A^{(1)}_{\alpha_1} A^{(2)}_{\alpha_2} A^{(3)}_{\alpha_3} A^{(4)}_{\alpha_4} \right\rangle_0 = \\
&= \left\langle C^{12} \right\rangle_0 \left\langle A^{(3)}_{\alpha_3} A^{(4)}_{\alpha_4} \right\rangle_0 + \varepsilon \left\langle C^{13} \right\rangle_0 \left\langle A^{(2)}_{\alpha_2} A^{(4)}_{\alpha_4} \right\rangle_0 + \left\langle C^{14} \right\rangle_0 \left\langle A^{(2)}_{\alpha_2} A^{(3)}_{\alpha_3} \right\rangle_0 , \qquad (7.30)
\end{aligned}
$$

and finally

$$
\left\langle A^{(1)}_{\alpha_1} A^{(2)}_{\alpha_2} A^{(3)}_{\alpha_3} A^{(4)}_{\alpha_4} \right\rangle_0
$$

$$= \frac{\langle C^{12}\rangle_0 \left\langle A^{(3)}_{\alpha_3} A^{(4)}_{\alpha_4}\right\rangle_0 + \varepsilon \langle C^{13}\rangle_0 \left\langle A^{(2)}_{\alpha_2} A^{(4)}_{\alpha_4}\right\rangle_0 + \langle C^{14}\rangle_0 \left\langle A^{(2)}_{\alpha_2} A^{(3)}_{\alpha_3}\right\rangle_0}{1 - \varepsilon e^{\delta_1 \beta \mathcal{E}_{\alpha_1}}}. \tag{7.31}$$

This expression explicitly shows how the thermal average of a four-operator product reduces to a combination of thermal contractions, in accordance with the generalized thermal Wick's theorem. Equation 7.31 represents the thermal Wick's theorem specialized to the case of four operators. □

Note that thermal Wick's theorem is referred to the thermal average of an even number of operators, since such an object is different from 0 only when the number of operators is half creation and half annihilation.

Now we show an equivalent form of thermal Wick's theorem 7.2.

Theorem 7.3 (Alternative Form of thermal Wick's Theorem for Time-Independent Operators (4-Operator Case)) *Starting from Eq. 7.31, therefore limiting to the 4-operator case, an equivalent formulation of the thermal Wick's theorem 7.2 follows*

$$\left\langle A^{(1)}_{\alpha_1} A^{(2)}_{\alpha_2} A^{(3)}_{\alpha_3} A^{(4)}_{\alpha_4}\right\rangle_0 = \left\langle A^{(1)}_{\alpha_1} A^{(2)}_{\alpha_2}\right\rangle_0 \left\langle A^{(3)}_{\alpha_3} A^{(4)}_{\alpha_4}\right\rangle_0 + \varepsilon \left\langle A^{(1)}_{\alpha_1} A^{(3)}_{\alpha_3}\right\rangle_0 \left\langle A^{(2)}_{\alpha_2} A^{(4)}_{\alpha_4}\right\rangle_0 +$$
$$+ \left\langle A^{(1)}_{\alpha_1} A^{(4)}_{\alpha_4}\right\rangle_0 \left\langle A^{(2)}_{\alpha_2} A^{(3)}_{\alpha_3}\right\rangle_0, \tag{7.32}$$

where ε is the statistical index given in Eq. 2.52.

Proof From Eq. 7.16, one finds

$$\left\langle A^{(1)}_{\alpha_1} A^{(2)}_{\alpha_2}\right\rangle_0 = \left\langle C^{12}\right\rangle_0 + \varepsilon \left\langle A^{(2)}_{\alpha_2} A^{(1)}_{\alpha_1}\right\rangle_0, \tag{7.33}$$

then

$$\begin{aligned}
\left\langle A^{(2)}_{\alpha_2} A^{(1)}_{\alpha_1}\right\rangle_0 &= \frac{1}{Z_0} \operatorname{Tr}\left(e^{-\beta \hat{\mathcal{H}}_0} A^{(2)}_{\alpha_2} A^{(1)}_{\alpha_1}\right) = \\
&= \frac{1}{Z_0} \operatorname{Tr}\left(A^{(1)}_{\alpha_1} e^{-\beta \hat{\mathcal{H}}_0} A^{(2)}_{\alpha_2}\right) = \\
&= \frac{1}{Z_0} \operatorname{Tr}\left(A^{(2)}_{\alpha_2} A^{(1)}_{\alpha_1} e^{-\beta \hat{\mathcal{H}}_0}\right) = \\
&= \frac{1}{Z_0} \operatorname{Tr}\left(A^{(2)}_{\alpha_2} e^{-\beta \hat{\mathcal{H}}_0} e^{\beta \hat{\mathcal{H}}_0} A^{(1)}_{\alpha_1} e^{-\beta \hat{\mathcal{H}}_0}\right) = \\
&= \frac{1}{Z_0} \operatorname{Tr}\left(A^{(2)}_{\alpha_2} e^{-\beta \hat{\mathcal{H}}_0} e^{\delta_1 \beta \mathcal{E}_{\alpha_1}} A^{(1)}_{\alpha_1}\right) = \\
&= \frac{e^{\delta_1 \beta \mathcal{E}_{\alpha_1}}}{Z_0} \operatorname{Tr}\left(A^{(2)}_{\alpha_2} e^{-\beta \hat{\mathcal{H}}_0} A^{(1)}_{\alpha_1}\right) =
\end{aligned}$$

$$
\begin{aligned}
&= \frac{e^{\delta_1 \beta \mathcal{E}_{\alpha_1}}}{Z_0} \operatorname{Tr}\left(A_{\alpha_1}^{(1)} A_{\alpha_2}^{(2)} e^{-\beta \hat{\mathcal{H}}_0}\right) = \\
&= \frac{e^{\delta_1 \beta \mathcal{E}_{\alpha_1}}}{Z_0} \operatorname{Tr}\left(e^{-\beta \hat{\mathcal{H}}_0} A_{\alpha_1}^{(1)} A_{\alpha_2}^{(2)}\right) = \\
&= e^{\delta_1 \beta \mathcal{E}_{\alpha_1}} \left\langle A_{\alpha_1}^{(1)} A_{\alpha_2}^{(2)} \right\rangle_0 ,
\end{aligned} \tag{7.34}
$$

where we have used the cyclic property of the trace (see Eq. J.3) and δ_1 is given in Eq. 2.243. It is important to note that the relation 7.34 represents a generalization of KMS relation for annihilation and creation operators, that is, Eq. 2.253. The general form of the KMS condition, however, allows one to treat any pair of operators $A_{\alpha_1}^{(1)}$ and $A_{\alpha_2}^{(2)}$, showing how the properties of periodicity in imaginary time extend beyond the particular case of bosons or fermions. One finds

$$
\left\langle A_{\alpha_1}^{(1)} A_{\alpha_2}^{(2)} \right\rangle_0 = \frac{\left\langle C^{12} \right\rangle_0}{1 - \varepsilon e^{\delta_1 \beta \mathcal{E}_{\alpha_1}}} . \tag{7.35}
$$

Now, given the thermal average of the product of the 4 operators (see Eq. 7.31), we use

$$
\left\langle C^{12} \right\rangle_0 = \left(1 - \varepsilon e^{\delta_1 \beta \mathcal{E}_{\alpha_1}}\right) \left\langle A_{\alpha_1}^{(1)} A_{\alpha_2}^{(2)} \right\rangle_0 \tag{7.36}
$$

for each $\langle C^{1j} \rangle_0$, then we finally arrive at the alternative form of thermal Wick's theorem, namely Eq. 7.32. □

Equation 7.32 represents the form most commonly employed in practice, where the basic building blocks are the thermal contractions

$$
\left\langle A_{\alpha_i}^{(i)} A_{\alpha_j}^{(j)} \right\rangle_0 . \tag{7.37}
$$

As mentioned earlier, such contractions yield non-vanishing contributions only when the two operators refer to the same quantum state and one of them is a creation operator while the other is an annihilation operator, in either order (see Eq. 2.252): in particular, for the number operator (see Eq. D.12), one recovers Eq. 2.251, which gives the average occupation number in terms of the Bose-Einstein or Fermi-Dirac statistics, that is, Eqs. E.3 or E.4, respectively.

The contraction terminology proves particularly convenient where it allows one to systematically organize terms when applying thermal Wick's theorem and facilitates the graphical representation of Dyson expansion through Feynman diagrams (see below).

However, when considering expressions such as the denominator in Eq. 7.5, the integrand of a generic contribution to the expansion involves a thermal average of the form

$$\left\langle \hat{T}_\tau \left\{ A_{\alpha_1}^{(1)(0)}(\tau_1) \dots A_{\alpha_{2N}}^{(2N)(0)}(\tau_{2N}) \right\} \right\rangle_0, \tag{7.38}$$

where each $A^{(0)}$ denotes either a creation or an annihilation operator in the thermal interaction picture (see Eq. 2.194), and $\hat{T}_\tau$ is the Wick time-ordering operator (see Eq. 2.200).

We shall not present the proof of the corresponding thermal Wick's theorem in this context. Such a proof can be obtained, for instance, by means of a generating functional approach or equivalent techniques. Here we limit ourselves to stating the result, which takes a form fully analogous to Theorem 7.3, suitably adapted to the case of time-dependent operators.

In particular, we shall present the statement of formulation that is fully analogous to Theorem 7.3, appropriately adapted to the case of time-dependent operators.

Theorem 7.4 (Thermal Wick's Theorem for Time-Dependent Operators) *Let $A_{\alpha_1}^{(1)(0)}(\tau_1)A_{\alpha_2}^{(2)(0)}(\tau_2)A_{\alpha_3}^{(3)(0)}(\tau_3)A_{\alpha_4}^{(4)(0)}(\tau_4)$ denote the product of creation or annihilation operators, either bosonic or fermionic, in the thermal interaction picture (see Eq. 2.194). Furthermore, consider their product placed under Wick time-ordering operator $\hat{T}_\tau$ (see Eq. 2.200). The thermal average $\langle \dots \rangle_0$ (see Eq. 1.379), taken with respect to the non-interacting Hamiltonian $\hat{\mathcal{H}}_0$ given in Eq. 2.181, is given by the sum over all possible pairwise Matsubara Green's functions (see Eq. 6.14), up to the factor $(-\hbar)^{-1}$ appearing in their definition, i.e.,*

$$\begin{aligned}
&\left\langle \hat{T}_\tau A_{\alpha_1}^{(1)(0)}(\tau_1)A_{\alpha_2}^{(2)(0)}(\tau_2)A_{\alpha_3}^{(3)(0)}(\tau_3)A_{\alpha_4}^{(4)(0)}(\tau_4) \right\rangle_0 = \\
&= \left\langle \hat{T}_\tau A_{\alpha_1}^{(1)(0)}(\tau_1)A_{\alpha_2}^{(2)(0)}(\tau_2) \right\rangle_0 \left\langle \hat{T}_\tau A_{\alpha_3}^{(3)(0)}(\tau_3)A_{\alpha_4}^{(4)(0)}(\tau_4) \right\rangle_0 + \\
&+ \varepsilon \left\langle \hat{T}_\tau A_{\alpha_1}^{(1)(0)}(\tau_1)A_{\alpha_3}^{(3)(0)}(\tau_3) \right\rangle_0 \left\langle \hat{T}_\tau A_{\alpha_2}^{(2)(0)}(\tau_2)A_{\alpha_4}^{(4)(0)}(\tau_4) \right\rangle_0 + \\
&+ \left\langle \hat{T}_\tau A_{\alpha_1}^{(1)(0)}(\tau_1)A_{\alpha_4}^{(4)(0)}(\tau_4) \right\rangle_0 \left\langle \hat{T}_\tau A_{\alpha_2}^{(2)(0)}(\tau_2)A_{\alpha_3}^{(3)(0)}(\tau_3) \right\rangle_0 .
\end{aligned} \tag{7.39}$$

We have established that thermal Wick's theorem 7.3 (or, equivalently, Theorem 7.2) admits a straightforward extension to time-dependent operators. The resulting expression has the same structure as in the time-independent case, with the only difference that each thermal contraction is replaced, up to numerical factors, by a Matsubara Green's function.

7.3 Dyson Equation for Thermal Green's Function of Field Operators in Diagrammatic and Integral Form

Here, our aim is to specialize the Dyson expansion of the thermal Green's function (see Eq. 7.5) to the case of field operators (see Eqs. 2.149 and 2.150), considering a

Hamiltonian composed of a free part (see Eq. 2.181) and a two-body interaction term (see Eq. 2.314), namely the Hamiltonian given in Eq. 2.327. From Eq. 6.4, consider the Matsubara Green's function of the field operators, i.e.,

$$G^{(m)}(x,\tau,x',0) = -\frac{1}{\hbar}\left\langle \hat{T}_\tau\, \hat{\psi}(x,\tau)\, \hat{\psi}^\dagger(x',0)\right\rangle, \tag{7.40}$$

then we adapt Theorem 7.1 for the field operators, that is, we aim to calculate the object

$$\begin{aligned} &G^{(m)}(x,\tau,x',0) \\ &= \frac{-\frac{1}{\hbar}\left\langle \hat{T}_\tau \hat{\psi}^{(0)}(x,\tau)\hat{\psi}^\dagger(x',0)\right\rangle_0}{1+\sum_{n=1}^{\infty}\left(-\frac{1}{\hbar}\right)^n \frac{1}{n!}\int_0^{\beta\hbar} d\tau_1 \ldots \int_0^{\beta\hbar} d\tau_n \left\langle \hat{T}_\tau \left\{\hat{\mathcal{H}}_I^{(0)}(\tau_1)\ldots\hat{\mathcal{H}}_I^{(0)}(\tau_n)\right\}\right\rangle_0} \\ &+ \frac{\begin{aligned}&-\frac{1}{\hbar}\sum_{n=1}^{\infty}\left(-\frac{1}{\hbar}\right)^n \frac{1}{n!}\int_0^{\beta\hbar} d\tau_1 \ldots \int_0^{\beta\hbar} d\tau_n \\ &\left\langle \hat{T}_\tau \left\{\hat{\mathcal{H}}_I^{(0)}(\tau_1)\ldots\hat{\mathcal{H}}_I^{(0)}(\tau_n)\right\}\hat{\psi}^{(0)}(x,\tau)\hat{\psi}^\dagger(x',0)\right\rangle_0\end{aligned}}{1+\sum_{n=1}^{\infty}\left(-\frac{1}{\hbar}\right)^n \frac{1}{n!}\int_0^{\beta\hbar} d\tau_1 \ldots \int_0^{\beta\hbar} d\tau_n \left\langle \hat{T}_\tau \left\{\hat{\mathcal{H}}_I^{(0)}(\tau_1)\ldots\hat{\mathcal{H}}_I^{(0)}(\tau_n)\right\}\right\rangle_0}. \end{aligned} \tag{7.41}$$

Note that the general form of an interaction Hamiltonian in second quantization always involves field operators, which are expressed as combinations of creation and annihilation operators. Therefore, since every interaction Hamiltonian is written in the thermal interaction picture (see Eq. 2.194), from thermal Wick's theorem for time-dependent operators (see Theorem 7.4) we are ultimately required to compute quantities of the form

$$\left\langle \hat{T}_\tau\, a_{\alpha_1}^{(0)}(\tau_1)\, a_{\alpha_2}^{\dagger(0)}(\tau_2)\right\rangle_0 = -\hbar\, G_{\alpha_1,\alpha_2}^{(m)(0)}(\tau_1,\tau_2), \tag{7.42}$$

which can be rewritten as

$$\left\langle \hat{T}_\tau\, a_{\alpha_2}^{\dagger(0)}(\tau_2)\, a_{\alpha_1}^{(0)}(\tau_1)\right\rangle_0 = -\hbar\, \varepsilon\, G_{\alpha_1,\alpha_2}^{(m)(0)}(\tau_1,\tau_2). \tag{7.43}$$

Consider the denominator of Eq. 7.41. For $n = 0$, it gives unity. For $n = 1$, the term of the series is

$$1 - \frac{1}{\hbar}\int_0^{\beta\hbar} d\tau_1 \left\langle \hat{T}_\tau\, \hat{\mathcal{H}}_I^{(0)}(\tau_1)\right\rangle_0, \tag{7.44}$$

where for an interaction Hamiltonian given by a two-body operator, that is, Eq. 2.314, one finds

$$1-\frac{1}{\hbar}\int_0^{\beta\hbar}d\tau_1\,\frac{1}{2}\int dx_1\int dx_2\,U(x_1,x_2)$$
$$\left\langle \hat{T}_\tau\hat{\psi}^{\dagger(0)}(x_1,\tau_1)\hat{\psi}^{\dagger(0)}(x_2,\tau_1)\hat{\psi}^{(0)}(x_2,\tau_1)\hat{\psi}^{(0)}(x_1,\tau_1)\right\rangle_0 . \tag{7.45}$$

We employ Theorem 7.4: $\hat{\psi}^{\dagger(0)}(x_1,\tau_1)$ can only contract with $\hat{\psi}^{(0)}(x_2,\tau_1)$ (there is one exchange and an ε appears) and with $\hat{\psi}^{(0)}(x_1,\tau_1)$ (there are two exchanges), therefore it follows

$$1-\frac{1}{\hbar}\int_0^{\beta\hbar}d\tau_1\,\frac{1}{2}\int dx_1\int dx_2\,U(x_1,x_2)$$
$$\Big(\left\langle \hat{T}_\tau\,\hat{\psi}^{\dagger(0)}(x_1,\tau_1)\,\hat{\psi}^{(0)}(x_1,\tau_1)\right\rangle_0\left\langle \hat{T}_\tau\,\hat{\psi}^{\dagger(0)}(x_2,\tau_1)\,\hat{\psi}^{(0)}(x_2,\tau_1)\right\rangle_0$$
$$+\,\varepsilon\left\langle \hat{T}_\tau\,\hat{\psi}^{\dagger(0)}(x_1,\tau_1)\,\hat{\psi}^{(0)}(x_2,\tau_1)\right\rangle_0\left\langle \hat{T}_\tau\,\hat{\psi}^{\dagger(0)}(x_2,\tau_1)\,\hat{\psi}^{(0)}(x_1,\tau_1)\right\rangle_0\Big). \tag{7.46}$$

Feynman introduced a diagrammatic representation, complemented by a corresponding set of rules, that enables one to visualize the individual contributions to the expansion. A wavy line (see Fig. 7.1) represents an interaction between particles. The endpoints of a wavy line are referred to as vertices: each interaction line carries two such vertices, corresponding to distinct spacetime points. For a Hamiltonian that includes two-body interaction, two annihilation field operators and two creation field operators are associated with the interaction line. An operator representing field annihilation (or, more generally, an annihilation operator) is depicted as a line entering a vertex (see Fig. 7.2a), whereas an operator representing field creation (or, more generally, a creation operator) is depicted as a line leaving a vertex (see Fig. 7.2b). At first order, the denominator of Eq. 7.41 contains a single insertion of the interaction, so a diagram of the type shown in Fig. 7.3 appears only once. Recall that according to the thermal Wick's theorem (see Theorems 7.2–7.4), the only nonvanishing contractions are of the form $\langle aa^\dagger\rangle_0$, $\langle a^\dagger a\rangle_0$. Consequently, the diagram can be completed in only two distinct ways:

1. Fig. 7.4 or equivalently Fig. 7.5: $\hat{\psi}^{\dagger(0)}(x_1)$ is contracted with $\hat{\psi}^{(0)}(x_1)$ and $\hat{\psi}^{\dagger(0)}(x_2)$ is contracted with $\hat{\psi}^{(0)}(x_2)$; briefly, $1 \leftrightarrow 1$ and $2 \leftrightarrow 2$. This configuration yields two loops.
2. Fig. 7.6 or equivalently Fig. 7.7: $\hat{\psi}^{\dagger(0)}(x_1)$ is contracted with $\hat{\psi}^{(0)}(x_2)$ and $\hat{\psi}^{\dagger(0)}(x_2)$ is contracted with $\hat{\psi}^{(0)}(x_1)$; briefly, $1\leftrightarrow 2$ and $2\leftrightarrow 1$.

Fig. 7.1 A horizontal wavy line illustrates an interaction in a Feynman diagram

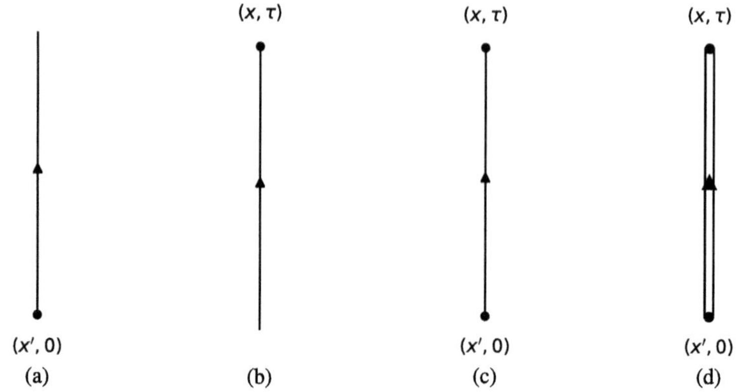

Fig. 7.2 The diagram shows, from left to right, the following elements in Feynman diagrammatics: (**a**) Annihilation operator: vertical line starting from a point at the bottom $(x', 0)$, representing the removal of a particle. (**b**) Creation operator: vertical line ending at a point at the top (x, τ), representing the addition of a particle. (**c**) Free (or bare) Green's function: vertical line connecting $(x', 0)$ and (x, τ), showing the propagation of a non-interacting particle. (**d**) Full (interacting) Green's function: two closely spaced vertical lines connecting $(x', 0)$ and (x, τ), illustrating propagation dressed by interactions. All lines have a black arrow in the middle indicating the direction of particle propagation

Due to thermal Wick's theorem, the contraction of operators is obtained by evaluating their thermal average with respect to the non-interacting Hamiltonian $\hat{\mathcal{H}}_0$, so that, in particular, we aim to compute the following free propagator

$$G^{(m)(0)}(x, \tau, x', 0) = -\frac{1}{\hbar}\left\langle \hat{T}_\tau \hat{\psi}^{(0)}(x, \tau)\hat{\psi}^{\dagger(0)}(x', 0)\right\rangle_0 . \tag{7.47}$$

Note that in Eq. 7.47 we have directly expressed both operators in the thermal interaction picture (see Eq. 2.194), unlike the Green's function defined in Eq. 6.14, where the thermal Heisenberg picture (see Eq. 6.12) is employed. This is because the latter Green's function refers to a non-interacting Hamiltonian.

We seek to express Eq. 7.47 in the language of Feynman diagrammatics. To this end, we consider two spacetime points—$(x', 0)$, where the particle is created, and (x, τ), where it is annihilated—as depicted in Fig. 7.2c. The corresponding line is interpreted as a creation—annihilation segment and represents the propagation from $(x', 0)$ to (x, τ). A solid line denotes a free propagator, i.e., a particle whose dynamics is governed by the non-interacting Hamiltonian, and it is conventionally read from top to bottom.

Regarding the first-order denominator of Eq. 7.41, the diagram can be closed in two distinct ways. In the first case, both solid lines are closed onto themselves, giving rise to two loops—that is, two factors of $G^{(m)(0)}$—and no particle exchange occurs. In the second case, the lines cross, corresponding to an exchange of particles. The first order of the denominator is shown in Fig. 7.8.

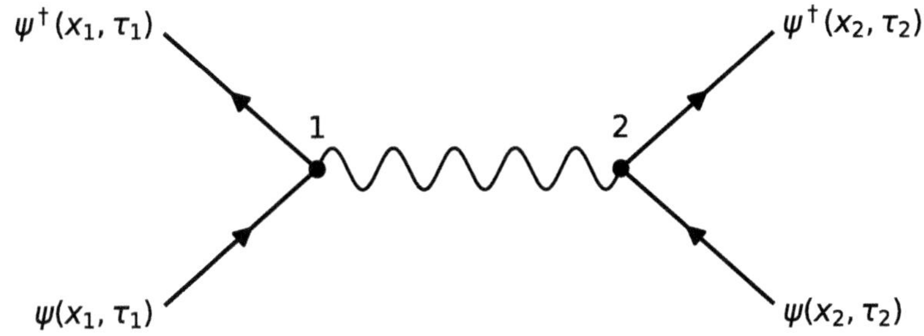

Fig. 7.3 Schematic representation of an interaction connecting two spacetime points, $1 \equiv (x_1, \tau_1)$ and $2 \equiv (x_2, \tau_2)$. At each vertex, one field is annihilated and another is created. The two vertices are connected by a horizontal wavy line, illustrating the propagation of the interacting field in standard Feynman diagrammatics

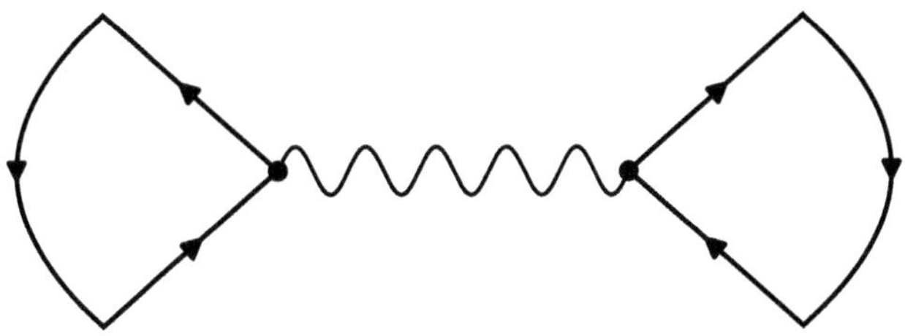

Fig. 7.4 Schematic representation of the first possible type of contractions contributing to the first-order term in the denominator of the many-body Green's function of field operators, obtained from the Dyson expansion (see Eq. 7.41). Two spacetime vertices are connected by a horizontal wavy line representing the interaction. Each vertex also features a curved contraction line, indicated by a semi-arc with an arrow pointing from the creation line (upper) to the annihilation line (lower)

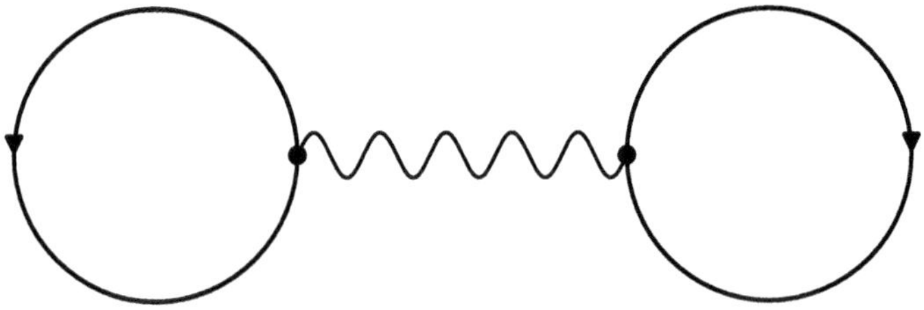

Fig. 7.5 Feynman diagram illustrating the contractions shown in Fig. 7.4

Now, consider the second order of the denominator of Eq. 7.41. Again, there are two possible ways of closing the diagram:

1. Fig. 7.9. One finds disconnected diagrams, meaning that the two interaction vertices are not linked and therefore remain independent; the corresponding contractions are analogous to those shown in Fig. 7.4.

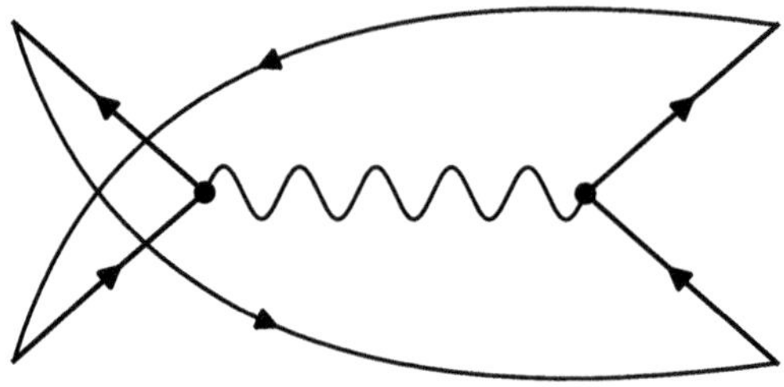

Fig. 7.6 Schematic representation of the second possible type of contractions contributing to the first-order term in the denominator of the many-body Green's function of field operators, obtained from the Dyson expansion (see Eq. 7.41). Two spacetime vertices are connected by a horizontal wavy line representing the interaction. In these contractions, the creation operator (upper line) at one vertex is contracted with the annihilation operator (lower line) at the other vertex

Fig. 7.7 Feynman diagram illustrating the contractions shown in Fig. 7.6

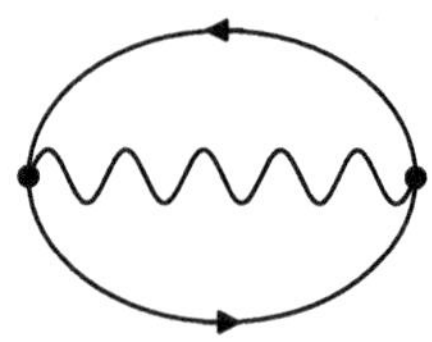

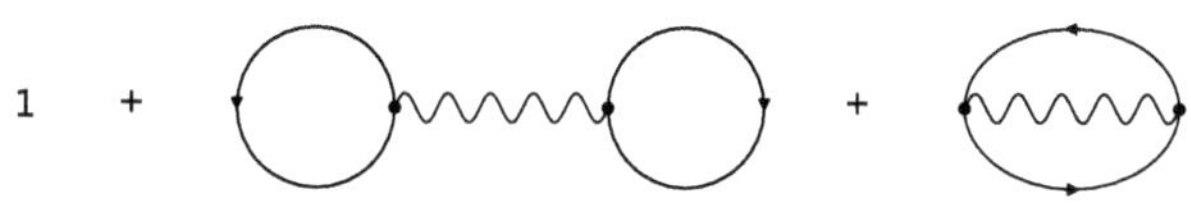

Fig. 7.8 Feynman diagrammatic representation of the first-order contribution to the denominator of the many-body Green's function of field operators, as obtained from the Dyson expansion (see Eq. 7.41). This contribution corresponds to summing the zeroth-order term (the identity, 1) with the two possible types of first-order contractions shown in Figs. 7.5 and 7.7

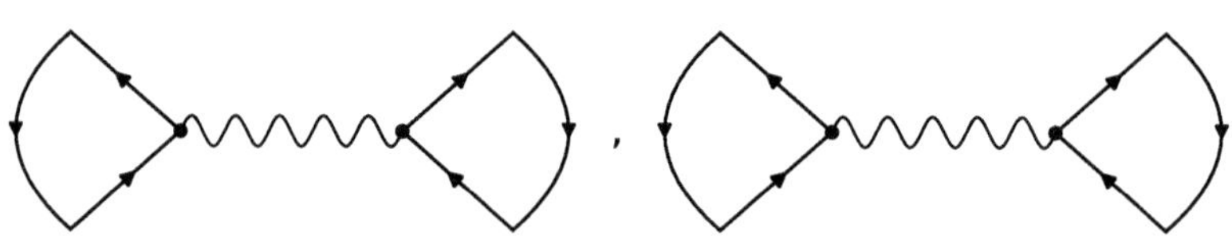

Fig. 7.9 Schematic representation of the first possible type of contractions contributing to the second-order term in the denominator of the many-body Green's function of field operators, as obtained from the Dyson expansion (see Eq. 7.41)

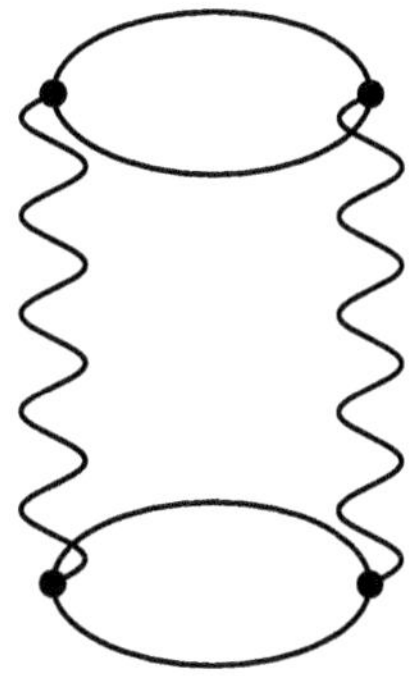

Fig. 7.10 Feynman diagrammatic representation of the second type of contractions contributing to the second-order term in the denominator of the many-body Green's function of field operators, as obtained from the Dyson expansion (see Eq. 7.41)

2. Fig. 7.10. One finds a connected diagram: in this case the lines may cross, and each vertex of one interaction is linked to a vertex of the other. Each ellipse corresponds to the contraction of two creation operators with the associated two annihilation operators, yielding four field operators per ellipse. In total, four vertices are present, arranged in pairs and connected by two vertical wavy lines that represent the interaction between the vertices.

The first-order contribution in Eq. 7.41 generates only connected diagrams, whereas the second-order term gives rise to both connected and disconnected ones.

Now, consider the numerator of Eq. 7.41. For $n = 0$ no integrations appear, and we simply obtain

$$-\frac{1}{\hbar}\left\langle \hat{T}_\tau\, \hat{\psi}^{(0)}(x,0)\hat{\psi}^{\dagger(0)}(x',\tau)\right\rangle_0 = G^{(m)(0)}(x,\tau,x',0), \tag{7.48}$$

which is the free thermal Green's function. In contrast, for $n = 0$ the denominator reduces to unity. This explicitly shows that Feynman's method is inherently perturbative, as the full expression is generated through a systematic expansion in powers of the interaction, starting from the free propagator. Note that the integrand of the n-th term in the Dyson series of the Matsubara Green's function for two generic operators A and B (see Theorem 7.1), that is

$$\left\langle \hat{T}_\tau \left\{\hat{\mathcal{H}}_I^{(0)}(\tau_1)\ldots\hat{\mathcal{H}}_I^{(0)}(\tau_n)\right\} A^{(0)}(\tau)B(0)\right\rangle_0 \tag{7.49}$$

can be written as

$$\left\langle \hat{T}_\tau \left\{A^{(0)}(\tau)\hat{\mathcal{H}}_I^{(0)}(\tau_1)\ldots\hat{\mathcal{H}}_I^{(0)}(\tau_n)B(0)\right\}\right\rangle_0. \tag{7.50}$$

Indeed, by inserting $A^{(0)}(\tau)$ at the beginning and $B(0)$ at the end of the product, the action of the Wick time-ordering operator $\hat{T}_\tau$ (see Eq. 2.200) is preserved. Since $B(0)$ is evaluated at $\tau = 0$, it naturally remains at the rightmost position in the time-ordered product. As for A, each exchange with $\hat{\mathcal{H}}_I$ involves an even number

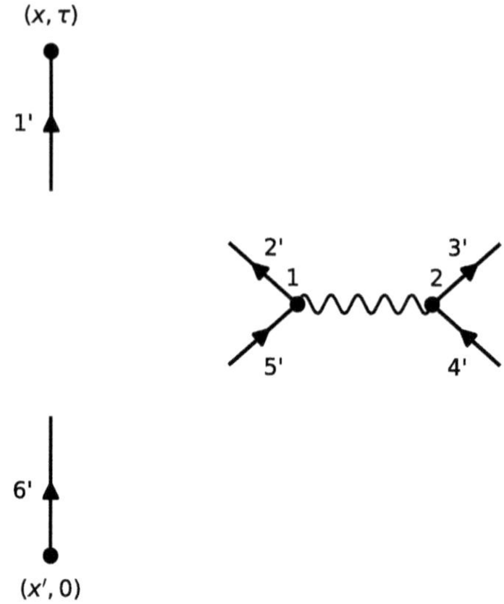

Fig. 7.11 Feynman diagrammatic representation of the first-order contribution to the numerator in the expansion of the many-body Green's function of field operators, as obtained from the Dyson expansion (see Eq. 7.41). This diagram is analogous to the first-order term in the denominator (as shown in Fig. 7.3), with the inclusion of an annihilation operator at (x, τ) and a creation operator at $(x', 0)$

of permutations, so its placement does not affect the ordering. Then, for $A^{(0)}(\tau) = \hat{\psi}^{(0)}(x, \tau)$ and $B(0) = \hat{\psi}^{\dagger}(x', 0)$, the first-order contribution to the numerator reads

$$G^{(m)(0)}(x, \tau, x', 0) - \frac{1}{\hbar}\left(\frac{1}{\hbar}\right)\int_0^{\beta\hbar} d\tau_1 \frac{1}{2}\int dx_1 \int dx_2 U(x_1, x_2)$$
$$\left\langle \hat{T}_\tau \hat{\psi}^{(0)}(x, \tau)\hat{\psi}^{\dagger(0)}(x_1, \tau_1)\hat{\psi}^{\dagger(0)}(x_2, \tau_1)\hat{\psi}^{(0)}(x_2, \tau_1)\hat{\psi}^{(0)}(x_1, \tau_1)\hat{\psi}^{\dagger}(x', 0)\right\rangle_0, \tag{7.51}$$

and we now apply the rules of diagrammatic perturbation theory. We use the identifications $1' \equiv \hat{\psi}^{(0)}(x, \tau)$, $2' \equiv \hat{\psi}^{\dagger(0)}(x_1, \tau_1)$, $3' \equiv \hat{\psi}^{\dagger(0)}(x_2, \tau_1)$, $4' \equiv \hat{\psi}^{(0)}(x_2, \tau_1)$, $5' \equiv \hat{\psi}^{(0)}(x_1, \tau_1)$, $6' \equiv \hat{\psi}^{\dagger}(x', 0)$. Here, operators $2'$, $3'$, $4'$, and $5'$ originate from the interaction, whereas $1'$ corresponds to the external annihilation at (x, τ) and $6'$ to the external creation at $(x', 0)$ (see Fig. 7.11). We now examine the possible contractions.

1. $1'$ is contracted with $2'$, $3'$ with $4'$, $5'$ with $6'$ as in Fig. 7.12, and we get Fig. 7.13, which represents a direct term, since there are no particle exchanges.
2. $1'$ is contracted with $3'$, $4'$ with $6'$, $2'$ with $5'$. The diagram we obtain is equivalent to that in case 1, two vertices of the interaction being equivalent to each other: since by the principle of indistinguishability $U(x_1, x_2) = U(x_2, x_1)$ holds, the direct term in Fig. 7.13 appears twice.
3. $1'$ is contracted with $6'$, the resulting contribution yields the free thermal Green's function $G^{(m)(0)}(x, \tau, x', 0)$. For the interaction part, the two contraction patterns previously discussed for the first-order denominator apply unchanged. Consequently, one obtains the two disconnected diagrams shown in Fig. 7.14.

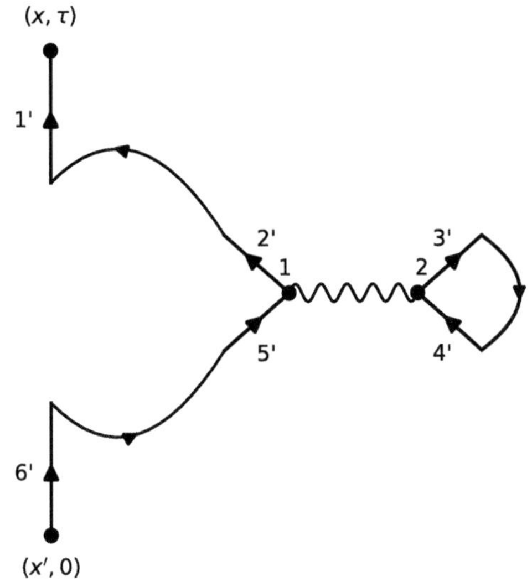

Fig. 7.12 Schematic representation of the first possible type of contractions (the second type is an equivalent diagram with different contractions) contributing to the first-order numerator of the many-body Green's function of field operators in the Dyson expansion (see Eq. 7.41). This contribution appears twice, originating from the two distinct sets of contractions $1' - 2', 3' - 4', 5' - 6'$ and $1' - 3', 2' - 5', 4' - 6'$, as shown in Fig. 7.11

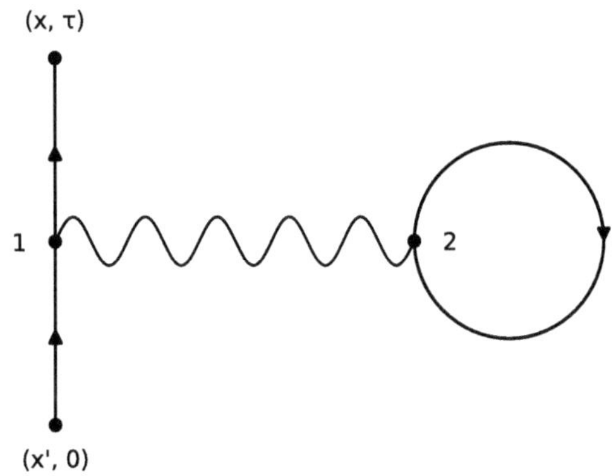

Fig. 7.13 Feynman diagram illustrating the contractions shown in Fig. 7.12

4. $1'$ is contracted with $2'$, $3'$ with $5'$ and $4'$ with $6'$. Figure 7.15 represents an exchange term, as operators 2 and 1 have been exchanged with each other. The particle is created in $(x', 0)$, propagates to 2, interacts, then propagates from 1 to (x, τ) (see Fig. 7.15).
5. $1'$ is contracted with $3'$, $2'$ with $4'$ and $5'$ with $6'$: 1 exchanges with 2 and by the principle of indistinguishability we get again the diagram in Fig. 7.15.

Finally, for the thermal Green's function of the field operators in Eq. 7.41, the structure of the denominator and numerator is summarized in Figs. 7.16 and 7.17, respectively. Inspection of these diagrams shows that certain numerator contributions cancel corresponding diagrams in the denominator. Moreover, only connected diagrams survive in the numerator. One can therefore demonstrate that the denominator cancels out entirely, provided that only connected diagrams are retained in the numerator.

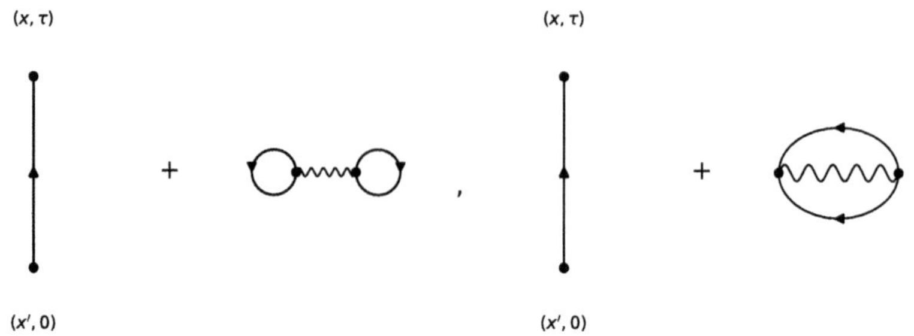

Fig. 7.14 Feynman diagrammatic representation of the third type of contractions contributing to the first-order numerator in the expansion of the many-body Green's function of field operators, as obtained from the Dyson expansion (see Eq. 7.41). Each diagram contains a free propagator (see Fig. 7.2c) combined in one case with a diagram like Fig. 7.5 and in another case with a diagram like Fig. 7.7

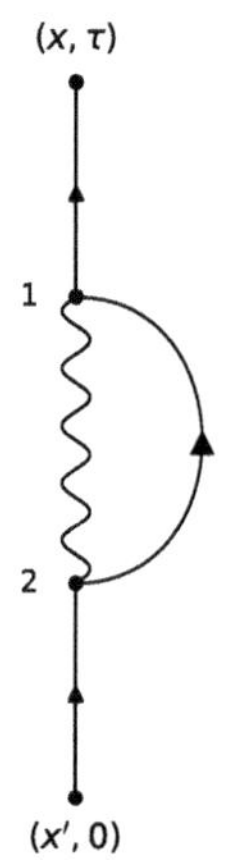

Fig. 7.15 Feynman diagrammatic representation of the fourth type of contractions (the fifth type is an equivalent diagram with different contractions) contributing to the first-order numerator in the expansion of the many-body Green's function of field operators, as obtained from the Dyson expansion (see Eq. 7.41). This contribution appears twice, originating from the two distinct sets of contractions $1' - 2'$, $3' - 5'$, $4' - 6'$ and $1' - 3'$, $2' - 4'$, $5' - 6'$, as shown in Fig. 7.11

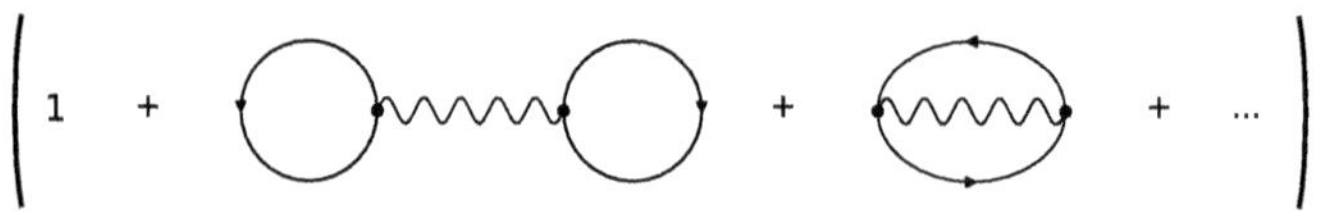

Fig. 7.16 Feynman diagrammatic representation of the denominator appearing in the Dyson expansion of the many-body Green's function of field operators (see Eq. 7.41)

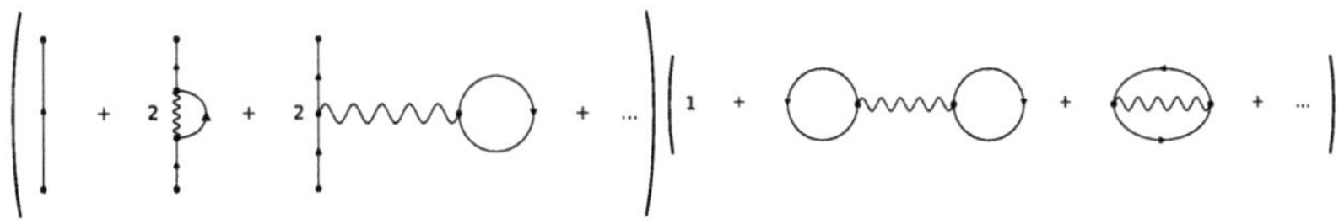

Fig. 7.17 Feynman diagrammatic representation of the numerator appearing in the Dyson expansion of the many-body Green's function of field operators (see Eq. 7.41)

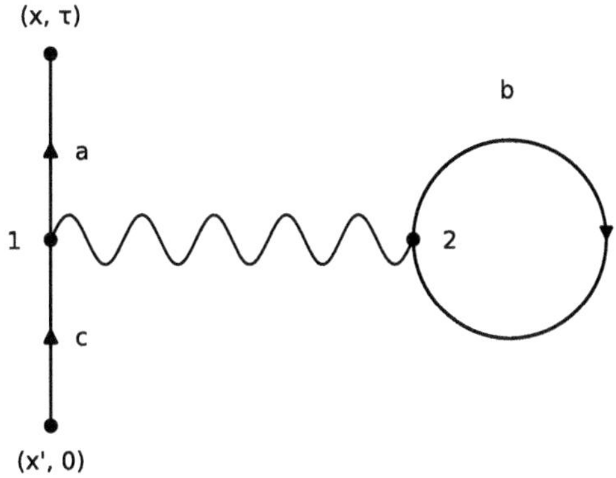

Fig. 7.18 Diagram identical to the one presented in Fig. 7.13, but shown here with explicit labels assigned to each element of the diagram. It represents the first type of contractions contributing to the first-order term in the perturbative expansion of the many-body Green's function of field operators (see Eq. 7.52), and it occurs twice in the expansion

Finally, the Feynman perturbative expansion of a thermal Green's function is given by

$$G^{(m)}_{A,B}(\tau,0) = G^{(m)(0)}_{A,B}(\tau,0) - \frac{1}{\hbar}\sum_{n=1}^{\infty}\left(-\frac{1}{\hbar}\right)^n \frac{1}{n!}\int_0^{\beta\hbar} d\tau_1 \cdots \int_0^{\beta\hbar} d\tau_n$$
$$\left\langle \hat{T}_\tau\, A^{(0)}(\tau)\,\hat{\mathcal{H}}^{(0)}_I(\tau_1)\cdots\hat{\mathcal{H}}^{(0)}_I(\tau_n) B(0)\right\rangle_{0,\mathrm{conn}}. \tag{7.52}$$

Thus, the Matsubara Green's function is expressed as a power series in the interaction contribution of the Hamiltonian. Since the expansion involves only the numerator—thanks to the cancellation of disconnected diagrams— the thermal Green's function is therefore a genuine perturbative series, rather than a ratio of two independent expansions.

In the following, we shall omit the subscript *conn*, which has been written here only to emphasize that the average includes connected contributions only.

7.3.1 First Order Analytic in Feynman's Perturbative Theory

Here, we turn to the analysis of the first-order contribution to Eq. 7.52.

1. Fig. 7.18. This diagram corresponds to the contractions $a \equiv 1' - 2'$, $b \equiv 3' - 4'$, and $c \equiv 5' - 6'$. a corresponds to $-\hbar G^{(m)(0)}(x,\tau,x_1,\tau_1)$ since, when reading the

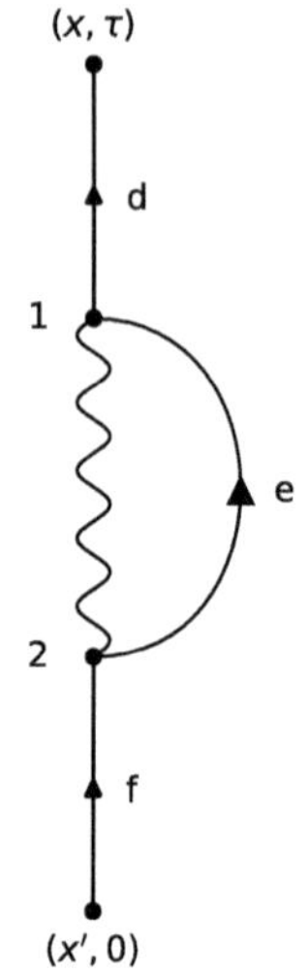

Fig. 7.19 Diagram identical to the one presented in Fig. 7.15, but shown here with explicit labels assigned to each element of the diagram. It represents the second type of contractions contributing to the first-order term in the perturbative expansion of the many-body Green's function of field operators (see Eq. 7.52), and it occurs twice in the expansion

operators from top to bottom, we encounter $\hat{\psi}^{(0)}(x,\tau)$ followed by $\hat{\psi}^{\dagger(0)}(x_1,\tau_1)$. Similarly, the contraction b corresponds to $-\hbar\varepsilon G^{(m)(0)}(x_2,\tau_1,x_2,\tau_1+\delta)$, with the limit $\delta\to 0^+$. The Green's function is evaluated at negative time differences, as the second time argument occurs after the first. Since the operators are taken at the same time, we consider the limit $\delta\to 0^+$. Finally, the contraction c corresponds to $-\hbar G^{(m)(0)}(x_1,\tau_1,x',0)$.

2. Fig. 7.19. This diagram involves the contractions $d\equiv 1'-2'$, $e\equiv 3'-5'$, and $f\equiv 4'-6'$. The contraction d corresponds to $-\hbar G^{(m)(0)}(x,\tau,x_1,\tau_1)$, while e corresponds to $-\hbar\varepsilon G^{(m)(0)}(x_1,\tau_1,x_2,\tau_1+\delta)$. Finally, the contraction f corresponds to $-\hbar G^{(m)(0)}(x_2,\tau_1,x',0)$.

Each of these diagrams appears twice, because of the principle of indistinguishability. The analytic first order in Eq. 7.52 evaluated at $\tau>0$ is

$$
\begin{aligned}
&G^{(m)(1)}(x,\tau,x',0)\\
&=G^{(m)(0)}(x,\tau,x',0)+\left(-\frac{1}{\hbar}\right)\left(-\frac{1}{\hbar}\right)\int_0^{\beta\hbar}d\tau_1\,\frac{1}{2}\int dx_1\int dx_2\,U(x_1,x_2)\\
&\quad\left\langle\hat{T}_\tau\hat{\psi}^{(0)}(x,\tau)\hat{\psi}^{\dagger(0)}(x_1,\tau_1)\hat{\psi}^{\dagger(0)}(x_2,\tau_1)\hat{\psi}^{(0)}(x_2,\tau_1)\hat{\psi}^{(0)}(x_1,\tau_1)\hat{\psi}^{\dagger}(x',0)\right\rangle_0=\\
&=G^{(m)(0)}(x,\tau,x',0)+\left(-\frac{1}{\hbar}\right)^2\int_0^{\beta\hbar}d\tau_1\,\frac{1}{2}\int dx_1\int dx_2\,U(x_1,x_2)\\
&\quad\Big\{2(-\hbar)G^{(m)(0)}(x,\tau,x_1,\tau_1)(-\hbar)\varepsilon G^{(m)(0)}(x_2,\tau_1,x_2,\tau_1+\delta)\\
&\quad\;(-\hbar)G^{(m)(0)}(x_1,\tau_1,x',0)
\end{aligned}
$$

$$+ 2\varepsilon(-\hbar)G^{(m)(0)}(x,\tau,x_1,\tau_1)(-\hbar)\varepsilon G^{(m)(0)}(x_1,\tau_1,x_2,\tau_1+\delta)$$
$$(-\hbar)G^{(m)(0)}(x_2,\tau_1,x',0)\Big\} =$$
$$= G^{(m)(0)}(x,\tau,x',0) + \left(-\frac{1}{\hbar}\right)^2 \int_0^{\beta\hbar} d\tau_1 \int dx_1 \int dx_2 U(x_1,x_2)$$
$$\Big\{(-\hbar)G^{(m)(0)}(x,\tau,x_1,\tau_1)(-\hbar)\varepsilon G^{(m)(0)}(x_2,\tau_1,x_2,\tau_1+\delta)$$
$$(-\hbar)G^{(m)(0)}(x_1,\tau_1,x',0)$$
$$+ \varepsilon(-\hbar)G^{(m)(0)}(x,\tau,x_1,\tau_1)\cdot(-\hbar)\varepsilon G^{(m)(0)}(x_1,\tau_1,x_2,\tau_1+\delta)$$
$$(-\hbar)G^{(m)(0)}(x_2,\tau_1,x',0)\Big\}. \tag{7.53}$$

Note that the spatial integrations are performed over the interaction vertices. Each vertex is assigned a spacetime coordinate, over which we integrate; thus, instead of the labels 1 and 2, we refer to the vertices as (x_1, τ_1) and (x_2, τ_2). The expression initially contains two spatial integrals and a single time integral. A second time integral may be introduced by multiplying by

$$\int_0^{\beta\hbar} d\tau_2 \delta(\tau_1 - \tau_2), \tag{7.54}$$

then the terms b, e, and f can be rewritten as

$$b \longrightarrow -\hbar\varepsilon G^{(m)(0)}(x_2,\tau_1,x_2,\tau_1+\delta) \longrightarrow -\hbar\varepsilon G^{(m)(0)}(x_2,\tau_2,x_2,\tau_2+\delta), \tag{7.55}$$
$$e \longrightarrow -\hbar\varepsilon G^{(m)(0)}(x_1,\tau_1,x_2,\tau_1+\delta) \longrightarrow -\hbar\varepsilon G^{(m)(0)}(x_1,\tau_1,x_2,\tau_2+\delta), \tag{7.56}$$
$$f \longrightarrow -\hbar G^{(m)(0)}(x_2,\tau_1,x',0) \longrightarrow -\hbar G^{(m)(0)}(x_2,\tau_2,x',0), \tag{7.57}$$

and finally

$$G^{(m)(1)}(x,\tau,x',0) =$$
$$= G^{(m)(0)}(x,\tau,x',0) + \left(-\frac{1}{\hbar}\right)^2 \int_0^{\beta\hbar} d\tau_1 \int_0^{\beta\hbar} d\tau_2$$
$$\int dx_1 \int dx_2\, U(x_1,x_2)\,\delta(\tau_1-\tau_2)$$
$$\Big\{(-\hbar)G^{(m)(0)}(x,\tau,x_1,\tau_1)(-\hbar)\varepsilon G^{(m)(0)}(x_2,\tau_2,x_2,\tau_2+\delta)$$
$$(-\hbar)G^{(m)(0)}(x_1,\tau_1,x',0)$$

$$+\,\varepsilon(-\hbar)G^{(m)(0)}(x,\tau,x_1,\tau_1)(-\hbar)\varepsilon G^{(m)(0)}(x_1,\tau_1,x_2,\tau_2+\delta)$$

$$(-\hbar)G^{(m)(0)}(x_2,\tau_2,x',0)\Big\} =$$

$$= G^{(m)(0)}(x,\tau,x',0) - \hbar\int_0^{\beta\hbar} d\tau_1 \int_0^{\beta\hbar} d\tau_2 \int dx_1 \int dx_2\, U(x_1,x_2)\,\delta(\tau_1-\tau_2)$$

$$\Big\{\varepsilon\, G^{(m)(0)}(x,\tau,x_1,\tau_1)\, G^{(m)(0)}(x_2,\tau_2,x_2,\tau_2+\delta)\, G^{(m)(0)}(x_1,\tau_1,x',0)$$

$$+\, G^{(m)(0)}(x,\tau,x_1,\tau_1)\, G^{(m)(0)}(x_1,\tau_1,x_2,\tau_2+\delta)\, G^{(m)(0)}(x_2,\tau_2,x',0)\Big\}. \tag{7.58}$$

We now turn to Eq. 7.58 in order to analyze the structure of the term at order n. At order n, the diagram contains $2n+1$ free propagators $G^{(m)(0)}$. This follows from the fact that each interaction line contributes four half-particle lines, which pair up to form two free propagators. The overall numerical prefactor can be identified as follows:

- The factor $(-\hbar)^{-1}$ that multiplies the entire series is included as a standard normalization convention.
- Each interaction vertex contributes a factor $(-\hbar)^{-1}$, so at order n we get $(-\hbar))^{-n}$.
- Every time a free propagator $G^{(m)(0)}$ appears, it contributes a factor $-\hbar$. Since there are $2n+1$ propagators, this gives $(-\hbar)^{2n+1}$.
- Finally, we include a factor ε (see Eq. 2.52) for each closed loop.

Putting everything together, the total numerical factor at order n becomes

$$\left(-\frac{1}{\hbar}\right)\left(-\frac{1}{\hbar}\right)^n (-\hbar)^{2n+1}\varepsilon^{\text{number of closed loops}} = (-\hbar)^n\, \varepsilon^{\text{number of closed loops}}. \tag{7.59}$$

7.4 Self-energy

The many-body Green's function $G^{(m)}(x,\tau,x',0)$ is depicted in Fig. 7.2d, while its perturbative expansion is shown in Fig. 7.20. As higher-order terms are included, the resulting diagrams exhibit a common structure: an *outgoing free propagator*, an *intermediate interaction structure*, and an *incoming free propagator*. This organization permits Feynman's perturbative expansion to be cast in the form shown in Fig. 7.21, where Σ denotes the proper self-energy. In this framework, the proper self-energy is defined in an operational sense: given any two-point diagram contributing to the Green's function, one removes, or amputates, the incoming and

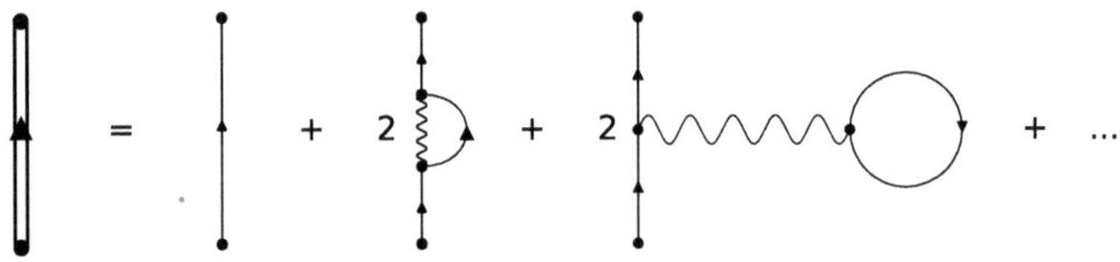

Fig. 7.20 Feynman diagrammatic representation of the perturbative expansion of the many-body Green's function of field operators (see Eq. 7.52). The diagram depicts the full (interacting) propagator on the left and the free (bare) propagator on the right, together with all connected diagrams at higher orders in the perturbation series, with ellipses indicating that the expansion continues

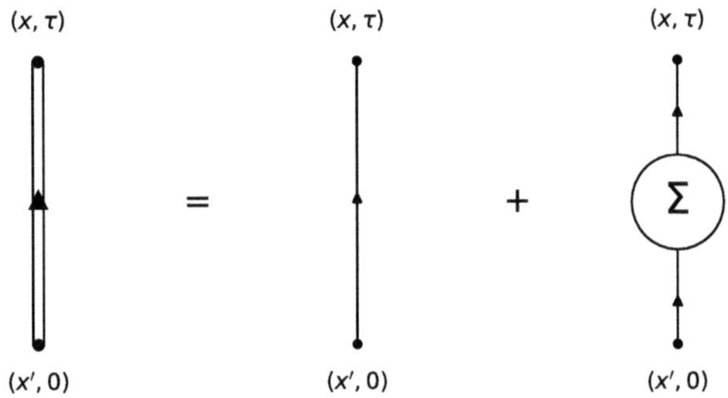

Fig. 7.21 Feynman diagrammatic representation of the many-body Green's function expressed in terms of the self-energy. The full (interacting) propagator is depicted as the free (bare) propagator with an insertion of the proper self-energy, Σ, between two free lines, illustrating how the interacting propagator is constructed from the bare propagator through self-energy corrections

outgoing free propagator lines, and the remaining structure is identified as the proper self-energy. Thus, the proper self-energy consists precisely of whatever lies between the incoming and outgoing propagators, independent of whether that internal structure is connected or appears disconnected prior to amputation. The proper self-energy is a function that depends on the interaction vertices x_1, x_2 and the times τ_1, τ_2. At first order, the self-energy can be obtained by cutting the incoming and outgoing propagator lines of a diagram; its diagrammatic representation is given in Fig. 7.22. Importantly, Fig. 7.21 corresponds to an equation that can be recast in integral form, namely

$$G^{(m)}(x,\tau,x',0) = G^{(m)(0)}(x,\tau,x',0) + \int_0^{\beta\hbar} d\tau_1 \int_0^{\beta\hbar} d\tau_2 \int dx_1 \int dx_2 \\ G^{(m)(0)}(x,\tau,x_1,\tau_1)\Sigma(x_1,\tau_1,x_2,\tau_2)G^{(m)(0)}(x_2,\tau_2,x',0), \tag{7.60}$$

which expresses the interacting Green's function $G^{(m)}(x,\tau,x',0)$ exactly in terms of the free propagators $G^{(m)(0)}$ and the proper self-energy $\Sigma(x_1,\tau_1,x_2,\tau_2)$. In

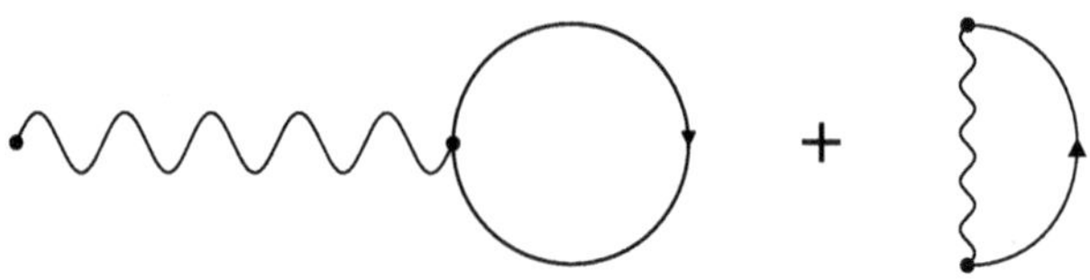

Fig. 7.22 Feynman diagrammatic representation of the first-order self-energy. This diagram corresponds to the analytical expression given in Eq. 7.61 and represents the leading-order contribution to the proper self-energy in the perturbative expansion of the many-body Green's function (see Eq. 7.52)

practice, the evaluation of these integrals is generally non-trivial, and one typically proceeds either numerically or by selecting an appropriate subset of higher-order diagrams. By comparing the integral equation above with the first-order perturbative expansion 7.58, one obtains the analytical expression for the first-order self-energy shown in Fig. 7.22. Explicitly,

$$\begin{aligned}\Sigma^{(1)}(x_1,\tau_1,x_2,\tau_2) = -\varepsilon\hbar\delta(x_1-x_2)\delta(\tau_1-\tau_2)\int dx_2' \\ \int_0^{\beta\hbar} d\tau_2' U(x_1,x_2')\delta(\tau_1-\tau_2')G^{(m)(0)}(x_2',\tau_2',x_2',\tau_2'+\delta)+ \\ -\hbar U(x_1,x_2)\delta(\tau_1-\tau_2)G^{(m)(0)}(x_1,\tau_1,x_2,\tau_2+\delta).\end{aligned} \tag{7.61}$$

We define a self-energy insertion Σ_i (or self-energy contribution) as the part of a Feynman diagram that results from removing one incoming and one outgoing propagator line. The proper self-energy Σ is then defined as the sum over all such self-energy insertions. A connected two-point diagram is said to be reducible if, after amputating the external propagator lines, it can be separated into two parts by cutting a single internal propagator line. Otherwise, it is irreducible (or proper). Note that all first-order diagrams are irreducible. As illustrative second-order examples, in Fig. 7.23 we show a reducible connected diagram, while in Fig. 7.24 we present an irreducible example. In general, at any order higher than one, both reducible and irreducible connected diagrams appear. Irreducible self-energy insertions are also referred to as proper, while reducible ones are called improper. Let Σ^* denote the sum of all improper self-energy insertions, as previously introduced in the context of the Anderson model (see Eq. 6.115). The full self-energy insertion can be decomposed into the proper part Σ and all improper (reducible) contributions (see Fig. 7.25). We can now express the diagram in Fig. 7.21 in terms of proper self-energy insertions. In Fig. 7.26, we isolate a single outgoing free propagator $G^{(m)(0)}$ and one proper self-energy insertion at each order, and sum over the remaining (excluded) diagrams. The resulting expression corresponds to the full propagator $G^{(m)}(x_2,\tau_2,x',0)$. This yields the diagrammatic form of the Dyson equation, shown in Fig. 7.27. In summary, the many-body Green's function can be

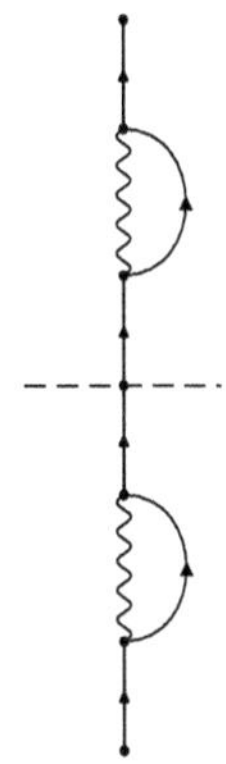

Fig. 7.23 Example of a second-order connected, reducible Feynman diagram. This diagram illustrates one possible configuration contributing to the second-order term in the perturbative expansion of the many-body Green's function (see Eq. 7.52)

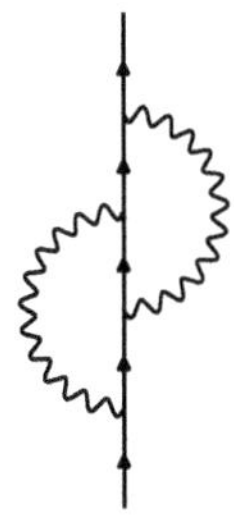

Fig. 7.24 Example of a second-order connected, irreducible Feynman diagram. This diagram illustrates one possible configuration contributing to the second-order term in the perturbative expansion of the many-body Green's function (see Eq. 7.52)

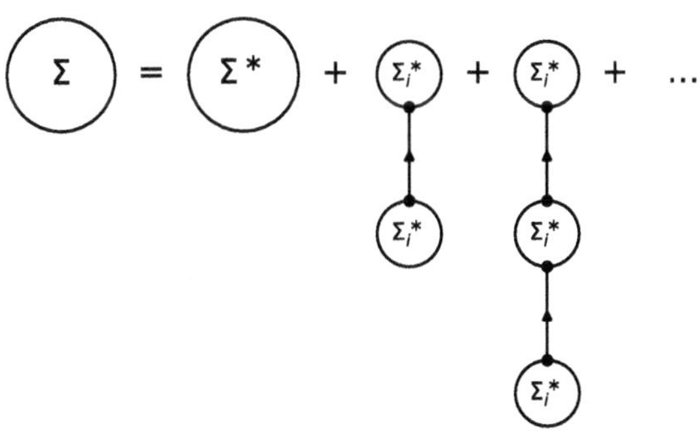

Fig. 7.25 Feynman diagrammatic representation of the decomposition of the proper self-energy, Σ, into the improper self-energy, Σ^*, and the reducible contributions generated by insertions of Σ_i^*. This illustrates how the full self-energy can be systematically expressed in terms of its improper components and their connected contributions

written either in terms of the proper self-energy Σ and two free propagators $G^{(m)(0)}$, or alternatively in terms of the improper self-energy Σ^* and the full (interacting) Green's function $G^{(m)}$.

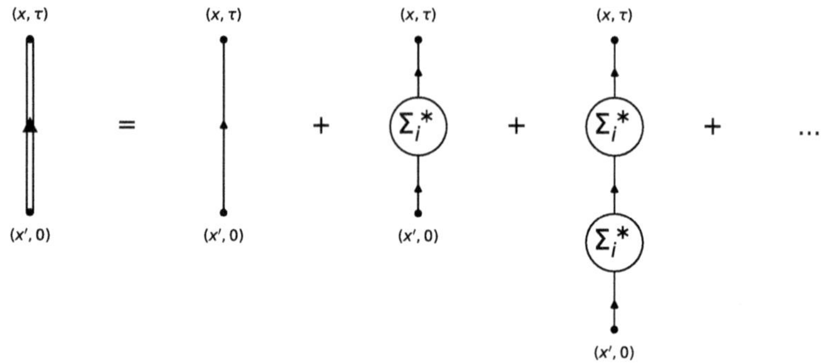

Fig. 7.26 Feynman diagrammatic representation of the perturbative series for the many-body Green's function, expressed as a sum of contributions containing an increasing number of improper self-energy insertions. The diagram illustrates how the full propagator is generated by repeatedly inserting Σ_i^* into the free propagator, giving rise to an infinite series of reducible terms

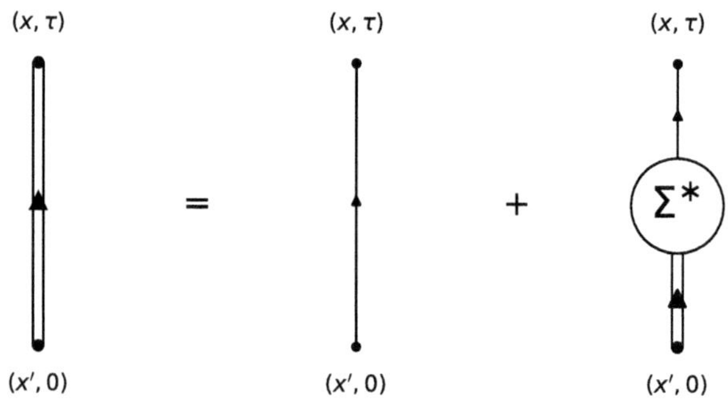

Fig. 7.27 Feynman diagrammatic representation of the Dyson equation for the many-body Green's function, illustrating how the full (interacting) propagator is expressed in terms of the bare propagator and the improper self-energy (see Eq. 7.62)

Since exact evaluations are not feasible, one typically resorts to perturbation theory. It should be emphasized that the two forms of the Dyson equation generate distinct perturbative expansions. In particular, the formulation involving Σ^* is often preferable. For instance, if Σ is computed to first order, as in Fig. 7.21, the resulting expansion of the Green's function consists of $G^{(m)(0)}$ supplemented only by first-order diagrams. In contrast, evaluating Σ^* to first order in the Dyson equation shown in Fig. 7.27 yields a Green's function that already embodies an infinite resummation of reducible diagrams. We may now express the Dyson equation in its integral form, namely,

$$G^{(m)}(x, \tau, x', 0) = G^{(m)(0)}(x, \tau, x', 0) + \int_0^{\beta\hbar} d\tau_1 \int_0^{\beta\hbar} d\tau_2 \int dx_1 \int dx_2$$

$$G^{(m)(0)}(x,\tau,x_1,\tau_1)\Sigma^*(x_1,\tau_1,x_2,\tau_2)G^{(m)}(x_2,\tau_2,x',0). \tag{7.62}$$

7.5 Dyson Algebraic Equation

The Dyson equation for field operators in its integral form (see Eq. 7.62) undergoes a substantial simplification for systems in thermodynamic equilibrium. In such cases, namely, when the system is invariant under time translations (see Theorem 6.1), the equation reduces to a more tractable form, i.e.,

$$G^{(m)}(x,x',\tau) = G^{(m)(0)}(x,x',\tau) + \int_0^{\beta\hbar} d\tau_1 \int_0^{\beta\hbar} d\tau_2 \int dx_1 \int dx_2$$
$$G^{(m)(0)}(x,x_1,\tau-\tau_1)\Sigma^*(x_1,x_2,\tau_1-\tau_2)G^{(m)}(x_2,x',\tau_2). \tag{7.63}$$

Recalling Eqs. 6.25 and 6.26, we multiply both members of Eq. 7.63 by $e^{i\omega_n\tau}$ and integrate with respect to time in the range $[0,\beta\hbar)$, i.e.,

$$G^{(m)}(x,x',i\omega_n) = G^{(m)(0)}(x,x',i\omega_n) + \int_0^{\beta\hbar} d\tau \int_0^{\beta\hbar} d\tau_1 \int_0^{\beta\hbar} d\tau_2$$
$$\int dx_1 \int dx_2\, e^{i\omega_n\tau}\left(\frac{1}{\beta\hbar}\right)^3$$
$$\sum_{n_1}\sum_{n_2}\sum_{n_3} G^{(m)(0)}(x,x_1,i\omega_{n_1})\Sigma^*(x_1,x_2,i\omega_{n_2})$$
$$G^{(m)}(x_2,x',i\omega_{n_3})e^{-i\omega_{n_1}(\tau-\tau_1)}e^{-i\omega_{n_2}(\tau_1-\tau_2)}e^{-i\omega_{n_3}\tau_2}. \tag{7.64}$$

With respect to the measure $d\tau$, the only nonzero contribution arises from the term $e^{(i\omega_n-i\omega_{n_1})\tau}$. Let us analyze the corresponding integral. Depending on whether the propagators are fermionic or bosonic, the Matsubara frequencies $i\omega_{n_1}$, $i\omega_{n_2}$, and $i\omega_{n_3}$ (see Eq. 6.27) are odd or even, respectively. As for the self-energy, in the case $n = 1$, it consists of two diagrams, each given by the product of an interaction line (bosonic) and a particle propagator (either bosonic or fermionic). Therefore, the frequency structure of the self-energy ultimately depends on the nature of the internal propagator. As a consequence, the quantity $\omega_n - \omega_{n_1} = \omega_m$ is always a bosonic frequency, i.e., it is always even. Hence,

$$\int_0^{\beta\hbar} e^{(i\omega_n-i\omega_{n_1})\tau}\,d\tau = \int_0^{\beta\hbar} e^{i\omega_m\tau}\,d\tau =$$

$$= \int_0^{\beta\hbar} e^{\frac{2\pi i m}{\beta\hbar}\tau} d\tau =$$

$$= \begin{cases} \beta\hbar, & m = 0 \\ \dfrac{e^{i\omega_m\beta\hbar} - 1}{i\omega_m}, & m \neq 0 \end{cases}. \tag{7.65}$$

Since $i\omega_m$ is a bosonic frequency, then $e^{i\omega_m\beta\hbar} = 1$ (see Eq. 6.28). As a result, the integral over τ yields a nonzero contribution only when $m = 0$. Ultimately, we obtain

$$\int_0^{\beta\hbar} e^{(i\omega_n - i\omega_{n_1})\tau} = \delta_{n,n_1}\beta\hbar, \tag{7.66}$$

and similarly

$$\int_0^{\beta\hbar} e^{(i\omega_{n_1} - i\omega_{n_2})\tau} = \delta_{n_1,n_2}\beta\hbar, \tag{7.67}$$

$$\int_0^{\beta\hbar} e^{(i\omega_{n_2} - i\omega_{n_3})\tau} = \delta_{n_2,n_3}\beta\hbar. \tag{7.68}$$

We simplify the prefactor $(\beta\hbar)^{-3}$ in the Dyson equation, and by summing over the indices n_1, n_2, and n_3, the Kronecker deltas enforce the conditions $n = n_1 = n_2 = n_3$. Consequently, the frequencies also coincide: $\omega_n = \omega_{n_1} = \omega_{n_2} = \omega_{n_3}$. This reflects the conservation of energy at each vertex of the corresponding Feynman diagram. The Dyson equation at thermodynamic equilibrium in frequency space can thus be rewritten as

$$G^{(m)}(x, x', i\omega_n) = G^{(m)(0)}(x, x', i\omega_n) + \int dx_1 \int dx_2 G^{(m)(0)}(x, x_1, i\omega_n)$$
$$\Sigma^*(x_1, x_2, i\omega_n) G^{(m)}(x_2, x', i\omega_n). \tag{7.69}$$

This form is already considerably simpler than the nonequilibrium Dyson equation, since time convolutions reduce to products in frequency space.

Moreover, if the system is translationally invariant, the Green's functions and the self-energy depend only on the spatial difference $\mathbf{r}_i - \mathbf{r}_j$. Consequently, we decompose the index x into its spatial coordinate $\mathbf{r}$ and spin component s, and apply a spatial Fourier transform to Eq. 7.69. In the thermodynamic limit, boundary conditions become irrelevant, and we adopt periodic boundary conditions to preserve translational invariance. In this limit, the discrete momentum values associated with the periodic boundary conditions become quasi-continuous, allowing sums over $\mathbf{k}$ to be replaced by integrals in momentum space. Given the standard forms of the

Fourier transform in real and frequency space, we obtain

$$G^{(m)}_{s,s'}(\mathbf{r}-\mathbf{r}', i\omega_n) = \int \frac{d^3\mathbf{k}}{(2\pi)^3} e^{-i\mathbf{k}\cdot(\mathbf{r}-\mathbf{r}')} G^{(m)}_{s,s'}(\mathbf{k}, i\omega_n), \tag{7.70}$$

$$G^{(m)}_{s,s'}(\mathbf{k}, i\omega_n) = \int d^3(\mathbf{r}-\mathbf{r}') e^{i\mathbf{k}\cdot(\mathbf{r}-\mathbf{r}')} G^{(m)}_{s,s'}(\mathbf{r}-\mathbf{r}', i\omega_n), \tag{7.71}$$

we multiply by $e^{i\mathbf{k}\cdot(\mathbf{r}-\mathbf{r}')}$ both members of Eq. 7.69 and integrate with respect to the measure $d^3(\mathbf{r}-\mathbf{r}')$ as follows

$$\begin{aligned}
G^{(m)}_{s,s'}(\mathbf{k}, i\omega_n) &= G^{(m)(0)}_{s,s'}(\mathbf{k}, i\omega_n) + \sum_{s_1,s_2} \int d^3(\mathbf{r}-\mathbf{r}') e^{i\mathbf{k}\cdot(\mathbf{r}-\mathbf{r}')} \int d^3\mathbf{r}_1 \int d^3\mathbf{r}_2 \\
&\left(\int \frac{d^3\mathbf{k}_1}{(2\pi)^3} e^{-i\mathbf{k}_1\cdot(\mathbf{r}-\mathbf{r}_1)} G^{(m)(0)}_{s,s_1}(\mathbf{k}_1, i\omega_n)\right) \\
&\left(\int \frac{d^3\mathbf{k}_2}{(2\pi)^3} e^{-i\mathbf{k}_2\cdot(\mathbf{r}_1-\mathbf{r}_2)} \Sigma^*_{s_1,s_2}(\mathbf{k}_2, i\omega_n)\right) \\
&\left(\int \frac{d^3\mathbf{k}_3}{(2\pi)^3} e^{-i\mathbf{k}_3\cdot(\mathbf{r}_2-\mathbf{r}')} G^{(m)}_{s_2,s'}(\mathbf{k}_3, i\omega_n)\right) = \\
&= G^{(m)(0)}_{s,s'}(\mathbf{k}, i\omega_n) + \sum_{s_1,s_2} \int d^3(\mathbf{r}-\mathbf{r}') \int d^3\mathbf{r}_1 \\
&\int d^3\mathbf{r}_2 \int \frac{d^3\mathbf{k}_1}{(2\pi)^3} \int \frac{d^3\mathbf{k}_2}{(2\pi)^3} \int \frac{d^3\mathbf{k}_3}{(2\pi)^3} \\
&G^{(m)(0)}_{s,s_1}(\mathbf{k}_1, i\omega_n) \Sigma^*_{s_1,s_2}(\mathbf{k}_2, i\omega_n) G^{(m)}_{s_2,s'}(\mathbf{k}_3, i\omega_n) \\
&e^{i\mathbf{k}\cdot\mathbf{r}} e^{-i\mathbf{k}\cdot\mathbf{r}'} e^{-i\mathbf{k}_1\cdot\mathbf{r}} e^{i\mathbf{k}_1\cdot\mathbf{r}_1} e^{-i\mathbf{k}_2\cdot\mathbf{r}_1} e^{i\mathbf{k}_2\cdot\mathbf{r}_2} e^{-i\mathbf{k}_3\cdot\mathbf{r}_2} e^{i\mathbf{k}_3\cdot\mathbf{r}'}.
\end{aligned} \tag{7.72}$$

If we integrate with respect to the measure $d\mathbf{r}_2$, we observe that

$$\int d^3\mathbf{r}_2 e^{i(\mathbf{k}_2-\mathbf{k}_3)\cdot\mathbf{r}_2} = (2\pi)^3 \delta(\mathbf{k}_2-\mathbf{k}_3), \tag{7.73}$$

which allows us to perform

$$\int \frac{d^3\mathbf{k}_3}{(2\pi)^3} (2\pi)^3 \delta(\mathbf{k}_2-\mathbf{k}_3) G^{(m)}_{s_2,s'}(\mathbf{k}_3, i\omega_n) e^{i\mathbf{k}_3\cdot\mathbf{r}'} = G^{(m)}_{s_2,s'}(\mathbf{k}_2, i\omega_n) e^{i\mathbf{k}_2\cdot\mathbf{r}'}, \tag{7.74}$$

and the Dyson equation is written as

$$G^{(m)}_{s,s'}(\mathbf{k}, i\omega_n) = G^{(m)(0)}_{s,s'}(\mathbf{k}, i\omega_n) + \sum_{s_1,s_2} \int d^3(\mathbf{r}-\mathbf{r}') \int d^3\mathbf{r}_1 \int \frac{d^3\mathbf{k}_1}{(2\pi)^3} \int \frac{d^3\mathbf{k}_2}{(2\pi)^3}$$
$$G^{(m)(0)}_{s,s_1}(\mathbf{k}_1, i\omega_n)\Sigma^*_{s_1,s_2}(\mathbf{k}_2, i\omega_n)G^{(m)}_{s_2,s'}(\mathbf{k}_2, i\omega_n)e^{i\mathbf{k}\cdot\mathbf{r}}e^{-i\mathbf{k}\cdot\mathbf{r}'}e^{-i\mathbf{k}_1\cdot\mathbf{r}}$$
$$e^{i\mathbf{k}_1\cdot\mathbf{r}_1}e^{-i\mathbf{k}_2\cdot\mathbf{r}_1}e^{i\mathbf{k}_2\cdot\mathbf{r}'}. \tag{7.75}$$

Similarly

$$\int d^3\mathbf{r}_1 e^{i(\mathbf{k}_1-\mathbf{k}_2)\cdot\mathbf{r}_1} = (2\pi)^3\delta(\mathbf{k}_1-\mathbf{k}_2), \tag{7.76}$$

$$\int \frac{d^3\mathbf{k}_2}{(2\pi)^3}(2\pi)^3\delta(\mathbf{k}_1-\mathbf{k}_2)e^{i\mathbf{k}_2\cdot\mathbf{r}'}\Sigma^*_{s_1,s_2}(\mathbf{k}_2, i\omega_n)G^{(m)}_{s_2,s'}(\mathbf{k}_2, i\omega_n)$$
$$= e^{i\mathbf{k}_1\cdot\mathbf{r}'}\Sigma^*_{s_1,s_2}(\mathbf{k}_1, i\omega_n)G^{(m)}_{s_2,s'}(\mathbf{k}_1, i\omega_n), \tag{7.77}$$

$$G^{(m)}_{s,s'}(\mathbf{k}, i\omega_n) = G^{(m)(0)}_{s,s'}(\mathbf{k}, i\omega_n) + \sum_{s_1,s_2} \int d^3(\mathbf{r}-\mathbf{r}') \int \frac{d^3\mathbf{k}_1}{(2\pi)^3}$$
$$G^{(m)(0)}_{s,s_1}(\mathbf{k}_1, i\omega_n)\Sigma^*_{s_1,s_2}(\mathbf{k}_1, i\omega_n)G^{(m)}_{s_2,s'}(\mathbf{k}_1, i\omega_n)e^{i\mathbf{k}\cdot\mathbf{r}}e^{-i\mathbf{k}\cdot\mathbf{r}'}$$
$$e^{-i\mathbf{k}_1\cdot\mathbf{r}}e^{i\mathbf{k}_1\cdot\mathbf{r}'}. \tag{7.78}$$

Finally, we use

$$\int d^3(\mathbf{r}-\mathbf{r}')e^{i\mathbf{k}\cdot(\mathbf{r}-\mathbf{r}')}e^{-i\mathbf{k}_1\cdot(\mathbf{r}-\mathbf{r}')} = (2\pi)^3\delta(\mathbf{k}-\mathbf{k}_1), \tag{7.79}$$

$$\int \frac{d^3\mathbf{k}_1}{(2\pi)^3}(2\pi)^3\delta(\mathbf{k}-\mathbf{k}_1)G^{(m)(0)}_{s,s_1}(\mathbf{k}_1, i\omega_n)\Sigma^*_{s_1,s_2}(\mathbf{k}_1, i\omega_n)G^{(m)}_{s_2,s'}(\mathbf{k}_1, i\omega_n) =$$
$$= G^{(m)(0)}_{s,s_1}(\mathbf{k}, i\omega_n)\Sigma^*_{s_1,s_2}(\mathbf{k}, i\omega_n)G^{(m)}_{s_2,s'}(\mathbf{k}, i\omega_n), \tag{7.80}$$

and we write the Dyson equation as

$$G^{(m)}_{s,s'}(\mathbf{k}, i\omega_n) = G^{(m)(0)}_{s,s'}(\mathbf{k}, i\omega_n)$$
$$+ \sum_{s_1,s_2} G^{(m)(0)}_{s,s_1}(\mathbf{k}, i\omega_n)\Sigma^*_{s_1,s_2}(\mathbf{k}, i\omega_n)G^{(m)}_{s_2,s'}(\mathbf{k}, i\omega_n), \tag{7.81}$$

which expresses the Dyson equation in frequency and momentum space, with full spin structure. This form reflects the conservation of momentum at each vertex of the diagram, as all propagators and self-energies share the same momentum.

An additional simplification is possible if the system is spin-rotation invariant or if spin-dependent interactions (such as spin–orbit coupling or external magnetic fields) are absent. Indeed, in this case the Green's functions and the self-energy become diagonal in spin space and identical for both spin projections, that is

$$G^{(m)(0)}_{s,s_1}(\mathbf{k}, i\omega_n) = \delta_{s,s_1} G^{(m)(0)}_{\sigma}(\mathbf{k}, i\omega_n), \tag{7.82}$$

$$\Sigma^{*}_{s_1,s_2}(\mathbf{k}, i\omega_n) = \delta_{s_1,s_2} \Sigma^{*}_{\sigma}(\mathbf{k}, i\omega_n), \tag{7.83}$$

$$G^{(m)}_{s_2,s'}(\mathbf{k}, i\omega_n) = \delta_{s_2,s'} G^{(m)}_{\sigma}(\mathbf{k}, i\omega_n). \tag{7.84}$$

Consequently, under this further assumption, the Dyson equation reduces to a scalar algebraic equation

$$G^{(m)}_{\sigma}(\mathbf{k}, i\omega_n) = G^{(m)(0)}_{\sigma}(\mathbf{k}, i\omega_n) + G^{(m)(0)}_{\sigma}(\mathbf{k}, i\omega_n)\Sigma^{*}_{\sigma}(\mathbf{k}, i\omega_n) G^{(m)}_{\sigma}(\mathbf{k}, i\omega_n), \tag{7.85}$$

for the Green's function corresponding to a given spin and momentum. Consider a system of non-interacting particles and let us compute $G^{(m)(0)}_{\sigma}(\mathbf{k}, i\omega_n)$ starting from

$$G^{(m)}_{\sigma}(\mathbf{k}, i\omega_n) = \frac{G^{(m)(0)}_{\sigma}(\mathbf{k}, i\omega_n)}{1 - G^{(m)(0)}_{\sigma}(\mathbf{k}, i\omega_n)\Sigma^{*}_{\sigma}(\mathbf{k}, i\omega_n)}. \tag{7.86}$$

Given

$$\begin{aligned} G^{(m)(0)}_{\sigma}(\mathbf{k}, \tau) &= -\frac{1}{\hbar}\left\langle C^{(0)}_{\mathbf{k},\sigma}(\tau) C^{\dagger(0)}_{\mathbf{k},\sigma}(0)\right\rangle_0 = \\ &= -\frac{1}{\hbar}\left[e^{-\frac{\mathcal{E}_{\mathbf{k},\sigma}\tau}{\hbar}} \left(1 + \varepsilon n_{\varepsilon}(\mathcal{E}_{\mathbf{k},\sigma})\right)\Theta(\tau) + \varepsilon e^{-\frac{\mathcal{E}_{\mathbf{k},\sigma}\tau}{\hbar}} n_{\varepsilon}(\mathcal{E}_{\mathbf{k},\sigma})\Theta(-\tau)\right], \end{aligned} \tag{7.87}$$

where we employed Eqs. D.8, 2.251, 2.236, and $n_{\varepsilon}(\mathcal{E}_{\mathbf{k},\sigma})$ is given in Eq. E.2. It follows

$$\begin{aligned} G^{(m)(0)}_{\sigma}(\mathbf{k}, i\omega_n) &= \int_0^{\beta\hbar} d\tau e^{i\omega_n\tau} G^{(m)(0)}_{\sigma}(\mathbf{k}, \tau) = \\ &= \frac{1}{\hbar\left[i\omega_n - \frac{\mathcal{E}_{\mathbf{k},\sigma}}{\hbar}\right]}, \end{aligned} \tag{7.88}$$

and the solution of the Dyson equation can then be written as follows

$$
\begin{aligned}
G_{\sigma}^{(m)}(\mathbf{k}, i\omega_n) &= \frac{G_{\sigma}^{(m)(0)}(\mathbf{k}, i\omega_n)}{1 - G_{\sigma}^{(m)(0)}(\mathbf{k}, i\omega_n)\Sigma_{\sigma}^{*}(\mathbf{k}, i\omega_n)} = \\
&= \frac{1}{\left[G_{\sigma}^{(m)(0)}(\mathbf{k}, i\omega_n)\right]^{-1} - \Sigma_{\sigma}^{*}(\mathbf{k}, i\omega_n)} = \\
&= \frac{1}{\hbar\left[i\omega_n - \frac{\mathcal{E}_{\mathbf{k},\sigma}}{\hbar}\right] - \Sigma_{\sigma}^{*}(\mathbf{k}, i\omega_n)} \equiv \\
&\equiv \frac{1}{\hbar\left[i\omega_n - \frac{\mathcal{E}_{\mathbf{k},\sigma}}{\hbar} - \frac{\Sigma_{\sigma}^{*}(\mathbf{k}, i\omega_n)}{\hbar}\right]}.
\end{aligned}
\tag{7.89}
$$

As previously discussed, we recover an expression of the form given in Eq. 6.116. For a system that is invariant under both spatial and time translations, the many-body Green's function takes a particularly simple form: it depends only on the Matsubara frequency $i\omega_n$ and the momentum $\mathbf{k}$ (reflecting the underlying time and space symmetries), and the improper self-energy Σ^*.

In the absence of interactions, namely, when the self-energy vanishes identically $\Sigma_{\sigma}^{*}(\mathbf{k}, i\omega_n) = 0$, the interacting Green's function $G_{\sigma}^{(m)}(\mathbf{k}, i\omega_n)$ in Eq. 7.89 reduces exactly to its non-interacting counterpart $G_{\sigma}^{(m)(0)}(\mathbf{k}, i\omega_n)$ given in Eq. 6.89, as expected. This confirms the internal consistency of the perturbative framework: interactions affect the system's dynamics exclusively through the self-energy. In turn, the self-energy leads to a renormalization of the energy levels and, possibly, to a finite broadening of the spectral features.

7.6 A Matsubara Sum

From the definition of time Fourier transform we can derive an equation for Matsubara frequencies (see Eq. 6.27). Indeed, the many-body Green's function 7.87 can be written in Fourier time transform as

$$
\begin{aligned}
G_{\sigma}^{(m)(0)}(\mathbf{k}, \tau) &= \frac{1}{\beta\hbar}\sum_{n} e^{-i\omega_n} G_{\sigma}^{(m)(0)}(\mathbf{k}, i\omega_n) = \\
&= \frac{1}{\beta\hbar}\sum_{n} \frac{e^{-i\omega_n}}{\hbar\left[i\omega_n - \frac{\mathcal{E}_{\mathbf{k},\sigma}}{\hbar}\right]},
\end{aligned}
\tag{7.90}
$$

and from comparison between Eqs. 7.87 and 7.90, one finds

$$\frac{1}{\beta}\sum_n \frac{e^{-i\omega_n}}{\hbar\left[i\omega_n - \frac{\mathcal{E}_{\mathbf{k},\sigma}}{\hbar}\right]} = \begin{cases} -e^{-\frac{\mathcal{E}_{\mathbf{k},\sigma}\tau}{\hbar}}\left(1+\varepsilon n_\varepsilon(\mathcal{E}_{\mathbf{k},\sigma})\right), & \tau > 0 \\ -\varepsilon e^{-\frac{\mathcal{E}_{\mathbf{k},\sigma}\tau}{\hbar}} n_\varepsilon(\mathcal{E}_{\mathbf{k},\sigma}), & \tau < 0 \end{cases}, \tag{7.91}$$

where $n_\varepsilon(\mathcal{E}_{\mathbf{k},\sigma})$ is given by E.2. Due to the discontinuity of the Heaviside theta function, we need to examine the limit as $\tau \to 0^+$ and $\tau \to 0^-$, we would like to interchange the sum and the limit operators. However, this is not allowed. In fact, we know that the bosonic and fermionic frequencies satisfy $\omega_n \sim n$, so for infinitesimal τ we have

$$\frac{1}{\beta}\sum_n \frac{e^{-i\omega_n}}{\hbar\left[i\omega_n - \frac{\mathcal{E}_{\mathbf{k},\sigma}}{\hbar}\right]} \sim \sum_n \frac{1}{n}, \tag{7.92}$$

which is divergent. We must therefore introduce a factor $e^{-i\omega_n\delta}$, with $\delta \in (0, \beta\hbar)$ arbitrarily small, in order to ensure the convergence of the series. Consequently, the series is replaced by

$$\frac{1}{\beta}\sum_n \frac{e^{-i\omega_n\delta}}{\hbar\left[i\omega_n - \frac{\mathcal{E}_{\mathbf{k},\sigma}}{\hbar}\right]}, \tag{7.93}$$

and from Eq. 7.91, it follows

Theorem 7.5 *The Matsubara frequencies, i.e., Eq. 6.27, satisfy*

$$\frac{1}{\beta}\sum_n \frac{e^{-i\omega_n\delta}}{\hbar\left[i\omega_n - \frac{\mathcal{E}_{\mathbf{k},\sigma}}{\hbar}\right]} = \begin{cases} -e^{-\frac{\mathcal{E}_{\mathbf{k},\sigma}\delta}{\hbar}}\left(1+\varepsilon n_\varepsilon(\mathcal{E}_{\mathbf{k},\sigma})\right), & \tau > 0 \\ -\varepsilon e^{-\frac{\mathcal{E}_{\mathbf{k},\sigma}\delta}{\hbar}} n_\varepsilon(\mathcal{E}_{\mathbf{k},\sigma}), & \tau < 0 \end{cases}. \tag{7.94}$$

where $n_\varepsilon(\mathcal{E}_{\mathbf{k},\sigma})$ is given in Eq. E.2.

Proof Here, we aim to compute the sum of the series 7.93 independently of the propagators and we use the residue theorem K.1 (see Fig. 7.28). We will consider the case of fermions, and the fermionic Matsubara frequencies ω_n are the poles of the Fermi-Dirac function (see Eq. E.4)

$$n_{-1}(z) = \frac{1}{e^{\beta z}+1}. \tag{7.95}$$

Such poles $z = i\omega_n$ lie on the imaginary axis and they are simple poles, indeed,

$$\begin{aligned} \mathrm{Res}_{i\omega_n}(n_{-1}(z)) &= \lim_{z\to i\omega_n}(z - i\omega_n)n_{-1}(z) = \\ &= \lim_{z\to i\omega_n}\frac{z - i\omega_n}{e^{\beta z}+1} = \end{aligned}$$

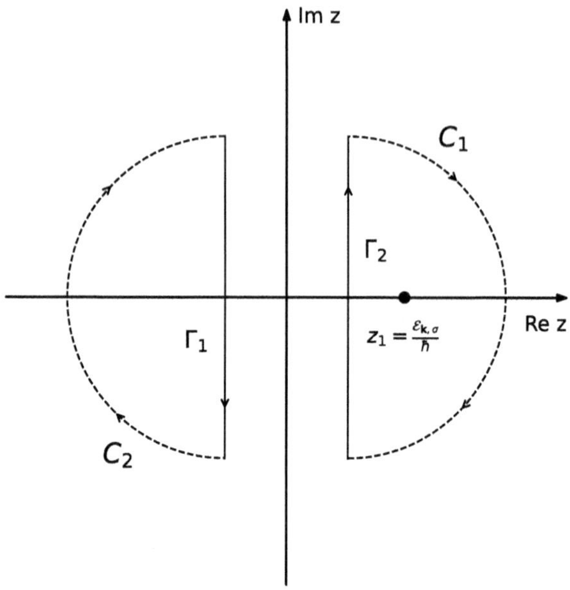

Fig. 7.28 Integration contour used to evaluate the Matsubara sum in Eq. 7.94 via the residue theorem. The horizontal axis corresponds to the real part of the complex variable z, representing the physical excitation energy, while the vertical axis corresponds to its imaginary part, associated with the discrete Matsubara frequencies (see Eq. 6.27). The contour is chosen to enclose the relevant singularities in the complex plane, allowing the corresponding residues to be computed and the summation over Matsubara frequencies to be performed

$$\begin{aligned} &= \lim_{z\to i\omega_n} \frac{0+1(z-i\omega_n)}{0+\beta e^{\beta z}(z-i\omega_n)} = \\ &= \lim_{z\to i\omega_n} \frac{1}{\beta e^{\beta z}} = \\ &= -\frac{1}{\beta}. \end{aligned} \tag{7.96}$$

In the complex frequency plane, the poles of the Fermi-Dirac function lie along the imaginary axis ($\operatorname{Im} z$), corresponding to the Matsubara frequencies, while the real axis ($\operatorname{Re} z$) represents the physical excitation energies. Around each pole, we draw small counterclockwise squares. Contributions from the horizontal sides cancel when summing over all poles. Since the sum runs over the set $\mathbb{Z}$, the resulting contour Γ consists of two lines parallel to the imaginary axis that close at infinity. We now show that the Matsubara sum can be computed via the integral

$$I = \int_\Gamma dz \left(-\frac{\beta}{2\pi i}\right) n_{-1}(z) g(z), \tag{7.97}$$

with

$$g(z) = \frac{e^{z\delta}}{z - \frac{\mathcal{E}_{\mathbf{k},\sigma}}{\hbar}}, \tag{7.98}$$

so that

$$\frac{1}{\beta}\sum_n \frac{e^{-i\omega_n\delta}}{\hbar\left[i\omega_n - \frac{\mathcal{E}_{\mathbf{k},\sigma}}{\hbar}\right]} = \int_\Gamma dz\left(-\frac{\beta}{2\pi i}\right)n_{-1}(z)\frac{e^{z\delta}}{z-\frac{\mathcal{E}_{\mathbf{k},\sigma}}{\hbar}}. \tag{7.99}$$

We fix the pole $z_1 = \frac{\mathcal{E}_{\mathbf{k},\sigma}}{\hbar}$ of the function $g(z)$ on the positive real semi-axis, assuming it to be positive. This pole lies on the real axis ($\mathrm{Re}\, z$), corresponding to the physical excitation energy of the system. We construct a contour consisting of two semicircles, C_1 and C_2, for each line. If we show that the contribution of the integrand along the arcs vanishes in the limit as the radius tends to infinity, we obtain the integral along the contour Γ. Let us assume $\mathrm{Re}\, z > 0$ and $z \to \infty$, i.e., we consider the curve C_1: in order for Jordan's lemma K.1 to apply, the integrand must vanish faster than z^{-1}. For $z \to \infty$, we have $e^{z\delta} \sim e^{\mathrm{Re}\, z\ \delta}$, and the Fermi-Dirac statistics $n_{-1}(z) \sim e^{-\beta\,\mathrm{Re}\, z}$ ensures convergence, since the exponential $e^{\mathrm{Re}\, z\delta}$ is well-defined for $\delta \in (0, \beta\hbar)$, with the endpoints excluded. Jordan's lemma K.1 can be applicated and the integral along C_1 vanishes. Now suppose $\mathrm{Re}\, z < 0$ and $z \to \infty$, i.e., we consider the curve C_2. In this case, we have $n_{-1}(z) \sim 1$. However, the exponential $e^{\mathrm{Re}\, z\ \delta}$ prevents the divergence of the integral, since we must perform the integral first and then take the limit $\delta \to 0$, exactly as for the series. We have shown that the integrals along the semicircles vanish, so the integrals over the two contours $\Gamma_1 + C_2$ and $\Gamma_2 + C_1$ reduce to the residue term only, and the only pole lies in the upper half-plane at $z_1 = \frac{\mathcal{E}_{\mathbf{k},\sigma}}{\hbar}$, and it lies outside the contour Γ. By the residue theorem K.1, we have

$$\frac{1}{\beta}\sum_n \frac{e^{-i\omega_n\delta}}{\hbar\left[i\omega_n - \frac{\mathcal{E}_{\mathbf{k},\sigma}}{\hbar}\right]} = e^{\frac{\mathcal{E}_{\mathbf{k},\sigma}\delta}{\hbar}}\, n_{-1}(\mathcal{E}_{\mathbf{k},\sigma}), \tag{7.100}$$

which is equal to Eq. 7.94 adapted for fermions ($\varepsilon = -1$) and for $\tau < 0$.

The proof for bosons proceeds similarly, with the substitutions $\varepsilon = +1$ and the Bose-Einstein function (see Eq. E.3)

$$n_{+1}(z) = \frac{1}{e^{\beta z} - 1}. \tag{7.101}$$

The poles of $n_{+1}(z)$ also lie along the imaginary axis, evenly spaced according to the bosonic Matsubara frequencies. In this representation, the imaginary axis captures the discrete Matsubara frequencies, while the real axis corresponds to the physical excitation energies. This separation allows the Matsubara sum to be evaluated efficiently via the residue theorem. □

7.7 Thermal Average of the Particle Density Operator via Green's Function

Here, we illustrate a useful application of the many-body Green's function by computing the thermal average of the particle density operator, see Eq. 2.261.

Theorem 7.6 (Thermal Average of the Particle Density Operator for a Non-interacting System via Green's Function) *Let $G^{(m)(0)}$ denote the non-interacting many-body Green's function. For a non-interacting particle system, the thermal average (see Eq. 1.379) of the particle density operator (see Eq. 2.261) is given by*

$$\langle \hat{\rho}(\mathbf{r}) \rangle_0 = \sum_{s=-S}^{S} \sum_j \left|\varphi_j(\mathbf{r}, s)\right|^2 n_\varepsilon(\mathcal{E}_j), \tag{7.102}$$

where ε is defined in Eq. 2.52, $n_\varepsilon(\mathcal{E}_j)$ is the unified distribution of Eq. E.2, and $\varphi_j(\mathbf{r}, s)$ and $\mathcal{E}_j$ denote, respectively, the eigenfunctions and eigenvalues of the corresponding non-interacting Hamiltonian.
As a consequence, the total particle number of the non-interacting system is

$$\left\langle \hat{N} \right\rangle_0 = \sum_j n_\varepsilon(\mathcal{E}_j). \tag{7.103}$$

Proof We start from the thermal average of the particle density operator,

$$\langle \hat{\rho}(\mathbf{r}) \rangle_0 = \sum_{s=-S}^{S} \left\langle \hat{\psi}^\dagger(\mathbf{r}, s) \hat{\psi}(\mathbf{r}, s) \right\rangle_0, \tag{7.104}$$

which can be manipulated as follows

$$\begin{aligned} \langle \hat{\rho}(\mathbf{r}) \rangle_0 &= \varepsilon \sum_{s=-S}^{S} \left\langle \hat{T}_\tau \, \hat{\psi}^{(0)}(\mathbf{r}, s, \tau) \, \hat{\psi}^{\dagger(0)}(\mathbf{r}, s, \tau + \delta) \right\rangle_0 = \\ &= -\hbar\varepsilon \sum_{s=-S}^{S} G^{(m)(0)}(\mathbf{r}, s, -\delta), \end{aligned} \tag{7.105}$$

where $G^{(m)(0)}$ denotes the non-interacting many-body Green's function. The small positive shift $\delta > 0$ ensures that the operators are arranged in ascending order with respect to time. Given a free propagator in the real space and imaginary time, i.e., of the form 6.59, we apply a Fourier transform to the particle density operator with respect to time

$$\left\langle \hat{\rho}(\mathbf{r}) \right\rangle_0 = -\hbar\varepsilon \sum_{s=-S}^{S} \frac{1}{\beta\hbar} \sum_n G^{(m)(0)}(\mathbf{r}, s, i\omega_n) e^{i\omega_n\delta}, \tag{7.106}$$

and we insert the explicit form of the free many-body Green's function, i.e.,

$$\left\langle \hat{\rho}(\mathbf{r}) \right\rangle_0 = -\hbar\varepsilon \sum_{s=-S}^{S} \frac{1}{\beta\hbar} \sum_n \sum_j \frac{\left|\varphi_j(\mathbf{r}, s)\right|^2}{\hbar\left[i\omega_n - \frac{\mathcal{E}_j}{\hbar}\right]} e^{i\omega_n\delta}. \tag{7.107}$$

From the Matsubara sum given in Eq. 7.94, it follows

$$\frac{1}{\beta\hbar} \sum_n \frac{e^{i\omega_n\delta}}{\left[i\omega_n - \frac{\mathcal{E}_j}{\hbar}\right]} = -\varepsilon n_\varepsilon(\mathcal{E}_j), \tag{7.108}$$

then

$$\begin{aligned} \left\langle \hat{\rho}(\mathbf{r}) \right\rangle_0 &= (-\varepsilon)^2 \sum_{s=-S}^{S} \sum_j \left|\varphi_j(\mathbf{r}, s)\right|^2 n_\varepsilon(\mathcal{E}_j) = \\ &= \sum_{s=-S}^{S} \sum_j \left|\varphi_j(\mathbf{r}, s)\right|^2 n_\varepsilon(\mathcal{E}_j), \end{aligned} \tag{7.109}$$

which shows explicitly that the thermal density is built from the free eigenfunctions and occupation numbers of the non-interacting many-body system. Finally, we can compute the average total number of particles as

$$\begin{aligned} \left\langle \hat{N} \right\rangle_0 &= \left\langle \int d\mathbf{r}\hat{\rho}(\mathbf{r}) \right\rangle_0 = \\ &= \int d\mathbf{r} \left\langle \hat{\rho}(\mathbf{r}) \right\rangle_0 = \\ &= \int d\mathbf{r} \sum_{s=-S}^{S} \sum_j \left|\varphi_j(\mathbf{r}, s)\right|^2 n_\varepsilon(\mathcal{E}_j) = \\ &= \sum_j n_\varepsilon(\mathcal{E}_j). \end{aligned} \tag{7.110}$$

□

The theorem makes explicit that, in a non-interacting system, the thermal particle density is entirely fixed by the one-particle spectrum together with the associated occupation factors. Once interactions are present, this simple structure no longer holds, and the density must be evaluated from the many-body Green's function of the interacting system.

Chapter 8
Finite-Temperature Hartree-Fock Method for Fermionic Systems

Abstract This chapter presents the finite-temperature Hartree-Fock method for fermionic systems, a key approach to describe interacting quantum systems in thermal equilibrium. We begin with the derivation of the finite-temperature Hartree-Fock equations, which incorporate both mean-field interactions and thermal statistical effects, and are applicable in the weak-coupling regime. The formalism is applied to a homogeneous paramagnetic system modeled using jellium, a simplified yet effective framework for studying electron-electron interactions. We solve the finite-temperature Hartree-Fock equations for this system, obtaining self-consistent solutions for the electron wavefunctions and energy levels, which reflect both exchange and direct interactions. From these, we compute the Hartree-Fock energy, offering an estimate of the system's total energy that includes thermal occupation and interaction effects. The chapter concludes with the calculation of the Hartree-Fock pair correlation function, which describes the conditional probability of finding two electrons at a given distance. This function reveals the impact of quantum statistics and Coulomb repulsion on the spatial distribution of particles, illustrating how interactions in the Hartree-Fock framework modify correlations compared to a non-interacting Fermi gas.

8.1 Hartree-Fock Equations at Finite Temperature

Here we apply the Hartree-Fock method at the finite temperature by means of Feynman's approach (see Chap. 7). Let us consider a quantized Hamiltonian of the form 1.85. According to Dyson's expansion (see Eq. 7.62), the propagator can be calculated from the self energy $\Sigma^*(x_1, \tau_1, x_2, \tau_2)$, which, however, includes a sum of infinite irreducible diagrams. We previously calculated the first-order self-energy analytically, that is, Eq. 7.61. The Hartree-Fock method consists of substituting the self-energy of the Dyson equation for its first-order analytically. We will treat the general case, but eventually apply the results to fermions, since bosons at low temperatures undergo the phenomenon of Bose-Einstein condensation (see Chap. 11). At thermodynamic equilibrium, Eq. 7.61 is invariant by time translation (see Theorem 6.1), so we can define $\tau = \tau_1 - \tau_2$ and apply a Fourier transform with

F. Tafuri et al., *Understanding Quantum Many-Body Theory*, UNITEXT for Physics, https://doi.org/10.1007/978-3-032-17849-7_8

respect to time as follows

$$\Sigma^{*(1)}(x_1, x_2, i\omega_n) = \int_0^{\beta\hbar} d\tau e^{i\omega_n \tau} \Sigma^{*(1)}(x_1, x_2, \tau). \tag{8.1}$$

By adapting Eq. 6.25 to the present case, we obtain

$$G^{(m)(0)}(x_1, x_2, -\tau) = \frac{1}{\beta\hbar} \sum_{n'} e^{i\omega_{n'}\tau} G^{(m)(0)}(x_1, x_2, i\omega_{n'}), \tag{8.2}$$

$$G^{(m)(0)}(x_2', x_2', -\tau) = \frac{1}{\beta\hbar} \sum_{n'} e^{i\omega_{n'}\tau} G^{(m)(0)}(x_2', x_2', i\omega_{n'}), \tag{8.3}$$

we have

$$\begin{aligned}
\Sigma^{*(1)}(x_1, x_2, i\omega_n) &= -\hbar U(x_1, x_2) \left(\int_0^{\beta\hbar} d\tau \delta(\tau) \right) \\
&\left(\frac{1}{\beta\hbar} \sum_{n'} e^{i\omega_{n'}\tau} G^{(m)(0)}(x_1, x_2, i\omega_{n'}) \right) + \\
&- \varepsilon\hbar\delta(x_1 - x_2) \left(\int_0^{\beta\hbar} d\tau \delta(\tau) \right) \left(\int_0^{\beta\hbar} d\tau_2' \delta(\tau_1 - \tau_2') \right) \\
&\int dx_2' U(x_1, x_2') \left(\frac{1}{\beta\hbar} \sum_{n'} e^{i\omega_{n'}\tau} G^{(m)(0)}(x_2', x_2', i\omega_{n'}) \right) = \\
&= -\hbar U(x_1, x_2) \left(\frac{1}{\beta\hbar} \sum_{n'} e^{i\omega_{n'}\tau} G^{(m)(0)}(x_1, x_2, i\omega_{n'}) \right) + \\
&- \varepsilon\hbar\delta(x_1 - x_2) \int dx_2' U(x_1, x_2') \left(\frac{1}{\beta\hbar} \sum_{n'} e^{i\omega_{n'}\tau} G^{(m)(0)}(x_2', x_2', i\omega_{n'}) \right),
\end{aligned} \tag{8.4}$$

that is, the first order self-energy does not depend on frequency but only on vertices x_1 and x_2. To include the effects of successive orders of self-energy, in Dyson's expansion we use the first order of self-energy and replace the free propagator with the propagator: we will see that this approach leads to the Hartree-Fock variational method. We apply a Fourier transform to Eq. 7.62 with respect to time, namely

$$\begin{aligned}
G^{(m)}(x, x', i\omega_n) &= G^{(m)(0)}(x, x', i\omega_n) \\
&+ \int dx_1 \int dx_2 G^{(m)(0)}(x, x_1, i\omega_n) \\
&\Sigma^*(x_1, x_2, i\omega_n) G^{(m)}(x_2, x', i\omega_n).
\end{aligned} \tag{8.5}$$

The Hartree-Fock method is a mean-field theory, which replaces the interaction between particles with an effective potential for independent particles. In general, different propagators correspond to different effective potentials, allowing us to estimate the form of the eigenfunctions of the mean-field Hamiltonian. We assume that the propagator has the same form as the free propagator given in Eq. 6.59. Consequently, the Ansatz for the Matsubara Green's function is given by

$$G^{(m)}(x, x', i\omega_n) = \sum_j \frac{\varphi_j^*(x')\varphi_j(x)}{\hbar\left[i\omega_n - \frac{\mathcal{E}_j}{\hbar}\right]}, \tag{8.6}$$

that is, it has the same structure as the free many-body function, that is, Eq. 6.59. If we prove that 8.6 satisfies Eq. 8.5 where $\{\varphi_j\}$ and $\{\mathcal{E}_j\}$ satisfy a self-consistent equation, then we have verified that solving the Dyson equation with first-order self-energy and the propagator in place of the free propagator is equivalent to solve a self-consistent mean-field equation. If the propagator is expressed as in Eq. 8.6, the first-order self-energy is accordingly modified as follows

$$\Sigma^{*(1)}(x_1, x_2, i\omega_n) \longrightarrow \tilde{\Sigma}^{*(1)}(x_1, x_2, i\omega_n), \tag{8.7}$$

where $\tilde{\Sigma}^{*(1)}(x_1, x_2, i\omega_n)$ is obtained by replacing $G^{(m)}$ with $G^{(m)(0)}$ in $\Sigma^{*(1)}(x_1, x_2, i\omega_n)$, i.e.,

$$\begin{aligned}
&\tilde{\Sigma}^{*(1)}(x_1, x_2, i\omega_n) \\
&\quad = -\hbar U(x_1, x_2)\left(\frac{1}{\beta\hbar}\sum_{n'} e^{i\omega_{n'}\tau} G^{(m)}(x_1, x_2, i\omega_{n'})\right) + \\
&\quad - \varepsilon\hbar\delta(x_1 - x_2)\int dx_2' U(x_1, x_2')\left(\frac{1}{\beta\hbar}\sum_{n'} e^{i\omega_{n'}\tau} G^{(m)}(x_2', x_2', i\omega_{n'})\right) = \\
&\quad = -\hbar U(x_1, x_2)\left(\frac{1}{\beta\hbar}\sum_{n'} e^{i\omega_{n'}\tau}\left(\sum_j \frac{\varphi_j^*(x_2)\varphi_j(x_1)}{\hbar\left[i\omega_{n'} - \frac{\mathcal{E}_j}{\hbar}\right]}\right)\right) + \\
&\quad - \varepsilon\hbar\delta(x_1 - x_2)\int dx_2' U(x_1, x_2')\left(\frac{1}{\beta\hbar}\sum_{n'} e^{i\omega_{n'}\tau}\left(\sum_j \frac{|\varphi_j(x_2')|^2}{\hbar\left[i\omega_{n'} - \frac{\mathcal{E}_j}{\hbar}\right]}\right)\right).
\end{aligned} \tag{8.8}$$

Since Eq. 8.6 has the same structure as Eq. 6.59, it also satisfies Eq. 7.91: in particular, we consider the case $\tau < 0$, where the exponential factor takes the form $e^{i\omega_n\tau}$ instead of $e^{-i\omega_n\tau}$. Combining these results, one finds

$$\tilde{\Sigma}^{*(1)}(x_1, x_2, i\omega_n)$$

$$= \varepsilon U(x_1, x_2) \sum_j \varphi_j^*(x_2)\varphi_j(x_1) n_\varepsilon(\mathcal{E}_j) + (-\varepsilon)^2 \hbar\delta(x_1 - x_2)$$

$$\int dx_2' U(x_1, x_2') \sum_j \left|\varphi_j(x_2')\right|^2 n_\varepsilon(\mathcal{E}_j) =$$

$$= \varepsilon U(x_1, x_2) \sum_j \varphi_j^*(x_2)\varphi_j(x_1) n_\varepsilon(\mathcal{E}_j) + \hbar\delta(x_1 - x_2)$$

$$\int dx_2' U(x_1, x_2') \sum_j \left|\varphi_j(x_2')\right|^2 n_\varepsilon(\mathcal{E}_j). \tag{8.9}$$

Theorem 8.1 (Finite-Temperature Hartree-Fock from the Matsubara Green's Function Ansatz) *Assume the interacting many-body system is described by the Hamiltonian 1.85, here expressed in the framework of second quantization. If the Matsubara Green's function is assumed to have the form 8.6, then the first-order self-energy evaluated with this Ansatz leads to a self-consistent mean-field theory. In particular, the eigenfunctions* $\{\varphi_j\}$ *and eigenvalues* $\{\mathcal{E}_j\}$ *satisfy the finite-temperature Hartree-Fock equations, so that solving the Dyson equation with the first-order self-energy and the propagator in place of the free propagator is equivalent to solving the Hartree-Fock variational problem for the system at finite temperature.*

Proof *We define the operator*

$$O_x = i\hbar\omega_n - \hat{\mathcal{H}}_1(x) \tag{8.10}$$

where

$$\hat{\mathcal{H}}_1(x) = \frac{\hat{\mathbf{p}}^2}{2m} + V(x) - \mu\hat{N} \tag{8.11}$$

is a grancanonical Hamiltonian, since it includes the chemical potential μ, *i.e., the number of particles is not fixed. Let us consider the non-interacting many-body Green's function* $G^{(m)(0)}(x, x', i\omega_n)$, *given by Eq. 6.59, and denote by* $\mathcal{E}_\alpha^{(0)}$ *and* $\varphi_\alpha^{(0)}(x)$ *the eigenvalues and eigenfunctions of* $\hat{\mathcal{H}}_1$, *respectively, in order to distinguish them from those used in the Ansatz presented here. By definition,* O_x *is a one-body operator and satisfies the following properties*

$$O_x\varphi_\alpha^{(0)}(x) = \left(i\hbar\omega_n - \mathcal{E}_\alpha^{(0)}\right)\varphi_\alpha^{(0)}(x), \tag{8.12}$$

and

$$
\begin{aligned}
O_x G^{(m)(0)}(x, x', i\omega_n) &= \left(i\hbar\omega_n - \hat{\mathcal{H}}_1(x)\right) \sum_\alpha \frac{\varphi_\alpha^{*(0)}(x')\varphi_\alpha^{(0)}(x)}{\hbar\left[i\omega_n - \frac{\mathcal{E}_\alpha^{(0)}}{\hbar}\right]} = \\
&= \left(i\hbar\omega_n - \mathcal{E}_\alpha^{(0)}\right) \sum_\alpha \frac{\varphi_\alpha^{*(0)}(x')\varphi_\alpha^{(0)}(x)}{\hbar\left[i\omega_n - \frac{\mathcal{E}_\alpha^{(0)}}{\hbar}\right]} = \\
&= \sum_\alpha \varphi_\alpha^{*(0)}(x')\varphi_\alpha^{(0)}(x) = \\
&= \delta(x - x').
\end{aligned}
\tag{8.13}
$$

where we used completeness 1.51. Relation 8.13 can be regarded as the many-body generalization, within the independent-particle approximation, of Eq. 1.282. By replacing the self-energy with its first-order correction and applying the operator O_x on both sides of Eq. 8.5, one finds

$$
\begin{aligned}
O_x G^{(m)}(x, x', i\omega_n) &= O_x G^{(m)(0)}(x, x', i\omega_n) \\
&+ O_x \int dx_1 \int dx_2 G^{(m)(0)}(x, x_1, i\omega_n)\tilde{\Sigma}^*(x_1, x_2, i\omega_n) G^{(m)}(x_2, x', i\omega_n),
\end{aligned}
\tag{8.14}
$$

$$
\begin{aligned}
O_x \sum_j \frac{\varphi_j^*(x')\varphi_j(x)}{\hbar\left[i\omega_n - \frac{\mathcal{E}_j}{\hbar}\right]} &= \delta(x - x') \\
&+ \int dx_1 \int dx_2 O_x G^{(m)(0)}(x, x_1, i\omega_n)\tilde{\Sigma}^*(x_1, x_2, i\omega_n) \sum_j \frac{\varphi_j^*(x')\varphi_j(x_2)}{\hbar\left[i\omega_n - \frac{\mathcal{E}_j}{\hbar}\right]},
\end{aligned}
\tag{8.15}
$$

$$
\begin{aligned}
\sum_j \frac{\varphi_j^*(x') O_x \varphi_j(x)}{\hbar\left[i\omega_n - \frac{\mathcal{E}_j}{\hbar}\right]} &= \delta(x - x') \\
&+ \int dx_1 \int dx_2 \delta(x - x_1)\tilde{\Sigma}^*(x_1, x_2, i\omega_n) \sum_j \frac{\varphi_j^*(x')\varphi_j(x_2)}{\hbar\left[i\omega_n - \frac{\mathcal{E}_j}{\hbar}\right]},
\end{aligned}
\tag{8.16}
$$

$$
\sum_j \frac{\varphi_j^*(x') O_x \varphi_j(x)}{\hbar\left[i\omega_n - \frac{\mathcal{E}_j}{\hbar}\right]} = \delta(x - x') + \int dx_2 \tilde{\Sigma}^*(x, x_2, i\omega_n) \sum_j \frac{\varphi_j^*(x')\varphi_j(x_2)}{\hbar\left[i\omega_n - \frac{\mathcal{E}_j}{\hbar}\right]}.
\tag{8.17}
$$

We multiply both members of the equation by $\varphi_h(x')$ and integrate with respect to the measure dx', to use the orthonormality of the eigenfunctions. We obtain

$$\frac{O_x\varphi_h(x)}{\hbar\left[i\omega_n - \frac{\mathcal{E}_h}{\hbar}\right]} = \varphi_h(x) + \int dx_2\tilde{\Sigma}^*(x, x_2, i\omega_n)\frac{\varphi_h(x_2)}{\hbar\left[i\omega_n - \frac{\mathcal{E}_h}{\hbar}\right]}, \tag{8.18}$$

$$O_x\varphi_h(x) = [i\hbar\omega_n - \mathcal{E}_h]\varphi_h(x) + \int dx_2\tilde{\Sigma}^*(x, x_2, i\omega_n)\varphi_h(x_2), \tag{8.19}$$

$$\left[i\hbar\omega_n - \hat{\mathcal{H}}_1(x)\right]\varphi_h(x) = [i\hbar\omega_n - \mathcal{E}_h]\varphi_h(x) + \int dx_2\tilde{\Sigma}^*(x, x_2, i\omega_n)\varphi_h(x_2), \tag{8.20}$$

$$\mathcal{E}_h\varphi_h(x) = \hat{\mathcal{H}}_1(x)\varphi_h(x) + \int dx_2\tilde{\Sigma}^*(x, x_2, i\omega_n)\varphi_h(x_2). \tag{8.21}$$

In the absence of interactions, $\tilde{\Sigma}^*(x, x_2, i\omega_n) = 0$*, and the above expression reduces to the standard eigenvalue equation. The presence of interactions modifies this equation through the self-energy term. We now show that Eq. 8.21 corresponds to the set of Hartree-Fock equations. From Eq. 8.9, it follows*

$$\tilde{\Sigma}^*(x, x_2, i\omega_n) = \varepsilon U(x, x_2)\sum_j \varphi_j^*(x_2)\varphi_j(x)n_\varepsilon(\mathcal{E}_j) + \\ + \hbar\delta(x - x_2)\int dx_2' U(x, x_2')\sum_j \left|\varphi_j(x_2')\right|^2 n_\varepsilon(\mathcal{E}_j), \tag{8.22}$$

then we insert into Eq. 8.21, i.e.,

$$\mathcal{E}_h\varphi_h(x) = \hat{\mathcal{H}}_1(x)\varphi_h(x) + \varphi_h(x)\left[\int dx_2\sum_j \left|\varphi_j(x_2)\right|^2 n_\varepsilon(\mathcal{E}_j)U(x, x_2)\right] + \\ + \varepsilon\int dx_2 U(x, x_2)\sum_j \varphi_j^*(x_2)\varphi_j(x)n_\varepsilon(\mathcal{E}_j)\varphi_h(x_2), \tag{8.23}$$

which are the Hartree-Fock equations.
Note that the temperature is contained in $n_\varepsilon(\mathcal{E}_j)$*, so in the limit* $T \to 0$*, Eq. 8.23 reduces to the Hartree-Fock equations for* $T = 0$*.* □

We show the meaning of Eq. 8.23. First of all, the object

$$\varphi_h(x)\left[\int dx_2\sum_j \left|\varphi_j(x_2)\right|^2 n_\varepsilon(\mathcal{E}_j)U(x, x_2)\right] \tag{8.24}$$

is called the *direct term*. If Eq. 8.23 contains only the direct term, that is, it takes the form

$$\mathcal{E}_h\varphi_h(x) = \hat{\mathcal{H}}_1(x)\varphi_h(x) + \varphi_h(x)\left[\int dx_2 \sum_j \left|\varphi_j(x_2)\right|^2 n_\varepsilon(\mathcal{E}_j)U(x,x_2)\right], \tag{8.25}$$

then the equation remains solvable, taking the form

$$\mathcal{E}_h\varphi_h(x) = \hat{\mathcal{H}}_1(x)\varphi_h(x) + \tilde{V}(x)\varphi_h(x), \tag{8.26}$$

and the eigenfunctions still satisfy a one-body problem, while the eigenvalues are shifted by the mean-field effective potential. The direct term represents an energy renormalization obtained by replacing each particle with its charge distribution $\left|\varphi_j(x_2)\right|^2 n_\varepsilon(\mathcal{E}_j)$, and it has a classical analogue. The object

$$\varepsilon \int dx_2 U(x,x_2) \sum_j \varphi_j^*(x_2)\varphi_j(x) n_\varepsilon(\mathcal{E}_j)\varphi_h(x_2) \tag{8.27}$$

is called the *exchange term*, since it includes the object ϵ (see Eq. 2.52), which depends on the statistical nature of the operators. This term has no classical equivalent; it makes sense only in quantum mechanics.

8.2 Solution of the Hartree-Fock Equations at Finite Temperature for Spatially Translationally Invariant Systems

Here, we introduce a set of simplifying assumptions that reduce the Hartree-Fock equations (see Eq. 8.23) to a more tractable form and ultimately permit an exact solution. First, assume that the two-body interaction is spin-independent, meaning it depends only on the spatial coordinates, i.e.,

$$U(x_1,x_2) = U(\mathbf{r}_1,\mathbf{r}_2). \tag{8.28}$$

Consequently, given a solution

$$\varphi_h(x) = \varphi_n(\mathbf{r})\chi_\sigma(s),\ h = (n,\sigma), \tag{8.29}$$

where $\chi_\sigma(s)$ are the standard eigenfunctions of the spin operator $\hat{S}_z$ (see Sect. 1.1), the direct and exchange terms (see Eqs. 8.24 and 8.27) become respectively

$$\varphi_h(x)\left[\int dx_2 \sum_j \left|\varphi_j(x_2)\right|^2 n_\varepsilon(\mathcal{E}_j) U(x,x_2)\right] =$$

$$= \varphi_n(\mathbf{r})\chi_\sigma(s)\left[\sum_{s_2}\int d^3\mathbf{r}_2 \sum_{n'}\sum_{\sigma'} |\varphi_{n'}(\mathbf{r}_2)|^2 \,|\chi_{\sigma'}(s_2)|^2\, n_\varepsilon(\mathcal{E}_{n',\sigma'}) U(\mathbf{r},\mathbf{r}_2)\right] =$$

$$= \varphi_n(\mathbf{r})\chi_\sigma(s)\left[\int d^3\mathbf{r}_2 \sum_{n'}\sum_{\sigma'} |\varphi_{n'}(\mathbf{r}_2)|^2 \sum_{s_2} |\chi_{\sigma'}(s_2)|^2\, n_\varepsilon(\mathcal{E}_{n',\sigma'}) U(\mathbf{r},\mathbf{r}_2)\right] =$$

$$= \varphi_n(\mathbf{r})\chi_\sigma(s)\left[\int d^3\mathbf{r}_2 \sum_{n'}\sum_{\sigma'} |\varphi_{n'}(\mathbf{r}_2)|^2\, n_\varepsilon(\mathcal{E}_{n',\sigma'}) U(\mathbf{r},\mathbf{r}_2)\right], \tag{8.30}$$

$$\varepsilon \int dx_2 U(x,x_2) \sum_j \varphi_j^*(x_2)\varphi_j(x)\varphi_h(x_2) n_\varepsilon(\mathcal{E}_j) =$$

$$= \varepsilon \sum_{s_2}\int d^3\mathbf{r}_2 U(\mathbf{r},\mathbf{r}_2) \sum_{n'}\sum_{\sigma'} \varphi_{n'}^*(\mathbf{r}_2)\varphi_{n'}(\mathbf{r})\varphi_n(\mathbf{r}_2)\chi_{\sigma'}^*(s_2)$$

$$\chi_{\sigma'}(s)\chi_\sigma(s_2) n_\varepsilon(\mathcal{E}_{n',\sigma'}) =$$

$$= \varepsilon \int d^3\mathbf{r}_2 U(\mathbf{r},\mathbf{r}_2) \sum_{n'}\sum_{\sigma'} \varphi_{n'}^*(\mathbf{r}_2)\varphi_{n'}(\mathbf{r})\varphi_n(\mathbf{r}_2)\chi_{\sigma'}(s) n_\varepsilon(\mathcal{E}_{n',\sigma'})$$

$$\left(\sum_{s_2} \chi_{\sigma'}^*(s_2)\chi_\sigma(s_2)\right) =$$

$$= \varepsilon \int d^3\mathbf{r}_2 U(\mathbf{r},\mathbf{r}_2) \sum_{n'}\sum_{\sigma'} \varphi_{n'}^*(\mathbf{r}_2)\varphi_{n'}(\mathbf{r})\varphi_n(\mathbf{r}_2)\chi_{\sigma'}(s) n_\varepsilon(\mathcal{E}_{n',\sigma'})\delta_{\sigma,\sigma'} =$$

$$= \varepsilon \int d^3\mathbf{r}_2 U(\mathbf{r},\mathbf{r}_2) \sum_{n'} \varphi_{n'}^*(\mathbf{r}_2)\varphi_{n'}(\mathbf{r})\varphi_n(\mathbf{r}_2)\chi_\sigma(s) n_\varepsilon(\mathcal{E}_{n',\sigma}), \tag{8.31}$$

and Hartree-Fock equations become

$$\begin{aligned}
\mathcal{E}_{n,\sigma}\varphi_n(\mathbf{r})\chi_\sigma(s) = \hat{\mathcal{H}}_1(\mathbf{r},s)\varphi_n(\mathbf{r})\chi_\sigma(s) + \\
+ \varphi_n(\mathbf{r})\chi_\sigma(s)\left[\int d^3\mathbf{r}_2 \sum_{n'}\sum_{\sigma'} |\varphi_{n'}(\mathbf{r}_2)|^2\, n_\varepsilon(\mathcal{E}_{n',\sigma'}) U(\mathbf{r},\mathbf{r}_2)\right] + \\
+ \varepsilon \int d^3\mathbf{r}_2 U(\mathbf{r},\mathbf{r}_2) \sum_{n'} \varphi_{n'}^*(\mathbf{r}_2)\varphi_{n'}(\mathbf{r})\varphi_n(\mathbf{r}_2)\chi_\sigma(s) n_\varepsilon(\mathcal{E}_{n',\sigma}).
\end{aligned} \tag{8.32}$$

Note that the direct term, Eq. 8.30, involves a summation over all spin states due to the sum over σ', whereas the exchange term, Eq. 8.31, depends exclusively on the spin component σ of the particle described by $\varphi_n(\mathbf{r})\chi_\sigma(s)$. Moreover, assuming that the one-body operator is independent of the spin variable, i.e.,

$$\hat{\mathcal{H}}_1(\mathbf{r}, s) = \hat{\mathcal{H}}_1(\mathbf{r}), \tag{8.33}$$

the spin eigenfunction can be factored out and simplified on both sides of the equation as follows

$$\begin{aligned} \mathcal{E}_{n,\sigma}\varphi_n(\mathbf{r}) &= \hat{\mathcal{H}}_1(\mathbf{r})\varphi_n(\mathbf{r}) + \\ &+ \varphi_n(\mathbf{r})\left[\int d^3\mathbf{r}_2 \sum_{n'}\sum_{\sigma'} |\varphi_{n'}(\mathbf{r}_2)|^2\, n_\varepsilon(\mathcal{E}_{n',\sigma'})U(\mathbf{r}, \mathbf{r}_2)\right] + \\ &+ \varepsilon \int d^3\mathbf{r}_2 U(\mathbf{r}, \mathbf{r}_2) \sum_{n'} \varphi^*_{n'}(\mathbf{r}_2)\varphi_{n'}(\mathbf{r})\varphi_n(\mathbf{r}_2) n_\varepsilon(\mathcal{E}_{n',\sigma}). \end{aligned} \tag{8.34}$$

However, the problem remains quite complex, as it still depends on both the two-body interaction U and the one-body potential V (in $\hat{H}_1$). The Hartree-Fock Eq. 8.34 can be solved exactly only under further simplifying assumptions, namely

- the one-body potential is zero, i.e.,

$$V = 0; \tag{8.35}$$

- the two-body interaction is invariant by spatial translation, i.e.,

$$U(\mathbf{r}_1, \mathbf{r}_2) = U(\mathbf{r}_1 - \mathbf{r}_2); \tag{8.36}$$

- the statistics and the energies do not depend on the spin variable, i.e.,

$$n_\varepsilon(\mathcal{E}_{n,\sigma}) = n_\varepsilon(\mathcal{E}_n), \tag{8.37}$$

$$\mathcal{E}_{n,\sigma} = \mathcal{E}_n, \tag{8.38}$$

  that is, the system is paramagnetic.

Consequently, the eigenvalue equation is modified as follows

$$\begin{aligned} \mathcal{E}_n\varphi_n(\mathbf{r}) &= \hat{\mathcal{H}}_1(\mathbf{r})\varphi_n(\mathbf{r}) + (2S+1)\,\varphi_n(\mathbf{r}) \\ &\left[\int d^3\mathbf{r}_2 \sum_{n'} |\varphi_{n'}(\mathbf{r}_2)|^2\, n_\varepsilon\,(\mathcal{E}_{n'})\,U(\mathbf{r} - \mathbf{r}_2)\right] + \end{aligned}$$

$$+\varepsilon \int d^3\mathbf{r}_2 U(\mathbf{r}-\mathbf{r}_2) \sum_{n'} \varphi^*_{n'}(\mathbf{r}_2)\varphi_{n'}(\mathbf{r})\varphi_n(\mathbf{r}_2) n_\varepsilon\left(\mathcal{E}_{n'}\right), \tag{8.39}$$

where in the direct term the sum over σ' provided the quantity $(2S+1)$, with S denoting the spin of the identical particles (see Sect. 1.1). Now, we show that the plane waves 1.77 satisfy 8.39. Indeed,

$$\begin{aligned}
\mathcal{E}_{\mathbf{k}} \frac{e^{i\mathbf{k}\cdot\mathbf{r}}}{\sqrt{V}} &= \hat{\mathcal{H}}_1(\mathbf{r}) \frac{e^{i\mathbf{k}\cdot\mathbf{r}}}{\sqrt{V}} + (2S+1)\frac{e^{i\mathbf{k}\cdot\mathbf{r}}}{\sqrt{V}}\left[\int d^3\mathbf{r}_2 \sum_{\mathbf{k}'} \frac{1}{V} n_\varepsilon\left(\mathcal{E}_{\mathbf{k}'}\right) U(\mathbf{r}-\mathbf{r}_2)\right] + \\
&+ \varepsilon \int d^3\mathbf{r}_2 U(\mathbf{r}-\mathbf{r}_2) \sum_{\mathbf{k}'} \frac{1}{V^{\frac{3}{2}}} e^{-i\mathbf{k}'\cdot\mathbf{r}_2} e^{i\mathbf{k}'\cdot\mathbf{r}} e^{i\mathbf{k}\cdot\mathbf{r}_2} n_\varepsilon\left(\mathcal{E}_{\mathbf{k}'}\right) = \\
&= \mathcal{E}^{(0)}_{\mathbf{k}} \frac{e^{i\mathbf{k}\cdot\mathbf{r}}}{\sqrt{V}} + (2S+1)\frac{e^{i\mathbf{k}\cdot\mathbf{r}}}{\sqrt{V}}\left[\int d^3\mathbf{r}_2 \sum_{\mathbf{k}'} \frac{1}{V} n_\varepsilon\left(\mathcal{E}_{\mathbf{k}'}\right) U(\mathbf{r}-\mathbf{r}_2)\right] + \\
&+ \varepsilon \int d^3\mathbf{r}_2 U(\mathbf{r}-\mathbf{r}_2) \sum_{\mathbf{k}'} \frac{1}{V^{\frac{3}{2}}} e^{-i\mathbf{k}'\cdot\mathbf{r}_2} e^{i\mathbf{k}'\cdot\mathbf{r}} e^{i\mathbf{k}\cdot\mathbf{r}_2} n_\varepsilon\left(\mathcal{E}_{\mathbf{k}'}\right).
\end{aligned} \tag{8.40}$$

To verify that the plane wave is a solution, we need to modify the exchange term to bring out the object $\frac{e^{i\mathbf{k}\cdot\mathbf{r}}}{\sqrt{V}}$. Here, the two-body operator satisfies

$$U(\mathbf{r}-\mathbf{r}_2) = U(\mathbf{r},\mathbf{r}_2) = U(\mathbf{r}_2,\mathbf{r}) = U(\mathbf{r}_2-\mathbf{r}), \tag{8.41}$$

then we set $\mathbf{r}_2 - \mathbf{r} \to \mathbf{r}_2$, $\mathbf{r}_2 \to \mathbf{r}_2 + \mathbf{r}$, then the direct term becomes

$$\begin{aligned}
&(2S+1)\frac{e^{i\mathbf{k}\cdot\mathbf{r}}}{\sqrt{V}} \sum_{\mathbf{k}'} \frac{1}{V} n_\varepsilon\left(\mathcal{E}_{\mathbf{k}'}\right) \left[\int d^3\mathbf{r}_2 U(\mathbf{r}_2)\right] \\
&= (2S+1)\frac{e^{i\mathbf{k}\cdot\mathbf{r}}}{\sqrt{V}} U(\mathbf{k}=\mathbf{0}) \sum_{\mathbf{k}'} \frac{1}{V} n_\varepsilon\left(\mathcal{E}_{\mathbf{k}'}\right) = \\
&= \frac{N(2S+1)}{V} \frac{e^{i\mathbf{k}\cdot\mathbf{r}}}{\sqrt{V}} U(\mathbf{k}=\mathbf{0}),
\end{aligned} \tag{8.42}$$

and the exchange term becomes

$$\begin{aligned}
&\varepsilon \int d^3\mathbf{r}_2 U(\mathbf{r}_2) \sum_{\mathbf{k}'} \frac{1}{V^{\frac{3}{2}}} e^{-i\mathbf{k}'\cdot(\mathbf{r}_2+\mathbf{r})} e^{i\mathbf{k}'\cdot\mathbf{r}} e^{i\mathbf{k}\cdot(\mathbf{r}_2+\mathbf{r})} n_\varepsilon\left(\mathcal{E}_{\mathbf{k}'}\right) = \\
&= \varepsilon \int d^3\mathbf{r}_2 U(\mathbf{r}_2) \sum_{\mathbf{k}'} \frac{1}{V^{\frac{3}{2}}} e^{-i\mathbf{k}'\cdot\mathbf{r}_2} e^{-i\mathbf{k}'\cdot\mathbf{r}} e^{i\mathbf{k}'\cdot\mathbf{r}} e^{i\mathbf{k}\cdot\mathbf{r}_2} e^{i\mathbf{k}\cdot\mathbf{r}} n_\varepsilon\left(\mathcal{E}_{\mathbf{k}'}\right) =
\end{aligned}$$

$$= \varepsilon \int d^3\mathbf{r}_2 U(\mathbf{r}_2) \sum_{\mathbf{k}'} \frac{1}{V^{\frac{3}{2}}} e^{-i\mathbf{k}'\cdot\mathbf{r}_2} e^{i\mathbf{k}\cdot\mathbf{r}_2} e^{i\mathbf{k}\cdot\mathbf{r}} n_{-1}(\mathcal{E}_{\mathbf{k}'}) =$$

$$= \varepsilon \frac{e^{i\mathbf{k}\cdot\mathbf{r}}}{\sqrt{V}} \sum_{\mathbf{k}'} \frac{1}{V} n_\varepsilon (\mathcal{E}_{\mathbf{k}'}) \left[\int d^3\mathbf{r}_2 U(\mathbf{r}_2) e^{i(\mathbf{k}-\mathbf{k}')\cdot\mathbf{r}_2} \right] =$$

$$= \varepsilon \sum_{\mathbf{k}'} \frac{1}{V} n_\varepsilon (\mathcal{E}_{\mathbf{k}'}) U(\mathbf{k}-\mathbf{k}') \frac{e^{i\mathbf{k}\cdot\mathbf{r}}}{\sqrt{V}}. \tag{8.43}$$

Finally, the plane waves are solutions of the Hartree-Fock equations, and the Hartree-Fock energies are given by

$$\mathcal{E}_{\mathbf{k}}^{\text{HF}} = \mathcal{E}_{\mathbf{k}}^{(0)} + \frac{N(2S+1)}{V} U(\mathbf{k}=\mathbf{0}) + \varepsilon \sum_{\mathbf{k}'} \frac{1}{V} n_\varepsilon \left(\mathcal{E}_{\mathbf{k}'}^{\text{HF}}\right) U(\mathbf{k}-\mathbf{k}'). \tag{8.44}$$

The direct term contributes a uniform energy shift that is independent of the momentum $\mathbf{k}$, while the exchange term introduces a nontrivial, momentum-dependent correction to the one-particle energy spectrum.

8.3 Hartree-Fock Method in the Jellium Model

Equation 8.44 can be specialized to the case of electrons in a solid by considering the grand-canonical Hamiltonian of the jellium model (see Eq. 2.393). In this specific application, the generic fermionic mass m is replaced by the electron mass m_e, while the interaction potential U takes its Coulomb form. Consequently, the one-particle energies read $\mathcal{E}_{\mathbf{k}}^{(0)} = \frac{\hbar^2\mathbf{k}^2}{2m_e} - \mu$, $\varepsilon = -1$ (see Eq. 2.52), and the unified statistics n_ε (see Eq. E.2) is replaced by the Fermi-Dirac statistics n_{-1} (see Eq. E.4). Moreover, since the direct term in Eq. 8.44 contains $U(\mathbf{k}=\mathbf{0})$ and the jellium model excludes the zero-momentum component from the sum over momenta, this term vanishes, and we have

$$\mathcal{E}_{\mathbf{k}}^{\text{HF}} = \mathcal{E}_{\mathbf{k}}^{(0)} - \sum_{\mathbf{k}'\neq\mathbf{k}} \frac{1}{V} n_{-1} \left(\mathcal{E}_{\mathbf{k}'}^{\text{HF}}\right) U(\mathbf{k}-\mathbf{k}') =$$

$$= \mathcal{E}_{\mathbf{k}}^{(0)} - \sum_{\mathbf{k}'\neq\mathbf{k}} \frac{1}{V} n_{-1} \left(\mathcal{E}_{\mathbf{k}'}^{\text{HF}}\right) \frac{4\pi q_e^2}{|\mathbf{k}-\mathbf{k}'|^2}, \tag{8.45}$$

that, by means of the thermodynamic limit C.11, becomes

$$\mathcal{E}_{\mathbf{k}}^{\text{HF}} = \mathcal{E}_{\mathbf{k}}^{(0)} - \text{P. V.} \int \frac{d^3\mathbf{k}'}{(2\pi)^3} n_{-1} \left(\mathcal{E}_{\mathbf{k}'}^{\text{HF}}\right) \frac{4\pi q_e^2}{|\mathbf{k}-\mathbf{k}'|^2}. \tag{8.46}$$

Previously, we showed that the Hartree-Fock propagator 8.6 admits plane-wave solutions under the specific conditions given in Eqs. 8.35–8.38. In this case, the Hartree-Fock propagator takes the explicit form

$$G^{(m)}(\mathbf{r}-\mathbf{r}', i\omega_n) = \sum_{\mathbf{k},\sigma} \frac{e^{-i\mathbf{k}\cdot\mathbf{r}}}{\sqrt{V}} \frac{e^{i\mathbf{k}\cdot\mathbf{r}'}}{\sqrt{V}} \frac{1}{\hbar\left[i\omega_n - \frac{\mathcal{E}^{\mathrm{HF}}_{\mathbf{k}}}{\hbar}\right]}. \tag{8.47}$$

Note that the system is translationally invariant, so all quantities depend only on the difference between two spatial coordinates. Moreover, there is no dependence on the spin component, as we are considering a paramagnetic system. By summing over the spin degrees of freedom, we obtain

$$G^{(m)}(\mathbf{r}-\mathbf{r}', i\omega_n) = \sum_{\mathbf{k}} \frac{e^{i\mathbf{k}\cdot(\mathbf{r}-\mathbf{r}')}}{V} \frac{1}{\hbar\left[i\omega_n - \frac{\mathcal{E}^{\mathrm{HF}}_{\mathbf{k}}}{\hbar}\right]}, \tag{8.48}$$

then we apply a spatial Fourier transform and we get

$$G^{(m)}(\mathbf{k}, i\omega_n) = \frac{1}{\hbar\left[i\omega_n - \frac{\mathcal{E}^{\mathrm{HF}}_{\mathbf{k}}}{\hbar}\right]}, \tag{8.49}$$

which corresponds to the typical form of the Green's function for systems that are invariant under spatial and time translations (see Eqs. 6.116 and 7.89).
We aim to compute the spectral function associated with the Hartree-Fock propagator: from Eq. 6.87, we have

$$A^{\mathrm{HF}}(\mathbf{k}, \omega) = 2\pi\delta\left[\omega - \frac{\mathcal{E}^{\mathrm{HF}}_{\mathbf{k}}}{\hbar}\right], \tag{8.50}$$

where we note that the spectral function is identical in form to that of free particles, and in particular the one-to-one correspondence between energies and momenta is preserved, although the energy is not that of a free particle. Since the true propagator is replaced by the Hartree-Fock one, it is natural to investigate whether the Hartree-Fock spectral function satisfies the same sum rules as the true spectral function $A(\mathbf{k}, \omega)$, i.e., whether it can be substituted for it while still providing the same results. It is straightforward to verify that $A^{\mathrm{HF}}(\mathbf{k}, \omega)$ satisfies 6.235, since the integral of the delta function yields 1, in agreement with the anticommutator between the destruction and creation operators associated with the Hartree-Fock plane-wave solution. We next consider the first moment of the Hartree-Fock spectral function, that is, Eq. 6.240. Regarding the left-hand side of 6.240, we have

$$\int_{-\infty}^{+\infty} \frac{d\omega}{2\pi} \omega A^{\mathrm{HF}}(\mathbf{k}, \omega) = \int_{-\infty}^{+\infty} \frac{d\omega}{2\pi} \omega 2\pi \delta \left[\omega - \frac{\mathcal{E}_{\mathbf{k}}^{\mathrm{HF}}}{\hbar} \right] =$$
$$= \frac{\mathcal{E}_{\mathbf{k}}^{\mathrm{HF}}}{\hbar} =$$
$$= \frac{\mathcal{E}_{\mathbf{k}}^{(0)}}{\hbar} - \frac{1}{\hbar} \mathrm{P.V.} \int \frac{d^3\mathbf{k}'}{(2\pi)^3} n_{-1}\left(\mathcal{E}_{\mathbf{k}'}^{\mathrm{HF}}\right) \frac{4\pi q_e^2}{|\mathbf{k} - \mathbf{k}'|^2}, \tag{8.51}$$

where in the last step we employed Eq. 8.46. According to Eq. 6.240, we need to compute the right-hand side of 6.240, that is,

$$\left\langle \left[\left[C_{\mathbf{k}}, \hat{\mathcal{H}} \right], C_{\mathbf{k}}^{\dagger} \right]^{(\varepsilon)} \right\rangle, \tag{8.52}$$

where $\varepsilon = -1$, then the brackets represent anticommutators (see Eq. 2.102), and the thermal average (see Eq. 1.379) is performed over a Hamiltonian with interaction; here we use the jellium Hamiltonian 2.366, adapted to the case with no spin dependence. We have

$$\left[C_{\mathbf{k}}, \hat{\mathcal{H}} \right] = \sum_{\mathbf{k}'} \mathcal{E}_{\mathbf{k}'}^{(0)} \left[C_{\mathbf{k}}, C_{\mathbf{k}'}^{\dagger} C_{\mathbf{k}'} \right] + \frac{1}{2V} \sum_{\substack{\mathbf{k}_1, \mathbf{k}_2, \\ \mathbf{q} \neq \mathbf{0}}} \frac{4\pi q_e^2}{q^2} \left[C_{\mathbf{k}}, C_{\mathbf{k}_1+\mathbf{q}}^{\dagger} C_{\mathbf{k}_2-\mathbf{q}}^{\dagger} C_{\mathbf{k}_2} C_{\mathbf{k}_1} \right], \tag{8.53}$$

from Eq. D.13 it follows

$$\sum_{\mathbf{k}'} \mathcal{E}_{\mathbf{k}'}^{(0)} \left[C_{\mathbf{k}}, C_{\mathbf{k}'}^{\dagger} C_{\mathbf{k}'} \right] = \sum_{\mathbf{k}'} \mathcal{E}_{\mathbf{k}'}^{(0)} \delta_{\mathbf{k},\mathbf{k}'} C_{\mathbf{k}'} =$$
$$= \mathcal{E}_{\mathbf{k}}^{(0)} C_{\mathbf{k}}, \tag{8.54}$$

Concerning the second commutator given in 8.53, we use the fermionic algebra D.4: first we use $C_{\mathbf{k}} C_{\mathbf{k}_1+\mathbf{q}}^{\dagger} = \delta_{\mathbf{k},\mathbf{k}_1+\mathbf{q}} - C_{\mathbf{k}_1+\mathbf{q}}^{\dagger} C_{\mathbf{k}}$ to move the operator $C_{\mathbf{k}}$ into one of the two products,

$$\left[C_{\mathbf{k}}, C_{\mathbf{k}_1+\mathbf{q}}^{\dagger} C_{\mathbf{k}_2-\mathbf{q}}^{\dagger} C_{\mathbf{k}_2} C_{\mathbf{k}_1} \right] = C_{\mathbf{k}} C_{\mathbf{k}_1+\mathbf{q}}^{\dagger} C_{\mathbf{k}_2-\mathbf{q}}^{\dagger} C_{\mathbf{k}_2} C_{\mathbf{k}_1} - C_{\mathbf{k}_1+\mathbf{q}}^{\dagger} C_{\mathbf{k}_2-\mathbf{q}}^{\dagger} C_{\mathbf{k}_2} C_{\mathbf{k}_1} C_{\mathbf{k}} =$$
$$= \delta_{\mathbf{k},\mathbf{k}_1+\mathbf{q}} C_{\mathbf{k}_2-\mathbf{q}}^{\dagger} C_{\mathbf{k}_2} C_{\mathbf{k}_1} - C_{\mathbf{k}_1+\mathbf{q}}^{\dagger} C_{\mathbf{k}} C_{\mathbf{k}_2-\mathbf{q}}^{\dagger} C_{\mathbf{k}_2} C_{\mathbf{k}_1} +$$
$$- C_{\mathbf{k}_1+\mathbf{q}}^{\dagger} C_{\mathbf{k}_2-\mathbf{q}}^{\dagger} C_{\mathbf{k}_2} C_{\mathbf{k}_1} C_{\mathbf{k}} =$$
$$= \delta_{\mathbf{k},\mathbf{k}_1+\mathbf{q}} C_{\mathbf{k}_2-\mathbf{q}}^{\dagger} C_{\mathbf{k}_2} C_{\mathbf{k}_1} - \delta_{\mathbf{k},\mathbf{k}_2-\mathbf{q}} C_{\mathbf{k}_1+\mathbf{q}}^{\dagger} C_{\mathbf{k}_2} C_{\mathbf{k}_1} +$$

$$+ C^{\dagger}_{\mathbf{k}_1+\mathbf{q}} C^{\dagger}_{\mathbf{k}_2-\mathbf{q}} C_{\mathbf{k}} C_{\mathbf{k}_2} C_{\mathbf{k}_1} - C^{\dagger}_{\mathbf{k}_1+\mathbf{q}} C^{\dagger}_{\mathbf{k}_2-\mathbf{q}} C_{\mathbf{k}_2} C_{\mathbf{k}_1} C_{\mathbf{k}}, \tag{8.55}$$

and from Eq. D.4 with $\alpha \neq \beta$, it follows $C_{\mathbf{k}} C_{\mathbf{k}_2} C_{\mathbf{k}_1} = -C_{\mathbf{k}_2} C_{\mathbf{k}} C_{\mathbf{k}_1} = C_{\mathbf{k}_2} C_{\mathbf{k}_1} C_{\mathbf{k}}$, then

$$\left[C_{\mathbf{k}}, C^{\dagger}_{\mathbf{k}_1+\mathbf{q}} C^{\dagger}_{\mathbf{k}_2-\mathbf{q}} C_{\mathbf{k}_2} C_{\mathbf{k}_1}\right] = \delta_{\mathbf{k},\mathbf{k}_1+\mathbf{q}} C^{\dagger}_{\mathbf{k}_2-\mathbf{q}} C_{\mathbf{k}_2} C_{\mathbf{k}_1} - \delta_{\mathbf{k},\mathbf{k}_2-\mathbf{q}} C^{\dagger}_{\mathbf{k}_1+\mathbf{q}} C_{\mathbf{k}_2} C_{\mathbf{k}_1}, \tag{8.56}$$

$$\begin{aligned}
&\frac{1}{2V} \sum_{\substack{\mathbf{k}_1,\mathbf{k}_2,\\ \mathbf{q}\neq\mathbf{0}}} \frac{4\pi q_e^2}{q^2} \left[C_{\mathbf{k}}, C^{\dagger}_{\mathbf{k}_1+\mathbf{q}} C^{\dagger}_{\mathbf{k}_2-\mathbf{q}} C_{\mathbf{k}_2} C_{\mathbf{k}_1}\right] = \\
&= \frac{1}{2V} \sum_{\substack{\mathbf{k}_1,\mathbf{k}_2,\\ \mathbf{q}\neq\mathbf{0}}} \frac{4\pi q_e^2}{q^2} \delta_{\mathbf{k},\mathbf{k}_1+\mathbf{q}} C^{\dagger}_{\mathbf{k}_2-\mathbf{q}} C_{\mathbf{k}_2} C_{\mathbf{k}_1} \\
&\quad - \frac{1}{2V} \sum_{\substack{\mathbf{k}_1,\mathbf{k}_2,\\ \mathbf{q}\neq\mathbf{0}}} \frac{4\pi q_e^2}{q^2} \delta_{\mathbf{k},\mathbf{k}_2-\mathbf{q}} C^{\dagger}_{\mathbf{k}_1+\mathbf{q}} C_{\mathbf{k}_2} C_{\mathbf{k}_1} = \\
&= \frac{1}{2V} \sum_{\substack{\mathbf{k}_2,\\ \mathbf{q}\neq\mathbf{0}}} \frac{4\pi q_e^2}{q^2} C^{\dagger}_{\mathbf{k}_2-\mathbf{q}} C_{\mathbf{k}_2} C_{\mathbf{k}-\mathbf{q}} - \frac{1}{2V} \sum_{\substack{\mathbf{k}_1,\\ \mathbf{q}\neq\mathbf{0}}} \frac{4\pi q_e^2}{q^2} C^{\dagger}_{\mathbf{k}_1+\mathbf{q}} C_{\mathbf{k}+\mathbf{q}} C_{\mathbf{k}_1} = \\
&= \frac{1}{2V} \sum_{\substack{\mathbf{k}_2,\\ \mathbf{q}\neq\mathbf{0}}} \frac{4\pi q_e^2}{q^2} C^{\dagger}_{\mathbf{k}_2-\mathbf{q}} C_{\mathbf{k}_2} C_{\mathbf{k}-\mathbf{q}} + \frac{1}{2V} \sum_{\substack{\mathbf{k}_1,\\ \mathbf{q}\neq\mathbf{0}}} \frac{4\pi q_e^2}{q^2} C^{\dagger}_{\mathbf{k}_1+\mathbf{q}} C_{\mathbf{k}_1} C_{\mathbf{k}+\mathbf{q}},
\end{aligned} \tag{8.57}$$

$$\left[C_{\mathbf{k}}, \hat{\mathcal{H}}\right] = \mathcal{E}^{(0)}_{\mathbf{k}} C_{\mathbf{k}} + \sum_{\substack{\mathbf{k}_2,\\ \mathbf{q}\neq\mathbf{0}}} \frac{4\pi q_e^2}{q^2} C^{\dagger}_{\mathbf{k}_2-\mathbf{q}} C_{\mathbf{k}_2} C_{\mathbf{k}-\mathbf{q}} + \sum_{\substack{\mathbf{k}_1,\\ \mathbf{q}\neq\mathbf{0}}} \frac{4\pi q_e^2}{q^2} C^{\dagger}_{\mathbf{k}_1+\mathbf{q}} C_{\mathbf{k}_1} C_{\mathbf{k}+\mathbf{q}}. \tag{8.58}$$

We now consider the anticommutator

$$\begin{aligned}
\left\{\left[C_{\mathbf{k}}, \hat{\mathcal{H}}\right], C^{\dagger}_{\mathbf{k}}\right\} &= \mathcal{E}^{(0)}_{\mathbf{k}} \left\{C_{\mathbf{k}}, C^{\dagger}_{\mathbf{k}}\right\} + \frac{1}{2V} \sum_{\substack{\mathbf{k}_2,\\ \mathbf{q}\neq\mathbf{0}}} \frac{4\pi q_e^2}{q^2} \left\{C^{\dagger}_{\mathbf{k}_2-\mathbf{q}} C_{\mathbf{k}_2} C_{\mathbf{k}-\mathbf{q}}, C^{\dagger}_{\mathbf{k}}\right\} + \\
&\quad + \frac{1}{2V} \sum_{\substack{\mathbf{k}_1,\\ \mathbf{q}\neq\mathbf{0}}} \frac{4\pi q_e^2}{q^2} \left\{C^{\dagger}_{\mathbf{k}_1+\mathbf{q}} C_{\mathbf{k}_1} C_{\mathbf{k}+\mathbf{q}}, C^{\dagger}_{\mathbf{k}}\right\}.
\end{aligned} \tag{8.59}$$

We now need to compute the remaining anticommutators. Regarding the first one, from the fermionic algebra D.4, it follows

$$
\begin{aligned}
\left\{C^{\dagger}_{\mathbf{k}_2-\mathbf{q}}C_{\mathbf{k}_2}C_{\mathbf{k}-\mathbf{q}},\, C^{\dagger}_{\mathbf{k}}\right\} &= C^{\dagger}_{\mathbf{k}_2-\mathbf{q}}C_{\mathbf{k}_2}C_{\mathbf{k}-\mathbf{q}}C^{\dagger}_{\mathbf{k}} + C^{\dagger}_{\mathbf{k}}C^{\dagger}_{\mathbf{k}_2-\mathbf{q}}C_{\mathbf{k}_2}C_{\mathbf{k}-\mathbf{q}} = \\
&= C^{\dagger}_{\mathbf{k}_2-\mathbf{q}}C_{\mathbf{k}_2}C_{\mathbf{k}-\mathbf{q}}C^{\dagger}_{\mathbf{k}} - C^{\dagger}_{\mathbf{k}_2-\mathbf{q}}C^{\dagger}_{\mathbf{k}}C_{\mathbf{k}_2}C_{\mathbf{k}-\mathbf{q}} = \\
&= C^{\dagger}_{\mathbf{k}_2-\mathbf{q}}C_{\mathbf{k}_2}C_{\mathbf{k}-\mathbf{q}}C^{\dagger}_{\mathbf{k}} - \delta_{\mathbf{k},\mathbf{k}_2}C^{\dagger}_{\mathbf{k}_2-\mathbf{q}}C_{\mathbf{k}-\mathbf{q}} + \\
&+ C^{\dagger}_{\mathbf{k}_2-\mathbf{q}}C_{\mathbf{k}_2}C^{\dagger}_{\mathbf{k}}C_{\mathbf{k}-\mathbf{q}} = \\
&= C^{\dagger}_{\mathbf{k}_2-\mathbf{q}}C_{\mathbf{k}_2}C_{\mathbf{k}-\mathbf{q}}C^{\dagger}_{\mathbf{k}} - \delta_{\mathbf{k},\mathbf{k}_2}C^{\dagger}_{\mathbf{k}_2-\mathbf{q}}C_{\mathbf{k}-\mathbf{q}} + \\
&+ \delta_{\mathbf{k},\mathbf{k}-\mathbf{q}}C^{\dagger}_{\mathbf{k}_2-\mathbf{q}}C_{\mathbf{k}_2} - C^{\dagger}_{\mathbf{k}_2-\mathbf{q}}C_{\mathbf{k}_2}C_{\mathbf{k}-\mathbf{q}}C^{\dagger}_{\mathbf{k}} = \\
&= -\delta_{\mathbf{k},\mathbf{k}_2}C^{\dagger}_{\mathbf{k}_2-\mathbf{q}}C_{\mathbf{k}-\mathbf{q}} + \delta_{\mathbf{k},\mathbf{k}-\mathbf{q}}C^{\dagger}_{\mathbf{k}_2-\mathbf{q}}C_{\mathbf{k}_2},
\end{aligned}
\tag{8.60}
$$

$$
\begin{aligned}
&\sum_{\substack{\mathbf{k}_2,\\ \mathbf{q}\neq\mathbf{0}}}\frac{4\pi q_e^2}{q^2}\left\{C^{\dagger}_{\mathbf{k}_2-\mathbf{q}}C_{\mathbf{k}_2}C_{\mathbf{k}-\mathbf{q}},\, C^{\dagger}_{\mathbf{k}}\right\} \\
&\quad = -\frac{4\pi q_e^2}{q^2}\sum_{\substack{\mathbf{k}_2,\\ \mathbf{q}\neq\mathbf{0}}}\delta_{\mathbf{k},\mathbf{k}_2}C^{\dagger}_{\mathbf{k}_2-\mathbf{q}}C_{\mathbf{k}-\mathbf{q}} + \sum_{\substack{\mathbf{k}_2,\\ \mathbf{q}\neq\mathbf{0}}}\frac{4\pi q_e^2}{q^2}\delta_{\mathbf{k},\mathbf{k}-\mathbf{q}}C^{\dagger}_{\mathbf{k}_2-\mathbf{q}}C_{\mathbf{k}_2},
\end{aligned}
\tag{8.61}
$$

note that in the second term, since $\mathbf{q}\neq\mathbf{0}$, the Kronecker delta $\delta_{\mathbf{k},\mathbf{k}-\mathbf{q}}$ is always zero, and therefore

$$
\begin{aligned}
\frac{1}{2V}\sum_{\substack{\mathbf{k}_2,\\ \mathbf{q}\neq\mathbf{0}}}\frac{4\pi q_e^2}{q^2}\left\{C^{\dagger}_{\mathbf{k}_2-\mathbf{q}}C_{\mathbf{k}_2}C_{\mathbf{k}-\mathbf{q}},\, C^{\dagger}_{\mathbf{k}}\right\} &= -\sum_{\substack{\mathbf{k}_2,\\ \mathbf{q}\neq\mathbf{0}}}\frac{4\pi q_e^2}{q^2}\delta_{\mathbf{k},\mathbf{k}_2}C^{\dagger}_{\mathbf{k}_2-\mathbf{q}}C_{\mathbf{k}-\mathbf{q}} = \\
&= -\frac{1}{2V}\sum_{\mathbf{q}\neq\mathbf{0}}\frac{4\pi q_e^2}{q^2}C^{\dagger}_{\mathbf{k}-\mathbf{q}}C_{\mathbf{k}-\mathbf{q}}.
\end{aligned}
\tag{8.62}
$$

Similarly, for the second anticommutator we have

$$
\begin{aligned}
\left\{C^{\dagger}_{\mathbf{k}_1+\mathbf{q}}C_{\mathbf{k}_1}C_{\mathbf{k}+\mathbf{q}},\, C^{\dagger}_{\mathbf{k}}\right\} &= C^{\dagger}_{\mathbf{k}_1+\mathbf{q}}C_{\mathbf{k}_1}C_{\mathbf{k}+\mathbf{q}}C^{\dagger}_{\mathbf{k}} + C^{\dagger}_{\mathbf{k}}C^{\dagger}_{\mathbf{k}_1+\mathbf{q}}C_{\mathbf{k}_1}C_{\mathbf{k}+\mathbf{q}} = \\
&= C^{\dagger}_{\mathbf{k}_1+\mathbf{q}}C_{\mathbf{k}_1}C_{\mathbf{k}+\mathbf{q}}C^{\dagger}_{\mathbf{k}} - C^{\dagger}_{\mathbf{k}_1+\mathbf{q}}C^{\dagger}_{\mathbf{k}}C_{\mathbf{k}_1}C_{\mathbf{k}+\mathbf{q}} = \\
&= C^{\dagger}_{\mathbf{k}_1+\mathbf{q}}C_{\mathbf{k}_1}C_{\mathbf{k}+\mathbf{q}}C^{\dagger}_{\mathbf{k}} - \delta_{\mathbf{k},\mathbf{k}_1}C^{\dagger}_{\mathbf{k}_1+\mathbf{q}}C_{\mathbf{k}+\mathbf{q}} + \\
&+ C^{\dagger}_{\mathbf{k}_1+\mathbf{q}}C_{\mathbf{k}_1}C^{\dagger}_{\mathbf{k}}C_{\mathbf{k}+\mathbf{q}} =
\end{aligned}
$$

$$= C^{\dagger}_{\mathbf{k}_1+\mathbf{q}} C_{\mathbf{k}_1} C_{\mathbf{k}+\mathbf{q}} C^{\dagger}_{\mathbf{k}} - \delta_{\mathbf{k},\mathbf{k}_1} C^{\dagger}_{\mathbf{k}_1+\mathbf{q}} C_{\mathbf{k}+\mathbf{q}} + \\ + \delta_{\mathbf{k},\mathbf{k}+\mathbf{q}} C^{\dagger}_{\mathbf{k}_1+\mathbf{q}} C_{\mathbf{k}_1} - C^{\dagger}_{\mathbf{k}_1+\mathbf{q}} C_{\mathbf{k}_1} C_{\mathbf{k}+\mathbf{q}} C^{\dagger}_{\mathbf{k}}, \tag{8.63}$$

$$\frac{1}{2V} \sum_{\substack{\mathbf{k}_1,\\ \mathbf{q}\neq\mathbf{0}}} \frac{4\pi q_e^2}{q^2} \left\{ C^{\dagger}_{\mathbf{k}_1+\mathbf{q}} C_{\mathbf{k}_1} C_{\mathbf{k}+\mathbf{q}}, C^{\dagger}_{\mathbf{k}} \right\} = -\frac{1}{2V} \sum_{\substack{\mathbf{k}_1,\\ \mathbf{q}\neq\mathbf{0}}} \frac{4\pi q_e^2}{q^2} \delta_{\mathbf{k},\mathbf{k}_1} C^{\dagger}_{\mathbf{k}_1+\mathbf{q}} C_{\mathbf{k}+\mathbf{q}} = \\ = -\frac{1}{2V} \sum_{\mathbf{q}\neq\mathbf{0}} \frac{4\pi q_e^2}{q^2} C^{\dagger}_{\mathbf{k}+\mathbf{q}} C_{\mathbf{k}+\mathbf{q}}. \tag{8.64}$$

Combining the results, we have

$$\left\{ \left[C_{\mathbf{k}}, \hat{\mathcal{H}} \right], C^{\dagger}_{\mathbf{k}} \right\} = \mathcal{E}^{(0)}_{\mathbf{k}} \mathbb{1} - \frac{1}{2V} \sum_{\mathbf{q}\neq\mathbf{0}} \frac{4\pi q_e^2}{q^2} \left(C^{\dagger}_{\mathbf{k}-\mathbf{q}} C_{\mathbf{k}-\mathbf{q}} + C^{\dagger}_{\mathbf{k}+\mathbf{q}} C_{\mathbf{k}+\mathbf{q}} \right), \tag{8.65}$$

where we employed fermionic algebra D.4. Then, we set $\mathbf{k}' = \mathbf{k} - \mathbf{q}$ in the first term of the sum, from which it follows that $\mathbf{q} = \mathbf{k} - \mathbf{k}'$, and we set $\mathbf{k}' = \mathbf{k} + \mathbf{q}$ in the second term of the sum, from which $\mathbf{q} = \mathbf{k}' - \mathbf{k}$, thus

$$-\frac{1}{2V} \sum_{\mathbf{q}\neq\mathbf{0}} \frac{4\pi q_e^2}{q^2} \left(C^{\dagger}_{\mathbf{k}-\mathbf{q}} C_{\mathbf{k}-\mathbf{q}} + C^{\dagger}_{\mathbf{k}+\mathbf{q}} C_{\mathbf{k}+\mathbf{q}} \right) = -\frac{1}{2V} 2 \sum_{\mathbf{k}'\neq\mathbf{k}} \frac{4\pi q_e^2}{|\mathbf{k} - \mathbf{k}'|^2} C^{\dagger}_{\mathbf{k}'} C_{\mathbf{k}'}, \tag{8.66}$$

and Eq. 8.65 can be rewritten as

$$\left\{ \left[C_{\mathbf{k}}, \hat{\mathcal{H}} \right], C^{\dagger}_{\mathbf{k}} \right\} = \mathcal{E}^{(0)}_{\mathbf{k}} \mathbb{1} - \frac{1}{V} \sum_{\mathbf{k}'\neq\mathbf{k}} \frac{4\pi q_e^2}{|\mathbf{k} - \mathbf{k}'|^2} C^{\dagger}_{\mathbf{k}'} C_{\mathbf{k}'}. \tag{8.67}$$

We apply a thermal average to Eq. 8.67 so that

$$\left\langle \left\{ \left[C_{\mathbf{k}}, \hat{\mathcal{H}} \right], C^{\dagger}_{\mathbf{k}} \right\} \right\rangle = \mathcal{E}^{(0)}_{\mathbf{k}} \langle \mathbb{1} \rangle - \frac{1}{V} \sum_{\mathbf{k}'\neq\mathbf{k}} \frac{4\pi q_e^2}{|\mathbf{k} - \mathbf{k}'|^2} \left\langle C^{\dagger}_{\mathbf{k}'} C_{\mathbf{k}'} \right\rangle = \\ = \mathcal{E}^{(0)}_{\mathbf{k}} - \frac{1}{V} \sum_{\mathbf{k}'\neq\mathbf{k}} n_{-1}\left(\mathcal{E}_{\mathbf{k}'}\right) \frac{4\pi q_e^2}{|\mathbf{k} - \mathbf{k}'|^2}, \tag{8.68}$$

where we employed Eq. 2.251 and we emphasize that the Fermi-Dirac distribution $n_{-1}(\mathcal{E}_{\mathbf{k}'})$ refers to the eigenvalues $\mathcal{E}_{\mathbf{k}'}$ of the jellium model. Ultimately, by applying the thermodynamic limit (see Eq. C.11), one finds

$$\left\langle\left\{\left[c_{\mathbf{k}}, \hat{\mathcal{H}}\right], c_{\mathbf{k}}^{\dagger}\right\}\right\rangle = \mathcal{E}_{\mathbf{k}}^{(0)} - \text{P.V.} \int \frac{d^3\mathbf{k}'}{(2\pi)^3} n_{-1}\left(\mathcal{E}_{\mathbf{k}'}\right) \frac{4\pi q_e^2}{|\mathbf{k}-\mathbf{k}'|^2}. \tag{8.69}$$

We divide Eq. 8.69 by the reduced Planck constant, i.e.,

$$\frac{1}{\hbar}\left\langle\left\{\left[c_{\mathbf{k}}, \hat{\mathcal{H}}\right], c_{\mathbf{k}}^{\dagger}\right\}\right\rangle = \frac{\mathcal{E}_{\mathbf{k}}^{(0)}}{\hbar} - \frac{1}{\hbar}\text{P.V.} \int \frac{d^3\mathbf{k}'}{(2\pi)^3} n_{-1}\left(\mathcal{E}_{\mathbf{k}'}\right) \frac{4\pi q_e^2}{|\mathbf{k}-\mathbf{k}'|^2}, \tag{8.70}$$

which is not equal to Eq. 8.51. Indeed, $n_{-1}\left(\mathcal{E}_{\mathbf{k}'}\right)$ given in 8.70 is the Fermi-Dirac distribution evaluated at the eigenvalues of the jellium Hamiltonian; on the other hand, Eq. 8.51 involves $n_{-1}\left(\mathcal{E}_{\mathbf{k}'}^{\text{HF}}\right)$, evaluated at the Hartree-Fock eigenvalues. Consequently, the first moment sum rule 6.240 is not satisfied by the Hartree-Fock spectral function $A_{\mathbf{k}}^{\text{HF}}(\omega)$, unlike the zeroth moment sum rule, and this discrepancy worsens as the order of the moments increases.

Let us then compute the energies 8.46 in a simple case at zero temperature, so that the Fermi-Dirac functions $n_{-1}\left(\mathcal{E}_{\mathbf{k}'}^{\text{HF}}\right)$ are replaced by step functions and the integral can be evaluated in spherical polar coordinates. We have

$$\begin{aligned}
\mathcal{E}_{\mathbf{k}}^{\text{HF}}\Big|_{T=0} &= \mathcal{E}_{\mathbf{k}}^{(0)} - \int_0^{k_F} \frac{2\pi k'^2}{(2\pi)^3} dk' \,\text{P.V.} \int_{-1}^{+1} d\cos(\theta) \frac{4\pi q_e^2}{k^2 + k'^2 - 2kk'\cos(\theta)} = \\
&= \mathcal{E}_{\mathbf{k}}^{(0)} - \frac{q_e^2}{\pi} \int_0^{k_F} dk' \frac{k'}{k} \ln\left|\frac{k+k'}{k-k'}\right| = \\
&= \mathcal{E}_{\mathbf{k}}^{(0)} - \frac{q_e^2}{\pi}\left[1 + \frac{1-y^2}{2y}\ln\left|\frac{1+y}{1-y}\right|\right] = \\
&= \mathcal{E}_{\mathbf{k}}^{(0)} + \frac{q_e^2}{\pi} H(y),
\end{aligned} \tag{8.71}$$

where we defined (see Fig. 8.1)

$$y = \frac{k}{k_F}, \tag{8.72}$$

$$H(y) = -\left[1 + \frac{1-y^2}{2y}\ln\left|\frac{1+y}{1-y}\right|\right], \tag{8.73}$$

The asymptotic behaviors of $H(y)$ for the regimes $y \ll 1$, $y = 1$, and $y \gg 1$ correspond, respectively, to points inside the Fermi sphere, on the Fermi surface, and outside the Fermi sphere (see Appendix O). Employing the series expansion

$$\ln\left|\frac{1+y}{1-y}\right| = 2\sum_{n=0}^{+\infty} \frac{y^{2n+1}}{2n+1}, \tag{8.74}$$

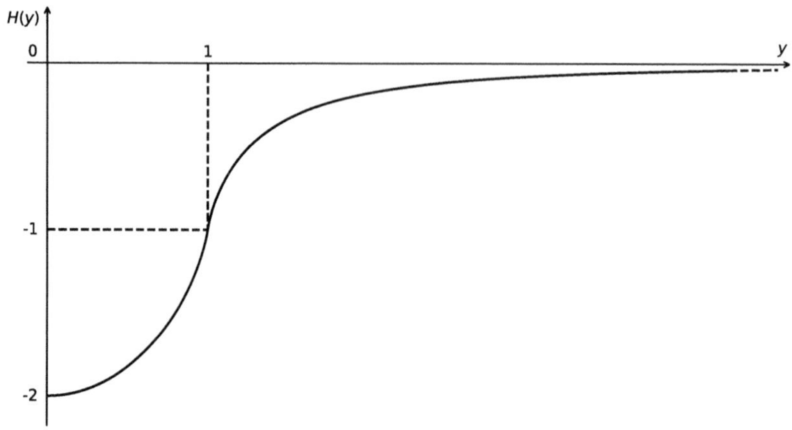

Fig. 8.1 Plot of the function $H(y)$ defined in Eq. 8.73, expressed in terms of the dimensionless variable $y = \frac{k}{k_F}$. This function arises in the evaluation of the Hartree-Fock energies of the jellium model at zero temperature (see Eq. 8.46)

we note that Eq. 8.73 satisfies

$$\lim_{y \to 0^+} H(y) = -2. \tag{8.75}$$

In addition,

$$\lim_{y \to +\infty} H(y) = 0, \tag{8.76}$$

$$\lim_{y \to 1} H(y) = -1, \tag{8.77}$$

as shown in Fig. 8.1. In addition, it can be readily verified that

$$\left. \frac{dS(y)}{dy} \right|_{y=1} \to +\infty \tag{8.78}$$

Specifically, the derivative of the function exhibits a logarithmic divergence at $y = 1$, corresponding to the Fermi surface, indicating a very slow divergence. The Hartree-Fock energies in the jellium model at zero temperature, as given by Eq. 8.71, can be expressed equivalently as

$$\left. \mathcal{E}_{\mathbf{k}}^{\mathrm{HF}} \right|_{T=0} = \mathcal{E}_{\mathbf{k}}^{(0)} + \Sigma(\mathbf{k}), \tag{8.79}$$

where

$$\Sigma(\mathbf{k}) = -\mathrm{P.\,V.} \int \frac{d^3\mathbf{k}'}{(2\pi)^3} n_{-1}\left(\mathcal{E}_{\mathbf{k}'}^{\mathrm{HF}}\right) \frac{4\pi q_e^2}{|\mathbf{k}-\mathbf{k}'|^2} \tag{8.80}$$

is the self-energy (see Sect. 7.4) in the Hartree-Fock model. It should be noted that Eq. 8.46 represents a static energy within the assumptions adopted here. In order to examine the relative behavior of the spectrum in the vicinity of the Fermi surface, it is convenient to consider the energy difference

$$\mathcal{E}_{\mathbf{k}}^{\mathrm{HF}} - \mathcal{E}_{\mathbf{k}_F}^{\mathrm{HF}} = \left[\mathcal{E}_{\mathbf{k}}^{(0)} - \mathcal{E}_{\mathbf{k}_F}^{(0)}\right] + \left[\Sigma(\mathbf{k}) - \Sigma(\mathbf{k}_F)\right]. \tag{8.81}$$

We have

$$\begin{aligned} \mathcal{E}_{\mathbf{k}}^{(0)} - \mathcal{E}_{\mathbf{k}_F}^{(0)} &= \frac{\hbar^2\mathbf{k}^2}{2m_e} - \frac{\hbar^2\mathbf{k}_F^2}{2m_e} = \\ &= \frac{\hbar^2}{2m_e}(\mathbf{k}+\mathbf{k}_F)\cdot(\mathbf{k}-\mathbf{k}_F) \end{aligned} \tag{8.82}$$

which, in the limit $\mathbf{k} \to \mathbf{k}_F$, can be approximated as

$$\begin{aligned} \mathcal{E}_{\mathbf{k}}^{(0)} - \mathcal{E}_{\mathbf{k}_F}^{(0)} &\sim \frac{\hbar^2}{2m_e} 2\mathbf{k}_F \cdot (\mathbf{k}-\mathbf{k}_F) = \\ &= \frac{\hbar\mathbf{k}_F}{m_e} \cdot (\hbar\mathbf{k} - \hbar\mathbf{k}_F) \equiv \\ &\equiv \mathbf{v}_F \cdot (\hbar\mathbf{k} - \hbar\mathbf{k}_F), \end{aligned} \tag{8.83}$$

where we have explicitly defined the Fermi velocity

$$\mathbf{v}_F = \frac{\hbar\mathbf{k}_F}{m_e}. \tag{8.84}$$

Furthermore, from

$$\Sigma(\mathbf{k}) - \Sigma(\mathbf{k}_F) \sim \nabla_{\mathbf{k}}\Sigma(\mathbf{k})|_{\mathbf{k}=\mathbf{k}_F} \cdot (\mathbf{k}-\mathbf{k}_F), \tag{8.85}$$

combining the results, we have

$$\left.\mathcal{E}_{\mathbf{k}}^{\mathrm{HF}}\right|_{T=0} \xrightarrow[\mathbf{k}\sim\mathbf{k}_F]{} \hbar(\mathbf{k}-\mathbf{k}_F)\cdot\mathbf{v}_F^{\mathrm{HF}}, \tag{8.86}$$

where

$$\mathbf{v}_F^{\mathrm{HF}} = \mathbf{v}_F + \nabla_{\mathbf{k}}\Sigma(\mathbf{k})|_{\mathbf{k}=\mathbf{k}_F} \tag{8.87}$$

is the Fermi velocity in the Hartree-Fock model. We have shown that near the Fermi surface, a linear dispersion relation holds. From Eq. 8.71, it follows

$$\nabla_{\mathbf{k}} \Sigma(\mathbf{k})|_{\mathbf{k}=\mathbf{k}_F} = \frac{q_e^2}{\pi} \left. \frac{dS(y)}{dy} \right|_{y=1}, \tag{8.88}$$

and thus

$$\mathbf{v}_F^{\mathrm{HF}} \xrightarrow[\mathbf{k}\to\mathbf{k}_F]{} \infty. \tag{8.89}$$

Since the effective mass of Hartree-Fock quasiparticles can be derived from

$$\mathbf{p}^{\mathrm{HF}} = m^{\mathrm{HF}} \mathbf{v}^{\mathrm{HF}}, \tag{8.90}$$

and the Fermi momentum must remain constant, it follows that

$$m^{\mathrm{HF}} \xrightarrow[\mathbf{k}\to\mathbf{k}_F]{} 0. \tag{8.91}$$

Consequently, the quasiparticles exhibit a mass tending to zero and a velocity tending to infinity near the Fermi surface. It can be shown that this divergence is a direct consequence of the long-range nature of the Coulomb interaction. In fact, if the two-body Coulomb potential U is replaced by a short-range interaction, the singular behavior disappears: the Fermi velocity remains finite, and the Hartree-Fock quasiparticle mass approaches a finite, non-zero value at the Fermi surface.

It should be noted that physical quantities calculated within the Hartree-Fock approximation, such as the specific heat, do not quantitatively reproduce experimental results. This discrepancy originates from the use of the bare Coulomb potential, which is long-ranged and gives rise to logarithmic divergences that are not observed experimentally. The underlying reason is that only first-order diagrams are taken into account, where the free propagators are replaced by the full Hartree-Fock propagators, while the interaction potential itself remains unrenormalized. To properly incorporate the modification of the interaction, one must include the response of the medium, equivalently, evaluate the dielectric function (see Chap. 10), so that the resulting effective potential becomes short-ranged, as will be demonstrated below.

8.4 Hartree-Fock Pair Correlation Function in the Jellium Model

Here we specialize the analysis of the pair correlation function 2.349 within the Hartree-Fock framework for the jellium model (see Eq. 2.393), explicitly expressing the field operators in the plane-wave basis tensored with the spin eigenstates of $\hat{S}_z$ (see Eqs. 2.163 and 2.164). The expectation value $\left\langle \hat{\psi}^\dagger(\mathbf{r}, s)\hat{\psi}(\mathbf{r}, s)\right\rangle$ corresponds to the local density of particles with spin projection s, and it quantifies the average number of particles per unit volume occupying the spin state s at position $\mathbf{r}$. Since the jellium model describes a spatially homogeneous electron gas, the system exhibits no preferential spin orientation. Consequently, the electronic population is equally shared between the two spin projections, i.e.,

$$\left\langle \hat{\psi}^\dagger(\mathbf{r}, s)\hat{\psi}(\mathbf{r}, s)\right\rangle = \frac{N}{2V}, \tag{8.92}$$

and the denominator of Eq. 2.349 is given by

$$\left\langle \hat{\psi}^\dagger(\mathbf{r}, s)\hat{\psi}(\mathbf{r}, s)\right\rangle\left\langle \hat{\psi}^\dagger(\mathbf{r}', s')\hat{\psi}(\mathbf{r}', s')\right\rangle = \frac{N^2}{4V^2}. \tag{8.93}$$

As for the numerator of Eq. 2.349, we have

$$\begin{aligned}
&\left\langle \hat{\psi}^\dagger(\mathbf{r}, s)\hat{\psi}^\dagger(\mathbf{r}', s')\hat{\psi}(\mathbf{r}', s')\hat{\psi}(\mathbf{r}, s)\right\rangle = \\
&= \sum_{\mathbf{k}_1,\sigma_1}\sum_{\mathbf{k}_2,\sigma_2}\sum_{\mathbf{k}_3,\sigma_3}\sum_{\mathbf{k}_4,\sigma_4} \frac{e^{-i\mathbf{k}_1\cdot\mathbf{r}}}{\sqrt{V}}\frac{e^{-i\mathbf{k}_2\cdot\mathbf{r}'}}{\sqrt{V}}\frac{e^{i\mathbf{k}_3\cdot\mathbf{r}'}}{\sqrt{V}}\frac{e^{i\mathbf{k}_4\cdot\mathbf{r}}}{\sqrt{V}}\chi^*_{\sigma_1}(s)\chi^*_{\sigma_2}(s')\chi_{\sigma_3}(s')\chi_{\sigma_4}(s) \\
&\quad \left\langle C^\dagger_{\mathbf{k}_1,\sigma_1}C^\dagger_{\mathbf{k}_2,\sigma_2}C_{\mathbf{k}_3,\sigma_3}C_{\mathbf{k}_4,\sigma_4}\right\rangle = \\
&= \frac{1}{V^2}\sum_{\mathbf{k}_1,\mathbf{k}_2,\mathbf{k}_3,\mathbf{k}_4}\sum_{\sigma_1,\sigma_2} e^{-i(\mathbf{k}_1-\mathbf{k}_4)\cdot\mathbf{r}}e^{-i(\mathbf{k}_2-\mathbf{k}_3)\cdot\mathbf{r}'}\delta_{\sigma_1,\sigma_4}\delta_{\sigma_2,\sigma_3} \\
&\quad \left\langle C^\dagger_{\mathbf{k}_1,\sigma_1}C^\dagger_{\mathbf{k}_2,\sigma_2}C_{\mathbf{k}_3,\sigma_3}C_{\mathbf{k}_4,\sigma_4}\right\rangle = \\
&= \frac{1}{V^2}\sum_{\mathbf{k}_1,\mathbf{k}_2,\mathbf{k}_3,\mathbf{k}_4} e^{-i(\mathbf{k}_1-\mathbf{k}_4)\cdot\mathbf{r}}e^{-i(\mathbf{k}_2-\mathbf{k}_3)\cdot\mathbf{r}'}\left\langle C^\dagger_{\mathbf{k}_1,\sigma_4}C^\dagger_{\mathbf{k}_2,\sigma_3}C_{\mathbf{k}_3,\sigma_3}C_{\mathbf{k}_4,\sigma_4}\right\rangle,
\end{aligned} \tag{8.94}$$

and by relabeling the dummy spin indices, we obtain

$$\begin{aligned}
\left\langle \hat{\psi}^\dagger(\mathbf{r}, s)\hat{\psi}^\dagger(\mathbf{r}', s')\hat{\psi}(\mathbf{r}', s')\hat{\psi}(\mathbf{r}, s)\right\rangle &= \frac{1}{V^2}\sum_{\mathbf{k}_1,\mathbf{k}_2,\mathbf{k}_3,\mathbf{k}_4} e^{-i(\mathbf{k}_1-\mathbf{k}_4)\cdot\mathbf{r}}e^{-i(\mathbf{k}_2-\mathbf{k}_3)\cdot\mathbf{r}'} \\
&\quad \left\langle C^\dagger_{\mathbf{k}_1,\sigma}C^\dagger_{\mathbf{k}_2,\sigma'}C_{\mathbf{k}_3,\sigma'}C_{\mathbf{k}_4,\sigma}\right\rangle.
\end{aligned} \tag{8.95}$$

Now, since the jellium model describes a system that is invariant under spatial translations, the pair correlation function depends only on the relative coordinate between particles, not on their individual positions, and total momentum is conserved. We therefore consider the term $e^{-i(\mathbf{k}_1-\mathbf{k}_4)\cdot\mathbf{r}}$, $e^{-i(\mathbf{k}_2-\mathbf{k}_3)\cdot\mathbf{r}'}$ and set $\mathbf{r} = \mathbf{r}' + \mathbf{r}''$, yielding

$$
\begin{aligned}
e^{-i(\mathbf{k}_1-\mathbf{k}_4)\cdot\mathbf{r}} e^{-i(\mathbf{k}_2-\mathbf{k}_3)\cdot\mathbf{r}'} &= e^{-i(\mathbf{k}_1-\mathbf{k}_4)\cdot(\mathbf{r}'+\mathbf{r}'')} e^{-i(\mathbf{k}_2-\mathbf{k}_3)\cdot\mathbf{r}'} = \\
&= e^{-i(\mathbf{k}_1-\mathbf{k}_4)\cdot\mathbf{r}'} e^{-i(\mathbf{k}_1-\mathbf{k}_4)\cdot\mathbf{r}''} e^{-i(\mathbf{k}_2-\mathbf{k}_3)\cdot\mathbf{r}'} = \\
&= e^{-i(\mathbf{k}_1-\mathbf{k}_4)\cdot\mathbf{r}''} e^{-i(\mathbf{k}_1+\mathbf{k}_2-\mathbf{k}_3-\mathbf{k}_4)\cdot\mathbf{r}'} . \qquad (8.96)
\end{aligned}
$$

Since total momentum must be conserved, we set $\mathbf{k}_1 = \mathbf{k}_4 + \mathbf{q}$ and $\mathbf{k}_2 = \mathbf{k}_3 - \mathbf{q}$, so that $\mathbf{k}_3$, $\mathbf{k}_4$, and $\mathbf{q}$ are independent variables, while $\mathbf{k}_1$ and $\mathbf{k}_2$ are determined by $\mathbf{k}_3$, $\mathbf{k}_4$, and $\mathbf{q}$. One finds

$$
\begin{aligned}
&\left\langle \hat{\psi}^\dagger(\mathbf{r}, s)\hat{\psi}^\dagger(\mathbf{r}', s')\hat{\psi}(\mathbf{r}', s')\hat{\psi}(\mathbf{r}, s)\right\rangle = \\
&= \frac{1}{V^2} \sum_{\mathbf{k}_1,\mathbf{k}_2,\mathbf{k}_3,\mathbf{k}_4} e^{-i(\mathbf{k}_1-\mathbf{k}_4)\cdot\mathbf{r}''} e^{-i(\mathbf{k}_1+\mathbf{k}_2-\mathbf{k}_3-\mathbf{k}_4)\cdot\mathbf{r}'} \left\langle C^\dagger_{\mathbf{k}_1,\sigma} C^\dagger_{\mathbf{k}_2,\sigma'} C_{\mathbf{k}_3,\sigma'} C_{\mathbf{k}_4,\sigma}\right\rangle = \\
&= \frac{1}{V^2} \sum_{\mathbf{k}_3,\mathbf{k}_4,\mathbf{q}} e^{-i\mathbf{q}\cdot(\mathbf{r}-\mathbf{r}')} \left\langle C^\dagger_{\mathbf{k}_4+\mathbf{q},\sigma} C^\dagger_{\mathbf{k}_3-\mathbf{q},\sigma'} C_{\mathbf{k}_3,\sigma'} C_{\mathbf{k}_4,\sigma}\right\rangle, \qquad (8.97)
\end{aligned}
$$

and by renaming the dummy momentum indices, we obtain

$$
\begin{aligned}
\left\langle \hat{\psi}^\dagger(\mathbf{r}, s)\hat{\psi}^\dagger(\mathbf{r}', s')\hat{\psi}(\mathbf{r}', s')\hat{\psi}(\mathbf{r}, s)\right\rangle &= \frac{1}{V^2} \sum_{\mathbf{k}_1,\mathbf{k}_2,\mathbf{q}} e^{-i\mathbf{q}\cdot(\mathbf{r}-\mathbf{r}')} \\
&\left\langle C^\dagger_{\mathbf{k}_1+\mathbf{q},\sigma} C^\dagger_{\mathbf{k}_2-\mathbf{q},\sigma'} C_{\mathbf{k}_2,\sigma'} C_{\mathbf{k}_1,\sigma}\right\rangle . \qquad (8.98)
\end{aligned}
$$

Dividing this numerator by the denominator, we obtain

$$
\begin{aligned}
g(\mathbf{r}-\mathbf{r}', \sigma, \sigma') &= \frac{4V^2}{N^2}\frac{1}{V^2} \sum_{\mathbf{k}_1,\mathbf{k}_2,\mathbf{q}} e^{-i\mathbf{q}\cdot(\mathbf{r}-\mathbf{r}')} \left\langle C^\dagger_{\mathbf{k}_1+\mathbf{q},\sigma} C^\dagger_{\mathbf{k}_2-\mathbf{q},\sigma'} C_{\mathbf{k}_2,\sigma'} C_{\mathbf{k}_1,\sigma}\right\rangle = \\
&= \frac{4}{N^2} \sum_{\mathbf{k}_1,\mathbf{k}_2,\mathbf{q}} e^{-i\mathbf{q}\cdot(\mathbf{r}-\mathbf{r}')} \left\langle C^\dagger_{\mathbf{k}_1+\mathbf{q},\sigma} C^\dagger_{\mathbf{k}_2-\mathbf{q},\sigma'} C_{\mathbf{k}_2,\sigma'} C_{\mathbf{k}_1,\sigma}\right\rangle, \qquad (8.99)
\end{aligned}
$$

where the thermal average (see Eq. 1.379) is to be taken over the jellium Hamiltonian. We adopt a perturbative approach and compute this average with respect to the free-fermion Hamiltonian, which allows us to apply thermal Wick's theorem (see Theorems 7.2–7.4). We make the identifications $1' \equiv C^\dagger_{\mathbf{k}_1+\mathbf{q},\sigma}$, $2' \equiv C^\dagger_{\mathbf{k}_2-\mathbf{q},\sigma'}$, $3' \equiv C_{\mathbf{k}_2,\sigma'}$, $4' \equiv C_{\mathbf{k}_1,\sigma}$. We begin by analyzing the term involving different spin components, $\sigma \neq \sigma'$, corresponding to the direct Hartree-Fock contribution previously computed: operator $1'$ can be contracted only with $4'$, and similarly $2'$ only with $3'$, yielding

$$
\begin{aligned}
g^{(0)}\left(\mathbf{r}-\mathbf{r}', \sigma \neq \sigma'\right) &= \frac{4}{N^2} \sum_{\mathbf{k}_1,\mathbf{k}_2,\mathbf{q}} e^{-i\mathbf{q}\cdot(\mathbf{r}-\mathbf{r}')} \delta_{\mathbf{q},\mathbf{0}} \left\langle \hat{N}_{\mathbf{k}_1,\sigma} \right\rangle \left\langle \hat{N}_{\mathbf{k}_2,\sigma'} \right\rangle = \\
&= \frac{4}{N^2} \frac{N}{2} \frac{N}{2} = \\
&= 1.
\end{aligned} \tag{8.100}
$$

Accordingly, to lowest order, two fermions with opposite spin projections are completely uncorrelated, as expected. This result follows directly from the fact that the reference Hamiltonian $\hat{\mathcal{H}}_0$ describes a system of non-interacting particles, where the absence of interactions implies that the connected part of the pair correlation function vanishes. We now evaluate the zeroth-order contribution to the pair correlation function for fermions with identical spin projections. This case corresponds to the exchange term in the Hartree-Fock approximation. For particles with the same spin, the antisymmetry of the fermionic wavefunction leads to nontrivial correlations even in the absence of interactions. In computing the four-point Green's function using thermal Wick's theorem, two distinct contractions contribute to the result. The first contraction, $1' - 4'$ and $2' - 3'$, reproduces the previously derived direct term. The second contraction, $1' - 3'$ and $2' - 4'$, yields an additional exchange contribution. This exchange term is accompanied by a minus sign, $\varepsilon = -1$ (see Eq. 2.52), reflecting the fermionic anticommutation relations. Therefore, the full zeroth-order contribution to the pair correlation function for particles with the same spin includes both the direct and exchange terms. The exchange term reduces the probability of finding two identical fermions at nearby positions, as required by the Pauli exclusion principle and the antisymmetric nature of the many-body wavefunction. We have

$$
\begin{aligned}
&g^{(0)}\left(\mathbf{r}-\mathbf{r}', \sigma = \sigma'\right) = \\
&= 1 - \frac{4}{N^2} \sum_{\mathbf{k}_1,\mathbf{k}_2,\mathbf{q}} e^{-i\mathbf{q}\cdot(\mathbf{r}-\mathbf{r}')} \delta_{\mathbf{k}_1+\mathbf{q},\mathbf{k}_2} \delta_{\mathbf{k}_2-\mathbf{q},\mathbf{k}_1} \left\langle C^\dagger_{\mathbf{k}_1+\mathbf{q},\sigma} C_{\mathbf{k}_2,\sigma} \right\rangle \left\langle C^\dagger_{\mathbf{k}_2-\mathbf{q},\sigma} C_{\mathbf{k}_1,\sigma} \right\rangle = \\
&= 1 - \frac{4}{N^2} \sum_{\mathbf{k}_1,\mathbf{k}_2} e^{-i(\mathbf{k}_2-\mathbf{k}_1)\cdot(\mathbf{r}-\mathbf{r}')} \left\langle C^\dagger_{\mathbf{k}_2,\sigma} C_{\mathbf{k}_2,\sigma} \right\rangle \left\langle C^\dagger_{\mathbf{k}_1,\sigma} C_{\mathbf{k}_1,\sigma} \right\rangle = \\
&= 1 - \frac{4}{N^2} \left(\sum_{\mathbf{k}_1} e^{i\mathbf{k}_1\cdot(\mathbf{r}-\mathbf{r}')} \left\langle C^\dagger_{\mathbf{k}_1,\sigma} C_{\mathbf{k}_1,\sigma} \right\rangle \right) \left(\sum_{\mathbf{k}_2} e^{-i\mathbf{k}_2\cdot(\mathbf{r}-\mathbf{r}')} \left\langle C^\dagger_{\mathbf{k}_2,\sigma} C_{\mathbf{k}_2,\sigma} \right\rangle \right) = \\
&= 1 - \frac{4}{N^2} \left(\sum_{\mathbf{k}_1} e^{-i\mathbf{k}_1\cdot(\mathbf{r}-\mathbf{r}')} \left\langle C^\dagger_{-\mathbf{k}_1,\sigma} C_{-\mathbf{k}_1,\sigma} \right\rangle \right) \left(\sum_{\mathbf{k}_2} e^{-i\mathbf{k}_2\cdot(\mathbf{r}-\mathbf{r}')} \left\langle C^\dagger_{\mathbf{k}_2,\sigma} C_{\mathbf{k}_2,\sigma} \right\rangle \right),
\end{aligned} \tag{8.101}
$$

where we have rewritten the sum over $-\mathbf{k}_1$ instead of $\mathbf{k}_1$. Since the jellium model satisfies the relation $C^{\dagger}_{-\mathbf{k}_1,\sigma}C_{-\mathbf{k}_1,\sigma} = C^{\dagger}_{\mathbf{k}_1,\sigma}C_{\mathbf{k}_1,\sigma}$, and the summation indices are dummy variables, we have

$$g^{(0)}\left(\mathbf{r}-\mathbf{r}',\sigma=\sigma'\right) = 1-\frac{4}{N^2}\left(\sum_{\mathbf{k}} e^{-i\mathbf{k}\cdot(\mathbf{r}-\mathbf{r}')}\left\langle \hat{N}_{\mathbf{k},\sigma}\right\rangle\right)^2$$
$$= 1-\frac{4}{N^2}\left(\sum_{\mathbf{k}} e^{-i\mathbf{k}\cdot(\mathbf{r}-\mathbf{r}')}n_{-1}\left(\mathcal{E}_{\mathbf{k},\sigma}\right)\right)^2, \quad (8.102)$$

where we employed Eq. 2.251 and the Fermi-Dirac statistics $n_{-1}(\mathcal{E}_{\mathbf{k},\sigma})$ refers to the eigenvalues $\mathcal{E}_{\mathbf{k},\sigma}$ of the jellium model. Now, the pair correlation function is generally less than 1, reflecting the consequences of Fermi-Dirac statistics. We now evaluate 8.102 in the thermodynamic limit C.11, in particular at zero temperature, where the Fermi-Dirac functions $n_{-1}\left(\mathcal{E}_{\mathbf{k},\sigma}\right)$ reduce to step functions with respect to the Fermi momentum $\mathbf{k}_F$. We have

$$g^{(0)}\left(\mathbf{r}-\mathbf{r}',\sigma=\sigma'\right)\Big|_{T=0} = 1-\frac{4}{N^2}\left(\frac{V}{(2\pi)^3}\int d^3\mathbf{k}e^{-i\mathbf{k}\cdot(\mathbf{r}-\mathbf{r}')}\Theta(\mathbf{k}-\mathbf{k}_F)\right)^2 =$$
$$= 1-\frac{4}{N^2}\left(\frac{V}{(2\pi)^3}\int_0^{k_F} 2\pi k^2 dk\int_{-1}^{+1} d\cos(\theta)e^{-ik|\mathbf{r}-\mathbf{r}'|\cos(\theta)}\right)^2 =$$
$$= 1-\frac{4}{N^2}\left(\frac{V}{4\pi^2}\int_0^{k_F} k^2 dk\frac{2\sin(k\left|\mathbf{r}-\mathbf{r}'\right|)}{k\left|\mathbf{r}-\mathbf{r}'\right|}\right)^2 =$$
$$= 1-\frac{4}{N^2}\left(\frac{V}{2\pi^2}\frac{1}{|\mathbf{r}-\mathbf{r}'|}\int_0^{k_F} k\sin\left(k\left|\mathbf{r}-\mathbf{r}'\right|\right)dk\right)^2, \quad (8.103)$$

then

$$g^{(0)}\left(\mathbf{r}-\mathbf{r}',\sigma=\sigma'\right)\Big|_{T=0} = g^{(0)}\left(\left|\mathbf{r}-\mathbf{r}'\right|,\sigma=\sigma'\right)\Big|_{T=0}. \quad (8.104)$$

From

$$\int_0^{k_F} k\sin\left(k\left|\mathbf{r}-\mathbf{r}'\right|\right)dk = \frac{\sin(k_F\left|\mathbf{r}-\mathbf{r}'\right|)}{|\mathbf{r}-\mathbf{r}'|^2} - \frac{k_F\cos(k_F\left|\mathbf{r}-\mathbf{r}'\right|)}{|\mathbf{r}-\mathbf{r}'|}, \quad (8.105)$$

it follows

$$g^{(0)}(\left|\mathbf{r}-\mathbf{r}'\right|,\sigma=\sigma')\Big|_{T=0} =$$

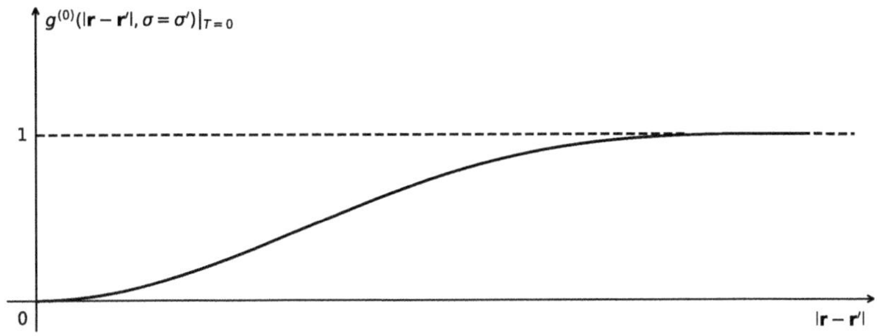

Fig. 8.2 Pair correlation function for electrons with same spin as a function of distance at zero temperature, calculated within the Hartree-Fock approximation (see Eq. 8.106)

$$\begin{aligned}
&= 1 - \frac{4}{N^2}\left\{\frac{V}{2\pi^2}\frac{1}{|\mathbf{r}-\mathbf{r}'|^3}\left[\sin\left(k_F\left|\mathbf{r}-\mathbf{r}'\right|\right) - k_F\left|\mathbf{r}-\mathbf{r}'\right|\cos\left(k_F\left|\mathbf{r}-\mathbf{r}'\right|\right)\right]\right\}^2 = \\
&= 1 - \frac{V^2}{N^2\pi^4}\frac{1}{|\mathbf{r}-\mathbf{r}'|^6}\left[\sin\left(k_F\left|\mathbf{r}-\mathbf{r}'\right|\right) - k_F\left|\mathbf{r}-\mathbf{r}'\right|\cos\left(k_F\left|\mathbf{r}-\mathbf{r}'\right|\right)\right]^2 = \\
&= 1 - \frac{9}{k_F^6\,|\mathbf{r}-\mathbf{r}'|^6}\left[\sin\left(k_F\left|\mathbf{r}-\mathbf{r}'\right|\right) - k_F\left|\mathbf{r}-\mathbf{r}'\right|\cos\left(k_F\left|\mathbf{r}-\mathbf{r}'\right|\right)\right]^2,
\end{aligned} \tag{8.106}$$

where in the last step we have explicitly expressed the density of non-interacting particles and employed Eq. O.8. The function 8.106 is illustrated in Fig. 8.2. We note that it satisfies

$$\begin{aligned}
&\lim_{|\mathbf{r}-\mathbf{r}'|\to+\infty} g^{(0)}\left(\left|\mathbf{r}-\mathbf{r}'\right|, \sigma=\sigma'\right)\Big|_{T=0} = \\
&= \lim_{|\mathbf{r}-\mathbf{r}'|\to+\infty} 1 - \frac{9}{k_F^6\,|\mathbf{r}-\mathbf{r}'|^6} \\
&\quad \left[\sin\left(k_F\left|\mathbf{r}-\mathbf{r}'\right|\right) - k_F\left|\mathbf{r}-\mathbf{r}'\right|\cos\left(k_F\left|\mathbf{r}-\mathbf{r}'\right|\right)\right]^2 = \\
&= 1,
\end{aligned} \tag{8.107}$$

as expected, two electrons with the same spin are uncorrelated at large distances. On the other hand,

$$\begin{aligned}
&\lim_{|\mathbf{r}-\mathbf{r}'|\to 0^+} g^{(0)}\left(\left|\mathbf{r}-\mathbf{r}'\right|, \sigma=\sigma'\right)\Big|_{T=0} \\
&= \lim_{|\mathbf{r}-\mathbf{r}'|\to 0^+}\left\{1 - \frac{9}{k_F^6\,|\mathbf{r}-\mathbf{r}'|^6}\left[\sin\left(k_F\left|\mathbf{r}-\mathbf{r}'\right|\right) - k_F\left|\mathbf{r}-\mathbf{r}'\right|\cos\left(k_F\left|\mathbf{r}-\mathbf{r}'\right|\right)\right]^2\right\},
\end{aligned} \tag{8.108}$$

for convenience, we define $x = k_F \left|\mathbf{r} - \mathbf{r}'\right|$ and we get

$$\lim_{x\to 0^+} \left\{1 - \frac{9}{x^6}\left[\sin(x) - x\cos(x)\right]^2\right\}$$

$$= \lim_{x\to 0^+} \left\{1 - \frac{9}{x^6}\left[x - \frac{x^3}{6} + o(x^3) - x\left(1 - \frac{x^2}{2} + o(x^2)\right)\right]^2\right\} =$$

$$= \lim_{x\to 0^+} \left\{1 - \frac{9}{x^6}\left[-\frac{x^3}{6} + \frac{x^3}{2} + o(x^3)\right]^2\right\} =$$

$$= \lim_{x\to 0^+} \left\{1 - \frac{9}{x^6}\left(\frac{x^6}{9} + o(x^6)\right)\right\}, \tag{8.109}$$

then

$$\lim_{|\mathbf{r}-\mathbf{r}'|\to 0^+} g^{(0)}\left(\left|\mathbf{r} - \mathbf{r}'\right|, \sigma = \sigma'\right)\Big|_{T=0} = 0, \tag{8.110}$$

and this result also follows from the Pauli exclusion principle, since it is impossible to find two particles at the same position with the same spin. This consistency was to be expected, given that we considered Slater determinants as eigenstates, and that we performed thermal averages on a Hamiltonian of non-interacting electrons.
From 8.99, we define

$$\tilde{g}(\mathbf{r} - \mathbf{r}') = \frac{1}{4}\sum_{\sigma,\sigma'} g(\mathbf{r} - \mathbf{r}', \sigma, \sigma') =$$

$$= \frac{1}{N^2}\sum_{\sigma,\sigma'}\sum_{\mathbf{k}_1,\mathbf{k}_2,\mathbf{q}} e^{-i\mathbf{q}\cdot(\mathbf{r}-\mathbf{r}')}\left\langle C^\dagger_{\mathbf{k}_1+\mathbf{q},\sigma} C^\dagger_{\mathbf{k}_2-\mathbf{q},\sigma'} C_{\mathbf{k}_2,\sigma'} C_{\mathbf{k}_1,\sigma}\right\rangle, \tag{8.111}$$

which is the probability of finding two electrons at a distance $\mathbf{r} - \mathbf{r}'$, regardless of their spin. We want to calculate this quantity with respect to the full Hamiltonian, including interactions. We separate the sum over $\mathbf{q}$ as follows

$$\tilde{g}(\mathbf{r} - \mathbf{r}') = \frac{1}{N^2}\sum_{\sigma,\sigma'}\sum_{\mathbf{k}_1,\mathbf{k}_2}\left\langle C^\dagger_{\mathbf{k}_1,\sigma} C^\dagger_{\mathbf{k}_2,\sigma'} C_{\mathbf{k}_2,\sigma'} C_{\mathbf{k}_1,\sigma}\right\rangle$$
$$+ \frac{1}{N^2}\sum_{\sigma,\sigma'}\sum_{\mathbf{k}_1,\mathbf{k}_2,\mathbf{q}\neq\mathbf{0}} e^{-i\mathbf{q}\cdot(\mathbf{r}-\mathbf{r}')}\left\langle C^\dagger_{\mathbf{k}_1+\mathbf{q},\sigma} C^\dagger_{\mathbf{k}_2-\mathbf{q},\sigma'} C_{\mathbf{k}_2,\sigma'} C_{\mathbf{k}_1,\sigma}\right\rangle, \tag{8.112}$$

and we exploit

$$
\begin{aligned}
C^{\dagger}_{\mathbf{k}_1,\sigma}C^{\dagger}_{\mathbf{k}_2,\sigma'}C_{\mathbf{k}_2,\sigma'}C_{\mathbf{k}_1,\sigma} &= -C^{\dagger}_{\mathbf{k}_1,\sigma}C^{\dagger}_{\mathbf{k}_2,\sigma'}C_{\mathbf{k}_1,\sigma}C_{\mathbf{k}_2,\sigma'} = \\
&= -C^{\dagger}_{\mathbf{k}_1,\sigma}\left(\delta_{\mathbf{k}_2,\mathbf{k}_1}\delta_{\sigma',\sigma} - C_{\mathbf{k}_1,\sigma}C^{\dagger}_{\mathbf{k}_2,\sigma'}\right)C_{\mathbf{k}_2,\sigma'}
\end{aligned}
\tag{8.113}
$$

in the first term, and we have

$$
\begin{aligned}
\tilde{g}(\mathbf{r}-\mathbf{r}') = &-\frac{1}{N^2}\sum_{\sigma,\sigma'}\sum_{\mathbf{k}_1,\mathbf{k}_2}\delta_{\mathbf{k}_2,\mathbf{k}_1}\delta_{\sigma',\sigma}\left\langle C^{\dagger}_{\mathbf{k}_1,\sigma}C_{\mathbf{k}_2,\sigma'}\right\rangle \\
&+\frac{1}{N^2}\sum_{\sigma,\sigma'}\sum_{\mathbf{k}_1,\mathbf{k}_2}\left\langle C^{\dagger}_{\mathbf{k}_1,\sigma}C_{\mathbf{k}_1,\sigma}C^{\dagger}_{\mathbf{k}_2,\sigma'}C_{\mathbf{k}_2,\sigma'}\right\rangle + \\
&+\frac{1}{N^2}\sum_{\sigma,\sigma'}\sum_{\mathbf{k}_1,\mathbf{k}_2,\mathbf{q}\neq\mathbf{0}} e^{-i\mathbf{q}\cdot(\mathbf{r}-\mathbf{r}')}\left\langle C^{\dagger}_{\mathbf{k}_1+\mathbf{q},\sigma}C^{\dagger}_{\mathbf{k}_2-\mathbf{q},\sigma'}C_{\mathbf{k}_2,\sigma'}C_{\mathbf{k}_1,\sigma}\right\rangle = \\
= &-\frac{1}{N^2}\sum_{\sigma}\sum_{\mathbf{k}_1}\left\langle C^{\dagger}_{\mathbf{k}_1,\sigma}C_{\mathbf{k}_1,\sigma}\right\rangle \\
&+\frac{1}{N^2}\sum_{\sigma,\sigma'}\sum_{\mathbf{k}_1,\mathbf{k}_2}\left\langle C^{\dagger}_{\mathbf{k}_1,\sigma}C_{\mathbf{k}_1,\sigma}C^{\dagger}_{\mathbf{k}_2,\sigma'}C_{\mathbf{k}_2,\sigma'}\right\rangle + \\
&+\frac{1}{N^2}\sum_{\sigma,\sigma'}\sum_{\mathbf{k}_1,\mathbf{k}_2,\mathbf{q}\neq\mathbf{0}} e^{-i\mathbf{q}\cdot(\mathbf{r}-\mathbf{r}')}\left\langle C^{\dagger}_{\mathbf{k}_1+\mathbf{q},\sigma}C^{\dagger}_{\mathbf{k}_2-\mathbf{q},\sigma'}C_{\mathbf{k}_2,\sigma'}C_{\mathbf{k}_1,\sigma}\right\rangle = \\
= &-\frac{1}{N}+\frac{1}{N^2}\sum_{\sigma,\sigma'}\sum_{\mathbf{k}_1,\mathbf{k}_2}\left\langle C^{\dagger}_{\mathbf{k}_1,\sigma}C_{\mathbf{k}_1,\sigma}C^{\dagger}_{\mathbf{k}_2,\sigma'}C_{\mathbf{k}_2,\sigma'}\right\rangle + \\
&+\frac{1}{N^2}\sum_{\sigma,\sigma'}\sum_{\mathbf{k}_1,\mathbf{k}_2,\mathbf{q}\neq\mathbf{0}} e^{-i\mathbf{q}\cdot(\mathbf{r}-\mathbf{r}')}\left\langle C^{\dagger}_{\mathbf{k}_1+\mathbf{q},\sigma}C^{\dagger}_{\mathbf{k}_2-\mathbf{q},\sigma'}C_{\mathbf{k}_2,\sigma'}C_{\mathbf{k}_1,\sigma}\right\rangle = \\
= &-\frac{1}{N}+\frac{1}{N^2}\sum_{\sigma,\sigma'}\sum_{\mathbf{k}_1,\mathbf{k}_2}\left\langle C^{\dagger}_{\mathbf{k}_1,\sigma}C_{\mathbf{k}_1,\sigma}\right\rangle\left\langle C^{\dagger}_{\mathbf{k}_2,\sigma'}C_{\mathbf{k}_2,\sigma'}\right\rangle + \\
&+\frac{1}{N^2}\sum_{\sigma,\sigma'}\sum_{\mathbf{k}_1,\mathbf{k}_2,\mathbf{q}\neq\mathbf{0}} e^{-i\mathbf{q}\cdot(\mathbf{r}-\mathbf{r}')}\left\langle C^{\dagger}_{\mathbf{k}_1+\mathbf{q},\sigma}C^{\dagger}_{\mathbf{k}_2-\mathbf{q},\sigma'}C_{\mathbf{k}_2,\sigma'}C_{\mathbf{k}_1,\sigma}\right\rangle,
\end{aligned}
\tag{8.114}
$$

where in the last step we used that the states are uncorrelated, allowing the factorization of the expectation value as

$$
\left\langle C^{\dagger}_{\mathbf{k}_1,\sigma}C_{\mathbf{k}_1,\sigma}C^{\dagger}_{\mathbf{k}_2,\sigma'}C_{\mathbf{k}_2,\sigma'}\right\rangle = \left\langle C^{\dagger}_{\mathbf{k}_1,\sigma}C_{\mathbf{k}_1,\sigma}\right\rangle\left\langle C^{\dagger}_{\mathbf{k}_2,\sigma'}C_{\mathbf{k}_2,\sigma'}\right\rangle . \tag{8.115}
$$

Integrating both sides over position, we obtain

$$\int d(\mathbf{r}-\mathbf{r}')\tilde{g}(\mathbf{r}-\mathbf{r}')$$
$$= -\frac{1}{N}\int d(\mathbf{r}-\mathbf{r}') + \frac{1}{N^2}\sum_{\sigma,\sigma'}\sum_{\mathbf{k}_1,\mathbf{k}_2}\left\langle C^{\dagger}_{\mathbf{k}_1,\sigma}C_{\mathbf{k}_1,\sigma}\right\rangle\left\langle C^{\dagger}_{\mathbf{k}_2,\sigma'}C_{\mathbf{k}_2,\sigma'}\right\rangle\int d(\mathbf{r}-\mathbf{r}')+$$
$$+\frac{1}{N^2}\sum_{\sigma,\sigma'}\sum_{\mathbf{k}_1,\mathbf{k}_2,\mathbf{q}\neq\mathbf{0}}\left(\int d(\mathbf{r}-\mathbf{r}')e^{-i\mathbf{q}\cdot(\mathbf{r}-\mathbf{r}')}\right)\left\langle C^{\dagger}_{\mathbf{k}_1+\mathbf{q},\sigma}C^{\dagger}_{\mathbf{k}_2-\mathbf{q},\sigma'}C_{\mathbf{k}_2,\sigma'}C_{\mathbf{k}_1,\sigma}\right\rangle, \tag{8.116}$$

and note that the last term vanishes, since it is given by the integral of an oscillatory function. Thus,

$$\int d(\mathbf{r}-\mathbf{r}')\tilde{g}(\mathbf{r}-\mathbf{r}') = -\frac{V}{N} + \frac{1}{N^2}N^2V, \tag{8.117}$$

$$\int d(\mathbf{r}-\mathbf{r}')\tilde{g}(\mathbf{r}-\mathbf{r}') = -\frac{1}{\rho_0} + V, \tag{8.118}$$

where ρ_0 denotes the particle density in the non-interacting case. From $V = \int d(\mathbf{r}-\mathbf{r}')$, we finally obtain

$$\int d(\mathbf{r}-\mathbf{r}')\left[1-\tilde{g}(\mathbf{r}-\mathbf{r}')\right] = \frac{1}{\rho_0}, \tag{8.119}$$

which constitutes a sum rule for the pair correlation function.

Chapter 9
Phonon Propagator

Abstract This chapter explores the Green's function for phonons, the quanta of lattice vibrations. It begins by introducing the free phonon propagator, describing phonon behaviour in an ideal harmonic lattice without interactions with electrons, other phonons, or defects. This propagator is derived in both real-time and imaginary-time formalisms, focusing on its mathematical properties and frequency dependence. Next, the chapter examines how interactions affect phonon propagation by introducing the phonon self-energy, which accounts for the corrections arising from scattering processes and the various coupling mechanisms present in the system. Using Feynman diagram techniques, the interaction effects are systematically analysed through perturbation theory. A key result discussed is the relationship between the phonon self-energy and the polarization propagator, connected to the particle density operator. This connection is fundamental for understanding critical solid-state phenomena such as phonon frequency shifts, energy dissipation, and symmetry breaking in strongly interacting materials. Overall, the chapter provides a comprehensive framework for describing phonon dynamics and interactions using Green's functions and diagrammatic methods.

9.1 Free Phonon Propagator and First-Order Self-energy for Electron-Phonon Interaction

In Chap. 7 we derived the Dyson series for the thermal Green's function of field operators (see Eq. 7.41), where the free Hamiltonian is given by 2.181 and the interaction Hamiltonian by 2.314, i.e., of the form 2.327. We now consider a full Hamiltonian composed of a free part given by

$$\hat{\mathcal{H}}_0 = \sum_{\mathbf{k},\mathbf{q}} \mathcal{E}_{\mathbf{k}} C^{\dagger}_{\mathbf{k},\sigma} C_{\mathbf{k},\sigma} + \sum_{\mathbf{q}} \hbar\omega_{\mathbf{q}} a^{\dagger}_{\mathbf{q}} a_{\mathbf{q}}, \tag{9.1}$$

and an interaction part given by Hamiltonian 5.37, describing a system of fermions or bosons interacting with phonons (see Chap. 4) in a finite volume V. In the free Hamiltonian above, the first term accounts for fermions or bosons with dispersion

F. Tafuri et al., *Understanding Quantum Many-Body Theory*, UNITEXT for Physics, https://doi.org/10.1007/978-3-032-17849-7_9

$\mathcal{E}_{\mathbf{k}}$, while the second term describes the free dynamics of bosonic field excitations (e.g., phonons, photons, or a thermal reservoir). A typical physical realization of such a model is that of a system coupled to an external thermal bath. The interaction Hamiltonians given in Eqs. 2.314 and 5.37 are among the most commonly used in the literature. In what follows, we demonstrate that Feynman's perturbative expansion retains its general structure regardless of the specific interaction Hamiltonian.

Let us analyze the meaning of Eq. 5.37. This equation describes a process in which a particle is annihilated at one vertex by the operator $C_{\mathbf{k},\sigma}$, and another particle is created at another vertex by the operator $C^{\dagger}_{\mathbf{k}+\mathbf{q},\sigma}$, mediated by bosonic annihilation and creation operators $a_{\mathbf{q}}$ and $a^{\dagger}_{\mathbf{q}}$, respectively.

Now, we proceed to write the Dyson expansion of the Matsubara Green's function for the field operators (see Eq. 7.41) for the interaction Hamiltonian given by Eq. 5.37. For $n = 0$, we recover the free propagator $G^{(m)(0)}$. For $n = 1$, the contribution to the Green's function is given by

$$\left\langle \hat{T}_{\tau}\, \hat{\psi}^{(0)}(x,\tau)\, \hat{\mathcal{H}}_{I}^{(0)}(\tau_1)\, \hat{\psi}^{\dagger}(x',0)\right\rangle_0, \tag{9.2}$$

where $\hat{T}_{\tau}$ is the Wick time-ordering operator given in 2.200 and $\hat{\mathcal{H}}_{I}^{(0)}(\tau_1)$ is the interaction Hamiltonian. Moreover, all operators, including the Hamiltonian, are evaluated in the thermal interaction picture (see Eq. 2.194) with respect to the free Hamiltonian $\hat{\mathcal{H}}_0$ defined in Eq. 9.1, while the operator on the far right, $\hat{\psi}^{\dagger}(x',0)$, is not evaluated in any picture, being taken at time zero (see Remark 7.1). Ultimately, the subscript 0 indicates that the thermal average (see Eq. 1.379) is taken with respect to this same non-interacting Hamiltonian. Such an average factorizes into the product of a fermionic and a bosonic thermal average. This follows from the fact that Eq. 9.1 describes independent particles: in the non-interacting Hamiltonian the fermionic and bosonic (phononic) spaces are decoupled, i.e.,

$$\left\langle \hat{T}_{\tau}\, \hat{\psi}^{(0)}(x,\tau)\, \hat{\mathcal{H}}_{I}^{(0)}(\tau_1)\, \hat{\psi}^{\dagger}(x',0)\right\rangle_0 = \sum_{\mathbf{q}_1} M_{\mathbf{q}_1}$$

$$\left\langle \hat{T}_{\tau}\left[\hat{\psi}^{(0)}(x,\tau)\left(\int dx_1\, \hat{\psi}^{\dagger(0)}(x_1,\tau_1)\, e^{i\mathbf{q}_1\cdot\mathbf{r}_1}\, \hat{\psi}^{(0)}(x_1,\tau_1)\right)\hat{\psi}^{\dagger(0)}(x',0)\right]\right\rangle_{f,0} \tag{9.3}$$

$$\left\langle \hat{T}_{\tau}\left(a^{(0)}_{\mathbf{q}_1}(\tau_1) + a^{\dagger(0)}_{-\mathbf{q}_1}(\tau_1)\right)\right\rangle_{b,0}. \tag{9.4}$$

Consider the bosonic thermal average. The operator $a_{\mathbf{q}}$ annihilates a phonon with momentum $\mathbf{q}$, thereby generating an $(N-1)$-particle state, while $a^{\dagger}_{-\mathbf{q}}$ creates a phonon with momentum $-\mathbf{q}$, producing an $(N+1)$-particle state. Since states with different particle numbers are orthonormal, both expectation values vanish. It follows that the first-order contribution in the perturbative expansion is identically zero. Consequently, the first-order contribution in the perturbative expansion is identically zero. We therefore proceed to the second-order term of the series, for an interaction Hamiltonian given by Eq. 5.37, i.e.,

$$G^{(m)(2)}(x,\tau,x',0) = G^{(m)(0)}(x,\tau,x',0) - \frac{1}{\hbar}\left(-\frac{1}{\hbar}\right)^2$$
$$\frac{1}{2}\int_0^{\beta\hbar} d\tau_1 \int_0^{\beta\hbar} d\tau_2 \int dx_1 \int dx_2 \sum_{q_1,q_2} M_{\mathbf{q}_1} M_{\mathbf{q}_2} e^{i\mathbf{q}_1\cdot\mathbf{r}_1} e^{i\mathbf{q}_2\cdot\mathbf{r}_2}$$
$$\Big\langle \hat{T}_\tau \hat{\psi}^{(0)}(x,\tau)\hat{\psi}^{\dagger(0)}(x_1,\tau_1)\hat{\psi}^{(0)}(x_1,\tau_1)\hat{\psi}^{\dagger(0)}$$
$$(x_2,\tau_2)\hat{\psi}^{(0)}(x_2,\tau_2)\hat{\psi}^{\dagger}(x',0)\Big\rangle_{f,0}$$
$$\left\langle \hat{T}_\tau \left(a_{\mathbf{q}_1} + a^\dagger_{-\mathbf{q}_1}\right)^{(0)}(\tau_1)\left(a_{\mathbf{q}_2} + a^\dagger_{-\mathbf{q}_2}\right)^{(0)}(\tau_2)\right\rangle_{b,0}, \tag{9.5}$$

where the two spatial integrations come from the operators $\hat{\mathcal{H}}_I^{(0)}(\tau_1)$, $\hat{\mathcal{H}}_I^{(0)}(\tau_2)$. We compute the bosonic term. Since the system is at thermodynamic equilibrium, it is invariant by time translation, i.e.,

$$\left\langle \hat{T}_\tau \left(a_{\mathbf{q}_1} + a^\dagger_{-\mathbf{q}_1}\right)^{(0)}(\tau_1-\tau_2)\left(a_{\mathbf{q}_2} + a^\dagger_{-\mathbf{q}_2}\right)^{(0)}(0)\right\rangle_{b,0}. \tag{9.6}$$

Assume $\tau_1 - \tau_2 > 0$, so that the operator $\hat{T}_\tau$ can be removed. The time evolution of the bosonic operators is given by Eqs. 2.236 and 2.241, which, for convenience, we rewrite as,

$$a_{\mathbf{q}}^{(0)}(\tau_1-\tau_2) = e^{-\omega_{\mathbf{q}}(\tau_1-\tau_2)}a_{\mathbf{q}}, \tag{9.7}$$

$$a_{-\mathbf{q}}^{\dagger(0)}(\tau_1-\tau_2) = e^{\omega_{\mathbf{q}}(\tau_1-\tau_2)}a^\dagger_{-\mathbf{q}}, \tag{9.8}$$

where we used the fact that, for a lattice with inversion symmetry, the phonon frequencies satisfy

$$\omega_{\mathbf{q}} = \omega_{-\mathbf{q}}, \tag{9.9}$$

allowing us to write $\omega_{\mathbf{q}}$ instead of $\omega_{-\mathbf{q}}$ in Eq. 9.8. To simplify the notation, we omit the subscript b, and note that the only non-vanishing thermal averages are those of the form $\left\langle a_{\mathbf{q}}a^\dagger_{\mathbf{q}'}\right\rangle_0$ and $\left\langle a^\dagger_{\mathbf{q}}a_{\mathbf{q}'}\right\rangle_0$. Therefore, we obtain

$$\left\langle \left(a_{\mathbf{q}_1} + a^\dagger_{-\mathbf{q}_1}\right)^{(0)}(\tau_1-\tau_2)\left(a_{\mathbf{q}_2} + a^\dagger_{-\mathbf{q}_2}\right)^{(0)}(0)\right\rangle_0 =$$

$$
\begin{aligned}
&= \left\langle a^{(0)}_{\mathbf{q}_1}(\tau_1-\tau_2)\, a^{\dagger(0)}_{-\mathbf{q}_2}(0)\right\rangle_0 + \left\langle a^{\dagger(0)}_{-\mathbf{q}_1}(\tau_1-\tau_2)\, a^{(0)}_{\mathbf{q}_2}(0)\right\rangle_0 = \\
&= e^{-\omega_{\mathbf{q}_1}(\tau_1-\tau_2)} \left\langle a_{\mathbf{q}_1}\, a^{\dagger}_{-\mathbf{q}_2}\right\rangle_0 + e^{\omega_{\mathbf{q}_1}(\tau_1-\tau_2)} \left\langle a^{\dagger}_{-\mathbf{q}_1}\, a_{\mathbf{q}_2}\right\rangle_0 = \\
&= e^{-\omega_{\mathbf{q}_1}(\tau_1-\tau_2)} \left\langle \mathbb{1} + a^{\dagger}_{-\mathbf{q}_2} a_{\mathbf{q}_1}\right\rangle_0 + e^{\omega_{\mathbf{q}_1}(\tau_1-\tau_2)} \left\langle a^{\dagger}_{-\mathbf{q}_1}\, a_{\mathbf{q}_2}\right\rangle_0 = \\
&= e^{-\omega_{\mathbf{q}_1}(\tau_1-\tau_2)} \left(\langle \mathbb{1}\rangle_0 + \delta_{\mathbf{q}_1,-\mathbf{q}_2} n_1(\hbar\omega_{\mathbf{q}_1})\right) + e^{\omega_{\mathbf{q}_1}(\tau_1-\tau_2)}\, \delta_{\mathbf{q}_1,-\mathbf{q}_2} n_1(\hbar\omega_{\mathbf{q}_1}) = \\
&= \delta_{\mathbf{q}_1,-\mathbf{q}_2} \left\{ e^{-\omega_{\mathbf{q}_1}(\tau_1-\tau_2)} \left[1 + \frac{1}{e^{\beta\hbar\omega_{\mathbf{q}_1}}-1}\right] + e^{\omega_{\mathbf{q}_1}(\tau_1-\tau_2)} \frac{1}{e^{\beta\hbar\omega_{\mathbf{q}_1}}-1} \right\},
\end{aligned}
\tag{9.10}
$$

where we employed Eq. 2.252 for Bose-Einstein statistics (see Eq. E.3), which has been adapted as follows

$$
n_1(\hbar\omega_{\pm\mathbf{q}}) = \frac{1}{e^{\beta\hbar\omega_{\pm\mathbf{q}}}-1}. \tag{9.11}
$$

We introduce

Definition 9.1 (Free Phonon Propagator) At thermodynamic equilibrium, given the phonon annihilation and creation operators (see Eqs. 4.88 and 4.90), we define the free phonon propagator as

$$
\begin{aligned}
D^{(0)}_{\mathbf{q}_1}(\tau_1,\tau_2) &= -\frac{1}{\hbar}\delta_{\mathbf{q}_1,-\mathbf{q}_2} \left\langle \hat{T}_\tau \left(a_{\mathbf{q}_1} + a^{\dagger}_{-\mathbf{q}_1}\right)^{(0)}(\tau_1) \left(a_{\mathbf{q}_2} + a^{\dagger}_{-\mathbf{q}_2}\right)^{(0)}(\tau_2)\right\rangle_0 \equiv \\
&\equiv -\frac{1}{\hbar} \left\langle \hat{T}_\tau \left(a_{\mathbf{q}_1} + a^{\dagger}_{-\mathbf{q}_1}\right)^{(0)}(\tau_1) \left(a_{-\mathbf{q}_1} + a^{\dagger}_{\mathbf{q}_1}\right)^{(0)}(\tau_2)\right\rangle_0,
\end{aligned}
\tag{9.12}
$$

where $\hat{T}_\tau$ is the Wick time-ordering operator given in Eq. 2.200.

In direct analogy with the fermionic or bosonic Matsubara Green's function of field operators, $G^{(m)}(x,\tau,x',0)$ (see Eq. 7.40), which describes the creation of a particle with spin s' at position $\mathbf{r}'$ and time 0 and its annihilation at position $\mathbf{r}$ with spin s at time τ, the displacement operator $\mathbf{S}_{\mathbf{n},\mu}$ (see Eq. 4.134) accounts for the annihilation of a boson with momentum $\mathbf{q}$ and the simultaneous creation of a boson with momentum $-\mathbf{q}$ within the unit cell indexed by the lattice vectors $\mathbf{R}_\mathbf{n}$ (see Eq. 3.2). Note that $G^{(m)}(x,\tau,x',0)$ involves two configuration indices, x and x', while $D^{(0)}_{\mathbf{q}_1}(\tau_1,\tau_2)$ depends on a single momentum variable, $\mathbf{q}$. This difference arises from spatial translation symmetry: the propagator depends only on the relative coordinate $\mathbf{r}-\mathbf{r}'$, and upon Fourier transforming, it reduces to a single momentum index.

We now turn to the second-order term in the Dyson series for $G^{(m)}(x,\tau,x',0)$ (see Eq. 7.41). In this step, we factor out a term $(-\hbar)^{-1}$ in order to reconstruct the phonon propagator, i.e.,

$$G^{(m)(2)}(x,\tau,x',0) = G^{(m)(0)}(x,\tau,x',0)+$$
$$+\left(-\frac{1}{\hbar}\right)^2 \frac{1}{2}\int_0^{\beta\hbar} d\tau_1 \int_0^{\beta\hbar} d\tau_2 \int dx_1 \int dx_2 \sum_{\mathbf{q}_1,\mathbf{q}_2} M_{\mathbf{q}_1} M_{\mathbf{q}_2} e^{i\mathbf{q}_1\cdot\mathbf{r}_1} e^{i\mathbf{q}_2\cdot\mathbf{r}_2}$$
$$\left\langle \hat{T}_\tau \hat{\psi}^{(0)}(x,\tau)\hat{\psi}^{\dagger(0)}(x_1,\tau_1)\hat{\psi}^{(0)}(x_1,\tau_1)\hat{\psi}^{\dagger(0)}(x_2,\tau_2)\hat{\psi}^{(0)}(x_2,\tau_2)\hat{\psi}^{\dagger}(x',0)\right\rangle_{f,0}$$
$$\delta_{\mathbf{q}_1,-\mathbf{q}_2} D^{(0)}_{\mathbf{q}_1}(\tau_1,\tau_2) =$$
$$= G^{(m)(0)}(x,\tau,x',0) + \left(-\frac{1}{\hbar}\right)^2 \frac{1}{2}\int_0^{\beta\hbar} d\tau_1 \int_0^{\beta\hbar} d\tau_2$$
$$\int dx_1 \int dx_2 \sum_{\mathbf{q}} |M_{\mathbf{q}}|^2 e^{i\mathbf{q}\cdot(\mathbf{r}_1-\mathbf{r}_2)}$$
$$\left\langle \hat{T}_\tau \hat{\psi}^{(0)}(x,\tau)\hat{\psi}^{\dagger(0)}(x_1,\tau_1)\hat{\psi}^{(0)}(x_1,\tau_1)\hat{\psi}^{\dagger(0)}(x_2,\tau_2)\hat{\psi}^{(0)}(x_2,\tau_2)\hat{\psi}^{\dagger}(x',0)\right\rangle_{f,0}$$
$$D^{(0)}_{\mathbf{q}}(\tau_1,\tau_2), \tag{9.13}$$

where we renamed the summation index $\mathbf{q}_1 = \mathbf{q}$ and we employed Eq. 5.36. To compare the resulting expression with the first-order Dyson expansion for the field operators, we focus on the structure of the fermionic thermal average. In particular, we aim to construct a *mirrored* operator arrangement of the form $\hat{\psi}^\dagger(x)\hat{\psi}^\dagger(x')\hat{\psi}(x')\hat{\psi}(x)$. Under the action of the time-ordering operator $\hat{T}_\tau$, the fermionic field operators can be reordered according to their anticommutation relations. Each exchange of two fermionic operators introduces a factor $\varepsilon = -1$ (see Eq. 2.52), reflecting their fermionic anticommutation relations. Consequently, we obtain

$$\hat{T}_\tau \left\{\hat{\psi}^{(0)}(x,\tau)\hat{\psi}^{\dagger(0)}(x_1,\tau_1)\hat{\psi}^{(0)}(x_1,\tau_1)\hat{\psi}^{\dagger(0)}(x_2,\tau_2)\hat{\psi}^{(0)}(x_2,\tau_2)\hat{\psi}^{\dagger}(x',0)\right\} =$$
$$= \hat{T}_\tau \left\{\varepsilon\, \hat{\psi}^{(0)}(x,\tau)\hat{\psi}^{\dagger(0)}(x_1,\tau_1)\hat{\psi}^{\dagger(0)}(x_2,\tau_2)\hat{\psi}^{(0)}(x_1,\tau_1)\hat{\psi}^{(0)}(x_2,\tau_2)\hat{\psi}^{\dagger}(x',0)\right\} =$$
$$= \hat{T}_\tau \left\{\varepsilon^2\, \hat{\psi}^{(0)}(x,\tau)\hat{\psi}^{\dagger(0)}(x_1,\tau_1)\hat{\psi}^{\dagger(0)}(x_2,\tau_2)\hat{\psi}^{(0)}(x_2,\tau_2)\hat{\psi}^{(0)}(x_1,\tau_1)\hat{\psi}^{\dagger}(x',0)\right\}, \tag{9.14}$$

and we get the mirror operator structure. For $n = 1$, we obtain the first nonzero contribution. Indeed, the constant

$$\frac{1}{2}\left(-\frac{1}{\hbar}\right)\left(-\frac{1}{\hbar}\right)\frac{1}{1!} = \frac{1}{2}\left(-\frac{1}{\hbar}\right)^2 \tag{9.15}$$

corresponds to the second-order term: the factor 2^{-1} originates from the expansion order. From the comparison between the first expansion of Eq. 2.314 and the second one of Eq. 5.37, one finds

$$U_{\text{eff}}(\mathbf{r}_1, \tau_1, \mathbf{r}_2, \tau_2) = \sum_{\mathbf{q}} |M_{\mathbf{q}}|^2 e^{i\mathbf{q}\cdot(\mathbf{r}_1-\mathbf{r}_2)} D_{\mathbf{q}}^{(0)}(\tau_1, \tau_2). \tag{9.16}$$

The particle interaction is replaced by an effective potential that is non-instantaneous, as it does not contain a $\delta(\tau_1 - \tau_2)$ function. The phonon term in Eq. 9.1 indicates that $\omega_{\mathbf{q}}$ corresponds to a frequency associated with a finite phonon propagation time. Moreover, the original two-body interaction is the repulsive electron-electron Coulomb potential, $U_{\text{Coulomb}} > 0$, whereas the effective potential satisfies $U_{\text{eff}} < 0$, since $D_{\mathbf{q}}^{(0)}(\tau_1, \tau_2) < 0$ (see Eq. 9.12). Consequently, the effective potential between electrons is attractive and retarded. Under thermodynamic equilibrium conditions, the free phonon propagator depends only on the time difference (see Theorem 6.1), i.e.,

$$D_{\mathbf{q}}^{(0)}(\tau_1, \tau_2) = D_{\mathbf{q}}^{(0)}(\tau_1 - \tau_2). \tag{9.17}$$

Moreover, being invariant under spatial translations, we can expand the field operators in the plane-wave basis (see Eq. 1.77). Consequently, the second-order spatial Fourier transform of the Green's function can be expressed as

$$\begin{aligned} G_{\sigma}^{(m)(2)}(\mathbf{k}, \tau) &= G_{\sigma}^{(m)(0)}(\mathbf{k}, \tau) + \\ &+ \left(-\frac{1}{\hbar}\right)^2 \frac{1}{2} \int_0^{\beta\hbar} d\tau_1 \int_0^{\beta\hbar} d\tau_2 \sum_{\mathbf{k}_1,\sigma_1} \sum_{\mathbf{k}_2,\sigma_2} \sum_{\mathbf{q}} |M_{\mathbf{q}}|^2 D_{\mathbf{q}}^{(0)}(\tau_1 - \tau_2) \\ &\left\langle \hat{T}_\tau C_{\mathbf{k},\sigma}^{(0)}(\tau) C_{\mathbf{k}_1+\mathbf{q},\sigma_1}^{\dagger(0)}(\tau_1) C_{\mathbf{k}_1,\sigma_1}^{(0)}(\tau_1) C_{\mathbf{k}_2-\mathbf{q},\sigma_2}^{\dagger(0)}(\tau_2) C_{\mathbf{k}_2,\sigma_2}^{(0)}(\tau_2) C_{\mathbf{k},\sigma}^{\dagger}(0) \right\rangle_{f,0}, \end{aligned} \tag{9.18}$$

where the spatial integral of the exponential factor has canceled the volume included in the plane waves. We apply thermal Wick's theorem (see Theorems 7.2–7.4) and compute the contractions, considering only the connected diagrams. We set $1' \equiv C_{\mathbf{k},\sigma}^{(0)}(\tau)$, $2' \equiv C_{\mathbf{k}_1+\mathbf{q},\sigma_1}^{\dagger(0)}(\tau_1)$, $3' \equiv C_{\mathbf{k}_1,\sigma_1}^{(0)}(\tau_1)$, $4' \equiv C_{\mathbf{k}_2-\mathbf{q},\sigma_2}^{\dagger(0)}(\tau_2)$, $5' \equiv C_{\mathbf{k}_2,\sigma_2}^{(0)}(\tau_2)$, $6' \equiv C_{\mathbf{k},\sigma}^{\dagger}(0)$, see Fig. 9.1. The contraction $1' - 6'$ produces disconnected diagrams, which must be discarded. Furthermore, we observe the presence of two diagrams with the structure *something in—something out—half interaction line*, which arises from the specific form of the Hamiltonian $\hat{\mathcal{H}}_I$. The contraction of the two half lines at times τ_1, τ_2 gives rise to the phonon propagator; in fact, the two half lines are represented by $\left(a_{\mathbf{q}} + a_{-\mathbf{q}}^{\dagger}\right)$ and $\left(a_{-\mathbf{q}} + a_{\mathbf{q}}^{\dagger}\right)$ (see Eq. 4.91), they compose the phonon propagator and represent the electron-phonon interaction. Ultimately, there are two diagrams, and both appear twice: the direct term diagram (see Fig. 9.2) from the contractions $1' - 2'$, $3' - 6'$, $4' - 5'$ and $1' - 4'$, $2' - 3'$, $5' - 6'$, and the exchange term diagram (see Fig. 9.3) from the contractions $1' - 2'$, $3' - 4'$, $5' - 6'$ and $1' - 4'$, $2' - 5'$, $3' - 6'$. In the direct diagram, due to the loop, an ε emerges, while in the exchange diagram, one ε arises from the half loop and one from thermal Wick's

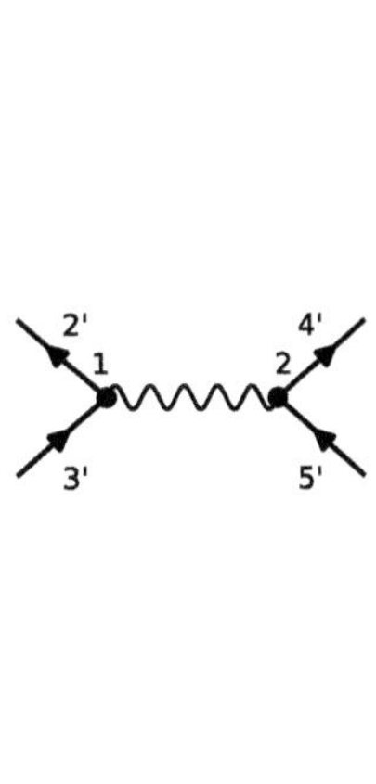

Fig. 9.1 Feynman diagram representing the fermionic operators involved in the electron–phonon interaction. The diagram corresponds to the second-order contribution to the fermionic many-body Green's function arising from the electron–phonon coupling, as given in Eq. 9.18

theorem, then $\varepsilon^2 = 1$. The analytical expression for the second-order correction to the many-body Green function is given by

$$G_\sigma^{(m)(2)}(\mathbf{k}, \tau) = G_\sigma^{(m)(0)}(\mathbf{k}, \tau) + \left(-\frac{1}{\hbar}\right)^2 \int_0^{\beta\hbar} d\tau_1$$

$$\int_0^{\beta\hbar} d\tau_2 \sum_{\mathbf{k}_1,\sigma_1} \sum_{\mathbf{k}_2,\sigma_2} \sum_{\mathbf{q}} |M_\mathbf{q}|^2 D_\mathbf{q}^{(0)}(\tau_1 - \tau_2)$$

$$\Bigg[(-\hbar) G_\sigma^{(m)(0)}(\mathbf{k}, \tau - \tau_1) \delta_{\mathbf{k}_1+\mathbf{q},\mathbf{k}} \delta_{\sigma_1,\sigma} (-\hbar)$$

$$G_\sigma^{(m)(0)}(\mathbf{k}, \tau_1) \delta_{\mathbf{q},\mathbf{0}} (-\hbar)(\varepsilon) G_{\sigma_2}^{(m)(0)}(\mathbf{k}_2, \tau_2, \tau_2 + \delta) +$$

$$+ (-\hbar) G_\sigma^{(m)(0)}(\mathbf{k}, \tau - \tau_1) \delta_{\mathbf{k}_1+\mathbf{q},\mathbf{k}} \delta_{\sigma_1,\sigma} (-\hbar) G_{\sigma_1}^{(m)(0)}(\mathbf{k}_1, \tau_1 - \tau_2)$$

$$\delta_{\sigma_1,\sigma_2} \delta_{\mathbf{k}_2-\mathbf{q},\mathbf{k}_1} (-\hbar) G_\sigma^{(m)(0)}(\mathbf{k}, \tau) \delta_{\mathbf{k}_2,\mathbf{k}} \delta_{\sigma_2,\sigma} \Bigg]. \tag{9.19}$$

We sum over momenta and spins: the delta functions imply $\mathbf{k}_1 = \mathbf{k}$, $\sigma_1 = \sigma$ for the direct term and $\mathbf{k}_1 = \mathbf{k} - \mathbf{q}$, $\mathbf{k}_2 = \mathbf{k}$, $\sigma_1 = \sigma_2 = \sigma$ for the exchange term, and we obtain

$$G_\sigma^{(m)(2)}(\mathbf{k}, \tau) = G_\sigma^{(m)(0)}(\mathbf{k}, \tau) - \hbar \int_0^{\beta\hbar} d\tau_1 \int_0^{\beta\hbar} d\tau_2 \sum_{\mathbf{q}} |M_\mathbf{q}|^2 D_\mathbf{q}^{(0)}(\tau_1 - \tau_2)$$

$$\Bigg[\varepsilon \delta_{\mathbf{q},\mathbf{0}} \sum_{\mathbf{k}_2,\sigma_2} G_\sigma^{(m)(0)}(\mathbf{k}, \tau - \tau_1) G_{\sigma_2}^{(m)(0)}(\mathbf{k}_2, \tau_2, \tau_2 + \delta) G_\sigma^{(m)(0)}(\mathbf{k}, \tau_1) +$$

$$+ G_\sigma^{(m)(0)}(\mathbf{k}, \tau - \tau_1) G_\sigma^{(m)(0)}(\mathbf{k} - \mathbf{q}, \tau_1 - \tau_2) G_\sigma^{(m)(0)}(\mathbf{k}, \tau_2) \Bigg]. \tag{9.20}$$

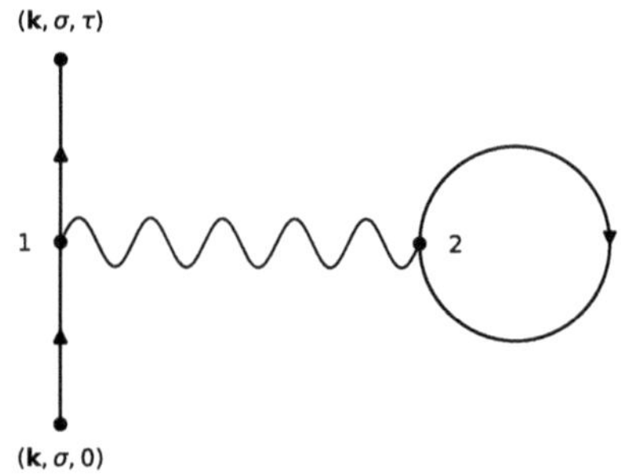

Fig. 9.2 Diagrammatic representation of the direct diagram arising from the electron-phonon interaction. This contribution appears twice, originating from the two distinct sets of contractions $1' - 2'$, $3' - 6'$, $4' - 5'$ and $1' - 4'$, $2' - 3'$, $5' - 6'$ (see Fig. 9.1). It contributes as a second-order term in the Dyson expansion of the fermionic many-body Green's function associated with the electron-phonon interaction (see Eq. 9.18)

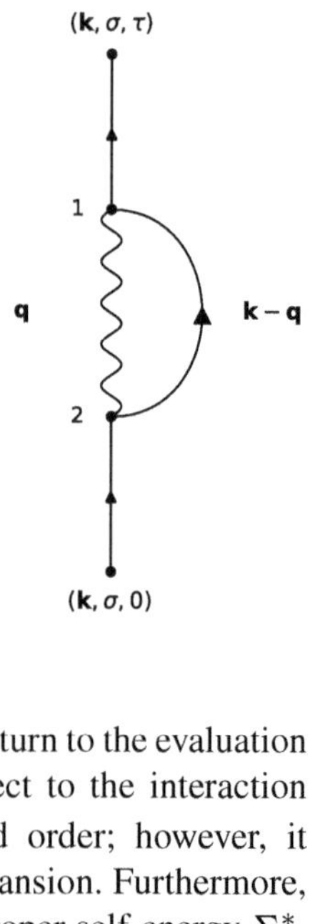

Fig. 9.3 Diagrammatic representation of the exchange diagram arising from the electron–phonon interaction. This contribution appears twice, originating from the two distinct sets of contractions $1' - 2'$, $3' - 4'$, $5' - 6'$ and $1' - 4'$, $2' - 5'$, $3' - 6'$, as shown in Fig. 9.1. It contributes as a second-order term in the Dyson expansion of the fermionic many-body Green's function associated with the electron-phonon interaction (see Eq. 9.18)

Recalling the Dyson algebraic equation (see Eq. 7.85), we now turn to the evaluation of the self-energy at the lowest non-trivial order. With respect to the interaction Hamiltonian $\hat{\mathcal{H}}_I$, the relevant contribution arises at second order; however, it constitutes the first non-vanishing term in the perturbative expansion. Furthermore, although the Dyson equation is expressed in terms of the improper self-energy Σ^*, at this order all contributing diagrams are already irreducible, and therefore we may identify

$$\Sigma^{*(1)} \equiv \Sigma^{(1)}. \tag{9.21}$$

From the comparison between the Dyson equation 7.85 and the explicit second-order expression for the Matsubara Green's function 9.20, one finds

$$\Sigma_\sigma^{(1)}(\mathbf{k}, \tau_1 - \tau_2) = -\varepsilon\hbar \sum_{\mathbf{q}} \delta_{\mathbf{q},\mathbf{0}} |M_{\mathbf{q}}|^2 \delta(\tau_1 - \tau_2) \int_0^{\beta\hbar} d\tau_2' D_{\mathbf{q}}^{(0)}(\tau_1 - \tau_2')$$
$$\sum_{\mathbf{k}_2,\sigma_2} G_{\sigma_2}^{(m)(0)}(\mathbf{k}_2, \tau_2', \tau_2' + \delta) +$$
$$- \hbar \sum_{\mathbf{q}} |M_{\mathbf{q}}|^2 D_{\mathbf{q}}^{(0)}(\tau_1 - \tau_2) G_\sigma^{(m)(0)}(\mathbf{k} - \mathbf{q}, \tau_1 - \tau_2). \tag{9.22}$$

Note that the contribution to the self-energy originating from the direct term is zero due to the delta $\delta_{\mathbf{q},\mathbf{0}}$, since $M_{\mathbf{q}=\mathbf{0}} = 0$. Indeed, the $\mathbf{q} = \mathbf{0}$ contribution does not lead to electronic scattering, corresponding to a uniform lattice displacement that preserves the electronic crystal momentum: for this reason, it is often neglected in transport calculations. Consequently, the second-order propagator reduces to

$$G_\sigma^{(m)(2)}(\mathbf{k}, \tau) = G_\sigma^{(m)(0)}(\mathbf{k}, \tau) - \hbar \int_0^{\beta\hbar} d\tau_1 \int_0^{\beta\hbar} d\tau_2$$
$$\sum_{\mathbf{q}} |M_{\mathbf{q}}|^2 D_{\mathbf{q}}^{(0)}(\tau_1 - \tau_2) G_\sigma^{(m)(0)}(\mathbf{k}, \tau - \tau_1)$$
$$G_\sigma^{(m)(0)}(\mathbf{k} - \mathbf{q}, \tau_1 - \tau_2) G_\sigma^{(m)(0)}(\mathbf{k}, \tau_2), \tag{9.23}$$

and the first-order self-energy is given by

$$\Sigma_\sigma^{(1)}(\mathbf{k}, \tau_1 - \tau_2) = -\hbar \sum_{\mathbf{q}} |M_{\mathbf{q}}|^2 D_{\mathbf{q}}^{(0)}(\tau_1 - \tau_2) G_\sigma^{(m)(0)}(\mathbf{k} - \mathbf{q}, \tau_1 - \tau_2). \tag{9.24}$$

Keeping in mind Eq. 6.25, we apply the time Fourier transform and obtain

$$\Sigma_\sigma^{(1)}(\mathbf{k}, i\omega_n) = \int_0^{\beta\hbar} d(\tau_1 - \tau_2) e^{i\omega_n(\tau_1 - \tau_2)} \Sigma_{\mathbf{k},\sigma}^{(1)}(\tau_1 - \tau_2) =$$
$$= -\hbar \sum_{\mathbf{q}} |M_{\mathbf{q}}|^2 \int_0^{\beta\hbar} d(\tau_1 - \tau_2) e^{i\omega_n(\tau_1 - \tau_2)}$$
$$\frac{1}{\beta\hbar} \sum_{n_1} D_{\mathbf{q}}^{(0)}(i\omega_{n_1}) e^{-i\omega_{n_1}(\tau_1 - \tau_2)}$$
$$\frac{1}{\beta\hbar} \sum_{n_2} G_\sigma^{(m)(0)}(\mathbf{k} - \mathbf{q}, i\omega_{n_2}) e^{-i\omega_{n_2}(\tau_1 - \tau_2)}. \tag{9.25}$$

Regarding the nature of the Matsubara frequencies (see Eq. 6.27), some clarifications are in order. Although we are dealing with electron-phonon interactions, n_1 is even, as it originates from a bosonic propagator. In contrast, n and n_2 are either both odd or both even, depending on whether the propagators $G^{(m)(0)}$ describe fermions or bosons, respectively. Consequently, the difference $n - n_2$ is always even, irrespective of the particle statistics, and thus the quantity $n_1 - (n - n_2)$ is also even. Therefore, the integral is zero unless the overall frequency vanishes, thus producing $\delta_{\omega_{n_2},\omega_n-\omega_{n_1}}$, which implies the presence of a factor $\beta\hbar$, and we have

$$\Sigma_\sigma^{(1)}(\mathbf{k}, i\omega_n) = -\frac{1}{\beta}\sum_{\mathbf{q}}|M_{\mathbf{q}}|^2\sum_{n_1} D_{\mathbf{q}}^{(0)}(i\omega_{n_1})G_\sigma^{(m)(0)}(\mathbf{k}-\mathbf{q}, i(\omega_n - \omega_{n_1})). \tag{9.26}$$

9.1.1 Free Phonon Propagator in Momentum and Frequency Space

We recall that, at thermodynamic equilibrium, the free phonon propagator $D_{\mathbf{q}}^{(0)}(\tau)$ is given in Eq. 9.12, which is defined with respect to the free Hamiltonian in Eq. 9.1. For simplicity, we restrict to the case $\tau > 0$, then

$$\begin{aligned} D_{\mathbf{q}}^{(0)}(\tau) &= -\frac{1}{\hbar}\left\{\left\langle a_{\mathbf{q}}^{(0)}(\tau)a_{\mathbf{q}}^{\dagger(0)}(0)\right\rangle_0 + \left\langle a_{-\mathbf{q}}^{\dagger(0)}(\tau)a_{-\mathbf{q}}^{(0)}(0)\right\rangle_0\right\} = \\ &= -\frac{1}{\hbar}\left\{e^{-\omega_{\mathbf{q}}\tau}\left\langle a_{\mathbf{q}}a_{\mathbf{q}}^\dagger\right\rangle_0 + e^{\omega_{-\mathbf{q}}\tau}\left\langle a_{-\mathbf{q}}^\dagger a_{-\mathbf{q}}\right\rangle_0\right\} = \\ &= -\frac{1}{\hbar}\left\{\left(1 + n_1(\hbar\omega_{\mathbf{q}})\right)e^{-\omega_{\mathbf{q}}\tau} + n_1(\hbar\omega_{-\mathbf{q}})e^{\omega_{-\mathbf{q}}\tau}\right\}, \end{aligned} \tag{9.27}$$

where we used Eqs. 2.251, 9.7 and 9.8, while n_1 denotes the Bose-Einstein statistics given in Eq. 9.11. From Eq. 9.9, it follows $n_1(\hbar\omega_{\mathbf{q}}) = n_1(\hbar\omega_{-\mathbf{q}})$, then

$$D_{\mathbf{q}}^{(0)}(\tau) = -\frac{1}{\hbar}\left\{\left(1 + n_1(\hbar\omega_{\mathbf{q}})\right)e^{-\omega_{\mathbf{q}}\tau} + n_1(\hbar\omega_{\mathbf{q}})e^{\omega_{\mathbf{q}}\tau}\right\}. \tag{9.28}$$

Now, the phonon propagator is evaluated at imaginary times and is therefore periodic with period $\beta\hbar$, as is the case for any bosonic Matsubara Green's function (see Eq. 6.17), i.e.,

$$D_{\mathbf{q}}^{(0)}(\tau - \beta\hbar) = D_{\mathbf{q}}^{(0)}(\tau), \ \tau \in (0, \beta\hbar]. \tag{9.29}$$

This periodicity allows us to expand it in a Fourier series (see Eq. 6.19) over the bosonic Matsubara frequencies (see Eq. 6.27) as follows

$$D^{(0)}_{\mathbf{q}}(\tau) = \frac{1}{\beta\hbar}\sum_{n\in\mathbb{Z}} D^{(0)}_{\mathbf{q}}(i\omega_n)e^{-i\omega_n\tau}. \tag{9.30}$$

Starting from the adapted form of Eq. 6.26, one finds

$$\begin{aligned}
&D^{(0)}_{\mathbf{q}}(i\omega_n) = \\
&= \int_0^{\beta\hbar} d\tau e^{i\omega_n\tau} D^{(0)}_{\mathbf{q}}(\tau) = \\
&= -\frac{1}{\hbar}\left\{\left(1+n_1(\hbar\omega_{\mathbf{q}})\right)\left(\int_0^{\beta\hbar} d\tau e^{i\omega_n\tau}e^{-\omega_{\mathbf{q}}\tau}\right) + n_1(\hbar\omega_{\mathbf{q}})\left(\int_0^{\beta\hbar} d\tau e^{i\omega_n\tau}e^{\omega_{\mathbf{q}}\tau}\right)\right\} = \\
&= -\frac{1}{\hbar}\left\{\frac{1+n_1(\hbar\omega_{\mathbf{q}})}{i\omega_n-\omega_{\mathbf{q}}}\left(e^{i\omega_n\beta\hbar}e^{-\beta\hbar\omega_{\mathbf{q}}}-1\right) + \frac{n_1(\hbar\omega_{\mathbf{q}})}{i\omega_n+\omega_{\mathbf{q}}}\left(e^{i\omega_n\beta\hbar}e^{\beta\hbar\omega_{\mathbf{q}}}-1\right)\right\} = \\
&= -\frac{1}{\hbar}\left\{\frac{1+n_1(\hbar\omega_{\mathbf{q}})}{i\omega_n-\omega_{\mathbf{q}}}\left(\varepsilon e^{-\beta\hbar\omega_{\mathbf{q}}}-1\right) + \frac{n_1(\hbar\omega_{\mathbf{q}})}{i\omega_n+\omega_{\mathbf{q}}}\left(\varepsilon e^{\beta\hbar\omega_{\mathbf{q}}}-1\right)\right\} = \\
&= -\frac{1}{\hbar}\left\{\frac{1+n_1(\hbar\omega_{\mathbf{q}})}{i\omega_n-\omega_{\mathbf{q}}}\left(e^{-\beta\hbar\omega_{\mathbf{q}}}-1\right) + \frac{n_1(\hbar\omega_{\mathbf{q}})}{i\omega_n+\omega_{\mathbf{q}}}\left(e^{\beta\hbar\omega_{\mathbf{q}}}-1\right)\right\},
\end{aligned} \tag{9.31}$$

and from (see Eq. 9.11)

$$\begin{aligned}
1+n_1(\hbar\omega_{\mathbf{q}}) &= 1+\frac{1}{e^{\beta\hbar\omega_{\mathbf{q}}}-1} = \\
&= \frac{e^{\beta\hbar\omega_{\mathbf{q}}}}{e^{\beta\hbar\omega_{\mathbf{q}}}-1},
\end{aligned} \tag{9.32}$$

we get

$$\begin{aligned}
D^{(0)}_{\mathbf{q}}(i\omega_n) &= -\frac{1}{\hbar}\left\{\frac{1}{i\omega_n-\omega_{\mathbf{q}}}\frac{e^{\beta\hbar\omega_{\mathbf{q}}}}{e^{\beta\hbar\omega_{\mathbf{q}}}-1}\left(e^{-\beta\hbar\omega_{\mathbf{q}}}-1\right)\right. \\
&\quad \left.+\frac{1}{i\omega_n+\omega_{\mathbf{q}}}\frac{1}{e^{\beta\hbar\omega_{\mathbf{q}}}-1}\left(e^{\beta\hbar\omega_{\mathbf{q}}}-1\right)\right\} = \\
&= -\frac{1}{\hbar}\left\{\frac{1}{i\omega_n-\omega_{\mathbf{q}}}\frac{1-e^{\beta\hbar\omega_q}}{e^{\beta\hbar\omega_{\mathbf{q}}}-1} + \frac{1}{i\omega_n+\omega_{\mathbf{q}}}\right\} = \\
&= -\frac{1}{\hbar}\left\{-\frac{1}{i\omega_n-\omega_{\mathbf{q}}} + \frac{1}{i\omega_n+\omega_{\mathbf{q}}}\right\} = \\
&= \frac{1}{\hbar}\left\{\frac{1}{i\omega_n-\omega_{\mathbf{q}}} - \frac{1}{i\omega_n+\omega_{\mathbf{q}}}\right\} \equiv \\
&\equiv \frac{1}{\hbar}\frac{2\omega_{\mathbf{q}}}{(i\omega_n)^2-\omega_{\mathbf{q}}^2}.
\end{aligned} \tag{9.33}$$

If we set $i\omega_n = \omega + i\delta$, it follows

$$D_{\mathbf{q}}^{(0)}(\omega + i\delta) = \frac{1}{\hbar}\left\{\frac{\omega - i\delta - \omega_{\mathbf{q}}}{(\omega + i\delta - \omega_{\mathbf{q}})(\omega - i\delta - \omega_{\mathbf{q}})} - \frac{\omega - i\delta + \omega_{\mathbf{q}}}{(\omega + i\delta + \omega_{\mathbf{q}})(\omega - i\delta + \omega_{\mathbf{q}})}\right\} =$$

$$= \frac{1}{\hbar}\left\{\frac{\omega - i\delta - \omega_{\mathbf{q}}}{\left(\omega - \omega_{\mathbf{q}}\right)^2 + \delta^2} - \frac{\omega - i\delta + \omega_{\mathbf{q}}}{\left(\omega + \omega_{\mathbf{q}}\right)^2 + \delta^2}\right\}, \tag{9.34}$$

then

$$-2\hbar \lim_{\delta \to 0^+} \mathrm{Im}\left(D_{\mathbf{q}}^{(0)}(\omega + i\delta)\right) = A_{\mathbf{q}}(\omega), \tag{9.35}$$

$$A_{\mathbf{q}}(\omega) = 2\pi\left\{\delta(\omega - \omega_{\mathbf{q}}) - \delta(\omega + \omega_{\mathbf{q}})\right\}, \tag{9.36}$$

that is, we have derived the spectral function of free phonons.

9.1.2 *Free Phonon Propagator at the High Frequency Limit*

Here, we examine the behavior of the free phonon propagator in the high-frequency limit. This regime is relevant when the phonon energy $\hbar\omega_{\mathbf{q}}$ greatly exceeds all other energy scales in the system. In this limit, the phonon propagator assumes a particularly simple form, as stated in the following theorem.

Theorem 9.1 (High-Frequency Limit of the Free Phonon Propagator) *The free phonon propagator given in Eq. 9.28 satisfies*

$$\lim_{\omega_{\mathbf{q}} \to +\infty}\left(-\hbar\omega_{\mathbf{q}} D_{\mathbf{q}}^{(0)}(\tau)\right) = 2\sum_{n \in \mathbb{Z}} \delta(\tau - n\beta\hbar). \tag{9.37}$$

Proof Let $\varphi(\tau)$ be a smooth $\beta\hbar$-periodic test function: $\varphi \in C^{\infty}(\mathbb{R})$ and $\varphi(\tau + \beta\hbar) = \varphi(\tau)$ for all τ. To prove distributional convergence, consider the quantity

$$I_{\omega}[\varphi] = \int_0^{\beta\hbar}\left(-\hbar\omega_{\mathbf{q}} D_{\mathbf{q}}^{(0)}(\tau)\right)\varphi(\tau)d\tau. \tag{9.38}$$

Substituting the explicit form of $D_{\mathbf{q}}^{(0)}(\tau)$ given in Eq. 9.28, one finds

$$I_\omega[\varphi] = \int_0^{\beta\hbar} \omega_{\mathbf{q}} \left\{(1 + n_1(\hbar\omega_{\mathbf{q}}))e^{-\omega_{\mathbf{q}}\tau} + n_1(\hbar\omega_{\mathbf{q}})e^{\omega_{\mathbf{q}}\tau}\right\} \varphi(\tau)d\tau. \tag{9.39}$$

For simplicity, we write ω for $\omega_{\mathbf{q}}$ and n for $n_1(\hbar\omega)$. We write

$$I_\omega[\varphi] = I_\omega^{(1)} + I_\omega^{(2)}, \tag{9.40}$$

with

$$I_\omega^{(1)} = \int_0^{\beta\hbar} \omega(1+n)e^{-\omega\tau}\varphi(\tau)d\tau, \tag{9.41}$$

$$I_\omega^{(2)} = \int_0^{\beta\hbar} \omega n e^{\omega\tau}\varphi(\tau)d\tau. \tag{9.42}$$

Compute the mass (integral) of each kernel over the period $[0, \beta\hbar]$. For the first kernel:

$$\int_0^{\beta\hbar} \omega(1+n)e^{-\omega\tau}\, d\tau = (1+n)(1-e^{-\beta\hbar\omega}). \tag{9.43}$$

From Eq. 9.32, we get

$$(1+n)(1-e^{-\beta\hbar\omega}) = 1. \tag{9.44}$$

Hence the first kernel is nonnegative and has unit mass on $[0, \beta\hbar]$. For the second kernel:

$$\int_0^{\beta\hbar} \omega n e^{\omega\tau}\, d\tau = \frac{\omega}{e^{\beta\hbar\omega}-1}\int_0^{\beta\hbar} e^{\omega\tau}\, d\tau = \\ = 1. \tag{9.45}$$

Thus the second kernel is also nonnegative and has mass 1. Consequently, for $\tau \in [0, \beta\hbar]$, we define the normalized kernels:

$$K_\omega^{(1)}(\tau) = \omega(1+n)e^{-\omega\tau}, \tag{9.46}$$

$$K_\omega^{(2)}(\tau) = \omega n e^{\omega\tau}. \tag{9.47}$$

They satisfy

$$\int_0^{\beta\hbar} K_\omega^{(1)}(\tau)\, d\tau = \int_0^{\beta\hbar} K_\omega^{(2)}(\tau)\, d\tau = 1. \tag{9.48}$$

Let $\varepsilon > 0$. Since φ is continuous on the compact $[0, \beta\hbar]$, it is uniformly continuous. There exists $\delta > 0$ such that

$$|\varphi(\tau) - \varphi(0)| < \varepsilon \quad \text{for } 0 \leq \tau < \delta, \tag{9.49}$$

$$|\varphi(\tau) - \varphi(\beta\hbar)| < \varepsilon \quad \text{for } \beta\hbar - \delta < \tau \leq \beta\hbar. \tag{9.50}$$

Concerning the first contribution,

$$\left| \int_0^{\beta\hbar} K_\omega^{(1)}(\tau)\varphi(\tau)\, d\tau - \varphi(0) \right| \leq \int_0^{\beta\hbar} K_\omega^{(1)}(\tau)|\varphi(\tau) - \varphi(0)|\, d\tau \tag{9.51}$$

$$\leq \int_0^{\delta} K_\omega^{(1)}(\tau)|\varphi(\tau) - \varphi(0)|\, d\tau + \int_\delta^{\beta\hbar} K_\omega^{(1)}(\tau)|\varphi(\tau) - \varphi(0)|\, d\tau \tag{9.52}$$

$$\leq \varepsilon \int_0^{\delta} K_\omega^{(1)}(\tau)\, d\tau + 2M \int_\delta^{\beta\hbar} K_\omega^{(1)}(\tau)\, d\tau, \tag{9.53}$$

where $M = \sup|\varphi|$. But

$$\int_\delta^{\beta\hbar} K_\omega^{(1)}(\tau)\, d\tau \leq (1+n)e^{-\omega\delta} \to 0 \quad \text{as } \omega \to \infty. \tag{9.54}$$

Hence for large ω,

$$\left| \int_0^{\beta\hbar} K_\omega^{(1)}(\tau)\varphi(\tau)\, d\tau - \varphi(0) \right| < 2\varepsilon, \tag{9.55}$$

so

$$\lim_{\omega\to\infty} \int_0^{\beta\hbar} K_\omega^{(1)}(\tau)\varphi(\tau)\, d\tau = \varphi(0). \tag{9.56}$$

Regarding the second contribution, we set $s = \beta\hbar - \tau$, and we obtain

$$I_\omega^{(2)} = \int_0^{\beta\hbar} \frac{\omega e^{-\omega s}}{1 - e^{-\beta\hbar\omega}} \varphi(\beta\hbar - s) ds. \tag{9.57}$$

We define

$$\widetilde{K}_\omega^{(2)}(s) = \frac{\omega e^{-\omega s}}{1 - e^{-\beta\hbar\omega}}, \tag{9.58}$$

then

$$\int_0^{\beta\hbar} \widetilde{K}_\omega^{(2)}(s)ds = 1. \tag{9.59}$$

Repeating the concentration argument near $s = 0$ gives

$$\lim_{\omega\to\infty} I_\omega^{(2)} = \varphi(\beta\hbar) = \varphi(0), \tag{9.60}$$

where the last step follows from the $\beta\hbar$-periodicity of φ. Combining both contributions, one finds

$$\begin{aligned}\lim_{\omega\to\infty} I_\omega[\varphi] = \lim_{\omega\to\infty}\left(I_\omega^{(1)} + I_\omega^{(2)}\right) &= \\ = \varphi(0) + \varphi(\beta\hbar) &= \\ = 2\varphi(0). \end{aligned}\tag{9.61}$$

Since this result holds for every smooth $\beta\hbar$-periodic test function φ, the statement follows. □

Physically, the high-frequency limit corresponds to a regime in which the characteristic phonon energy is much larger than all other relevant energy scales in the system, particularly those associated with the electronic dynamics. In this regime, the phonons oscillate so rapidly that, on the timescale relevant for the electrons, their response can be considered instantaneous. Mathematically, this is reflected in the propagator reducing to a Dirac delta function in imaginary time, which encodes a perfectly local response at each instant. Consequently, the phonon-mediated electron-electron interaction becomes effectively instantaneous, or static, along the imaginary-time axis. This observation provides a rigorous justification for replacing the full retarded interaction with a time-local approximation in certain effective theories. In particular, it underpins the derivation of simplified models such as attractive Hubbard-type Hamiltonians, where the phonon-mediated attraction between electrons is treated as an instantaneous on-site interaction.

9.2 Phonon Propagator Self-energy

After introducing the free phonon propagator, it is natural to extend the definition to the full interacting case. In particular, the phonon propagator describes the time-ordered correlation of the phonon displacement operators $\mathbf{S}_{\mathbf{n},\mu}$ (see Eq. 4.134), accounting for interaction effects.

Definition 9.2 (Phonon Propagator) At thermodynamic equilibrium, given the phonon annihilation and creation operators (see Eqs. 4.88 and 4.90), the phonon propagator is defined as

$$D_{\mathbf{q}}(\tau) = -\frac{1}{\hbar}\left\langle \hat{T}_\tau \left(a_{\mathbf{q}} + a^{\dagger}_{-\mathbf{q}}\right)(\tau)\left(a_{-\mathbf{q}} + a^{\dagger}_{\mathbf{q}}\right)(0)\right\rangle, \tag{9.62}$$

where the thermal average (see Eq. 1.379) is taken with respect to the Hamiltonian 5.37, and the operators at time τ are evaluated in the thermal Heisenberg picture (see Eq. 6.12).

From Dyson expansion (see Theorem 7.1), we can expand phonon propagator using the free propagators and the non-interacting Hamiltonian, i.e., we have to compute

$$D_{\mathbf{q}}(\tau) = -\frac{1}{\hbar}\left\langle \hat{T}_\tau \left(a_{\mathbf{q}} + a^{\dagger}_{-\mathbf{q}}\right)^{(0)}(\tau)\left(a_{-\mathbf{q}} + a^{\dagger}_{\mathbf{q}}\right)(0)\right\rangle_0, \tag{9.63}$$

where the operators at time τ are evaluated in the thermal interaction picture (see Eq. 2.194). For convenience, we define the operator

$$A_{\mathbf{q}} = a_{\mathbf{q}} + a^{\dagger}_{-\mathbf{q}}. \tag{9.64}$$

From Theorem 7.1 and from Eq. 9.64, the phonon propagator 9.63 can be written as

$$D_{\mathbf{q}}(\tau) = -\frac{1}{\hbar}\left\langle \hat{T}_\tau A^{(0)}_{\mathbf{q}}(\tau) A^{\dagger}_{\mathbf{q}}(0)\right\rangle_0, \tag{9.65}$$

where note that the operator on the far right, $A^{\dagger}_{\mathbf{q}}(0)$, is not evaluated in any picture, being taken at time zero (see Remark 7.1). Consider Theorem 7.1: at zero order, the phonon propagator coincides with the free phonon propagator. At order $n = 1$, the interaction appears only once, but the term is zero since the interaction Hamiltonian $\hat{\mathcal{H}}_I$ appears only once and the thermal average

$$\left\langle a_{\mathbf{q}} + a^{\dagger}_{-\mathbf{q}}\right\rangle_0 \tag{9.66}$$

vanishes. The first non-zero order is for $n = 2$ and the phonon propagator at second order is written as

$$\begin{aligned} D^{(2)}_{\mathbf{q}}(\tau) = D^{(0)}_{\mathbf{q}}(\tau) - \frac{1}{\hbar}\left(\frac{1}{\hbar}\right)^2 \frac{1}{2}\int_0^{\beta\hbar} d(\tau_1)\int_0^{\beta\hbar} d(\tau_2) \sum_{\mathbf{q}_1,\mathbf{k}_1,\sigma_1}\sum_{\mathbf{q}_2,\mathbf{k}_2,\sigma_2} M_{\mathbf{q}_1} M_{\mathbf{q}_2} \\ \left\langle \hat{T}_\tau A^{(0)}_{\mathbf{q}}(\tau) A^{\dagger(0)}_{-\mathbf{q}_1}(\tau_1) A^{(0)}_{\mathbf{q}_2}(\tau_2) A^{\dagger}_{\mathbf{q}}(0)\right\rangle_0 \\ \left\langle \hat{T}_\tau C^{\dagger(0)}_{\mathbf{k}_1+\mathbf{q}_1,\sigma_1}(\tau_1) C^{(0)}_{\mathbf{k}_1,\sigma_1}(\tau_1) C^{\dagger(0)}_{\mathbf{k}_2+\mathbf{q}_2,\sigma_2}(\tau_2) C^{(0)}_{\mathbf{k}_2,\sigma_2}(\tau_2)\right\rangle_0. \end{aligned} \tag{9.67}$$

From Eqs. 2.194 and 4.91, it follows

$$
\begin{aligned}
A_{\mathbf{q}1}^{\dagger(0)}(\tau_1) &= \left(e^{\frac{\hat{\mathcal{H}}_0}{\hbar}\tau}\right)^{\dagger} A_{\mathbf{q}_1}^{\dagger} \left(e^{-\frac{\hat{\mathcal{H}}_0}{\hbar}\tau}\right)^{\dagger} = \\
&= e^{\frac{\hat{\mathcal{H}}_0}{\hbar}\tau} A_{-\mathbf{q}_1} e^{-\frac{\hat{\mathcal{H}}_0}{\hbar}\tau} = \\
&= A_{-\mathbf{q}_1}^{(0)}(\tau),
\end{aligned}
\tag{9.68}
$$

and in particular, we make use of Eq. 9.68 in the form $A_{-\mathbf{q}_1}^{\dagger(0)}(\tau_1) = A_{\mathbf{q}_1}^{(0)}(\tau_1)$, and we write

$$
\begin{aligned}
D_{\mathbf{q}}^{(2)}(\tau) = D_{\mathbf{q}}^{(0)}(\tau) - \frac{1}{\hbar}\left(\frac{1}{\hbar}\right)^2 \frac{1}{2} \int_0^{\beta\hbar} d(\tau_1) \int_0^{\beta\hbar} d(\tau_2) \sum_{\mathbf{q}_1,\mathbf{k}_1,\sigma_1} \sum_{\mathbf{q}_2,\mathbf{k}_2,\sigma_2} M_{\mathbf{q}_1} M_{\mathbf{q}_2} \\
\left\langle \hat{T}_\tau A_{\mathbf{q}}^{(0)}(\tau) A_{\mathbf{q}_1}^{(0)}(\tau_1) A_{\mathbf{q}_2}^{(0)}(\tau_2) A_{\mathbf{q}}^{\dagger}(0) \right\rangle_0 \\
\left\langle \hat{T}_\tau C_{\mathbf{k}_1+\mathbf{q}_1,\sigma_1}^{\dagger(0)}(\tau_1) C_{\mathbf{k}_1,\sigma_1}^{(0)}(\tau_1) C_{\mathbf{k}_2+\mathbf{q}_2,\sigma_2}^{\dagger(0)}(\tau_2) C_{\mathbf{k}_2,\sigma_2}^{(0)}(\tau_2) \right\rangle_0 .
\end{aligned}
\tag{9.69}
$$

We set $1' \equiv A_{\mathbf{q}}^{(0)}(\tau)$, $2' \equiv A_{\mathbf{q}_1}^{(0)}(\tau_1)$, $3' \equiv A_{\mathbf{q}_2}^{(0)}(\tau_2)$, $4' \equiv A_{\mathbf{q}}^{\dagger}(0)$, $5' \equiv C_{\mathbf{k}_1+\mathbf{q}_1,\sigma_1}^{\dagger(0)}(\tau_1)$, $6' \equiv C_{\mathbf{k}_1,\sigma_1}^{(0)}(\tau_1)$, $7' \equiv C_{\mathbf{k}_2+\mathbf{q}_2,\sigma_2}^{\dagger(0)}(\tau_2)$, $8' \equiv C_{\mathbf{k}_2,\sigma_2}^{(0)}(\tau_2)$ and we apply the thermal Wick's theorem (see Theorems 7.2–7.4). The operators $A_{\mathbf{q}}^{(0)}(\tau)$ and $A_{\mathbf{q}}^{\dagger}(0)$ serve as the annihilation and creation field operators, evaluated at the final time τ and the initial time 0, respectively: they are represented by a bosonic line that originates at time 0, $A_{\mathbf{q}}^{\dagger}(0)$ (see Fig. 9.4a) or terminates at time τ, i.e., $A_{\mathbf{q}}^{(0)}(\tau)$ (see Fig. 9.4b). Consequently, similarly to the free and interacting propagators of the field operators, the free phonon propagator is represented as in Fig. 9.4c, and the phonon propagator as in Fig. 9.4d. The interaction vertices are two, τ_1 and τ_2, since we are at second order, and each vertex includes an incoming particle line (C) and an outgoing one ($C^{\dagger}$), and a half bosonic line ($2'$ and $3'$), as in Fig. 9.5. We cannot contract $1'$ with $4'$ since that would yield disconnected diagrams, which are excluded within the framework of Feynman's perturbative theory of the thermal Green's function (see Chap. 7). We must contract $1'$ with one of the two half interaction lines, and both these contractions give the same contribution since both vertices are integrated from 0 to $\beta\hbar$. We choose to contract $1' - 2'$ and $3' - 4'$, and this double diagram removes the factor $(2)^{-1}$ in the second order expansion. The contraction $1' - 2'$ produces a free phonon propagator with $\mathbf{q}_1 = -\mathbf{q}$ up to a factor $(-\hbar)^{-1}$, and the contraction $3' - 4'$ produces a free phonon propagator with $\mathbf{q}_2 = \mathbf{q}$, up to a factor $(-\hbar)^{-1}$: the numerical factor $(-\hbar)^{-2}$ in the second order expansion facilitates this algebraic manipulation. Furthermore, the contractions $5' - 6'$ and $7' - 8'$ are not admissible, as they imply $\mathbf{q}_1 = 0$ and $\mathbf{q}_2 = 0$, respectively, where the interaction identically vanishes. Therefore, we must contract $5' - 8'$ and $6' - 7'$. Equation 9.69 becomes

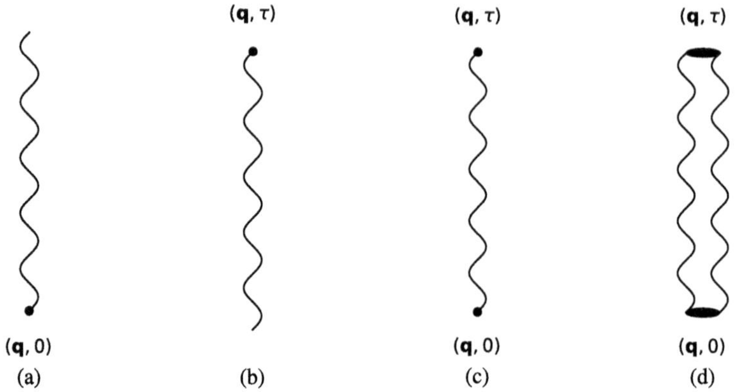

Fig. 9.4 The diagram shows, from left to right, the following elements in Feynman diagrammatics: (**a**) Bosonic annihilation operator: a wavy line with a vertex at the bottom labeled $(\mathbf{q}, 0)$, corresponding to the absorption of a boson in a process. (**b**) Bosonic creation operator: a wavy line with a vertex at the top labeled $(\mathbf{q}, \tau)$, corresponding to the emission of a boson in a process. (**c**) Free (or bare) phonon propagator: a wavy line connecting a vertex at the bottom labeled $(\mathbf{q}, 0)$ and a vertex at the top labeled $(\mathbf{q}, \tau)$, describing the propagation of a phonon in the absence of interactions. (**d**) Phonon propagator: two parallel wavy lines connecting a vertex at the bottom labeled $(\mathbf{q}, 0)$ and a vertex at the top labeled $(\mathbf{q}, \tau)$, describing the propagation of a phonon dressed by interactions

Fig. 9.5 Feynman diagram representing four fermionic and two bosonic operators in the electron-phonon interaction, corresponding to the second-order term in the Dyson expansion of the phonon propagator (see Eq. 9.69)

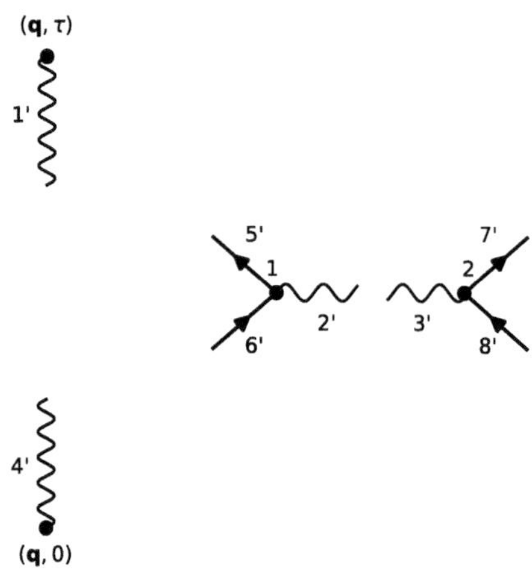

$$
\begin{aligned}
D_{\mathbf{q}}^{(2)}(\tau) = & \\
= D_{\mathbf{q}}^{(0)}(\tau) - \frac{1}{\hbar}\int_0^{\beta\hbar} d\tau_1 \int_0^{\beta\hbar} d\tau_2 \sum_{\mathbf{k}_1,\sigma_1}\sum_{\mathbf{k}_2,\sigma_2} \left|M_{\mathbf{q}}\right|^2 & \\
(-\hbar) G_{\sigma_1}^{(m)(0)}(\mathbf{k}_1, \tau_1 - \tau_2)\delta_{\mathbf{k}_2,\mathbf{k}_1-\mathbf{q}}\delta_{\sigma_1,\sigma_2}\varepsilon(-\hbar) & \\
G_{\sigma_1}^{(m)(0)}(\mathbf{k}_1 - \mathbf{q}, \tau_2 - \tau_1) D_{\mathbf{q}}^{(0)}(\tau - \tau_1) D_{\mathbf{q}}^{(0)}(\tau_2), & \qquad (9.70)
\end{aligned}
$$

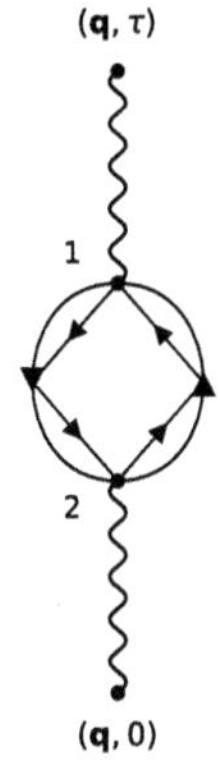

Fig. 9.6 Schematic representation of the electron-phonon interaction corresponding to the second-order term in the Dyson expansion of the phonon propagator (see Eq. 9.69). This contribution originates from the contractions $1' - 2'$, $3' - 4'$, $5' - 8'$, and $6' - 7'$, as shown in Fig. 9.5

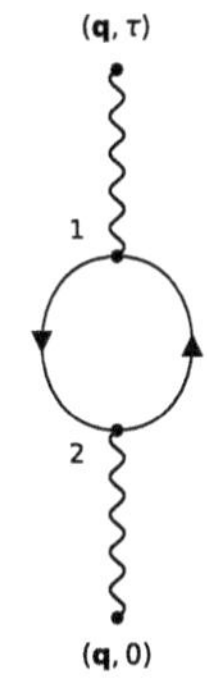

Fig. 9.7 Feynman diagram illustrating the contractions shown in Fig. 9.6

where $(-\hbar)G_{\sigma_1}^{(m)(0)}(\mathbf{k}_1, \tau_1 - \tau_2)\delta_{\mathbf{k}_2,\mathbf{k}_1-\mathbf{q}}\delta_{\sigma_1,\sigma_2}$ comes from the contraction $5' - 8'$ and $\varepsilon(-\hbar)G_{\sigma_1}^{(m)(0)}(\mathbf{k}_1 - \mathbf{q}, \tau_2 - \tau_1)$ comes from the contraction $6' - 7'$. The resulting diagram is Fig. 9.6: the contraction $1' - 2'$ produces a free phonon propagator with vertices at τ and τ_1, and in particular at τ_1, there is an incoming and an outgoing line ($C^\dagger$ and C); similarly, the contraction $3' - 4'$ produces a free phonon propagator with vertices at τ_2 and 0, and in particular at τ_2 there is an incoming and an outgoing line ($C^\dagger$ and C); finally, $G_{\sigma_1}^{(m)(0)}(\mathbf{k}_1 - \mathbf{q}, \tau_2 - \tau_1)$ is the propagator of an electron propagating from the internal time τ_1 to the time τ_2, while $G_{\sigma_1}^{(m)(0)}(\mathbf{k}_1, \tau_1 - \tau_2)$ is the propagator of an electron propagating from time τ_2 to time τ_1, and together the two propagators form a closed electron loop into which the two pairs of incoming and outgoing lines for each vertex are inserted. In the Feynman diagrammatics, the contractions in Fig. 9.6 are schematized as in Fig. 9.7, where the four fermionic lines are absorbed into the loop. From Eq. 9.69, summing up we have

$$
\begin{aligned}
D_{\mathbf{q}}^{(2)}(\tau) = D_{\mathbf{q}}^{(0)}(\tau) - \varepsilon\hbar \int_0^{\beta\hbar} d(\tau_1) \int_0^{\beta\hbar} d(\tau_2) \\
\sum_{\mathbf{k}_1,\sigma_1} \left|M_{\mathbf{q}}\right|^2 G_{\sigma_1}^{(m)(0)}(\mathbf{k}_1 - \mathbf{q}, \tau_2 - \tau_1) G_{\sigma_1}^{(m)(0)}(\mathbf{k}_1, \tau_1 - \tau_2) \\
D_{\mathbf{q}}^{(0)}(\tau - \tau_1) D_{\mathbf{q}}^{(0)}(\tau_2).
\end{aligned}
\tag{9.71}
$$

To remove the integration with respect to time, we apply a Fourier transform, i.e.,

$$D_{\mathbf{q}}^{(2)}(i\omega_n) =$$
$$= \int_0^{\beta\hbar} d\tau e^{i\omega_n \tau} D_{\mathbf{q}}^{(2)}(\tau) =$$
$$= D_{\mathbf{q}}^{(0)}(i\omega_n) - \varepsilon\hbar \int_0^{\beta\hbar} d\tau e^{i\omega_n \tau} \int_0^{\beta\hbar} d(\tau_1) \int_0^{\beta\hbar} d(\tau_2) \sum_{\mathbf{k}_1,\sigma_1} \left|M_{\mathbf{q}}\right|^2 \frac{1}{(\beta\hbar)^4}$$
$$\sum_{n_1}\sum_{n_2}\sum_{n_3}\sum_{n_4} G_{\sigma_1}^{(m)(0)}(\mathbf{k}_1 - \mathbf{q}, i\omega_{n_1}) G_{\sigma_1}^{(m)(0)}(\mathbf{k}_1, i\omega_{n_2}) D_{\mathbf{q}}^{(0)}(i\omega_{n_3}) D_{\mathbf{q}}^{(0)}(i\omega_{n_4})$$
$$e^{i\omega_{n_1}(\tau_2-\tau_1)} e^{i\omega_{n_2}(\tau_1-\tau_2)} e^{i\omega_{n_3}(\tau-\tau_1)} e^{i\omega_{n_4}(\tau_2)}. \tag{9.72}$$

From the integrals with respect to τ, τ_1, τ_2 we respectively have: $n = n_3$, $n_2 = n_1 + n$, $n_2 = n_4 + n_1$ and combining, we get $n_4 = n$, furthermore the three integrals provide a numerical factor $\beta\hbar$ each, that is,

$$D_{\mathbf{q}}^{(2)}(i\omega_n) = D_{\mathbf{q}}^{(0)}(i\omega_n) - \frac{\varepsilon}{\beta} \left|M_{\mathbf{q}}\right|^2 \sum_{\mathbf{k}_1,\sigma_1,n_1} G_{\sigma_1}^{(m)(0)}(\mathbf{k}_1 - \mathbf{q}, i\omega_{n_1})$$
$$G_{\sigma_1}^{(m)(0)}(\mathbf{k}_1, i(\omega_{n_1} + \omega_n)) D_{\mathbf{q}}^{(0)}(i\omega_n) D_{\mathbf{q}}^{(0)}(i\omega_n). \tag{9.73}$$

By comparing Eqs. 7.85 and 9.73, remembering that at the first nonzero order the self-energies Σ and Σ^* coincide, we have

$$\Sigma_{\sigma}^{(1)}(\mathbf{k}, i\omega_n) = -\frac{\varepsilon}{\beta} \left|M_{\mathbf{q}}\right|^2 \sum_{\mathbf{k}_1,\sigma_1,n_1} G_{\sigma_1}^{(m)(0)}(\mathbf{k}_1 - \mathbf{q}, i\omega_{n_1}) G_{\sigma_1}^{(m)(0)}(\mathbf{k}_1, i(\omega_{n_1} + \omega_n)). \tag{9.74}$$

Such self-energy is actually the lowest order of another type of propagator, as we will now verify.

Definition 9.3 (Polarization Propagator (in Momentum Space and Imaginary Time)) Given the particle density operator in second quantization and in momentum space (see Eq. 2.262), the polarization propagator in momentum space and imaginary time is defined as the following thermal average (see Eq. 1.379)

$$\Pi_{\mathbf{q}}(\tau) = -\frac{i}{\hbar} \left\langle \left[\hat{\rho}_{\mathbf{q}}(\tau), \hat{\rho}_{\mathbf{q}}^{\dagger}(0) \right] \right\rangle, \tag{9.75}$$

where the operators at time τ are evaluated in the thermal Heisenberg picture (see Eq. 6.12). The propagator encodes the correlations between particle density fluctuations at different times.

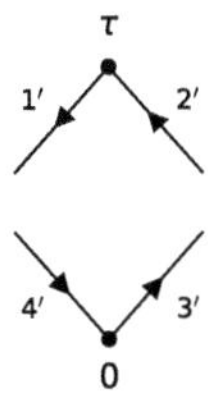

Fig. 9.8 Schematic representation of the first-order polarization propagator, also called the polarization bubble, corresponding to Eq. 9.77. In this diagram, the two time vertices are connected by particle-hole lines, forming a closed loop (the polarization bubble) that represents a particle-hole excitation. At each vertex, a particle is created and a particle is annihilated, corresponding to the two Matsubara Green's functions that appear in the expression for $\Pi_{\mathbf{q}}^{(0)}(\tau)$

Definition 9.4 (Free Polarization Propagator) In the non-interacting case, i.e., with respect to the free Hamiltonian 9.1, the polarization propagator given in Eq. 9.75 takes the form

$$\Pi_{\mathbf{q}}^{(0)}(\tau) = -\frac{1}{\hbar}\left\langle \hat{T}_\tau C_{\mathbf{k}_1+\mathbf{q},\sigma_1}^{\dagger(0)}(\tau) C_{\mathbf{k}_1,\sigma_1}^{(0)}(\tau) C_{\mathbf{k}_2-\mathbf{q},\sigma_2}^{\dagger}(0) C_{\mathbf{k}_2,\sigma_2}(0) \right\rangle_0, \tag{9.76}$$

where the operators at time τ are evaluated in the thermal interaction picture (see Eq. 2.194). The momentum transfer $\mathbf{q}$ is assumed to be non-zero, since $M_{\mathbf{q}=\mathbf{0}} = 0$.

Note that, in contrast to the one-particle Green's function associated with single field operators, the object considered here is a two-particle Green's function, describing correlations between pairs of operators and thus capturing higher-order interaction effects. This is because the particle density operators involve two field operators each, so the thermal average at first order involves a four-operator expectation value. At each vertex, located at times 0 and τ, one particle is created and one is annihilated, leading to the contraction of four operators in total. With the identifications $1' \equiv C_{\mathbf{k}_1+\mathbf{q},\sigma_1}^{\dagger(0)}(\tau)$, $2' \equiv C_{\mathbf{k}_1,\sigma_1}^{(0)}(\tau)$, $3' \equiv C_{\mathbf{k}_2-\mathbf{q},\sigma_2}^{\dagger}(0)$, $4' \equiv C_{\mathbf{k}_2,\sigma_2}(0)$, the only possible contractions are $1'-4'$, $2'-3'$, see Fig. 9.8. The contraction $1'-2'$ would be a contraction between two operators acting at the same space-time point, and it would be physically untenable in this context. This is because it would imply a coupling between the same particle being created and annihilated at the same point and at the same instant, which leads to the condition $\mathbf{q} = 0$. This would mean there is no momentum transfer (i.e., no interaction between particles with different momenta), making this contraction physically irrelevant for the interaction we are studying. The only permissible diagram is the one shown in Fig. 9.9, commonly referred to as the *polarization bubble*. The first-order polarization propagator is then expressed analytically as follows

$$\Pi_{\mathbf{q}}^{(0)}(\tau) = -\varepsilon\hbar \sum_{\mathbf{k}_1,\sigma_1} G_{\sigma_1}^{(m)(0)}(\mathbf{k}_1+\mathbf{q}, -\tau) G_{\sigma_1}^{(m)(0)}(\mathbf{k}_1, \tau). \tag{9.77}$$

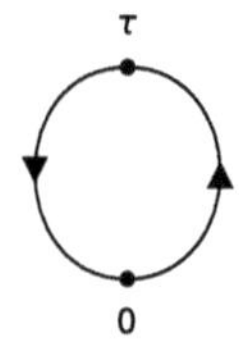

Fig. 9.9 Feynman diagram illustrating the contractions shown in Fig. 9.8

We apply the time Fourier transform and we have

$$\Pi_{\mathbf{q}}^{(0)}(i\omega_n) =$$
$$= \int_0^{\beta\hbar} d\tau e^{i\omega_n\tau}\Pi_{\mathbf{q}}^{(0)}(\tau) =$$
$$= -\varepsilon\hbar\int_0^{\beta\hbar} d\tau e^{i\omega_n\tau}\sum_{\mathbf{k}_1,\sigma_1}\frac{1}{(\beta\hbar)^2}\sum_{n_1,n_2} G_{\sigma_1}^{(m)(0)}(\mathbf{k}_1+\mathbf{q}, i\omega_{n_1})$$
$$G_{\sigma_1}^{(m)(0)}(\mathbf{k}_1, i\omega_{n_2})e^{-i\omega_{n_1}(-\tau)}e^{-i\omega_{n_2}\tau} =$$
$$= -\varepsilon\hbar\int_0^{\beta\hbar} d\tau e^{-i(\omega_n+\omega_{n_1}-\omega_{n_2})\tau}$$
$$\sum_{\mathbf{k}_1,\sigma_1}\frac{1}{(\beta\hbar)^2}\sum_{n_1,n_2} G_{\sigma_1}^{(m)(0)}(\mathbf{k}_1+\mathbf{q}, i\omega_{n_1})G_{\sigma_1}^{(m)(0)}(\mathbf{k}_1, i\omega_{n_2}). \quad (9.78)$$

The integral with respect to τ is equal to $\beta\hbar\delta_{n+n_1,n_2}$, then

$$\Pi_{\mathbf{q}}^{(0)}(i\omega_n) = -\frac{\varepsilon}{\beta}\sum_{\mathbf{k}_1,\sigma_1,n_1} G_{\sigma_1}^{(m)(0)}(\mathbf{k}_1+\mathbf{q}, i\omega_{n_1})G_{\sigma_1}^{(m)(0)}(\mathbf{k}_1, i(\omega_{n_1}+\omega_n)). \quad (9.79)$$

By comparing Eqs. 9.74 and 9.79, we note that the two propagators would be proportional, apart from the sign of the momentum $\mathbf{q}$; however, the following theorems show that this sign difference is irrelevant for the polarization propagator object. We have

Theorem 9.2 (A Matsubara Sum for Bosonic and Fermionic Frequencies) *Let ω_n, ω_{n_1} be Matsubara frequencies (see Eq. 6.27), respectively bosonic and fermionic, then they satisfy*

$$\frac{1}{\beta}\sum_{n_1}\frac{1}{\hbar\left[i\omega_{n_1}-\frac{\mathcal{E}_{\mathbf{k}}}{\hbar}\right]}\frac{1}{\hbar\left[i\omega_{n_1}+i\omega_n-\frac{\mathcal{E}_{\mathbf{p}}}{\hbar}\right]} = \frac{1}{\hbar}\frac{n_{-1}(\mathcal{E}_{\mathbf{p}})-n_{-1}(\mathcal{E}_{\mathbf{k}})}{i\omega_n+\frac{\mathcal{E}_{\mathbf{p}}-\mathcal{E}_{\mathbf{k}}}{\hbar}}, \quad (9.80)$$

where n_{-1} denotes the Fermi-Dirac statistics E.4.

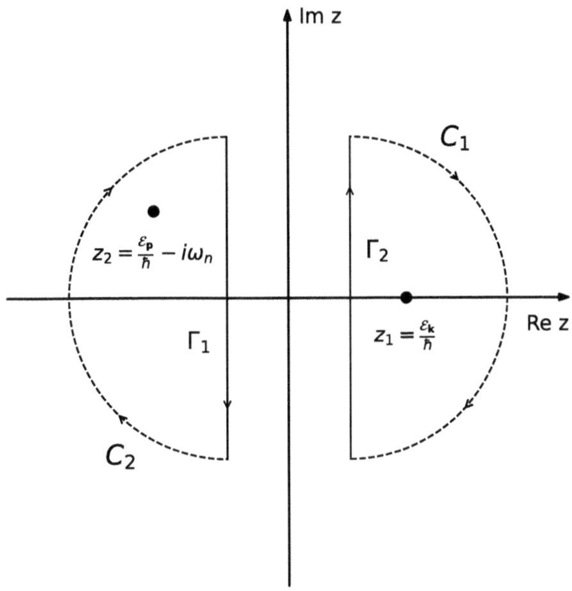

Fig. 9.10 Integration contour used to evaluate the Matsubara sum in Eq. 9.80 via the residue theorem. The horizontal axis corresponds to the real part of the complex variable z, representing the physical excitation energy, while the vertical axis corresponds to its imaginary part, associated with the discrete Matsubara frequencies (see Eq. 6.27). The contour is chosen to enclose the relevant singularities in the complex plane, allowing the corresponding residues to be computed and the summation over Matsubara frequencies to be performed

Proof First note that in Eq. 9.80 the factor $e^{i\omega_n\delta}$ is absent, unlike the series in 7.93, since here the general term in the Matsubara sum is proportional to

$$\sum_{n_1} \frac{1}{(i\omega_{n_1})^2} \sim \sum_{n_1} \frac{1}{n_1^2}, \tag{9.81}$$

which is already convergent. In other words, the summand here decays sufficiently fast due to the $(n_1)^{-2}$ behavior. Therefore, no additional damping is required, and the series converges naturally. Now, we proceed by applying the residue method, following steps similar to those used for the Matsubara sum in Eq. 9.12 (see Fig. 9.10). The frequencies ω_{n_1} are fermionic, hence n_1 is odd, and the frequency ω_n is bosonic, hence n is even; consequently, $n_1 + n$ is odd. The Matsubara sum can be computed using the integral

$$I = \int dz \left(-\frac{\beta}{2\pi i}\right) n_{-1}(z) g(z), \tag{9.82}$$

with

$$g(z) = \frac{1}{\hbar\left[z - \frac{\mathcal{E}_{\mathbf{k}}}{\hbar}\right]} \frac{1}{\hbar\left[z + i\omega_n - \frac{\mathcal{E}_{\mathbf{p}}}{\hbar}\right]}. \tag{9.83}$$

The poles of $g(z)$ are two: $z_1 = \frac{\mathcal{E}_\mathbf{k}}{\hbar}$ (which we assume to be positive and on the real axis), and $z_2 = \frac{\mathcal{E}_\mathbf{p}}{\hbar} - i\omega_n$, both lying outside the contour Γ. By applying the residue theorem K.1, we obtain

$$I = \frac{1}{\hbar}\left[\frac{n_{-1}(\mathcal{E}_\mathbf{k})}{\frac{\mathcal{E}_\mathbf{k}}{\hbar} + i\omega_n - \frac{\mathcal{E}_\mathbf{p}}{\hbar}} + \frac{n_{-1}(\mathcal{E}_\mathbf{p} - i\hbar\omega_n)}{\frac{\mathcal{E}_\mathbf{p}}{\hbar} - i\omega_n - \frac{\mathcal{E}_\mathbf{k}}{\hbar}}\right], \tag{9.84}$$

where the first term corresponds to the clockwise propagation around the contour, while the second term corresponds to the counterclockwise propagation, reflecting the two possible time-orderings of the internal lines in the Matsubara diagram. Note that the following holds

$$\begin{aligned} n_{-1}(\mathcal{E}_\mathbf{p} - i\hbar\omega_n) &= \frac{1}{e^{\beta(\mathcal{E}_\mathbf{p} - i\hbar\omega_n)} + 1} = \\ &= \frac{1}{e^{\beta\mathcal{E}_\mathbf{p}} e^{-i\beta\hbar\omega_n} + 1} = \\ &= \frac{1}{e^{\beta\mathcal{E}_\mathbf{p}} + 1} = \\ &= n_{-1}(\mathcal{E}_\mathbf{p}), \end{aligned} \tag{9.85}$$

where we used that ω_n is a bosonic frequency, then $e^{-i2\pi n} = 1$ (see Eq. 6.28). We get

$$I = \frac{1}{\hbar}\left[\frac{n_{-1}(\mathcal{E}_\mathbf{k})}{\frac{\mathcal{E}_\mathbf{k}}{\hbar} + i\omega_n - \frac{\mathcal{E}_\mathbf{p}}{\hbar}} - \frac{n_{-1}(\mathcal{E}_\mathbf{p})}{\frac{\mathcal{E}_\mathbf{k}}{\hbar} + i\omega_n - \frac{\mathcal{E}_\mathbf{p}}{\hbar}}\right], \tag{9.86}$$

from which the statement follows. □

From Eq. 9.80, we write the propagator given in Eq. 9.79 as

$$\Pi^{(0)}_\mathbf{q}(i\omega_n) = \frac{1}{\hbar}\sum_{\mathbf{k},\sigma} \frac{n_{-1}(\mathcal{E}_{\mathbf{k}+\mathbf{q}}) - n_{-1}(\mathcal{E}_\mathbf{k})}{i\omega_n + \frac{\mathcal{E}_{\mathbf{k}+\mathbf{q}} - \mathcal{E}_\mathbf{k}}{\hbar}}. \tag{9.87}$$

We are now able to show the following

Theorem 9.3 *The first-order polarization propagator (see Eqs. 9.79 and 9.87) satisfies*

$$\Pi^{(0)}_\mathbf{q}(i\omega_n) = \Pi^{(0)}_\mathbf{q}(-i\omega_n), \tag{9.88}$$

$$\Pi^{(0)}_\mathbf{q}(i\omega_n) = \Pi^{(0)}_{-\mathbf{q}}(i\omega_n). \tag{9.89}$$

Proof Let us prove 9.88. We add and subtract $n_{-1}(\mathcal{E}_{\mathbf{k+q}})n_{-1}(\mathcal{E}_{\mathbf{k}})$ in the numerator of Eq. 9.87 and we write

$$\begin{aligned}
&\Pi^{(0)}_{\mathbf{q}}(i\omega_n)\\
&=\frac{1}{\hbar}\sum_{\mathbf{k},\sigma}\frac{n_{-1}(\mathcal{E}_{\mathbf{k+q}})-n_{-1}(\mathcal{E}_{\mathbf{k+q}})+n_{-1}(\mathcal{E}_{\mathbf{k+q}})n_{-1}(\mathcal{E}_{\mathbf{k}})-n_{-1}(\mathcal{E}_{\mathbf{k+q}})n_{-1}(\mathcal{E}_{\mathbf{k}})}{i\omega_n+\frac{\mathcal{E}_{\mathbf{k+q}}-\mathcal{E}_{\mathbf{k}}}{\hbar}}\\
&=\frac{1}{\hbar}\sum_{\mathbf{k},\sigma}\frac{n_{-1}(\mathcal{E}_{\mathbf{k+q}})(1-n_{-1}(\mathcal{E}_{\mathbf{k}}))}{i\omega_n+\frac{\mathcal{E}_{\mathbf{k+q}}-\mathcal{E}_{\mathbf{k}}}{\hbar}}-\frac{1}{\hbar}\sum_{\mathbf{k},\sigma}\frac{n_{-1}(\mathcal{E}_{\mathbf{k}})(1-n_{-1}(\mathcal{E}_{\mathbf{k+q}}))}{i\omega_n+\frac{\mathcal{E}_{\mathbf{k+q}}-\mathcal{E}_{\mathbf{k}}}{\hbar}},
\end{aligned}\tag{9.90}$$

so we rename the dummy index in the second term as $-\mathbf{k}'=\mathbf{k}+\mathbf{q}$ and we have

$$\begin{aligned}
\Pi^{(0)}_{\mathbf{q}}(i\omega_n)=&\frac{1}{\hbar}\sum_{\mathbf{k},\sigma}\frac{n_{-1}(\mathcal{E}_{\mathbf{k+q}})(1-n_{-1}(\mathcal{E}_{\mathbf{k}}))}{i\omega_n+\frac{\mathcal{E}_{\mathbf{k+q}}-\mathcal{E}_{\mathbf{k}}}{\hbar}}\\
&-\frac{1}{\hbar}\sum_{\mathbf{k}',\sigma}\frac{n_{-1}(\mathcal{E}_{-\mathbf{k}'-\mathbf{q}})(1-n_{-1}(\mathcal{E}_{-\mathbf{k}'}))}{i\omega_n+\frac{\mathcal{E}_{-\mathbf{k}'}-\mathcal{E}_{-\mathbf{k}'-\mathbf{q}}}{\hbar}}.
\end{aligned}\tag{9.91}$$

We now recall that we are working within the jellium model in the grand canonical ensemble, that is Eq. 2.393. In this framework, the one-particle energies take the form $\mathcal{E}_{\mathbf{k}}=\frac{\hbar^2\mathbf{k}^2}{2m}-\mu$. It immediately follows that $\mathcal{E}_{-\mathbf{k}}=\mathcal{E}_{\mathbf{k}}$, $n_{-1}(\mathcal{E}_{-\mathbf{k}})=n_{-1}(\mathcal{E}_{\mathbf{k}})$. Using these identities, and by relabeling the dummy summation index $\mathbf{k}'\to\mathbf{k}$ in the second term, we obtain

$$\begin{aligned}
\Pi^{(0)}_{\mathbf{q}}(i\omega_n)&=\frac{1}{\hbar}\sum_{\mathbf{k},\sigma}\frac{n_{-1}(\mathcal{E}_{\mathbf{k+q}})(1-n_{-1}(\mathcal{E}_{\mathbf{k}}))}{i\omega_n+\frac{\mathcal{E}_{\mathbf{k+q}}-\mathcal{E}_{\mathbf{k}}}{\hbar}}-\frac{1}{\hbar}\sum_{\mathbf{k},\sigma}\frac{n_{-1}(\mathcal{E}_{\mathbf{k+q}})(1-n_{-1}(\mathcal{E}_{\mathbf{k}}))}{i\omega_n+\frac{\mathcal{E}_{\mathbf{k}}-\mathcal{E}_{\mathbf{k+q}}}{\hbar}}=\\
&=\frac{1}{\hbar}\sum_{\mathbf{k},\sigma}\frac{n_{-1}(\mathcal{E}_{\mathbf{k+q}})(1-n_{-1}(\mathcal{E}_{\mathbf{k}}))}{i\omega_n+\frac{\mathcal{E}_{\mathbf{k+q}}-\mathcal{E}_{\mathbf{k}}}{\hbar}}-\frac{1}{\hbar}\sum_{\mathbf{k},\sigma}\frac{n_{-1}(\mathcal{E}_{\mathbf{k+q}})(1-n_{-1}(\mathcal{E}_{\mathbf{k}}))}{i\omega_n-\frac{\mathcal{E}_{\mathbf{k+q}}-\mathcal{E}_{\mathbf{k}}}{\hbar}}=\\
&=\frac{1}{\hbar}\sum_{\mathbf{k},\sigma}\frac{n_{-1}(\mathcal{E}_{\mathbf{k+q}})(1-n_{-1}(\mathcal{E}_{\mathbf{k}}))}{(i\omega_n)^2+\left(\frac{\mathcal{E}_{\mathbf{k+q}}-\mathcal{E}_{\mathbf{k}}}{\hbar}\right)^2}2\left[\frac{\mathcal{E}_{\mathbf{k}}-\mathcal{E}_{\mathbf{k+q}}}{\hbar}\right].
\end{aligned}\tag{9.92}$$

That is, we have shown that $\Pi^{(0)}_{\mathbf{q}}(i\omega_n)$ depends on $i\omega_n$ through its square power, so 9.88 has been proven.

Let us prove 9.89. Consider the last step of Eq. 9.92 and set $-\mathbf{k}'=\mathbf{k}+\mathbf{q}$, so that

$$\Pi^{(0)}_{\mathbf{q}}(i\omega_n)=\frac{1}{\hbar}\sum_{\mathbf{k}',\sigma}\frac{n_{-1}(\mathcal{E}_{-\mathbf{k}'})(1-n_{-1}(\mathcal{E}_{-\mathbf{k}'-\mathbf{q}}))}{(i\omega_n)^2+\left(\frac{\mathcal{E}_{-\mathbf{k}'}-\mathcal{E}_{-\mathbf{k}'-\mathbf{q}}}{\hbar}\right)^2}2\left[\frac{\mathcal{E}_{-\mathbf{k}'-\mathbf{q}}-\mathcal{E}_{-\mathbf{k}'}}{\hbar}\right],\tag{9.93}$$

and we rewrite the energies and Fermi-Dirac distributions for the opposite momenta, then rename the dummy index $\mathbf{k}' \to \mathbf{k}$ as follows

$$\Pi^{(0)}_{\mathbf{q}}(i\omega_n) = \frac{1}{\hbar}\sum_{\mathbf{k},\sigma} \frac{n_{-1}(\mathcal{E}_{\mathbf{k}})(1 - n_{-1}(\mathcal{E}_{\mathbf{k+q}})}{(i\omega_n)^2 + \left(\frac{\mathcal{E}_{\mathbf{k}}-\mathcal{E}_{\mathbf{k+q}}}{\hbar}\right)^2} 2\left[\frac{\mathcal{E}_{\mathbf{k+q}} - \mathcal{E}_{\mathbf{k}}}{\hbar}\right]. \tag{9.94}$$

From the previous equation, it follows

$$\Pi^{(0)}_{-\mathbf{q}}(i\omega_n) = \frac{1}{\hbar}\sum_{\mathbf{k},\sigma} \frac{n_{-1}(\mathcal{E}_{\mathbf{k}})(1 - n_{-1}(\mathcal{E}_{\mathbf{k-q}})}{(i\omega_n)^2 + \left(\frac{\mathcal{E}_{\mathbf{k}}-\mathcal{E}_{\mathbf{k-q}}}{\hbar}\right)^2} 2\left[\frac{\mathcal{E}_{\mathbf{k-q}} - \mathcal{E}_{\mathbf{k}}}{\hbar}\right], \tag{9.95}$$

and if we set $\mathbf{k}' = \mathbf{k} - \mathbf{q}$, we have

$$\Pi^{(0)}_{-\mathbf{q}}(i\omega_n) = \frac{1}{\hbar}\sum_{\mathbf{k}',\sigma} \frac{n_{-1}(\mathcal{E}_{\mathbf{k'+q}})(1 - n_{-1}(\mathcal{E}_{\mathbf{k'}}))}{(i\omega_n)^2 + \left(\frac{\mathcal{E}_{\mathbf{k'+q}}-\mathcal{E}_{\mathbf{k'}}}{\hbar}\right)^2} 2\left[\frac{\mathcal{E}_{\mathbf{k'}} - \mathcal{E}_{\mathbf{k'+q}}}{\hbar}\right], \tag{9.96}$$

and renaming the dummy index $\mathbf{k}' \to \mathbf{k}$, we obtain Eq. 9.92. □

Theorem 9.3 shows that the first-order self-energy given in Eq. 9.74 and the first-order polarization propagator satisfy

$$\Sigma^{(1)}_{\sigma}(\mathbf{k}, i\omega_n) = \left|M_{\mathbf{q}}\right|^2 \Pi^{(0)}_{\mathbf{q}}(i\omega_n), \tag{9.97}$$

and this equality can be shown to hold at every order, that is,

$$\Sigma_{\sigma}(\mathbf{k}, i\omega_n) = \left|M_{\mathbf{q}}\right|^2 \Pi_{\mathbf{q}}(i\omega_n), \tag{9.98}$$

$$\Sigma^{*}_{\sigma}(\mathbf{k}, i\omega_n) = \left|M_{\mathbf{q}}\right|^2 \Pi^{*}_{\mathbf{q}}(i\omega_n), \tag{9.99}$$

Starting from the algebraic Dyson equation (cf. Eq. 7.85), here suitably adapted to describe the electron–phonon interaction, namely

$$D_{\mathbf{q}}(i\omega_n) = D^{(0)}_{\mathbf{q}}(i\omega_n) + D^{(0)}_{\mathbf{q}}(i\omega_n)\left|M_{\mathbf{q}}\right|^2 \Pi_{\mathbf{q}}(i\omega_n) D^{(0)}_{\mathbf{q}}(i\omega_n), \tag{9.100}$$

$$D_{\mathbf{q}}(i\omega_n) = D^{(0)}_{\mathbf{q}}(i\omega_n) + D^{(0)}_{\mathbf{q}}(i\omega_n)\left|M_{\mathbf{q}}\right|^2 \Pi^{*}_{\mathbf{q}}(i\omega_n) D_{\mathbf{q}}(i\omega_n), \tag{9.101}$$

we multiply both sides by $\left|M_{\mathbf{q}}\right|^2$, then

$$\left|M_{\mathbf{q}}\right|^2 D_{\mathbf{q}}(i\omega_n) = \left|M_{\mathbf{q}}\right|^2 D^{(0)}_{\mathbf{q}}(i\omega_n) + \left|M_{\mathbf{q}}\right|^2 D^{(0)}_{\mathbf{q}}(i\omega_n)\Pi_{\mathbf{q}}(i\omega_n)\left|M_{\mathbf{q}}\right|^2 D^{(0)}_{\mathbf{q}}(i\omega_n), \tag{9.102}$$

$$\left|M_{\mathbf{q}}\right|^2 D_{\mathbf{q}}(i\omega_n) = \left|M_{\mathbf{q}}\right|^2 D_{\mathbf{q}}^{(0)}(i\omega_n) + \left|M_{\mathbf{q}}\right|^2 D_{\mathbf{q}}^{(0)}(i\omega_n)\Pi_{\mathbf{q}}^{*}(i\omega_n) \left|M_{\mathbf{q}}\right|^2 D_{\mathbf{q}}(i\omega_n). \tag{9.103}$$

Note that the quantity

$$U_{\text{eff},\mathbf{q}}(i\omega_n) = \left|M_{\mathbf{q}}\right|^2 D_{\mathbf{q}}(i\omega_n) \tag{9.104}$$

it is the effective potential between two particles renormalized by the electron-phonon interaction, and it includes all orders of the perturbative expansion; similarly,

$$U_{\text{eff},\mathbf{q}}^{(0)}(i\omega_n) = \left|M_{\mathbf{q}}\right|^2 D_{\mathbf{q}}^{(0)}(i\omega_n) \tag{9.105}$$

is the effective potential without renormalization. The Dyson equations can then be rewritten in a more suggestive form as

$$U_{\text{eff},\mathbf{q}}(i\omega_n) = U_{\text{eff},\mathbf{q}}^{(0)}(i\omega_n) + U_{\text{eff},\mathbf{q}}^{(0)}(i\omega_n)\Pi_{\mathbf{q}}(i\omega_n)U_{\text{eff},\mathbf{q}}^{(0)}(i\omega_n), \tag{9.106}$$

$$U_{\text{eff},\mathbf{q}}(i\omega_n) = U_{\text{eff},\mathbf{q}}^{(0)}(i\omega_n) + U_{\text{eff},\mathbf{q}}^{(0)}(i\omega_n)\Pi_{\mathbf{q}}^{*}(i\omega_n)U_{\text{eff},\mathbf{q}}(i\omega_n). \tag{9.107}$$

Equation 9.107 can be written as

$$U_{\text{eff},\mathbf{q}}(i\omega_n) = \frac{U_{\text{eff},\mathbf{q}}^{(0)}(i\omega_n)}{1 - U_{\text{eff},\mathbf{q}}^{(0)}(i\omega_n)\Pi_{\mathbf{q}}^{*}(i\omega_n)}. \tag{9.108}$$

In conclusion, we note that the same reasoning developed above can be applied to instantaneous interactions, such as the Coulomb interaction. In this case, one simply replaces the phonon potentials $D_{\mathbf{q}}(i\omega_n)$ and $D_{\mathbf{q}}^{(0)}(i\omega_n)$ in Eqs. 9.104 and 9.105 with the corresponding Coulomb potentials. This leads to effective interactions that are frequency-independent, i.e., they occur instantaneously. In particular, the effective potential calculated in the absence of interactions for the Coulomb case is given by

$$U_{\text{eff},\mathbf{q}}^{(0)} = \frac{1}{V}\frac{4\pi q_e^2}{q^2}. \tag{9.109}$$

Chapter 10
Linear Response Theory

Abstract This chapter presents the theory of linear response, which describes how quantum systems react to weak external perturbations. In most condensed-matter applications, electromagnetic fields can be treated classically, while the microscopic degrees of freedom of the material require a quantum-mechanical description. A fully quantum treatment of the electromagnetic field will be introduced in a later chapter, where field quantization is discussed. The chapter begins by defining response functions in terms of many-body Green's functions and correlation operators in space-time and frequency domains. Particular attention is devoted to the dielectric function, magnetic susceptibility, and optical conductivity, as they characterize the response of materials to electric and magnetic fields and determine how electromagnetic energy propagates and dissipates in matter. Sum rules for the dielectric function are examined, showing how frequency integrals constrain physical observables. The role of dielectric-function zeros is highlighted, since they signal possible instabilities or collective excitations such as plasmons. Three main theoretical approximations for calculating the dielectric function are introduced: the Hartree-Fock method, the Thomas-Fermi approximation, and the Random Phase Approximation (R.P.A.). Overall, the chapter provides the formal tools needed to analyze dynamic and response properties in condensed-matter systems.

10.1 General Formalism

Let $\hat{\mathcal{H}}$ denote a time-independent Hamiltonian describing a system of particles interacting both among themselves and with a static external field. The solution to the corresponding Schrödinger equation (see Eq. 1.86) is given by $\psi(t) = e^{-i\frac{\hat{\mathcal{H}}}{\hbar}t}\,\tilde{\psi}(0)$ (see Eq. 1.87). At a particular instant, say t_0, a time-dependent external perturbation is introduced, so that the total Hamiltonian becomes

$$\hat{\mathcal{H}}(t) = \hat{\mathcal{H}} + \hat{\mathcal{H}}_{\text{ext}}(t), \tag{10.1}$$

F. Tafuri et al., *Understanding Quantum Many-Body Theory*, UNITEXT for Physics, https://doi.org/10.1007/978-3-032-17849-7_10

with

$$\hat{\mathcal{H}}_{\text{ext}}(t) = 0, \; \forall\, t \leq t_0. \tag{10.2}$$

In the following, we consider a solution to the Schrödinger equation of the form

$$\psi(t) = e^{-i\frac{\hat{\mathcal{H}}}{\hbar}t} O(t)\tilde{\psi}(0), \tag{10.3}$$

$$O(t) = \mathbb{1}, \; \forall\, t \leq t_0, \tag{10.4}$$

where the operator $O(t)$ is the evolution operator generated by $\hat{\mathcal{H}}_{\text{ext}}$. We substitute $\psi(t)$ into Schrödinger's equation and we get

$$i\hbar\partial_t \left(e^{-i\frac{\hat{\mathcal{H}}}{\hbar}t} O(t)\tilde{\psi}(0)\right) = \left(\hat{\mathcal{H}} + \hat{\mathcal{H}}_{\text{ext}}(t)\right)\psi(t), \tag{10.5}$$

$$i\hbar\left(-\frac{i}{\hbar}\hat{\mathcal{H}}e^{-i\frac{\hat{\mathcal{H}}}{\hbar}t} O(t)\tilde{\psi}(0) + e^{-i\frac{\hat{\mathcal{H}}}{\hbar}t}\frac{dO(t)}{dt}\tilde{\psi}(0)\right) = \hat{\mathcal{H}}\psi(t) + \hat{\mathcal{H}}_{\text{ext}}(t)\psi(t), \tag{10.6}$$

$$\hat{\mathcal{H}}e^{-i\frac{\hat{\mathcal{H}}}{\hbar}t} O(t)\tilde{\psi}(0) + i\hbar\, e^{-i\frac{\hat{\mathcal{H}}}{\hbar}t}\frac{dO(t)}{dt}\tilde{\psi}(0) = \hat{\mathcal{H}}\psi(t) + \hat{\mathcal{H}}_{\text{ext}}(t)\psi(t), \tag{10.7}$$

$$\hat{\mathcal{H}}\psi(t) + i\hbar\, e^{-i\frac{\hat{\mathcal{H}}}{\hbar}t}\frac{dO(t)}{dt}\tilde{\psi}(0) = \hat{\mathcal{H}}\psi(t) + \hat{\mathcal{H}}_{\text{ext}}(t)\psi(t), \tag{10.8}$$

$$i\hbar\, e^{-i\frac{\hat{\mathcal{H}}}{\hbar}t}\frac{dO(t)}{dt}\tilde{\psi}(0) = \hat{\mathcal{H}}_{\text{ext}}(t)\psi(t), \tag{10.9}$$

$$i\hbar\, e^{-i\frac{\hat{\mathcal{H}}}{\hbar}t}\frac{dO(t)}{dt}\tilde{\psi}(0) = \hat{\mathcal{H}}_{\text{ext}}(t)\, e^{-i\frac{\hat{\mathcal{H}}}{\hbar}t} O(t)\tilde{\psi}(0), \tag{10.10}$$

$$i\hbar e^{-i\frac{\hat{\mathcal{H}}}{\hbar}t}\frac{dO(t)}{dt} = \hat{\mathcal{H}}_{\text{ext}}(t)e^{-i\frac{\hat{\mathcal{H}}}{\hbar}t} O(t), \tag{10.11}$$

and we multiply to the left both members by $e^{i\frac{\hat{\mathcal{H}}}{\hbar}t}$, then

$$i\hbar\frac{dO(t)}{dt} = e^{i\frac{\hat{\mathcal{H}}}{\hbar}t}\hat{\mathcal{H}}_{\text{ext}}(t)e^{-i\frac{\hat{\mathcal{H}}}{\hbar}t} O(t), \tag{10.12}$$

$$i\hbar\frac{dO(t)}{dt} = \hat{\mathcal{H}}^{(0)}_{\text{ext}}(t)O(t). \tag{10.13}$$

This formulation makes explicit that we are working in the interaction picture (see Eq. 1.107) defined with respect to the full many-body Hamiltonian of the unperturbed system. The following Remark provides a detailed explanation of the superscript (0) and its role in the corresponding time evolution.

Remark 10.1 (Clarification on the Superscript 0 **in Time Evolution)** It is important to emphasize that, although we are working in the interaction picture, the Hamiltonian that generates the time evolution is not a non-interacting Hamiltonian, but rather the full many-body Hamiltonian of the system, with all internal interactions included. In other words, the decomposition introduced in Eq. 1.85 is entirely analogous to the standard linear-response decomposition in Eq. 10.1, where $\hat{\mathcal{H}}_0$ is identified with the full Hamiltonian $\hat{\mathcal{H}}$ of the system (including all internal interactions), while $\hat{\mathcal{H}}_I$ is identified with the external perturbation $\hat{\mathcal{H}}_{\text{ext}}$.

In general, Eq. 10.13 is an operator equation, which we can formally integrate between t_0 and t to obtain

$$i\hbar\left[O(t) - O(t_0)\right] = \int_{t_0}^{t} dt' \hat{\mathcal{H}}_{\text{ext}}^{(0)}(t') O(t'), \tag{10.14}$$

$$i\hbar\left[O(t) - \mathbb{1}\right] = \int_{t_0}^{t} dt' \hat{\mathcal{H}}_{\text{ext}}^{(0)}(t') O(t'), \tag{10.15}$$

$$O(t) = \mathbb{1} + \frac{1}{i\hbar}\int_{t_0}^{t} dt' \hat{\mathcal{H}}_{\text{ext}}^{(0)}(t') O(t'), \tag{10.16}$$

which we solve by iterating and we get

$$O(t) = \mathbb{1} + \frac{1}{i\hbar}\int_{t_0}^{t} dt' \hat{\mathcal{H}}_{\text{ext}}^{(0)}(t') + \frac{1}{i\hbar}\int_{t_0}^{t} dt' \hat{\mathcal{H}}_{\text{ext}}^{(0)}(t') \left[\frac{1}{i\hbar}\int_{t_0}^{t'} dt'' \hat{\mathcal{H}}_{\text{ext}}^{(0)}(t'') O(t'')\right]. \tag{10.17}$$

In the linear response theory approach, if the magnitude of the perturbation is negligible with respect to the initial Hamiltonian, we neglect terms of higher than first order in $\hat{\mathcal{H}}_{\text{ext}}^{(0)}$, i.e.,

$$O(t) \simeq \mathbb{1} + \frac{1}{i\hbar}\int_{t_0}^{t} dt' \hat{\mathcal{H}}_{\text{ext}}^{(0)}(t'), \tag{10.18}$$

$$\psi(t) = e^{-i\frac{\hat{\mathcal{H}}}{\hbar}t}\tilde{\psi}(0) + e^{-i\frac{\hat{\mathcal{H}}}{\hbar}t}\frac{1}{i\hbar}\int_{t_0}^{t} dt' \hat{\mathcal{H}}_{\text{ext}}^{(0)}(t')\tilde{\psi}(0), \tag{10.19}$$

where, in the context of linear response theory, the wave function is expressed as the sum of the unperturbed wave function and a correction induced by the external perturbation. If we neglect the second-order term in $\hat{\mathcal{H}}_{\text{ext}}^{(0)}$, the expectation value of a generic observable can be evaluated using the perturbed wave function, and we obtain

$$\langle A(t)\rangle = \left\langle \tilde{\psi}^*(0) A^{(0)}(t)\tilde{\psi}(0)\right\rangle - \frac{1}{i\hbar}\left\langle \tilde{\psi}^*(0)\int_{t_0}^{t} dt'\hat{\mathcal{H}}_{\text{ext}}^{(0)}(t')A^{(0)}(t)\tilde{\psi}(0)\right\rangle + \\ + \frac{1}{i\hbar}\left\langle \tilde{\psi}^*(0)\int_{t_0}^{t} dt' A^{(0)}(t)\hat{\mathcal{H}}_{\text{ext}}^{(0)}(t')\tilde{\psi}(0)\right\rangle, \tag{10.20}$$

which is equivalent to

$$\langle A\rangle(t) = \langle A\rangle_0(t) - \frac{1}{i\hbar}\int_{t_0}^{t} dt' \left\langle \left[\hat{\mathcal{H}}_{\text{ext}}^{(0)}(t'), A^{(0)}(t)\right]\right\rangle_0, \tag{10.21}$$

where $\langle\ldots\rangle_0$ denotes the expectation value taken with respect to the initial unperturbed state $\tilde{\psi}(t=0)$. We define

$$\begin{aligned} \delta\langle A\rangle(t) &= \langle A\rangle(t) - \langle A\rangle_0(t) = \\ &= -\frac{1}{i\hbar}\int_{t_0}^{t} dt' \left\langle \left[\hat{\mathcal{H}}_{\text{ext}}^{(0)}(t'), A^{(0)}(t)\right]\right\rangle_0 = \\ &= \frac{i}{\hbar}\int_{t_0}^{t} dt' \left\langle \left[\hat{\mathcal{H}}_{\text{ext}}^{(0)}(t'), A^{(0)}(t)\right]\right\rangle_0 = \\ &= -\frac{i}{\hbar}\int_{t_0}^{t} dt' \left\langle \left[A^{(0)}(t), \hat{\mathcal{H}}_{\text{ext}}^{(0)}(t')\right]\right\rangle_0, \end{aligned} \tag{10.22}$$

which quantifies the variation of $\langle A\rangle$ at time t due to the external perturbation. It is important to note that, regardless of whether the particles obey Bose-Einstein or Fermi-Dirac statistics (see Eqs. E.3 and E.4), the integrand must involve a commutator. We now rewrite $\delta\langle A\rangle(t)$ by extending the integration limits to infinity. Since we assume $\hat{\mathcal{H}}_{\text{ext}}(t_0) = 0$, we can safely take $t_0 \to -\infty$ without affecting the value of the integral. For the upper limit, we introduce a Heaviside function $\Theta(t-t')$ and take $t \to \infty$, in accordance with the principle of causality: the system cannot respond to a perturbation before it is applied. We then obtain

$$\delta\langle A\rangle(t) = -\frac{i}{\hbar}\int_{-\infty}^{+\infty} dt'\Theta(t-t')\left\langle \left[A^{(0)}(t), \hat{\mathcal{H}}_{\text{ext}}^{(0)}(t')\right]\right\rangle_0. \tag{10.23}$$

Now, assume that the perturbation takes the form $\hat{\mathcal{H}}_{\text{ext}}(t) = h(t)B(t)$, where $h(t)$ is an external classical field, that is, a prescribed function of time, which represents the perturbation, and $B(t)$ is the quantum observable to which the field is coupled. We then obtain

$$\begin{aligned} \delta\langle A\rangle(t) &= -\frac{i}{\hbar}\int_{-\infty}^{+\infty} dt'\Theta(t-t')\left\langle \left[A^{(0)}(t), h(t')B^{(0)}(t')\right]\right\rangle_0 = \\ &= -\frac{i}{\hbar}\int_{-\infty}^{+\infty} dt' h(t')\,\Theta(t-t')\left\langle \left[A^{(0)}(t), B^{(0)}(t')\right]\right\rangle_0. \end{aligned} \tag{10.24}$$

We get the Kubo's formula (see Definition 6.2)

$$\delta\langle A\rangle(t) = \int_{-\infty}^{+\infty} dt' h(t') G^{(r)}_{A,B}(t,t'), \tag{10.25}$$

which is a manifestation of the fluctuation-dissipation theorem: within the framework of linear response theory, the dissipation of a physical system due to an external perturbation is governed by time-correlated fluctuations of operators. At thermodynamic equilibrium, the many-body Green's function serves as the response function, as it quantifies how the system reacts to external perturbations. The retarded many-body Green's function $G^{(r)}_{A,B}(t,t')$ encapsulates the causal response of the observable A to a perturbation conjugate to B. Note that $h(t')$ comes out of the commutator, being a function and not an operator.

Remark 10.2 (Clarification on the Subscript 0 in Thermal Average) It is crucial to emphasize that the retarded Green's function obtained in Eq. 10.25 is computed in the absence of the external perturbation. However, this does not imply a non-interacting system. The thermal average (see Eq. 1.379) denoted by the subscript 0 (see Eq. 10.24) is evaluated with respect to the unperturbed Hamiltonian, which still contains all particle interactions. This distinction is essential to avoid confusion: the 0 subscript does not mean that interactions are absent, it simply refers to the absence of the external perturbation. Thanks to this, we are justified in introducing the many-body retarded Green's function, which incorporates all internal interactions exactly, while treating the external field as a small perturbation. Note that this is consistent with the fact that the retarded Green's function, which already includes all internal interactions, is not labeled with a superscript (0), unlike in the case of free particles.

If the external perturbation oscillates with a well-defined frequency ω, we can express the response in the frequency domain via a Fourier transform

$$\delta\langle A\rangle(\omega) = \int_{-\infty}^{+\infty} dt e^{i\omega t} \int_{-\infty}^{+\infty} dt' h(t') G^{(r)}_{A,B}(t,t'), \tag{10.26}$$

At zero temperature, the system is no longer described by a statistical mixture of eigenstates of the unperturbed Hamiltonian, but rather by a pure state, namely the ground state $|\psi_{gs}\rangle$ (see Eqs. 1.380 and 1.381). Assume the system is at thermodynamic equilibrium at $T = 0$, before the perturbation; then, Eq. 10.26 becomes a convolution (see Theorem 6.1); the wave function of the system is the ground state ψ_{gs} of $\hat{\mathcal{H}}$, which includes interactions; the retarded Green's function is no longer a thermal average but an expected value of the operator $\left[A^{(0)}(t), B^{(0)}(t')\right]$ with respect to the state ψ_{gs}, that is,

$$\begin{aligned} G^{(r)}_{A,B}(t,t') &= -\frac{i}{\hbar}\Theta(t-t')\left\langle \psi_{gs} \left| \left[A^{(0)}(t), B^{(0)}(t')\right] \right| \psi_{gs} \right\rangle \equiv \\ &\equiv -\frac{i}{\hbar}\Theta(t-t')\left\langle \left[A^{(0)}(t), B^{(0)}(t')\right] \right\rangle_{T=0}. \end{aligned} \tag{10.27}$$

Note that in Eq. 10.27 the subscript 0 is omitted, since it no longer denotes a statistical average but an expectation value over a single pure state. In particular, consider the first term of the commutator in Eq. 10.27, i.e.,

$$A^{(0)}(t)B^{(0)}(t') = e^{i\frac{\hat{\mathcal{H}}}{\hbar}t}Ae^{-i\frac{\hat{\mathcal{H}}}{\hbar}(t-t')}Be^{-i\frac{\hat{\mathcal{H}}}{\hbar}t'}. \tag{10.28}$$

With respect to the fundamental state ψ_{gs} of $\hat{\mathcal{H}}$, the operator $e^{i\frac{\hat{\mathcal{H}}}{\hbar}t}$ acts as $e^{i\frac{\mathcal{E}_{gs}}{\hbar}t}\mathbb{1}$, i.e.,

$$\left\langle \psi_{gs} \left| A^{(0)}(t)B^{(0)}(t') \right| \psi_{gs} \right\rangle = e^{i\frac{\mathcal{E}_{gs}}{\hbar}(t-t')} \left\langle \psi_{gs} \left| A\, e^{-i\frac{\hat{\mathcal{H}}}{\hbar}(t-t')} B \right| \psi_{gs} \right\rangle. \tag{10.29}$$

Now, the quantity in Eq. 10.26 represents a convolution in the time domain, consequently, in the frequency domain it takes the simple multiplicative form

$$\delta\langle A\rangle(\omega) = h(\omega)G^{(r)}_{A,B}(\omega). \tag{10.30}$$

However, $T = 0$ cannot be reached experimentally. Assuming that, prior to the perturbation, the system is in contact with a thermal bath at a finite temperature $T \neq 0$, the ground state ψ_{gs} is replaced by an ensemble of excited states ψ_r. Accordingly, the expectation value of any observable must be evaluated with respect to the statistical mixture of states accessible to the system. The coefficients p_r, representing the statistical weights of these states, are generally altered by the perturbation, reflecting the resulting changes in the system's probability distribution. Note that, as in the $T = 0$ case, since the expectation value considered here is taken with respect to a single (unperturbed) eigenstate of $\hat{\mathcal{H}}$, denoted by $|\psi_r\rangle$, the average does not carry the subscript 0. From

$$\delta\langle A\rangle(t) = -\frac{i}{\hbar}\int_{t_0}^{t} dt' \left\langle \psi_r \left| \left[A^{(0)}(t), \hat{\mathcal{H}}^{(0)}_{\text{ext}}(t') \right] \right| \psi_r \right\rangle, \tag{10.31}$$

we note that if the modified statistical weights were to incorporate the external perturbation $\hat{\mathcal{H}}_{\text{ext}}$, higher-order terms would arise in the expansion, which are inconsistent with the assumptions of linear response theory. Consequently, the only admissible choice for the statistical weights is given by Maxwell-Boltzmann weights 1.355 corresponding to the unperturbed Hamiltonian. Moreover, since experimental temperatures are strictly positive ($T > 0$), many-body Green's functions are naturally defined as thermal averages and are therefore always evaluated over excited states, rather than over the ground state.

In the following sections, we will demonstrate that applying linear response theory to Hamiltonians involving the particle density, spin density, or current density operators leads, respectively, to the dielectric function, the magnetic susceptibility, and the optical conductivity.

Remark 10.3 (Omission of Subscript 0 in Thermal Averages) For notational convenience, in the remainder of this chapter, thermal averages computed with respect to the unperturbed Hamiltonian, i.e., without the external perturbation but still including all particle interactions, will be denoted simply as $\langle \ldots \rangle$, omitting the 0 subscript. It is understood that these averages always refer to the interacting system in the absence of the external field.

10.2 Dielectric Function

10.2.1 Derivation of the Dielectric Function

Let us consider a system of interacting electrons, each carrying a charge q_e, coupled to an external longitudinal electric field. The interaction between the electrons and the external field is described by an interaction Hamiltonian that involves the particle density operator $\hat{\rho}$ (see Eq. 2.261) and the scalar potential φ_{ext} associated with the external electric field. In other words, the Hamiltonian couples the local charge density of the system to the external potential, thus accounting for how the field perturbs the electrons. The potential is a fixed, time-dependent function that is not subject to quantum fluctuations, meaning that it does not evolve dynamically within the quantum system and is unaffected by the time-evolution operators acting on quantum states. Since the scalar potential is a classical function of space and time, it commutes with quantum operators and can be factored out of time-ordered or commutator expressions. Accordingly, in the interaction picture (see Eq. 1.107), the external perturbation takes the form

$$\hat{\mathcal{H}}^{(0)}_{\text{ext}}(t) = \int d^3\mathbf{r} q_e \hat{\rho}^{(0)}(\mathbf{r}, t)\varphi_{\text{ext}}(\mathbf{r}, t). \tag{10.32}$$

From Eq. 10.24, with $A = \hat{\rho}(\mathbf{r})$, we compute the variation in the particle number density as follows

$$\begin{aligned}
&\delta\left\langle\hat{\rho}(\mathbf{r})\right\rangle(t) = \\
&= \frac{i}{\hbar}\int_{-\infty}^{+\infty} dt'\,\Theta(t-t')\int d^3\mathbf{r}'\,q_e\,\varphi_{\text{ext}}(\mathbf{r}',t')\left\langle\left[\hat{\rho}^{(0)}(\mathbf{r}',t'),\hat{\rho}^{(0)}(\mathbf{r},t)\right]\right\rangle = \\
&= -\frac{i}{\hbar}\int_{-\infty}^{+\infty} dt'\Theta(t-t')\int d^3\mathbf{r}' q_e\varphi_{\text{ext}}(\mathbf{r}',t')\left\langle\left[\hat{\rho}^{(0)}(\mathbf{r},t),\hat{\rho}^{(0)}(\mathbf{r}',t')\right]\right\rangle.
\end{aligned} \tag{10.33}$$

Keeping in mind the polarization propagator in momentum space and imaginary time, given in Eq. 9.75, we now introduce

Definition 10.1 (Polarization Propagator (in Real Space and Real Time)) Given the particle density operator in second quantization (see Eq. 2.261), the polarization propagator in real space and real time is defined as

$$\Pi(\mathbf{r}, t, \mathbf{r}', t') = -\frac{i}{\hbar}\Theta(t-t')\left\langle\left[\hat{\rho}^{(0)}(\mathbf{r}, t), \hat{\rho}^{(0)}(\mathbf{r}', t')\right]\right\rangle, \tag{10.34}$$

where $\hat{\rho}$ denotes the particle density operator in second quantization given in Eq. 2.261, and the superscript (0) indicates time evolution under the unperturbed many-body Hamiltonian (see Remark 10.1).

This definition in real space and real time will serve as the starting point for the derivation of the dielectric function, a central physical quantity that can be experimentally measured. We write

$$\delta\left\langle\hat{\rho}(\mathbf{r})\right\rangle(t) = -\frac{i}{\hbar}\int_{-\infty}^{+\infty} dt'\Theta(t-t')\int d^3\mathbf{r}' q_e \varphi_{\text{ext}}(\mathbf{r}', t')\Pi(\mathbf{r}, t, \mathbf{r}', t'), \tag{10.35}$$

that is, the induced polarization in the medium, meaning the variation of the charged particle density in the medium due to the external perturbation, is related to the polarization Green's function, which does not contain the external electric field, since the densities evolve according to the Hamiltonian $\hat{\mathcal{H}}_0$ that describes only the electron system. Let us assume the system is in thermal equilibrium (see Theorem 6.1) and invariant under spatial translations, then $\Pi(\mathbf{r}, t, \mathbf{r}', t') = \Pi(\mathbf{r}-\mathbf{r}', t-t')$. By applying a Fourier transform in space and time, Eq. 10.35 becomes a convolution. In particular, the polarization Green's function is a retarded function, due to the principle of causality. If we substitute the particle density operator in second quantization, that is Eq. 2.261, into Eq. 10.34, we obtain

$$\begin{aligned}\Pi(\mathbf{r}-\mathbf{r}', t-t') = {} & -\frac{i}{\hbar}\Theta(t-t') \\ & \sum_{s,s'}\left\langle\left[\hat{\psi}^{\dagger(0)}(\mathbf{r}, s, t)\hat{\psi}^{(0)}(\mathbf{r}, s, t), \hat{\psi}^{\dagger(0)}(\mathbf{r}', s', t')\hat{\psi}^{(0)}(\mathbf{r}', s', t')\right]\right\rangle,\end{aligned} \tag{10.36}$$

and we expand the field operators in the tensor-product basis of plane-wave states and spin eigenstates (see Eq. 1.81) as follows

$$\begin{aligned}\Pi(\mathbf{r}-\mathbf{r}', t-t') = {} & -\frac{i}{\hbar}\Theta(t-t')\frac{1}{V^2}\sum_{s,s'}\sum_{\mathbf{k}_1,\mathbf{k}_2}\sum_{\mathbf{k}_3,\mathbf{k}_4}\sum_{\sigma_1,\sigma_2}\sum_{\sigma_3,\sigma_4} e^{-i(\mathbf{k}_1-\mathbf{k}_2)\cdot\mathbf{r}} e^{-i(\mathbf{k}_3-\mathbf{k}_4)\cdot\mathbf{r}'} \\ & \left\langle\left[\left(C^{\dagger}_{\mathbf{k}_1,\sigma_1}C_{\mathbf{k}_2,\sigma_2}\chi^*_{\sigma_1}\chi_{\sigma_2}\right)^{(0)}(t), \left(C^{\dagger}_{\mathbf{k}_3,\sigma_3}C_{\mathbf{k}_4,\sigma_4}\chi^*_{\sigma_3}\chi_{\sigma_4}\right)^{(0)}(t')\right]\right\rangle.\end{aligned} \tag{10.37}$$

For notational convenience, we factor out all explicit time dependence into a function $d(t,t')$, allowing us to focus on the relevant operator structure, that is,

$$\Pi(\mathbf{r}-\mathbf{r}',t-t')=-\frac{i}{\hbar}\Theta(t-t')\frac{1}{V^2}\sum_{s,s'}\sum_{\mathbf{k}_1,\mathbf{k}_2}\sum_{\mathbf{k}_3,\mathbf{k}_4}\sum_{\sigma_1,\sigma_2}\sum_{\sigma_3,\sigma_4}e^{-i(\mathbf{k}_1-\mathbf{k}_2)\cdot\mathbf{r}}e^{-i(\mathbf{k}_3-\mathbf{k}_4)\cdot\mathbf{r}'}$$

$$d(t,t')\left\langle\left[C^{\dagger}_{\mathbf{k}_1,\sigma_1}C_{\mathbf{k}_2,\sigma_2}\,\chi^*_{\sigma_1}\chi_{\sigma_2},\,C^{\dagger}_{\mathbf{k}_3,\sigma_3}C_{\mathbf{k}_4,\sigma_4}\,\chi^*_{\sigma_3}\chi_{\sigma_4}\right]\right\rangle. \tag{10.38}$$

From orthonormality relation 1.23, that is,

$$\sum_{s}\chi^*_{\sigma_1}(s)\chi_{\sigma_2}(s)=\delta_{\sigma_1,\sigma_2}, \tag{10.39}$$

$$\sum_{s'}\chi^*_{\sigma_3}(s')\chi_{\sigma_4}(s')=\delta_{\sigma_3,\sigma_4}, \tag{10.40}$$

one finds

$$\Pi(\mathbf{r}-\mathbf{r}',t-t')=-\frac{i}{\hbar}\Theta(t-t')\frac{1}{V^2}\sum_{\mathbf{k}_1,\mathbf{k}_2}\sum_{\mathbf{k}_3,\mathbf{k}_4}\sum_{\sigma_1,\sigma_3}e^{-i(\mathbf{k}_1-\mathbf{k}_2)\cdot\mathbf{r}}e^{-i(\mathbf{k}_3-\mathbf{k}_4)\cdot\mathbf{r}'}$$

$$d(t,t')\left\langle\left[C^{\dagger}_{\mathbf{k}_1,\sigma_1}C_{\mathbf{k}_2,\sigma_1},\,C^{\dagger}_{\mathbf{k}_3,\sigma_3}C_{\mathbf{k}_4,\sigma_3}\right]\right\rangle. \tag{10.41}$$

The polarization propagator involves four independent sums over momenta vectors $\mathbf{k}_1$, $\mathbf{k}_2$, $\mathbf{k}_3$, and $\mathbf{k}_4$. By introducing momentum transfers, we set

$$\mathbf{k}_1=\mathbf{k}_2+\mathbf{q}_1, \tag{10.42}$$

$$\mathbf{k}_3=\mathbf{k}_4+\mathbf{q}_2. \tag{10.43}$$

Translational invariance of the system enforces conservation of total linear momentum at each interaction vertex. As a consequence, the two momentum transfers must be equal and opposite, namely

$$\mathbf{q}_1=-\mathbf{q}_2. \tag{10.44}$$

This condition reflects the fact that the polarization propagator describes the creation of a particle-hole pair with momentum transfer $\mathbf{q}_1$, which must then be annihilated with the opposite transfer $\mathbf{q}_2$, ensuring overall momentum conservation. Moreover, we rename the dummy spin index σ_3 as σ_2. We therefore obtain

$$\Pi(\mathbf{r}-\mathbf{r}',t-t')=-\frac{i}{\hbar}\,\Theta(t-t')\,\frac{1}{V^2}\sum_{\mathbf{k}_1,\sigma_1,\mathbf{q}_1}\sum_{\mathbf{k}_2,\sigma_2}e^{-i\mathbf{q}_1\cdot(\mathbf{r}-\mathbf{r}')}$$

$$d(t,t')\left\langle\left[C^{\dagger}_{\mathbf{k}_1+\mathbf{q}_1,\sigma_1}C_{\mathbf{k}_1,\sigma_1},C^{\dagger}_{\mathbf{k}_2-\mathbf{q}_1,\sigma_2}C_{\mathbf{k}_2,\sigma_2}\right]\right\rangle. \tag{10.45}$$

We apply a Fourier transform in space and time as follows

$$\Pi_{\mathbf{q}}(\omega+i\delta)=\int d^3\mathbf{r}e^{i\mathbf{q}\cdot\mathbf{r}}\int_{-\infty}^{+\infty}dte^{i(\omega+i\delta)t}\Pi(\mathbf{r},t), \tag{10.46}$$

where we have renamed $\mathbf{r}-\mathbf{r}'\to\mathbf{r}$, $t-t'\to t$, and the factor $\delta>0$ allows us to integrate in the upper half-plane; thus, the integration is performed first, and the limit $\delta\to 0^+$ is taken afterwards. The spatial integral yields

$$\int d^3\mathbf{r}e^{i(\mathbf{q}-\mathbf{q}_1)\cdot\mathbf{r}}=V\delta_{\mathbf{q},\mathbf{q}_1} \tag{10.47}$$

and by summing over $\mathbf{q}_1$ we have $\mathbf{q}=\mathbf{q}_1$. In the commutator, on the other hand, we obtain the spatial Fourier transforms of the particle density (see Eq. 2.262), then

$$\Pi_{\mathbf{q}}(\omega+i\delta)=-\frac{i}{\hbar}\frac{1}{V}\int_{-\infty}^{+\infty}dte^{i(\omega+i\delta)t}\Theta(t)\left\langle\left[\hat{\rho}^{(0)}_{\mathbf{q}}(t),\hat{\rho}^{\dagger}_{\mathbf{q}}(0)\right],\right\rangle, \tag{10.48}$$

and the variation of the particle density in Fourier space is a convolution, that is

$$\delta\langle\hat{\rho}\rangle(\mathbf{q},\omega)=\frac{1}{V}q_e\varphi_{\text{ext},\mathbf{q}}(\omega)\Pi_{\mathbf{q}}(\omega). \tag{10.49}$$

Now we can write the dielectric function from the variation in the particle density. The potential φ_{ext} is generated by a charge distribution ρ_{ext}, which polarizes a system of electrons with charge q_e. To describe electric interactions, we make use of the electric field $\mathbf{E}$ and the electric displacement field $\mathbf{D}$, which satisfy Maxwell's equations (see Sect. 13.1). In this regard, we recall the full set of Maxwell's equations in the CGS-Gaussian system, that is

$$\begin{cases}\nabla\cdot\mathbf{E}=4\pi\rho\\ \nabla\cdot\mathbf{B}=0\\ \nabla\times\mathbf{E}=-\dfrac{1}{c}\dfrac{\partial\mathbf{B}}{\partial t}\\ \nabla\times\mathbf{B}=\dfrac{1}{c}\dfrac{\partial\mathbf{E}}{\partial t}+\dfrac{4\pi}{c}\mathbf{J}\end{cases}. \tag{10.50}$$

Here, ρ and $\mathbf{J}$ represent the total charge and current densities. When considering electromagnetic fields in a medium, it is often useful to distinguish between external and induced sources. In this context, the displacement field $\mathbf{D}$ satisfies

$$\nabla\cdot\mathbf{D}=4\pi\rho_{\text{ext}}, \tag{10.51}$$

where ρ_{ext} is the external charge density. Meanwhile, the electric field $\mathbf{E}$ satisfies

$$\nabla \cdot \mathbf{E} = 4\pi(\rho_{\text{ext}} + \rho_i), \tag{10.52}$$

where ρ_i is the induced charge density that accounts for the response of the medium. Here, we consider the fields $\mathbf{D}$ and $\mathbf{E}$ to be slowly varying in time, so that the effects induced by the magnetic field can be neglected. We apply the spatial and time Fourier transform and we obtain

$$-i\mathbf{q} \cdot \mathbf{D}_{\mathbf{q}}(\omega) = 4\pi \rho_{\text{ext},\mathbf{q}}(\omega), \tag{10.53}$$

$$-i\mathbf{q} \cdot \mathbf{E}_{\mathbf{q}}(\omega) = 4\pi \left(\rho_{\text{ext},\mathbf{q}}(\omega) + \rho_{i,\mathbf{q}}(\omega)\right). \tag{10.54}$$

Given the relations between the fields and the potentials, namely $\mathbf{D} = -\nabla\varphi_{\text{ext}}$, $\mathbf{E} = -\nabla\varphi$, where φ is generated by the total charge distribution $\rho_{\text{ext}} + \rho_i$, we apply a Fourier transform and obtain

$$\mathbf{D}_{\mathbf{q}}(\omega) = i\varphi_{\text{ext},\mathbf{q}}(\omega)\mathbf{q}, \tag{10.55}$$

$$\mathbf{E}_{\mathbf{q}}(\omega) = i\varphi_{\mathbf{q}}(\omega)\mathbf{q}. \tag{10.56}$$

By comparing the equations for $\mathbf{D}_{\mathbf{q}}(\omega)$ and $\mathbf{E}_{\mathbf{q}}(\omega)$, we have

$$\varphi_{\text{ext},\mathbf{q}}(\omega) = \frac{4\pi}{q^2}\rho_{\text{ext},\mathbf{q}}(\omega), \tag{10.57}$$

$$\varphi_{\mathbf{q}}(\omega) = \frac{4\pi}{q^2}\left(\rho_{\text{ext},\mathbf{q}}(\omega) + \rho_{i,\mathbf{q}}(\omega)\right). \tag{10.58}$$

By definition, the dielectric function $\epsilon(\omega)$ relates the electric displacement field $\mathbf{D}$ to the electric field $\mathbf{E}$ through the relation

$$\mathbf{E}(\omega) = \frac{\mathbf{D}(\omega)}{\epsilon(\omega)}, \tag{10.59}$$

then

$$\varphi_{\mathbf{q}}(\omega) = \frac{\varphi_{\text{ext},\mathbf{q}}(\omega)}{\epsilon_{\mathbf{q}}(\omega)}, \tag{10.60}$$

$$\frac{1}{\epsilon_{\mathbf{q}}(\omega)} = \frac{\rho_{\text{ext},\mathbf{q}}(\omega) + \rho_{i,\mathbf{q}}(\omega)}{\rho_{\text{ext},\mathbf{q}}(\omega)}, \tag{10.61}$$

$$\frac{1}{\epsilon_{\mathbf{q}}(\omega)} = 1 + \frac{\rho_{i,\mathbf{q}}(\omega)}{\rho_{\text{ext},\mathbf{q}}(\omega)}. \tag{10.62}$$

Ultimately, to calculate the dielectric function, it is necessary to determine the induced charge distribution $\rho_{i,\mathbf{q}}(\omega)$, which can be computed using linear response theory. Note that it must be

$$\rho_i = q_e \delta \langle \hat{\rho} \rangle, \tag{10.63}$$

in other words, the induced charge distribution in the material medium is proportional to the electron charge and to the fluctuation in the number of electron, that is why we computed the variation in particle density. We substitute these quantities into Eq. 10.62 and obtain

$$\frac{1}{\epsilon_{\mathbf{q}}(\omega)} = 1 + \frac{1}{V} \frac{q_e^2 \varphi_{\text{ext},\mathbf{q}}(\omega) \Pi_{\mathbf{q}}(\omega)}{\rho_{\text{ext},\mathbf{q}}(\omega)}, \tag{10.64}$$

then (see Eq. 10.57)

$$\frac{1}{\epsilon_{\mathbf{q}}(\omega)} = 1 + \frac{1}{V} \frac{4\pi q_e^2}{q^2} \Pi_{\mathbf{q}}(\omega). \tag{10.65}$$

Note that if we use the equivalent definition of the dielectric function

$$\frac{1}{\epsilon_{\mathbf{q}}(\omega)} = \frac{U_{\text{eff},\mathbf{q}}(\omega)}{U^{(0)}_{\text{eff},\mathbf{q}}(\omega)}, \tag{10.66}$$

from Eq. 9.108, we have

$$\frac{1}{\epsilon_{\mathbf{q}}(\omega)} = \frac{U^{(0)}_{\text{eff},\mathbf{q}}(\omega)}{U^{(0)}_{\text{eff},\mathbf{q}}(\omega)\left(1 - U^{(0)}_{\text{eff},\mathbf{q}}(\omega)\Pi^*_{\mathbf{q}}(\omega)\right)}, \tag{10.67}$$

$$\epsilon_{\mathbf{q}}(\omega) = 1 - U^{(0)}_{\text{eff},\mathbf{q}}(\omega)\Pi^*_{\mathbf{q}}(\omega), \tag{10.68}$$

and using Eq. 9.109, we get

$$\epsilon_{\mathbf{q}}(\omega) = 1 - \frac{1}{V} \frac{4\pi q_e^2}{q^2} \Pi^*_{\mathbf{q}}(\omega). \tag{10.69}$$

In the following sections, we will investigate both the general properties of the dielectric function $\epsilon_{\mathbf{q}}(\omega)$ and various theoretical models used to approximate it in many-body systems.

Remark 10.4 (Dielectric Function and Its Experimental Probing via EELS) Understanding the behavior of $\epsilon_{\mathbf{q}}(\omega)$ is essential, as it governs how electric fields are screened in a material and determines the effective interactions between charged particles. The dielectric function $\epsilon_{\mathbf{q}}(\omega)$ encapsulates important information about the material's response to external perturbations, such as electric fields, and its ability to shield the interactions between charges. Moreover, the study of the dielectric function is not limited to theoretical interest: it is also directly accessible through experimental techniques. One of the most prominent examples is Electron Energy Loss Spectroscopy (EELS), which measures the energy lost by fast electrons as they pass through a material. This energy loss is associated with the excitation of collective modes, such as plasmons (see below), and is directly related to the imaginary part of the inverse dielectric function,

$$\text{EELS} \propto \text{Im}\left[-\frac{1}{\epsilon_{\mathbf{q}}(\omega)}\right]. \tag{10.70}$$

As such, EELS provides valuable momentum- and frequency-resolved information about the screening properties and charge dynamics of quantum systems. By analyzing the energy loss spectrum, one can gain insight into both the static and dynamic properties of the material, including the nature of the collective excitations that occur at different energy scales. In a typical EELS experiment, a beam of high-energy electrons (in the range of 10–300 keV) is directed through a thin sample. As the electrons interact with the material, they may lose energy by exciting internal degrees of freedom such as plasmons or interband transitions. By analyzing the energy distribution of the transmitted electrons with high resolution, one obtains the energy loss spectrum. Moreover, by varying the scattering angle, it is possible to determine the momentum transfer $\mathbf{q}$, thereby mapping out the full dispersion $\omega(\mathbf{q})$ of collective excitations. This makes EELS a powerful tool to probe the dielectric response over a broad range of energies and wavevectors, providing a direct link between microscopic theoretical models and experimental data.

10.2.2 Zeros of the Dielectric Function

In this section, we investigate the physical meaning of the zeros of the dielectric function. Note that from this point onward we explicitly include the functional dependence on $\mathbf{q}$ in the particle density fluctuation operator, writing it in parentheses. From Eq. 10.49, it follows

$$q_e \delta\langle\hat{\rho}\rangle(\mathbf{q}, \omega) = \frac{1}{V} q_e^2 \varphi_{\text{ext},\mathbf{q}}(\omega) \Pi_{\mathbf{q}}(\omega), \tag{10.71}$$

furthermore, by inverting Eqs. 10.65 and 10.69, we obtain, respectively,

$$\Pi_{\mathbf{q}}(\omega) = \frac{q^2 V}{4\pi q_e^2}\left(\frac{1}{\epsilon_{\mathbf{q}}(\omega)} - 1\right), \tag{10.72}$$

$$\Pi^*_{\mathbf{q}}(\omega) = \frac{q^2 V}{4\pi q_e^2}\left(\epsilon_{\mathbf{q}}(\omega) - 1\right). \tag{10.73}$$

Equation 10.71 can be written from Eq. 10.72 as

$$\begin{aligned} q_e \delta\langle\hat{\rho}\rangle(\mathbf{q}, \omega) &= \frac{1}{V} q_e^2 \varphi_{\text{ext},\mathbf{q}}(\omega) \Pi_{\mathbf{q}}(\omega) = \\ &= \frac{1}{V} q_e^2 \varphi_{\text{ext},\mathbf{q}}(\omega) \frac{q^2 V}{4\pi q_e^2}\left(\frac{1}{\epsilon_{\mathbf{q}}(\omega)} - 1\right) = \\ &= \frac{q^2}{4\pi} \varphi_{\text{ext},\mathbf{q}}(\omega)\left(\frac{1}{\epsilon_{\mathbf{q}}(\omega)} - 1\right), \end{aligned} \tag{10.74}$$

or equivalently, if we use Eq. 10.73 into Eq. 10.74, then Eq. 10.71 can be written as

$$\begin{aligned} q_e \delta\langle\hat{\rho}\rangle(\mathbf{q}, \omega) &= \frac{q^2}{4\pi} \varphi_{\text{ext},\mathbf{q}}(\omega)\left(\frac{1}{\epsilon_{\mathbf{q}}(\omega)} - 1\right) = \\ &= \frac{q^2}{4\pi} \varphi_{\text{ext},\mathbf{q}}(\omega)\left(\frac{1 - \epsilon_{\mathbf{q}}(\omega)}{\epsilon_{\mathbf{q}}(\omega)}\right) = \\ &= \frac{q^2}{4\pi} \varphi_{\text{ext},\mathbf{q}}(\omega)\left(-\frac{\epsilon_{\mathbf{q}}(\omega) - 1}{\epsilon_{\mathbf{q}}(\omega)}\right) = \\ &= \frac{q^2}{4\pi} \varphi_{\text{ext},\mathbf{q}}(\omega) \frac{\frac{1}{V}\frac{4\pi q_e^2}{q^2}\Pi^*_{\mathbf{q}}(\omega)}{\epsilon_{\mathbf{q}}(\omega)} = \\ &= \frac{q_e^2}{V} \varphi_{\text{ext},\mathbf{q}}(\omega) \frac{\Pi^*_{\mathbf{q}}(\omega)}{\epsilon_{\mathbf{q}}(\omega)}. \end{aligned} \tag{10.75}$$

Here, we aim to study the meaning of the zeros of the dielectric function. Let us consider an impulsive external perturbation, that is of the form

$$\varphi_{\text{ext}}(\mathbf{r}, t) = \varphi_0 e^{-i\mathbf{q}\cdot\mathbf{r}} \delta(t), \tag{10.76}$$

and we compute the spatial and time Fourier transform, that is

$$\begin{aligned} \varphi_{\text{ext},\mathbf{k}}(\omega) &= \int d^3\mathbf{r} e^{i\mathbf{k}\cdot\mathbf{r}} \int dt e^{i\omega t} \varphi_0 e^{-i\mathbf{q}\cdot\mathbf{r}} = \\ &= \delta(\mathbf{k} - \mathbf{q})\varphi_0. \end{aligned} \tag{10.77}$$

Then, we consider the spatial and time Fourier transform of Eq. 10.75, namely

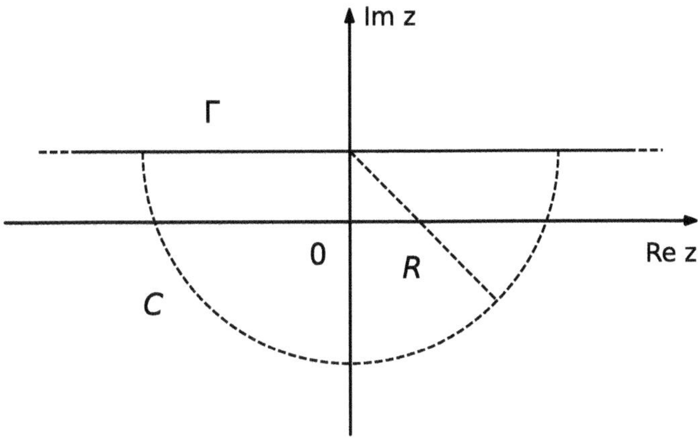

Fig. 10.1 Integration contour used to evaluate the Matsubara sum in Eq. 10.79 via the residue theorem

$$q_e \delta \langle \hat{\rho} \rangle (\mathbf{r}, t) = \int \frac{d^3\mathbf{k}}{(2\pi)^3} e^{-i\mathbf{k}\cdot\mathbf{r}} \int \frac{d\omega}{2\pi} e^{-i\omega t} \frac{q_e^2}{V} \varphi_{\text{ext},\mathbf{k}}(\omega) \frac{\Pi^*_{\mathbf{k}}(\omega)}{\epsilon_{\mathbf{k}}(\omega)}, \tag{10.78}$$

and we compute it in the case of the impulsive field described above, i.e.,

$$\begin{aligned} q_e \delta \langle \hat{\rho} \rangle (\mathbf{r}, t) &= \int \frac{d^3\mathbf{k}}{(2\pi)^3} e^{-i\mathbf{k}\cdot\mathbf{r}} \int \frac{d\omega}{2\pi} e^{-i\omega t} \frac{q_e^2}{V} \delta(\mathbf{k} - \mathbf{q}) \varphi_0 \frac{\Pi^*_{\mathbf{k}}(\omega)}{\epsilon_{\mathbf{k}}(\omega)} = \\ &= \frac{e^{-i\mathbf{q}\cdot\mathbf{r}}}{(2\pi)^3} \varphi_0 \int_\Gamma dz e^{-izt} \frac{\Pi^*_{\mathbf{q}}(z)}{\epsilon_{\mathbf{q}}(z)}, \end{aligned} \tag{10.79}$$

where the integral is over a line Γ parallel to the real axis contained in the upper half-plane. Consider the contour $\Gamma + C$ in Fig. 10.1. The factor e^{-izt} ensures convergence along the semicircle. Applying Jordan's lemma, in the limit $R \to \infty$, the contribution is given by the poles of the integrand function in the lower half-plane, namely by the zeros of the function $\left[\epsilon_{\mathbf{q}}(z)\right]^{-1}$. Suppose that a zero is of the form $z_{\mathbf{q}} = \Omega_{\mathbf{q}} - i\gamma_{\mathbf{q}}$, with $\gamma_{\mathbf{q}} > 0$. From the residue Theorem K.1, we obtain, up to constants, that the variation of particle density, following an impulsive perturbation, evolves as

$$q_e \delta \langle \hat{\rho} \rangle (\mathbf{r}, t) \propto e^{-i\mathbf{q}\cdot\mathbf{r}} e^{-i\Omega_{\mathbf{q}} t} e^{-i\omega_{\mathbf{q}} t}. \tag{10.80}$$

We have shown that the Fourier transform of an impulsive perturbation includes all the oscillation frequencies of the material medium, with a decay time that directly depends on the zeros of the function $\left[\epsilon_{\mathbf{q}}(\omega)\right]^{-1}$.

We now proceed to compute these frequencies within a semiclassical framework. After the impulsive perturbation, the particle density takes the form

$$\rho(\mathbf{r}, t) = \rho_0 + \delta\rho(\mathbf{r}, t), \tag{10.81}$$

where ρ_0 is the particle density at equilibrium and the quantity $\delta\rho(\mathbf{r}, t)$ measures the induced variation. At thermodynamic equilibrium, the density ρ_0 is balanced by that of the positive ions. The external perturbation implies electron motion, and thus charge imbalances, which in turn create an electric field that satisfies

$$\nabla \cdot \mathbf{E} = 4\pi q_e \delta\rho. \tag{10.82}$$

We will apply Newton's law $\mathbf{F} = m\mathbf{a}$, following an Eulerian approach to motion. We are interested in studying the quantities related to a fixed infinitesimal volume, such as the acting force and velocity, as time passes. Since we work within a continuous model for the material medium, the mass contained in an infinitesimal macroscopical volume is given by

$$\begin{aligned} dm &= \rho_m dV = \\ &= m_e \rho dV, \end{aligned} \tag{10.83}$$

where ρ_m is the electronic mass density, which is obviously related to the electronic particle (or number) density ρ through $\rho_m = m_e\rho$. With respect to the infinitesimal volume dV, Newton's equation is written as

$$q_e \mathbf{E} \rho(\mathbf{r}, t) dV = m_e dV \frac{d\mathbf{v}}{dt} \rho(\mathbf{r}, t), \tag{10.84}$$

$$q_e \mathbf{E} \rho(\mathbf{r}, t) = m_e \frac{d\mathbf{v}}{dt} \rho(\mathbf{r}, t), \tag{10.85}$$

where the Eulerian velocity $\mathbf{v}(x, y, z, t)$ indicates the velocity of the infinitesimal volume. At time $t + dt$ the volume has moved, and the velocity must be calculated at the new point, that is, it is a velocity $\mathbf{v}(x+v_x dt, y+v_y dt, z+v_z dt, t+dt)$. From the Taylor expansion we have

$$\begin{aligned} \mathbf{v}(x + v_x dt, y + v_y dt, z + v_z dt, t + dt) = \mathbf{v}(x, y, z, t) + \frac{\partial \mathbf{v}}{\partial t} dt + \frac{\partial \mathbf{v}_x}{\partial x} v_x dt \\ + \frac{\partial \mathbf{v}_y}{\partial y} v_y dt + \frac{\partial \mathbf{v}_z}{\partial z} v_z dt, \end{aligned} \tag{10.86}$$

thus, the acceleration is given by

$$\begin{aligned} \mathbf{a} &= \frac{\mathbf{v}(x + v_x dt, y + v_y dt, z + v_z dt, t + dt) - \mathbf{v}(x, y, z, t)}{dt} = \\ &= \frac{\partial \mathbf{v}}{\partial t} + (\mathbf{v} \cdot \nabla)\, \mathbf{v}. \end{aligned} \tag{10.87}$$

Within the framework of linear response theory, we retain only terms up to first order in the velocity, so that the total material derivative (i.e., the total derivative following the motion of the fluid element) of the velocity may be approximated by the partial derivative with respect to time. Under the assumption of linear theory, moreover, $\rho_e \sim \rho_0$, then

$$-q_e \mathbf{E} \rho_0 = m_e \frac{\partial \mathbf{v}}{\partial t} \rho_0, \tag{10.88}$$

$$-q_e \mathbf{E} = m_e \frac{\partial \mathbf{v}}{\partial t}. \tag{10.89}$$

Now, from the continuity equation

$$\frac{d\rho}{dt} + \rho \nabla \cdot \mathbf{v} = 0, \tag{10.90}$$

and considering a first-order approximation, i.e., linearizing the equation around a uniform equilibrium state, we obtain

$$\begin{aligned} \frac{d\rho}{dt} \sim \frac{\partial \rho}{\partial t} &= \\ &= \frac{\partial \rho_0}{\partial t} + \frac{\partial \delta\rho}{\partial t} = \\ &= \frac{\partial \delta\rho}{\partial t}, \end{aligned} \tag{10.91}$$

and $\rho \sim \rho_0$, and Eq. 10.90 becomes

$$\frac{\partial \delta\rho}{\partial t} + \rho_0 \nabla \cdot \mathbf{v} = 0. \tag{10.92}$$

We differentiate Eq. 10.92 with respect to time, i.e.,

$$\frac{\partial^2 \delta\rho}{\partial t^2} + \rho_0 \nabla \cdot \frac{\partial \mathbf{v}}{\partial t} = 0, \tag{10.93}$$

from Eq. 10.89, one finds

$$\frac{\partial^2 \delta\rho}{\partial t^2} + \frac{q_e \rho_0}{m_e} \nabla \cdot \mathbf{E} = 0, \tag{10.94}$$

and from Eq. 10.82, it follows

$$\frac{\partial^2 \delta\rho}{\partial t^2} + \frac{4\pi q_e^2 \rho_0}{m_e} \delta\rho = 0, \tag{10.95}$$

which is a harmonic motion equation for $\delta\rho$. We have derived the frequency at which the induced charge density oscillates, namely the plasma frequency

$$\omega_p^2 = \frac{4\pi q_e^2 \rho_0}{m_e}, \tag{10.96}$$

which is associated with quasiparticles called *plasmons*. In particular, such quasiparticles satisfy bosonic statistics, despite originating from a system of fermions. Indeed, the dielectric function is related to the polarization propagator, which involves commutators of particle density operators. According to Remark 2.7, these operators are not genuinely bosonic, but they effectively exhibit bosonic behavior. This allows one to treat the corresponding excitations, i.e., plasmons, as bosonic in this context.

10.2.3 Sum Rules for the Dielectric Function

Here, we introduce some fundamental properties of the dielectric function, known as sum rules, which directly arise from the spectral structure of the polarization propagator. These sum rules relate moments of the spectral function to microscopic properties of the system. From Eq. 6.87, the polarization propagator must satisfy

$$\lim_{\delta\to 0^+} -2\hbar\, \mathrm{Im}\, \Pi_{\mathbf{q}}(\omega + i\delta) = A_{\mathbf{q}}(\omega), \tag{10.97}$$

where $A_{\mathbf{q}}(\omega)$ is the spectral function of the polarization propagator, which from 6.43 can be written in terms of the eigenstates of the complete Hamiltonian as

$$A_{\mathbf{q}}(\omega) = \frac{2\pi}{Z}\sum_{n,m}\left(e^{-\beta\mathcal{E}_n} - e^{-\beta\mathcal{E}_m}\right)\delta\left[\omega - \frac{\mathcal{E}_m - \mathcal{E}_n}{\hbar}\right]\left|\langle n|\hat{\rho}_{\mathbf{q}}|m\rangle\right|^2. \tag{10.98}$$

Theorem 10.1 *The spectral function (see Definition 6.5) of the polarization propagator (see Eq. 10.98) satisfies*

1.

$$A_{\mathbf{q}}(-\omega) = -A_{\mathbf{q}}(\omega); \tag{10.99}$$

2.

$$\int_{-\infty}^{+\infty}\frac{d\omega}{2\pi}A_{\mathbf{q}}(\omega) = 0; \tag{10.100}$$

3.

$$\int_{-\infty}^{+\infty} \frac{d\omega}{2\pi} \omega A_{\mathbf{q}}(\omega) = \frac{\hbar q^2 N}{m_e}, \tag{10.101}$$

where N denotes the total number of electrons in the system.

Proof

1. From Eq. 10.98, it follows

$$A_{\mathbf{q}}(-\omega) = \frac{2\pi}{Z} \sum_{n,m} \left(e^{-\beta\mathcal{E}_n} - e^{-\beta\mathcal{E}_m}\right) \delta\left[-\omega - \frac{\mathcal{E}_m - \mathcal{E}_n}{\hbar}\right] \left|\langle n|\hat{\rho}_{\mathbf{q}}|m\rangle\right|^2, \tag{10.102}$$

then we use the property $\delta(x) = \delta(-x)$ and we write

$$A_{\mathbf{q}}(-\omega) = \frac{2\pi}{Z} \sum_{n,m} \left(e^{-\beta\mathcal{E}_n} - e^{-\beta\mathcal{E}_m}\right) \delta\left[\omega + \frac{\mathcal{E}_m - \mathcal{E}_n}{\hbar}\right] \left|\langle n|\hat{\rho}_{\mathbf{q}}|m\rangle\right|^2, \tag{10.103}$$

and we invert the order of the indices n, m as follows

$$A_{\mathbf{q}}(-\omega) = \frac{2\pi}{Z} \sum_{n,m} \left(e^{-\beta\mathcal{E}_m} - e^{-\beta\mathcal{E}_n}\right) \delta\left[\omega + \frac{\mathcal{E}_n - \mathcal{E}_m}{\hbar}\right] \left|\langle m|\hat{\rho}_{\mathbf{q}}|n\rangle\right|^2. \tag{10.104}$$

Note that the delta can be rewritten as

$$\delta\left[\omega + \frac{\mathcal{E}_n - \mathcal{E}_m}{\hbar}\right] = \delta\left[\omega - \frac{\mathcal{E}_m - \mathcal{E}_n}{\hbar}\right]. \tag{10.105}$$

Let us define

$$X \equiv \langle m|\hat{\rho}_{\mathbf{q}}|n\rangle. \tag{10.106}$$

Its complex conjugate reads

$$\begin{aligned} X^* &= \langle n|\hat{\rho}_{\mathbf{q}}^{\dagger}|m\rangle = \\ &= \langle n|\hat{\rho}_{-\mathbf{q}}|m\rangle, \end{aligned} \tag{10.107}$$

where we used Eq. 2.268. Therefore, the squared modulus is

$$|X|^2 = \langle m|\hat{\rho}_{\mathbf{q}}|n\rangle \, \langle n|\hat{\rho}_{-\mathbf{q}}|m\rangle. \tag{10.108}$$

Analogously, by exchanging $m \leftrightarrow n$, we obtain

$$\left|\langle n|\hat{\rho}_{\mathbf{q}}|m\rangle\right|^2 = \langle n|\hat{\rho}_{\mathbf{q}}|m\rangle\langle m|\hat{\rho}_{-\mathbf{q}}|n\rangle. \tag{10.109}$$

Since the two expressions coincide, one finds

$$\left|\langle m|\hat{\rho}_{\mathbf{q}}|n\rangle\right|^2=\left|\langle n|\hat{\rho}_{\mathbf{q}}|m\rangle\right|^2. \tag{10.110}$$

It follows

$$A_{\mathbf{q}}(-\omega)=\frac{2\pi}{Z}\sum_{n,m}\left(e^{-\beta\mathcal{E}_m}-e^{-\beta\mathcal{E}_n}\right)\delta\left[\omega-\frac{\mathcal{E}_m-\mathcal{E}_n}{\hbar}\right]\left|\langle n|\hat{\rho}_{\mathbf{q}}|m\rangle\right|^2. \tag{10.111}$$

which is the thesis.

2. From 6.235 with $A=\hat{\rho}_{\mathbf{q}}$, $B=\hat{\rho}^{\dagger}_{\mathbf{q}}$, we have

$$\int_{-\infty}^{+\infty}\frac{d\omega}{2\pi}A_{\mathbf{q}}(\omega)=\left\langle\left[\hat{\rho}_{\mathbf{q}},\hat{\rho}^{\dagger}_{\mathbf{q}}\right]\right\rangle, \tag{10.112}$$

and from Eq. 2.275, the thesis follows.

3. From Eq. 6.240 with $A=\hat{\rho}_{\mathbf{q}}$, $B=\hat{\rho}^{\dagger}_{\mathbf{q}}$, we have

$$\int_{-\infty}^{+\infty}\frac{d\omega}{2\pi}\omega A_{\mathbf{q}}(\omega)=\frac{1}{\hbar}\left\langle\left[\left[\hat{\rho}_{\mathbf{q}},\hat{\mathcal{H}}\right],\hat{\rho}^{\dagger}_{\mathbf{q}}\right]\right\rangle, \tag{10.113}$$

where $\hat{\mathcal{H}}$ is the jellium Hamiltonian; in particular, we employ the form given in Eq. 2.388 of the jellium model, with the obvious identifications

$$\hat{\mathcal{H}}_0=\sum_{\mathbf{k},\sigma}\mathcal{E}^{(0)}_{\mathbf{k}}C^{\dagger}_{\mathbf{k},\sigma}C_{\mathbf{k},\sigma}, \tag{10.114}$$

$$\hat{\mathcal{H}}_I=\frac{1}{2V}\sum_{\mathbf{q}\neq\mathbf{0}}\frac{4\pi q_e^2}{q^2}\left(\hat{N}-\hat{\rho}_{\mathbf{q}}\hat{\rho}^{\dagger}_{\mathbf{q}}\right), \tag{10.115}$$

where $\hat{N}$ is the total number operator and $\mathcal{E}^{(0)}_{\mathbf{k}}=\frac{\hbar^2\mathbf{k}^2}{2m_e}-\mu$, and we compute

$$\left[\hat{\rho}_{\mathbf{q}},\hat{\mathcal{H}}\right]=\left[\hat{\rho}_{\mathbf{q}},\hat{\mathcal{H}}_0\right]+\left[\hat{\rho}_{\mathbf{q}},\hat{\mathcal{H}}_I\right]. \tag{10.116}$$

The commutator $\left[\hat{\rho}_{\mathbf{q}},\hat{\mathcal{H}}_0\right]$ is obtained from 2.276, with $\mathcal{E}_{\mathbf{k}}=\mathcal{E}^{(0)}_{\mathbf{k}}$, while the second term decomposes as

$$\begin{aligned}\left[\hat{\rho}_{\mathbf{q}},\hat{\mathcal{H}}_I\right]&=\frac{1}{2V}\sum_{\mathbf{q}'\neq\mathbf{0}}\frac{4\pi q_e^2}{q'^2}\left[\hat{\rho}_{\mathbf{q}},\left(\hat{N}-\hat{\rho}_{\mathbf{q}'}\hat{\rho}^{\dagger}_{\mathbf{q}'}\right)\right]=\\&=\frac{1}{2V}\sum_{\mathbf{q}'\neq\mathbf{0}}\frac{4\pi q_e^2}{q'^2}\left(\left[\hat{\rho}_{\mathbf{q}},\hat{N}\right]-\left[\hat{\rho}_{\mathbf{q}},\hat{\rho}_{\mathbf{q}'}\hat{\rho}^{\dagger}_{\mathbf{q}'}\right]\right).\end{aligned} \tag{10.117}$$

and from 2.278 and 2.279, it vanishes, and the first moment can be rewritten as

$$\int_{-\infty}^{+\infty} \frac{d\omega}{2\pi} \omega A_{\mathbf{q}}(\omega) = \frac{1}{\hbar} \left\langle \left[\left[\hat{\rho}_{\mathbf{q}}, \hat{\mathcal{H}}_0 \right], \hat{\rho}_{\mathbf{q}}^{\dagger} \right] \right\rangle. \tag{10.118}$$

From 2.277, with $\mathcal{E}_{\mathbf{k}} = \mathcal{E}_{\mathbf{k}}^{(0)}$, we have

$$\left\langle \left[\left[\hat{\rho}_{\mathbf{q}}, \hat{\mathcal{H}}_0 \right], \hat{\rho}_{\mathbf{q}}^{\dagger} \right] \right\rangle = \sum_{\mathbf{k},\sigma} \left\langle \left(\mathcal{E}_{\mathbf{k}+\mathbf{q}}^{(0)} - 2\mathcal{E}_{\mathbf{k}}^{(0)} + \mathcal{E}_{\mathbf{k}-\mathbf{q}}^{(0)} \right) C_{\mathbf{k},\sigma}^{\dagger} C_{\mathbf{k},\sigma} \right\rangle, \tag{10.119}$$

and from

$$\begin{aligned} \mathcal{E}_{\mathbf{k}+\mathbf{q}}^{(0)} - 2\mathcal{E}_{\mathbf{k}}^{(0)} + \mathcal{E}_{\mathbf{k}-\mathbf{q}}^{(0)} &= \frac{\hbar^2(\mathbf{k}+\mathbf{q})^2}{2m_e} - 2\frac{\hbar^2\mathbf{k}^2}{2m_e} + \frac{\hbar^2(\mathbf{k}-\mathbf{q})^2}{2m_e} = \\ &= \hbar^2 \frac{\mathbf{k}^2 + 2\mathbf{k}\cdot\mathbf{q} + \mathbf{q}^2 - 2\mathbf{k}^2 + \mathbf{k}^2 - 2\mathbf{k}\cdot\mathbf{q} + \mathbf{q}^2}{2m_e} = \\ &= \frac{\hbar^2 q^2}{2m_e}, \end{aligned} \tag{10.120}$$

it follows

$$\begin{aligned} \left\langle \left[\left[\hat{\rho}_{\mathbf{q}}, \hat{\mathcal{H}}_0 \right], \hat{\rho}_{\mathbf{q}}^{\dagger} \right] \right\rangle &= \frac{\hbar^2 q^2}{m_e} \sum_{\mathbf{k},\sigma} \left\langle C_{\mathbf{k},\sigma}^{\dagger} C_{\mathbf{k},\sigma} \right\rangle = \\ &= \frac{\hbar^2 q^2 N}{m_e}, \end{aligned} \tag{10.121}$$

from which the thesis follows.

Thanks to Eq. 10.101, we are now able to prove a sum rule for the dielectric function.

Theorem 10.2 (The f-Sum Rule for the Dielectric Function) *The dielectric function (see Eq. 10.65) satisfies the f-sum rule*

$$\int_{-\infty}^{+\infty} \frac{d\omega}{2\pi} \omega \operatorname{Im} \frac{1}{\varepsilon_{\mathbf{q}}(\omega)} = -\frac{\omega_p^2}{2}, \tag{10.122}$$

where ω_p is the plasma frequency given in Eq. 10.96.

Proof *We multiply the imaginary part of Eq. 10.65 by $(-2\hbar)$, that is*

$$-2\hbar \operatorname{Im} \frac{1}{\epsilon_{\mathbf{q}}(\omega)} = \frac{1}{V} \frac{4\pi q_e^2}{q^2} \operatorname{Im}\left[-2\hbar \Pi_{\mathbf{q}}(\omega) \right], \tag{10.123}$$

and given that the imaginary part of the polarization propagator is related to the spectral function (see Eq. 6.87), we have

$$-2\hbar\,\mathrm{Im}\,\frac{1}{\epsilon_{\mathbf{q}}(\omega)}=\frac{1}{V}\frac{4\pi q_e^2}{q^2}A_{\mathbf{q}}(\omega), \tag{10.124}$$

$$A_{\mathbf{q}}(\omega)=-2\hbar\frac{Vq^2}{4\pi q_e^2}\,\mathrm{Im}\,\frac{1}{\epsilon_{\mathbf{q}}(\omega)}. \tag{10.125}$$

We substitute such equation into Eq. 10.101, and we have

$$\int_{-\infty}^{+\infty}\frac{d\omega}{2\pi}\omega\left(-2\hbar\frac{Vq^2}{4\pi q_e^2}\,\mathrm{Im}\,\frac{1}{\epsilon_{\mathbf{q}}(\omega)}\right)=\frac{\hbar q^2 N}{m_e}, \tag{10.126}$$

$$\int_{-\infty}^{+\infty}\frac{d\omega}{2\pi}\omega\,\mathrm{Im}\,\frac{1}{\epsilon_{\mathbf{q}}(\omega)}=-\frac{4\pi q_e^2 N}{2m_e V}, \tag{10.127}$$

which is the thesis.

We will now derive another sum rule for the dielectric function. From Eq. 10.34, we define the polarization propagator in momentum and real-time space as

$$\Pi_{\mathbf{q}}(t)=-\frac{i}{\hbar}\Theta(t)\left\langle\left[\hat{\rho}_{\mathbf{q}}^{(0)}(t),\hat{\rho}_{\mathbf{q}}^{\dagger}(0)\right]\right\rangle. \tag{10.128}$$

Assuming that the eigenstates of the interacting Hamiltonian are known, and following calculations analogous to those in Sect. 6.3, the propagator $\Pi_{\mathbf{q}}(t)$ can be expressed as

$$\Pi_{\mathbf{q}}(t)=\frac{1}{\hbar}\sum_{n,m}\frac{e^{-\beta\mathcal{E}_n}-e^{-\beta\mathcal{E}_m}}{Z}\left|\langle n|\hat{\rho}_{\mathbf{q}}|m\rangle\right|^2. \tag{10.129}$$

We apply a Fourier transform in frequency space, inserting as usual an arbitrarily small factor δ to ensure convergence, and we have

$$\begin{aligned}\Pi_{\mathbf{q}}(\omega+i\delta)&=\int_{-\infty}^{+\infty}dze^{i(\omega+i\delta)t}\,\Pi_{\mathbf{q}}(t)=\\&=\frac{1}{\hbar}\sum_{n,m}\frac{e^{-\beta\mathcal{E}_n}-e^{-\beta\mathcal{E}_m}}{Z}\left|\langle n|\hat{\rho}_{\mathbf{q}}|m\rangle\right|^2\frac{1}{\omega+i\delta+\frac{\mathcal{E}_n-\mathcal{E}_m}{\hbar}},\end{aligned} \tag{10.130}$$

consequently, from 10.65 the dielectric function can be written as

$$\frac{1}{\epsilon_{\mathbf{q}}(\omega+i\delta)}=1+\frac{1}{V}\frac{4\pi q_e^2}{q^2}\frac{1}{\hbar}\sum_{n,m}\frac{e^{-\beta\mathcal{E}_n}-e^{-\beta\mathcal{E}_m}}{Z}\left|\langle n|\hat{\rho}_{\mathbf{q}}|m\rangle\right|^2\frac{1}{\omega+i\delta+\frac{\mathcal{E}_n-\mathcal{E}_m}{\hbar}}. \tag{10.131}$$

If we rewrite the previous equation as

$$\frac{1}{\epsilon_{\mathbf{q}}(\omega+i\delta)}=1+\frac{1}{V}\frac{4\pi q_e^2}{q^2}\frac{1}{\hbar\omega}\sum_{n,m}\frac{e^{-\beta\mathcal{E}_n}-e^{-\beta\mathcal{E}_m}}{Z}\left|\langle n|\hat{\rho}_{\mathbf{q}}|m\rangle\right|^2\frac{1}{1+\frac{i\delta}{\omega}+\frac{\mathcal{E}_n-\mathcal{E}_m}{\hbar\omega}}, \tag{10.132}$$

and if we write

$$\frac{1}{1+\frac{i\delta}{\omega}+\frac{\mathcal{E}_n-\mathcal{E}_m}{\hbar\omega}}=\frac{1}{1+\frac{i\delta+\frac{\mathcal{E}_n-\mathcal{E}_m}{\hbar}}{\omega}}, \tag{10.133}$$

we observe that in the high-frequency limit $\omega\to+\infty$, we can use the following approximation

$$\frac{1}{1+\frac{i\delta+\frac{\mathcal{E}_n-\mathcal{E}_m}{\hbar}}{\omega}}=1-\frac{i\delta+\frac{\mathcal{E}_n-\mathcal{E}_m}{\hbar}}{\omega}, \tag{10.134}$$

then

$$\left.\frac{1}{\epsilon_{\mathbf{q}}(\omega+i\delta)}\right|_{\omega\to\infty}=1+\frac{1}{V}\frac{4\pi q_e^2}{q^2}\frac{1}{\hbar\omega}\sum_{n,m}\frac{e^{-\beta\mathcal{E}_n}-e^{-\beta\mathcal{E}_m}}{Z}\left|\langle n|\hat{\rho}_{\mathbf{q}}|m\rangle\right|^2 \times\left(1-\frac{i\delta}{\omega}-\frac{\mathcal{E}_n-\mathcal{E}_m}{\hbar\omega}\right). \tag{10.135}$$

As for the sum

$$\left(1-\frac{i\delta}{\omega}\right)\sum_{n,m}\frac{e^{-\beta\mathcal{E}_n}-e^{-\beta\mathcal{E}_m}}{Z}\left|\langle n|\hat{\rho}_{\mathbf{q}}|m\rangle\right|^2 -\sum_{n,m}\frac{e^{-\beta\mathcal{E}_n}-e^{-\beta\mathcal{E}_m}}{Z}\left|\langle n|\hat{\rho}_{\mathbf{q}}|m\rangle\right|^2\frac{\mathcal{E}_n-\mathcal{E}_m}{\hbar\omega}, \tag{10.136}$$

we show that the first term vanishes: indeed, the general term of the series is antisymmetric, as can be seen by swapping the indices n and m,

$$A_{mn}=\frac{e^{-\beta\mathcal{E}_m}-e^{-\beta\mathcal{E}_n}}{Z}\left|\langle m\left|\hat{\rho}_{\mathbf{q}}\right|n\rangle\right|^2=-\frac{e^{-\beta\mathcal{E}_n}-e^{-\beta\mathcal{E}_m}}{Z}\left|\langle n\left|\hat{\rho}_{\mathbf{q}}\right|m\rangle\right|^2=-A_{nm}, \tag{10.137}$$

then it is a sum of antisymmetric terms over a pair of symmetric indices, so it is zero, and the dielectric function in the limit $\omega\to+\infty$ can be rewritten as

$$\left.\frac{1}{\epsilon_{\mathbf{q}}(\omega)}\right|_{\omega\to\infty} = 1 - \frac{1}{V}\frac{4\pi q_e^2}{q^2}\frac{1}{(\hbar\omega)^2}\sum_{n,m}\frac{e^{-\beta\mathcal{E}_n} - e^{-\beta\mathcal{E}_m}}{Z}\left|\langle n|\hat{\rho}_{\mathbf{q}}|m\rangle\right|^2(\mathcal{E}_n - \mathcal{E}_m)\,. \tag{10.138}$$

From

$$\begin{aligned}
\left|\langle n\left|\hat{\rho}_{\mathbf{q}}\right|m\rangle\right|^2(\mathcal{E}_n - \mathcal{E}_m) &= \langle n\left|\hat{\rho}_{\mathbf{q}}\right|m\rangle\left\langle m\left|\hat{\rho}_{\mathbf{q}}^{\dagger}\right|n\right\rangle(\mathcal{E}_n - \mathcal{E}_m) = \\
&= \langle n\left|(\mathcal{E}_n - \mathcal{E}_m)\,\hat{\rho}_{\mathbf{q}}\right|m\rangle\left\langle m\left|\hat{\rho}_{\mathbf{q}}^{\dagger}\right|n\right\rangle = \\
&= \left\langle n\left|\hat{\mathcal{H}}_0\hat{\rho}_{\mathbf{q}} - \hat{\rho}_{\mathbf{q}}\hat{\mathcal{H}}_0\right|m\right\rangle\left\langle m\left|\hat{\rho}_{\mathbf{q}}^{\dagger}\right|n\right\rangle = \\
&= \left\langle n\left|\left[\hat{\mathcal{H}}_0, \hat{\rho}_{\mathbf{q}}\right]\right|m\right\rangle\left\langle m\left|\hat{\rho}_{\mathbf{q}}^{\dagger}\right|n\right\rangle = \\
&= -\left\langle n\left|\left[\hat{\rho}_{\mathbf{q}}, \hat{\mathcal{H}}_0\right]\right|m\right\rangle\left\langle m\left|\hat{\rho}_{\mathbf{q}}^{\dagger}\right|n\right\rangle,
\end{aligned} \tag{10.139}$$

we get

$$\left.\frac{1}{\epsilon_{\mathbf{q}}(\omega)}\right|_{\omega\to\infty} = 1 + \frac{1}{V}\frac{4\pi q_e^2}{q^2}\frac{1}{(\hbar\omega)^2}\sum_{n,m}\frac{e^{-\beta\mathcal{E}_n} - e^{-\beta\mathcal{E}_m}}{Z}\left\langle n\right|\left[\hat{\rho}_{\mathbf{q}}, \hat{\mathcal{H}}_0\right]\left|m\right\rangle\left\langle m\right|\hat{\rho}_{\mathbf{q}}^{\dagger}\left|n\right\rangle, \tag{10.140}$$

then, we manipulate the sum as follows

$$\begin{aligned}
&\sum_{n,m}\frac{e^{-\beta\mathcal{E}_n} - e^{-\beta\mathcal{E}_m}}{Z}\left\langle n\left|\left[\hat{\rho}_{\mathbf{q}}, \hat{\mathcal{H}}_0\right]\right|m\right\rangle\left\langle m\left|\hat{\rho}_{\mathbf{q}}^{\dagger}\right|n\right\rangle = \\
&\quad = \sum_{n,m}\frac{e^{-\beta\mathcal{E}_n}}{Z}\left\langle n\left|\left[\hat{\rho}_{\mathbf{q}}, \hat{\mathcal{H}}_0\right]\right|m\right\rangle\left\langle m\left|\hat{\rho}_{\mathbf{q}}^{\dagger}\right|n\right\rangle \\
&\qquad - \sum_{n,m}\frac{e^{-\beta\mathcal{E}_m}}{Z}\left\langle n\left|\left[\hat{\rho}_{\mathbf{q}}, \hat{\mathcal{H}}_0\right]\right|m\right\rangle\left\langle m\left|\hat{\rho}_{\mathbf{q}}^{\dagger}\right|n\right\rangle = \\
&\quad = \sum_{n,m}\frac{e^{-\beta\mathcal{E}_n}}{Z}\left\langle m\left|\hat{\rho}_{\mathbf{q}}^{\dagger}\right|n\right\rangle\left\langle n\left|\left[\hat{\rho}_{\mathbf{q}}, \hat{\mathcal{H}}_0\right]\right|m\right\rangle \\
&\qquad - \sum_{n,m}\frac{e^{-\beta\mathcal{E}_m}}{Z}\left\langle n\left|\left[\hat{\rho}_{\mathbf{q}}, \hat{\mathcal{H}}_0\right]\right|m\right\rangle\left\langle m\left|\hat{\rho}_{\mathbf{q}}^{\dagger}\right|n\right\rangle = \\
&= \sum_{m}\left\langle m\left|\hat{\rho}_{\mathbf{q}}^{\dagger}\left(\sum_{n}\frac{e^{-\beta\mathcal{E}_n}}{Z}\left|n\right\rangle\left\langle n\right|\right)\left[\hat{\rho}_{\mathbf{q}}, \hat{\mathcal{H}}_0\right]\right|m\right\rangle \\
&\quad - \sum_{n}\left\langle n\left|\left[\hat{\rho}_{\mathbf{q}}, \hat{\mathcal{H}}_0\right]\left(\sum_{m}\frac{e^{-\beta\mathcal{E}_m}}{Z}\left|m\right\rangle\left\langle m\right|\right)\hat{\rho}_{\mathbf{q}}^{\dagger}\right|n\right\rangle =
\end{aligned}$$

$$= \sum_m \left\langle m \left| \hat{\rho}_{\mathbf{q}}^\dagger \left[\hat{\rho}_{\mathbf{q}}, \hat{\mathcal{H}}_0 \right] \right| m \right\rangle - \sum_n \left\langle n \left| \left[\hat{\rho}_{\mathbf{q}}, \hat{\mathcal{H}}_0 \right] \hat{\rho}_{\mathbf{q}}^\dagger \right| n \right\rangle, \tag{10.141}$$

and since the summation indices are dummy indices, we have constructed a thermal average (see Eq. 1.379)

$$\left\langle \left[\left[\hat{\rho}_{\mathbf{q}}, \hat{\mathcal{H}}_0 \right], \hat{\rho}_{\mathbf{q}}^\dagger \right] \right\rangle, \tag{10.142}$$

from which it follows

$$\left. \frac{1}{\epsilon_{\mathbf{q}}(\omega)} \right|_{\omega \to \infty} = 1 + \frac{1}{V} \frac{4\pi q_e^2}{q^2} \frac{1}{(\hbar\omega)^2} \left\langle \left[\left[\hat{\rho}_{\mathbf{q}}, \hat{\mathcal{H}}_0 \right], \hat{\rho}_{\mathbf{q}}^\dagger \right] \right\rangle. \tag{10.143}$$

From 10.121 we have

$$\begin{aligned} \left. \frac{1}{\epsilon_{\mathbf{q}}(\omega)} \right|_{\omega \to \infty} &= 1 + \frac{1}{V} 4\pi q_e^2 \frac{1}{\omega^2} \frac{N}{m_e} = \\ &= 1 + \frac{\omega_p^2}{\omega^2}. \end{aligned} \tag{10.144}$$

If we write

$$\left. \frac{1}{\epsilon_{\mathbf{q}}(\omega)} \right|_{\omega \to \infty} = \frac{\omega^2 + \omega_p^2}{\omega^2}, \tag{10.145}$$

we equivalently have

$$\begin{aligned} \left. \epsilon_{\mathbf{q}}(\omega) \right|_{\omega \to \infty} &= \frac{\omega^2}{\omega^2 + \omega_p^2} = \\ &= \frac{1}{1 + \frac{\omega_p^2}{\omega^2}}, \end{aligned} \tag{10.146}$$

and in the limit $\omega \to \infty$ that we are considering here, we can use the following approximation

$$\left. \epsilon_{\mathbf{q}}(\omega) \right|_{\omega \to \infty} = 1 - \frac{\omega_p^2}{\omega^2}. \tag{10.147}$$

In other words, the dielectric function has a zero at the plasma frequency, meaning a resonance phenomenon occurs at the plasma frequency.

Remark 10.5 (Landau's Interpretation of the Zeros of the Dielectric Function) Starting from expression 10.69, we consider a frequency of the form $\omega = \Omega_{\mathbf{q}} - i\gamma_{\mathbf{q}}$, with $\gamma_{\mathbf{q}} > 0$, that is, we consider the lower complex half-plane, where we expect the dielectric function to have zeros, as they correspond to poles of the polarization propagator. We set $\epsilon_{\mathbf{q}}(\Omega_{\mathbf{q}} - i\gamma_{\mathbf{q}}) = 0$, from Eq. 10.69, it follows

$$1 - \frac{1}{V}\frac{4\pi q_e^2}{q^2}\left[\operatorname{Re}\Pi_{\mathbf{q}}^*\left(\Omega_{\mathbf{q}} - i\gamma_{\mathbf{q}}\right) + i\operatorname{Im}\Pi_{\mathbf{q}}^*\left(\Omega_{\mathbf{q}} - i\gamma_{\mathbf{q}}\right)\right] = 0. \tag{10.148}$$

Landau considers a Taylor expansion of the polarization propagator for $\gamma_{\mathbf{q}} \to 0$. In particular, since the propagator is analytic at points that are zeros of the dielectric function, the complex derivative can be taken along any direction; thus, we differentiate the propagator with respect to the real axis and write

$$1 - \frac{1}{V}\frac{4\pi q_e^2}{q^2}\left[\Pi_{\mathbf{q}}^*\left(\Omega_{\mathbf{q}}\right) - i\gamma_{\mathbf{q}}\left.\frac{\partial\Pi_{\mathbf{q}}^*(\omega)}{\partial\omega}\right|_{\omega=\Omega_{\mathbf{q}}}\right] = 0, \tag{10.149}$$

which implies two equations for the real and imaginary parts. Regarding the real part of Eq. 10.149, neglecting terms of order higher than first in $\gamma_{\mathbf{q}}$, we have

$$1 - \frac{1}{V}\frac{4\pi q_e^2}{q^2}\operatorname{Re}\Pi_{\mathbf{q}}^*\left(\Omega_{\mathbf{q}}\right) = 0. \tag{10.150}$$

From the imaginary part equation, we have

$$\gamma_{\mathbf{q}} = \frac{\operatorname{Im}\Pi_{\mathbf{q}}^*\left(\Omega_{\mathbf{q}}\right)}{\left.\dfrac{\partial\operatorname{Re}\Pi_{\mathbf{q}}^*(\omega)}{\partial\omega}\right|_{\omega=\Omega_{\mathbf{q}}}}, \tag{10.151}$$

That is, the imaginary part of the polarization propagator is responsible for the damping, similarly to what we showed for the many-body propagator. If the imaginary part is zero, the damping is zero and the associated particles have an infinite lifetime.

10.2.4 Hartree-Fock Approximation

A first approximation for calculating the dielectric function is the Hartree-Fock approximation, which consists of replacing the polarization propagator (see Eq. 10.128) with its first-order term, that is

$$\left[q_e\delta\langle\hat{\rho}\rangle(\mathbf{q},\omega)\right]\Big|_{\mathrm{HF}} = \frac{1}{V}q_e^2\varphi_{\mathrm{ext},\mathbf{q}}(\omega)\Pi_{\mathbf{q}}^{(0)}(\omega), \tag{10.152}$$

That is, from Eq. 10.65, this means performing a first order expansion in terms of $\left[\epsilon_{\mathbf{q}}(\omega)\right]^{-1}$. Since the dielectric function is written as in Eq. 10.131, the Hartree-Fock approach consists of replacing the eigenfunctions of the complete Hamiltonian with those corresponding to the non-interacting Hamiltonian. Now, considering the case $T = 0$ and recalling Eqs. 1.380 and 1.381, we need to compute quantities of the form

$$\left|\langle n \left|\hat{\rho}_{\mathbf{q}}\right| 0\rangle\right|^2 . \tag{10.153}$$

Note that the ground state is a Slater determinant of plane waves with maximum momentum equal to the Fermi momentum $\mathbf{k}_F$ (see Appendix O), and the particle density operator destroys the plane wave of momentum $\mathbf{k}$ and creates one of momentum $\mathbf{k}+\mathbf{q}$. The operator $C_{\mathbf{k},\sigma}$ contained in $\hat{\rho}_{\mathbf{q}}$ can act on the ground state $|0\rangle$ only if the new momentum $\mathbf{k}$ has magnitude less than $\mathbf{k}_F$, while the operator $C^{\dagger}_{\mathbf{k}+\mathbf{q},\sigma}$ can act on the state $|n\rangle$ only if the new momentum is $\mathbf{k}+\mathbf{q}$. In other words, in the Hartree-Fock approximation, hole-particle processes occur, which are fixed-particle-number excitations. Let $\mathcal{E}_m$ be the energy of the initial state, which in our case is the ground state, so $\mathcal{E}_m = 0$, and $\mathcal{E}_n$ the energy of the final state, with the energy difference given by

$$\begin{aligned}
\Delta\mathcal{E}(\mathbf{q}) &= \mathcal{E}^{(0)}_{\mathbf{k}+\mathbf{q}} - \mu - \mathcal{E}^{(0)}_{\mathbf{k}} + \mu = \\
&= \frac{\hbar^2(\mathbf{k}+\mathbf{q})^2}{2m_e} - \frac{\hbar^2\mathbf{k}^2}{2m_e} = \\
&= \frac{\hbar^2}{2m_e}\left[\mathbf{q}^2 + 2\mathbf{k}\cdot\mathbf{q}\right],
\end{aligned} \tag{10.154}$$

where we used energies of the grand canonical jellium model, with the Hamiltonian given in Eq. 2.393. Such energy is the energy required to create a hole-particle pair and occurs at a fixed transferred momentum $\mathbf{q}$, and for the constraints discussed previously, it must hold that

$$\max\left\{0, \frac{\hbar^2}{2m_e}\left(\mathbf{q}^2 - 2\mathbf{k}_F\cdot\mathbf{q}\right)\right\} \le \Delta\mathcal{E}(|\mathbf{q}|) \le \frac{\hbar^2}{2m_e}\left(\mathbf{q}^2 + 2\mathbf{k}_F\cdot\mathbf{q}\right). \tag{10.155}$$

The graph of the energy as a function of the modulus of $\mathbf{q}$ is shown in Fig. 10.2; in particular, it is the planar region enclosed between arcs (b) and (c), including segment (a). Arc (c) corresponds to the parabola $\hbar\omega(q) = \frac{\hbar^2}{2m_e}\left(\mathbf{q}^2 + 2\mathbf{k}_F\cdot\mathbf{q}\right)$, which has zeros at $\mathbf{q} = 0$ and $\mathbf{q} = -2\mathbf{k}_F$; arc (b) corresponds to the parabola $\hbar\omega(q) = \frac{\hbar^2}{2m_e}\left(\mathbf{q}^2 - 2\mathbf{k}_F\cdot\mathbf{q}\right)$, which has zeros at $\mathbf{q} = 0$ and $\mathbf{q} = 2\mathbf{k}_F$; segment (a) corresponds to processes in which the energy variation is zero because the particle is annihilated on the Fermi surface and recreated with the same momentum, also on the Fermi surface. Note that the excitations produced in the Hartree-Fock approxi-

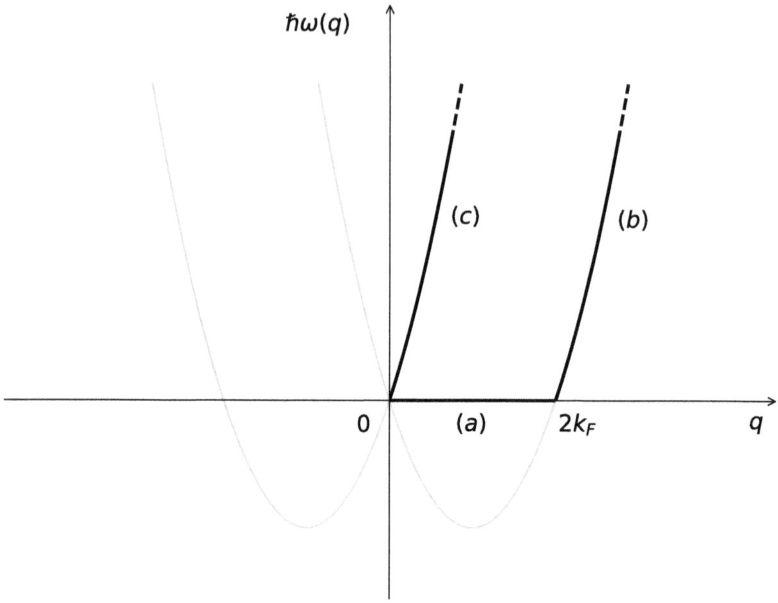

Fig. 10.2 The hole-particle processes contributing to the dielectric function within the Hartree-Fock approximation occur for momentum values lying in the region enclosed by curves (**a**), (**b**), and (**c**), as defined in Eq. 10.155. Curves (**b**) and (**c**) correspond to the parabolic energy boundaries associated with hole-particle excitations, while segment (**a**) denotes the zero-energy processes involving states on the Fermi surface

mation are one-particle excitations and not collective. Moreover, the Hartree-Fock approximation contains no singularities, whereas we have seen previously that the zeros of the dielectric function correspond to resonance frequencies, the plasma frequencies, which are frequencies of collective excitations.

10.2.5 Thomas-Fermi Approximation

Here we study the Thomas-Fermi approximation for the dielectric function, which consists of considering an external perturbation constant in time, that is, of the form

$$\varphi_{\text{ext},\mathbf{q}}(\omega) = 2\pi\delta(\omega)\varphi_{\text{ext},\mathbf{q}}, \tag{10.156}$$

and we aim to study the variation of particle density, that is, the polarization propagator. If we divide Eq. 10.74 by the charge q_e, we obtain

$$\delta\langle\hat{\rho}\rangle(\mathbf{q},\omega) = \frac{q^2}{4\pi q_e}\varphi_{\text{ext},\mathbf{q}}(\omega)\left(\frac{1}{\epsilon_{\mathbf{q}}(\omega)} - 1\right), \tag{10.157}$$

and by substituting the Thomas-Fermi perturbation, we have

$$\delta\langle\hat{\rho}\rangle(\mathbf{q},\omega) = \frac{q^2}{2q_e}\delta(\omega)\varphi_{\text{ext},\mathbf{q}}\left(\frac{1}{\epsilon_{\mathbf{q}}(\omega)} - 1\right). \tag{10.158}$$

We aim to compute this density variation in the time domain, so we apply a time Fourier transform as follows

$$\begin{aligned}\delta\langle\hat{\rho}\rangle(\mathbf{q},t) &= \int_{-\infty}^{+\infty}\frac{d\omega}{2\pi}e^{-i\omega t}\delta\langle\hat{\rho}\rangle(\mathbf{q},\omega) = \\ &= \int_{-\infty}^{+\infty}\frac{d\omega}{2\pi}e^{-i\omega t}\frac{q^2}{2q_e}\delta(\omega)\varphi_{\text{ext},\mathbf{q}}\left(\frac{1}{\epsilon_{\mathbf{q}}(\omega)} - 1\right),\end{aligned} \tag{10.159}$$

then

$$\delta\langle\hat{\rho}\rangle(\mathbf{q}) = \frac{q^2}{4\pi q_e}\varphi_{\text{ext},\mathbf{q}}\left(\frac{1}{\epsilon_{\mathbf{q}}(0)} - 1\right). \tag{10.160}$$

As expected, this variation is time-independent, reflecting the fact that it results from a static (i.e., zero-frequency) perturbation in the frequency domain. The response of the medium is therefore stationary. Since the explicit form of the static dielectric function $\epsilon_{\mathbf{q}}(0)$ is not known, Thomas and Fermi proposed an Ansatz for the variation of the particle density, i.e., $\delta\langle\hat{\rho}\rangle(\mathbf{q})$. In the absence of perturbation, the particle density is calculated as

$$\begin{aligned}\rho_0 &= \frac{1}{V}\sum_{\mathbf{k},\sigma}\left\langle C^{\dagger}_{\mathbf{k},\sigma}C_{\mathbf{k},\sigma}\right\rangle_0 = \\ &= \frac{1}{V}\sum_{\mathbf{k},\sigma}n_{-1}\left(\mathcal{E}^{(0)}_{\mathbf{k}}\right),\end{aligned} \tag{10.161}$$

where we employed Eq. 2.251 and $n_{-1}(\mathcal{E}^{(0)}_{\mathbf{k}})$ corresponds to Fermi-Dirac statistics given in Eq. E.4, with energies $\mathcal{E}^{(0)}_{\mathbf{k}} = \frac{\hbar^2\mathbf{k}^2}{2m_e} - \mu$, and represents a zeroth-order approximation for the particle density, as expected. The quantity ρ_0 is a scalar function (not an operator), since it results from the thermal expectation value of the number operator per unit volume, and thus represents a measurable macroscopic property rather than a quantum operator. Again, our analysis is performed within the grand canonical jellium model, whose Hamiltonian is given in Eq. 2.393. When an external perturbation φ_{ext} is applied, it induces a charge density within the material. As a result, the system becomes polarized and produces its own electric field, which contributes to the total electrostatic potential φ.

Thomas and Fermi proposed an Ansatz for this perturbation in real space, considering a Hamiltonian for charged particles of the form

$$\hat{\mathcal{H}} = \frac{\hat{\mathbf{p}}^2}{2m_e} + q_e\varphi(\mathbf{r}), \tag{10.162}$$

and it is assumed that the new particle density becomes (note that the operator symbol on ρ is omitted here, as we will explain shortly)

$$\begin{aligned}\rho(\mathbf{r}) &= \frac{1}{V}\sum_{\mathbf{k},\sigma} n_{-1}\left(\mathcal{E}_{\mathbf{k}}^{(0)} + q_e\varphi(\mathbf{r})\right) = \\ &= \frac{1}{V}\sum_{\mathbf{k},\sigma} \frac{1}{e^{\beta\left(\mathcal{E}_{\mathbf{k}}^{(0)} + q_e\varphi(\mathbf{r}) - \mu\right)} + 1}.\end{aligned} \tag{10.163}$$

It is important to recognize that this approximation assumes knowledge of the particle density at every point in space and for each corresponding conjugate momentum. In a strict sense, such an assumption is inconsistent with the fundamental principles of quantum mechanics. However, since the Thomas-Fermi model is a semiclassical approach, it is appropriate to treat the particle densities as continuous functions rather than as quantum mechanical operators. Furthermore, since the system is at thermodynamic equilibrium, the chemical potential must be constant within each macroscopic volume element; otherwise, there would be charge flows from one point to another. Since we are interested in the variation of particle density, we expand 10.163 around $\varphi = 0$ using a first-order Taylor polynomial, that is

$$\rho(\mathbf{r}) = \rho_0 + \frac{1}{V}\sum_{\mathbf{k},\sigma} \frac{\partial n_{-1}\left(\mathcal{E}_{\mathbf{k}}^{(0)}, \varphi(\mathbf{r})\right)}{\partial \mathcal{E}_{\mathbf{k}}^{(0)}} q_e\varphi(\mathbf{r}), \tag{10.164}$$

then

$$\delta\langle\rho\rangle(\mathbf{r}) = \frac{1}{V}\sum_{\mathbf{k},\sigma} \frac{n_{-1}\left(\mathcal{E}_{\mathbf{k}}^{(0)}, \varphi(\mathbf{r})\right)}{\partial \mathcal{E}_{\mathbf{k}}^{(0)}} q_e\varphi(\mathbf{r}). \tag{10.165}$$

We apply a Fourier transform in momentum space and obtain

$$\begin{aligned}\delta\langle\rho\rangle(\mathbf{q}) &= \frac{1}{V}\sum_{\mathbf{k},\sigma} \frac{\partial n_{-1}\left(\mathcal{E}_{\mathbf{k}}^{(0)}, \varphi(\mathbf{r})\right)}{\partial \mathcal{E}_{\mathbf{k}}^{(0)}} q_e\varphi_{\mathbf{q}} = \\ &= \frac{1}{V}\sum_{\mathbf{k},\sigma} \frac{\partial n_{-1}\left(\mathcal{E}_{\mathbf{k}}^{(0)}, \varphi(\mathbf{r})\right)}{\partial \mathcal{E}_{\mathbf{k}}^{(0)}} q_e \frac{\varphi_{\mathrm{ext},\mathbf{q}}}{\epsilon_{\mathbf{q}}(0)},\end{aligned} \tag{10.166}$$

where we have substituted Eq. 10.60 in the case where the external potential does not depend on frequency and the dielectric function is considered only for $\omega = 0$, that is, in the static case we are addressing. Equating 10.160 and 10.166, we have

$$\frac{q^2}{4\pi q_e}\varphi_{\text{ext},\mathbf{q}}\left(\frac{1}{\epsilon_{\mathbf{q}}(0)}-1\right)=\frac{1}{V}\sum_{\mathbf{k},\sigma}\frac{\partial n_{-1}\left(\mathcal{E}_{\mathbf{k}}^{(0)},\varphi(\mathbf{r})\right)}{\partial\mathcal{E}_{\mathbf{k}}^{(0)}}q_e\frac{\varphi_{\text{ext},\mathbf{q}}}{\epsilon_{\mathbf{q}}(0)}, \tag{10.167}$$

$$\frac{q^2}{4\pi q_e}\left(\frac{1-\epsilon_{\mathbf{q}}(0)}{\epsilon_{\mathbf{q}}(0)}\right)=\frac{1}{V}\sum_{\mathbf{k},\sigma}\frac{\partial n_{-1}\left(\mathcal{E}_{\mathbf{k}}^{(0)},\varphi(\mathbf{r})\right)}{\partial\mathcal{E}_{\mathbf{k}}^{(0)}}q_e\frac{1}{\epsilon_{\mathbf{q}}(0)}, \tag{10.168}$$

$$\frac{q^2}{4\pi q_e^2}\left(1-\epsilon_{\mathbf{q}}(0)\right)=\frac{1}{V}\sum_{\mathbf{k},\sigma}\frac{\partial n_{-1}\left(\mathcal{E}_{\mathbf{k}}^{(0)},\varphi(\mathbf{r})\right)}{\partial\mathcal{E}_{\mathbf{k}}^{(0)}}, \tag{10.169}$$

$$1-\epsilon_{\mathbf{q}}(0)=\frac{4\pi q_e^2}{q^2}\frac{1}{V}\sum_{\mathbf{k},\sigma}\frac{\partial n_{-1}\left(\mathcal{E}_{\mathbf{k}}^{(0)},\varphi(\mathbf{r})\right)}{\partial\mathcal{E}_{\mathbf{k}}^{(0)}}, \tag{10.170}$$

and we get

$$\epsilon_{\mathbf{q}}(0)=1-\frac{4\pi q_e^2}{q^2}\frac{1}{V}\sum_{\mathbf{k},\sigma}\frac{\partial n_{-1}\left(\mathcal{E}_{\mathbf{k}}^{(0)},\varphi(\mathbf{r})\right)}{\partial\mathcal{E}_{\mathbf{k}}^{(0)}}. \tag{10.171}$$

Ultimately, to compute Eq. 10.171, we need to estimate the quantity

$$\frac{1}{V}\sum_{\mathbf{k},\sigma}\frac{\partial n_{-1}\left(\mathcal{E}_{\mathbf{k}}^{(0)},\varphi(\mathbf{r})\right)}{\partial\mathcal{E}_{\mathbf{k}}^{(0)}}. \tag{10.172}$$

First of all, note that the energies do not depend on spin, so the sum over σ yields the spin degeneracy $g=2$,

$$\frac{2}{V}\sum_{\mathbf{k}}\frac{\partial n_{-1}(\mathbf{k})}{\partial\mathcal{E}_{\mathbf{k}}^{(0)}}. \tag{10.173}$$

We will compute this quantity in the limits of low and high temperatures relative to the Fermi temperature

$$T_F=\frac{\mathcal{E}_F}{k_B}, \tag{10.174}$$

where $\mathcal{E}_F$ is the Fermi energy and k_B denotes the Boltzmann constant. At low temperatures, the Fermi-Dirac distribution can be approximated by a Heaviside theta function $\Theta(\mathbf{k}_F - \mathbf{k}) = \Theta(\mathcal{E}_{\mathbf{k}_F} - \mathcal{E}_{\mathbf{k}}^{(0)})$, which equals 1 below the Fermi level and 0 otherwise, and its derivative with respect to energy is $-\delta\left(\mathcal{E}_{\mathbf{k}}^{(0)}\right)$. In the high-temperature limit, the Fermi function is replaced by the Maxwell-Boltzmann distribution $e^{-\beta\mathcal{E}_{\mathbf{k}}^{(0)}}$ (see Eq. 1.355), whose derivative with respect to energy is $-\beta e^{-\beta\mathcal{E}_{\mathbf{k}}^{(0)}}$. We have

$$\frac{2}{V}\sum_{\mathbf{k}}\frac{\partial n_{-1}\left(\mathcal{E}_{\mathbf{k}}^{(0)},\varphi(\mathbf{r})\right)}{\partial\mathcal{E}_{\mathbf{k}}^{(0)}} = -\frac{2}{V}\sum_{\mathbf{k}}\begin{cases}\delta\left(\mathcal{E}_{\mathbf{k}}^{(0)}\right), & T \ll T_F\\ \beta e^{-\beta\mathcal{E}_{\mathbf{k}}^{(0)}}, & T \gg T_F\end{cases}. \tag{10.175}$$

For $T \ll T_F$, it follows

$$\begin{aligned}-\frac{2}{V}\sum_{\mathbf{k}}\delta\left(\mathcal{E}_{\mathbf{k}}^{(0)}\right) &\simeq 2\int\frac{d^3\mathbf{k}}{(2\pi)^3}\delta\left(\mathcal{E}_{\mathbf{k}}^{(0)}\right) =\\ &= 2\int_{-\infty}^{+\infty}\frac{4\pi k^2 dk}{(2\pi)^3}\delta\left[\frac{\hbar^2k^2}{2m_e}-\mathcal{E}_F\right],\end{aligned} \tag{10.176}$$

where we have used the thermodynamic limit approximation C.11 and spherical polar coordinates. We observe that within the integral appears a composite function of the momentum, the function

$$f(k) = \frac{\hbar^2k^2}{2m_e} - \mathcal{E}_F, \tag{10.177}$$

which has zeros at

$$k_F = \pm\sqrt{\frac{2m_e\mathcal{E}_F}{\hbar^2}}, \tag{10.178}$$

and whose first derivative is

$$f'(k) = \frac{\hbar^2k}{m_e}. \tag{10.179}$$

By the property of the delta function of a composite argument, that is,

$$\begin{aligned}\delta(k) &= \frac{1}{|f'(k_F)|}\delta(k-k_F) + \frac{1}{|f'(-k_F)|}\delta(k+k_F) =\\ &= \frac{m_e}{\hbar^2k_F}\left(\delta(k-k_F)+\delta(k+k_F)\right),\end{aligned} \tag{10.180}$$

we have

$$
\begin{aligned}
-\frac{2}{V}\sum_{\mathbf{k}}\delta\left(\mathcal{E}_{\mathbf{k}}^{(0)}\right) &= 2\int_{-\infty}^{+\infty}\frac{4\pi k^2\,dk}{(2\pi)^3}\frac{m_e}{\hbar^2 k_F}\left(\delta(k-k_F)+\delta(k+k_F)\right) = \\
&= \frac{m_e}{\pi^2\hbar^2 k_F}\left(\int k^2\delta(k-k_F)\,dk + \int k^2\delta(k+k_F)dk\right) = \\
&= \frac{2m_e k_F^2}{\pi^2\hbar^2 k_F} = \\
&= \frac{2m_e k_F}{\pi^2\hbar^2}.
\end{aligned}
\tag{10.181}
$$

Now, since we are considering low temperatures, a first-order approximation for the particle density variation, and working in models where electronic interactions are neglected, such as the jellium model, we can use Eq. O.8 and proceed as follows

$$
\begin{aligned}
-\frac{2}{V}\sum_{\mathbf{k}}\delta\left(\mathcal{E}_{\mathbf{k}}^{(0)}\right) &= \frac{2m_e k_F}{\pi^2\hbar^2}\frac{3k_F^2}{3k_F^2} = \\
&= 3\frac{2m_e}{\hbar^2 k_F^2}\frac{k_F^2}{3\pi^2} = \\
&= \frac{3\rho_0}{2\mathcal{E}_F}.
\end{aligned}
\tag{10.182}
$$

For $T \gg T_F$, the Maxwell-Boltzmann distribution 1.355, including the spin degeneracy $g = 2$, yields the number of particles N, thus

$$
-\frac{2}{V}\sum_{\mathbf{k}}\beta e^{-\beta\mathcal{E}_{\mathbf{k}}^{(0)}} = -\beta\rho_0. \tag{10.183}
$$

The dielectric function in the Thomas-Fermi model satisfies

$$
\epsilon_{\mathbf{q}}(0) = 1 + \frac{6\pi q_e^2}{q^2}\frac{\rho_0}{\mathcal{E}_F},\ T \ll T_F, \tag{10.184}
$$

$$
\epsilon_{\mathbf{q}}(0) = 1 + \frac{4\pi q_e^2}{q^2}\beta\rho_0,\ T \gg T_F, \tag{10.185}
$$

or equivalently we can write in compact form as follows

$$
\epsilon_{\mathbf{q}}(0)\big|_{\mathrm{TF}} = 1 + \frac{q_{\mathrm{TF}}^2}{q^2}, \tag{10.186}
$$

with

$$q_{\mathrm{TF}}^2 = \begin{cases} \dfrac{6\pi q_e^2 \rho_0}{\mathcal{E}_F}, & T \ll T_F \\ 4\pi q_e^2 \beta \rho_0, & T \gg T_F \end{cases} . \tag{10.187}$$

Consider as the external perturbation the potential produced by a charged particle Q placed at a distance from the material medium, that is,

$$\varphi_{\mathrm{ext}}(\mathbf{r}) = \frac{Q}{\mathbf{r}}, \tag{10.188}$$

$$\varphi_{\mathrm{ext},\mathbf{q}} = \frac{4\pi Q}{q^2}, \tag{10.189}$$

which is a stationary perturbation that induces a redistribution of charges and consequently it implies an induced total potential $\varphi_{\mathbf{q}}$ that we want to determine. From Eqs. 10.60 and 10.186, we have

$$\begin{aligned} \varphi_{\mathbf{q}}(\omega) &= \frac{\varphi_{\mathrm{ext},\mathbf{q}}(\omega)}{\epsilon_{\mathbf{q}}(\omega)} = \\ &= \frac{4\pi Q}{q^2} \frac{1}{1 + \frac{q_{\mathrm{TF}}^2}{q^2}} = \\ &= \frac{4\pi Q}{q^2 + q_{\mathrm{TF}}^2}. \end{aligned} \tag{10.190}$$

We note that from 2.378, the total potential is a Yukawa potential,

$$\varphi_{\mathrm{ext}}(r) = \frac{Q}{r} e^{-q_{\mathrm{TF}} r}, \tag{10.191}$$

that is, the Thomas-Fermi approximation transforms a long-range external perturbation (Coulomb potential) into a short-range interaction (Yukawa potential).

10.2.6 Random Phase Approximation

Unlike the Hartree-Fock approximation, which performs an expansion in terms of $\left[\epsilon_{\mathbf{q}}(\omega)\right]^{-1}$, the Random Phase Approximation (R.P.A.) method performs an expansion in terms of $\epsilon_{\mathbf{q}}(\omega)$. In particular, given 10.69, the random phase approximation consists in replacing the polarization propagator with the lowest order, that is

$$\epsilon_{\mathbf{q}}(\omega)\Big|_{\mathrm{RPA}} = 1 - \frac{1}{V} \frac{4\pi q_e^2}{q^2} \Pi_{\mathbf{q}}^{(0)}(\omega), \tag{10.192}$$

where $\Pi^{(0)}_{\mathbf{q}}(\omega)$ is the polarization bubble and is given by 9.87, which we write as

$$\Pi^{(0)}_{\mathbf{q}}(\omega + i\delta) = \frac{1}{\hbar}\sum_{\mathbf{k},\sigma} \frac{n_{-1}(\mathcal{E}_{\mathbf{k}+\mathbf{q}}) - n_{-1}(\mathcal{E}_{\mathbf{k}})}{\omega + i\delta + \frac{\mathcal{E}_{\mathbf{k}+\mathbf{q}} - \mathcal{E}_{\mathbf{k}}}{\hbar}}, \tag{10.193}$$

that is, we have introduced a convergence factor $\delta > 0$. Again, our discussion is conducted within the framework of the grand canonical jellium model, as given in Eq. 2.393, then, the energies are of the form $\mathcal{E}_{\mathbf{k}} = \frac{\hbar^2\mathbf{k}^2}{2m_e} - \mu$. Moreover, we can eliminate the sum over spin σ and introduce a spin degeneracy $g = 2$ as follows

$$\Pi^{(0)}_{\mathbf{q}}(\omega + i\delta) = \frac{2}{\hbar}\sum_{\mathbf{k}} \frac{n_{-1}(\mathcal{E}_{\mathbf{k}+\mathbf{q}}) - n_{-1}(\mathcal{E}_{\mathbf{k}})}{\omega + i\delta + \frac{\mathcal{E}_{\mathbf{k}+\mathbf{q}} - \mathcal{E}_{\mathbf{k}}}{\hbar}}. \tag{10.194}$$

Here we are interested in the static case. Since the spectral function is antisymmetric with respect to ω, see Eq. 10.99, the imaginary part of the polarization propagator is also antisymmetric, and therefore it vanishes at $\omega = 0$, that is

$$\mathrm{Im}\, \Pi^{(0)}_{\mathbf{q}}(\omega = 0) = 0. \tag{10.195}$$

The polarization propagator for $\omega = 0$ has only a real part, and in the thermodynamic limit the sum is replaced by a principal value integral as follows

$$\mathrm{Re}\, \Pi^{(0)}_{\mathbf{q}}(\omega = 0) = 2V\, \mathrm{P.\,V.} \int \frac{d^3\mathbf{k}}{(2\pi)^3} \frac{n_{-1}(\mathcal{E}_{\mathbf{k}+\mathbf{q}}) - n_{-1}(\mathcal{E}_{\mathbf{k}})}{\mathcal{E}_{\mathbf{k}+\mathbf{q}} - \mathcal{E}_{\mathbf{k}}}, \tag{10.196}$$

from which the R.P.A. approximation of dielectric function is

$$\epsilon_{\mathbf{q}}(0)\big|_{\mathrm{RPA}} = 1 - \frac{4\pi q_e^2}{q^2} 2\, \mathrm{P.\,V.} \int \frac{d^3\mathbf{k}}{(2\pi)^3} \frac{n_{-1}(\mathcal{E}_{\mathbf{k}+\mathbf{q}}) - n_{-1}(\mathcal{E}_{\mathbf{k}})}{\mathcal{E}_{\mathbf{k}+\mathbf{q}} - \mathcal{E}_{\mathbf{k}}}. \tag{10.197}$$

Note that in the limit $\mathbf{q} \to \mathbf{0}$, the integrand function becomes $\frac{\partial n_{-1}(\mathcal{E}_{\mathbf{k}})}{\partial \mathcal{E}_{\mathbf{k}}}$, that is, in the limit of infinitesimal momenta, the R.P.A. dielectric function reduces to the Thomas-Fermi result. We then need to quantify this function for finite $\mathbf{q}$. A famous approximation in the scientific literature at zero temperature is the following. Let us first rewrite the first term of the integral as

$$I_1(\mathbf{q}) = \mathrm{P.\,V.} \int \frac{d^3\mathbf{k}}{(2\pi)^3} \frac{n_{-1}(\mathcal{E}_{\mathbf{k}+\mathbf{q}})}{\mathcal{E}_{\mathbf{k}+\mathbf{q}} - \mathcal{E}_{\mathbf{k}}}. \tag{10.198}$$

By performing the change of variable $\mathbf{k}' = \mathbf{k} + \mathbf{q}$, one obtains

$$I_1(\mathbf{q}) = \mathrm{P.\,V.} \int \frac{d^3\mathbf{k}}{(2\pi)^3} \frac{n_{-1}(\mathcal{E}_{\mathbf{k}})}{\mathcal{E}_{\mathbf{k}} - \mathcal{E}_{\mathbf{k}-\mathbf{q}}}. \tag{10.199}$$

Therefore, the total integral can be expressed in the symmetric form

$$I(\mathbf{q}) = \text{P. V.} \int \frac{d^3\mathbf{k}}{(2\pi)^3} n_{-1}(\mathcal{E}_\mathbf{k}) \left[\frac{1}{\mathcal{E}_\mathbf{k} - \mathcal{E}_{\mathbf{k}-\mathbf{q}}} - \frac{1}{\mathcal{E}_{\mathbf{k}+\mathbf{q}} - \mathcal{E}_\mathbf{k}} \right]. \tag{10.200}$$

Combining the two terms inside the brackets gives

$$\frac{1}{\mathcal{E}_\mathbf{k} - \mathcal{E}_{\mathbf{k}-\mathbf{q}}} - \frac{1}{\mathcal{E}_{\mathbf{k}+\mathbf{q}} - \mathcal{E}_\mathbf{k}} = \frac{\mathcal{E}_{\mathbf{k}+\mathbf{q}} + \mathcal{E}_{\mathbf{k}-\mathbf{q}} - 2\mathcal{E}_\mathbf{k}}{\left(\mathcal{E}_\mathbf{k} - \mathcal{E}_{\mathbf{k}-\mathbf{q}}\right)\left(\mathcal{E}_{\mathbf{k}+\mathbf{q}} - \mathcal{E}_\mathbf{k}\right)}. \tag{10.201}$$

Thus, the integral can be rewritten as

$$I(\mathbf{q}) = \text{P. V.} \int \frac{d^3\mathbf{k}}{(2\pi)^3} n_{-1}(\mathcal{E}_\mathbf{k}) \frac{\mathcal{E}_{\mathbf{k}+\mathbf{q}} + \mathcal{E}_{\mathbf{k}-\mathbf{q}} - 2\mathcal{E}_\mathbf{k}}{\left(\mathcal{E}_\mathbf{k} - \mathcal{E}_{\mathbf{k}-\mathbf{q}}\right)\left(\mathcal{E}_{\mathbf{k}+\mathbf{q}} - \mathcal{E}_\mathbf{k}\right)}. \tag{10.202}$$

If the denominators are assumed to have similar magnitudes and opposite signs, namely

$$\mathcal{E}_\mathbf{k} - \mathcal{E}_{\mathbf{k}-\mathbf{q}} \simeq -\left(\mathcal{E}_{\mathbf{k}+\mathbf{q}} - \mathcal{E}_\mathbf{k}\right), \tag{10.203}$$

then the product of the denominators can be approximated as

$$\left(\mathcal{E}_\mathbf{k} - \mathcal{E}_{\mathbf{k}-\mathbf{q}}\right)\left(\mathcal{E}_{\mathbf{k}+\mathbf{q}} - \mathcal{E}_\mathbf{k}\right) \simeq -\left(\mathcal{E}_{\mathbf{k}+\mathbf{q}} - \mathcal{E}_\mathbf{k}\right)^2. \tag{10.204}$$

Consequently, the fraction simplifies to

$$\frac{\mathcal{E}_{\mathbf{k}+\mathbf{q}} + \mathcal{E}_{\mathbf{k}-\mathbf{q}} - 2\mathcal{E}_\mathbf{k}}{\left(\mathcal{E}_\mathbf{k} - \mathcal{E}_{\mathbf{k}-\mathbf{q}}\right)\left(\mathcal{E}_{\mathbf{k}+\mathbf{q}} - \mathcal{E}_\mathbf{k}\right)} \simeq -\frac{2}{\mathcal{E}_{\mathbf{k}+\mathbf{q}} - \mathcal{E}_\mathbf{k}}. \tag{10.205}$$

Substituting this result into $I(\mathbf{q})$ finally gives

$$I(\mathbf{q}) \simeq -2\,\text{P. V.} \int \frac{d^3\mathbf{k}}{(2\pi)^3} \frac{n_{-1}(\mathcal{E}_\mathbf{k})}{\mathcal{E}_{\mathbf{k}+\mathbf{q}} - \mathcal{E}_\mathbf{k}}, \tag{10.206}$$

which, in compact form, corresponds to the approximation

$$\left[n_{-1}(\mathcal{E}_{\mathbf{k}+\mathbf{q}}) - n_{-1}(\mathcal{E}_\mathbf{k})\right] \simeq -2\,n_{-1}(\mathcal{E}_\mathbf{k}), \tag{10.207}$$

which implies

$$\epsilon_\mathbf{q}(\omega)\big|_{\text{RPA}, T=0} = 1 + \frac{4\pi q_e^2}{q^2} 2\,\text{P. V.} \int \frac{d^3\mathbf{k}}{(2\pi)^3} \frac{2n_{-1}(\mathcal{E}_\mathbf{k})}{\mathcal{E}_{\mathbf{k}+\mathbf{q}} - \mathcal{E}_\mathbf{k}}, \tag{10.208}$$

where the Fermi-Dirac distribution is replaced by a Heaviside function $\Theta(\mathbf{k}_F - \mathbf{k})$. We have

$$\begin{aligned}
\epsilon_{\mathbf{q}}(\omega)\big|_{\text{RPA},T=0} &= 1 - \frac{4\pi q_e^2}{q^2} 4 \int_0^{k_F} \frac{2\pi k^2 dk}{(2\pi)^3} \text{P.V.} \\
&\int_{-1}^{+1} d\cos(\theta) \frac{1}{\dfrac{\hbar^2 q^2}{2m_e} + \dfrac{2\hbar^2 kq\cos(\theta)}{2m_e}} = \\
&= 1 + \frac{4\pi q_e^2}{q^2} \frac{4}{(2\pi)^2} \frac{2m_e}{\hbar^2} \int_0^{k_F} \frac{2\pi k^2 dk}{(2\pi)^3} \text{P.V.} \\
&\int_{-1}^{+1} d\cos(\theta) \frac{1}{q^2 + 2kq\cos(\theta)}.
\end{aligned} \tag{10.209}$$

By evaluating the well-known integral, we find

$$\epsilon_{\mathbf{q}}(\omega)\big|_{\text{RPA},T=0} = 1 + \frac{q_{\text{TF}}^2(T=0)}{2q^2} S(y), \tag{10.210}$$

with

$$y = \frac{q}{2k_F}, \tag{10.211}$$

$$S(y) = 1 + \frac{1-y^2}{2y} \ln\left|\frac{1+y}{1-y}\right|, \tag{10.212}$$

which is the opposite of function given in Eq. 8.73. Consequently, from Eqs. 8.75, 8.76 and 8.77, Eq. 10.212 satisfies

$$\lim_{y\to 0^+} S(y) = 2, \tag{10.213}$$

$$\lim_{y\to +\infty} S(y) = 0, \tag{10.214}$$

$$\lim_{y\to 1} S(y) = 1, \tag{10.215}$$

as shown in Fig. 10.3. In particular, for $q \to 0^+$, i.e., $y \to 0^+$, from Eqs. 10.213 and 10.210, it follows that

$$\begin{aligned}
\epsilon_{\mathbf{q}}(0)\big|_{\text{RPA},T=0,\ q\to 0^+} &= 1 + \frac{q_{\text{TF}}^2(T=0)}{q^2} = \\
&= \epsilon_{\mathbf{q}}(0)\big|_{\text{TF}}\,.
\end{aligned} \tag{10.216}$$

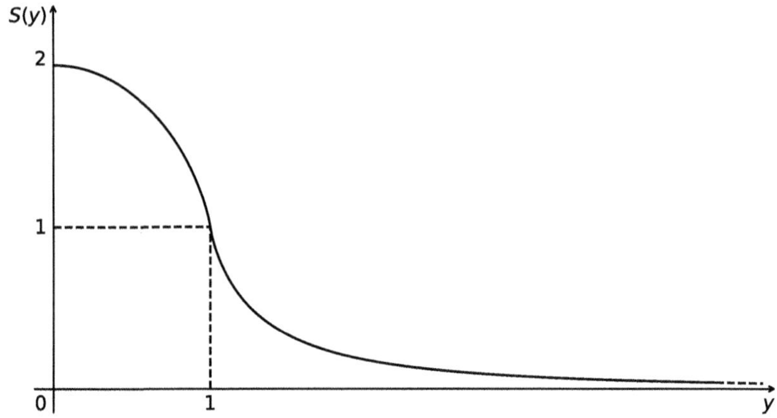

Fig. 10.3 The function $S(y)$ defined in Eq. 10.212 appears in the R.P.A. approximation of the dielectric function for $\omega = 0$ and $T = 0$. It corresponds to the negative of the function defined in Eq. 8.73, shown in Fig. 8.1

Now, let us consider an external perturbation produced by a charged particle at a distance. Within the R.P.A. approximation, the problem reduces to determining the induced total potential. By applying the residue Theorem K.1, one finds that this potential in real space takes the form

$$\varphi(r) = \frac{\cos(2k_F r)}{r^3}, \tag{10.217}$$

which is the inverse Fourier transform of

$$\varphi(q) = \frac{4\pi Q}{q^2}\left(1 + \frac{q_{\mathrm{TF}}^2(T=0)}{2q^2} S(y)\right)^{-1}. \tag{10.218}$$

Note that, unlike the Thomas-Fermi approximation where the potential decays exponentially, here the induced potential decays following a power law and includes oscillations, known as Friedel oscillations. Physically, these oscillations reflect the tendency of electrons to arrange themselves in oscillating densities around the perturbing charge, alternating regions of excess and deficit of electronic density. They are connected to the fact that scattering processes near the Fermi surface involve characteristic momenta of the order of $2k_F$. Friedel oscillations are therefore a direct fingerprint of the existence of a well-defined Fermi surface and play an important role in several solid-state phenomena. A more detailed analysis of these oscillations goes beyond the scope of the present discussion. Let us now consider arbitrary frequencies, i.e., finite frequencies. The imaginary part of Eq. 10.197 is equal to the imaginary part of the integral, which can be easily shown to be antisymmetric with respect to the frequency ω. Furthermore, this imaginary part

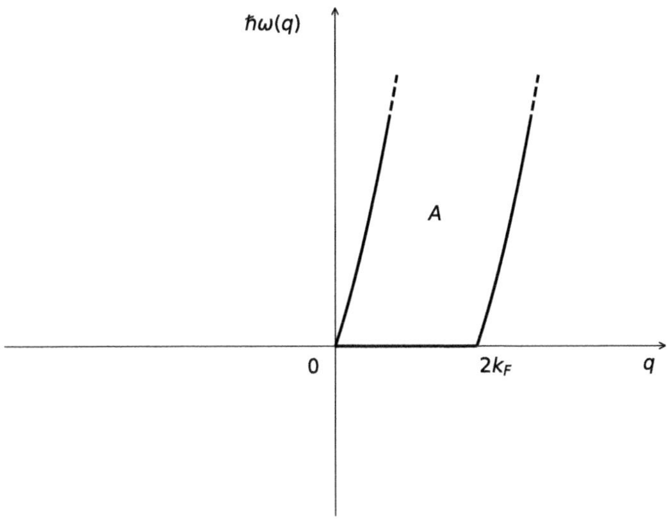

Fig. 10.4 The hole-particle processes in the Hartree-Fock approximation for the dielectric function occur for momentum and energy values belonging to the A region, that is, in the region of the plane enclosed by the segment and the two arcs of the parabola (bold curve)

is responsible for the hole-particle processes, which are allowed for energies within the region A of Fig. 10.4, as seen in the Hartree-Fock approximation. It follows that the imaginary part of the dielectric function in the R.P.A. model must be zero for energies within the region $\mathbb{R}^2 \setminus A$ of Fig. 10.4. Thus, we focus on energies belonging to this region, where the dielectric function has only a real part. Since hole-particle excitations are not allowed, we are investigating whether the R.P.A. approximation generates new phenomena. We write 10.197 for arbitrary ω, that is, from 10.194 and 10.192, we have

$$\epsilon_{\mathbf{q}}(\omega + i\delta)\big|_{\text{RPA}} = 1 - \frac{4\pi q_e^2}{q^2}\frac{2}{\hbar}\,\text{P.V.}\int \frac{d^3\mathbf{k}}{(2\pi)^3}\frac{n_{-1}(\mathcal{E}_{\mathbf{k}+\mathbf{q}}) - n_{-1}(\mathcal{E}_{\mathbf{k}})}{\omega + i\delta + \frac{\mathcal{E}_{\mathbf{k}+\mathbf{q}} - \mathcal{E}_{\mathbf{k}}}{\hbar}}. \tag{10.219}$$

and we manipulate as follows

$$\epsilon_{\mathbf{q}}(\omega + i\delta)\big|_{\text{RPA}} = 1 - \frac{4\pi q_e^2}{q^2}\frac{2}{\hbar}\left[\text{P.V.}\int \frac{d^3\mathbf{k}}{(2\pi)^3}\frac{n_{-1}(\mathcal{E}_{\mathbf{k}+\mathbf{q}})}{\omega + i\delta + \frac{\mathcal{E}_{\mathbf{k}+\mathbf{q}} - \mathcal{E}_{\mathbf{k}}}{\hbar}}\right.$$
$$\left. - \text{P.V.}\int \frac{d^3\mathbf{k}}{(2\pi)^3}\frac{n_{-1}(\mathcal{E}_{\mathbf{k}})}{\omega + i\delta + \frac{\mathcal{E}_{\mathbf{k}+\mathbf{q}} - \mathcal{E}_{\mathbf{k}}}{\hbar}}\right] =$$

$$= 1 - \frac{4\pi q_e^2}{q^2}\frac{2}{\hbar}\left[\text{P. V.}\int \frac{d^3\mathbf{k}'}{(2\pi)^3}\frac{n_{-1}(\mathcal{E}_{\mathbf{k}'})}{\omega + i\delta + \frac{\mathcal{E}_{\mathbf{k}'}-\mathcal{E}_{\mathbf{k}'-\mathbf{q}}}{\hbar}}\right.$$

$$\left. - \text{P. V.}\int \frac{d^3\mathbf{k}}{(2\pi)^3}\frac{n_{-1}(\mathcal{E}_{\mathbf{k}})}{\omega + i\delta + \frac{\mathcal{E}_{\mathbf{k}+\mathbf{q}}-\mathcal{E}_{\mathbf{k}}}{\hbar}}\right], \tag{10.220}$$

from which we rename the dummy variable and factorize as follows

$$\epsilon_{\mathbf{q}}(\omega + i\delta)\big|_{\text{RPA}} = 1 - \frac{4\pi q_e^2}{q^2}\frac{2}{\hbar}\text{P. V.}\int \frac{d^3\mathbf{k}}{(2\pi)^3}n_{-1}(\mathcal{E}_{\mathbf{k}})$$

$$\left[\frac{1}{\omega + i\delta + \frac{\mathcal{E}_{\mathbf{k}}-\mathcal{E}_{\mathbf{k}-\mathbf{q}}}{\hbar}} - \frac{1}{\omega + i\delta + \frac{\mathcal{E}_{\mathbf{k}+\mathbf{q}}-\mathcal{E}_{\mathbf{k}}}{\hbar}}\right] =$$

$$= 1 - \frac{4\pi q_e^2}{q^2}\frac{2}{\hbar\omega}\text{P. V.}\int \frac{d^3\mathbf{k}}{(2\pi)^3}n_{-1}(\mathcal{E}_{\mathbf{k}})$$

$$\left[\frac{1}{1 + \frac{i\delta}{\omega} + \frac{\mathcal{E}_{\mathbf{k}}-\mathcal{E}_{\mathbf{k}-\mathbf{q}}}{\hbar\omega}} - \frac{1}{1 + \frac{i\delta}{\omega} + \frac{\mathcal{E}_{\mathbf{k}+\mathbf{q}}-\mathcal{E}_{\mathbf{k}}}{\hbar\omega}}\right]. \tag{10.221}$$

Now, consider the limit of small $\mathbf{q}$. If $\mathbf{q}$ is small, we can expand the energy differences

$$\mathcal{E}_{\mathbf{k}+\mathbf{q}} - \mathcal{E}_{\mathbf{k}} = \frac{\hbar^2}{m_e}\mathbf{k}\cdot\mathbf{q} + \frac{\hbar^2 q^2}{2m_e}, \tag{10.222}$$

$$\mathcal{E}_{\mathbf{k}} - \mathcal{E}_{\mathbf{k}-\mathbf{q}} = \frac{\hbar^2}{m_e}\mathbf{k}\cdot\mathbf{q} - \frac{\hbar^2 q^2}{2m_e}, \tag{10.223}$$

to first order in $\mathbf{q}$. We introduce the dimensionless variables

$$X = \frac{\hbar}{m_e}\frac{\mathbf{k}\cdot\mathbf{q}}{\omega}, \tag{10.224}$$

$$Y = \frac{\hbar q^2}{2m_e\omega}. \tag{10.225}$$

With these notations, the denominators in the integral become

$$1 + \frac{i\delta}{\omega} + \frac{\mathcal{E}_{\mathbf{k}} - \mathcal{E}_{\mathbf{k}-\mathbf{q}}}{\hbar\omega} = 1 + \frac{i\delta}{\omega} + X - Y, \tag{10.226}$$

$$1 + \frac{i\delta}{\omega} + \frac{\mathcal{E}_{\mathbf{k}+\mathbf{q}} - \mathcal{E}_{\mathbf{k}}}{\hbar\omega} = 1 + \frac{i\delta}{\omega} + X + Y. \tag{10.227}$$

Then, expanding to first order in Y,

$$\frac{1}{1 + X - Y + \frac{i\delta}{\omega}} - \frac{1}{1 + X + Y + \frac{i\delta}{\omega}} \simeq \frac{2Y}{(1+X)^2}. \tag{10.228}$$

Substituting this approximation into Eq. 10.221, and assuming $T = 0$, so that the Fermi-Dirac distributions can be replaced by step functions, then

$$\epsilon_{\mathbf{q}}(\omega) \simeq 1 - \frac{4\pi q_e^2}{m_e\omega^2} g \int \frac{d^3\mathbf{k}}{(2\pi)^3} \Theta\left(\mathcal{E}_F - \mathcal{E}_{\mathbf{k}}\right) \frac{1}{(1+X)^2}, \tag{10.229}$$

where g denotes the spin degeneracy factor ($g = 2$ for electrons), which multiplies the integral because up to this point we have considered a single spin state. For very small q, the denominator can be expanded as follows

$$\frac{1}{(1+X)^2} \simeq 1 - 2X + 3X^2. \tag{10.230}$$

The linear term in X vanishes due to the symmetry of the integral, then

$$g \int \frac{d^3\mathbf{k}}{(2\pi)^3} \Theta\left(\mathcal{E}_F - \mathcal{E}_{\mathbf{k}}\right) \frac{1}{(1+X)^2} \simeq g \int \frac{d^3\mathbf{k}}{(2\pi)^3} \Theta\left(\mathcal{E}_F - \mathcal{E}_{\mathbf{k}}\right) \left(1 + 3X^2\right). \tag{10.231}$$

The first term yields the particle density

$$\begin{aligned} g \int \frac{d^3\mathbf{k}}{(2\pi)^3} \Theta(\mathcal{E}_F - \mathcal{E}_{\mathbf{k}}) &= g \frac{1}{(2\pi)^3} \int_{|\mathbf{k}| \le k_F} d^3\mathbf{k} = \\ &= g \frac{1}{(2\pi)^3} \int_0^{2\pi} d\phi \int_0^{\pi} d\theta \, \sin\theta \int_0^{k_F} dk \, k^2 = \\ &= g \frac{4\pi}{(2\pi)^3} \frac{k_F^3}{3} = \\ &= \rho_0, \end{aligned} \tag{10.232}$$

where Eq. O.8 with $g = 2$ has been used. For the X^2 term, we obtain

$$3g \int \frac{d^3\mathbf{k}}{(2\pi)^3} \Theta\left(\mathcal{E}_F - \mathcal{E}_{\mathbf{k}}\right) X^2 = 3g \left(\frac{\hbar q}{m_e\omega}\right)^2 \int \frac{d^3\mathbf{k}}{(2\pi)^3} \Theta\left(\mathcal{E}_F - \mathcal{E}_{\mathbf{k}}\right) k^2. \tag{10.233}$$

Regarding the integral involving k^2, we have

$$\int \frac{d^3\mathbf{k}}{(2\pi)^3}\,\Theta(\mathcal{E}_F-\mathcal{E}_{\mathbf{k}})\,k^2 = \frac{1}{(2\pi)^3}\int_{|\mathbf{k}|\le k_F} d^3\mathbf{k}\,k^2 =$$
$$= \frac{1}{(2\pi)^3}\int_0^{2\pi} d\phi \int_0^{\pi} d\theta\,\sin\theta \int_0^{k_F} dk\,k^4 =$$
$$= \frac{1}{2\pi^2}\frac{k_F^5}{5}, \tag{10.234}$$

and again from Eq. O.8, we can write

$$\frac{1}{\rho_0}3g\int \frac{d^3\mathbf{k}}{(2\pi)^3}\Theta\left(\mathcal{E}_F-\mathcal{E}_{\mathbf{k}}\right)X^2 = 3g\left(\frac{\hbar q}{m_e\omega}\right)^2\frac{1}{10}k_F^2 =$$
$$= \frac{9}{5}\frac{\hbar^2 q^2 k_F^2}{m_e^2\omega^2}. \tag{10.235}$$

Substituting the results above into Eq. 10.221, we finally obtain

$$\epsilon_{\mathbf{q}}(\omega)\big|_{\mathrm{RPA},T=0,\ \omega\neq 0} = 1-\frac{4\pi q_e^2\rho_0}{m_e\omega^2}\left(1+\frac{9}{5}\frac{\hbar^2}{m_e^2}\frac{q^2k_F^2}{\omega^2}\right). \tag{10.236}$$

If we set this dielectric function equal to zero, we obtain

$$\frac{4\pi q_e^2\rho_0}{m_e\omega^2}\left(1+\frac{9}{5}\frac{\hbar^2}{m_e^2}\frac{q^2k_F^2}{\omega^2}\right) = 1. \tag{10.237}$$

We set

$$A = \frac{4\pi q_e^2\rho_0}{m_e}, \tag{10.238}$$

$$B = \frac{9}{5}\frac{\hbar^2}{m_e^2}q^2k_F^2, \tag{10.239}$$

then

$$\frac{A}{\omega^2}\left(1+\frac{B}{\omega^2}\right) = 1, \tag{10.240}$$

$$\frac{A}{\omega^2}+\frac{AB}{\omega^4} = 1, \tag{10.241}$$

$$A\omega^2 + AB = \omega^4, \tag{10.242}$$

$$\omega^4 - A\omega^2 - AB = 0. \tag{10.243}$$

Whose solution is given by

$$\omega^2 = \frac{A + \sqrt{A^2 + 4AB}}{2}, \tag{10.244}$$

then

$$\begin{aligned}
\omega &= \sqrt{\frac{A + \sqrt{A^2 + 4AB}}{2}} = \\
&= \sqrt{\frac{\frac{4\pi q_e^2 \rho_0}{m_e} + \sqrt{\left(\frac{4\pi q_e^2 \rho_0}{m_e}\right)^2 + 4\frac{4\pi q_e^2 \rho_0}{m_e}\frac{9}{5}\frac{\hbar^2}{m_e^2}q^2 k_F^2}}{2}} = \\
&= \sqrt{\frac{\omega_p^2 + \sqrt{\omega_p^4 + \frac{36}{5}\omega_p^2\frac{\hbar^2}{m_e^2}q^2 k_F^2}}{2}},
\end{aligned} \tag{10.245}$$

where we have made explicit the plasma frequency 10.96. We now compute a first-order expansion of the frequency 10.245 in q^2. We have

$$\delta = \frac{36}{5}\omega_p^2 \frac{\hbar^2}{m_e^2} q^2 k_F^2 \ll \omega_p^2, \tag{10.246}$$

$$\begin{aligned}
\sqrt{\omega_p^4 + \omega_p^2 \delta} &= \omega_p^2 \sqrt{1 + \frac{\delta}{\omega_p^2}} = \\
&\approx \omega_p^2 \left(1 + \frac{1}{2}\frac{\delta}{\omega_p^2}\right) = \\
&= \omega_p^2 + \frac{1}{2}\delta,
\end{aligned} \tag{10.247}$$

$$\begin{aligned}
\omega &\approx \sqrt{\frac{\omega_p^2 + \left(\omega_p^2 + \frac{1}{2}\delta\right)}{2}} = \\
&= \sqrt{\frac{2\omega_p^2 + \frac{1}{2}\delta}{2}} =
\end{aligned}$$

$$
\begin{aligned}
&= \sqrt{\omega_p^2 + \frac{1}{4}\delta} = \\
&\approx \omega_p \left(1 + \frac{1}{2}\frac{1}{4}\frac{\delta}{\omega_p^2}\right) = \\
&= \omega_p \left(1 + \frac{1}{8}\frac{\delta}{\omega_p^2}\right) = \\
&= \omega_p \left(1 + \frac{1}{8}\frac{36}{5}\frac{\hbar^2}{m_e^2}\frac{q^2 k_F^2}{\omega_p^2}\right) = \\
&= \omega_p \left(1 + \frac{9}{10}\frac{\hbar^2}{m_e^2 \omega_p^2} q^2 k_F^2\right).
\end{aligned} \tag{10.248}
$$

If we invert the first of Eq. 10.187 as follows

$$
\rho_0 = \frac{\mathcal{E}_F q_{\mathrm{TF}}^2 (T = 0)}{6\pi q_e^2}, \tag{10.249}
$$

and we manipulate as follows

$$
\begin{aligned}
\omega &= \omega_p \left(1 + \frac{9}{10}\frac{\hbar^2 k_F^2}{2m_e}\frac{1}{m_e \omega_p^2} q^2\right) = \\
&= \omega_p \left(1 + \frac{9}{10}\mathcal{E}_F \frac{1}{m_e}\frac{1}{\omega_p^2} q^2\right) = \\
&= \omega_p \left(1 + \frac{9}{10}\mathcal{E}_F \frac{1}{m_e}\frac{m}{4\pi q_e^2 \rho_0} q^2\right) = \\
&= \omega_p \left(1 + \frac{9}{10}\mathcal{E}_F \frac{1}{4\pi q_e^2 \rho_0} q^2\right) = \\
&= \omega_p \left(1 + \frac{9}{10}\mathcal{E}_F \frac{1}{4\pi q_e^2 \frac{\mathcal{E}_F q_{\mathrm{TF}}^2 (T=0)}{6\pi q_e^2}} q^2\right),
\end{aligned} \tag{10.250}
$$

and we finally get

$$
\omega = \omega_p \left(1 + \frac{27}{20}\frac{q^2}{q_{\mathrm{TF}}^2}\right). \tag{10.251}
$$

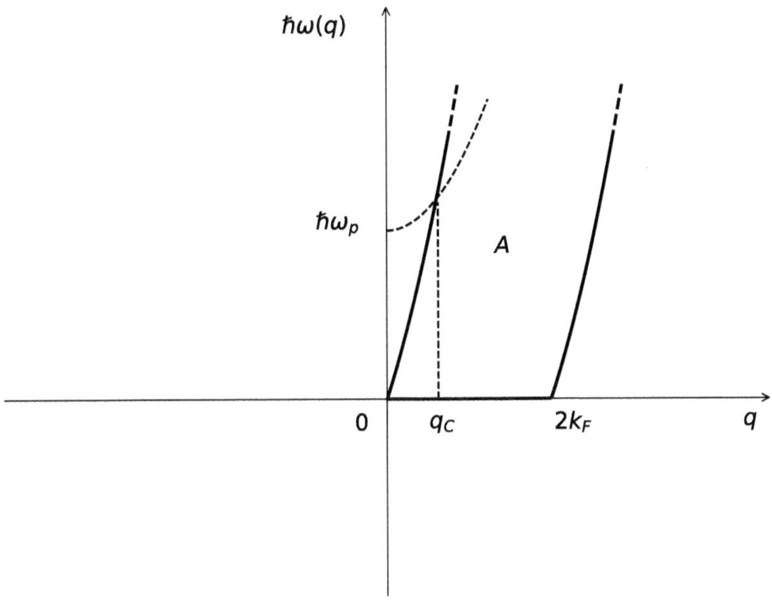

Fig. 10.5 The hole-particle processes in the Hartree-Fock approximation for the dielectric function occur for momentum and energy values belonging to the A region, while the dashed curve corresponds to Eq. 10.251, i.e., the plasmon dispersion relation

As we anticipated, the semiclassical and quantum calculations of the plasma frequency coincide, but the quantum treatment has additionally provided Eq. 10.251, which is a dispersion relation at small momenta q for quasiparticles, the plasmons, which are collective excitations of an electron gas. Equation 10.251 is shown in Fig. 10.5, together with the allowed region for hole-particle processes. In the region where the imaginary part of the dielectric function is zero, hole-particle processes do not occur, and the plasmons correspond to a dielectric function that has only a nonzero real part: from the discussion on Landau damping, we deduce that in this region the plasmons have infinite lifetime. Note that the plasmon dispersion relation intersects with the energy region of hole-particle processes; consequently, for sufficiently large values of q, starting from a critical momentum q_C, plasmons are no longer stable because they can decay into hole-particle processes.

10.3 Magnetic Susceptibility

Here we aim to study the response of a material medium to a weak magnetic field. The following Hamiltonian is written in the framework of first quantization, and describes the coupling between a single spin and a magnetic field directed along the z-axis, that is,

$$\mathcal{H}_I = -\frac{g\mu_B}{\hbar}\hat{S}_z H_z, \tag{10.252}$$

where

$$\mu_B = \frac{\hbar|q_e|}{2m_e c} \tag{10.253}$$

is the Bohr magneton, and g is a constant known as the gyromagnetic factor. For a system of N non-interacting spins, we generalize as follows

$$\mathcal{H}_I = -\frac{g\mu_B}{\hbar}\sum_{i=1}^{N}\hat{S}_{i,z} H_z. \tag{10.254}$$

In the thermodynamic limit the interaction hamiltonian in first quantization takes the form

$$\mathcal{H}_I = -\frac{g\mu_B}{\hbar}\int d^3\mathbf{r} s_z(\mathbf{r}) H_z(\mathbf{r}), \tag{10.255}$$

where $s_z(\mathbf{r})$ is a spin density, that is it has dimensions of spin per unit volume, and the discrete and continuous expressions are equivalent only if

$$s_z(\mathbf{r}) = \sum_{i=1}^{N}\delta(\mathbf{r}-\mathbf{r}_i)\hat{S}_z(\mathbf{r}), \tag{10.256}$$

and we express such a density in second quantization. Since it is a one-body operator, from Theorem 2.9, one finds

$$\hat{s}_z(\mathbf{r}) = \int dx' \hat{\psi}^\dagger(x')\delta(\mathbf{r}-\mathbf{r}')\hat{S}_z\hat{\psi}(x') = \\ = \sum_s \hat{\psi}^\dagger(\mathbf{r},s)\hat{S}_z\hat{\psi}(\mathbf{r},s). \tag{10.257}$$

Let us assume that the material medium is spatially translationally invariant. In this case, we employ Eqs. 2.163 and 2.164, and from Eq. 1.63, it follows

$$\hat{s}_z(\mathbf{r}) = \sum_s \hat{\psi}^\dagger(\mathbf{r},s)\delta(\mathbf{r}-\mathbf{r}')\hat{S}_z\hat{\psi}(\mathbf{r},s) = \\ = \frac{1}{V}\sum_s \sum_{\mathbf{k}_1,\mathbf{k}_2,\sigma_1,\sigma_2} e^{i(\mathbf{k}_2-\mathbf{k}_1)\cdot\mathbf{r}} C^\dagger_{\mathbf{k}_1,\sigma_1}\chi_{\sigma_1}(s)\hat{S}_z\chi_{\sigma_2}(s)C_{\mathbf{k}_2,\sigma_2} =$$

$$= \frac{1}{V}\sum_{s}\sum_{\mathbf{k}_1,\mathbf{k}_2,\sigma_1,\sigma_2} e^{i(\mathbf{k}_2-\mathbf{k}_1)\cdot\mathbf{r}} C^{\dagger}_{\mathbf{k}_1,\sigma_1}\chi_{\sigma_1}(s)\hbar\sigma_2\chi_{\sigma_2}(s)C_{\mathbf{k}_2,\sigma_2} =$$

$$= \frac{1}{V}\sum_{\mathbf{k}_1,\mathbf{k}_2,\sigma_1,\sigma_2} e^{i(\mathbf{k}_2-\mathbf{k}_1)\cdot\mathbf{r}}\hbar\sigma_2 C^{\dagger}_{\mathbf{k}_1,\sigma_1}\delta_{\sigma_1,\sigma_2}C_{\mathbf{k}_2,\sigma_2} =$$

$$= \frac{1}{V}\sum_{\mathbf{k}_1,\mathbf{k}_2,\sigma} e^{i(\mathbf{k}_2-\mathbf{k}_1)\cdot\mathbf{r}}\hbar\sigma C^{\dagger}_{\mathbf{k}_1,\sigma}C_{\mathbf{k}_2,\sigma}, \tag{10.258}$$

where we used the completeness relation 1.24, and then performed a sum over σ_2 renaming the dummy index $\sigma_1 \to \sigma$. Now that the spin density is expressed in second quantization, we compute its spatial Fourier transform, that is

$$\hat{s}_z(\mathbf{q}) = \int d^3\mathbf{r} e^{i\mathbf{q}\cdot\mathbf{r}}\hat{s}_z(\mathbf{r}) =$$

$$= \frac{1}{V}\sum_{\mathbf{k}_1,\mathbf{k}_2,\sigma}\int d^3\mathbf{r} e^{i(\mathbf{q}+\mathbf{k}_2-\mathbf{k}_1)\cdot\mathbf{r}}\hbar\sigma C^{\dagger}_{\mathbf{k}_1,\sigma}C_{\mathbf{k}_2,\sigma} =$$

$$= \frac{1}{V}\hbar\sum_{\mathbf{k}_1,\mathbf{k}_2,\sigma} V\delta_{\mathbf{q}+\mathbf{k}_2-\mathbf{k}_1}\sigma C^{\dagger}_{\mathbf{k}_1,\sigma}C_{\mathbf{k}_2,\sigma} =$$

$$= \hbar\sum_{\mathbf{k},\sigma}\sigma C^{\dagger}_{\mathbf{k}+\mathbf{q},\sigma}C_{\mathbf{k},\sigma}, \tag{10.259}$$

where we used

$$\int d^3\mathbf{r} e^{i(\mathbf{q}+\mathbf{k}_2-\mathbf{k}_1)} = V\delta_{\mathbf{q}+\mathbf{k}_2-\mathbf{k}_1}, \tag{10.260}$$

$$\mathbf{k}_1 = \mathbf{k}_2 + \mathbf{q}, \tag{10.261}$$

and finally, we renamed the dummy index $\mathbf{k}_2 \to \mathbf{k}$. We observe that $\hat{s}_z(\mathbf{q})$ has a structure similar to the Fourier transform of the number particle density (see Eq. 2.262). We sum over $\sigma = \mp\frac{1}{2}$, and we obtain

$$\hat{s}_z(\mathbf{q}) = \frac{\hbar}{2}\sum_{\mathbf{k}}\left[C^{\dagger}_{\mathbf{k}+\mathbf{q},\frac{1}{2}}C_{\mathbf{k},\frac{1}{2}} - C^{\dagger}_{\mathbf{k}+\mathbf{q},-\frac{1}{2}}C_{\mathbf{k},-\frac{1}{2}}\right], \tag{10.262}$$

which has an immediate interpretation: an electron with momentum $\mathbf{k}$ and spin up is annihilated and created with unchanged spin and momentum $\mathbf{k}+\mathbf{q}$, and the term is preceded by a positive sign; alternatively, an electron with momentum $\mathbf{k}$ and spin down is annihilated and created with unchanged spin and momentum $\mathbf{k}+\mathbf{q}$, and the term is preceded by a negative sign.

Remark 10.6 (Spin Volume Densities Along the x and y Axes) Let us write the spatial spin densities corresponding to the x and y components. The difference compared to the z component case treated above lies in the action of the operators $\hat{S}_x$ and $\hat{S}_y$ on the eigenfunction $\chi_\sigma(s)$, (see Eqs. 1.66 and 1.69). The spin volume density along x in second quantization is then given by

$$\begin{aligned}
\hat{s}_x(\mathbf{r}) &= \sum_s \hat{\psi}^\dagger(\mathbf{r}, s)\delta(\mathbf{r}-\mathbf{r}')\hat{S}_x\hat{\psi}(\mathbf{r}, s) = \\
&= \frac{1}{V}\sum_s \sum_{\mathbf{k}_1,\mathbf{k}_2,\sigma_1,\sigma_2} e^{i(\mathbf{k}_2-\mathbf{k}_1)\cdot\mathbf{r}} C^\dagger_{\mathbf{k}_1,\sigma_1}\chi_{\sigma_1}(s)\hat{S}_x\chi_{\sigma_2}(s)C_{\mathbf{k}_2,\sigma_2} = \\
&= \frac{1}{V}\sum_s \sum_{\mathbf{k}_1,\mathbf{k}_2,\sigma_1,\sigma_2} e^{i(\mathbf{k}_2-\mathbf{k}_1)\cdot\mathbf{r}} C^\dagger_{\mathbf{k}_1,\sigma_1}\chi_{\sigma_1}(s)\hbar\,|\sigma_2|\,\chi_{-\sigma_2}(s)C_{\mathbf{k}_2,\sigma_2} = \\
&= \frac{1}{V}\sum_{\mathbf{k}_1,\mathbf{k}_2,\sigma_1,\sigma_2} e^{i(\mathbf{k}_2-\mathbf{k}_1)\cdot\mathbf{r}}\hbar\,|\sigma_2|\,C^\dagger_{\mathbf{k}_1,\sigma_1}\delta_{\sigma_1,-\sigma_2}C_{\mathbf{k}_2,\sigma_2} = \\
&= \frac{1}{V}\sum_{\mathbf{k}_1,\mathbf{k}_2,\sigma} e^{i(\mathbf{k}_2-\mathbf{k}_1)\cdot\mathbf{r}}\hbar\,|\sigma|\,C^\dagger_{\mathbf{k}_1,\sigma}C_{\mathbf{k}_2,-\sigma},
\end{aligned} \tag{10.263}$$

and its Fourier transform is

$$\begin{aligned}
\hat{s}_x(\mathbf{q}) &= \int d^3\mathbf{r} e^{i\mathbf{q}\cdot\mathbf{r}}\hat{s}_x(\mathbf{r}) = \\
&= \frac{1}{V}\sum_{\mathbf{k}_1,\mathbf{k}_2,\sigma}\int d^3\mathbf{r} e^{i(\mathbf{q}+\mathbf{k}_2-\mathbf{k}_1)\cdot\mathbf{r}}\hbar\,|\sigma|\,C^\dagger_{\mathbf{k}_1,\sigma}C_{\mathbf{k}_2,-\sigma} = \\
&= \frac{1}{V}\hbar\sum_{\mathbf{k}_1,\mathbf{k}_2,\sigma} V\delta_{\mathbf{q}+\mathbf{k}_2-\mathbf{k}_1}\,|\sigma|\,C^\dagger_{\mathbf{k}_1,\sigma}C_{\mathbf{k}_2,\sigma} = \\
&= \hbar\sum_{\mathbf{k},\sigma}|\sigma|\,C^\dagger_{\mathbf{k}+\mathbf{q},\sigma}C_{\mathbf{k},-\sigma},
\end{aligned} \tag{10.264}$$

that is

$$\hat{s}_x(\mathbf{q}) = \frac{\hbar}{2}\sum_{\mathbf{k}}\left[C^\dagger_{\mathbf{k}+\mathbf{q},-\frac{1}{2}}C_{\mathbf{k},\frac{1}{2}} + C^\dagger_{\mathbf{k}+\mathbf{q},\frac{1}{2}}C_{\mathbf{k},-\frac{1}{2}}\right]. \tag{10.265}$$

The spin volume density along y in second quantization is then given by

$$\hat{s}_y(\mathbf{r}) = \sum_s \hat{\psi}^\dagger(\mathbf{r}, s)\delta(\mathbf{r}-\mathbf{r}')\hat{S}_y\hat{\psi}(\mathbf{r}, s) =$$

$$= \frac{1}{V}\sum_{s}\sum_{\mathbf{k}_1,\mathbf{k}_2,\sigma_1,\sigma_2} e^{i(\mathbf{k}_2-\mathbf{k}_1)\cdot\mathbf{r}} C^{\dagger}_{\mathbf{k}_1,\sigma_1}\chi_{\sigma_1}(s)\hat{S}_y\chi_{\sigma_2}(s)C_{\mathbf{k}_2,\sigma_2} =$$

$$= \frac{1}{V}\sum_{s}\sum_{\mathbf{k}_1,\mathbf{k}_2,\sigma_1,\sigma_2} e^{i(\mathbf{k}_2-\mathbf{k}_1)\cdot\mathbf{r}} C^{\dagger}_{\mathbf{k}_1,\sigma_1}\chi_{\sigma_1}(s)i\hbar\sigma_2\chi_{-\sigma_2}(s)C_{\mathbf{k}_2,\sigma_2} =$$

$$= \frac{1}{V}i\hbar\sum_{\mathbf{k}_1,\mathbf{k}_2,\sigma_1,\sigma_2} e^{i(\mathbf{k}_2-\mathbf{k}_1)\cdot\mathbf{r}}\sigma_2 C^{\dagger}_{\mathbf{k}_1,\sigma_1}\delta_{\sigma_1,-\sigma_2}C_{\mathbf{k}_2,\sigma_2} =$$

$$= -\frac{1}{V}i\hbar\sum_{\mathbf{k}_1,\mathbf{k}_2,\sigma} e^{i(\mathbf{k}_2-\mathbf{k}_1)\cdot\mathbf{r}}\sigma C^{\dagger}_{\mathbf{k}_1,\sigma}C_{\mathbf{k}_2,-\sigma}, \tag{10.266}$$

and its Fourier transform is

$$\hat{s}_y(\mathbf{q}) = \int d^3\mathbf{r} e^{i\mathbf{q}\cdot\mathbf{r}}\hat{s}_y(\mathbf{r}) =$$

$$= -\frac{1}{V}i\hbar\sum_{\mathbf{k}_1,\mathbf{k}_2,\sigma}\int d^3\mathbf{r} e^{i(\mathbf{q}+\mathbf{k}_2-\mathbf{k}_1)\cdot\mathbf{r}}\sigma C^{\dagger}_{\mathbf{k}_1,\sigma}C_{\mathbf{k}_2,-\sigma} =$$

$$= -\frac{1}{V}i\hbar\sum_{\mathbf{k}_1,\mathbf{k}_2,\sigma} V\delta_{\mathbf{q}+\mathbf{k}_2-\mathbf{k}_1}\sigma C^{\dagger}_{\mathbf{k}_1,\sigma}C_{\mathbf{k}_2,\sigma} =$$

$$= -i\hbar\sum_{\mathbf{k},\sigma}\sigma C^{\dagger}_{\mathbf{k}+\mathbf{q},\sigma}C_{\mathbf{k},-\sigma}, \tag{10.267}$$

that is

$$\hat{s}_y(\mathbf{q}) = i\frac{\hbar}{2}\sum_{\mathbf{k}}\left[C^{\dagger}_{\mathbf{k}+\mathbf{q},-\frac{1}{2}}C_{\mathbf{k},\frac{1}{2}} - C^{\dagger}_{\mathbf{k}+\mathbf{q},\frac{1}{2}}C_{\mathbf{k},-\frac{1}{2}}\right]. \tag{10.268}$$

Note that, unlike the spin volume density along x, in second quantization the spin volume densities along y and z annihilate an electron with momentum $\mathbf{k}$ and spin down (up) and create one with momentum $\mathbf{k}+\mathbf{q}$ and spin up (down), and vice versa: these densities flip the spin, as in first quantization.

Starting from $\hat{s}_z(\mathbf{r})$, we define the magnetization density as

$$\hat{M}_z(\mathbf{r}) = \frac{g\mu_B}{\hbar}\hat{s}_z(\mathbf{r}), \tag{10.269}$$

and its Fourier transform, obviously proportional to $\hat{s}_z(\mathbf{q})$ (see Eq. 10.262), is

$$\hat{M}_z(\mathbf{q}) = \frac{g\mu_B}{2}\sum_{\mathbf{k}}\left[C^{\dagger}_{\mathbf{k}+\mathbf{q},\frac{1}{2}}C_{\mathbf{k},\frac{1}{2}} - C^{\dagger}_{\mathbf{k}+\mathbf{q},-\frac{1}{2}}C_{\mathbf{k},-\frac{1}{2}}\right]. \tag{10.270}$$

Here we aim to apply linear response theory. As a first step, we recall that in this framework the relevant Hamiltonian is the one describing the unperturbed system. In our case, this Hamiltonian is expressed in the formalism of second quantization and, using the definition of the magnetization volume density, it takes the form

$$\hat{\mathcal{H}}_I = -\int d^3\mathbf{r}\hat{M}_z(\mathbf{r})H_z(\mathbf{r}). \tag{10.271}$$

Now we aim to quantify the change in magnetization density, that is $\delta\langle\hat{M}_z\rangle(\mathbf{q},\omega)$, as a function of the external magnetic field. Through calculations analogous to those performed to calculate $\delta\langle\hat{\rho}\rangle(\mathbf{q},\omega)$ in the Sect. 10.2, Eq. 10.24 takes the form

$$\delta\langle\hat{M}_z\rangle(\mathbf{q},\omega) = -\frac{1}{V}H_z(\mathbf{q},\omega)G_M(\mathbf{q},\omega), \tag{10.272}$$

where

$$G_M(\mathbf{q},\omega+i\delta) = -\frac{i}{\hbar}\frac{1}{V}\int_{-\infty}^{+\infty} dt e^{i(\omega+i\delta)t}\Theta(t)\left\langle\left[\hat{M}_z^{(0)}(\mathbf{q},t),\hat{M}_z^\dagger(\mathbf{q},0)\right]\right\rangle \tag{10.273}$$

is an equivalent of the polarization propagator (see Eqs. 10.34 and 10.128), and it is called the magnetization propagator. Note that the variable δ is introduced to compute the integral in the upper complex half-plane and then take the limit $\delta \to 0^+$. We define the magnetic susceptibility as

$$\chi(\mathbf{q},\omega) = -G_M(\mathbf{q},\omega), \tag{10.274}$$

which includes a thermal average (see Eq. 1.379) over the Hamiltonian without interaction with the field, but in which the particles interact among themselves. Recall that at the lowest order of the density-density Green's function expansion one obtains the polarization bubble, similarly, the lowest order of the spin-spin Green's function produces the so-called magnetization bubble. The thermal average contained in the magnetic susceptibility is of the form

$$\left\langle C^\dagger_{\mathbf{k}+\mathbf{q},\sigma}C_{\mathbf{k},\sigma}C^\dagger_{\mathbf{k}'-\mathbf{q},\sigma'}C_{\mathbf{k}',\sigma'}\right\rangle, \tag{10.275}$$

note that from thermal Wick's theorem (see Theorems 7.2–7.4), the contraction $C^\dagger_{\mathbf{k}+\mathbf{q},\sigma}C_{\mathbf{k},\sigma}$ implies $\mathbf{q} = \mathbf{0}$, which is not admissible, so the unique contraction is $C^\dagger_{\mathbf{k}+\mathbf{q},\sigma}C_{\mathbf{k}',\sigma'}$, which implies $\mathbf{k}' = \mathbf{k}+\mathbf{q}$, $\sigma = \sigma'$, and therefore the lowest order of the spin-spin Green's function is equal to the polarization bubble. In other words, the polarization bubble at lowest order describes both the dielectric response of a system to a weak longitudinal electric field and the magnetic response to a weak external magnetic field.

10.4 Optical Conductivity

Here we aim to derive Ohm's law at the microscopic level

$$\mathbf{J} = \sigma \mathbf{E}, \tag{10.276}$$

or equivalently, to derive the optical conductivity σ. For now, we begin with a classical treatment of the problem. This will allow us to develop the physical intuition and identify the key quantities involved. We will then proceed to reformulate the discussion within the frameworks of first and second quantization. At this stage, we adopt a classical approach: let q_e denote the charge of an electron in an electromagnetic field; its Hamiltonian is of the form

$$\mathcal{H} = \frac{1}{2m_e}\left(\mathbf{p} - \frac{q_e}{c}\mathbf{A}\right)^2 + q_e\varphi + V, \tag{10.277}$$

where φ and $\mathbf{A}$ are respectively the scalar and vector potentials. The fields are related to the potentials by

$$\mathbf{E} = -\frac{1}{c}\frac{\partial \mathbf{A}}{\partial t} - \nabla\varphi, \tag{10.278}$$

$$\mathbf{B} = \nabla \times \mathbf{A}. \tag{10.279}$$

Moreover, any other scalar and vector potentials must satisfy

$$\begin{cases} \mathbf{A}' = \mathbf{A} + \nabla f \\ \varphi' = \varphi - \dfrac{1}{c}\dfrac{\partial f}{\partial t} \end{cases} . \tag{10.280}$$

where f is the gauge function. Among the possible gauge functions, we choose one that enforces the Coulomb gauge condition, which ensures that the vector potential is purely transverse, i.e., $\mathbf{A} = \mathbf{A}_T$ (for a detailed discussion, see Sect. 13.1). In this gauge, the equations for the scalar potential φ are decoupled from the vector potential, and one solution is the Coulomb potential, which depends instantaneously on the charge distribution, that is

$$\varphi(\mathbf{r}, t) = \int d^3\mathbf{r}\frac{\rho(\mathbf{r}, t)}{|\mathbf{r} - \mathbf{r}'|}, \tag{10.281}$$

with $\rho(\mathbf{r}, t)$ the charge density. Here, we aim to study the response of a material medium to a weak transverse electromagnetic field. Under the assumption that the medium is far from the sources of the field, the following approximations hold

- the electromagnetic wave can be approximated as a plane wave;

- the Coulomb scalar potential is negligible, since $|\mathbf{r}-\mathbf{r}'| \to \infty$.

  Under the above assumptions, it follows

$$\mathbf{E} = -\frac{1}{c}\frac{\partial \mathbf{A}}{\partial t}, \tag{10.282}$$

$$\mathcal{H} = \frac{1}{2m_e}\left(\mathbf{p} - \frac{q_e}{c}\mathbf{A}\right)^2 + V. \tag{10.283}$$

We also recall that the velocity of the particle, as a consequence of the electromagnetic field, becomes

$$\mathbf{v} = \frac{1}{m_e}\left(\mathbf{p} - \frac{q_e}{c}\mathbf{A}\right), \tag{10.284}$$

then the current density vector can be written as

$$\mathbf{J} = q_e \rho \mathbf{v}, \tag{10.285}$$

with ρ the particle density. In quantum mechanics, for an N-particle Hilbert space, the current density vector is written as

$$\mathbf{J}_{\mathbf{A}} = \frac{q_e}{m_e}\sum_{i=1}^{N}\left(\mathbf{p}_i - \frac{q_e}{c}\mathbf{A}(\mathbf{r}_i)\right)\delta(\mathbf{r}-\mathbf{r}_i), \tag{10.286}$$

where we have made explicit the dependence of the current density on the vector potential. We write the current density as the sum of the contribution with zero vector potential and the contribution with nonzero vector potential, that is,

$$\mathbf{J}_{\mathbf{A}} = \mathbf{J}_{\mathbf{A}=\mathbf{0}} - \frac{q_e^2}{m_e c}\mathbf{A}(\mathbf{r})\rho(\mathbf{r}), \tag{10.287}$$

where we have explicitly written the particle density operator in first quantization (see Eq. 2.259). We now apply linear response theory to the current density. To this end, we explicitly switch to the formalism of first quantization, rewriting the Hamiltonian as in Eq. 1.85, where

$$\mathcal{H}_0 = \frac{\hat{\mathbf{p}}^2}{2m_e} + V, \tag{10.288}$$

$$\mathcal{H}_I = -\frac{q_e^2}{2m_e c^2}\mathbf{A}^2 - \frac{q_e}{2m_e c}\left(\hat{\mathbf{p}}\cdot\mathbf{A} + \mathbf{A}\cdot\hat{\mathbf{p}}\right). \tag{10.289}$$

Under the assumption of a weak field, we can neglect the term $\mathbf{A}^2$; moreover, in general the operators $\mathbf{p}$ and $\mathbf{A}$ do not commute, but in the Coulomb gauge they do commute, so

$$\begin{aligned}\mathcal{H}_I &= -\frac{q_e}{m_e c}\hat{\mathbf{p}} \cdot \mathbf{A} = \\ &= -\frac{q_e}{c}\mathbf{v}_{\mathbf{A}=\mathbf{0}} \cdot \mathbf{A}.\end{aligned} \tag{10.290}$$

We generalize this interaction term between the particles and the external field to the case of N particles, as follows

$$\mathcal{H}_I = -\frac{1}{c}\int d^3\mathbf{r}\mathbf{J}_{\mathbf{A}=\mathbf{0}}(\mathbf{r}) \cdot \mathbf{A}(\mathbf{r}). \tag{10.291}$$

Here we have replaced the current density evaluated in the absence of the external perturbation. Recall that, within linear response theory, one must employ the interaction picture (see Eq. 1.107) with respect to the unperturbed Hamiltonian, denoted here by $\hat{\mathcal{H}}_0$. It is important to stress that $\hat{\mathcal{H}}_0$ may still include interaction terms, but it does not contain the external perturbation (see Remark 10.1). It follows

$$\mathcal{H}_I^{(0)}(t) = -\frac{1}{c}\int d^3\mathbf{r}\mathbf{J}^{(0)}_{\mathbf{A}=\mathbf{0}}(\mathbf{r},t) \cdot \mathbf{A}(\mathbf{r},t), \tag{10.292}$$

where we emphasize that, since the vector potential $\mathbf{A}$ is a function rather than an operator, the interaction picture does not act on it. Note that the interaction Hamiltonian contains the perturbation, the potential $\mathbf{A}$, at the lowest order, consistent with linear response theory. From Eq. 10.24, with $A = \mathbf{J}_{\mathbf{A}}(\mathbf{r})$, one finds

$$\begin{aligned}&\langle\mathbf{J}_{\mathbf{A}}(\mathbf{r})\rangle(t) = \langle\mathbf{J}_{\mathbf{A}}(\mathbf{r})\rangle(0)+ \\ &+\frac{i}{\hbar}\int_{-\infty}^{+\infty} dt'\Theta(t-t')\left\langle\left[-\frac{1}{c}\int d^3\mathbf{r}'\mathbf{J}^{(0)}_{\mathbf{A}=\mathbf{0}}(\mathbf{r}',t') \cdot \mathbf{A}(\mathbf{r}',t'), \mathbf{J}^{(0)}_{\mathbf{A}=\mathbf{0}}(\mathbf{r},t)\right]\right\rangle.\end{aligned} \tag{10.293}$$

Note that the time evolution of the current density,

$$\mathbf{J}^{(0)}_{\mathbf{A}}(\mathbf{r},t) = \mathbf{J}^{(0)}_{\mathbf{A}=\mathbf{0}}(\mathbf{r},t) - \frac{q_e^2}{m_e c}\mathbf{A}(\mathbf{r})\rho(\mathbf{r}), \tag{10.294}$$

has been replaced in the commutator by $\mathbf{J}^{(0)}_{\mathbf{A}=\mathbf{0}}(\mathbf{r},t)$. The reason is that the additional term gives rise to

$$\left[\mathbf{J}^{(0)}_{\mathbf{A}=\mathbf{0}}(\mathbf{r}',t') \cdot \mathbf{A}(\mathbf{r}',t'), \mathbf{A}(\mathbf{r})\rho(\mathbf{r})\right], \tag{10.295}$$

which, up to constant prefactors, generates contributions of second order in $\mathbf{A}$. Since we are working within the linear response regime, such terms are neglected. The thermal average (see Eq. 1.379) of the current density 10.287 is given by

$$\begin{aligned}\langle \mathbf{J_A} \rangle (0) &= \langle \mathbf{J_{A=0}} \rangle - \left\langle \frac{q_e^2}{m_e c} \mathbf{A}(\mathbf{r}, t=0) \rho(\mathbf{r}) \right\rangle = \\ &= -\frac{q_e^2}{m_e c} \langle \mathbf{A}(\mathbf{r}, t=0) \rho(\mathbf{r}) \rangle = \\ &= -\frac{q_e^2}{m_e c} \mathbf{A}(\mathbf{r}, t=0) \rho(\mathbf{r}). \end{aligned} \tag{10.296}$$

In the absence of any external perturbation, the current density within the material medium vanishes, and therefore its expectation value is also zero. Let us consider a field of the form

$$\mathbf{E}(t) = E_0 e^{-i\omega t} \hat{\mathbf{i}}, \tag{10.297}$$

that is, a plane wave traveling along the $\hat{\mathbf{i}}$ axis. Note that the dependence of $\mathbf{E}$ on $\mathbf{r}$ has been omitted; we will explain why. We have implicitly assumed $\mathbf{q} = \mathbf{0}$, then from $q = \frac{2\pi}{\lambda}$, we are assuming that the wavelength of the incident wave is very large: in solid-state physics, the system excitations are relevant for energies associated with wavelengths on the order of 10^3 Angstrom, whereas the lattice constant in the material medium is on the order of 1 Angstrom. In other words, the photon momentum plays no role. From Eq. 10.282, one finds

$$\mathbf{A}(t) = \frac{c}{i\omega} \mathbf{E}(t), \tag{10.298}$$

and since the vector potential also does not depend on position, it is irrotational, and thus $\mathbf{B} = \mathbf{0}$. In other words, we are studying the case of a transverse electric field perturbing a material medium. We integrate with respect to position and divide both sides of Eq. 10.293, adapted for a transverse field, by the volume, that is,

$$\begin{aligned}&\frac{1}{V} \int d^3\mathbf{r} \langle \mathbf{J_A}(\mathbf{r}) \rangle (t) = \frac{1}{V} \int d^3\mathbf{r} \langle \mathbf{J_A}(\mathbf{r}) \rangle (0) + \\ &+ \frac{i}{\hbar} \frac{1}{V} \int d^3\mathbf{r} \int_{-\infty}^{+\infty} dt' \Theta(t-t') \left\langle \left[-\frac{1}{c} \int d^3\mathbf{r}' \mathbf{J}^{(0)}_{\mathbf{A=0}}(\mathbf{r}', t') \cdot \mathbf{A}(t'), \mathbf{J}^{(0)}_{\mathbf{A=0}}(\mathbf{r}, t) \right] \right\rangle . \end{aligned} \tag{10.299}$$

We have

$$\frac{1}{V} \int d^3\mathbf{r} \langle \mathbf{J_A}(\mathbf{r}) \rangle (t) = \frac{1}{V} \langle \mathbf{J_{q=0}} \rangle (t), \tag{10.300}$$

and from Eq. 10.296 it follows

$$\frac{1}{V}\int d^3\mathbf{r}\langle \mathbf{J}_{\mathbf{A}}(\mathbf{r})\rangle(0) = -\frac{1}{V}\frac{q_e^2}{m_e c}\int d^3\mathbf{r}\mathbf{A}(\mathbf{r}, t=0)\rho(\mathbf{r}) =$$
$$= -\frac{q_e^2}{i m_e \omega}\mathbf{E}(t)\int d^3\mathbf{r}\frac{\rho(\mathbf{r})}{V} =$$
$$= -\frac{q_e^2}{i m_e \omega}\frac{N}{V}\mathbf{E}(t), \tag{10.301}$$

where we have made explicit that $N = \int d^3\mathbf{r}\rho(\mathbf{r})$. Moreover, in the last term, only $\mathbf{J}^{(0)}_{\mathbf{A}=\mathbf{0}}(\mathbf{r}, t)$ depends on $\mathbf{r}$, and only $\mathbf{J}^{(0)}_{\mathbf{A}=\mathbf{0}}(\mathbf{r}', t')$ depends on $\mathbf{r}'$. Therefore, by performing these integrals, we obtain their respective Fourier transforms in momentum space for $\mathbf{q} = \mathbf{0}$. Ultimately, Eq. 10.299 becomes

$$\frac{1}{V}\langle \mathbf{J}_{\mathbf{q}=\mathbf{0}}\rangle(t)$$
$$= -\frac{q_e^2}{i m_e \omega}\frac{N}{V}\mathbf{E}(t)$$
$$+ \frac{i}{\hbar}\frac{1}{V}\int_{-\infty}^{+\infty} dt'\Theta(t-t')\left(-\frac{1}{c}\right)\left\langle\left[\mathbf{J}^{(0)}_{\mathbf{q}=\mathbf{0},\mathbf{A}=\mathbf{0}}(t')\cdot\frac{c}{i\omega}\mathbf{E}(t'), \mathbf{J}^{(0)}_{\mathbf{q}=\mathbf{0},\mathbf{A}=\mathbf{0}}(t)\right]\right\rangle =$$
$$= -\frac{q_e^2}{i m_e \omega}\frac{N}{V}\mathbf{E}(t)$$
$$- \frac{1}{\hbar\omega}\frac{1}{V}\int_{-\infty}^{+\infty} dt'\Theta(t-t')\left\langle\left[\mathbf{J}^{(0)}_{\mathbf{q}=\mathbf{0},\mathbf{A}=\mathbf{0}}(t')\cdot E_0 e^{-i\omega t'}\hat{\mathbf{i}}, \mathbf{J}^{(0)}_{\mathbf{q}=\mathbf{0},\mathbf{A}=\mathbf{0}}(t)\right]\right\rangle =$$
$$= -\frac{q_e^2}{i m_e \omega}\frac{N}{V}\mathbf{E}(t)$$
$$- \frac{1}{\hbar\omega}\frac{1}{V}\int_{-\infty}^{+\infty} dt'\Theta(t-t')e^{-i\omega t'}\left\langle\left[\mathbf{J}^{(0)}_{\mathbf{q}=\mathbf{0},\mathbf{A}=\mathbf{0}}(t'), \mathbf{J}^{(0)}_{\mathbf{q}=\mathbf{0},\mathbf{A}=\mathbf{0}}(t)\right]\right\rangle\cdot E_0\hat{\mathbf{i}}. \tag{10.302}$$

We observe that $e^{-i\omega t'} = e^{i\omega(t-t')}e^{-i\omega t}$, which allows us to factor out the electric field and rewrite the expression as

$$\frac{1}{V}\langle \mathbf{J}_{\mathbf{q}=\mathbf{0}}\rangle(t)$$
$$= -\frac{q_e^2}{i m_e \omega}\frac{N}{V}\mathbf{E}(t)$$
$$- \frac{1}{\hbar\omega}\frac{1}{V}\int_{-\infty}^{+\infty} dt'\Theta(t-t')e^{i\omega(t-t')}\left\langle\left[\mathbf{J}^{(0)}_{\mathbf{q}=\mathbf{0},\mathbf{A}=\mathbf{0}}(t'), \mathbf{J}^{(0)}_{\mathbf{q}=\mathbf{0},\mathbf{A}=\mathbf{0}}(t)\right]\right\rangle\cdot\mathbf{E}(t), \tag{10.303}$$

after which we invert the commutator,

$$\frac{1}{V}\langle \mathbf{J}_{\mathbf{q}=\mathbf{0}}\rangle(t)$$
$$= -\frac{q_e^2}{im_e\omega}\frac{N}{V}\mathbf{E}(t)$$
$$+\frac{1}{\hbar\omega}\frac{1}{V}\int_{-\infty}^{+\infty} dt'\Theta(t-t')e^{i\omega(t-t')}\left\langle\left[\mathbf{J}^{(0)}_{\mathbf{q}=\mathbf{0},\mathbf{A}=\mathbf{0}}(t),\mathbf{J}^{(0)}_{\mathbf{q}=\mathbf{0},\mathbf{A}=\mathbf{0}}(t')\right]\right\rangle\cdot\mathbf{E}(t), \tag{10.304}$$

we exploit the condition of thermodynamic equilibrium (see Theorem 6.1),

$$\frac{1}{V}\langle \mathbf{J}_{\mathbf{q}=\mathbf{0}}\rangle(t)$$
$$= -\frac{q_e^2}{im_e\omega}\frac{N}{V}\mathbf{E}(t)$$
$$+\frac{1}{\hbar\omega}\frac{1}{V}\int_{-\infty}^{+\infty} dt'\Theta(t-t')e^{i\omega(t-t')}\left\langle\left[\mathbf{J}^{(0)}_{\mathbf{q}=\mathbf{0},\mathbf{A}=\mathbf{0}}(t-t'),\mathbf{J}^{(0)}_{\mathbf{q}=\mathbf{0},\mathbf{A}=\mathbf{0}}(0)\right]\right\rangle\cdot\mathbf{E}(t), \tag{10.305}$$

we rename the dummy variable in the integral as $t-t'\to t''$,

$$\frac{1}{V}\langle \mathbf{J}_{\mathbf{q}=\mathbf{0}}\rangle(t)$$
$$= -\frac{q_e^2}{im_e\omega}\frac{N}{V}\mathbf{E}(t)$$
$$-\frac{1}{\hbar\omega}\frac{1}{V}\int_{-\infty}^{+\infty} (-dt'')\Theta(t'')e^{i\omega t''}\left\langle\left[\mathbf{J}^{(0)}_{\mathbf{q}=\mathbf{0},\mathbf{A}=\mathbf{0}}(t''),\mathbf{J}^{(0)}_{\mathbf{q}=\mathbf{0},\mathbf{A}=\mathbf{0}}(0)\right]\right\rangle\cdot\mathbf{E}(t), \tag{10.306}$$

and finally we set $t''=t$, then we obtain

$$\frac{1}{V}\langle \mathbf{J}_{\mathbf{q}=\mathbf{0}}\rangle(t)$$
$$= -\frac{q_e^2}{im_e\omega}\frac{N}{V}\mathbf{E}(t)$$
$$+\frac{1}{\hbar\omega}\frac{1}{V}\int_{-\infty}^{+\infty} dt\Theta(t)e^{i\omega t}\left\langle\left[\mathbf{J}^{(0)}_{\mathbf{q}=\mathbf{0},\mathbf{A}=\mathbf{0}}(t),\mathbf{J}^{(0)}_{\mathbf{q}=\mathbf{0},\mathbf{A}=\mathbf{0}}(0)\right]\right\rangle\cdot\mathbf{E}(t). \tag{10.307}$$

Ohm's law 10.276 has been derived in frequency space and for $\mathbf{q}=\mathbf{0}$, that is,

$$\frac{1}{V}\mathbf{J}_{\mathbf{q}=\mathbf{0}}(\omega) = \sigma(\omega)\mathbf{E}(\omega), \tag{10.308}$$

where we set

$$\sigma(\omega) = \frac{q_e^2\rho}{im_e\omega} + \frac{i}{\omega}G_{\mathbf{J}}(\mathbf{q}=\mathbf{0}, \omega + i\delta), \tag{10.309}$$

$$G_{\mathbf{J}}(\mathbf{q}=\mathbf{0}, \omega+i\delta) = -\frac{i}{\hbar}\frac{1}{V}\int_{-\infty}^{+\infty} dt\Theta(t)e^{-i(\omega+i\delta)t}\left\langle\left[\mathbf{J}^{(0)}_{\mathbf{q}=\mathbf{0},\mathbf{A}=\mathbf{0}}(t), \mathbf{J}^{(0)}_{\mathbf{q}=\mathbf{0},\mathbf{A}=\mathbf{0}}(0)\right]\right\rangle. \tag{10.310}$$

Again, the integral is evaluated in the upper complex half-plane, followed by taking the limit $\delta \rightarrow 0^+$. If the electric field is weak, the optical conductivity is an intrinsic property of a material medium and is calculated as the current density autocorrelation in the absence of external perturbation. Since calculating the optical conductivity is very complex, dedicated models have been developed for insulating and conducting materials, using semiclassical approaches. In particular, in the following, we present two simple models of linear optical response: the Lorentz and Drude models. These provide intuitive insight into the behavior of the conductivity function. A fully microscopic calculation of the optical conductivity is considerably more complex and requires the Mori-Zwanzig formalism, which lies beyond the scope of this text.

10.4.1 Lorentz and Drude Models of Optical Conductivity

A basic model for describing an insulator is provided by the Lorentz model, in which atoms oscillate around their equilibrium positions and the electrons are bound to their respective atomic nuclei by elastic forces characterized by a constant k_{el}. In addition, the electrons move within a medium that exerts a viscous damping force, which, according to Stokes' law, is proportional to the electron velocity with a damping coefficient γ. When subjected to an external electric field of the type introduced previously, the motion of an electron is governed by the following differential equation

$$m_e\ddot{\mathbf{r}} = -k_{\text{el}}\mathbf{r} - m_e\gamma\dot{\mathbf{r}} + q_eE_0e^{-i\omega t}\hat{\mathbf{i}}, \tag{10.311}$$

whose solution is given by the sum of the general integral of the associated homogeneous equation and a particular solution of the full equation. The solution of the associated homogeneous equation is transient, meaning it tends to zero at large times, and since we are interested in stationary solutions, we directly consider a particular solution. Let us assume that this particular solution is of the form

$$\mathbf{r}(t) = \mathbf{r}(0)e^{-i\omega t}, \tag{10.312}$$

that is, it has the same oscillation frequency as the external perturbation, and we substitute this solution into the differential equation. We obtain

$$-m_e\omega^2\mathbf{r}(0)e^{-i\omega t} = -k_{\mathrm{el}}\mathbf{r}(0)e^{-i\omega t} + im_e\omega\gamma\mathbf{r}(0)e^{-i\omega t} + q_e E_0 e^{-i\omega t}\hat{\mathbf{i}}, \qquad (10.313)$$

which implies

$$\mathbf{r}(0) = \frac{q_e E_0 \hat{\mathbf{i}}}{k_{\mathrm{el}} - im_e\omega\gamma - m_e\omega^2}. \qquad (10.314)$$

The current density in the Lorentz model is given by

$$\mathbf{J} = nq_e(-i\omega)\mathbf{r}(0)e^{-i\omega t} \qquad (10.315)$$

which, from Ohm's law 10.276, implies

$$nq_e(-i\omega)\mathbf{r}(0)e^{-i\omega t} = \sigma E_0 e^{-i\omega t}\hat{\mathbf{i}}, \qquad (10.316)$$

and by substituting $\mathbf{r}(0)$, we have

$$nq_e(-i\omega)\frac{q_e E_0 \hat{\mathbf{i}}}{k_{\mathrm{el}} - im_e\omega\gamma - m_e\omega^2} = \sigma E_0 \hat{\mathbf{i}}, \qquad (10.317)$$

and

$$\begin{aligned}
\sigma(\omega) &= -inq_e^2 \frac{\omega}{k_{\mathrm{el}} - im_e\omega\gamma - m_e\omega^2} = \\
&= -i\frac{nq_e^2}{m_e}\frac{\omega}{\omega_0^2 - i\omega\gamma - \omega^2} = \\
&= i\frac{nq_e^2}{m_e}\frac{\omega}{\omega^2 - \omega_0^2 + i\omega\gamma} = \\
&= i\frac{nq_e^2}{m_e\gamma}\frac{\gamma\frac{\omega}{\omega_0^2}}{\frac{\omega^2}{\omega_0^2} - 1 + i\frac{\omega\gamma}{\omega_0^2}} = \\
&= i\frac{nq_e^2}{m_e\gamma}\frac{\gamma\frac{\omega}{\omega_0^2}}{\left(\frac{\omega^2}{\omega_0^2} - 1\right)^2 + \frac{\omega^2\gamma^2}{\omega_0^4}}\left(\frac{\omega^2}{\omega_0^2} - 1 - i\frac{\omega\gamma}{\omega_0^2}\right),
\end{aligned} \qquad (10.318)$$

where we have defined the electron's natural frequency as

$$\omega_0^2 = \frac{k_{\mathrm{el}}}{m_e}. \qquad (10.319)$$

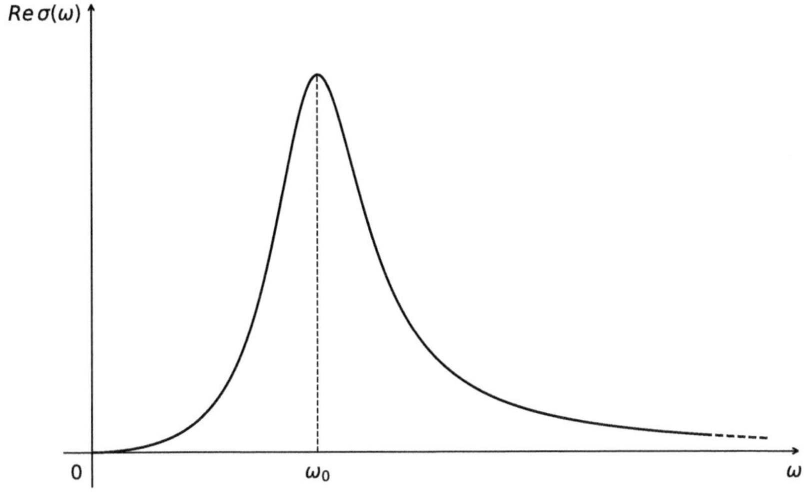

Fig. 10.6 Real part of the optical conductivity in the Lorentz model, see Eq. 10.320

Let us consider the real part of the Lorentz optical conductivity,

$$\mathrm{Re}\,\sigma(\omega) = \frac{nq_e^2}{m_e\gamma}\frac{\gamma^2\frac{\omega^2}{\omega_0^4}}{\left[\frac{\omega^2}{\omega_0^2}-1\right]^2+\frac{\omega^2\gamma^2}{\omega_0^4}} \tag{10.320}$$

as shown in Fig. 10.6. This function vanishes both at $\omega = 0$ and in the high-frequency limit. It reaches its maximum at $\omega = \omega_0$, where its value is given by

$$\mathrm{Re}\,\sigma(\omega_0) = \frac{nq_e^2}{m_e\gamma}. \tag{10.321}$$

Let us now consider the imaginary part of the Lorentz optical conductivity,

$$\mathrm{Im}\,\sigma(\omega) = \frac{nq_e^2}{m_e\gamma}\frac{\gamma\frac{\omega}{\omega_0^2}}{\left[\frac{\omega^2}{\omega_0^2}-1\right]^2+\frac{\omega^2\gamma^2}{\omega_0^4}}\left(\frac{\omega^2}{\omega_0^2}-1\right), \tag{10.322}$$

as shown in Fig. 10.7. As expected, $\mathrm{Im}\,\sigma(\omega)$ vanishes at $\omega = 0$. This behavior is characteristic of an insulator: a static electric field ($\omega = 0$) cannot induce any current density. Moreover, the imaginary part is negative for $\omega < \omega_0$, positive for $\omega > \omega_0$, crosses zero at $\omega = \omega_0$, and tends to zero in the high-frequency limit.

Now, let us consider the case where the material medium is a conductor. The simplest model is the Drude model, which is very similar to the Lorentz model,

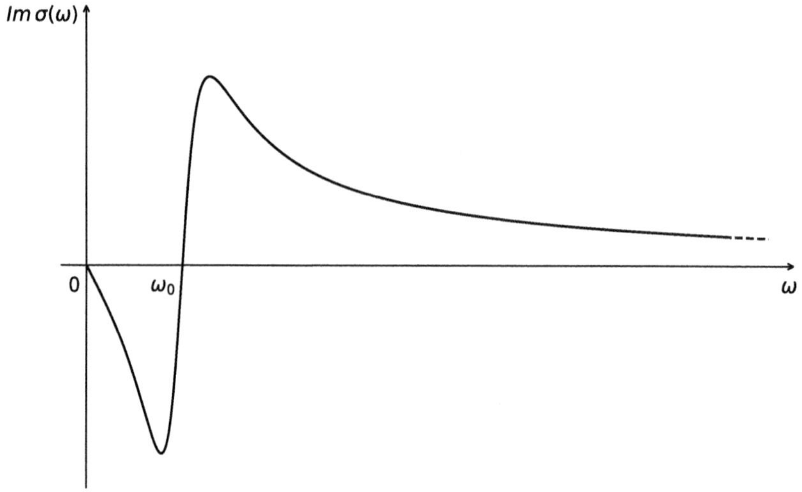

Fig. 10.7 Imaginary part of the optical conductivity in the Lorentz model, see Eq. 10.322

except that the electrons are free to move. Consequently, the differential equation to study is similar to Eq. 10.311, but with $k_{\text{el}} = 0$. By performing calculations analogous to those in the Lorentz model, we obtain the expression for the optical conductivity in the Drude model

$$\sigma(\omega) = i \frac{nq_e^2}{m_e \gamma} \frac{\gamma}{\omega^2 + \gamma^2} (\omega - i\gamma) . \tag{10.323}$$

Let us consider the real part of the Drude optical conductivity,

$$\text{Re}\, \sigma(\omega) = \frac{nq_e^2}{m_e \gamma} \frac{\gamma^2}{\omega^2 + \gamma^2}, \tag{10.324}$$

as shown in Fig. 10.8. This function is a Lorentzian with parameter γ; in particular, it is nonzero at $\omega = 0$, which is due to the resistance that the material medium opposes to the moving electrons. This resistance arises from electron-phonon interactions, meaning that the Hamiltonian on which we want to calculate the optical conductivity in a conducting material must include the electron-phonon interaction. A maximum at zero frequency is a distinctive feature of a conducting material.

Ultimately, let us consider the imaginary part of the Drude optical conductivity,

$$\text{Im}\, \sigma(\omega) = \frac{nq_e^2}{m_e \gamma} \frac{\gamma\omega}{\omega^2 + \gamma^2}, \tag{10.325}$$

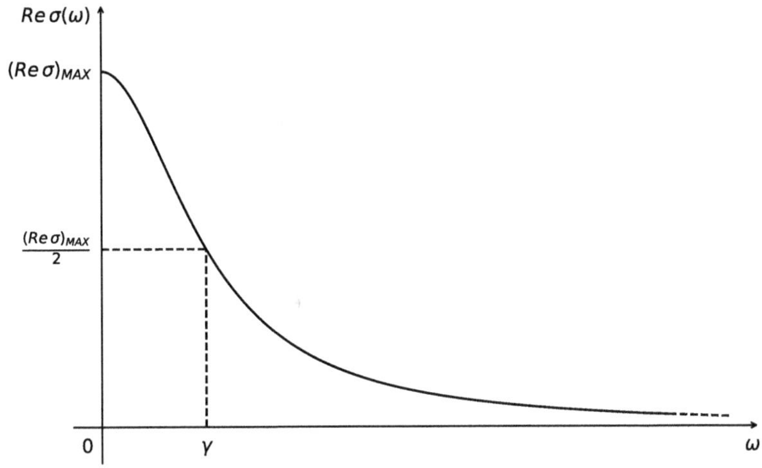

Fig. 10.8 Real part of the optical conductivity in the Drude model, see Eq. 10.324

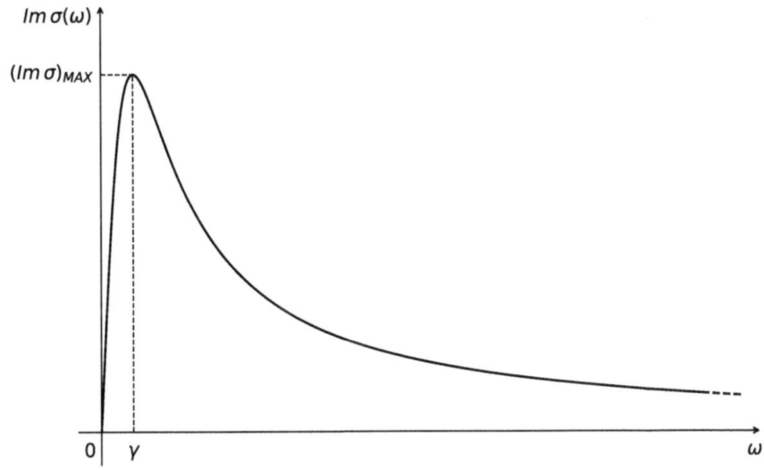

Fig. 10.9 Imaginary part of the optical conductivity in the Drude model, see Eq. 10.325

as shown in Fig. 10.9. This function vanishes at $\omega = 0$, remains positive for all $\omega > 0$, reaches its maximum at $\omega = \gamma$, and decreases monotonically as the frequency increases further.

This formal analogy is well known in solid state optics and provides a direct mapping to familiar electrical circuits. Importantly, in formal terms, the Lorentz model for bound electrons in an insulator is mathematically equivalent to a driven series RLC electrical circuit. In this analogy, the electron mass m_e plays the role of the inductance L, accounting for the inertial response; the elastic restoring force with constant $k_{\rm el}$ corresponds to the reciprocal of the capacitance C^{-1}; and the damping due to viscous friction, characterized by the coefficient γ, is identified with the resistance R. The external oscillating electric field acts like a time-dependent voltage source applied to the circuit. As a result, the frequency-dependent optical conductivity $\sigma(\omega)$ mirrors the complex admittance of the circuit. Importantly, just as in an RLC circuit, only the real part of the optical conductivity corresponds to energy dissipation, it reflects the irreversible conversion of electromagnetic energy into heat due to friction. The imaginary part, by contrast, is associated with the reactive (non-dissipative) response of the system, related to energy temporarily stored in the bound motion of the electrons or in the electric and magnetic fields of the circuit. This formal correspondence provides a powerful physical intuition for interpreting optical responses in solids in terms of familiar circuit elements.

On the other hand, the Drude model is formally equivalent to that of a series RL circuit driven by an alternating voltage $V(t)$. In this analogy, the electron mass m_e plays the role of the inductance L, representing the electron's inertia which resists changes in current (electron velocity). The damping coefficient γ, which quantifies the friction due to scattering, corresponds directly to the resistance R, responsible for dissipating electrical energy into heat. The external electric field $\mathbf{E}(t)$, which drives the electrons, is analogous to the voltage source $V(t)$ in the circuit, supplying energy to maintain oscillations. The frequency-dependent optical conductivity $\sigma(\omega)$ of the Drude model corresponds to the complex admittance of the RL circuit, where the real part $\text{Re}[\sigma(\omega)]$ represents the dissipative conductance that converts electromagnetic energy irreversibly into heat, and the imaginary part $\text{Im}[\sigma(\omega)]$ represents the reactive susceptance, associated with energy temporarily stored in the electrons' kinetic motion and released back to the field. This analogy not only provides a clear physical picture of how electrons respond to time-varying electric fields but also offers a useful framework to understand how resistive losses and inertial effects shape the frequency-dependent response of metals, just as resistance and inductance govern the behavior of an RL circuit.

Finally, to better understand the physical significance of the imaginary parts of the Lorentz and Drude optical conductivities, let us consider their phase relationship with the driving electric field. Ohm's law can be rewritten as

$$
\begin{aligned}
\mathbf{J} &= \text{Re}\,[\sigma]\,\mathbf{E} + i\,\text{Im}\,[\sigma]\,\mathbf{E} = \\
&= \text{Re}\,[\sigma]\,\mathbf{E} + e^{i\frac{\pi}{2}\,\text{sgn}(\text{Im}\,\sigma)}\,|\text{Im}\,\sigma|\,\mathbf{E},
\end{aligned}
\tag{10.326}
$$

showing that the current density consists of two components: an in-phase part proportional to the real part of the conductivity, and an out-of-phase part proportional to the imaginary part, which leads or lags the electric field by $\pm\frac{\pi}{2}$. This phase difference is characteristic of reactive behavior, corresponding to energy temporarily stored and released rather than dissipated. The sign of $\text{Im}\,\sigma$ thus indicates whether the system behaves more inductively or capacitively, which directly connects back to the effective circuit analogies of the Lorentz (RLC) and Drude (RL) models.

Part IV
Further Bosonic Quantum Fields

Chapter 11
Bose-Einstein Condensation of Non-Interacting Bosons

Abstract In this chapter, we explore Bose-Einstein condensation (BEC), a macroscopic quantum phase transition occurring when a dilute gas of indistinguishable bosons is cooled near absolute zero. At this temperature, a large fraction of particles occupies the ground state, leading to the emergence of a coherent quantum phase. The chapter covers theoretical frameworks for BEC, including thermodynamic methods using the grand canonical ensemble, chemical potential, and density of states, as well as quantum field-theoretic approaches based on second quantization. Here, bosonic field operators acquire macroscopic eigenvalues, reflecting condensation. BECs are fundamental in fields such as quantum gases, superfluidity, quantum simulation, and precision metrology. Their tunability and coherence make them powerful platforms for studying quantum many-body physics, symmetry breaking, phase transitions, and developing quantum technologies.

11.1 Qualitative Study at the Low and High Temperature Limit for Non-Interacting Bosons

In this section, we study the behavior of a non-interacting Bose gas in the low- and high-temperature limits, working within the framework of the grand canonical ensemble. Given a system of non-interacting bosons, let us consider the average number $\left\langle \hat{N}_i \right\rangle$ of bosons in a given state i, where $\hat{N}_i$ denotes the number operator (see Sect. 2.7.3). From Eq. 2.251, it follows

$$\begin{aligned} \left\langle \hat{N}_i \right\rangle &= n_1\left(\mathcal{E}_i\right) = \\ &= \frac{1}{e^{\beta \mathcal{E}_i} e^{-\beta\mu} - 1} = \\ &= \frac{e^{\beta\mu} e^{-\beta \mathcal{E}_i}}{1 - e^{\beta\mu} e^{-\beta \mathcal{E}_i}}, \end{aligned} \tag{11.1}$$

F. Tafuri et al., *Understanding Quantum Many-Body Theory*, UNITEXT for Physics, https://doi.org/10.1007/978-3-032-17849-7_11

where $n_1(\mathcal{E}_i)$ is the Bose-Einstein statistics given in E.3. Since $\left\langle \hat{N}_i \right\rangle$ is nonnegative, it follows

$$e^{\beta\mathcal{E}_i}e^{-\beta\mu} - 1 \geq 0, \ \forall\, i \tag{11.2}$$

$$\mu \leq \mathcal{E}_i, \ \forall\, i, \tag{11.3}$$

which coincides with condition E.1 as stated in Theorem E.1. Without loss of generality, we set the ground-state energy to zero, $\mathcal{E}_0 = 0$, so that

$$\mathcal{E}_i > 0, \ \forall\, i > 0. \tag{11.4}$$

As $T \to 0^+$, consider any excited state $i > 0$, then

$$\left.\left\langle \hat{N}_i \right\rangle\right|_{T=0} = 0, \ \forall\, i > 0. \tag{11.5}$$

The total average number of particles $\langle \hat{N} \rangle$ is conserved, then it is equal to the average number of bosons in the fundamental state, i.e., $\langle \hat{N}_0 \rangle$, and we have

$$\begin{aligned}\left\langle \hat{N} \right\rangle = \left\langle \hat{N}_0 \right\rangle &= \\ &= \frac{e^{\beta\mu}e^{-\beta\mathcal{E}_0}}{1 - e^{\beta\mu}e^{-\beta\mathcal{E}_0}} = \\ &= \frac{e^{\beta\mu}}{1 - e^{\beta\mu}} = \\ &= \frac{\alpha}{1-\alpha},\end{aligned} \tag{11.6}$$

where we defined the *fugacity* as

$$\alpha = e^{\beta\mu}. \tag{11.7}$$

From the fugacity we get μ as follows

$$\left\langle \hat{N} \right\rangle - \alpha \left\langle \hat{N} \right\rangle = \alpha, \tag{11.8}$$

$$\alpha\left(\left\langle \hat{N} \right\rangle + 1\right) = \left\langle \hat{N} \right\rangle, \tag{11.9}$$

$$\alpha = \frac{\langle \hat{N} \rangle}{\langle \hat{N} \rangle + 1}, \tag{11.10}$$

$$\beta\mu = \ln \frac{\langle \hat{N} \rangle}{\langle \hat{N} \rangle + 1}, \tag{11.11}$$

$$\mu = k_B T \ln \frac{\langle \hat{N} \rangle}{\langle \hat{N} \rangle + 1}. \tag{11.12}$$

At the thermodynamic limit, the chemical potential tends to 0, but since it is defined as non-positive, at most null, we deduce that

$$\lim_{T \to 0^+} \lim_{\langle \hat{N} \rangle \to \infty} \mu = 0. \tag{11.13}$$

This reflects the macroscopic occupation of the ground state at low temperature, i.e., the onset of BEC.

In the opposite regime $T \to \infty$, the Bose-Einstein distribution E.3 approaches the Maxwell-Boltzmann distribution 1.355. Given that $\beta \to 0^+$, we require

$$e^{\beta(\mathcal{E}_i - \mu)} \gg 1, \tag{11.14}$$

which implies

$$\beta\mu \to -\infty. \tag{11.15}$$

In this limit, the average occupation number becomes

$$\begin{aligned} \left.\langle \hat{N}_i \rangle\right|_{T \to +\infty} &\sim e^{\beta\mu} e^{-\beta\mathcal{E}_i} \sim \\ &\sim e^{\beta\mu}, \end{aligned} \tag{11.16}$$

since

$$\lim_{\beta \to 0^+} e^{-\beta\mathcal{E}_i} = 1. \tag{11.17}$$

This shows that at infinite temperature, the particles are evenly distributed over the available excited states, and quantum statistical effects are suppressed.

11.2 Quantitative Study at the Low and High Temperature Limit for Non-Interacting Bosons

11.2.1 Average Number of Bosons in a Box

Previously, we qualitatively discussed the behavior of a system of non-interacting bosons in the low-temperature (quantum) and high-temperature (classical) regimes. We now aim to quantitatively analyze these results. To begin, consider the total average number of non-interacting bosons (see Eq. 11.7)

$$\langle \hat{N} \rangle = \sum_i \frac{e^{\beta\mu} e^{-\beta\mathcal{E}_i}}{1 - e^{\beta\mu} e^{-\beta\mathcal{E}_i}} = \\ = \sum_i \frac{\alpha e^{-\beta\mathcal{E}_i}}{1 - \alpha e^{-\beta\mathcal{E}_i}}. \tag{11.18}$$

From $\mu \leq 0$, we have $\alpha \leq 1$. Moreover, if $\mathcal{E}_i > 0$, then $e^{-\beta\mathcal{E}_i} < 1$, so in particular $\alpha e^{-\beta\mathcal{E}_i} < 1$. The function

$$\frac{1}{1 - \alpha e^{-\beta\mathcal{E}_i}} \tag{11.19}$$

can be seen as the sum of the geometric series

$$\sum_{j=0}^{N} \left(\alpha e^{-\beta\mathcal{E}_i}\right)^j, \tag{11.20}$$

and the average number of particles is written as

$$\langle \hat{N} \rangle = \sum_i \sum_{j=0}^{\infty} \alpha e^{-\beta\mathcal{E}_i} \left(\alpha e^{-\beta\mathcal{E}_i}\right)^j = \\ = \sum_i \sum_{j=0}^{\infty} \alpha^{j+1} e^{-(j+1)\beta\mathcal{E}_i} = \\ = \sum_i \sum_{j=1}^{\infty} \alpha^j e^{-j\beta\mathcal{E}_i}. \tag{11.21}$$

Since the eigenvalues of the bosonic Hamiltonian in a box are independent of the spin degree of freedom, a spin degeneracy factor g arises, and we get

$$\langle \hat{N} \rangle = g \sum_{\mathbf{k}} \sum_{j=1}^{\infty} \alpha^j e^{-j\beta\mathcal{E}_{\mathbf{k}}}. \tag{11.22}$$

Recall that the chemical potential is contained in the definition of the fugacity (see Eq. 11.7). Now, we aim to rewrite the sum over momentum states as an integral over energy, introducing the density of states $\rho(\mathcal{E})$. Accordingly, the sum over $\mathbf{k}$ is replaced by an integral with respect to the measure $d\mathcal{E}$, namely

$$\left\langle \hat{N} \right\rangle = g \sum_{j=1}^{\infty} \int d\mathcal{E} \rho(\mathcal{E}) \alpha^j e^{-j\beta\mathcal{E}}, \tag{11.23}$$

where the density of states must be defined as follows

$$\rho(\mathcal{E}) = \sum_{\mathbf{k}} \delta(\mathcal{E} - \mathcal{E}_{\mathbf{k}}). \tag{11.24}$$

In the thermodynamic limit C.11, the density of states can be rewritten in integral form as

$$\begin{aligned} \rho(\mathcal{E}) &= \frac{V}{(2\pi)^3} \int d^3\mathbf{k} \delta(\mathcal{E} - \mathcal{E}_{\mathbf{k}}) = \\ &= \frac{V}{(2\pi)^3} \int d^3\mathbf{k} \delta \left(\mathcal{E} - \frac{\hbar^2 \mathbf{k}^2}{2m} \right), \end{aligned} \tag{11.25}$$

which in spherical polar coordinates becomes

$$\begin{aligned} \rho(\mathcal{E}) &= \frac{V}{(2\pi)^3} 4\pi \int_0^{\infty} dk k^2 \delta \left(\mathcal{E} - \frac{\hbar^2 k^2}{2m} \right) = \\ &= \frac{V}{2\pi^2} \int_0^{\infty} dk k^2 \delta \left(\mathcal{E} - \frac{\hbar^2 k^2}{2m} \right). \end{aligned} \tag{11.26}$$

Now, given a decomposition

$$\delta(f(x)) = \sum_i \frac{\delta(x - x_i)}{|f'(x_i)|}, \tag{11.27}$$

where x_i is a zero of f but not of f', we recognize the identifications

$$f(x_i) \longrightarrow \mathcal{E}(k_i), \tag{11.28}$$

$$x_i \longrightarrow k_{\pm} = \pm \sqrt{\frac{2m\mathcal{E}}{\hbar^2}}, \tag{11.29}$$

$$\left| f'(x) \right| \longrightarrow \left| \frac{d\mathcal{E}}{dk} \right| = \frac{\hbar^2 k}{m}, \tag{11.30}$$

$$\left|f'(x_i)\right| \longrightarrow \left|\frac{d\mathcal{E}}{dk}(k_i)\right| = \frac{\hbar^2}{m}\sqrt{\frac{2m\mathcal{E}}{\hbar^2}}. \tag{11.31}$$

Since the integration is with respect to the positive semi-axis, we are only interested in k_+ and the density of states transforms as follows

$$\begin{aligned}\rho(\mathcal{E}) &= \frac{V}{2\pi^2}\int_0^\infty dk k^2 \frac{\delta\left(k - \sqrt{\frac{2m\mathcal{E}}{\hbar^2}}\right)}{\frac{\hbar^2}{m}\sqrt{\frac{2m\mathcal{E}}{\hbar^2}}} = \\ &= \frac{V}{2\pi^2}\frac{2m\mathcal{E}}{\hbar^2}\frac{1}{\frac{\hbar^2}{m}\sqrt{\frac{2m\mathcal{E}}{\hbar^2}}} = \\ &= \frac{V}{2\pi^2}\sqrt{\frac{2m\mathcal{E}}{\hbar^2}\frac{m}{\hbar^2}}.\end{aligned} \tag{11.32}$$

The average number of bosons becomes

$$\left\langle\hat{N}\right\rangle = g\sum_{j=1}^{\infty}\alpha^j\int_0^\infty d\mathcal{E}\frac{V}{2\pi^2}\sqrt{\frac{2m\mathcal{E}}{\hbar^2}\frac{m}{\hbar^2}}e^{-j\beta\mathcal{E}}. \tag{11.33}$$

We set $x = j\beta\mathcal{E}$, then

$$\left\langle\hat{N}\right\rangle = g\sum_{j=1}^{\infty}\alpha^j\int_0^\infty \frac{dx}{j\beta}\frac{V}{2\pi^2}\sqrt{\frac{2m\mathcal{E}}{\hbar^2}\frac{m}{\hbar^2}}e^{-x}, \tag{11.34}$$

and multiply and divide by $j\beta$ as follows

$$\begin{aligned}\left\langle\hat{N}\right\rangle &= g\sum_{j=1}^{\infty}\alpha^j\int_0^\infty \frac{dx}{(j\beta)^{\frac{3}{2}}}\frac{V}{2\pi^2}\sqrt{\frac{2mj\beta\mathcal{E}}{\hbar^2}\frac{m}{\hbar^2}}e^{-x} = \\ &= g\sum_{j=1}^{\infty}\frac{\alpha^j}{j^{\frac{3}{2}}}\int_0^\infty dx(k_BT)^{\frac{3}{2}}\frac{V}{2\pi^2}\sqrt{\frac{2m}{\hbar^2}\frac{m}{\hbar^2}}\sqrt{x}e^{-x} = \\ &= g\sum_{j=1}^{\infty}\frac{\alpha^j}{j^{\frac{3}{2}}}\left(\frac{V}{\sqrt{2}}\frac{1}{\pi^2}\left(\frac{mk_BT}{\hbar^2}\right)^{\frac{3}{2}}\right)\left[\int_0^\infty dx\sqrt{x}e^{-x}\right].\end{aligned} \tag{11.35}$$

Clearly, $\left\langle\hat{N}\right\rangle$ is a dimensionless quantity, all physical dimensions are instead contained in

$$V\left(\frac{mk_BT}{\hbar^2}\right)^{\frac{3}{2}}, \tag{11.36}$$

then the quantity

$$\left[\frac{\hbar^2}{mk_BT}\right] \tag{11.37}$$

has the dimensions of a length squared. We recall de Broglie thermal wavelength

$$\begin{aligned}\lambda_T &\equiv \sqrt{\frac{2\pi\hbar^2}{mk_BT}} \equiv \\ &\equiv \sqrt{\frac{h^2}{2\pi mk_BT}},\end{aligned} \tag{11.38}$$

we multiply and divide $\left\langle\hat{N}\right\rangle$ by $(2\pi)^{\frac{3}{2}}$ and obtain

$$\begin{aligned}\left\langle\hat{N}\right\rangle &= g\sum_{j=1}^{\infty}\frac{\alpha^j}{j^{\frac{3}{2}}}\left(\frac{V}{\sqrt{2}}\frac{1}{\pi^2}(2\pi)^{\frac{3}{2}}\left(\frac{mk_BT}{2\pi\hbar^2}\right)^{\frac{3}{2}}\right)\left[\int_0^{\infty}dx\sqrt{x}e^{-x}\right] = \\ &= g\sum_{j=1}^{\infty}\frac{\alpha^j}{j^{\frac{3}{2}}}\frac{V}{\lambda_T^3}\frac{2}{\sqrt{\pi}}\left[\int_0^{\infty}dx\sqrt{x}e^{-x}\right] = \\ &= g\sum_{j=1}^{\infty}\frac{\alpha^j}{j^{\frac{3}{2}}}\frac{V}{\lambda_T^3}\frac{2}{\sqrt{\pi}}\Gamma\left(\frac{3}{2}\right),\end{aligned} \tag{11.39}$$

where we recognize the Gamma function (see Appendix M). In particular, from

$$\begin{aligned}\Gamma\left(\frac{3}{2}\right) &= \int_0^{\infty}dx\sqrt{x}e^{-x} = \\ &= \frac{\sqrt{\pi}}{2},\end{aligned} \tag{11.40}$$

we get

$$\left\langle\hat{N}\right\rangle = g\sum_{j=1}^{\infty}\frac{\alpha^j}{j^{\frac{3}{2}}}\frac{V}{\lambda_T^3}. \tag{11.41}$$

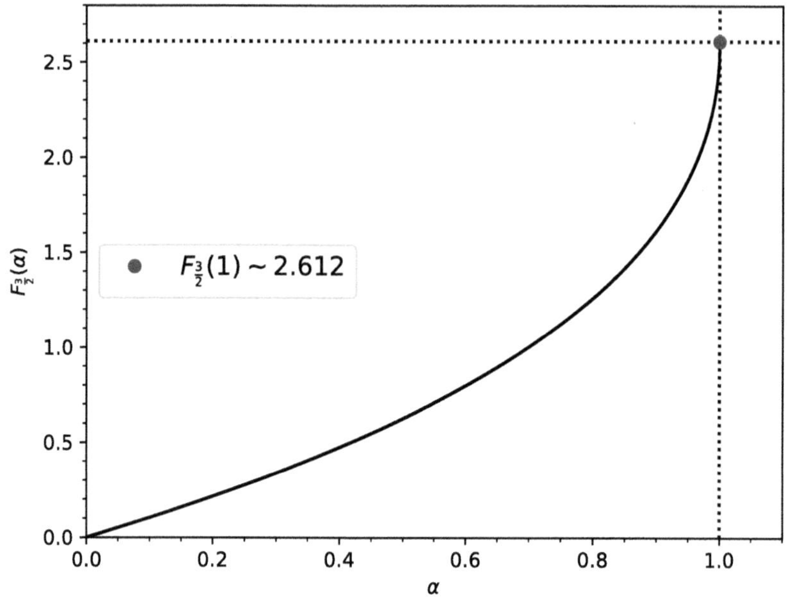

Fig. 11.1 The polylogarithmic function $F_{\frac{3}{2}}$ as a function of the fugacity α, defined in Eq. 11.7, over the interval [0, 1], including its maximum. This function appears in the expression for the average number of non-interacting bosons, expressed in terms of the thermal de Broglie wavelength, i.e., Eq. 11.41

The function

$$F_{\frac{3}{2}}(\alpha) = \sum_{j=1}^{\infty} \frac{\alpha^j}{j^{\frac{3}{2}}}, \ \alpha \in [0, 1] \tag{11.42}$$

is known in the literature as polylogarithmic function, see Fig. 11.1, and it is monotonically increasing with respect to fugacity, reaching its maximum (2.612 approximately) at $\alpha = 1$. Here, we are interested in emphasizing that for any physically allowed value of the chemical potential, the polylogarithmic function is a finite quantity, so the average number of bosons is a power law of the form

$$\left\langle \hat{N} \right\rangle \propto \frac{V}{\lambda_T^3}. \tag{11.43}$$

Indeed, at the low temperature limit, the thermal de Broglie wavelength becomes large, i.e.,

$$\frac{1}{\lambda_T^3} \sim T^{\frac{3}{2}}. \tag{11.44}$$

That is, below a certain temperature there is no longer any value of α for which Eq. 11.41 can be satisfied. This would appear to imply a violation of particle-number conservation. The issue lies in having replaced the sum over **k** with an integral. This leads to

$$\rho(\mathcal{E}) \propto \sqrt{\mathcal{E}}, \tag{11.45}$$

where the weight associated with $\mathcal{E}_0$ is seen to vanish; however, we know that the single-particle ground state is the only one that is macroscopically occupied at $T = 0$. Equation 11.41 describes the population of excited states, so the replacement of the series with the integral in the calculation of $\rho(\mathcal{E})$ is allowed for $\mathcal{E} > 0$. If we denote by $\langle \hat{N}_e \rangle$ the number of bosons in the excited states, i.e.,

$$\langle \hat{N}_e \rangle = \sum_{i \neq 0} \langle \hat{N}_i \rangle, \tag{11.46}$$

then Eq. 11.41 must be corrected as

$$\langle \hat{N}_e \rangle = g F_{\frac{3}{2}}(\alpha) \frac{V}{\lambda_T^3}, \quad T \geq T_C, \tag{11.47}$$

with

$$\langle \hat{N} \rangle = \langle \hat{N}_0 \rangle + \langle \hat{N}_e \rangle. \tag{11.48}$$

The temperature T_C is the lowest temperature at which Eq. 11.41 can be satisfied without introducing the correction factor $\langle \hat{N}_0 \rangle$. At this critical temperature, α attains its maximum value; that is, $\mu = 0$. Assuming that the chemical potential μ remains constant and strictly negative, let us consider two temperatures T and T', with $T' < T$, then

$$\beta' > \beta, \tag{11.49}$$

$$\beta' \mu < \beta \mu, \tag{11.50}$$

$$\alpha' = e^{\beta' \mu} < e^{\beta \mu} = \alpha, \tag{11.51}$$

$$F_{\frac{3}{2}}(\alpha') < F_{\frac{3}{2}}(\alpha). \tag{11.52}$$

Together with $\lambda_{T'} > \lambda_T$ (see Eqs. 11.38 and 11.41) provides

$$\langle \hat{N}' \rangle = g F_{\frac{3}{2}}(\alpha') \frac{V}{\lambda_{T'}^3} < g F_{\frac{3}{2}}(\alpha) \frac{V}{\lambda_T^3} = \langle \hat{N} \rangle, \tag{11.53}$$

which implies that the number of particles accommodated in the excited states decreases as the temperature is lowered, eventually saturating at a finite value. We manipulate the expression of the average number of bosons to bring out the dependence on the critical temperature. For $T = T_C$, Eq. 11.47 can be written as

$$\frac{\left\langle \hat{N} \right\rangle \lambda_{T_C}^3}{gV} = F_{\frac{3}{2}}(1). \tag{11.54}$$

A comparison of Eqs. 11.48 and 11.54 yields

$$\left\langle \hat{N} \right\rangle - \left\langle \hat{N}_0 \right\rangle = \left(\frac{T}{T_C}\right)^{\frac{3}{2}} \left\langle \hat{N} \right\rangle, \tag{11.55}$$

$$\left\langle \hat{N}_0 \right\rangle = \left[1 - \left(\frac{T}{T_C}\right)^{\frac{3}{2}}\right] \left\langle \hat{N} \right\rangle, \tag{11.56}$$

which is the average number of bosons in the fundamental state as a function of temperature. Obviously, the number of bosons in the fundamental state increases as the temperature decreases, and BEC is complete when $T = 0$. Finally, if we write Eq. 11.54 as

$$\frac{\left\langle \hat{N} \right\rangle \lambda_{T_C}^3}{V} = gF_{\frac{3}{2}}(1), \tag{11.57}$$

that is, we note that BEC becomes observable when the De Broglie thermal wavelength is comparable to the average spatial distance between bosons.

Remark 11.1 (Absence of Bose-Einstein Condensation in One and Two Dimensions) The possibility of Bose-Einstein condensation crucially depends on the low-energy behavior of the density of states. For non-interacting bosons confined in a box, it is straightforward to show that the density of states in one and two spatial dimensions reads

$$\rho_{1\mathrm{D}}(\mathcal{E}) = \frac{L}{\pi}\sqrt{\frac{m}{2\hbar^2}}\frac{1}{\sqrt{\mathcal{E}}}, \tag{11.58}$$

$$\rho_{2\mathrm{D}}(\mathcal{E}) = \rho_{2\mathrm{D}} = \frac{A\,m}{2\pi\hbar^2}. \tag{11.59}$$

Repeating the same steps leading to Eq. 11.47, one finds that the integral in Eq. 11.23 diverges for both $d = 1$ and $d = 2$, for $\alpha = 1$ (see Eq. 11.7). As a consequence, the number of particles that can be accommodated in the excited states is not bounded from above, and no macroscopic occupation of the ground state occurs at any finite temperature. Therefore, Bose-Einstein condensation is not allowed in one and two dimensions for an ideal homogeneous Bose gas.

11.2.2 Average Energy

Here, we calculate the grand potential and the average energy of a system of non-interacting bosons. These quantities allow us to derive thermodynamic properties such as pressure and internal energy, providing insight into the statistical behavior of the bosonic gas. For convenience, and only within a part of this section, we make use of

$$\mathcal{E}'_{\mathbf{k}} = \mathcal{E}_{\mathbf{k}} - \mu, \tag{11.60}$$

where μ is the chemical potential. In the grand canonical ensemble, the grand partition function Z is obtained as a trace over all Fock space states:

$$Z = \sum_{N=0}^{\infty} \sum_{\{N_{\mathbf{k},\sigma}\}} \left\langle N_0, N_1, \ldots \left| e^{-\beta \hat{\mathcal{H}}} \right| N_0, N_1, \ldots \right\rangle, \tag{11.61}$$

where the second sum runs over all occupation-number states $\{N_{\mathbf{k},\sigma}\}$ (see Sect. 2.7.3) with a fixed number of particles N. The Hamiltonian acts diagonally on these states, yielding

$$\begin{aligned}\hat{\mathcal{H}}|N_0, N_1, \ldots\rangle &= \sum_{\mathbf{k},\sigma} \mathcal{E}'_{\mathbf{k}} \hat{N}_{\mathbf{k},\sigma} |N_0, N_1, \ldots\rangle = \\ &= \sum_{\mathbf{k},\sigma} \mathcal{E}'_{\mathbf{k}} N_{\mathbf{k},\sigma} |N_0, N_1, \ldots\rangle,\end{aligned} \tag{11.62}$$

so that the grand partition function becomes

$$\begin{aligned}Z &= \sum_{N_0=0}^{\infty} \sum_{N_1=0}^{\infty} \cdots e^{-\beta \sum_{\mathbf{k},\sigma} \mathcal{E}'_{\mathbf{k}} N_{\mathbf{k},\sigma}} = \\ &= \prod_{\mathbf{k},\sigma} \sum_{N=0}^{\infty} e^{-\beta \mathcal{E}'_{\mathbf{k}} N}.\end{aligned} \tag{11.63}$$

Since for all excited states we have $\mathcal{E}'_{\mathbf{k}} > 0$ (see Eqs. 11.3 and 11.4), the geometric series converges (see Eq. E.11), i.e.,

$$\sum_{N=0}^{\infty} e^{-\beta \mathcal{E}'_{\mathbf{k}} N} = \frac{1}{1 - e^{-\beta \mathcal{E}'_{\mathbf{k}}}}. \tag{11.64}$$

Hence, the grand partition function can be written as

$$Z = \prod_{\mathbf{k},\sigma} \frac{1}{1 - e^{-\beta \mathcal{E}'_{\mathbf{k}}}}. \tag{11.65}$$

From the statistical mechanics, the grand potential $\Omega(V, T, \mu)$ is related to the grand partition function via $\Omega = -k_B T \ln Z$, then the grand potential describing non-interacting bosons is given by

$$\begin{aligned}\Omega &= -k_B T \ln \prod_{\mathbf{k},\sigma} \frac{1}{1 - e^{-\beta \mathcal{E}'_{\mathbf{k}}}} = \\ &= -k_B T \sum_{\mathbf{k},\sigma} \ln \frac{1}{1 - e^{-\beta \mathcal{E}'_{\mathbf{k}}}} = \\ &= -g k_B T \sum_{\mathbf{k}} \ln \frac{1}{1 - e^{-\beta \mathcal{E}'_{\mathbf{k}}}} = \\ &= g k_B T \sum_{\mathbf{k}} \ln \left(1 - e^{-\beta \mathcal{E}'_{\mathbf{k}}}\right) = \\ &= g k_B T \ln \left(1 - e^{\beta \mu}\right) + g k_B T \sum_{\mathbf{k} \neq \mathbf{0}} \ln \left(1 - e^{-\beta \mathcal{E}'_{\mathbf{k}}}\right),\end{aligned} \tag{11.66}$$

where in the last step we separated the ground state from the excited states. In the thermodynamic limit, the terms of excited states can be written in a integral form by means of the density of states (see Eq. 11.32) as follows

$$\begin{aligned}g k_B T \sum_{\mathbf{k} \neq \mathbf{0}} \ln \left(1 - e^{-\beta \mathcal{E}'_{\mathbf{k}}}\right) &\simeq g k_B T \int_0^\infty d\mathcal{E} \rho(\mathcal{E}) \ln \left(1 - e^{-\beta(\mathcal{E} - \mu)}\right) = \\ &= g k_B T \frac{V}{2\pi^2} \sqrt{\frac{2m}{\hbar^2} \frac{m}{\hbar^2}} \int_0^\infty d\mathcal{E} \sqrt{\mathcal{E}} \ln \left(1 - e^{-\beta(\mathcal{E} - \mu)}\right),\end{aligned} \tag{11.67}$$

where we made explicit $\mathcal{E}'_{\mathbf{k}} = \mathcal{E}_{\mathbf{k}} - \mu$, and the discrete variable $\mathcal{E}_{\mathbf{k}}$ is replaced by the continuous measure $d\mathcal{E}$. For simplicity, we omit the numerical constants and perform integration by parts, i.e.,

$$\begin{aligned}&\frac{2}{3} \int_0^\infty d\mathcal{E} \frac{3}{2} \sqrt{\mathcal{E}} \ln \left(1 - e^{-\beta(\mathcal{E} - \mu)}\right) = \\ &= \frac{2}{3} \left[\mathcal{E}^{\frac{3}{2}} \ln \left(1 - e^{-\beta(\mathcal{E} - \mu)}\right)\right]_0^\infty - \frac{2}{3} \int_0^\infty d\mathcal{E} \frac{\mathcal{E}^{\frac{3}{2}} (-e^{-\beta \mathcal{E}} e^{\beta \mu})(-\beta)}{1 - e^{-\beta \mathcal{E}} e^{\beta \mu}} = \\ &= -\frac{2}{3} \beta \int_0^\infty d\mathcal{E} \frac{\mathcal{E}^{\frac{3}{2}} e^{-\beta \mathcal{E}} e^{\beta \mu}}{1 - e^{-\beta \mathcal{E}} e^{\beta \mu}} = \\ &= -\frac{2}{3} \beta \int_0^\infty d\mathcal{E} \frac{\mathcal{E}^{\frac{3}{2}}}{e^{\beta(\mathcal{E} - \mu)} - 1}.\end{aligned} \tag{11.68}$$

Now we insert the numerical constants and multiply and divide by 2

$$
\begin{aligned}
& - g k_B T \frac{V}{2\pi^2} \sqrt{\frac{2m}{\hbar^2}} \frac{m}{\hbar^2} \frac{2}{3} \beta \int_0^{\infty} d\mathcal{E} \frac{\mathcal{E}^{\frac{3}{2}}}{e^{\beta(\mathcal{E}-\mu)} - 1} \\
& = -\frac{2}{3} g \frac{V}{(2\pi)^2} \left(\frac{2m}{\hbar^2} \right)^{\frac{3}{2}} \int_0^{\infty} d\mathcal{E} \frac{\mathcal{E}^{\frac{3}{2}}}{e^{\beta(\mathcal{E}-\mu)} - 1},
\end{aligned}
\tag{11.69}
$$

then the grand potential of a system of non-interacting bosons is given by

$$
\Omega = g k_B T \ln \left(1 - e^{\beta\mu}\right) - \frac{2}{3} g \frac{V}{(2\pi)^2} \left(\frac{2m}{\hbar^2} \right)^{\frac{3}{2}} \int_0^{\infty} d\mathcal{E} \frac{\mathcal{E}^{\frac{3}{2}}}{e^{\beta(\mathcal{E}-\mu)} - 1}. \tag{11.70}
$$

The pressure of the system is defined as

$$
\begin{aligned}
p &= - \left. \frac{\partial \Omega}{\partial V} \right|_{T,\mu} = \\
&= \frac{2}{3} g \frac{1}{(2\pi)^2} \left(\frac{2m}{\hbar^2} \right)^{\frac{3}{2}} \int_0^{\infty} d\mathcal{E} \frac{\mathcal{E}^{\frac{3}{2}}}{e^{\beta(\mathcal{E}-\mu)} - 1},
\end{aligned}
\tag{11.71}
$$

which does not depend on the condensate term: in other words, the condensate exerts no pressure on the walls of the box. Now we compute the average value of the internal energy U, which is related to the grand partition function as

$$
U = -\frac{\partial}{\partial \beta} \ln Z \,|_{\beta\mu}, \tag{11.72}
$$

with

$$
\ln Z = -g \ln \left(1 - e^{\beta\mu}\right) + g \sum_{\mathbf{k}\neq\mathbf{0}} \ln \left(1 - e^{-\beta \mathcal{E}'_{\mathbf{k}}}\right). \tag{11.73}
$$

It follows

$$
\begin{aligned}
U &= -g \sum_{\mathbf{k}\neq\mathbf{0}} \frac{(-\mathcal{E}_{\mathbf{k}}) e^{-\beta\mathcal{E}_{\mathbf{k}}} e^{\beta\mu}}{1 - e^{-\beta\mathcal{E}_{\mathbf{k}}} e^{\beta\mu}} = \\
&= g \sum_{\mathbf{k}\neq\mathbf{0}} \frac{\mathcal{E}_{\mathbf{k}} e^{-\beta\mathcal{E}_{\mathbf{k}}} e^{\beta\mu}}{1 - e^{-\beta\mathcal{E}_{\mathbf{k}}} e^{\beta\mu}} = \\
&= g \sum_{\mathbf{k}\neq\mathbf{0}} \frac{\mathcal{E}_{\mathbf{k}}}{e^{\beta(\mathcal{E}_{\mathbf{k}}-\mu)} - 1},
\end{aligned}
\tag{11.74}
$$

and by means of the density of states 11.32, we have

$$U = g\frac{V}{(2\pi)^2}\left(\frac{2m}{\hbar^2}\right)^{\frac{3}{2}}\int_0^\infty d\mathcal{E}\frac{\mathcal{E}^{\frac{3}{2}}}{e^{\beta(\mathcal{E}-\mu)}-1}. \tag{11.75}$$

Upon comparing the expressions for the internal energy U and the pressure p, one finds that the gas of non-interacting bosons obeys the relation

$$pV = \frac{2}{3}U. \tag{11.76}$$

11.3 A Comparison with Classical Physics Scenario

We analyze the high-temperature limit of a non-interacting Bose gas, where quantum effects become negligible. In this regime, the classical expression for the average energy is recovered, thereby illustrating the principle of energy equipartition.

11.3.1 Average Number of Bosons at High Temperatures

At high temperatures, Bose-Einstein statistics E.3 can be replaced by Maxwell-Boltzmann statistics 1.355, namely

$$\begin{aligned}\left\langle \hat{N}_{\mathbf{k}} \right\rangle &= \frac{1}{e^{\beta\mathcal{E}_{\mathbf{k}}}e^{-\beta\mu}-1} \approx \\ &\approx e^{-\beta\mathcal{E}_{\mathbf{k}}}e^{\beta\mu},\end{aligned} \tag{11.77}$$

then

$$\begin{aligned}\left\langle \hat{N} \right\rangle &= \sum_{\mathbf{k},\sigma}\left\langle \hat{N}_{\mathbf{k}} \right\rangle \approx \\ &\approx g e^{\beta\mu}\sum_{\mathbf{k}} e^{-\beta\mathcal{E}_{\mathbf{k}}},\end{aligned} \tag{11.78}$$

where we employed Eq. 2.251 in the thermodynamic limit (see Eq. C.11), we get

$$\begin{aligned}\left\langle \hat{N} \right\rangle &\approx g\frac{V}{(2\pi)^3}e^{\beta\mu}\int d^3\mathbf{k}e^{-\beta\mathcal{E}_{\mathbf{k}}} = \\ &= g\frac{V}{(2\pi)^3}4\pi e^{\beta\mu}\int_0^\infty dk k^2 e^{-\beta\mathcal{E}_{\mathbf{k}}}.\end{aligned} \tag{11.79}$$

We define the dimensionless variable

$$x = \hbar k \sqrt{\frac{\beta}{2m}}, \tag{11.80}$$

then

$$x^3 = \hbar^3 k^3 \left(\frac{\beta}{2m}\right)^{\frac{3}{2}}, \tag{11.81}$$

$$3x^2 dx = 3\hbar^3 k^2 \left(\frac{\beta}{2m}\right)^{\frac{3}{2}} dk, \tag{11.82}$$

$$x^2 dx = \hbar^3 k^2 \left(\frac{\beta}{2m}\right)^{\frac{3}{2}} dk. \tag{11.83}$$

We multiply and divide the integral by

$$\frac{(2mk_B T)^{\frac{3}{2}}}{\hbar^3} \equiv \left(\frac{2mk_B T}{\hbar^2}\right)^{\frac{3}{2}}, \tag{11.84}$$

and we get

$$\begin{aligned}
\left\langle \hat{N} \right\rangle &= g \frac{V}{(2\pi)^3} 4\pi e^{\beta\mu} \left(\frac{2mk_B T}{\hbar^2}\right)^{\frac{3}{2}} \int_0^\infty dk e^{-\beta \frac{\hbar^2 k^2}{2m}} \hbar^3 k^2 \left(\frac{1}{2mk_B T}\right)^{\frac{3}{2}} \equiv \\
&\equiv g \frac{V}{(2\pi)^3} 4\pi e^{\beta\mu} \left(\frac{2mk_B T}{\hbar^2}\right)^{\frac{3}{2}} \int_0^\infty dx e^{-x^2} x^2 = \\
&= g \frac{V}{(2\pi)^3} 4\pi \frac{\sqrt{\pi}}{4} e^{\beta\mu} \left(\frac{2mk_B T}{\hbar^2}\right)^{\frac{3}{2}} = \\
&= g \frac{V}{2^3 \pi^{\frac{3}{2}}} e^{\beta\mu} \left(\frac{2mk_B T}{\hbar^2}\right)^{\frac{3}{2}} = \\
&= g V e^{\beta\mu} \left(\frac{mk_B T}{2\pi\hbar^2}\right)^{\frac{3}{2}}.
\end{aligned} \tag{11.85}$$

It follows

$$e^{\beta\mu} = \frac{\left\langle \hat{N} \right\rangle}{V} \frac{1}{g} \left(\frac{2\pi\hbar^2}{mk_B T}\right)^{\frac{3}{2}}, \tag{11.86}$$

$$e^{\beta\mu} = \frac{\langle \hat{N} \rangle}{V} \frac{\lambda_T^3}{g}, \tag{11.87}$$

$$\beta\mu = \ln \left[\frac{\langle \hat{N} \rangle}{V} \frac{\lambda_T^3}{g} \right]. \tag{11.88}$$

By recalling the definition of the thermal de Broglie wavelength, one obtains the following limiting expressions

$$\lim_{T \to +\infty} \beta\mu = -\infty, \tag{11.89}$$

$$\lim_{T \to 0^+} \beta\mu = +\infty. \tag{11.90}$$

At lower temperatures, quantum effects become significant, and the classical approximation fails to capture the emergence of collective phenomena such as BEC.

11.3.2 Average Energy in the High-Temperature Regime

In the high-temperature regime and in the thermodynamic limit (see Eq. C.11), we get (see Eq. 2.251)

$$\begin{aligned} U &= \sum_{\mathbf{k},\sigma} \mathcal{E}_{\mathbf{k}} \left\langle \hat{N}_{\mathbf{k}} \right\rangle = \\ &= g \sum_{\mathbf{k}} \mathcal{E}_{\mathbf{k}} \frac{1}{e^{\beta \mathcal{E}_{\mathbf{k}}} e^{-\beta\mu} - 1} \approx \\ &\approx g \sum_{\mathbf{k}} \mathcal{E}_{\mathbf{k}} e^{-\beta \mathcal{E}_{\mathbf{k}}} e^{\beta\mu} \approx \\ &\approx g \frac{V}{(2\pi)^3} e^{\beta\mu} \int d^3\mathbf{k} \frac{\hbar^2 k^2}{2m} e^{-\beta \frac{\hbar^2 k^2}{2m}} = \\ &= g \frac{V}{(2\pi)^3} 4\pi e^{\beta\mu} \int_0^\infty dk \frac{\hbar^2 k^4}{2m} e^{-\beta \frac{\hbar^2 k^2}{2m}}. \end{aligned} \tag{11.91}$$

Again, using the variable x from Eq. 11.80, we get

$$x^5 = \hbar^5 k^5 \left(\frac{\beta}{2m} \right)^{\frac{5}{2}}, \tag{11.92}$$

$$5x^4 dx = 5\hbar^5 k^4 \left(\frac{\beta}{2m}\right)^{\frac{5}{2}} dk, \tag{11.93}$$

$$x^4 dx = \hbar^5 k^4 \left(\frac{\beta}{2m}\right)^{\frac{5}{2}} dk. \tag{11.94}$$

Multiplying and dividing the integral by $\frac{\hbar^3 \beta^{\frac{5}{2}}}{(2m)^{\frac{3}{2}}}$, we get

$$\begin{aligned} U &= g\frac{V}{(2\pi)^3} 4\pi e^{\beta\mu} \frac{(2m)^{\frac{3}{2}}}{\hbar^3 \beta^{\frac{5}{2}}} \int_0^\infty dk \hbar^5 k^4 \left(\frac{\beta}{2m}\right)^{\frac{5}{2}} e^{-\beta\frac{\hbar^2 k^2}{2m}} = \\ &= g\frac{V}{(2\pi)^3} 4\pi e^{\beta\mu} \frac{(2m)^{\frac{3}{2}}}{\hbar^3 \beta^{\frac{5}{2}}} \int_0^\infty dx x^4 e^{-x^2} \equiv \\ &\equiv g\frac{V}{(2\pi)^3} 4\pi e^{\beta\mu} \frac{(2m)^{\frac{3}{2}} (k_B T)^{\frac{5}{2}}}{\hbar^3} \int_0^\infty dx x^4 e^{-x^2} = \\ &= g\frac{V}{(2\pi)^3} 4\pi \frac{3\sqrt{\pi}}{8} e^{\beta\mu} \frac{(2m)^{\frac{3}{2}} (k_B T)^{\frac{5}{2}}}{\hbar^3}. \end{aligned} \tag{11.95}$$

Now, we manipulate physical and numerical quantities, excluding V and g, as follows

$$\begin{aligned} \frac{m^{\frac{3}{2}} (k_B T)^{\frac{5}{2}}}{\hbar^3} &= k_B T \frac{m^{\frac{3}{2}} (k_B T)^{\frac{3}{2}}}{\hbar^3} = \\ &= k_B T \left(\frac{m k_B T}{\hbar^2}\right)^{\frac{3}{2}}, \end{aligned} \tag{11.96}$$

$$\begin{aligned} \frac{(2^{\frac{3}{2}})(4)(3)}{(8)(8)} \frac{\pi^{\frac{1}{2}} \pi}{\pi^2} &= 3\frac{2^{\frac{3}{2}}}{2^4} \frac{1}{\pi^{\frac{3}{2}}} = \\ &= 3\frac{2^3}{(2^4)(2^{\frac{3}{2}})} \frac{1}{\pi^{\frac{3}{2}}} = \\ &= \frac{3}{2} \frac{1}{(2\pi)^{\frac{3}{2}}}, \end{aligned} \tag{11.97}$$

and we write

$$
\begin{aligned}
U &= \frac{3}{2} g k_B T V e^{\beta\mu} \left(\frac{m k_B T}{2\pi\hbar^2} \right)^{\frac{3}{2}} = \\
&= \frac{3}{2} g k_B T V e^{\beta\mu} \frac{1}{\lambda_T^3} = \\
&= \frac{3}{2} \langle \hat{N} \rangle k_B T,
\end{aligned} \tag{11.98}
$$

where we included the average number of bosons at high temperatures given in Eq. 11.85, and we get the energy equipartition principle.

11.4 The Specific Heat of a Non-Interacting Boson System

Here, we analyze the temperature dependence of the specific heat of a non-interacting Bose gas. We focus on the distinct behaviors in the low- and high-temperature regimes, highlighting the influence of Bose-Einstein condensation on the system's thermodynamic properties. From energy given in Eq. 11.75, derived within a quantum framework and valid at all temperatures, we now focus on its behavior in the low-temperature regime. As previously established, BEC occurs below a critical temperature T_C, at which point the bosonic gas undergoes a phase transition and a macroscopic number of particles occupies the ground state. We have

$$
\begin{aligned}
U &= g \frac{V}{(2\pi)^2} \left(\frac{2m}{\hbar^2} \right)^{\frac{3}{2}} \int_0^\infty d\mathcal{E} \frac{\mathcal{E}^{\frac{3}{2}}}{e^{\beta(\mathcal{E}-\mu)} - 1} = \\
&= g \frac{V}{(2\pi)^2} \left(\frac{2m}{\hbar^2} \right)^{\frac{3}{2}} \int_0^\infty d\mathcal{E} \frac{\mathcal{E}^{\frac{3}{2}}}{e^{\beta\mathcal{E}} - 1},
\end{aligned} \tag{11.99}
$$

given that for temperatures much lower than the critical temperature, $T < T_C$, it follows that $e^{-\beta\mu} \to 1$. If we rewrite the variable x from Eq. 11.80 as

$$
x = \sqrt{\beta\mathcal{E}}, \tag{11.100}
$$

from

$$
x^5 = \beta^{\frac{5}{2}} \mathcal{E}^{\frac{5}{2}}, \tag{11.101}
$$

$$
5x^4 dx = \frac{5}{2} \beta^{\frac{5}{2}} \mathcal{E}^{\frac{3}{2}} d\mathcal{E}, \tag{11.102}
$$

$$x^4 dx = \frac{1}{2}\beta^{\frac{5}{2}}\mathcal{E}^{\frac{3}{2}} d\mathcal{E}, \tag{11.103}$$

we multiply and divide U by $\frac{\beta^{\frac{5}{2}}}{2}$ and we get

$$\begin{aligned} U &= \frac{2}{\beta^{\frac{5}{2}}} g \frac{V}{(2\pi)^2}\left(\frac{2m}{\hbar^2}\right)^{\frac{3}{2}} \int_0^\infty dx \frac{x^4}{e^{\sqrt{x}}-1} \equiv \\ &\equiv 2g \frac{V}{(2\pi)^2}\left(\frac{2m}{\hbar^2}\right)^{\frac{3}{2}} (k_B T)^{\frac{5}{2}} \int_0^\infty dx \frac{x^4}{e^{\sqrt{x}}-1}. \end{aligned} \tag{11.104}$$

It is now clear that the internal energy U of a system of non-interacting bosons follows a power-law dependence on temperature in the low-temperature regime, that is,

$$U \propto T^{\frac{5}{2}}, \quad T \ll T_C. \tag{11.105}$$

In the high-temperature limit, the Bose-Einstein statistics (see Eq. E.3) approaches the Maxwell-Boltzmann statistics (see Eq. 1.355). According to the equipartition theorem, the internal energy then scales linearly with temperature

$$U \propto T, \quad T \gg T_C. \tag{11.106}$$

As a result, the specific heat at constant volume, C_V, behaves as follows

$$C_V = \frac{\partial U}{\partial T} \propto \begin{cases} T^{\frac{3}{2}}, & T \ll T_C, \\ \frac{3}{2}R, & T \gg T_C, \end{cases} \tag{11.107}$$

where R is the ideal gas constant. According to this simplified model, $C_V(T)$ features a cusp at the critical temperature $T = T_C$.

It is important to stress that the above result is a direct consequence of the assumption of a non-interacting Bose gas. While Bose–Einstein condensation correctly captures the macroscopic occupation of the ground state below T_C, the ideal-gas model fails to reproduce the experimentally observed behavior of the specific heat at low temperatures. In particular, measurements in superfluid helium and in weakly interacting ultracold Bose gases show a low-temperature specific heat dominated by phonon excitations, leading to a $C_V \propto T^3$ behavior in three dimensions, rather than the $T^{\frac{3}{2}}$ law obtained here. In addition, at the critical temperature the noninteracting model predicts a cusp, whereas measurements on liquid helium indicate a singularity.

11.5 Landau's Theory of Superfluidity

The failure of the ideal Bose gas to reproduce the low-temperature thermodynamics observed experimentally indicates that Bose-Einstein condensation alone is not sufficient to account for superfluid behavior. In particular, a correct description requires a theory of the elementary excitations of the condensed phase and of their role in the energetic stability of a flowing state. This is precisely the problem addressed by Landau's theory of superfluidity. Superfluidity is closely related to Bose-Einstein condensation, although the two notions are not identical. Superfluidity is an emergent dynamical property of a quantum fluid: it refers to the ability of the system to sustain a macroscopic flow without dissipation below a certain critical velocity. When one states that a superfluid has no friction, this assertion requires a careful interpretation. Friction is not an intrinsic or absolute property of a physical system, but rather refers to the dissipation of kinetic energy associated with motion relative to an external reference, typically the confining walls or the surrounding environment. Without specifying such a reference frame, the statement lacks a precise physical meaning.

Landau's theory of superfluidity finds its primary physical realization in liquid helium. The atoms of helium-4 are bosons and are characterized by a small mass and relatively weak interatomic interactions. These properties enhance quantum effects at low temperatures and prevent crystallization at ordinary pressures, so that helium does not solidify even near absolute zero. For these reasons, the system may be regarded, at least qualitatively, as a dense quantum fluid. Upon cooling, liquid helium undergoes a phase transition signaled by a pronounced anomaly in the specific heat, marking the onset of a qualitatively new low-temperature phase. Above this temperature the fluid behaves as an ordinary liquid (helium I), whereas below it enters a distinct phase (helium II) characterized by superfluid behavior. The existence of this phase provides the paradigmatic example of superfluidity and historically motivated Landau's theoretical formulation.

Consider a fluid flowing through a pipe. Saying that the fluid slows down means that its velocity decreases with respect to the pipe. There is no notion of absolute slowing down: by changing the reference frame, one could equally say that the fluid is at rest while the pipe moves in the opposite direction. What matters physically is whether the energy associated with the relative motion between the fluid and the walls decreases over time. In an ordinary fluid this relative kinetic energy is dissipated because many processes are available to convert it into heat: random collisions, vortices, surface excitations, density fluctuations. Through these processes the fluid can transfer energy to the walls or to the environment, and in the laboratory frame the flow eventually comes to rest. The situation is fundamentally different in a superfluid, but not because the walls cease to interact with the fluid, nor because the fluid is somehow isolated. Interactions with the environment are still present. The crucial point is that, below a certain velocity, there exists no physically allowed process that can convert the energy of the relative motion into internal excitations of the fluid. The energy is present, but it has nowhere to go.

This is the essence of Landau's idea. Dissipation is not forbidden by a general principle; it is forbidden because, when viewed in the reference frame of the walls, all possible excitations of the fluid cost energy rather than release it. Creating them would increase the total energy instead of lowering it, so such processes do not occur. As a consequence, the relative motion between the fluid and the walls is stable. This is why superfluidity should not be understood as the absence of friction in absolute terms. In a reference frame moving with the fluid, the fluid is at rest and the very notion of friction of the fluid is meaningless. The physical content of superfluidity lies entirely in the fact that, in the frame of the walls, the relative flow cannot dissipate energy. In this sense, a superfluid is not a fluid that does not interact with its surroundings, but a fluid that, as long as it flows slowly enough, has no available energetic channel to lose its velocity relative to the walls. This is precisely what Landau made quantitative through his criterion for superfluidity.

Having clarified what superfluidity means, one can now address its connection with Bose-Einstein condensation. Landau's criterion, by itself, states only that a fluid is superfluid if the spectrum of its elementary excitations has a suitable form, such that dissipation is impossible below a finite critical velocity. However, the criterion does not explain why the excitation spectrum should have this form. This is where Bose-Einstein condensation becomes relevant. In a Bose-Einstein condensate, a macroscopic fraction of the particles occupies the same quantum state. As a result, the system can no longer be viewed as a collection of independent particles moving individually. The motion of the fluid acquires a collective character. The ground state of the system is described by a macroscopic wave function with a well-defined phase over large distances. Small perturbations of this state do not correspond to single particles changing their momentum, but rather to collective oscillations of the condensate as a whole. At this stage, no assumption is made on the detailed form of the excitation spectrum.

The formulation of Landau's criterion is based on the Galilean transformation properties of energy and momentum between the rest frame of the fluid and the laboratory frame. Consider a homogeneous quantum fluid whose elementary excitations are characterized, in the reference frame in which the fluid is at rest, by an energy–momentum relation $\mathcal{E}(\mathbf{p})$. In this frame, the ground state of the system has energy $\mathcal{E}_0$ and zero total momentum. We consider an elementary excitation of the fluid with momentum $\mathbf{p}$, defined in the reference frame where the fluid is at rest. A state containing a single such excitation then has energy

$$\mathcal{E} = \mathcal{E}_0 + \mathcal{E}(\mathbf{p}). \tag{11.108}$$

We now consider the same system as observed in a reference frame in which the fluid moves with a constant velocity $\mathbf{v}$ relative to the confining walls. In the rest frame of the fluid, the total momentum of the ground state vanishes. The creation of a single elementary excitation carrying momentum $\mathbf{p}$ changes the total momentum of the system by $\mathbf{p}$, so that the total momentum of the fluid-excitation system is $\mathbf{P} = \mathbf{p}$. The total mass of the system is denoted by M. Since the system is Galilean invariant, the energy and momentum in the laboratory frame are related to those in the rest frame of the fluid by the standard Galilean transformation. Consequently, the energy transforms as

$$\mathcal{E}' = \mathcal{E} + \mathbf{v} \cdot \mathbf{P} + \frac{1}{2} M v^2. \tag{11.109}$$

Applying this transformation to the ground state, whose momentum vanishes in the rest frame, one finds that in the laboratory frame its energy is simply

$$\mathcal{E}_0' = \mathcal{E}_0 + \frac{1}{2} M v^2, \tag{11.110}$$

which corresponds to the macroscopic kinetic energy of the fluid flowing with velocity $\mathbf{v}$. Let us now consider a state in which an excitation of momentum $\mathbf{p}$ is present. In the laboratory frame, its energy is given by

$$\mathcal{E}' = \mathcal{E}_0 + \mathcal{E}(\mathbf{p}) + \mathbf{v} \cdot \mathbf{p} + \frac{1}{2} M v^2. \tag{11.111}$$

The energy difference between this state and the flowing ground state is therefore

$$\Delta\mathcal{E}' = \mathcal{E}(\mathbf{p}) + \mathbf{v} \cdot \mathbf{p}. \tag{11.112}$$

The physical meaning of this expression is immediately apparent. In the reference frame of the walls, the fluid can dissipate energy only if it is possible to create an excitation such that the total energy of the system decreases. This requires the existence of at least one momentum $\mathbf{p}$ for which $\Delta\mathcal{E}' < 0$. If no such excitation exists, the flow is energetically stable and dissipation cannot occur. For a given excitation momentum $\mathbf{p}$, the quantity $\mathbf{v} \cdot \mathbf{p}$ is minimized when $\mathbf{p}$ is antiparallel to the flow velocity. In this case, the condition for stability becomes

$$\mathcal{E}(p) - vp \geq 0, \tag{11.113}$$

where $p = |\mathbf{p}|$ and $v = |\mathbf{v}|$. The flow is therefore stable provided that

$$v \leq \frac{\mathcal{E}(p)}{p} \tag{11.114}$$

for all possible excitation momenta. This leads directly to Landau's criterion for superfluidity. The critical velocity is defined as

$$v_c = \min_p \frac{\mathcal{E}(p)}{p}. \tag{11.115}$$

If this minimum is finite, then for all velocities $v < v_c$ no excitation can be created that lowers the energy in the laboratory frame, and the flow is dissipationless.

Conversely, if the minimum vanishes, excitations with arbitrarily small energy cost exist and the critical velocity is zero: dissipation is unavoidable, no matter how slow the flow.

Landau's criterion immediately reveals a fundamental limitation of the ideal Bose-Einstein condensate. Although condensation implies macroscopic occupation of a single quantum state, the quadratic low-momentum dispersion of the ideal gas leads to a vanishing critical velocity. Superfluidity therefore does not follow from condensation alone, but requires a qualitative modification of the excitation spectrum. Such a modification arises only when interparticle interactions are taken into account, a point that will be addressed in the following chapter.

Chapter 12
Bogoljubov's Theory for Interacting Bosonic Systems

Abstract In this chapter, we present a detailed development of Bogoljubov's theory for weakly interacting bosonic systems, offering a self-contained and rigorous formulation from physical assumptions to experimentally relevant predictions. The core idea is to exploit the macroscopic occupation of the ground state in a Bose-Einstein condensate, thereby approximating the many-body Hamiltonian in a quadratic, diagonalizable form. Starting from the second-quantized Hamiltonian with two-body interactions, we apply Bogoljubov's key assumption, treating the zero-momentum component of the field as a classical variable, and derive the conditions for its validity. This leads to a simplified Hamiltonian expressible via bosonic creation and annihilation operators. We then introduce the Bogoljubov canonical transformations, designed to preserve bosonic commutation relations, enabling the approximate diagonalization of the Hamiltonian and identification of quasiparticle excitations. The resulting quasiparticle spectrum is derived explicitly, showing a linear dispersion at low momenta that corresponds to phonon-like collective excitations and confirms the microscopic origin of superfluidity. Finally, we analyse the ground state, computing both the condensate depletion and quantum correction to the vacuum energy, thereby establishing a quantitative link between microscopic theory and macroscopic properties of interacting bosonic systems.

12.1 Bogoljubov Assumption

Previously (see Chap. 11) we demonstrated that a quantitatively accurate description of the specific heat of a bosonic system necessitates accounting for inter-particle interactions, particularly in the vicinity of the condensation transition. In the present chapter, we initiate the analysis of the interacting Bose gas by considering the simplest regime, namely the zero-temperature limit. At $T = 0$, one expects that a large macroscopic part of bosons within a Bose-Einstein condensate occupy a one-particle quantum state. We denote by $\varphi_0 = |0\rangle$ the ground-state eigenfunction of the one-particle Hamiltonian, which defines the condensate mode. Consequently, the many-body wavefunction $|\phi_N\rangle$ of a condensate comprising N bosons can be expressed as (see Eq. 2.142)

F. Tafuri et al., *Understanding Quantum Many-Body Theory*, UNITEXT for Physics, https://doi.org/10.1007/978-3-032-17849-7_12

$$|\phi_N\rangle = \frac{\left(a_0^\dagger\right)^N}{\sqrt{N!}}|0\rangle, \tag{12.1}$$

where $a_0^\dagger$ creates a boson in the state φ_0. Recall that the number operator satisfies (see Sect. 1.3)

$$a_0^\dagger a_0|\phi_N\rangle = N|\phi_N\rangle. \tag{12.2}$$

Instead, in the thermodynamic limit, the operator $a_0 a_0^\dagger$ satisfies

$$\begin{aligned} a_0 a_0^\dagger|\phi_N\rangle &= (N+1)|\phi_N\rangle \simeq \\ &\simeq N|\phi_N\rangle, \end{aligned} \tag{12.3}$$

and similarly for the annihilation operator

$$\begin{aligned} a_0|\phi_N\rangle &= \sqrt{N}|\phi_{N-1}\rangle \simeq \\ &\simeq \sqrt{N}|\phi_N\rangle. \end{aligned} \tag{12.4}$$

In other words, Bogoljubov assumes that the bosonic operators describing the condensate commute. Considering the operators as classical functions, we can replace them by their eigenvalues, that is, by $\sqrt{N}$. This is first assumption of Bogoljubov's theory. In addition, given an Hamiltonian of interacting boson particles of the form given in Eq. 2.327, with

$$\mathcal{H}_0 = \frac{\hat{\mathbf{p}}^2}{2m} + V(x) - \mu, \tag{12.5}$$

Bogoljubov considered a system composed of ultracold atoms that is strongly dilute. The ultracold condition is essential for the realization and study of BEC, while the diluteness implies that the characteristic range of the interatomic interactions is much smaller than the average interparticle spacing. Consequently, the pairwise interaction potential can be effectively approximated by a contact potential, i.e.,

$$U(x, x') = U\delta(|x - x'|), \tag{12.6}$$

and our starting point is the second-quantized Hamiltonian

$$\hat{\mathcal{H}} = \int dx \hat{\psi}^\dagger(x) \left[\frac{\hat{\mathbf{p}}^2}{2m} + V(x) - \mu\right] \hat{\psi}(x) + \frac{U}{2}\int dx \hat{\psi}^\dagger(x)\hat{\psi}^\dagger(x)\hat{\psi}(x)\hat{\psi}(x), \tag{12.7}$$

where we employed Theorems 2.9 and 2.16, with $\hat{\psi}(x)$ and $\hat{\psi}^\dagger(x)$ being the fields operators (see Eqs. 2.149 and 2.150). Bogoljubov assumed

$$\hat{\psi}(x) = \varphi_0(x)a_0 + \hat{\phi}(x), \tag{12.8}$$

where $\hat{\phi}(x)$ denotes the excited states and the BEC state $\varphi_0(x)$ must satisfy

$$\int dx \varphi_0^*(x)\hat{\phi}(x) = 0, \tag{12.9}$$

$$\int dx\, |\varphi_0(x)|^2 = 1. \tag{12.10}$$

Our goal is to determine the ground state $\varphi_0(x)$. Since the ground state corresponds to a stable equilibrium configuration, the Hamiltonian is expanded around the ground state and truncated at quadratic order in the creation and annihilation operators. At zero temperature, in the case of non-interacting bosons, the number of particles in the condensate would be such that $N_0 = N$. However, due to the interactions between the bosons, we expect $N_0 \neq N$, although both quantities remain of the same order of magnitude. Therefore, we replace the operator a_0 with the variable $\sqrt{N_0}$, and we introduce the Bogoljubov field operator

$$\hat{\psi}(x) = \sqrt{N_0}\varphi_0(x) + \hat{\phi}(x). \tag{12.11}$$

where $\sqrt{N_0}\varphi_0(x)$ is a c-number, i.e., in particular, it is not an operator. In other words, the macroscopic part of the field is treated as a number, and only the fluctuations are quantized.

12.2 Bogoljubov's Hamiltonian

Here, we derive the quadratic Bogoljubov Hamiltonian by performing a second-order expansion in the fluctuation operators about the condensate. To achieve this, we introduce the Bogoljubov field operator and keep only the terms that are second-order or lower in $\hat{\phi}$, $\hat{\phi}^\dagger$, namely

$$\begin{aligned}\hat{\mathcal{H}} = &\int dx \Big(\sqrt{N_0}\varphi_0^*(x) + \hat{\phi}^\dagger(x)\Big)\left[\frac{\hat{\mathbf{p}}^2}{2m} + V(x) - \mu\right]\Big(\sqrt{N_0}\varphi_0(x) + \hat{\phi}(x)\Big) + \\ &+ \frac{U}{2}\int dx \Big(\sqrt{N_0}\varphi_0^*(x) + \hat{\phi}^\dagger(x)\Big)\Big(\sqrt{N_0}\varphi_0^*(x) + \hat{\phi}^\dagger(x)\Big)\Big(\sqrt{N_0}\varphi_0(x) + \hat{\phi}(x)\Big) \\ &\times \Big(\sqrt{N_0}\varphi_0(x) + \hat{\phi}(x)\Big).\end{aligned} \tag{12.12}$$

The contributions in the first integral are

$$N_0 \int dx \varphi_0^*(x) \left[\frac{\hat{\mathbf{p}}^2}{2m} + V(x) - \mu \right] \varphi_0(x), \tag{12.13}$$

$$\int dx \hat{\phi}^\dagger(x) \left[\frac{\hat{\mathbf{p}}^2}{2m} + V(x) - \mu \right] \hat{\phi}(x), \tag{12.14}$$

$$\sqrt{N_0} \int dx \varphi_0^*(x) \left[\frac{\hat{\mathbf{p}}^2}{2m} + V(x) - \mu \right] \hat{\phi}(x), \tag{12.15}$$

$$\sqrt{N_0} \int dx \hat{\phi}^\dagger(x) \left[\frac{\hat{\mathbf{p}}^2}{2m} + V(x) - \mu \right] \varphi_0(x). \tag{12.16}$$

On the other hand, the contributions in the second integral can be simplified by recognizing the following identification

$$\int dx (a+b)(c+d)(e+f)(g+h), \tag{12.17}$$

where

$$aceg \equiv \frac{U N_0^2}{2} \int dx \varphi_0^*(x) \varphi_0^*(x) \varphi_0(x) \varphi_0(x), \tag{12.18}$$

$$aceh \equiv \frac{U N_0^{\frac{3}{2}}}{2} \int dx \varphi_0^*(x) \varphi_0^*(x) \varphi_0(x) \hat{\phi}(x), \tag{12.19}$$

$$acfg \equiv \frac{U N_0^{\frac{3}{2}}}{2} \int dx \varphi_0^*(x) \varphi_0^*(x) \hat{\phi}(x) \varphi_0(x), \tag{12.20}$$

$$acfh \equiv \frac{U N_0}{2} \int dx \varphi_0^*(x) \varphi_0^*(x) \hat{\phi}(x) \hat{\phi}(x), \tag{12.21}$$

$$adeg \equiv \frac{U N_0^{\frac{3}{2}}}{2} \int dx \varphi_0^*(x) \hat{\phi}^\dagger(x) \varphi_0(x) \varphi_0(x), \tag{12.22}$$

$$adeh \equiv \frac{U N_0}{2} \int dx \varphi_0^*(x) \hat{\phi}^\dagger(x) \varphi_0(x) \hat{\phi}(x), \tag{12.23}$$

$$adfg \equiv \frac{U \sqrt{N_0}}{2} \int dx \varphi_0^*(x) \hat{\phi}^\dagger(x) \hat{\phi}(x) \varphi_0(x), \tag{12.24}$$

$$adfh \equiv \frac{U\sqrt{N_0}}{2}\int dx\varphi_0^*(x)\hat{\phi}^\dagger(x)\hat{\phi}(x)\hat{\phi}(x), \tag{12.25}$$

$$bdfh \equiv \frac{U}{2}\int dx\hat{\phi}^\dagger(x)\hat{\phi}^\dagger(x)\hat{\phi}(x)\hat{\phi}(x), \tag{12.26}$$

$$bdfg \equiv \frac{U\sqrt{N_0}}{2}\int dx\hat{\phi}^\dagger(x)\hat{\phi}^\dagger(x)\hat{\phi}(x)\varphi_0(x), \tag{12.27}$$

$$bdeg \equiv \frac{UN_0}{2}\int dx\hat{\phi}^\dagger(x)\hat{\phi}^\dagger(x)\varphi_0(x)\varphi_0(x), \tag{12.28}$$

$$bdeh \equiv \frac{U\sqrt{N_0}}{2}\int dx\hat{\phi}^\dagger(x)\hat{\phi}^\dagger(x)\varphi_0(x)\hat{\phi}(x), \tag{12.29}$$

$$bceg \equiv \frac{UN_0^{\frac{3}{2}}}{2}\int dx\hat{\phi}^\dagger(x)\varphi_0^*(x)\varphi_0(x)\varphi_0(x), \tag{12.30}$$

$$bceh \equiv \frac{UN_0}{2}\int dx\hat{\phi}^\dagger(x)\varphi_0^*(x)\varphi_0(x)\hat{\phi}(x), \tag{12.31}$$

$$bcfg \equiv \frac{UN_0}{2}\int dx\hat{\phi}^\dagger(x)\varphi_0^*(x)\hat{\phi}(x)\varphi_0(x), \tag{12.32}$$

$$bcfh \equiv \frac{U\sqrt{N_0}}{2}\int dx\hat{\phi}^\dagger(x)\varphi_0^*(x)\hat{\phi}(x)\hat{\phi}(x). \tag{12.33}$$

We neglect all terms beyond second order: this involves disregarding contributions such as $adfh$, $bdfh$, $bdfg$, $bdeh$ and $bcfh$. As a result, the contributions to the second integral are given by

$$aceg \equiv \frac{UN_0^2}{2}\int dx\,|\varphi_0(x)|^2\,|\varphi_0(x)|^2, \tag{12.34}$$

$$aceh \equiv \frac{UN_0^{\frac{3}{2}}}{2}\int dx\varphi_0^*(x)\,|\varphi_0(x)|^2\,\hat{\phi}(x), \tag{12.35}$$

$$acfg \equiv \frac{UN_0^{\frac{3}{2}}}{2}\int dx\varphi_0^*(x)\,|\varphi_0(x)|^2\,\hat{\phi}(x), \tag{12.36}$$

$$acfh \equiv \frac{UN_0}{2}\int dx\left[\varphi_0^*(x)\right]^2\hat{\phi}(x)\hat{\phi}(x), \tag{12.37}$$

$$adeg \equiv \frac{UN_0^{\frac{3}{2}}}{2} \int dx\varphi_0(x) \left|\varphi_0(x)\right|^2 \hat{\phi}^\dagger(x), \tag{12.38}$$

$$adeh \equiv \frac{UN_0}{2} \int dx \left|\varphi_0(x)\right|^2 \hat{\phi}^\dagger(x)\hat{\phi}(x), \tag{12.39}$$

$$adfg \equiv \frac{U\sqrt{N_0}}{2} \int dx \left|\varphi_0(x)\right|^2 \hat{\phi}^\dagger(x)\hat{\phi}(x), \tag{12.40}$$

$$bdeg \equiv \frac{UN_0}{2} \int dx \left[\varphi_0(x)\right]^2 \hat{\phi}^\dagger(x)\hat{\phi}^\dagger(x), \tag{12.41}$$

$$bceg \equiv \frac{UN_0^{\frac{3}{2}}}{2} \int dx\varphi_0(x) \left|\varphi_0(x)\right|^2 \hat{\phi}^\dagger(x), \tag{12.42}$$

$$bceh \equiv \frac{UN_0}{2} \int dx \left|\varphi_0(x)\right|^2 \hat{\phi}^\dagger(x)\hat{\phi}(x), \tag{12.43}$$

$$bcfg \equiv \frac{UN_0}{2} \int dx \left|\varphi_0(x)\right|^2 \hat{\phi}^\dagger(x)\hat{\phi}(x), \tag{12.44}$$

and from $aceh \equiv acfg$, $adeg \equiv bceg$, $adeh \equiv adfg \equiv bceh \equiv bcfg$, we have

$$\begin{aligned}\int dx(a+b)(c+d)(e+f)(g+h) \simeq \int dx\ aceg + 2\int dx\ aceh \\ + \int dx\ acfh + 2\int dx\ adeg + \\ +4\int dx\ adeh + \int dx\ bdeg.\end{aligned} \tag{12.45}$$

By combining the results obtained from the integrals contained in the Hamiltonian, we have

$$\begin{aligned}\hat{\mathcal{H}} = N_0 \int dx\varphi_0^*(x)\left[\frac{\hat{\mathbf{p}}^2}{2m} + V(x) - \mu\right]\varphi_0(x) \\ + \int dx\hat{\phi}^\dagger(x)\left[\frac{\hat{\mathbf{p}}^2}{2m} + V(x) - \mu\right]\hat{\phi}(x) + \\ + \sqrt{N_0}\int dx\varphi_0^*(x)\left[\frac{\hat{\mathbf{p}}^2}{2m} + V(x) - \mu\right]\hat{\phi}(x) + \sqrt{N_0}\int dx\hat{\phi}^\dagger(x)\left[\frac{\hat{\mathbf{p}}^2}{2m}\right.\end{aligned}$$

$$
\begin{aligned}
&+ V(x) - \mu\Big]\varphi_0(x)+ \\
&+ \frac{UN_0^2}{2}\int dx\,|\varphi_0(x)|^2\,|\varphi_0(x)|^2 + UN_0^{\frac{3}{2}}\int dx\varphi_0^*(x)\,|\varphi_0(x)|^2\,\hat{\phi}(x) \\
&+ \frac{UN_0}{2}\int dx\,\left[\varphi_0^*(x)\right]^2\hat{\phi}(x)\hat{\phi}(x)+ \\
&+ UN_0^{\frac{3}{2}}\int dx\varphi_0(x)\,|\varphi_0(x)|^2\,\hat{\phi}^\dagger(x) + 2UN_0\int dx\,|\varphi_0(x)|^2\,\hat{\phi}^\dagger(x)\phi(x) \\
&+ \frac{UN_0}{2}\int dx\,[\varphi_0(x)]^2\,\hat{\phi}^\dagger(x)\hat{\phi}^\dagger(x).
\end{aligned}
\tag{12.46}
$$

The state φ_0 represents the ground state, corresponding to a minimum of the system's energy. Consequently, the sum of all first-order terms in $\hat{\phi}$ and $\hat{\phi}^\dagger$ must vanish, thereby allowing the Hamiltonian to be expanded to second order as follows

$$
\begin{aligned}
\hat{\mathcal{H}} = N_0\int dx\varphi_0^*(x)&\left[\frac{\hat{\mathbf{p}}^2}{2m} + V(x) - \mu\right]\varphi_0(x) \\
&+ \int dx\hat{\phi}^\dagger(x)\left[\frac{\hat{\mathbf{p}}^2}{2m} + V(x) - \mu\right]\hat{\phi}(x)+ \\
&+ \frac{UN_0^2}{2}\int dx\,|\varphi_0(x)|^2\,|\varphi_0(x)|^2 + \frac{UN_0}{2}\int dx\,\left[\varphi_0^*(x)\right]^2\hat{\phi}(x)\hat{\phi}(x)+ \\
&+ 2UN_0\int dx\,|\varphi_0(x)|^2\,\hat{\phi}^\dagger(x)\hat{\phi}(x) + \frac{UN_0}{2}\int dx\,[\varphi_0(x)]^2\,\hat{\phi}^\dagger(x)\hat{\phi}^\dagger(x),
\end{aligned}
\tag{12.47}
$$

which holds if and only if

$$
\begin{aligned}
\sqrt{N_0}\int dx\varphi_0^*(x)&\left[\frac{\hat{\mathbf{p}}^2}{2m} + V(x) - \mu\right]\hat{\phi}(x) \\
&+ \sqrt{N_0}\int dx\hat{\phi}^\dagger(x)\left[\frac{\hat{\mathbf{p}}^2}{2m} + V(x) - \mu\right]\varphi_0(x)+ \\
&+ UN_0^{\frac{3}{2}}\int dx\varphi_0^*(x)\,|\varphi_0(x)|^2\,\hat{\phi}(x) + UN_0^{\frac{3}{2}}\int dx\varphi_0(x)\,|\varphi_0(x)|^2\,\hat{\phi}^\dagger(x) = 0.
\end{aligned}
\tag{12.48}
$$

Note that the equation above is of the form $z + z^* = 0$, with

$$
z \equiv \sqrt{N_0}\int dx\hat{\phi}^\dagger(x)\left[\frac{\hat{\mathbf{p}}^2}{2m}+V(x)-\mu\right]\varphi_0(x)+UN_0^{\frac{3}{2}}\int dx\varphi_0(x)\,|\varphi_0(x)|^2\,\hat{\phi}^\dagger(x)=
$$

$$= \sqrt{N_0} \int dx \hat{\phi}^\dagger(x) \left[\frac{\hat{\mathbf{p}}^2}{2m} + V(x) - \mu + U N_0 |\varphi_0(x)|^2 \right] \varphi_0(x). \tag{12.49}$$

From

$$0 = z + z^* = 2 \operatorname{Re} z, \tag{12.50}$$

the real part of z is required to vanish. In our case, the Hamiltonian is generally complex, so an alternative particular solution is given by

$$\int dx \hat{\phi}^\dagger(x) \left[\frac{\hat{\mathbf{p}}^2}{2m} + V(x) - \mu + U N_0 |\varphi_0(x)|^2 \right] \varphi_0(x) = 0, \ \forall \, \hat{\phi}^\dagger(x), \tag{12.51}$$

then

$$\left[\frac{\hat{\mathbf{p}}^2}{2m} + V(x) - \mu + U N_0 |\varphi_0(x)|^2 \right] \varphi_0(x) = 0, \tag{12.52}$$

which is known as Gross-Pitaevskij equation and must be satisfied. Bogoljubov demonstrated that a Hamiltonian of the form 12.47 can be exactly diagonalized in a particular case. Under the assumptions of ultracold atoms interacting via a two-body contact potential, and further assuming that $V(x) = 0$, the condensate state corresponds to a plane wave with zero momentum, that is

$$\varphi_0 = \frac{1}{\sqrt{V}}, \tag{12.53}$$

which satisfies Eq. 12.52. Indeed,

$$\left[-\mu + U N_0 \left| \frac{1}{\sqrt{V}} \right|^2 \right] \frac{1}{\sqrt{V}} = 0, \tag{12.54}$$

$$-\mu + \frac{U N_0}{V} = 0, \tag{12.55}$$

$$\mu = \frac{U N_0}{V}. \tag{12.56}$$

Note that we have omitted the momentum operator, since when acting on φ_0 it yields the zero eigenvalue. Given the form of the solution in Eq. 12.53, we expand the excitation field operator in terms of plane waves given in Eq. 1.77 as follows

$$\hat{\phi}(\mathbf{r}) = \sum_{\mathbf{k} \neq \mathbf{0}} \frac{e^{i \mathbf{k} \cdot \mathbf{r}}}{\sqrt{V}} a_{\mathbf{k}}. \tag{12.57}$$

Note that we have disregarded the spin degree of freedom, as it only contributes a numerical degeneracy factor to each term in the equations. The Bogoljubov field operator given in Eq. 12.11 becomes

$$\hat{\psi}(\mathbf{r}) = \sqrt{N_0}\varphi_0 + \hat{\phi}(\mathbf{r}) = \\ = \sqrt{\frac{N_0}{V}} + \sum_{\mathbf{k}\neq\mathbf{0}} \frac{e^{i\mathbf{k}\cdot\mathbf{r}}}{\sqrt{V}} a_{\mathbf{k}}, \tag{12.58}$$

where the variable x (see Eq. 1.43) reduces to $\mathbf{r}$, since there is no spin dependence, and the Hamiltonian in Eq. 12.47 becomes

$$\begin{aligned}
\hat{\mathcal{H}} = &-N_0\mu \int d\mathbf{r}\frac{1}{V} + \int d\mathbf{r}\left(\sum_{\mathbf{k}\neq\mathbf{0}} \frac{e^{-i\mathbf{k}\cdot\mathbf{r}}}{\sqrt{V}} a_{\mathbf{k}}^{\dagger}\right)\left[\frac{\hat{\mathbf{p}}^2}{2m} - \mu\right]\left(\sum_{\mathbf{k}'\neq\mathbf{0}} \frac{e^{i\mathbf{k}'\cdot\mathbf{r}}}{\sqrt{V}} a_{\mathbf{k}'}\right) + \\
&+ \frac{UN_0^2}{2}\int d\mathbf{r}\frac{1}{V^2} + \frac{UN_0}{2}\int d\mathbf{r}\frac{1}{V}\left[\sum_{\mathbf{k}\neq\mathbf{0}} \frac{e^{i\mathbf{k}\cdot\mathbf{r}}}{\sqrt{V}} a_{\mathbf{k}}\right]^2 + \\
&+ 2UN_0\int d\mathbf{r}\frac{1}{V}\left|\sum_{\mathbf{k}\neq\mathbf{0}} \frac{e^{i\mathbf{k}\cdot\mathbf{r}}}{\sqrt{V}} a_{\mathbf{k}}\right|^2 + \frac{UN_0}{2}\int d\mathbf{r}\frac{1}{V}\left[\sum_{\mathbf{k}\neq\mathbf{0}} \frac{e^{-i\mathbf{k}\cdot\mathbf{r}}}{\sqrt{V}} a_{\mathbf{k}}^{\dagger}\right]^2 = \\
= &-\frac{UN_0^2}{V} + \sum_{\mathbf{k}\neq\mathbf{0}}\sum_{\mathbf{k}'\neq\mathbf{0}} \int d\mathbf{r}\frac{e^{-i\mathbf{k}\cdot\mathbf{r}}}{\sqrt{V}}\frac{\hbar^2\mathbf{k}'^2}{2m}\frac{e^{i\mathbf{k}'\cdot\mathbf{r}}}{\sqrt{V}} a_{\mathbf{k}}^{\dagger}a_{\mathbf{k}'} \\
&- \frac{UN_0}{V}\int d\mathbf{r}\left(\sum_{\mathbf{k}\neq\mathbf{0}} \frac{e^{-i\mathbf{k}\cdot\mathbf{r}}}{\sqrt{V}} a_{\mathbf{k}}^{\dagger}\right)\left(\sum_{\mathbf{k}'\neq\mathbf{0}} \frac{e^{i\mathbf{k}'\cdot\mathbf{r}}}{\sqrt{V}} a_{\mathbf{k}'}\right) + \frac{UN_0^2}{2V} + \\
&+ \frac{UN_0}{2V}\int d\mathbf{r}\left(\sum_{\mathbf{k}\neq\mathbf{0}} \frac{e^{i\mathbf{k}\cdot\mathbf{r}}}{\sqrt{V}} a_{\mathbf{k}}\right)\left(\sum_{\mathbf{k}'\neq\mathbf{0}} \frac{e^{i\mathbf{k}'\cdot\mathbf{r}}}{\sqrt{V}} a_{\mathbf{k}'}\right) \\
&+ \frac{2UN_0}{V}\int d\mathbf{r}\left(\sum_{\mathbf{k}\neq\mathbf{0}} \frac{e^{-i\mathbf{k}\cdot\mathbf{r}}}{\sqrt{V}} a_{\mathbf{k}}^{\dagger}\right)\left(\sum_{\mathbf{k}'\neq\mathbf{0}} \frac{e^{i\mathbf{k}'\cdot\mathbf{r}}}{\sqrt{V}} a_{\mathbf{k}'}\right) + \\
&+ \frac{UN_0}{2V}\int d\mathbf{r}\left(\sum_{\mathbf{k}\neq\mathbf{0}} \frac{e^{-i\mathbf{k}\cdot\mathbf{r}}}{\sqrt{V}} a_{\mathbf{k}}^{\dagger}\right)\left(\sum_{\mathbf{k}'\neq\mathbf{0}} \frac{e^{-i\mathbf{k}'\cdot\mathbf{r}}}{\sqrt{V}} a_{\mathbf{k}'}^{\dagger}\right) = \\
= &-\frac{UN_0^2}{2V} + \sum_{\mathbf{k}\neq\mathbf{0}}\sum_{\mathbf{k}'\neq\mathbf{0}} \int d\mathbf{r}\frac{e^{-i(\mathbf{k}-\mathbf{k}')\cdot\mathbf{r}}}{V}\frac{\hbar^2\mathbf{k}'^2}{2m} a_{\mathbf{k}}^{\dagger}a_{\mathbf{k}'}
\end{aligned}$$

$$
\begin{aligned}
& -\frac{UN_0}{V}\sum_{\mathbf{k}\neq\mathbf{0}}\sum_{\mathbf{k}'\neq\mathbf{0}}\int d\mathbf{r}\frac{e^{-i(\mathbf{k}-\mathbf{k}')\cdot\mathbf{r}}}{V}a^{\dagger}_{\mathbf{k}}a_{\mathbf{k}'}+ \\
& +\frac{UN_0}{2V}\sum_{\mathbf{k}\neq\mathbf{0}}\sum_{\mathbf{k}'\neq\mathbf{0}}\int d\mathbf{r}\frac{e^{i(\mathbf{k}+\mathbf{k}')\cdot\mathbf{r}}}{V}a_{\mathbf{k}}a_{\mathbf{k}'}+\frac{2UN_0}{V}\sum_{\mathbf{k}\neq\mathbf{0}}\sum_{\mathbf{k}'\neq\mathbf{0}}\int d\mathbf{r}\frac{e^{-i(\mathbf{k}-\mathbf{k}')\cdot\mathbf{r}}}{V}a^{\dagger}_{\mathbf{k}}a_{\mathbf{k}'}+ \\
& +\frac{UN_0}{2V}\sum_{\mathbf{k}\neq\mathbf{0}}\sum_{\mathbf{k}'\neq\mathbf{0}}\int d\mathbf{r}\frac{e^{-i(\mathbf{k}+\mathbf{k}')\cdot\mathbf{r}}}{V}a^{\dagger}_{\mathbf{k}}a^{\dagger}_{\mathbf{k}'}= \\
& =-\frac{UN_0^2}{2V}+\sum_{\mathbf{k}\neq\mathbf{0}}\sum_{\mathbf{k}'\neq\mathbf{0}}\delta_{\mathbf{k},\mathbf{k}'}\frac{\hbar^2\mathbf{k}'^2}{2m}a^{\dagger}_{\mathbf{k}}a_{\mathbf{k}'}-\frac{UN_0}{V}\sum_{\mathbf{k}\neq\mathbf{0}}\sum_{\mathbf{k}'\neq\mathbf{0}}\delta_{\mathbf{k},\mathbf{k}'}a^{\dagger}_{\mathbf{k}}a_{\mathbf{k}'} \\
& +\frac{UN_0}{2V}\sum_{\mathbf{k}\neq\mathbf{0}}\sum_{\mathbf{k}'\neq\mathbf{0}}\delta_{\mathbf{k},-\mathbf{k}'}a_{\mathbf{k}}a_{\mathbf{k}'}+ \\
& +\frac{2UN_0}{V}\sum_{\mathbf{k}\neq\mathbf{0}}\sum_{\mathbf{k}'\neq\mathbf{0}}\delta_{\mathbf{k},\mathbf{k}'}a^{\dagger}_{\mathbf{k}}a_{\mathbf{k}'}+\frac{UN_0}{2V}\sum_{\mathbf{k}\neq\mathbf{0}}\sum_{\mathbf{k}'\neq\mathbf{0}}\delta_{\mathbf{k},-\mathbf{k}'}a^{\dagger}_{\mathbf{k}}a^{\dagger}_{\mathbf{k}'}= \\
& =-\frac{UN_0^2}{2V}+\sum_{\mathbf{k}\neq\mathbf{0}}\frac{\hbar^2\mathbf{k}^2}{2m}a^{\dagger}_{\mathbf{k}}a_{\mathbf{k}}-\frac{UN_0}{V}\sum_{\mathbf{k}\neq\mathbf{0}}a^{\dagger}_{\mathbf{k}}a_{\mathbf{k}}+\frac{UN_0}{2V}\sum_{\mathbf{k}'\neq\mathbf{0}}a_{-\mathbf{k}'}a_{\mathbf{k}'} \\
& +\frac{2UN_0}{V}\sum_{\mathbf{k}\neq\mathbf{0}}a^{\dagger}_{\mathbf{k}}a_{\mathbf{k}}+\frac{UN_0}{2V}\sum_{\mathbf{k}\neq\mathbf{0}}a^{\dagger}_{\mathbf{k}}a^{\dagger}_{-\mathbf{k}}.
\end{aligned}
\tag{12.59}
$$

After appropriately renaming the dummy indices, we finally obtain

$$
\hat{\mathcal{H}}=-\frac{UN_0^2}{2V}+\sum_{\mathbf{k}\neq\mathbf{0}}\left(\frac{\hbar^2\mathbf{k}^2}{2m}+\frac{UN_0}{V}\right)a^{\dagger}_{\mathbf{k}}a_{\mathbf{k}}+\frac{UN_0}{2V}\sum_{\mathbf{k}\neq\mathbf{0}}\left(a^{\dagger}_{\mathbf{k}}a^{\dagger}_{-\mathbf{k}}+a_{-\mathbf{k}}a_{\mathbf{k}}\right),
\tag{12.60}
$$

which is known as the Bogoljubov's Hamiltonian. The quadratic terms of the form $a^{\dagger}_{\mathbf{k}}a^{\dagger}_{-\mathbf{k}}$ and $a_{-\mathbf{k}}a_{\mathbf{k}}$ play a central role in the excitation spectrum and will be analyzed in detail in the following section.

12.3 Bogoljubov's Bosonic Transformations

Here, we derive the transformations necessary to diagonalize the bosonic Hamiltonian 12.60, commonly known as the *Bogoljubov bosonic transformations*. We make use of the Jacobi identity (see Eq. F.1) and we exploit bosonic algebra (see Eq. D.1), and we have

$$\left[a_{\mathbf{k}}, a_{\mathbf{k}'}^{\dagger}a_{-\mathbf{k}'}^{\dagger}\right] = \left[a_{\mathbf{k}}, a_{\mathbf{k}'}^{\dagger}\right]a_{-\mathbf{k}'}^{\dagger} + a_{\mathbf{k}'}^{\dagger}\left[a_{\mathbf{k}}, a_{-\mathbf{k}'}^{\dagger}\right] = \\ = \delta_{\mathbf{k},\mathbf{k}'}a_{-\mathbf{k}'}^{\dagger} + \delta_{\mathbf{k},-\mathbf{k}'}a_{\mathbf{k}'}^{\dagger}, \tag{12.61}$$

$$\left[a_{\mathbf{k}}, \sum_{\mathbf{k}'\neq\mathbf{0}} a_{\mathbf{k}'}^{\dagger}a_{-\mathbf{k}'}^{\dagger}\right] = \sum_{\mathbf{k}'\neq\mathbf{0}}\left[a_{\mathbf{k}}, a_{\mathbf{k}'}^{\dagger}a_{-\mathbf{k}'}^{\dagger}\right] = \\ = \sum_{\mathbf{k}'\neq\mathbf{0}}\left(\delta_{\mathbf{k},\mathbf{k}'}a_{-\mathbf{k}'}^{\dagger} + \delta_{\mathbf{k},-\mathbf{k}'}a_{\mathbf{k}'}^{\dagger}\right) = \\ = 2a_{-\mathbf{k}}^{\dagger}, \tag{12.62}$$

and

$$\left[a_{\mathbf{k}}^{\dagger}, a_{-\mathbf{k}'}a_{\mathbf{k}'}\right] = \left[a_{\mathbf{k}}^{\dagger}, a_{-\mathbf{k}'}\right]a_{\mathbf{k}'} + a_{-\mathbf{k}'}\left[a_{\mathbf{k}}^{\dagger}, a_{\mathbf{k}'}\right] = \\ = -\delta_{\mathbf{k},-\mathbf{k}'}a_{\mathbf{k}'} - \delta_{\mathbf{k},\mathbf{k}'}a_{-\mathbf{k}'}, \tag{12.63}$$

$$\left[a_{\mathbf{k}}^{\dagger}, \sum_{\mathbf{k}'\neq\mathbf{0}} a_{-\mathbf{k}'}a_{\mathbf{k}'}\right] = \sum_{\mathbf{k}'\neq\mathbf{0}}\left[a_{\mathbf{k}}^{\dagger}, a_{-\mathbf{k}'}a_{\mathbf{k}'}\right] = \\ = \sum_{\mathbf{k}'\neq\mathbf{0}}\left(-\delta_{\mathbf{k},-\mathbf{k}'}a_{\mathbf{k}'} - \delta_{\mathbf{k},\mathbf{k}'}a_{-\mathbf{k}'}\right) = \\ = -2a_{-\mathbf{k}}. \tag{12.64}$$

Bogoljubov approach involves considering transformations that map $a_{\mathbf{k}}$ and $a_{-\mathbf{k}}^{\dagger}$ into a linear combination of new creation and annihilation operators related to a momentum and its opposite. We define

$$b_{\mathbf{k}} = u_{\mathbf{k}}a_{\mathbf{k}} + v_{\mathbf{k}}a_{-\mathbf{k}}^{\dagger}, \ \mathbf{k} \neq \mathbf{0}, \tag{12.65}$$

$$b_{-\mathbf{k}}^{\dagger} = v_{-\mathbf{k}}^{*}a_{\mathbf{k}} + u_{-\mathbf{k}}^{*}a_{-\mathbf{k}}^{\dagger}, \ \mathbf{k} \neq \mathbf{0}, \tag{12.66}$$

or equivalently in matrix form

$$\begin{pmatrix} b_{\mathbf{k}} \\ b_{-\mathbf{k}}^{\dagger} \end{pmatrix} = \begin{pmatrix} u_{\mathbf{k}} & v_{\mathbf{k}} \\ v_{-\mathbf{k}}^{*} & u_{-\mathbf{k}}^{*} \end{pmatrix} \begin{pmatrix} a_{\mathbf{k}} \\ a_{-\mathbf{k}}^{\dagger} \end{pmatrix}, \ \mathbf{k} \neq \mathbf{0}, \tag{12.67}$$

where

$$\begin{pmatrix} u_{\mathbf{k}} & v_{\mathbf{k}} \\ v^*_{-\mathbf{k}} & u^*_{-\mathbf{k}} \end{pmatrix}, \quad \mathbf{k} \neq \mathbf{0} \tag{12.68}$$

is the Bogoljubov's bosonic matrix. Then, the goal is to compute the complex coefficients $u_{\mathbf{k}}$, $u^*_{-\mathbf{k}}$, $v_{\mathbf{k}}$, $v^*_{-\mathbf{k}}$. As a first assumption, Bogoljubov requires that transformations defined in Eqs. 12.65 and 12.66, be canonical, i.e., that the corresponding operators satisfy the bosonic commutation relations. Then, we set

$$\left[b_{\mathbf{k}}, b^\dagger_{\mathbf{k}'}\right] = \delta_{\mathbf{k},\mathbf{k}'}, \tag{12.69}$$

which implies

$$\left[u_{\mathbf{k}} a_{\mathbf{k}} + v_{\mathbf{k}} a^\dagger_{-\mathbf{k}}, v^*_{\mathbf{k}'} a_{-\mathbf{k}'} + u^*_{\mathbf{k}'} a^\dagger_{\mathbf{k}'}\right] = \delta_{\mathbf{k},\mathbf{k}'}, \tag{12.70}$$

$$\left[u_{\mathbf{k}'} a_{\mathbf{k}'}, v^*_{\mathbf{k}} a_{-\mathbf{k}}\right] + \left[u_{\mathbf{k}'} a_{\mathbf{k}'}, u^*_{\mathbf{k}} a^\dagger_{\mathbf{k}}\right] + \left[v_{\mathbf{k}'} a^\dagger_{-\mathbf{k}'}, v^*_{\mathbf{k}} a_{-\mathbf{k}}\right] + \left[v_{\mathbf{k}'} a^\dagger_{-\mathbf{k}'}, u^*_{\mathbf{k}} a^\dagger_{\mathbf{k}}\right] = \delta_{\mathbf{k}',\mathbf{k}}, \tag{12.71}$$

$$u_{\mathbf{k}} u^*_{\mathbf{k}'} \delta_{\mathbf{k},\mathbf{k}'} - v_{\mathbf{k}} v^*_{\mathbf{k}'} \delta_{-\mathbf{k},-\mathbf{k}'} = \delta_{\mathbf{k},\mathbf{k}'}, \tag{12.72}$$

$$u_{\mathbf{k}} u^*_{\mathbf{k}'} \delta_{\mathbf{k},\mathbf{k}'} - v_{\mathbf{k}} v^*_{\mathbf{k}'} \delta_{\mathbf{k},\mathbf{k}'} = \delta_{\mathbf{k},\mathbf{k}'}, \tag{12.73}$$

and for $\mathbf{k} = \mathbf{k}'$, we have

$$|u_{\mathbf{k}}|^2 - |v_{\mathbf{k}}|^2 = 1. \tag{12.74}$$

As a second assumption, Bogoljubov requires that the Hamiltonian 12.60 becomes diagonal with respect the new operators, i.e.,

$$\hat{\mathcal{H}} = \tilde{\mathcal{E}} + \sum_{\mathbf{k}\neq\mathbf{0}} \mathcal{E}_{\mathbf{k}} b^\dagger_{\mathbf{k}} b_{\mathbf{k}}, \tag{12.75}$$

then according to Eq. D.13, Eq. 12.60 must satisfy

$$\left[\hat{\mathcal{H}}, b_{\mathbf{k}}\right] = -\mathcal{E}_{\mathbf{k}} b_{\mathbf{k}}. \tag{12.76}$$

Obviously, involving the commutator of the Hamiltonian with the creation operator $b^\dagger_{-\mathbf{k}}$ is equally correct. From the definition of $b_{\mathbf{k}}$, Eq. 12.76 implies the study of the commutators between 12.60 and $a_{\mathbf{k}}$, $a^\dagger_{-\mathbf{k}}$. From Eqs. D.13 and D.14, it follows

$$\left[a^\dagger_{\mathbf{k}} a_{\mathbf{k}}, a_{\mathbf{k}'}\right] = -\delta_{\mathbf{k},\mathbf{k}'} a_{\mathbf{k}'}, \tag{12.77}$$

$$\left[a^\dagger_{\mathbf{k}} a_{\mathbf{k}}, a^\dagger_{\mathbf{k}'}\right] = \delta_{\mathbf{k},\mathbf{k}'} a^\dagger_{\mathbf{k}'}, \tag{12.78}$$

respectively. Moreover, employing the bosonic commutation relations D.1 together with the Jacobi identity F.1, we obtain

$$\left[a^{\dagger}_{-\mathbf{k}}, a_{-\mathbf{k}'}a_{\mathbf{k}'}\right] = \left[a^{\dagger}_{-\mathbf{k}}, a_{-\mathbf{k}'}\right]a_{\mathbf{k}'} + a_{-\mathbf{k}'}\left[a^{\dagger}_{-\mathbf{k}}, a_{\mathbf{k}'}\right] = \\ = -\delta_{-\mathbf{k},-\mathbf{k}'}a_{\mathbf{k}'} - \delta_{-\mathbf{k},\mathbf{k}'}a_{-\mathbf{k}'}. \tag{12.79}$$

Now let us compute 12.76, that is

$$\left[\sum_{\mathbf{k}'\neq\mathbf{0}}\left(\frac{\hbar^2\mathbf{k}'^2}{2m}+\frac{UN_0}{V}\right)a^{\dagger}_{\mathbf{k}'}a_{\mathbf{k}'} + \frac{UN_0}{2V}\sum_{\mathbf{k}'\neq\mathbf{0}}\left(a^{\dagger}_{\mathbf{k}'}a^{\dagger}_{-\mathbf{k}'} + a_{-\mathbf{k}'}a_{\mathbf{k}'}\right), b_{\mathbf{k}}\right] = -\mathcal{E}_{\mathbf{k}}b_{\mathbf{k}}, \tag{12.80}$$

$$\left[\sum_{\mathbf{k}'\neq\mathbf{0}}\left(\frac{\hbar^2\mathbf{k}'^2}{2m}+\frac{UN_0}{V}\right)a^{\dagger}_{\mathbf{k}'}a_{\mathbf{k}'} + \frac{UN_0}{2V}\sum_{\mathbf{k}'\neq\mathbf{0}}\left(a^{\dagger}_{\mathbf{k}'}a^{\dagger}_{-\mathbf{k}'} + a_{-\mathbf{k}'}a_{\mathbf{k}'}\right), u_{\mathbf{k}}a_{\mathbf{k}} + v_{\mathbf{k}}a^{\dagger}_{-\mathbf{k}}\right] \\ = -\mathcal{E}_{\mathbf{k}}\left(u_{\mathbf{k}}a_{\mathbf{k}} + v_{\mathbf{k}}a^{\dagger}_{-\mathbf{k}}\right), \tag{12.81}$$

$$\sum_{\mathbf{k}'\neq\mathbf{0}}\left(\frac{\hbar^2\mathbf{k}'^2}{2m}+\frac{UN_0}{V}\right)\left[a^{\dagger}_{\mathbf{k}'}a_{\mathbf{k}'}, u_{\mathbf{k}}a_{\mathbf{k}}\right] + \sum_{\mathbf{k}'\neq\mathbf{0}}\left(\frac{\hbar^2\mathbf{k}'^2}{2m}+\frac{UN_0}{V}\right)\left[a^{\dagger}_{\mathbf{k}'}a_{\mathbf{k}'}, v_{\mathbf{k}}a^{\dagger}_{-\mathbf{k}}\right] + \\ + \frac{UN_0}{2V}\sum_{\mathbf{k}'\neq\mathbf{0}}\left[a^{\dagger}_{\mathbf{k}'}a^{\dagger}_{-\mathbf{k}'}, u_{\mathbf{k}}a_{\mathbf{k}}\right] + \frac{UN_0}{2V}\sum_{\mathbf{k}'\neq\mathbf{0}}\left[a_{-\mathbf{k}'}a_{\mathbf{k}'}, v_{\mathbf{k}}a^{\dagger}_{-\mathbf{k}}\right] \\ = -\mathcal{E}_{\mathbf{k}}\left(u_{\mathbf{k}}a_{\mathbf{k}} + v_{\mathbf{k}}a^{\dagger}_{-\mathbf{k}}\right), \tag{12.82}$$

$$-\sum_{\mathbf{k}'\neq\mathbf{0}}\left(\frac{\hbar^2\mathbf{k}'^2}{2m}+\frac{UN_0}{V}\right)u_{\mathbf{k}}\delta_{\mathbf{k}',\mathbf{k}}a_{\mathbf{k}} + \sum_{\mathbf{k}'\neq\mathbf{0}}\left(\frac{\hbar^2\mathbf{k}'^2}{2m}+\frac{UN_0}{V}\right)v_{\mathbf{k}}\delta_{\mathbf{k}',\mathbf{k}}a^{\dagger}_{-\mathbf{k}} \\ + \frac{UN_0}{2V}\sum_{\mathbf{k}'\neq\mathbf{0}}u_{\mathbf{k}}\left(-\delta_{\mathbf{k},\mathbf{k}'}a^{\dagger}_{-\mathbf{k}'} - \delta_{\mathbf{k},-\mathbf{k}'}a^{\dagger}_{\mathbf{k}'}\right) + \\ + \frac{UN_0}{2V}\sum_{\mathbf{k}'\neq\mathbf{0}}v_{\mathbf{k}}\left(\delta_{-\mathbf{k},-\mathbf{k}'}a_{\mathbf{k}'} + \delta_{-\mathbf{k},\mathbf{k}'}a_{-\mathbf{k}'}\right) = -\mathcal{E}_{\mathbf{k}}\left(u_{\mathbf{k}}a_{\mathbf{k}} + v_{\mathbf{k}}a^{\dagger}_{-\mathbf{k}}\right), \tag{12.83}$$

$$-\left(\frac{\hbar^2\mathbf{k}^2}{2m}+\frac{UN_0}{V}\right)u_{\mathbf{k}}a_{\mathbf{k}}+\left(\frac{\hbar^2\mathbf{k}^2}{2m}+\frac{UN_0}{V}\right)v_{\mathbf{k}}a^\dagger_{-\mathbf{k}}-\frac{UN_0}{2V}u_{\mathbf{k}}\left(a^\dagger_{-\mathbf{k}}+a^\dagger_{-\mathbf{k}}\right)$$
$$+\frac{UN_0}{2V}v_{\mathbf{k}}\left(a_{\mathbf{k}}+a_{\mathbf{k}}\right)=-\mathcal{E}_{\mathbf{k}}\left(u_{\mathbf{k}}a_{\mathbf{k}}+v_{\mathbf{k}}a^\dagger_{-\mathbf{k}}\right), \tag{12.84}$$

$$-\left(\frac{\hbar^2\mathbf{k}^2}{2m}+\frac{UN_0}{V}\right)u_{\mathbf{k}}a_{\mathbf{k}}+\left(\frac{\hbar^2\mathbf{k}^2}{2m}+\frac{UN_0}{V}\right)v_{\mathbf{k}}a^\dagger_{-\mathbf{k}}-\frac{UN_0}{V}u_{\mathbf{k}}a^\dagger_{-\mathbf{k}}+\frac{UN_0}{V}v_{\mathbf{k}}a_{\mathbf{k}}$$
$$=-\mathcal{E}_{\mathbf{k}}\left(u_{\mathbf{k}}a_{\mathbf{k}}+v_{\mathbf{k}}a^\dagger_{-\mathbf{k}}\right). \tag{12.85}$$

By comparing the coefficients of $a_{\mathbf{k}}$ and $a^\dagger_{-\mathbf{k}}$, we obtain

$$\begin{cases}-\left(\dfrac{\hbar^2\mathbf{k}^2}{2m}+\dfrac{UN_0}{V}\right)u_{\mathbf{k}}+\dfrac{UN_0}{V}v_{\mathbf{k}}=-\mathcal{E}_{\mathbf{k}}u_{\mathbf{k}}\\ \left(\dfrac{\hbar^2\mathbf{k}^2}{2m}+\dfrac{UN_0}{V}\right)v_{\mathbf{k}}-\dfrac{UN_0}{V}u_{\mathbf{k}}=-\mathcal{E}_{\mathbf{k}}v_{\mathbf{k}}\end{cases}, \tag{12.86}$$

$$\begin{cases}\left(\dfrac{\hbar^2\mathbf{k}^2}{2m}+\dfrac{UN_0}{V}\right)u_{\mathbf{k}}-\dfrac{UN_0}{V}v_{\mathbf{k}}=\mathcal{E}_{\mathbf{k}}u_{\mathbf{k}}\\ -\dfrac{UN_0}{V}u_{\mathbf{k}}+\left(\dfrac{\hbar^2\mathbf{k}^2}{2m}+\dfrac{UN_0}{V}\right)v_{\mathbf{k}}=-\mathcal{E}_{\mathbf{k}}v_{\mathbf{k}}\end{cases}, \tag{12.87}$$

or in matrix form

$$\begin{pmatrix}\dfrac{\hbar^2\mathbf{k}^2}{2m}+\dfrac{UN_0}{V} & -\dfrac{UN_0}{V}\\ -\dfrac{UN_0}{V} & \dfrac{\hbar^2\mathbf{k}^2}{2m}+\dfrac{UN_0}{V}\end{pmatrix}\begin{pmatrix}u_{\mathbf{k}}\\ v_{\mathbf{k}}\end{pmatrix}=\begin{pmatrix}\mathcal{E}_{\mathbf{k}} & 0\\ 0 & -\mathcal{E}_{\mathbf{k}}\end{pmatrix}\begin{pmatrix}u_{\mathbf{k}}\\ v_{\mathbf{k}}\end{pmatrix}, \tag{12.88}$$

which is known as a *Bogoljubov-de Gennes eigenvalue problem.* We set

$$\begin{cases}u_{\mathbf{k}}+v_{\mathbf{k}}=s_{\mathbf{k}}\\ u_{\mathbf{k}}-v_{\mathbf{k}}=d_{\mathbf{k}}\end{cases}, \tag{12.89}$$

we then add the equations in 12.87 term by term, namely

$$\left(\frac{\hbar^2\mathbf{k}^2}{2m}+\frac{UN_0}{V}\right)(u_{\mathbf{k}}+v_{\mathbf{k}})-\frac{UN_0}{V}(u_{\mathbf{k}}+v_{\mathbf{k}})=\mathcal{E}_{\mathbf{k}}(u_{\mathbf{k}}-v_{\mathbf{k}}), \tag{12.90}$$

$$\frac{\hbar^2\mathbf{k}^2}{2m}(u_{\mathbf{k}}+v_{\mathbf{k}})=\mathcal{E}_{\mathbf{k}}(u_{\mathbf{k}}-v_{\mathbf{k}}), \tag{12.91}$$

and we subtract the equations in 12.87 term by term as follows

$$\left(\frac{\hbar^2\mathbf{k}^2}{2m}+\frac{UN_0}{V}\right)(u_{\mathbf{k}}-v_{\mathbf{k}})+\frac{UN_0}{V}(u_{\mathbf{k}}-v_{\mathbf{k}})=\mathcal{E}_{\mathbf{k}}(u_{\mathbf{k}}+v_{\mathbf{k}}), \tag{12.92}$$

$$\left(\frac{\hbar^2\mathbf{k}^2}{2m}+2\frac{UN_0}{V}\right)(u_{\mathbf{k}}-v_{\mathbf{k}})=\mathcal{E}_{\mathbf{k}}(u_{\mathbf{k}}+v_{\mathbf{k}}). \tag{12.93}$$

Thanks to Eq. 12.89, Eqs. 12.91 and 12.93 become

$$\frac{\hbar^2\mathbf{k}^2}{2m}s_{\mathbf{k}}=\mathcal{E}_{\mathbf{k}}d_{\mathbf{k}}, \tag{12.94}$$

$$\left(\frac{\hbar^2\mathbf{k}^2}{2m}+2\frac{UN_0}{V}\right)d_{\mathbf{k}}=\mathcal{E}_{\mathbf{k}}s_{\mathbf{k}}. \tag{12.95}$$

We multiply both members of Eq. 12.94 by

$$\frac{\hbar^2\mathbf{k}^2}{2m}+2\frac{UN_0}{V}, \tag{12.96}$$

and we have

$$\left(\frac{\hbar^2\mathbf{k}^2}{2m}+2\frac{UN_0}{V}\right)\frac{\hbar^2\mathbf{k}^2}{2m}s_{\mathbf{k}}=\left(\frac{\hbar^2\mathbf{k}^2}{2m}+2\frac{UN_0}{V}\right)\mathcal{E}_{\mathbf{k}}d_{\mathbf{k}}, \tag{12.97}$$

then we make use of Eq. 12.95, and we get

$$\left(\frac{\hbar^2\mathbf{k}^2}{2m}+2\frac{UN_0}{V}\right)\frac{\hbar^2\mathbf{k}^2}{2m}s_{\mathbf{k}}=\mathcal{E}_{\mathbf{k}}^2s_{\mathbf{k}}. \tag{12.98}$$

A possible solution of the above equation is $s_{\mathbf{k}}=0$, but this must be discarded, as it would imply

$$u_{\mathbf{k}}=-v_{\mathbf{k}}, \tag{12.99}$$

$$u_{\mathbf{k}}^2=v_{\mathbf{k}}^2, \tag{12.100}$$

$$u_{\mathbf{k}}^2-v_{\mathbf{k}}^2=0, \tag{12.101}$$

which contradicts the assumption given in Eq. 12.74. Since $s_{\mathbf{k}} \neq 0$, it must be

$$\mathcal{E}_{\mathbf{k}}^2 = \left(\frac{\hbar^2 \mathbf{k}^2}{2m} + 2\frac{UN_0}{V} \right) \frac{\hbar^2 \mathbf{k}^2}{2m}, \tag{12.102}$$

$$\mathcal{E}_{\mathbf{k}} = \sqrt{\mathcal{E}_{\mathbf{k}}^0 \left(\mathcal{E}_{\mathbf{k}}^0 + 2\frac{UN_0}{V} \right)}. \tag{12.103}$$

Moreover, given (see Eq. 12.74)

$$\begin{aligned} 1 &= u_{\mathbf{k}}^2 - v_{\mathbf{k}}^2 = \\ &= (u_{\mathbf{k}} + v_{\mathbf{k}})(u_{\mathbf{k}} - v_{\mathbf{k}}) = \\ &= s_{\mathbf{k}} d_{\mathbf{k}}, \end{aligned} \tag{12.104}$$

we insert $d_{\mathbf{k}} = s_{\mathbf{k}}^{-1}$ into Eq. 12.94, and we get

$$\frac{\hbar^2 \mathbf{k}^2}{2m} s_{\mathbf{k}}^2 = \mathcal{E}_{\mathbf{k}}, \tag{12.105}$$

$$s_{\mathbf{k}} = \sqrt{\frac{\mathcal{E}_{\mathbf{k}}}{\mathcal{E}_{\mathbf{k}}^0}}, \tag{12.106}$$

$$d_{\mathbf{k}} = \sqrt{\frac{\mathcal{E}_{\mathbf{k}}^0}{\mathcal{E}_{\mathbf{k}}}}, \tag{12.107}$$

$$2u_{\mathbf{k}} = s_{\mathbf{k}} + d_{\mathbf{k}}, \tag{12.108}$$

$$2v_{\mathbf{k}} = s_{\mathbf{k}} - d_{\mathbf{k}}, \tag{12.109}$$

which imply

$$u_{\mathbf{k}} = \frac{1}{2} \left(\sqrt{\frac{\mathcal{E}_{\mathbf{k}}}{\mathcal{E}_{\mathbf{k}}^0}} + \sqrt{\frac{\mathcal{E}_{\mathbf{k}}^0}{\mathcal{E}_{\mathbf{k}}}} \right), \tag{12.110}$$

$$v_{\mathbf{k}} = \frac{1}{2} \left(\sqrt{\frac{\mathcal{E}_{\mathbf{k}}}{\mathcal{E}_{\mathbf{k}}^0}} - \sqrt{\frac{\mathcal{E}_{\mathbf{k}}^0}{\mathcal{E}_{\mathbf{k}}}} \right), \tag{12.111}$$

that is, we have a solution for the coefficients $u_{\mathbf{k}}$, $v_{\mathbf{k}}$. Note that such solutions satisfy

$$u_{-\mathbf{k}} = u_{\mathbf{k}}, \tag{12.112}$$

$$v_{-\mathbf{k}} = v_{\mathbf{k}}. \tag{12.113}$$

Now we want to write the old operators as a function of Bogoljubov bosonic operators. For this purpose, it is necessary to compute the inverse Bogoljubov's bosonic transformation, that is, the inverse of the matrix 12.68. We have

Theorem 12.1 *If the coefficients of the Bogoljubov's bosonic matrix 12.68 satisfy*

$$|u_{\mathbf{k}}|^2 - |v_{-\mathbf{k}}|^2 = 1, \tag{12.114}$$

then such a matrix is invertible and its inverse is given by the matrix

$$\begin{pmatrix} u_{\mathbf{k}}^* & -v_{-\mathbf{k}} \\ -v_{\mathbf{k}}^* & u_{-\mathbf{k}} \end{pmatrix}, \ \mathbf{k} \neq \mathbf{0}. \tag{12.115}$$

Proof From

$$\left[b_{\mathbf{k}}, b_{-\mathbf{k}}\right] = 0, \tag{12.116}$$

$$\left[u_{\mathbf{k}}a_{\mathbf{k}} + v_{\mathbf{k}}a_{-\mathbf{k}}^{\dagger}, u_{-\mathbf{k}}a_{-\mathbf{k}} + v_{-\mathbf{k}}a_{\mathbf{k}}^{\dagger}\right] = 0, \tag{12.117}$$

$$u_{\mathbf{k}}v_{-\mathbf{k}}\left[a_{\mathbf{k}}, a_{\mathbf{k}}^{\dagger}\right] + v_{\mathbf{k}}u_{-\mathbf{k}}\left[a_{-\mathbf{k}}^{\dagger}, a_{-\mathbf{k}}\right] = 0, \tag{12.118}$$

$$u_{\mathbf{k}}v_{-\mathbf{k}} - v_{\mathbf{k}}u_{-\mathbf{k}} = 0, \tag{12.119}$$

we define

$$r_{\mathbf{k}} = \frac{v_{\mathbf{k}}}{u_{\mathbf{k}}}, \tag{12.120}$$

which satisfies

$$r_{\mathbf{k}} = r_{-\mathbf{k}}, \tag{12.121}$$

thanks to

$$\frac{v_{\mathbf{k}}}{u_{\mathbf{k}}} = \frac{v_{-\mathbf{k}}}{u_{-\mathbf{k}}}. \tag{12.122}$$

We divide Eq. 12.74 by $|u_{\mathbf{k}}|^2$, obtaining

$$\frac{|u_{\mathbf{k}}|^2}{|u_{\mathbf{k}}|^2} - \frac{|v_{\mathbf{k}}|^2}{|u_{\mathbf{k}}|^2} = \frac{1}{|u_{\mathbf{k}}|^2}, \tag{12.123}$$

then

$$1 - |r_{\mathbf{k}}|^2 = \frac{1}{|u_{\mathbf{k}}|^2}, \tag{12.124}$$

$$|u_{\mathbf{k}}|^2 \left(1 - |r_{\mathbf{k}}|^2\right) = 1, \tag{12.125}$$

$$|u_{\mathbf{k}}|^2 = \frac{1}{1 - |r_{\mathbf{k}}|^2}. \tag{12.126}$$

Since $u_{\mathbf{k}} \in \mathbb{C}$, its solutions are of the form

$$u_{\mathbf{k}} = \frac{e^{i\gamma_{\mathbf{k}}}}{\sqrt{1 - |r_{\mathbf{k}}|^2}}, \tag{12.127}$$

$$u_{-\mathbf{k}} = \frac{e^{i\gamma_{-\mathbf{k}}}}{\sqrt{1 - |r_{\mathbf{k}}|^2}}, \tag{12.128}$$

with $\gamma_{\mathbf{k}},\ \gamma_{-\mathbf{k}} \in \mathbb{R}$. We substitute the solutions $|u_{\mathbf{k}}|^2$ into Eq. 12.74, i.e.,

$$\frac{1}{1 - |r_{\mathbf{k}}|^2} - |v_{\mathbf{k}}|^2 = 1, \tag{12.129}$$

$$\frac{1 - |v_{\mathbf{k}}|^2 + |v_{\mathbf{k}}|^2 |r_{\mathbf{k}}|^2}{1 - |r_{\mathbf{k}}|^2} = 1, \tag{12.130}$$

$$1 - |v_{\mathbf{k}}|^2 + |v_{\mathbf{k}}|^2 |r_{\mathbf{k}}|^2 = 1 - |r_{\mathbf{k}}|^2, \tag{12.131}$$

$$- |v_{\mathbf{k}}|^2 + |v_{\mathbf{k}}|^2 |r_{\mathbf{k}}|^2 = - |r_{\mathbf{k}}|^2, \tag{12.132}$$

$$|v_{\mathbf{k}}|^2 \left(1 - |r_{\mathbf{k}}|^2\right) = |r_{\mathbf{k}}|^2. \tag{12.133}$$

$$|v_{\mathbf{k}}|^2 = \frac{|r_{\mathbf{k}}|^2}{\left(1 - |r_{\mathbf{k}}|^2\right)}. \tag{12.134}$$

We choose for the solution of $v_{\mathbf{k}}$ the one having the same phase as $u_{\mathbf{k}}$, i.e.,

$$v_{\mathbf{k}} = \frac{e^{i\gamma_{\mathbf{k}}} r_{\mathbf{k}}}{\sqrt{1 - |r_{\mathbf{k}}|^2}}, \tag{12.135}$$

$$v_{-\mathbf{k}} = \frac{e^{i\gamma_{-\mathbf{k}}} r_{\mathbf{k}}}{\sqrt{1 - |r_{\mathbf{k}}|^2}}. \tag{12.136}$$

Now, we multiply Eq. 12.65 by $u^*_{-\mathbf{k}}$ and Eq. 12.66 by $v_{\mathbf{k}}$, i.e.,

$$u^*_{-\mathbf{k}} b_{\mathbf{k}} = u^*_{-\mathbf{k}} u_{\mathbf{k}} a_{\mathbf{k}} + u^*_{-\mathbf{k}} v_{\mathbf{k}} a^\dagger_{-\mathbf{k}}, \tag{12.137}$$

$$v_{\mathbf{k}} b^\dagger_{-\mathbf{k}} = v_{\mathbf{k}} v^*_{-\mathbf{k}} a_{\mathbf{k}} + v_{\mathbf{k}} u^*_{-\mathbf{k}} a^\dagger_{-\mathbf{k}}, \tag{12.138}$$

and we subtract the second equation from the first equation as follows

$$u^*_{-\mathbf{k}} b_{\mathbf{k}} - v_{\mathbf{k}} b^\dagger_{-\mathbf{k}} = \left(u^*_{-\mathbf{k}} u_{\mathbf{k}} - v_{\mathbf{k}} v^*_{-\mathbf{k}}\right) a_{\mathbf{k}}, \tag{12.139}$$

$$a_{\mathbf{k}} = \frac{u^*_{-\mathbf{k}} b_{\mathbf{k}} - v_{\mathbf{k}} b^\dagger_{-\mathbf{k}}}{u^*_{-\mathbf{k}} u_{\mathbf{k}} - v_{\mathbf{k}} v^*_{-\mathbf{k}}}. \tag{12.140}$$

Now, the denominator of the above equation can be manipulated as

$$\begin{aligned} u^*_{-\mathbf{k}} u_{\mathbf{k}} - v^*_{-\mathbf{k}} v_{\mathbf{k}} &= \frac{e^{-i\gamma_{-\mathbf{k}}}}{\sqrt{1-|r_{\mathbf{k}}|^2}} \frac{e^{i\gamma_{\mathbf{k}}}}{\sqrt{1-|r_{\mathbf{k}}|^2}} - \frac{e^{-i\gamma_{-\mathbf{k}}} r^*_{\mathbf{k}}}{\sqrt{1-|r_{\mathbf{k}}|^2}} \frac{e^{i\gamma_{\mathbf{k}}} r_{\mathbf{k}}}{\sqrt{1-|r_{\mathbf{k}}|^2}} = \\ &= \frac{e^{i(\gamma_{\mathbf{k}}-\gamma_{-\mathbf{k}})}}{1-|r_{\mathbf{k}}|^2} - \frac{e^{i(\gamma_{\mathbf{k}}-\gamma_{-\mathbf{k}})}|r_{\mathbf{k}}|^2}{1-|r_{\mathbf{k}}|^2} = \\ &= \frac{e^{i(\gamma_{\mathbf{k}}-\gamma_{-\mathbf{k}})}\left(1-|r_{\mathbf{k}}|^2\right)}{1-|r_{\mathbf{k}}|^2} = \\ &= e^{i(\gamma_{\mathbf{k}}-\gamma_{-\mathbf{k}})}, \end{aligned} \tag{12.141}$$

then

$$\begin{aligned} \frac{u^*_{-\mathbf{k}}}{u^*_{-\mathbf{k}} u_{\mathbf{k}} - v^*_{-\mathbf{k}} v_{\mathbf{k}}} &= \frac{e^{-i\gamma_{-\mathbf{k}}}}{\sqrt{1-|r_{\mathbf{k}}|^2} e^{i(\gamma_{\mathbf{k}}-\gamma_{-\mathbf{k}})}} = \\ &= \frac{e^{-i\gamma_{\mathbf{k}}}}{\sqrt{1-|r_{\mathbf{k}}|^2}} = \\ &= u^*_{\mathbf{k}}, \end{aligned} \tag{12.142}$$

$$\begin{aligned} \frac{v_{\mathbf{k}}}{u^*_{-\mathbf{k}} u_{\mathbf{k}} - v^*_{-\mathbf{k}} v_{\mathbf{k}}} &= \frac{e^{i\gamma_{\mathbf{k}}} r_{\mathbf{k}}}{\sqrt{1-|r_{\mathbf{k}}|^2} e^{i(\gamma_{\mathbf{k}}-\gamma_{-\mathbf{k}})}} = \\ &= \frac{e^{i\gamma_{-\mathbf{k}}} r_{\mathbf{k}}}{\sqrt{1-|r_{\mathbf{k}}|^2}} = \\ &= v_{-\mathbf{k}}. \end{aligned} \tag{12.143}$$

We insert the two results above into Eq. 12.140, i.e.,

$$a_{\mathbf{k}} = \frac{u^*_{-\mathbf{k}}}{u^*_{-\mathbf{k}}u_{\mathbf{k}} - v_{\mathbf{k}}v^*_{-\mathbf{k}}} b_{\mathbf{k}} - \frac{v_{\mathbf{k}}}{u^*_{-\mathbf{k}}u_{\mathbf{k}} - v_{\mathbf{k}}v^*_{-\mathbf{k}}} b^\dagger_{-\mathbf{k}} = \\ = u^*_{\mathbf{k}} b_{\mathbf{k}} - v_{-\mathbf{k}} b^\dagger_{-\mathbf{k}}, \tag{12.144}$$

which together with its complex conjugate provide

$$\begin{cases} a_{\mathbf{k}} = u^*_{\mathbf{k}} b_{\mathbf{k}} - v_{-\mathbf{k}} b^\dagger_{-\mathbf{k}} \\ a^\dagger_{\mathbf{k}} = u_{\mathbf{k}} b^\dagger_{\mathbf{k}} - v^*_{-\mathbf{k}} b_{-\mathbf{k}} \end{cases}, \tag{12.145}$$

or equivalently

$$\begin{cases} a_{\mathbf{k}} = u^*_{\mathbf{k}} b_{\mathbf{k}} - v_{-\mathbf{k}} b^\dagger_{-\mathbf{k}} \\ a^\dagger_{-\mathbf{k}} = u_{-\mathbf{k}} b^\dagger_{-\mathbf{k}} - v^*_{\mathbf{k}} b_{\mathbf{k}} \end{cases}, \tag{12.146}$$

where the momentum in the creation operator $a^\dagger_{\mathbf{k}}$ has been inverted. Such transformations are given in matrix form as

$$\begin{pmatrix} a_{\mathbf{k}} \\ a^\dagger_{-\mathbf{k}} \end{pmatrix} = \begin{pmatrix} u^*_{\mathbf{k}} & -v_{-\mathbf{k}} \\ -v^*_{\mathbf{k}} & u_{-\mathbf{k}} \end{pmatrix} \begin{pmatrix} b_{\mathbf{k}} \\ b^\dagger_{-\mathbf{k}} \end{pmatrix}, \quad \mathbf{k} \neq \mathbf{0}. \tag{12.147}$$

Let us verify that operators 12.145 satisfy a bosonic algebra D.1, indeed

$$\left[a_{\mathbf{k}}, a^\dagger_{\mathbf{k}'}\right] = \left[u^*_{\mathbf{k}} b_{\mathbf{k}} - v_{-\mathbf{k}} b^\dagger_{-\mathbf{k}}, u_{\mathbf{k}'} b^\dagger_{\mathbf{k}'} - v^*_{-\mathbf{k}'} b_{-\mathbf{k}'}\right] = \\ = \left[u^*_{\mathbf{k}} b_{\mathbf{k}}, u_{\mathbf{k}'} b^\dagger_{\mathbf{k}'}\right] + \left[v_{-\mathbf{k}} b^\dagger_{-\mathbf{k}}, v^*_{-\mathbf{k}'} b_{-\mathbf{k}'}\right] = \\ = u^*_{\mathbf{k}} u_{\mathbf{k}'} \delta_{\mathbf{k},\mathbf{k}'} - v_{-\mathbf{k}} v^*_{-\mathbf{k}'} \delta_{-\mathbf{k},-\mathbf{k}'}, \tag{12.148}$$

that for $\mathbf{k} \neq \mathbf{k}'$ is null, while for $\mathbf{k} = \mathbf{k}'$ given that $\delta_{-\mathbf{k}',-\mathbf{k}} = \delta_{\mathbf{k}',\mathbf{k}}$ becomes

$$\left[a_{\mathbf{k}}, a^\dagger_{\mathbf{k}}\right] = u^*_{\mathbf{k}} u_{\mathbf{k}} - v_{-\mathbf{k}} v^*_{-\mathbf{k}} = \\ = |u_{\mathbf{k}}|^2 - |v_{-\mathbf{k}}|^2, \tag{12.149}$$

which is equal to the identity if Eq. 12.114 holds. Note that Eq. 12.114 is similar to Eq. 12.74 except for a sign of the momentum of v. However, an Ansatz for coefficients $u_{\mathbf{k}}$ and $v_{\mathbf{k}}$ will be set in this section (see Eqs. 12.152 and 12.153), according to which 12.74 and 12.114 become equal. □

Now, let us insert the above solutions into Eq. 12.60.

$$
\begin{aligned}
\hat{\mathcal{H}} = & -\frac{UN_0^2}{2V} + \sum_{\mathbf{k}\neq\mathbf{0}} \left(\frac{\hbar^2 \mathbf{k}^2}{2m} + \frac{UN_0}{V} \right) \left[\left(u_{\mathbf{k}} b_{\mathbf{k}}^{\dagger} - v_{-\mathbf{k}}^{*} b_{-\mathbf{k}} \right) \left(u_{\mathbf{k}}^{*} b_{\mathbf{k}} - v_{-\mathbf{k}} b_{-\mathbf{k}}^{\dagger} \right) \right] + \\
& + \frac{UN_0}{2V} \sum_{\mathbf{k}\neq\mathbf{0}} \left[\left(u_{\mathbf{k}} b_{\mathbf{k}}^{\dagger} - v_{-\mathbf{k}}^{*} b_{-\mathbf{k}} \right) \left(u_{-\mathbf{k}} b_{-\mathbf{k}}^{\dagger} - v_{\mathbf{k}}^{*} b_{\mathbf{k}} \right) \right. \\
& \qquad \left. + \left(u_{-\mathbf{k}}^{*} b_{-\mathbf{k}} - v_{\mathbf{k}} b_{\mathbf{k}}^{\dagger} \right) \left(u_{\mathbf{k}}^{*} b_{\mathbf{k}} - v_{-\mathbf{k}} b_{-\mathbf{k}}^{\dagger} \right) \right] = \\
= & -\frac{UN_0^2}{2V} + \sum_{\mathbf{k}\neq\mathbf{0}} \left(\frac{\hbar^2 \mathbf{k}^2}{2m} + \frac{UN_0}{V} \right) \left[|u_{\mathbf{k}}|^2 \, b_{\mathbf{k}}^{\dagger} b_{\mathbf{k}} + |v_{-\mathbf{k}}|^2 \, b_{-\mathbf{k}} b_{-\mathbf{k}}^{\dagger} \right. \\
& \qquad \left. - u_{\mathbf{k}} v_{-\mathbf{k}} b_{\mathbf{k}}^{\dagger} b_{-\mathbf{k}}^{\dagger} - u_{\mathbf{k}}^{*} v_{-\mathbf{k}}^{*} b_{-\mathbf{k}} b_{\mathbf{k}} \right] + \\
& + \frac{UN_0}{2V} \sum_{\mathbf{k}\neq\mathbf{0}} \left[\left(u_{\mathbf{k}} u_{-\mathbf{k}} b_{\mathbf{k}}^{\dagger} b_{-\mathbf{k}}^{\dagger} - u_{\mathbf{k}} v_{\mathbf{k}}^{*} b_{\mathbf{k}}^{\dagger} b_{\mathbf{k}} - u_{-\mathbf{k}} v_{-\mathbf{k}}^{*} b_{-\mathbf{k}} b_{-\mathbf{k}}^{\dagger} + v_{-\mathbf{k}}^{*} v_{\mathbf{k}}^{*} b_{-\mathbf{k}} b_{\mathbf{k}} \right) + \right. \\
& \qquad \left. + \left(u_{-\mathbf{k}}^{*} u_{\mathbf{k}}^{*} b_{-\mathbf{k}} b_{\mathbf{k}} - u_{-\mathbf{k}}^{*} v_{-\mathbf{k}} b_{-\mathbf{k}} b_{-\mathbf{k}}^{\dagger} - u_{\mathbf{k}}^{*} v_{\mathbf{k}} b_{\mathbf{k}}^{\dagger} b_{\mathbf{k}} + v_{\mathbf{k}} v_{-\mathbf{k}} b_{\mathbf{k}}^{\dagger} b_{-\mathbf{k}}^{\dagger} \right) \right].
\end{aligned}
\tag{12.150}
$$

In order to simplify the form of the Hamiltonian introduced above, we now assume that

$$u_{\mathbf{k}}, v_{\mathbf{k}} \in \mathbb{R}, \tag{12.151}$$

$$u_{-\mathbf{k}} = u_{\mathbf{k}}, \tag{12.152}$$

$$v_{-\mathbf{k}} = v_{\mathbf{k}}, \tag{12.153}$$

and write

$$
\begin{aligned}
\hat{\mathcal{H}} = & -\frac{UN_0^2}{2V} + \sum_{\mathbf{k}\neq\mathbf{0}} \left(\frac{\hbar^2 \mathbf{k}^2}{2m} + \frac{UN_0}{V} \right) \left[u_{\mathbf{k}}^2 b_{\mathbf{k}}^{\dagger} b_{\mathbf{k}} + v_{\mathbf{k}}^2 b_{-\mathbf{k}} b_{-\mathbf{k}}^{\dagger} \right. \\
& \qquad \left. - u_{\mathbf{k}} v_{\mathbf{k}} \left(b_{\mathbf{k}}^{\dagger} b_{-\mathbf{k}}^{\dagger} + b_{-\mathbf{k}} b_{\mathbf{k}} \right) \right] + \\
& + \frac{UN_0}{2V} \sum_{\mathbf{k}\neq\mathbf{0}} \left[\left(u_{\mathbf{k}}^2 b_{\mathbf{k}}^{\dagger} b_{-\mathbf{k}}^{\dagger} - u_{\mathbf{k}} v_{\mathbf{k}} b_{\mathbf{k}}^{\dagger} b_{\mathbf{k}} - u_{\mathbf{k}} v_{\mathbf{k}} b_{-\mathbf{k}} b_{-\mathbf{k}}^{\dagger} + v_{\mathbf{k}}^2 b_{-\mathbf{k}} b_{\mathbf{k}} \right) + \right. \\
& \qquad \left. + \left(u_{\mathbf{k}}^2 b_{-\mathbf{k}} b_{\mathbf{k}} - u_{\mathbf{k}} v_{\mathbf{k}} b_{-\mathbf{k}} b_{-\mathbf{k}}^{\dagger} - u_{\mathbf{k}} v_{\mathbf{k}} b_{\mathbf{k}}^{\dagger} b_{\mathbf{k}} + v_{\mathbf{k}}^2 b_{\mathbf{k}}^{\dagger} b_{-\mathbf{k}}^{\dagger} \right) \right].
\end{aligned}
\tag{12.154}
$$

Using Eq. D.8 adapted for bosonic particles, and redefining the index $-\mathbf{k} \to \mathbf{k}$ in operators $b_{-\mathbf{k}} b^{\dagger}_{-\mathbf{k}}$, $b^{\dagger}_{-\mathbf{k}} b_{-\mathbf{k}}$, we obtain

$$
\begin{aligned}
\hat{\mathcal{H}} = & -\frac{UN_0^2}{2V} + \sum_{\mathbf{k}\neq\mathbf{0}} \left(\frac{\hbar^2\mathbf{k}^2}{2m} + \frac{UN_0}{V} \right) \Big[\left(u_{\mathbf{k}}^2 + v_{\mathbf{k}}^2 \right) b^{\dagger}_{\mathbf{k}} b_{\mathbf{k}} + v_{\mathbf{k}}^2 \\
& - u_{\mathbf{k}} v_{\mathbf{k}} \left(b^{\dagger}_{\mathbf{k}} b^{\dagger}_{-\mathbf{k}} + b_{-\mathbf{k}} b_{\mathbf{k}} \right) \Big] + \\
& + \frac{UN_0}{2V} \sum_{\mathbf{k}\neq\mathbf{0}} \Big[\left(u_{\mathbf{k}}^2 b^{\dagger}_{\mathbf{k}} b^{\dagger}_{-\mathbf{k}} - u_{\mathbf{k}} v_{\mathbf{k}} b^{\dagger}_{\mathbf{k}} b_{\mathbf{k}} - u_{\mathbf{k}} v_{\mathbf{k}} b^{\dagger}_{\mathbf{k}} b_{\mathbf{k}} - u_{\mathbf{k}} v_{\mathbf{k}} + v_{\mathbf{k}}^2 b_{-\mathbf{k}} b_{\mathbf{k}} \right) + \\
& + \left(u_{\mathbf{k}}^2 b_{-\mathbf{k}} b_{\mathbf{k}} - u_{\mathbf{k}} v_{\mathbf{k}} b^{\dagger}_{\mathbf{k}} b_{\mathbf{k}} - u_{\mathbf{k}} v_{\mathbf{k}} - u_{\mathbf{k}} v_{\mathbf{k}} b^{\dagger}_{\mathbf{k}} b_{\mathbf{k}} + v_{\mathbf{k}}^2 b^{\dagger}_{\mathbf{k}} b^{\dagger}_{-\mathbf{k}} \right) \Big] = \\
= & -\frac{UN_0^2}{2V} + \sum_{\mathbf{k}\neq\mathbf{0}} \left(\frac{\hbar^2\mathbf{k}^2}{2m} + \frac{UN_0}{V} \right) \Big[\left(u_{\mathbf{k}}^2 + v_{\mathbf{k}}^2 \right) b^{\dagger}_{\mathbf{k}} b_{\mathbf{k}} - u_{\mathbf{k}} v_{\mathbf{k}} \left(b^{\dagger}_{\mathbf{k}} b^{\dagger}_{-\mathbf{k}} + b_{-\mathbf{k}} b_{\mathbf{k}} \right) \\
& + v_{\mathbf{k}}^2 \Big] + \\
& + \frac{UN_0}{2V} \sum_{\mathbf{k}\neq\mathbf{0}} \Big[\left(u_{\mathbf{k}}^2 + v_{\mathbf{k}}^2 \right) \left(b^{\dagger}_{\mathbf{k}} b^{\dagger}_{-\mathbf{k}} + b_{-\mathbf{k}} b_{\mathbf{k}} \right) - 4 u_{\mathbf{k}} v_{\mathbf{k}} b^{\dagger}_{\mathbf{k}} b_{\mathbf{k}} - 2 u_{\mathbf{k}} v_{\mathbf{k}} \Big] .
\end{aligned}
\tag{12.155}
$$

We define the constant

$$
\tilde{\mathcal{E}} = -\frac{UN_0^2}{2V} + \sum_{\mathbf{k}\neq\mathbf{0}} \left[\left(\frac{\hbar^2\mathbf{k}^2}{2m} + \frac{UN_0}{V} \right) v_{\mathbf{k}}^2 - \frac{UN_0}{V} u_{\mathbf{k}} v_{\mathbf{k}} \right], \tag{12.156}
$$

and the Hamiltonian becomes

$$
\begin{aligned}
\hat{\mathcal{H}} = \tilde{\mathcal{E}} + & \sum_{\mathbf{k}\neq\mathbf{0}} \left[\left(\frac{\hbar^2\mathbf{k}^2}{2m} + \frac{UN_0}{V} \right) \left(u_{\mathbf{k}}^2 + v_{\mathbf{k}}^2 \right) - 2 \frac{UN_0}{V} u_{\mathbf{k}} v_{\mathbf{k}} \right] b^{\dagger}_{\mathbf{k}} b_{\mathbf{k}} + \\
- & \sum_{\mathbf{k}\neq\mathbf{0}} \left[\left(\frac{\hbar^2\mathbf{k}^2}{2m} + \frac{UN_0}{V} \right) u_{\mathbf{k}} v_{\mathbf{k}} - \frac{UN_0}{2V} \left(u_{\mathbf{k}}^2 + v_{\mathbf{k}}^2 \right) \right] \left(b^{\dagger}_{\mathbf{k}} b^{\dagger}_{-\mathbf{k}} + b_{-\mathbf{k}} b_{\mathbf{k}} \right) .
\end{aligned}
\tag{12.157}
$$

If we set

$$
F_{\mathbf{k}} \equiv \frac{\hbar^2\mathbf{k}^2}{2m} + \frac{UN_0}{V}, \tag{12.158}
$$

$$T \equiv \frac{UN_0}{V}, \tag{12.159}$$

then the Hamiltonian

$$\hat{\mathcal{H}} = \tilde{\mathcal{E}} + \sum_{\mathbf{k}\neq\mathbf{0}} \left[F_{\mathbf{k}} \left(u_{\mathbf{k}}^2 + v_{\mathbf{k}}^2 \right) - T \left(2u_{\mathbf{k}}v_{\mathbf{k}} \right) \right] b_{\mathbf{k}}^\dagger b_{\mathbf{k}} - \sum_{\mathbf{k}\neq\mathbf{0}} \left[F_{\mathbf{k}} u_{\mathbf{k}} v_{\mathbf{k}} - \frac{T}{2} \left(u_{\mathbf{k}}^2 + v_{\mathbf{k}}^2 \right) \right]$$
$$\left(b_{\mathbf{k}}^\dagger b_{-\mathbf{k}}^\dagger + b_{-\mathbf{k}} b_{\mathbf{k}} \right) \tag{12.160}$$

shows a similarity between the coefficient of $b_{\mathbf{k}}^\dagger b_{\mathbf{k}}$ and that of $\left(b_{\mathbf{k}}^\dagger b_{-\mathbf{k}}^\dagger + b_{-\mathbf{k}} b_{\mathbf{k}} \right)$. Indeed, first we multiply and divide by 2 the number operator and manipulate as follows

$$\frac{1}{2} \sum_{\mathbf{k}\neq\mathbf{0}} 2 b_{\mathbf{k}}^\dagger b_{\mathbf{k}} = \frac{1}{2} \sum_{\mathbf{k}\neq\mathbf{0}} \left(b_{\mathbf{k}}^\dagger b_{\mathbf{k}} + b_{-\mathbf{k}}^\dagger b_{-\mathbf{k}} \right), \tag{12.161}$$

and we write

$$\hat{\mathcal{H}} = \tilde{\mathcal{E}} + \frac{1}{2} \sum_{\mathbf{k}\neq\mathbf{0}} \left[F_{\mathbf{k}} \left(u_{\mathbf{k}}^2 + v_{\mathbf{k}}^2 \right) - T \left(2u_{\mathbf{k}}v_{\mathbf{k}} \right) \right] \left(b_{\mathbf{k}}^\dagger b_{\mathbf{k}} + b_{-\mathbf{k}}^\dagger b_{-\mathbf{k}} \right)$$
$$- \frac{1}{2} \sum_{\mathbf{k}\neq\mathbf{0}} \left[F_{\mathbf{k}} \left(2u_{\mathbf{k}}v_{\mathbf{k}} \right) - T \left(u_{\mathbf{k}}^2 + v_{\mathbf{k}}^2 \right) \right] \left(b_{\mathbf{k}}^\dagger b_{-\mathbf{k}}^\dagger + b_{-\mathbf{k}} b_{\mathbf{k}} \right). \tag{12.162}$$

Consequently, the hamiltonian is diagonal if

$$F_{\mathbf{k}} \left(2u_{\mathbf{k}}v_{\mathbf{k}} \right) - T \left(u_{\mathbf{k}}^2 + v_{\mathbf{k}}^2 \right) \equiv \left(\frac{\hbar^2 \mathbf{k}^2}{2m} + \frac{UN_0}{V} \right) 2u_{\mathbf{k}}v_{\mathbf{k}} - \frac{UN_0}{V} \left(u_{\mathbf{k}}^2 + v_{\mathbf{k}}^2 \right) =$$
$$= 0, \ \forall\, \mathbf{k} \neq \mathbf{0}. \tag{12.163}$$

From Eq. 12.74, we set the Ansatz

$$u_{\mathbf{k}} = \cosh(\theta_{\mathbf{k}}), \tag{12.164}$$

$$v_{\mathbf{k}} = \sinh(\theta_{\mathbf{k}}), \tag{12.165}$$

and we insert

$$2\cosh(\theta_{\mathbf{k}}) \sinh(\theta_{\mathbf{k}}) = \sinh(2\theta_{\mathbf{k}}), \tag{12.166}$$

$$\cosh^2(\theta_{\mathbf{k}}) + \sinh^2(\theta_{\mathbf{k}}) = \cosh(2\theta_{\mathbf{k}}), \tag{12.167}$$

into Eq. 12.163, then

$$F_{\mathbf{k}} \sinh(2\theta_{\mathbf{k}}) = T \cosh(2\theta_{\mathbf{k}}), \tag{12.168}$$

$$F_{\mathbf{k}} \sinh(2\theta_{\mathbf{k}}) = T\sqrt{1 + \sinh^2(2\theta_{\mathbf{k}})}, \tag{12.169}$$

$$F_{\mathbf{k}}^2 \sinh^2(2\theta_{\mathbf{k}}) = T^2 \left(1 + \sinh^2(2\theta_{\mathbf{k}})\right), \tag{12.170}$$

$$\sinh^2(2\theta_{\mathbf{k}}) = \frac{T^2}{F_{\mathbf{k}}^2 - T^2}, \tag{12.171}$$

$$\begin{aligned}\cosh^2(2\theta_{\mathbf{k}}) &= 1 + \sinh^2(2\theta_{\mathbf{k}}) = \\ &= \frac{F_{\mathbf{k}}^2 - T^2 + T^2}{F_{\mathbf{k}}^2 - T^2} = \\ &= \frac{F_{\mathbf{k}}^2}{F_{\mathbf{k}}^2 - T^2},\end{aligned} \tag{12.172}$$

$$\begin{cases} \sinh(2\theta_{\mathbf{k}}) = \dfrac{T}{\sqrt{F_{\mathbf{k}}^2 - T^2}} \\ \cosh(2\theta_{\mathbf{k}}) = \dfrac{F_{\mathbf{k}}}{\sqrt{F_{\mathbf{k}}^2 - T^2}} \end{cases} . \tag{12.173}$$

In conclusion,

$$\begin{aligned}\hat{\mathcal{H}} &= \tilde{\mathcal{E}} + \frac{1}{2}\sum_{\mathbf{k}\neq\mathbf{0}} \left[F_{\mathbf{k}}\left(u_{\mathbf{k}}^2 + v_{\mathbf{k}}^2\right) - T\,(2u_{\mathbf{k}}v_{\mathbf{k}})\right]\left(b_{\mathbf{k}}^\dagger b_{\mathbf{k}} + b_{-\mathbf{k}}^\dagger b_{-\mathbf{k}}\right) = \\ &= \tilde{\mathcal{E}} + \frac{1}{2}\sum_{\mathbf{k}\neq\mathbf{0}} \left[F_{\mathbf{k}} \cosh(2\theta_{\mathbf{k}}) - T \sinh(2\theta_{\mathbf{k}})\right]\left(b_{\mathbf{k}}^\dagger b_{\mathbf{k}} + b_{-\mathbf{k}}^\dagger b_{-\mathbf{k}}\right) = \\ &= \tilde{\mathcal{E}} + \frac{1}{2}\sum_{\mathbf{k}\neq\mathbf{0}} \left[F_{\mathbf{k}}\frac{F_{\mathbf{k}}}{\sqrt{F_{\mathbf{k}}^2 - T^2}} - T\frac{T}{\sqrt{F_{\mathbf{k}}^2 - T^2}}\right]\left(b_{\mathbf{k}}^\dagger b_{\mathbf{k}} + b_{-\mathbf{k}}^\dagger b_{-\mathbf{k}}\right) = \\ &= \tilde{\mathcal{E}} + \frac{1}{2}\sum_{\mathbf{k}\neq\mathbf{0}} \left[\frac{F_{\mathbf{k}}^2}{\sqrt{F_{\mathbf{k}}^2 - T^2}} - \frac{T^2}{\sqrt{F_{\mathbf{k}}^2 - T^2}}\right]\left(b_{\mathbf{k}}^\dagger b_{\mathbf{k}} + b_{-\mathbf{k}}^\dagger b_{-\mathbf{k}}\right) =\end{aligned}$$

$$
\begin{aligned}
&= \tilde{\mathcal{E}} + \frac{1}{2}\sum_{\mathbf{k}\neq\mathbf{0}} \left[\sqrt{F_{\mathbf{k}}^2 - T^2}\right] \left(b_{\mathbf{k}}^{\dagger} b_{\mathbf{k}} + b_{-\mathbf{k}}^{\dagger} b_{-\mathbf{k}}\right) = \\
&= \tilde{\mathcal{E}} + \frac{1}{2}\sum_{\mathbf{k}\neq\mathbf{0}} \left[\sqrt{\left(\frac{\hbar^2\mathbf{k}^2}{2m} + \frac{UN_0}{V}\right)^2 - \left(\frac{UN_0}{V}\right)^2}\right] \left(b_{\mathbf{k}}^{\dagger} b_{\mathbf{k}} + b_{-\mathbf{k}}^{\dagger} b_{-\mathbf{k}}\right) = \\
&= \tilde{\mathcal{E}} + \frac{1}{2}\sum_{\mathbf{k}\neq\mathbf{0}} \left[\sqrt{\left(\frac{\hbar^2\mathbf{k}^2}{2m}\right)^2 + \frac{U^2N_0^2}{V^2} + 2\frac{\hbar^2\mathbf{k}^2}{2m}\frac{UN_0}{V} - \frac{U^2N_0^2}{V^2}}\right] \\
&\quad \times \left(b_{\mathbf{k}}^{\dagger} b_{\mathbf{k}} + b_{-\mathbf{k}}^{\dagger} b_{-\mathbf{k}}\right) = \\
&= \tilde{\mathcal{E}} + \frac{1}{2}\sum_{\mathbf{k}\neq\mathbf{0}} \left[\sqrt{\frac{\hbar^2\mathbf{k}^2}{2m}\left(\frac{\hbar^2\mathbf{k}^2}{2m} + 2\frac{UN_0}{V}\right)}\right] \left(b_{\mathbf{k}}^{\dagger} b_{\mathbf{k}} + b_{-\mathbf{k}}^{\dagger} b_{-\mathbf{k}}\right) = \\
&= \tilde{\mathcal{E}} + \frac{1}{2}\sum_{\mathbf{k}\neq\mathbf{0}} \mathcal{E}_{\mathbf{k}} \left(b_{\mathbf{k}}^{\dagger} b_{\mathbf{k}} + b_{-\mathbf{k}}^{\dagger} b_{-\mathbf{k}}\right) = \\
&= \tilde{\mathcal{E}} + \frac{1}{2}\sum_{\mathbf{k}\neq\mathbf{0}} \mathcal{E}_{\mathbf{k}} \left(2 b_{\mathbf{k}}^{\dagger} b_{\mathbf{k}}\right),
\end{aligned}
\tag{12.174}
$$

and we finally obtain the diagonalized Bogoljubov Hamiltonian

$$
\hat{\mathcal{H}} = \tilde{\mathcal{E}} + \sum_{\mathbf{k}\neq\mathbf{0}} \mathcal{E}_{\mathbf{k}} b_{\mathbf{k}}^{\dagger} b_{\mathbf{k}}, \tag{12.175}
$$

with

$$
\mathcal{E}_{\mathbf{k}} = \sqrt{\frac{\hbar^2\mathbf{k}^2}{2m}\left(\frac{\hbar^2\mathbf{k}^2}{2m} + 2\frac{UN_0}{V}\right)}. \tag{12.176}
$$

Bogoljubov's bosonic transformations (see Eq. 12.67) are given by

$$
\begin{aligned}
\begin{pmatrix} b_{\mathbf{k}} \\ b_{-\mathbf{k}}^{\dagger} \end{pmatrix} &= \begin{pmatrix} u_{\mathbf{k}} & v_{\mathbf{k}} \\ v_{\mathbf{k}} & u_{\mathbf{k}} \end{pmatrix} \begin{pmatrix} a_{\mathbf{k}} \\ a_{-\mathbf{k}}^{\dagger} \end{pmatrix} \equiv \\
&\equiv \begin{pmatrix} \cosh(\theta_{\mathbf{k}}) & \sinh(\theta_{\mathbf{k}}) \\ \sinh(\theta_{\mathbf{k}}) & \cosh(\theta_{\mathbf{k}}) \end{pmatrix} \begin{pmatrix} a_{\mathbf{k}} \\ a_{-\mathbf{k}}^{\dagger} \end{pmatrix}, \quad \mathbf{k} \neq \mathbf{0},
\end{aligned}
\tag{12.177}
$$

$$
\begin{pmatrix} a_{\mathbf{k}} \\ a^{\dagger}_{-\mathbf{k}} \end{pmatrix} = \begin{pmatrix} u_{\mathbf{k}} & -v_{\mathbf{k}} \\ -v_{\mathbf{k}} & u_{\mathbf{k}} \end{pmatrix} \begin{pmatrix} b_{\mathbf{k}} \\ b^{\dagger}_{-\mathbf{k}} \end{pmatrix} \equiv
$$

$$
\equiv \begin{pmatrix} \cosh(\theta_{\mathbf{k}}) & -\sinh(\theta_{\mathbf{k}}) \\ -\sinh(\theta_{\mathbf{k}}) & \cosh(\theta_{\mathbf{k}}) \end{pmatrix} \begin{pmatrix} b_{\mathbf{k}} \\ b^{\dagger}_{-\mathbf{k}} \end{pmatrix}, \quad \mathbf{k} \neq \mathbf{0}. \tag{12.178}
$$

Note that

$$
\begin{pmatrix} u_{\mathbf{k}} & v_{\mathbf{k}} \\ v_{\mathbf{k}} & u_{\mathbf{k}} \end{pmatrix} \begin{pmatrix} u_{\mathbf{k}} & -v_{\mathbf{k}} \\ -v_{\mathbf{k}} & u_{\mathbf{k}} \end{pmatrix} = \begin{pmatrix} u_{\mathbf{k}}^2 - v_{\mathbf{k}}^2 & -u_{\mathbf{k}} v_{\mathbf{k}} \\ u_{\mathbf{k}} v_{\mathbf{k}} - u_{\mathbf{k}} v_{\mathbf{k}} & -v_{\mathbf{k}}^2 + u_{\mathbf{k}}^2 \end{pmatrix} =
$$

$$
= \begin{pmatrix} 1 & 0 \\ 0 & 1 \end{pmatrix}, \tag{12.179}
$$

$$
\begin{pmatrix} \cosh(\theta_{\mathbf{k}}) & \sinh(\theta_{\mathbf{k}}) \\ \sinh(\theta_{\mathbf{k}}) & \cosh(\theta_{\mathbf{k}}) \end{pmatrix} \begin{pmatrix} \cosh(\theta_{\mathbf{k}}) & -\sinh(\theta_{\mathbf{k}}) \\ -\sinh(\theta_{\mathbf{k}}) & \cosh(\theta_{\mathbf{k}}) \end{pmatrix} =
$$

$$
= \begin{pmatrix} \cosh^2(\theta_{\mathbf{k}}) - \sinh^2(\theta_{\mathbf{k}}) & -\cosh(\theta_{\mathbf{k}}) \sinh(\theta_{\mathbf{k}}) + \sinh(\theta_{\mathbf{k}}) \cosh(\theta_{\mathbf{k}}) \\ \sinh(\theta_{\mathbf{k}}) \cosh(\theta_{\mathbf{k}}) - \cosh(\theta_{\mathbf{k}}) \sinh(\theta_{\mathbf{k}}) & -\sinh^2(\theta_{\mathbf{k}}) + \cosh^2(\theta_{\mathbf{k}}) \end{pmatrix} =
$$

$$
= \begin{pmatrix} 1 & 0 \\ 0 & 1 \end{pmatrix}, \tag{12.180}
$$

as it must be.

12.3.1 Bogoljubov Excitation Spectrum and Landau's Criterion

Let us examine the behavior of the eigenvalues 12.103 in the limits of high and low momenta. Note that for low momenta we have

$$
\mathcal{E}_{\mathbf{k}} \underset{\mathbf{k} \to \mathbf{0}}{\sim} 2 \frac{U N_0}{V} \mathcal{E}^0_{\mathbf{k}} =
$$

$$
= \sqrt{\frac{U N_0 \hbar^2 k^2}{m V}} =
$$

$$
= \hbar k \sqrt{\frac{U N_0}{m V}}, \tag{12.181}
$$

i.e., in Bogoljubov's theory, for small momenta, the interactions between bosons imply a linear dispersion relation. From

$$\omega = kc, \tag{12.182}$$

$$\mathcal{E} = \hbar kc, \tag{12.183}$$

we derive the dispersion relation for small momenta, where c is the sound velocity associated with the low-energy collective excitations,

$$c = \sqrt{\frac{UN_0}{mV}}, \tag{12.184}$$

$$\begin{aligned} \left[\frac{UN_0}{mV}\right] = \frac{[E]}{[m]\,[L]^3} = \\ = \frac{[m]\,[L]^2}{[m]\,[T]^2} = \\ = \frac{[L]^2}{[T]^2}. \end{aligned} \tag{12.185}$$

Conversely, at high momenta, the bosons behave effectively as non-interacting particles and the spectrum approaches the quadratic one-particle form. However, high-momentum excitations are irrelevant for the stability of a slowly moving fluid. The Landau's theory of superfluidity (see Sect. 11.5) is determined by the low-energy, long-wavelength part of the spectrum. According to Landau's criterion, the critical velocity v_c is given by the minimum value of the ratio between the excitation energy and its momentum (see Eq. 11.115). Here $p = \hbar k$ denotes the momentum associated with the excitation. For a linear dispersion relation of the form $\mathcal{E}(p) = pc$, this ratio is constant and equal to c (see Eq. 12.184), which coincides with the sound velocity of the system. The critical velocity is therefore finite and given by

$$v_c = c. \tag{12.186}$$

Bogoljubov's theory shows that interactions play a crucial role in determining the nature of the low-energy excitations. By transforming the quadratic one-particle spectrum into a linear, phononic one at small momenta, interactions qualitatively reorganize the low-energy excitation structure. In this way, they ensure a finite Landau critical velocity, equal to the speed of sound, and provide the microscopic mechanism underlying superfluidity.

Nevertheless, the sound velocity given in Eq. 12.184 generally provides only an upper estimate of the Landau critical velocity. This limitation reflects the fact that Bogoljubov theory describes solely the phononic branch of the excitation spectrum. At higher densities, additional excitations, such as rotons, can emerge and lead to a smaller critical velocity.

12.4 Fundamental State of the Bogoljubov's Bosonic Hamiltonian

In order to characterize the ground state of the Bogoljubov Hamiltonian, it is necessary to determine the state annihilated by all Bogoljubov quasiparticle operators. This can be achieved by drawing an analogy with a single harmonic oscillator (see Sect. 1.3), where a suitable canonical transformation allows the diagonalization of the Hamiltonian and the identification of its vacuum state, denoted by $|\Psi\rangle$. We thus consider the transformation generated by the operator

$$b_{\mathbf{k}}|\Psi\rangle = 0, \ \forall \mathbf{k}. \tag{12.187}$$

To do this, consider a one-dimensional harmonic oscillator Hamiltonian $\mathcal{H}_1$ as given in Eq. 1.138, and define the operator

$$S = z\left(aa - a^{\dagger}a^{\dagger}\right), \ z \in \mathbb{R}. \tag{12.188}$$

The parameter z is chosen such that the transformation e^S is unitary, which requires $S^{\dagger} = -S$.

Remark 12.1 The operator 12.188 exhibits a formal resemblance to the exponential operator employed in the construction of coherent states, namely Eq. 1.215, as both involve exponentials of linear combinations of a and $a^{\dagger}$. Nevertheless, in the Bogoljubov transformation the operator acts on pairs of opposite momenta ($\mathbf{k}$, $-\mathbf{k}$), so that the structure of S differs substantially from that of the operator generating coherent states.

In order to analyze the transformed operator $e^S\mathcal{H}_1e^{-S}$, we define

$$h(z) = e^{S(z)}ae^{-S(z)}, \tag{12.189}$$

and we compute the first derivative of $h(z)$ with respect to the z, i.e.,

$$\begin{aligned}
\frac{dh(z)}{dz} &= \frac{dS(z)}{dz}e^{S(z)}ae^{-S(z)} - e^{S(z)}ae^{-S(z)}\frac{dS(z)}{dz} = \\
&= e^{S(z)}\frac{dS(z)}{dz}ae^{-S(z)} - e^{S(z)}a\frac{dS(z)}{dz}e^{-S(z)} = \\
&= e^{S(z)}\left(\frac{dS(z)}{dz}a - a\frac{dS(z)}{dz}\right)e^{-S(z)} = \\
&= e^{S(z)}\left\{\left(aa - a^{\dagger}a^{\dagger}\right)a - a\left(aa - a^{\dagger}a^{\dagger}\right)\right\}e^{-S(z)} = \\
&= e^{S(z)}\left(aaa - a^{\dagger}a^{\dagger}a - aaa + aa^{\dagger}a^{\dagger}\right)e^{-S(z)} = \\
&= e^{S(z)}\left(-a^{\dagger}a^{\dagger}a + aa^{\dagger}a^{\dagger}\right)e^{-S(z)},
\end{aligned} \tag{12.190}$$

and we use the bosonic algebra D.1 as follows

$$\begin{aligned}\frac{dh(z)}{dz} &= e^{S(z)}\left(-a^\dagger\left(-\mathbb{1}+aa^\dagger\right)+\left(a^\dagger a+\mathbb{1}\right)a^\dagger\right)e^{-S(z)} = \\ &= e^{S(z)}\left(a^\dagger - a^\dagger aa^\dagger + a^\dagger aa^\dagger + a^\dagger\right)e^{-S(z)},\end{aligned} \tag{12.191}$$

then

$$\frac{dh(z)}{dz} = 2e^{S(z)}a^\dagger e^{-S(z)}. \tag{12.192}$$

Now we compute the second derivative of $h(z)$ with respect to z, i.e.,

$$\begin{aligned}\frac{d^2h(z)}{dz^2} &= 2e^{S(z)}\left(\frac{dS(z)}{dz}a^\dagger - a^\dagger\frac{dS(z)}{dz}\right)e^{-S(z)} = \\ &= 2e^{S(z)}\left\{\left(aa-a^\dagger a^\dagger\right)a^\dagger - a^\dagger\left(aa-a^\dagger a^\dagger\right)\right\}e^{-S(z)} = \\ &= 2e^{S(z)}\left(aaa^\dagger - a^\dagger a^\dagger a^\dagger - a^\dagger aa + a^\dagger a^\dagger a^\dagger\right)e^{-S(z)} = \\ &= 2e^{S(z)}\left(aaa^\dagger - a^\dagger aa\right)e^{-S(z)} = \\ &= 2e^{S(z)}\left(a\left(\mathbb{1}+a^\dagger a\right)-\left(-\mathbb{1}+aa^\dagger\right)a\right)e^{-S(z)} = \\ &= 2e^{S(z)}\left(a+aa^\dagger a+a-aa^\dagger a\right)e^{-S(z)},\end{aligned} \tag{12.193}$$

where we use the bosonic algebra. We get

$$\begin{aligned}\frac{d^2h(z)}{dz^2} &= 4e^{S(z)}ae^{-S(z)} \equiv \\ &\equiv 4h(z),\end{aligned} \tag{12.194}$$

which is a second-order differential equation and has solution

$$h(z) = c_1e^{2z} + c_2e^{-2z}. \tag{12.195}$$

The system of initial conditions

$$\begin{cases} h(0) = a \\ h'(0) = 2a^\dagger \end{cases}, \tag{12.196}$$

implies

$$\begin{cases} c_1 + c_2 = a \\ 2c_1 - 2c_2 = 2a^\dagger \end{cases}, \tag{12.197}$$

$$\begin{cases} c_1 + c_2 = a \\ c_1 - c_2 = a^\dagger \end{cases}. \tag{12.198}$$

We add the equations and have

$$c_1 = \frac{a + a^\dagger}{2}, \tag{12.199}$$

then we subtract the equations and have

$$c_2 = \frac{a - a^\dagger}{2}. \tag{12.200}$$

We insert c_1 and c_2 in $h(z)$ as follows

$$\begin{aligned} h(z) &= \frac{a + a^\dagger}{2} e^{2z} + \frac{a - a^\dagger}{2} e^{-2z} = \\ &= \frac{e^{2z} a}{2} + \frac{e^{2z} a^\dagger}{2} + \frac{e^{-2z} a}{2} - \frac{e^{-2z} a^\dagger}{2} = \\ &= \frac{e^{2z} + e^{-2z}}{2} a + \frac{e^{2z} - e^{-2z}}{2} a^\dagger = \\ &= \cosh(2z) a + \sinh(2z) a^\dagger \equiv \\ &\equiv b. \end{aligned} \tag{12.201}$$

Similarly, we define the operator

$$l(z) = e^{S(z)} a^\dagger e^{-S(z)}, \tag{12.202}$$

and we get

$$\begin{aligned} \frac{dl(z)}{dz} &= \frac{dS(z)}{dz} e^{S(z)} a^\dagger e^{-S(z)} - e^{S(z)} a^\dagger e^{-S(z)} \frac{dS(z)}{dz} = \\ &= e^{S(z)} \frac{dS(z)}{dz} a^\dagger e^{-S(z)} - e^{S(z)} a^\dagger \frac{dS(z)}{dz} e^{-S(z)} = \\ &= e^{S(z)} \left(\frac{dS(z)}{dz} a^\dagger - a^\dagger \frac{dS(z)}{dz} \right) e^{-S(z)} = \end{aligned}$$

$$
\begin{aligned}
&= e^{S(z)} \left\{ \left(aa - a^\dagger a^\dagger \right) a^\dagger - a^\dagger \left(aa - a^\dagger a^\dagger \right) \right\} e^{-S(z)} = \\
&= e^{S(z)} \left(aaa^\dagger - a^\dagger a^\dagger a^\dagger - a^\dagger aa + a^\dagger a^\dagger a^\dagger \right) e^{-S(z)} = \\
&= e^{S(z)} \left(aaa^\dagger - a^\dagger aa \right) e^{-S(z)} = \\
&= e^{S(z)} \left(a \left(\mathbb{1} + a^\dagger a \right) - \left(-\mathbb{1} + aa^\dagger \right) a \right) e^{-S(z)} = \\
&= e^{S(z)} \left(a + aa^\dagger a + a - aa^\dagger a \right) e^{-S(z)} = \\
&= 2e^{S(z)} a e^{-S(z)},
\end{aligned}
\tag{12.203}
$$

$$
\frac{dl(z)}{dz} = 2e^{S(z)} a e^{-S(z)}. \tag{12.204}
$$

$$
\begin{aligned}
\frac{d^2 l(z)}{dz^2} &= 2e^{S(z)} \left(\frac{dS(z)}{dz} a - a \frac{dS(z)}{dz} \right) e^{-S(z)} = \\
&= 2e^{S(z)} \left\{ \left(aa - a^\dagger a^\dagger \right) a - a \left(aa - a^\dagger a^\dagger \right) \right\} e^{-S(z)} = \\
&= 2e^{S(z)} \left(aaa - a^\dagger a^\dagger a - aaa + aa^\dagger a^\dagger \right) e^{-S(z)} = \\
&= 2e^{S(z)} \left(-a^\dagger a^\dagger a + aa^\dagger a^\dagger \right) e^{-S(z)} = \\
&= 2e^{S(z)} \left(-a^\dagger \left(-\mathbb{1} + aa^\dagger \right) + \left(\mathbb{1} + a^\dagger a \right) a^\dagger \right) e^{-S(z)} = \\
&= 2e^{S(z)} \left(a^\dagger - a^\dagger aa^\dagger + a^\dagger + a^\dagger aa^\dagger \right) e^{-S(z)},
\end{aligned}
\tag{12.205}
$$

$$
\begin{aligned}
\frac{d^2 l(z)}{dz^2} &= 4e^{S(z)} a^\dagger e^{-S(z)} \equiv \\
&\equiv 4l(z),
\end{aligned}
\tag{12.206}
$$

$$
l(z) = c_1 e^{2z} + c_2 e^{-2z}, \tag{12.207}
$$

$$
\begin{cases} l(0) = a^\dagger \\ l'(0) = 2a \end{cases}, \tag{12.208}
$$

$$
\begin{cases} c_1 + c_2 = a^\dagger \\ 2c_1 - 2c_2 = 2a \end{cases}, \tag{12.209}
$$

$$\begin{cases} c_1 + c_2 = a^\dagger \\ c_1 - c_2 = a \end{cases}, \tag{12.210}$$

$$c_1 = \frac{a + a^\dagger}{2}, \tag{12.211}$$

$$c_2 = \frac{a^\dagger - a}{2}. \tag{12.212}$$

We insert c_1 and c_2 in $l(z)$ as follows

$$\begin{aligned} l(z) &= \frac{a + a^\dagger}{2} e^{2z} + \frac{a^\dagger - a}{2} e^{-2z} = \\ &= \frac{e^{2z} a}{2} + \frac{e^{2z} a^\dagger}{2} + \frac{e^{-2z} a^\dagger}{2} - \frac{e^{-2z} a}{2} = \\ &= \frac{e^{2z} - e^{-2z}}{2} a + \frac{e^{2z} + e^{-2z}}{2} a^\dagger = \\ &= \sinh(2z) a + \cosh(2z) a^\dagger \equiv \\ &\equiv b^\dagger, \end{aligned} \tag{12.213}$$

given that $l(z)$ is the adjoint of $h(z)$. Note that if we set $u = \cosh(2z)$ and $v = \sinh(2z)$, as we assumed for the Bogoljubov's bosonic transformations, then b and $b^\dagger$ become Eqs. 12.65 and 12.66, respectively. Let us verify that such a transformation is canonical, that is,

$$\begin{aligned} \left[b, b^\dagger\right] &= \left[\cosh(2z) a + \sinh(2z) a^\dagger, \sinh(2z) a + \cosh(2z) a^\dagger\right] = \\ &= \cosh(2z)^2 \left[a, a^\dagger\right] + \sinh^2(2z) \left[a^\dagger, a\right] = \\ &= \cosh^2(2z) \mathbb{1} - \sinh^2(2z) \mathbb{1} = \\ &= \mathbb{1}. \end{aligned} \tag{12.214}$$

We now compute how e^{-S} acts on the position and momentum operators in one spatial dimension, that is, on $\hat{x}$ and $\hat{p}$. We have

$$\begin{aligned} \hat{x}' &= e^{S(z)} \hat{x} e^{-S(z)} = \\ &= \sqrt{\frac{\hbar}{2m\omega}} e^{S(z)} \left(a + a^\dagger\right) e^{-S(z)} = \\ &= \sqrt{\frac{\hbar}{2m\omega}} \left(\cosh(2z) a + \sinh(2z) a^\dagger + \cosh(2z) a^\dagger + \sinh(2z) a\right) = \end{aligned}$$

$$
\begin{aligned}
&= \sqrt{\frac{\hbar}{2m\omega}} \left(\cosh(2z) + \sinh(2z)\right) \left(a + a^\dagger\right) = \\
&= \sqrt{\frac{\hbar}{2m\omega}} \left(\frac{e^{2z} + e^{-2z}}{2} + \frac{e^{2z} - e^{-2z}}{2}\right) \left(a + a^\dagger\right) = \\
&= \sqrt{\frac{\hbar}{2m\omega}} e^{2z} \left(a + a^\dagger\right) = \\
&= e^{2z} \hat{x},
\end{aligned}
\tag{12.215}
$$

$$
\begin{aligned}
\hat{p}' &= e^{S(z)} \hat{p} e^{-S(z)} = \\
&= i\sqrt{\frac{m\hbar\omega}{2}} e^{S(z)} \left(a^\dagger - a\right) e^{-S(z)} = \\
&= i\sqrt{\frac{m\hbar\omega}{2}} \left(\cosh(2z)a^\dagger + \sinh(2z)a - \cosh(2z)a - \sinh(2z)a^\dagger\right) = \\
&= i\sqrt{\frac{m\hbar\omega}{2}} \left(\cosh(2z) - \sinh(2z)\right) \left(a^\dagger - a\right) = \\
&= i\sqrt{\frac{m\hbar\omega}{2}} \left(\frac{e^{2z} + e^{-2z}}{2} - \frac{e^{2z} - e^{-2z}}{2}\right) \left(a^\dagger - a\right) = \\
&= i\sqrt{\frac{m\hbar\omega}{2}} e^{-2z} \left(a^\dagger - a\right) = \\
&= e^{-2z} \hat{p}.
\end{aligned}
\tag{12.216}
$$

Recall that the ground state of one-dimensional harmonic oscillator has a Gaussian form in both position and momentum representations. When new operators are introduced through a linear canonical transformation, the corresponding ground state preserves its Gaussian nature. Such transformations modify the variances of the conjugate variables in a reciprocal manner: the spatial width increases, while the width in momentum space decreases, consistently with the uncertainty principle. This result will not be demonstrated here. It is important to note that the operators $\hat{x}'$ and $\hat{p}'$ still satisfy the uncertainty principle. Next, let us examine the transformed vacuum state

$$
\begin{aligned}
|\tilde{0}\rangle &\equiv e^{-S(z)}|0\rangle = \\
&= e^{-z\left(aa - a^\dagger a^\dagger\right)}|0\rangle.
\end{aligned}
\tag{12.217}
$$

From Jacobi identity (see Eq. F.1), we get

$$
\begin{aligned}
\left[aa, a^\dagger a^\dagger\right] &= a\left[a, a^\dagger a^\dagger\right] + \left[a, a^\dagger a^\dagger\right] a = \\
&= aa^\dagger\left[a, a^\dagger\right] + a\left[a, a^\dagger\right] a^\dagger + a^\dagger\left[a, a^\dagger\right] a + \left[a, a^\dagger\right] a^\dagger a = \\
&= 2aa^\dagger + 2a^\dagger a = \\
&= 2\mathbb{1} + 2a^\dagger a + 2a^\dagger a = \\
&= 2\mathbb{1} + 4a^\dagger a. \qquad (12.218)
\end{aligned}
$$

Accordingly, we recall that by definition the fundamental state $|\tilde{0}\rangle$ of the transformed Hamiltonian must satisfy (see Eq. 1.174)

$$b|\tilde{0}\rangle = 0, \qquad (12.219)$$

which implies

$$\left(\cosh(2z)a + \sinh(2z)a^\dagger\right)|\tilde{0}\rangle = 0. \qquad (12.220)$$

Now, we want to write the eigenstate of the fundamental state of the transformed Hamiltonian as a function of the eigenfunctions of the original Hamiltonian by means of a series expansion. Since the new eigenstates are even functions in the space of positions (they are Gaussians), only even powers of the creation operator $a^\dagger$ can be considered in the expansion of $|\tilde{0}\rangle$. We therefore expect a solution of the form

$$|\tilde{0}\rangle = c_0|0\rangle + c_2 a^\dagger a^\dagger|0\rangle + c_4 a^\dagger a^\dagger a^\dagger a^\dagger|0\rangle + \dots . \qquad (12.221)$$

In place of the series we consider only a quadratic term $a^\dagger a^\dagger$ and impose for the transformed fundamental state the following Ansatz

$$|\tilde{0}\rangle = \mathcal{N}(z)\exp\left\{-\frac{1}{2}f(z)a^\dagger a^\dagger\right\}|0\rangle, \qquad (12.222)$$

then we must compute $\mathcal{N}(z)$ and $f(z)$. Equation 12.220 gives

$$\left(\cosh(2z)a + \sinh(2z)a^\dagger\right)\exp\left\{-\frac{1}{2}f(z)a^\dagger a^\dagger\right\}|0\rangle = 0, \qquad (12.223)$$

$$\cosh(2z)a\exp\left\{-\frac{1}{2}f(z)a^\dagger a^\dagger\right\}|0\rangle + \sinh(2z)a^\dagger\exp\left\{-\frac{1}{2}f(z)a^\dagger a^\dagger\right\}|0\rangle = 0. \qquad (12.224)$$

Since the action of the annihilation operator a on the vacuum state is the null state, the argument of the first exponential can be replaced with a commutator as follows

$$\cosh(2z)\left[a, \exp\left\{-\frac{1}{2}f(z)a^\dagger a^\dagger\right\}\right]|0\rangle + \sinh(2z)a^\dagger \exp\left\{-\frac{1}{2}f(z)a^\dagger a^\dagger\right\}|0\rangle = 0. \tag{12.225}$$

Note that by choosing $A = a$ and $B = -\frac{1}{2}f(z)a^\dagger a^\dagger$, the condition H.7 is fulfilled, then the truncated conjugation formula H.8 can be applied, then

$$\cosh(2z)\left[a, -\frac{1}{2}f(z)a^\dagger a^\dagger\right]\exp\left\{-\frac{1}{2}f(z)a^\dagger a^\dagger\right\}|0\rangle$$
$$+ \sinh(2z)a^\dagger \exp\left\{-\frac{1}{2}f(z)a^\dagger a^\dagger\right\}|0\rangle = 0, \tag{12.226}$$

$$\left(\cosh(2z)\left[a, -\frac{1}{2}f(z)a^\dagger a^\dagger\right] + \sinh(2z)a^\dagger\right)\exp\left\{-\frac{1}{2}f(z)a^\dagger a^\dagger\right\}|0\rangle = 0. \tag{12.227}$$

From

$$\begin{aligned}\left[a, -\frac{1}{2}f(z)a^\dagger a^\dagger\right] &= -\frac{1}{2}f(z)\left[a, a^\dagger a^\dagger\right] = \\ &= -\frac{1}{2}f(z)\left(a^\dagger\left[a, a^\dagger\right] + \left[a, a^\dagger\right]a^\dagger\right) = \\ &= -f(z)a^\dagger,\end{aligned} \tag{12.228}$$

we have

$$\left(-\cosh(2z)f(z) + \sinh(2z)\right)a^\dagger \exp\left\{-\frac{1}{2}f(z)a^\dagger a^\dagger\right\}|0\rangle = 0, \tag{12.229}$$

$$-\cosh(2z)f(z) + \sinh(2z) = 0, \tag{12.230}$$

$$f(z) = \tanh(2z), \tag{12.231}$$

then

$$|\tilde{0}\rangle = \mathcal{N}(z)\exp\left\{-\frac{1}{2}\tanh(2z)a^\dagger a^\dagger\right\}|0\rangle. \tag{12.232}$$

First we multiply both members of the above equation by $\langle 0|$ and we then expand the exponential operator explicitly as a power series as follows

$$\begin{aligned}\left\langle 0\middle|\tilde{0}\right\rangle &= \mathcal{N}(z)\left\langle 0\left|\exp\left\{-\frac{1}{2}\tanh(2z)\,a^\dagger a^\dagger\right\}\right|0\right\rangle = \\ &= \mathcal{N}(z)\lim_{N\to\infty}\left\langle 0\left|\left(1 - \frac{1}{2}\tanh(2z)\frac{a^\dagger a^\dagger}{N}\right)^N\right|0\right\rangle =\end{aligned}$$

$$
\begin{aligned}
&= \mathcal{N}(z) \lim_{N\to\infty} \left\langle 0 \left| \left(1 - \frac{1}{2}\tanh(2z)\frac{aa}{N}\right)^N \right| 0 \right\rangle = \\
&= \mathcal{N}(z),
\end{aligned}
\tag{12.233}
$$

then, by integrating over the one-dimensional spatial coordinate x, we obtain

$$
\begin{aligned}
\mathcal{N}(z) &= \langle 0|\tilde{0}\rangle = \\
&= \int_{-\infty}^{+\infty} dx \langle 0|x\rangle\langle x|\tilde{0}\rangle = \\
&= \int_{-\infty}^{+\infty} dx \, (\psi_0(x))^* \, \tilde{\psi}_0(x),
\end{aligned}
\tag{12.234}
$$

where the fundamental state of the harmonic oscillator is given by

$$
\psi_0(x) = N_0 \exp\left(-\frac{m\omega}{2\hbar}x^2\right),
\tag{12.235}
$$

with

$$
N_0 = \left(\frac{m\omega}{\pi\hbar}\right)^{\frac{1}{4}}.
\tag{12.236}
$$

We have

$$
\begin{aligned}
\mathcal{N}(z) &= \left(\frac{m\omega}{\pi\hbar}\right)^{\frac{1}{4}} \left(\frac{\tilde{m}\tilde{\omega}}{\pi\hbar}\right)^{\frac{1}{4}} \int_{-\infty}^{+\infty} dx \exp\left(-\frac{m\omega + \tilde{m}\tilde{\omega}}{2\hbar}x^2\right) = \\
&= (m\omega\tilde{m}\tilde{\omega})^{\frac{1}{4}} \frac{1}{\sqrt{\pi\hbar}} \frac{\sqrt{2\pi\hbar}}{\sqrt{m\omega + \tilde{m}\tilde{\omega}}} = \\
&= \left(\frac{1}{2}\frac{m\omega + \tilde{m}\tilde{\omega}}{\sqrt{m\omega\tilde{m}\tilde{\omega}}}\right)^{-\frac{1}{2}} = \\
&= \left(\frac{1}{2}\left[\sqrt{\frac{m\omega}{\tilde{m}\tilde{\omega}}} + \sqrt{\frac{\tilde{m}\tilde{\omega}}{m\omega}}\right]\right)^{-\frac{1}{2}},
\end{aligned}
\tag{12.237}
$$

where $\tilde{m}$ and $\tilde{\omega}$ are the mass and frequency of the quasiparticle of the transformed Hamiltonian, respectively. We define

$$
e^{2z} = \sqrt{\frac{m\omega}{\tilde{m}\tilde{\omega}}},
\tag{12.238}
$$

then

$$\mathcal{N}(z) = \left(\frac{1}{2} \left[e^{2z} + e^{-2z} \right] \right)^{-\frac{1}{2}} = \\ = \frac{1}{\sqrt{\cosh(2z)}}, \tag{12.239}$$

$$|\tilde{0}\rangle = \frac{1}{\sqrt{\cosh(2z)}} \exp\left\{ -\frac{1}{2} \tanh(2z) a^\dagger a^\dagger \right\} |0\rangle. \tag{12.240}$$

We have obtained the operatorial identity

$$e^{-z\left(aa - a^\dagger a^\dagger\right)} |0\rangle = \frac{1}{\sqrt{\cosh(2z)}} \exp\left\{ -\frac{1}{2} \tanh(2z) a^\dagger a^\dagger \right\} |0\rangle. \tag{12.241}$$

Regarding the Bogoljubov transformed bosonic Hamiltonian, the property

$$b_{\mathbf{k}} |\psi\rangle = 0, \; \forall \mathbf{k} \neq \mathbf{0} \tag{12.242}$$

is verified if the fundamental state has the structure

$$|\psi\rangle = \frac{\left(a_0^\dagger\right)^{N_0}}{\sqrt{N_0!}} |0\rangle \exp\left\{ - \sum_{\substack{\mathbf{q}_1 \neq \mathbf{0}, \\ \mathbf{q}_2 \neq \mathbf{0}}} \frac{M_{\mathbf{q}_1, \mathbf{q}_2}}{2} a_{\mathbf{q}_1}^\dagger a_{\mathbf{q}_2}^\dagger \right\} |0\rangle, \tag{12.243}$$

where (see Eq. 2.142)

$$\frac{\left(a_0^\dagger\right)^{N_0}}{\sqrt{N_0!}} |0\rangle \tag{12.244}$$

represents the eigenfunction associated with the condensate (see Eq. 12.1). Now we want to prove that Eq. 12.243 satisfies 12.242. Given

$$\left(u_{\mathbf{k}} a_{\mathbf{k}} + v_{\mathbf{k}} a_{-\mathbf{k}}^\dagger \right) \frac{\left(a_0^\dagger\right)^{N_0}}{\sqrt{N_0!}} |0\rangle \exp\left\{ - \sum_{\substack{\mathbf{q}_1 \neq \mathbf{0}, \\ \mathbf{q}_2 \neq \mathbf{0}}} \frac{M_{\mathbf{q}_1, \mathbf{q}_2}}{2} a_{\mathbf{q}_1}^\dagger a_{\mathbf{q}_2}^\dagger \right\} |0\rangle, \tag{12.245}$$

note that the operator $a_{-\mathbf{k}}^\dagger$ commutes with both $\left(a_0^\dagger\right)^{N_0}$ and $a_{\mathbf{q}_1}^\dagger a_{\mathbf{q}_2}^\dagger$. On the other hand, $a_{\mathbf{k}}$ commutes with $\left(a_0^\dagger\right)^{N_0}$, but not with the exponential operator, due to the presence of terms like $a_{\mathbf{q}_1}^\dagger a_{\mathbf{q}_2}^\dagger$ in the series expansion. Then, given

$$a_{\mathbf{k}} \exp\left\{-\sum_{\substack{\mathbf{q}_1\neq\mathbf{0},\\ \mathbf{q}_2\neq\mathbf{0}}} \frac{M_{\mathbf{q}_1,\mathbf{q}_2}}{2} a^\dagger_{\mathbf{q}_1} a^\dagger_{\mathbf{q}_2}\right\}|0\rangle, \tag{12.246}$$

we insert the identity

$$\mathbb{1} = \exp\left\{-\sum_{\substack{\mathbf{q}_1\neq\mathbf{0},\\ \mathbf{q}_2\neq\mathbf{0}}} \frac{M_{\mathbf{q}_1,\mathbf{q}_2}}{2} a^\dagger_{\mathbf{q}_1} a^\dagger_{\mathbf{q}_2}\right\} \exp\left\{\sum_{\substack{\mathbf{q}_1\neq\mathbf{0},\\ \mathbf{q}_2\neq\mathbf{0}}} \frac{M_{\mathbf{q}_1,\mathbf{q}_2}}{2} a^\dagger_{\mathbf{q}_1} a^\dagger_{\mathbf{q}_2}\right\} \tag{12.247}$$

to the left of $a_{\mathbf{k}}$ as follows

$$\begin{aligned}
&\exp\left\{-\sum_{\substack{\mathbf{q}_1\neq\mathbf{0},\\ \mathbf{q}_2\neq\mathbf{0}}} \frac{M_{\mathbf{q}_1,\mathbf{q}_2}}{2} a^\dagger_{\mathbf{q}_1} a^\dagger_{\mathbf{q}_2}\right\} \exp\left\{\sum_{\substack{\mathbf{q}_1\neq\mathbf{0},\\ \mathbf{q}_2\neq\mathbf{0}}} \frac{M_{\mathbf{q}_1,\mathbf{q}_2}}{2} a^\dagger_{\mathbf{q}_1} a^\dagger_{\mathbf{q}_2}\right\} a_{\mathbf{k}} \\
&\quad\times \exp\left\{-\sum_{\substack{\mathbf{q}_1\neq\mathbf{0},\\ \mathbf{q}_2\neq\mathbf{0}}} \frac{M_{\mathbf{q}_1,\mathbf{q}_2}}{2} a^\dagger_{\mathbf{q}_1} a^\dagger_{\mathbf{q}_2}\right\},
\end{aligned} \tag{12.248}$$

and we study the transformed of $a_{\mathbf{k}}$, i.e.,

$$\begin{aligned}
\tilde{a}_{\mathbf{k}} &\equiv \exp\left\{\sum_{\substack{\mathbf{q}_1\neq\mathbf{0},\\ \mathbf{q}_2\neq\mathbf{0}}} \frac{M_{\mathbf{q}_1,\mathbf{q}_2}}{2} a^\dagger_{\mathbf{q}_1} a^\dagger_{\mathbf{q}_2}\right\} a_{\mathbf{k}} \exp\left\{-\sum_{\substack{\mathbf{q}_1\neq\mathbf{0},\\ \mathbf{q}_2\neq\mathbf{0}}} \frac{M_{\mathbf{q}_1,\mathbf{q}_2}}{2} a^\dagger_{\mathbf{q}_1} a^\dagger_{\mathbf{q}_2}\right\} \equiv \\
&\equiv e^T a_{\mathbf{k}} e^{-T},
\end{aligned} \tag{12.249}$$

with

$$T = \sum_{\substack{\mathbf{q}_1\neq\mathbf{0},\\ \mathbf{q}_2\neq\mathbf{0}}} \frac{M_{\mathbf{q}_1,\mathbf{q}_2}}{2} a^\dagger_{\mathbf{q}_1} a^\dagger_{\mathbf{q}_2}. \tag{12.250}$$

From the conjugation formula H.5, we compute

$$[T, a_{\mathbf{k}}] = \frac{1}{2}\left(\sum_{\mathbf{q}_1\neq\mathbf{0}} M_{\mathbf{q}_1,\mathbf{k}}\left[a^\dagger_{\mathbf{q}_1} a^\dagger_{\mathbf{k}}, a_{\mathbf{k}}\right] + \sum_{\mathbf{q}_2\neq\mathbf{0}} M_{\mathbf{k},\mathbf{q}_2}\left[a^\dagger_{\mathbf{k}} a^\dagger_{\mathbf{q}_2}, a_{\mathbf{k}}\right]\right) =$$

$$
\begin{aligned}
&= \frac{1}{2}\Bigg(\sum_{\mathbf{q}_1 \neq \mathbf{0}} M_{\mathbf{q}_1,\mathbf{k}} a^\dagger_{\mathbf{q}_1} \left[a^\dagger_{\mathbf{k}}, a_{\mathbf{k}} \right] + \sum_{\mathbf{q}_1 \neq \mathbf{0}} M_{\mathbf{q}_1,\mathbf{k}} \left[a^\dagger_{\mathbf{q}_1}, a_{\mathbf{k}} \right] a^\dagger_{\mathbf{k}} + \\
&+ \sum_{\mathbf{q}_2 \neq \mathbf{0}} M_{\mathbf{k},\mathbf{q}_2} a^\dagger_{\mathbf{k}} \left[a^\dagger_{\mathbf{q}_2}, a_{\mathbf{k}} \right] + \sum_{\mathbf{q}_2 \neq \mathbf{0}} M_{\mathbf{k},\mathbf{q}_2} \left[a^\dagger_{\mathbf{k}}, a_{\mathbf{k}} \right] a^\dagger_{\mathbf{q}_2} \Bigg) = \\
&= \frac{1}{2}\left(\sum_{\mathbf{q}_1 \neq \mathbf{0}} M_{\mathbf{q}_1,\mathbf{k}} \left(-a^\dagger_{\mathbf{q}_1} - \delta_{\mathbf{q}_1,\mathbf{k}} a^\dagger_{\mathbf{k}} \right) + \sum_{\mathbf{q}_2 \neq \mathbf{0}} M_{\mathbf{k},\mathbf{q}_2} \left(-\delta_{\mathbf{q}_2,\mathbf{k}} a^\dagger_{\mathbf{k}} - a^\dagger_{\mathbf{q}_2} \right) \right) = \\
&= \frac{1}{2}\left(- \sum_{\mathbf{q}_1 \neq \mathbf{0}} M_{\mathbf{q}_1,\mathbf{k}} a^\dagger_{\mathbf{k}} - M_{\mathbf{k},\mathbf{k}} a^\dagger_{\mathbf{k}} - \sum_{\mathbf{q}_2 \neq \mathbf{0}} M_{\mathbf{k},\mathbf{q}_2} a^\dagger_{\mathbf{q}_2} - M_{\mathbf{k},\mathbf{k}} a^\dagger_{\mathbf{k}} \right).
\end{aligned} \tag{12.251}
$$

Since $M_{\mathbf{q}_1,\mathbf{q}_2}$ quantifies the interaction between two particles with momenta $\mathbf{q}_1$ and $\mathbf{q}_2$, we assume that M is a symmetric matrix with vanishing diagonal elements, i.e.,

$$
M_{\mathbf{k},\mathbf{k}} = 0, \ \forall \mathbf{k} \neq \mathbf{0}, \tag{12.252}
$$

$$
M_{\mathbf{k},\mathbf{q}_2} \equiv M_{\mathbf{q}_2,\mathbf{k}}, \tag{12.253}
$$

then we rename the dummy indices $\mathbf{q}_1 \equiv \mathbf{q}_2 \equiv \mathbf{q}$ and write

$$
[T, a_{\mathbf{k}}] = - \sum_{\mathbf{q} \neq \mathbf{0}} M_{\mathbf{q},\mathbf{k}} a^\dagger_{\mathbf{q}}. \tag{12.254}
$$

Since the other terms in $e^T a_{\mathbf{k}} e^{-T}$ (see conjugation formula H.5) involve commutators of $[T, a_{\mathbf{k}}]$ with T itself, the only non-zero term is the one we calculated, i.e.,

$$
\begin{aligned}
\tilde{a}_{\mathbf{k}} &= e^T a_{\mathbf{k}} e^{-T} = \\
&= a_{\mathbf{k}} + [T, a_{\mathbf{k}}] = \\
&= a_{\mathbf{k}} - \sum_{\mathbf{q} \neq \mathbf{0}} M_{\mathbf{q},\mathbf{k}} a^\dagger_{\mathbf{q}}.
\end{aligned} \tag{12.255}
$$

From Eq. 12.245, it follows

$$
\frac{\left(a^\dagger_0\right)^{N_0}}{\sqrt{N_0!}} |0\rangle \exp\left\{ - \sum_{\substack{\mathbf{q}_1 \neq \mathbf{0}, \\ \mathbf{q}_2 \neq \mathbf{0}}} \frac{M_{\mathbf{q}_1,\mathbf{q}_2}}{2} a^\dagger_{\mathbf{q}_1} a^\dagger_{\mathbf{q}_2} \right\} \left(u_{\mathbf{k}} \tilde{a}_{\mathbf{k}} + v_{\mathbf{k}} a^\dagger_{-\mathbf{k}} \right) |0\rangle = 0, \tag{12.256}
$$

$$\frac{\left(a_0^\dagger\right)^{N_0}}{\sqrt{N_0!}}|0\rangle \exp\left\{-\sum_{\substack{\mathbf{q}_1\neq\mathbf{0},\\ \mathbf{q}_2\neq\mathbf{0}}}\frac{M_{\mathbf{q}_1,\mathbf{q}_2}}{2}a_{\mathbf{q}_1}^\dagger a_{\mathbf{q}_2}^\dagger\right\}\left(u_{\mathbf{k}}a_{\mathbf{k}}-u_{\mathbf{k}}\sum_{\mathbf{q}\neq\mathbf{0}}M_{\mathbf{q},\mathbf{k}}a_{\mathbf{q}}^\dagger+v_{\mathbf{k}}a_{-\mathbf{k}}^\dagger\right)|0\rangle = 0, \tag{12.257}$$

$$\frac{\left(a_0^\dagger\right)^{N_0}}{\sqrt{N_0!}}|0\rangle \exp\left\{-\sum_{\substack{\mathbf{q}_1\neq\mathbf{0},\\ \mathbf{q}_2\neq\mathbf{0}}}\frac{M_{\mathbf{q}_1,\mathbf{q}_2}}{2}a_{\mathbf{q}_1}^\dagger a_{\mathbf{q}_2}^\dagger\right\}\left(-u_{\mathbf{k}}\sum_{\mathbf{q}\neq\mathbf{0}}M_{\mathbf{q},\mathbf{k}}a_{\mathbf{q}}^\dagger+v_{\mathbf{k}}a_{-\mathbf{k}}^\dagger\right)|0\rangle = 0, \tag{12.258}$$

where in the last step the annihilation operator $a_{\mathbf{k}}$ has been removed, since it acts on the vacuum state. A possible solution for the above equation is given by

$$-u_{\mathbf{k}}\sum_{\mathbf{q}\neq\mathbf{0}}M_{\mathbf{q},\mathbf{k}}a_{\mathbf{q}}^\dagger+v_{\mathbf{k}}a_{-\mathbf{k}}^\dagger = 0, \tag{12.259}$$

which is verified if and only if

$$M_{\mathbf{q},\mathbf{k}} = \frac{v_{\mathbf{k}}}{u_{\mathbf{k}}}\delta_{\mathbf{q},-\mathbf{k}}. \tag{12.260}$$

We have shown that Eq. 12.243 satisfies 12.242 with $M_{\mathbf{q},\mathbf{q}} = 0$, $M_{\mathbf{q}_1,\mathbf{q}_2} = M_{\mathbf{q},-\mathbf{q}}$, then the fundamental state of a Bogoljubov's bosonic Hamiltonian is

$$|\psi\rangle = \frac{\left(a_0^\dagger\right)^{N_0}}{\sqrt{N_0!}}|0\rangle \exp\left\{-\sum_{\mathbf{q}\neq\mathbf{0}}\frac{M_{\mathbf{q},-\mathbf{q}}}{2}a_{\mathbf{q}}^\dagger a_{-\mathbf{q}}^\dagger\right\}|0\rangle. \tag{12.261}$$

Note that the object

$$M_{\mathbf{q},-\mathbf{q}} = \frac{v_{\mathbf{q}}}{u_{\mathbf{q}}} \tag{12.262}$$

explicitly incorporates the effects of interactions, as expected.

12.5 Number Density of Bosons

Given the annihilation field operator of the Bogoljubov's theory (see Eqs. 12.58, 12.146, 12.151 and 12.153), i.e.,

$$\hat{\psi}(\mathbf{r}) = \sqrt{\frac{N_0}{V}}+\sum_{\mathbf{k}\neq\mathbf{0}}\frac{e^{i\mathbf{k}\cdot\mathbf{r}}}{\sqrt{V}}\left(u_{\mathbf{k}}b_{\mathbf{k}}-v_{\mathbf{k}}b_{-\mathbf{k}}^\dagger\right), \tag{12.263}$$

let us compute its thermal average (see Eq. 1.379) with respect to the diagonalized Hamiltonian 12.175. Note that the interaction term does not contribute to the thermal average, as it does not conserve particle number. In particular,

$$\left\langle u_{\mathbf{k}} b_{\mathbf{k}} - v_{\mathbf{k}} b^{\dagger}_{-\mathbf{k}} \right\rangle = 0, \tag{12.264}$$

since the operator $u_{\mathbf{k}} b_{\mathbf{k}} - v_{\mathbf{k}} b^{\dagger}_{-\mathbf{k}}$ changes the total particle number. Its expectation value therefore vanishes when evaluated in states with a fixed number of particles. Then

$$\langle \hat{\psi}(\mathbf{r}) \rangle = \sqrt{\frac{N_0}{V}}, \tag{12.265}$$

that is, the Bogoljubov field operator is the sum of its mean and a fluctuation term. The particle number density (see Eq. 2.261) of bosons within an infinitesimal volume centered at $\mathbf{r}$ provides

$$\begin{aligned}\left\langle \hat{\rho}(\mathbf{r}) \right\rangle &= \langle \hat{\psi}^{\dagger}(\mathbf{r}) \hat{\psi}(\mathbf{r}) \rangle = \\ &= \left\langle \left(\sqrt{\frac{N_0}{V}} + \sum_{\mathbf{k} \neq \mathbf{0}} \frac{e^{-i\mathbf{k}\cdot\mathbf{r}}}{\sqrt{V}} \left(u_{\mathbf{k}} b^{\dagger}_{\mathbf{k}} - v_{\mathbf{k}} b_{-\mathbf{k}} \right) \right) \right. \\ &\quad \left. \times \left(\sqrt{\frac{N_0}{V}} + \sum_{\mathbf{k}' \neq \mathbf{0}} \frac{e^{i\mathbf{k}'\cdot\mathbf{r}}}{\sqrt{V}} \left(u_{\mathbf{k}'} b_{\mathbf{k}'} - v_{\mathbf{k}'} b^{\dagger}_{-\mathbf{k}'} \right) \right) \right\rangle, \end{aligned} \tag{12.266}$$

and from Eq. 12.264, we get

$$\left\langle \hat{\rho}(\mathbf{r}) \right\rangle = \frac{N_0}{V} + \sum_{\mathbf{k} \neq \mathbf{0}} \sum_{\mathbf{k}' \neq \mathbf{0}} \left\langle \frac{e^{-i\mathbf{k}\cdot\mathbf{r}}}{\sqrt{V}} \frac{e^{i\mathbf{k}'\cdot\mathbf{r}}}{\sqrt{V}} \left(u_{\mathbf{k}} b^{\dagger}_{\mathbf{k}} - v_{\mathbf{k}} b_{-\mathbf{k}} \right) \left(u_{\mathbf{k}'} b_{\mathbf{k}'} - v_{\mathbf{k}'} b^{\dagger}_{-\mathbf{k}'} \right) \right\rangle. \tag{12.267}$$

Moreover, the thermal average of the product of two generic wave functions associated with different momenta vanishes, due to the presence of an oscillatory term in the integrand. Therefore,

$$\left\langle \hat{\rho}(\mathbf{r}) \right\rangle = \frac{N_0}{V} + \frac{1}{V} \sum_{\mathbf{k} \neq \mathbf{0}} \left\langle \left(u_{\mathbf{k}} b^{\dagger}_{\mathbf{k}} - v_{\mathbf{k}} b_{-\mathbf{k}} \right) \left(u_{\mathbf{k}} b_{\mathbf{k}} - v_{\mathbf{k}} b^{\dagger}_{-\mathbf{k}} \right) \right\rangle. \tag{12.268}$$

The thermal averages of the products of two creation operators and two annihilation operators are zero, then the number density operator becomes

$$\left\langle \hat{\rho}(\mathbf{r}) \right\rangle = \frac{N_0}{V} + \frac{1}{V} \sum_{\mathbf{k} \neq \mathbf{0}} \left\langle u^2_{\mathbf{k}} b^{\dagger}_{\mathbf{k}} b_{\mathbf{k}} + v^2_{\mathbf{k}} b_{-\mathbf{k}} b^{\dagger}_{-\mathbf{k}} \right\rangle =$$

$$
\begin{aligned}
&= \frac{N_0}{V} + \frac{1}{V}\sum_{\mathbf{k}\neq\mathbf{0}} u_{\mathbf{k}}^2 \left\langle b_{\mathbf{k}}^\dagger b_{\mathbf{k}}\right\rangle + \frac{1}{V}\sum_{\mathbf{k}\neq\mathbf{0}} v_{\mathbf{k}}^2 \left\langle b_{\mathbf{k}} b_{\mathbf{k}}^\dagger\right\rangle = \\
&= \frac{N_0}{V} + \frac{1}{V}\sum_{\mathbf{k}\neq\mathbf{0}} u_{\mathbf{k}}^2 \left\langle b_{\mathbf{k}}^\dagger b_{\mathbf{k}}\right\rangle + \frac{1}{V}\sum_{\mathbf{k}\neq\mathbf{0}} v_{\mathbf{k}}^2 \left\langle \mathbb{1} + b_{\mathbf{k}}^\dagger b_{\mathbf{k}}\right\rangle = \\
&= \frac{N_0}{V} + \frac{1}{V}\sum_{\mathbf{k}\neq\mathbf{0}} u_{\mathbf{k}}^2 \langle \hat{N}_{\mathbf{k}}\rangle + \frac{1}{V}\sum_{\mathbf{k}\neq\mathbf{0}} v_{\mathbf{k}}^2 (1 + \langle \hat{N}_{\mathbf{k}}\rangle) \equiv \\
&\equiv \frac{N_0}{V} + \frac{1}{V}\sum_{\mathbf{k}\neq\mathbf{0}} \left(u_{\mathbf{k}}^2 + v_{\mathbf{k}}^2\right) \langle \hat{N}_{\mathbf{k}}\rangle + \frac{1}{V}\sum_{\mathbf{k}\neq\mathbf{0}} v_{\mathbf{k}}^2 = \\
&= \frac{N_0}{V} + \frac{1}{V}\sum_{\mathbf{k}\neq\mathbf{0}} \left(u_{\mathbf{k}}^2 + v_{\mathbf{k}}^2\right) n_1(\mathcal{E}_{\mathbf{k}}) + \frac{1}{V}\sum_{\mathbf{k}\neq\mathbf{0}} v_{\mathbf{k}}^2,
\end{aligned}
\tag{12.269}
$$

where we renamed the dummy index $-\mathbf{k} \rightarrow \mathbf{k}$, we applied the bosonic commutation relations, and we employed Eq. 2.251 and the Bose-Einstein distribution $n_1(\mathcal{E}_{\mathbf{k}})$, Eq. E.3. At finite temperature, the number density of bosons includes two additional contributions beyond the condensate density. However, at low temperatures, we have the following behavior

$$
\begin{aligned}
\lim_{T\to 0^+} n_1(\mathcal{E}_{\mathbf{k}}) &= \lim_{T\to 0^+} \frac{1}{e^{\beta\mathcal{E}_{\mathbf{k}}} - 1} = \\
&= 0,
\end{aligned}
\tag{12.270}
$$

$$
\lim_{T\to 0^+} \frac{1}{V}\sum_{\mathbf{k}\neq\mathbf{0}} \left(u_{\mathbf{k}}^2 + v_{\mathbf{k}}^2\right) n_1(\mathcal{E}_{\mathbf{k}}) = 0,
\tag{12.271}
$$

then

$$
\lim_{T\to 0^+} \left\langle \hat{\rho}(\mathbf{r})\right\rangle = \frac{N_0}{V} + \frac{1}{V}\sum_{\mathbf{k}\neq\mathbf{0}} v_{\mathbf{k}}^2,
\tag{12.272}
$$

that is, at low temperatures the number density of bosons includes also an additional term. Given (see Eq. 12.74)

$$
\begin{cases}
u_{\mathbf{k}}^2 - v_{\mathbf{k}}^2 = 1 \\
u_{\mathbf{k}}^2 + v_{\mathbf{k}}^2 = (u_{\mathbf{k}} + v_{\mathbf{k}})^2 - 2u_{\mathbf{k}}v_{\mathbf{k}}
\end{cases},
\tag{12.273}
$$

from (see Eqs. 12.110 and 12.111)

$$(u_{\mathbf{k}} + v_{\mathbf{k}})^2 = \left[\frac{1}{2} \left(\sqrt{\frac{\mathcal{E}_{\mathbf{k}}}{\mathcal{E}^0_{\mathbf{k}}}} + \sqrt{\frac{\mathcal{E}^0_{\mathbf{k}}}{\mathcal{E}_{\mathbf{k}}}} \right) + \frac{1}{2} \left(\sqrt{\frac{\mathcal{E}_{\mathbf{k}}}{\mathcal{E}^0_{\mathbf{k}}}} - \sqrt{\frac{\mathcal{E}^0_{\mathbf{k}}}{\mathcal{E}_{\mathbf{k}}}} \right) \right]^2 = \\
= \left[\frac{1}{2} 2 \sqrt{\frac{\mathcal{E}_{\mathbf{k}}}{\mathcal{E}^0_{\mathbf{k}}}} \right]^2 = \\
= \frac{\mathcal{E}_{\mathbf{k}}}{\mathcal{E}^0_{\mathbf{k}}}, \tag{12.274}$$

$$2u_{\mathbf{k}} v_{\mathbf{k}} = 2 \frac{1}{4} \left(\sqrt{\frac{\mathcal{E}_{\mathbf{k}}}{\mathcal{E}^0_{\mathbf{k}}}} + \sqrt{\frac{\mathcal{E}^0_{\mathbf{k}}}{\mathcal{E}_{\mathbf{k}}}} \right) \left(\sqrt{\frac{\mathcal{E}_{\mathbf{k}}}{\mathcal{E}^0_{\mathbf{k}}}} - \sqrt{\frac{\mathcal{E}^0_{\mathbf{k}}}{\mathcal{E}_{\mathbf{k}}}} \right) = \\
= \frac{1}{2} \left(\frac{\mathcal{E}_{\mathbf{k}}}{\mathcal{E}^0_{\mathbf{k}}} - \frac{\mathcal{E}^0_{\mathbf{k}}}{\mathcal{E}_{\mathbf{k}}} \right) = \\
= \frac{1}{2} \frac{\mathcal{E}^2_{\mathbf{k}} - \mathcal{E}^0_{\mathbf{k}}}{\mathcal{E}^0_{\mathbf{k}} \mathcal{E}_{\mathbf{k}}} = \\
= \frac{1}{2} \frac{\mathcal{E}^0_{\mathbf{k}} \left(\mathcal{E}^0_{\mathbf{k}} + 2 \frac{U N_0}{V} \right) - \left(\mathcal{E}^0_{\mathbf{k}} \right)^2}{\mathcal{E}^0_{\mathbf{k}} \mathcal{E}_{\mathbf{k}}} = \\
= \frac{1}{2} \frac{\mathcal{E}^0_{\mathbf{k}} + 2 \frac{U N_0}{V} - \mathcal{E}^0_{\mathbf{k}}}{\mathcal{E}_{\mathbf{k}}} = \\
= \frac{U N_0}{\mathcal{E}_{\mathbf{k}} V}, \tag{12.275}$$

we have

$$(u_{\mathbf{k}} + v_{\mathbf{k}})^2 - 2u_{\mathbf{k}} v_{\mathbf{k}} = \frac{\mathcal{E}_{\mathbf{k}}}{\mathcal{E}^0_{\mathbf{k}}} - \frac{U N_0}{\mathcal{E}_{\mathbf{k}} V} = \\
= \frac{\mathcal{E}^2_{\mathbf{k}} - \mathcal{E}^0_{\mathbf{k}} \frac{U N_0}{V}}{\mathcal{E}^0_{\mathbf{k}} \mathcal{E}_{\mathbf{k}}} = \\
= \frac{\mathcal{E}^0_{\mathbf{k}} \left(\mathcal{E}^0_{\mathbf{k}} + 2 \frac{U N_0}{V} \right) - \mathcal{E}^0_{\mathbf{k}} \frac{U N_0}{V}}{\mathcal{E}^0_{\mathbf{k}} \mathcal{E}_{\mathbf{k}}} = \\
= \frac{\mathcal{E}^0_{\mathbf{k}} + 2 \frac{U N_0}{V} - \frac{U N_0}{V}}{\mathcal{E}_{\mathbf{k}}} = \\
= \frac{\mathcal{E}^0_{\mathbf{k}} + \frac{U N_0}{V}}{\mathcal{E}_{\mathbf{k}}}, \tag{12.276}$$

and the system becomes

$$\begin{cases} u_{\mathbf{k}}^2 - v_{\mathbf{k}}^2 = 1 \\ u_{\mathbf{k}}^2 + v_{\mathbf{k}}^2 = \dfrac{\mathcal{E}_{\mathbf{k}}^0 + \frac{UN_0}{V}}{\mathcal{E}_{\mathbf{k}}} \end{cases} . \tag{12.277}$$

By subtracting the second equation from the first equation, we get

$$2v_{\mathbf{k}}^2 = \frac{\mathcal{E}_{\mathbf{k}}^0 + \frac{UN_0}{V}}{\mathcal{E}_{\mathbf{k}}} - 1, \tag{12.278}$$

$$v_{\mathbf{k}}^2 = \frac{1}{2}\left(\frac{\mathcal{E}_{\mathbf{k}}^0 + \frac{UN_0}{V}}{\mathcal{E}_{\mathbf{k}}} - 1\right), \tag{12.279}$$

which we insert in the number density of bosons at low temperatures, i.e.,

$$\lim_{T\to 0^+} \langle \hat{\rho}(\mathbf{r}) \rangle = \frac{N_0}{V} + \frac{1}{2V}\sum_{\mathbf{k}\neq\mathbf{0}}\left(\frac{\mathcal{E}_{\mathbf{k}}^0 + \frac{UN_0}{V}}{\mathcal{E}_{\mathbf{k}}} - 1\right). \tag{12.280}$$

We modify the dispersion relation as follows

$$\begin{aligned} \mathcal{E}_{\mathbf{k}} &= \sqrt{\left(\frac{\hbar^2 k^2}{2m}\right)^2 + \frac{\hbar^2 k^2}{2m} 2\frac{UN_0}{V}} \equiv \\ &\equiv \sqrt{\frac{\hbar^2 k^2 U N_0}{mV}}\sqrt{\frac{\hbar^2 k^2 V}{4mUN_0} + 1}, \end{aligned} \tag{12.281}$$

and we note that

$$\left[\frac{\hbar^2 V}{4mUN_0}\right] = \left[\text{length}\right]^2, \tag{12.282}$$

consequently, the object

$$\begin{aligned} l &= \frac{\hbar}{m}\sqrt{\frac{mV}{2UN_0}} \equiv \\ &\equiv \frac{\hbar}{\sqrt{2}mc} \end{aligned} \tag{12.283}$$

has the dimensions of a length. The dispersion relation becomes

$$\mathcal{E}_{\mathbf{k}} = \hbar k c \sqrt{\frac{l^2 k^2}{2} + 1}, \tag{12.284}$$

where we used the velocity 12.184. From the identity

$$\begin{aligned} l^2 k^2 &= k^2 \frac{\hbar^2 V}{2mUN_0} = \\ &= \frac{\mathcal{E}^0_{\mathbf{k}} V}{UN_0}, \end{aligned} \tag{12.285}$$

we write the non-condensate term as

$$\begin{aligned} \frac{1}{2V}\sum_{\mathbf{k}\neq\mathbf{0}}\left(\frac{\mathcal{E}^0_{\mathbf{k}} + \frac{UN_0}{V}}{\mathcal{E}_{\mathbf{k}}} - 1\right) &= \frac{1}{2V}\sum_{\mathbf{k}\neq\mathbf{0}}\left(\frac{\frac{UN_0}{V}\left(\frac{\mathcal{E}^0_{\mathbf{k}} V}{UN_0} + 1\right)}{\hbar k c\sqrt{\frac{l^2k^2}{2} + 1}} - 1\right) = \\ &= \frac{1}{2V}\sum_{\mathbf{k}\neq\mathbf{0}}\left(\frac{\frac{UN_0}{V}\left(l^2k^2 + 1\right)}{\hbar k\sqrt{\frac{UN_0}{mV}}\sqrt{\frac{l^2k^2+2}{2}}} - 1\right) = \\ &= \frac{1}{2V}\sum_{\mathbf{k}\neq\mathbf{0}}\left(\frac{\sqrt{\frac{2mUN_0}{\hbar^2k^2V}}\left(l^2k^2 + 1\right)}{\sqrt{l^2k^2+2}} - 1\right) = \\ &= \frac{1}{2V}\sum_{\mathbf{k}\neq\mathbf{0}}\left(\frac{l^2k^2 + 1}{lk\sqrt{l^2k^2+2}} - 1\right). \end{aligned} \tag{12.286}$$

In the thermodynamic limit C.11, given that the series excludes the zero-momentum term, the integral must be interpreted as a Cauchy principal value, i.e.,

$$\begin{aligned} \frac{1}{2V}\sum_{\mathbf{k}\neq\mathbf{0}}\left(\frac{\mathcal{E}^0_{\mathbf{k}} + \frac{UN_0}{V}}{\mathcal{E}_{\mathbf{k}}} - 1\right) &\simeq \frac{1}{2V}\frac{V}{(2\pi)^3}\,\text{P. V.}\int d^3\mathbf{k}\left(\frac{l^2k^2 + 1}{lk\sqrt{l^2k^2+2}} - 1\right) = \\ &= \frac{1}{2\,(2\pi)^3}\,\text{P. V.}\int d^3\mathbf{k}\left(\frac{l^2k^2 + 1}{lk\sqrt{l^2k^2+2}} - 1\right). \end{aligned} \tag{12.287}$$

For convenience, we multiply and divide by l^3, and rewrite as

$$\frac{1}{l^3}\frac{1}{2\,(2\pi)^3}\,\text{P. V.}\int d^3\,(l\mathbf{k})\left(\frac{l^2k^2 + 1}{lk\sqrt{l^2k^2+2}} - 1\right). \tag{12.288}$$

Since the integrand depends only on the modulus of $\mathbf{k}$, we switch to spherical coordinates, and we set $y = lk$, then

$$\begin{aligned}
&\frac{1}{l^3} \frac{1}{2\,(2\pi)^3} \,\mathrm{P.V.} \int d^3y \left(\frac{y^2+1}{y\sqrt{y^2+2}} - 1 \right) \\
&= \frac{2^{\frac{3}{2}} m^3 c^3}{\hbar^3} \frac{1}{2^4 \pi^3} 4\pi \,\mathrm{P.V.} \int_0^{+\infty} dy y^2 \left(\frac{y^2+1}{y\sqrt{y^2+2}} - 1 \right) = \\
&= \frac{m^3 c^3}{\sqrt{2}\hbar^3 \pi^2} \,\mathrm{P.V.} \int_0^{+\infty} dy y^2 \left(\frac{y^2+1}{y\sqrt{y^2+2}} - 1 \right) .
\end{aligned} \tag{12.289}$$

The number density of bosons at low temperatures does not depend on the position and it is given by

$$\langle \hat{\rho} \rangle = \frac{N_0}{V} + \frac{m^3 c^3}{\sqrt{2}\hbar^3 \pi^2} \,\mathrm{P.V.} \int_0^{+\infty} dy\; y^2 \left(\frac{y^2+1}{y\sqrt{y^2+2}} - 1 \right) . \tag{12.290}$$

In the low-temperature regime, the total number density is given by the sum of the condensate and non-condensate contributions, with the latter described through a principal value integral. Importantly, this latter term depends on the interaction strength through the sound velocity c, with a characteristic power-law behavior

$$\left\langle \hat{\rho}_{\text{non-condensate}} \right\rangle \propto c^3 \propto U^{\frac{3}{2}} . \tag{12.291}$$

This non-integer dependence on the interaction parameter U reveals a key insight: Bogoljubov's prediction goes beyond standard perturbation theory, which would only yield integer powers of U. In other words, the result embodies the effect of an infinite resummation of Feynman diagrams, accounting for collective many-body phenomena that cannot be captured by expanding to finite order in U. This highlights the non-perturbative nature of the Bogoljubov approximation and its ability to capture long-wavelength quantum correlations in an interacting Bose gas.

Chapter 13
Quantization of the Electromagnetic Field: Photons

Abstract In this chapter, we introduce a framework for quantizing the electromagnetic field, where the relativistic field is treated quantum mechanically, while matter remains in a nonrelativistic regime. This approach is particularly suitable for studying quantum systems in which photon dynamics is essential, but relativistic effects on particles can be neglected. We begin with the classical electromagnetic field expressed through scalar and vector potentials, emphasizing gauge invariance. The fields and potentials are decomposed into longitudinal and transverse components. The longitudinal component is absorbed into the particle degrees of freedom, while only the transverse part contributes to the fields. This separation holds independently of the chosen gauge. The next step involves deriving the Hamiltonian for a system of charged particles interacting with the electromagnetic field. This includes light-matter interaction terms, the field energy, and the particle kinetic energy. A careful constraint analysis is performed to justify the gauge choice and construct a consistent Hamiltonian. Finally, the transverse vector potential is quantized by expanding it into normal modes associated with creation and annihilation operators. This leads to a photon-based description of the field and forms the basis for understanding light-matter interactions, including absorption, emission, and collective quantum phenomena in many-body systems. This chapter closely follows the framework and notation introduced in [15]. While the structure and symbols are aligned with that source, the presentation, explanations, and context have been independently developed for this work.

13.1 Electromagnetic Field and Potentials

We begin by revisiting some fundamental properties of the electromagnetic (e.m.) field, which provide the foundation for its subsequent quantization. In particular, the distinction between longitudinal and transverse components of a vector field will play a central role.

F. Tafuri et al., *Understanding Quantum Many-Body Theory*, UNITEXT for Physics, https://doi.org/10.1007/978-3-032-17849-7_13

Definition 13.1 (Helmholtz Decomposition) Let $\mathbf{V}(\mathbf{r})$ be a sufficiently regular vector field. We define its longitudinal and transverse components, $\mathbf{V}^{(L)}(\mathbf{r})$ and $\mathbf{V}^{(T)}(\mathbf{r})$, through

$$\mathbf{V}(\mathbf{r}) = \mathbf{V}^{(L)}(\mathbf{r}) + \mathbf{V}^{(T)}(\mathbf{r}), \tag{13.1}$$

with

$$\nabla \cdot \mathbf{V}^{(T)}(\mathbf{r}) = 0, \tag{13.2}$$

$$\nabla \times \mathbf{V}^{(L)}(\mathbf{r}) = \mathbf{0}. \tag{13.3}$$

That is, $\mathbf{V}^{(L)}$ is irrotational (curl-free), whereas $\mathbf{V}^{(T)}$ is solenoidal (divergence-free). This decomposition, known as the Helmholtz decomposition, is unique under suitable boundary conditions.

Given a vector field $\mathbf{V}(\mathbf{r})$, we now determine its longitudinal and transverse components as defined in Definition 13.1. Let $f(\mathbf{r})$ and $\mathbf{u}(\mathbf{r})$ be the divergence and the curl of the field $\mathbf{V}(\mathbf{r})$, respectively. From Definition 13.1, it follows

$$\begin{aligned}\nabla \cdot \mathbf{V}(\mathbf{r}) &= \nabla \cdot \mathbf{V}^{(L)}(\mathbf{r}) = \\ &= f(\mathbf{r}),\end{aligned} \tag{13.4}$$

$$\begin{aligned}\nabla \times \mathbf{V}(\mathbf{r}) &= \nabla \times \mathbf{V}^{(T)}(\mathbf{r}) = \\ &= \mathbf{u}(\mathbf{r}),\end{aligned} \tag{13.5}$$

and by applying the Fourier transform, we obtain

$$-i\mathbf{q} \cdot \mathbf{V}^{(L)}(\mathbf{q}) = f(\mathbf{q}), \tag{13.6}$$

$$-i\mathbf{q} \times \mathbf{V}^{(T)}(\mathbf{q}) = \mathbf{u}(\mathbf{q}). \tag{13.7}$$

Furthermore, from Eqs. 13.2 and 13.3, the Fourier transforms of the components $\mathbf{V}^{(T)}(\mathbf{r})$ and $\mathbf{V}^{(L)}(\mathbf{r})$ are given by

$$-i\mathbf{q} \cdot \mathbf{V}^{(T)}(\mathbf{q}) = 0, \tag{13.8}$$

$$-i\mathbf{q} \times \mathbf{V}^{(L)}(\mathbf{q}) = \mathbf{0}, \tag{13.9}$$

then

$$\mathbf{V}^{(L)}(\mathbf{q}) \parallel \mathbf{q}, \tag{13.10}$$

$$\mathbf{V}^{(T)}(\mathbf{q}) \perp \mathbf{q}. \tag{13.11}$$

From Eq. 13.10, we can set

$$\mathbf{V}^{(L)}(\mathbf{q}) = a\mathbf{q}, \tag{13.12}$$

and from Eq. 13.6, we derive

$$f(\mathbf{q}) = -iaq^2, \tag{13.13}$$

$$a = i\frac{f(\mathbf{q})}{q^2}, \tag{13.14}$$

then

$$\mathbf{V}^{(L)}(\mathbf{q}) = i\frac{f(\mathbf{q})}{q^2}\mathbf{q}. \tag{13.15}$$

From Eq. 13.6, it follows

$$\begin{aligned}\mathbf{V}^{(L)}(\mathbf{q}) &= i\frac{f(\mathbf{q})}{q^2}\mathbf{q} = \\ &= i\frac{(-i\mathbf{q}\cdot\mathbf{V}(\mathbf{q}))}{q^2}\mathbf{q} = \\ &= \frac{\mathbf{q}\cdot\mathbf{V}(\mathbf{q})}{q^2}\mathbf{q}.\end{aligned} \tag{13.16}$$

The transverse component $V^{(T)}(\mathbf{q})$ can be obtained by taking the cross product of both sides of Eq. 13.7 with $\mathbf{q}$ and employing the vector triple product identity

$$\mathbf{A}\times(\mathbf{B}\times\mathbf{C}) = (\mathbf{A}\cdot\mathbf{C})\mathbf{B} - (\mathbf{A}\cdot\mathbf{B})\mathbf{C} \tag{13.17}$$

as follows

$$\begin{aligned}-i\mathbf{q}\times\left(\mathbf{q}\times\mathbf{V}^{(T)}(\mathbf{q})\right) &= -i\left[\left(\mathbf{q}\cdot\mathbf{V}^{(T)}\right)\mathbf{q} - q^2\mathbf{V}^{(T)}\right] = \\ &= iq^2\mathbf{V}^{(T)}(\mathbf{q}) = \\ &= \mathbf{q}\times\mathbf{u}(\mathbf{q}),\end{aligned} \tag{13.18}$$

then

$$\mathbf{V}^{(T)}(\mathbf{q}) = -i\frac{\mathbf{q}\times\mathbf{u}(\mathbf{q})}{q^2}. \tag{13.19}$$

From Eq. 13.3, we have

$$\begin{aligned}\mathbf{u}(\mathbf{q}) &= -i\mathbf{q} \times \mathbf{V}^{(T)}(\mathbf{q}) = \\ &= -i\mathbf{q} \times \mathbf{V}(\mathbf{q}),\end{aligned} \tag{13.20}$$

then

$$\begin{aligned}\mathbf{V}^{(T)}(\mathbf{q}) &= -\frac{i}{q^2}\mathbf{q} \times (-i\mathbf{q} \times \mathbf{V}(\mathbf{q})) = \\ &= -\frac{\mathbf{q} \times (\mathbf{q} \times \mathbf{V}(\mathbf{q}))}{q^2}.\end{aligned} \tag{13.21}$$

Ultimately, we wrote the Fourier transforms of the longitudinal and transverse components of a vector field as a function of the Fourier transform of the field itself.

We next apply the Helmholtz decomposition introduced in Definition 13.1 to the electromagnetic fields. To this purpose, let us recall Maxwell's equations expressed in the CGS-Gaussian system of units (see Eq. 10.50). This formulation allows one to separate explicitly the longitudinal and transverse components of the fields, thereby clarifying their respective dynamical roles. In particular, from $\nabla \cdot \mathbf{E}^{(T)} = 0$, $\nabla \times \mathbf{E}^{(L)} = \mathbf{0}$, $\nabla \cdot \mathbf{B}^{(T)} = 0$, $\nabla \times \mathbf{B}^{(L)} = \mathbf{0}$, we get

$$\begin{cases}\nabla \cdot \mathbf{E}^{(L)} = 4\pi\rho \\ \nabla \cdot \mathbf{B}^{(L)} = 0 \\ \nabla \times \mathbf{E}^{(T)} = -\frac{1}{c}\frac{\partial \mathbf{B}}{\partial t} \\ \nabla \times \mathbf{B}^{(T)} = \frac{1}{c}\frac{\partial \mathbf{E}}{\partial t} + \frac{4\pi}{c}\mathbf{J}\end{cases} . \tag{13.22}$$

The divergence and curl of the longitudinal component of the magnetic field vanish. As a result, this component must be zero, implying that the magnetic field is purely transverse, i.e.,

$$\mathbf{B} = \mathbf{B}^{(T)}. \tag{13.23}$$

In contrast, the electric field contains both longitudinal and transverse components. Noting that the divergence of a curl is identically zero, we apply the divergence operator to Maxwell's fourth equation in Eq. 10.50, yielding

$$\begin{aligned}0 &= \nabla \cdot (\nabla \times \mathbf{B}) = \\ &= \frac{1}{c}\frac{\partial \nabla \cdot \mathbf{E}}{\partial t} + \frac{4\pi}{c}\nabla \cdot \mathbf{J} = \\ &= \frac{1}{c}\frac{\partial \rho}{\partial t} + \frac{4\pi}{c}\nabla \cdot \mathbf{J},\end{aligned} \tag{13.24}$$

then continuity law 10.90 has been derived. We aim to show that the longitudinal component of the right-hand side of the fourth Maxwell equation in Eq. 10.50 vanishes. Indeed, by definition, the curl of the longitudinal component is zero; moreover, the divergence of the longitudinal component is

$$\begin{aligned}\nabla \cdot \left(\frac{1}{c}\frac{\partial \mathbf{E}^{(L)}}{\partial t} + \frac{4\pi}{c}\mathbf{J}^{(L)}\right) &= \frac{4\pi}{c}\left(\frac{\partial \rho}{\partial t} + \nabla \cdot \mathbf{J}\right) = \\ &= 0,\end{aligned} \tag{13.25}$$

where we used $\nabla \cdot \mathbf{J}^{(L)} = \nabla \cdot \mathbf{J}$, since $\nabla \cdot \mathbf{J}^{(T)} = 0$. Consequently, both the curl and the divergence of the longitudinal component of the considered field are zero, so the considered field is transverse, and we can rewrite Maxwell's equations in 13.22 as

$$\begin{cases}\nabla \cdot \mathbf{E}^{(L)} = 4\pi\rho \\ \nabla \cdot \mathbf{B}^{(L)} = 0 \\ \nabla \times \mathbf{E}^{(T)} = -\dfrac{1}{c}\dfrac{\partial \mathbf{B}^{(T)}}{\partial t} \\ \nabla \times \mathbf{B}^{(T)} = \dfrac{1}{c}\dfrac{\partial \mathbf{E}^{(T)}}{\partial t} + \dfrac{4\pi}{c}\mathbf{J}^{(T)}.\end{cases} \tag{13.26}$$

Thus, the first and second Maxwell equations correspond to the longitudinal components of the fields, while the third and fourth Maxwell equations correspond to the transverse components.

13.1.1 Scalar and Vector Potentials

We now introduce the scalar and vector potentials. Noting that the magnetic field is divergence-free (second equation in 10.50) and recalling that the divergence of a curl always vanishes, it follows that the magnetic field can be represented as the curl of a vector field $\mathbf{A}$ (see Eq. 10.279), and in particular

$$\begin{aligned}\mathbf{B} &= \nabla \times \left(\mathbf{A}^{(L)} + \mathbf{A}^{(T)}\right) = \\ &= \nabla \times \mathbf{A}^{(T)}.\end{aligned} \tag{13.27}$$

From Maxwell's third equation of Eq. 10.50, we derive

$$\begin{aligned}\nabla \times \mathbf{E} &= -\frac{1}{c}\frac{\partial \mathbf{B}}{\partial t} = \\ &= -\frac{1}{c}\frac{\partial\left(\nabla \times \mathbf{A}\right)}{\partial t},\end{aligned} \tag{13.28}$$

then

$$\nabla \times \left(\mathbf{E} + \frac{1}{c} \frac{\partial \mathbf{A}}{\partial t} \right) = \mathbf{0}. \tag{13.29}$$

Since the curl of a gradient is zero, we can write the quantity

$$\mathbf{E} + \frac{1}{c} \frac{\partial \mathbf{A}}{\partial t} \tag{13.30}$$

as the gradient of a scalar function φ, which we call a scalar potential, and we get Eq. 10.278. In particular, since the gradient is a longitudinal vector and the curl of a gradient is always zero, we obtain

$$\mathbf{E}^{(L)} = -\frac{1}{c} \frac{\partial \mathbf{A}^{(L)}}{\partial t} - \nabla \varphi, \tag{13.31}$$

$$\mathbf{E}^{(T)} = -\frac{1}{c} \frac{\partial \mathbf{A}^{(T)}}{\partial t}. \tag{13.32}$$

By introducing the potentials $\mathbf{A}$, φ, Maxwell's second and third equations are automatically satisfied, so it is necessary to determine the potentials from the first and fourth equations. From Eq. 10.50, by substituting the fields $\mathbf{E}$, $\mathbf{B}$ as a function of the potentials $\mathbf{A}$, φ into Maxwell's first and fourth equations, we find

$$\nabla^2 \varphi = -\frac{1}{c} \frac{\partial \nabla \cdot \mathbf{A}}{\partial t} - 4\pi \rho, \tag{13.33}$$

$$\left(\frac{1}{c^2} \frac{\partial^2}{\partial t^2} - \nabla^2 \right) \mathbf{A} = \frac{4\pi}{c} \mathbf{J} - \nabla \left(\nabla \cdot \mathbf{A} \right) - \frac{1}{c} \nabla \left(\frac{\partial \varphi}{\partial t} \right). \tag{13.34}$$

Since the curl of a gradient is identically zero, the vector potential $\mathbf{A}$ can be transformed by adding the gradient of a scalar function without altering the magnetic field. Under the gauge transformation

$$\mathbf{A} \to \mathbf{A}' = \mathbf{A} + \nabla f, \tag{13.35}$$

the magnetic field remains invariant. To ensure that the electric field is also unchanged under the transformation $(\mathbf{A}, \varphi) \to (\mathbf{A}', \varphi')$, we derive

$$\begin{aligned} \mathbf{E}' &= -\frac{1}{c} \frac{\partial \mathbf{A}'}{\partial t} - \nabla \varphi' = \\ &= -\frac{1}{c} \frac{\partial \mathbf{A}}{\partial t} - \frac{1}{c} \frac{\partial \nabla f}{\partial t} - \nabla \varphi', \end{aligned} \tag{13.36}$$

and from Eq. 10.278, we have

$$\varphi' = \varphi - \frac{1}{c}\frac{\partial f}{\partial t}. \tag{13.37}$$

Recall, the gauge transformation that leaves the electromagnetic field unchanged between two reference systems is described by means of 10.280. The curl of the vector potential is determined by Eq. 10.279: to uniquely specify the gauge, or equivalently to define the vector potential, it is necessary to assign $\nabla \cdot \mathbf{A}$. In particular, we note that the gauge transformation of the vector potential involves adding the gradient of the gauge function, which is a longitudinal field. Consequently, the transverse component of the vector potential remains unaffected by the gauge transformation. If we set $\nabla \cdot \mathbf{A} = 0$, we get Coulomb gauge. In this case, we have $\nabla \cdot \mathbf{A} = \nabla \cdot \mathbf{A}^{(L)} = 0$, $\nabla \times \mathbf{A}^{(L)} = \mathbf{0}$, so the vector potential becomes a purely transverse field. Under the Coulomb gauge, Maxwell's first equation in Eq. 13.26 takes the form

$$\begin{aligned} \nabla \cdot \mathbf{E}^{(L)} &= \nabla \cdot (-\nabla\varphi) = \\ &= 4\pi\rho, \end{aligned} \tag{13.38}$$

then

$$\nabla^2\varphi = -4\pi\rho, \tag{13.39}$$

which is the Poisson equation for the scalar potential, which has as its solution an instantaneous scalar potential, that is, $\varphi(t)$ depends on $\rho(t)$. It is important to note that the physically relevant quantity is the electromagnetic field itself, meaning there is no contradiction in allowing the scalar potential to propagate instantaneously. However, the electromagnetic field must remain retarded. In the Coulomb gauge, from $\mathbf{E}^{(L)} = -\nabla\varphi$ and from Eq. 13.32, it follows that the longitudinal component of the electric field is determined solely by the scalar potential, while the transverse component is determined solely by the vector potential. It should be emphasized that the identification of the longitudinal component with the instantaneous Coulomb interaction is strictly valid in the Coulomb gauge. In other gauges, such as the Lorenz gauge (see below), both ϕ and $\mathbf{A}$ obey wave equations and propagate at the speed of light, but only the transverse components of the fields correspond to physical, radiative degrees of freedom.

If we set

$$\nabla \cdot \mathbf{A} + \frac{1}{c}\frac{\partial\varphi}{\partial t} = 0, \tag{13.40}$$

we get Lorenz gauge. In this case the equations of $\mathbf{A}$, φ are decoupled, and in particular we have

$$\begin{cases} \left(\frac{1}{c^2}\frac{\partial^2}{\partial t^2} - \nabla^2\right)\varphi = 4\pi\rho \\ \left(\frac{1}{c^2}\frac{\partial^2}{\partial t^2} - \nabla^2\right)\mathbf{A} = \frac{4\pi}{c}\mathbf{J} \end{cases} . \tag{13.41}$$

13.1.2 Fourier Transform of the Electromagnetic Fields

Here, we compute the Fourier transforms of the electric and magnetic fields.

Remark 13.1 For convenience, throughout this chapter we will denote Fourier-transformed quantities using the wave vector $\mathbf{q}$ as a subscript, even though $\mathbf{q}$ is a continuous index. This convention will be useful later when we move to quantization, as it allows for a more compact and consistent notation.

The Fourier transforms of Maxwell's equations 10.50 are given by

$$\begin{cases} -i\mathbf{q}\cdot\mathbf{E_q} = 4\pi\rho_\mathbf{q} \\ -i\mathbf{q}\cdot\mathbf{B_q} = 0 \\ -i\mathbf{q}\times\mathbf{E_q} = -\frac{1}{c}\frac{\partial\mathbf{B_q}}{\partial t} \\ -i\mathbf{q}\times\mathbf{B_q} = \frac{1}{c}\frac{\partial\mathbf{E_q}}{\partial t} + \frac{4\pi}{c}\mathbf{J_q} \end{cases} , \tag{13.42}$$

and the Fourier transform of the continuity equation 10.90 is given by

$$-i\mathbf{q}\cdot\mathbf{J_q} + \frac{\partial\rho_\mathbf{q}}{\partial t} = 0. \tag{13.43}$$

Since $\mathbf{E}(\mathbf{r})$, $\mathbf{B}(\mathbf{r})$ are real, their Fourier transforms are complex fields. In particular,

$$\mathbf{E}(\mathbf{r}) = \frac{1}{(2\pi)^3}\int d\mathbf{q}\mathbf{E_q}e^{-i\mathbf{q}\cdot\mathbf{r}}, \tag{13.44}$$

$$\mathbf{E}^*(\mathbf{r}) = \frac{1}{(2\pi)^3}\int d\mathbf{q}\mathbf{E}^*_\mathbf{q}e^{i\mathbf{q}\cdot\mathbf{r}}, \tag{13.45}$$

then $\mathbf{E}(\mathbf{r}) = \mathbf{E}^*(\mathbf{r})$ implies

$$\mathbf{E}^*_\mathbf{q} = \mathbf{E}_{-\mathbf{q}}. \tag{13.46}$$

Similarly, we get

$$\mathbf{B}^*_\mathbf{q} = \mathbf{B}_{-\mathbf{q}}. \tag{13.47}$$

From Eqs. 13.46 and 13.47, we get

$$\mathbf{A}_{\mathbf{q}}^* = \mathbf{A}_{-\mathbf{q}}, \tag{13.48}$$

$$\varphi_{\mathbf{q}}^* = \varphi_{-\mathbf{q}}. \tag{13.49}$$

From Eqs. 10.278 and 10.279, in wave-vector space, electric and magnetic fields are related to potentials by

$$\mathbf{E}_{\mathbf{q}} = -\frac{1}{c}\frac{\partial \mathbf{A}_{\mathbf{q}}}{\partial t} + i\varphi_{\mathbf{q}}\mathbf{q}, \tag{13.50}$$

$$\mathbf{B}_{\mathbf{q}} = -i\mathbf{q} \times \mathbf{A}_{\mathbf{q}}, \tag{13.51}$$

respectively. Moreover, from Eq. 13.16, we get

$$\mathbf{E}_{\mathbf{q}}^{(L)} = \frac{\mathbf{E}\cdot\mathbf{q}}{q^2}\mathbf{q}. \tag{13.52}$$

Combining first equation in 13.42 and Eq. 13.52, one finds

$$\mathbf{E}_{\mathbf{q}}^{(L)} = \frac{i4\pi\rho_{\mathbf{q}}}{q^2}\mathbf{q}. \tag{13.53}$$

Now, by the convolution theorem, the Fourier transform of $\int d\mathbf{r} f(\mathbf{r})g(\mathbf{r}-\mathbf{r}')$ is given by $f(\mathbf{q})g(\mathbf{q})$. On the other hand, from Parseval identity (in the context of Fourier transforms)

$$\begin{aligned}\int d\mathbf{r} f^*(\mathbf{r})g(\mathbf{r}) &= \int d\mathbf{r}\int \frac{d\mathbf{q}}{(2\pi)^3} f_{\mathbf{q}}^* e^{i\mathbf{q}\cdot\mathbf{r}} \int \frac{d\mathbf{q}'}{(2\pi)^3} g_{\mathbf{q}'} e^{-i\mathbf{q}'\cdot\mathbf{r}} = \\ &= \int \frac{d\mathbf{q}}{(2\pi)^3}\int d\mathbf{q}'\delta(\mathbf{q}-\mathbf{q}') f_{\mathbf{q}}^* g_{\mathbf{q}'} = \\ &= \int \frac{d\mathbf{q}}{(2\pi)^3} f_{\mathbf{q}}^* g_{\mathbf{q}},\end{aligned} \tag{13.54}$$

we can infer that the integral of the product of two functions in real space is equal to the integral of the product of their Fourier transforms in momentum space. This relation is fundamental for understanding the behavior of the fields in Fourier space. In particular, it allows us to decompose and analyze the longitudinal and transverse components of the electromagnetic field more easily by transforming to momentum space.

Next, we consider the Fourier transform of the longitudinal electric field. To do so, recall that the Fourier transform of $\frac{1}{r}$ is given by

$$\frac{4\pi}{q^2} = \int d\mathbf{r} \frac{e^{i\mathbf{q}\cdot\mathbf{r}}}{r}. \tag{13.55}$$

In addition, recall that given a field $\mathbf{V}(\mathbf{r}) = \nabla h(\mathbf{r})$, its Fourier transform is given by $\mathbf{V}_\mathbf{q} = -ih_\mathbf{q}\mathbf{q}$. Consequently, the quantity

$$i\frac{4\pi}{q^2}\mathbf{q} \tag{13.56}$$

is the Fourier transform of

$$-\nabla\left(\frac{1}{r}\right) = \frac{\mathbf{r}}{r^3}. \tag{13.57}$$

From the convolution theorem and from Eq. 13.53, we derive

$$\mathbf{E}^{(L)}(\mathbf{r}, t) = \int d\mathbf{r}'\rho(\mathbf{r}', t)\frac{\mathbf{r} - \mathbf{r}'}{|\mathbf{r} - \mathbf{r}'|^3}. \tag{13.58}$$

Thus, we have solved Maxwell's first equation independently of the potentials, that is, without making reference to a specific gauge. The solution to the equation shows that the longitudinal component of the electric field is instantaneous, representing the electrostatic field generated by the charge distribution $\rho(\mathbf{r}, t)$. This result does not present a contradiction, as the physically meaningful quantity is the total electric field, which, in its entirety, is retarded. In particular, the transverse component of the electric field compensates for the instantaneous nature of the longitudinal component, ensuring the overall field remains retarded.

Next, let us consider the energy of the electromagnetic field, which is given by the spatial integral of the energy density, that is

$$U = \frac{1}{8\pi}\int d\mathbf{r}\left[\mathbf{E}(\mathbf{r}) \cdot \mathbf{E}(\mathbf{r}) + \mathbf{B}(\mathbf{r}) \cdot \mathbf{B}(\mathbf{r})\right]. \tag{13.59}$$

The fields $\mathbf{E}(\mathbf{r})$, $\mathbf{B}(\mathbf{r})$ are real, then we can write

$$U = \frac{1}{8\pi}\int d\mathbf{r}\left[\mathbf{E}^*(\mathbf{r}) \cdot \mathbf{E}(\mathbf{r}) + \mathbf{B}^*(\mathbf{r}) \cdot \mathbf{B}(\mathbf{r})\right]. \tag{13.60}$$

Now we can use the Parseval identity 13.54 and write the energy of the electromagnetic field as a function of the Fourier transforms of the electric and magnetic fields as

$$U = \frac{1}{8\pi}\int \frac{d\mathbf{q}}{(2\pi)^3}\left[\mathbf{E}^*_\mathbf{q} \cdot \mathbf{E}_\mathbf{q} + \mathbf{B}^*_\mathbf{q} \cdot \mathbf{B}_\mathbf{q}\right]. \tag{13.61}$$

Recalling that the magnetic field is purely transverse, while the electric field possesses both longitudinal and transverse components, we get

$$U = \frac{1}{8\pi} \int \frac{d\mathbf{q}}{(2\pi)^3} \left[\left| \mathbf{E}_{\mathbf{q}}^{(L)} \right|^2 + \left| \mathbf{E}_{\mathbf{q}}^{(T)} \right|^2 + \left| \mathbf{B}_{\mathbf{q}}^{(T)} \right|^2 \right]. \tag{13.62}$$

Specifically, regarding the contribution to the energy of the longitudinal component of the electric field, we have

$$\begin{aligned} U^{(L)} &= \frac{1}{8\pi} \int \frac{d\mathbf{q}}{(2\pi)^3} \mathbf{E}_{\mathbf{q}}^{(L)*} \cdot \mathbf{E}_{\mathbf{q}}^{(L)} = \\ &= \frac{1}{8\pi} \int \frac{d\mathbf{q}}{(2\pi)^3} (4\pi)^2 \frac{\rho_{\mathbf{q}}^* \rho_{\mathbf{q}}}{q^2}. \end{aligned} \tag{13.63}$$

From

$$\frac{4\pi}{q^2} \rho(\mathbf{q}) = \int d\mathbf{r}' \frac{\rho(\mathbf{r}')}{|\mathbf{r} - \mathbf{r}'|}, \tag{13.64}$$

and from Parseval identity (see Eq. 13.54), we get

$$U^{(L)} = \frac{1}{2} \int d\mathbf{r} \rho(\mathbf{r}) \int d\mathbf{r}' \frac{\rho(\mathbf{r}')}{|\mathbf{r} - \mathbf{r}'|}. \tag{13.65}$$

Consequently, regardless of the gauge, the contribution to the energy of $\mathbf{E}^{(L)}$ is equal to the electrostatic energy of the charge distribution.

13.2 Lagrangian of a System of Charged Particles in an Electromagnetic Field

To proceed with the quantization of the electromagnetic field, we must first examine the Lagrangian of a system of charges in an electromagnetic field, and subsequently derive the corresponding Hamiltonian, which will lead us to the quantization process.

The Lagrangian for a particle in an electromagnetic field is given by

$$\mathcal{L} = \frac{1}{2} m \mathbf{v}^2 + Q \left(\frac{\mathbf{A} \cdot \mathbf{v}}{c} - \varphi \right). \tag{13.66}$$

where m is the mass of the particle, $\mathbf{v}$ is its velocity, Q is its charge, $\mathbf{A}$ is the vector potential, φ is the scalar potential, and c is the speed of light. For a system consisting of multiple charges interacting with the electromagnetic field, the total Lagrangian is assumed to be

$$\mathcal{L} = \frac{1}{2}\sum_{\alpha} m_\alpha \mathbf{v}_\alpha^2 + \frac{1}{8\pi}\int d\mathbf{r}\,(\mathbf{E}^* \cdot \mathbf{E} - \mathbf{B}^* \cdot \mathbf{B}) + \sum_{\alpha} Q_\alpha \left(\frac{\mathbf{A}(\mathbf{r}_\alpha, t)\cdot \mathbf{v}}{c} - \varphi(\mathbf{r}_\alpha, t)\right). \tag{13.67}$$

Since the charge density and current density are respectively written as

$$\rho(\mathbf{r}) = \sum_{\alpha} Q_\alpha \delta(\mathbf{r} - \mathbf{r}_\alpha), \tag{13.68}$$

$$\mathbf{J}(\mathbf{r}) = \sum_{\alpha} Q_\alpha \mathbf{v}_\alpha \delta(\mathbf{r} - \mathbf{r}_\alpha), \tag{13.69}$$

we can write the Lagrangian as

$$\mathcal{L} = \frac{1}{2}\sum_{\alpha} m_\alpha \mathbf{v}_\alpha^2 + \frac{1}{8\pi}\int d\mathbf{r}\,(\mathbf{E}^* \cdot \mathbf{E} - \mathbf{B}^* \cdot \mathbf{B}) + \frac{1}{c}\int d\mathbf{r}\,(\mathbf{J}\cdot\mathbf{A} - c\rho\varphi). \tag{13.70}$$

In order to establish the validity of the Lagrangian reported in Eq. 13.70, it is necessary to demonstrate that the corresponding Euler-Lagrange equations (see Eq. 4.8) consistently reproduce the fundamental dynamical laws of the system. Specifically, one must verify that, on the one hand, the equations of motion for the charged particles reduce to the Lorentz force law, and, on the other hand, that the variation with respect to the field degrees of freedom gives rise to the Maxwell equations. We focus primarily on the fields. We can express Lagrangian 13.70 in terms of Fourier transforms as follows

$$\mathcal{L} = \frac{1}{2}\sum_{\alpha} m_\alpha \mathbf{v}_\alpha^2 + \frac{1}{8\pi}\int \frac{d\mathbf{q}}{(2\pi)^3}\left(\mathbf{E}_\mathbf{q}^* \cdot \mathbf{E}_\mathbf{q} - \mathbf{B}_\mathbf{q}^* \cdot \mathbf{B}_\mathbf{q}\right) + \frac{1}{c}\int \frac{d\mathbf{q}}{(2\pi)^3}\left(\mathbf{J}_\mathbf{q}^* \cdot \mathbf{A}_\mathbf{q} - c\rho_\mathbf{q}^*\varphi_\mathbf{q}\right), \tag{13.71}$$

where, since charge density and current density are real quantities, we have substituted $\rho(\mathbf{r}) \to \rho^*(\mathbf{r})$, $\mathbf{J}(\mathbf{r}) \to \mathbf{J}^*(\mathbf{r})$ and then we used the Parseval identity 13.54. Note that, by expressing the fields in terms of the potentials, Maxwell's second and third equations are automatically satisfied. Therefore, we need only verify that the Lagrange equations derived from this Lagrangian lead to the first and fourth Maxwell equations. The Lagrangian is written as a function of the variables $\mathbf{A}_\mathbf{q}(t)$, $\varphi_\mathbf{q}(t)$, so let us treat $\mathbf{A}_\mathbf{q}^*(t)$, $\varphi_\mathbf{q}^*(t)$ as Lagrangian variables. By taking the derivative of the Lagrangian with respect to $\varphi_\mathbf{q}^*$, and recalling that $\mathcal{L}$ exhibits no dependence on $\dot{\varphi}_\mathbf{q}^*$, we obtain

$$\left(-\frac{1}{c}\dot{\mathbf{A}}_\mathbf{q} + i\varphi_\mathbf{q}\mathbf{q}\right)\frac{-i\mathbf{q}}{4\pi} - \rho_\mathbf{q} = 0, \tag{13.72}$$

which is the Maxwell's first equation in wave vector space, that is first equation in 13.42. It is important to note that when performing the derivative, we must remember that the integral $\int d\mathbf{q}$ is invariant under the change of variable $\mathbf{q} \to -\mathbf{q}$.

To derive Maxwell's fourth equation, we need to treat $\mathbf{A}_\mathbf{q}^*$ as a Lagrangian variable. In this case, the Lagrangian depends on $\dot{\mathbf{A}}_\mathbf{q}^*$, since the electric field depends on the time derivative of the vector potential. Specifically, the canonical momentum conjugate to $\mathbf{A}_\mathbf{q}^*$ is given by

$$\frac{\partial \mathcal{L}}{\partial \dot{\mathbf{A}}_\mathbf{q}^*} = -\frac{1}{4\pi c}\dot{\mathbf{E}}_\mathbf{q}. \tag{13.73}$$

This equation will play a fundamental role in the quantization of the electromagnetic field. From Eq. 13.51, and from identity

$$(\mathbf{X} \times \mathbf{Y}) \cdot \mathbf{Z} = \mathbf{X} \cdot (\mathbf{Y} \times \mathbf{Z}), \tag{13.74}$$

with

$$\mathbf{X} = \mathbf{q}, \tag{13.75}$$

$$\mathbf{Y} = \mathbf{A}_\mathbf{q}^*, \tag{13.76}$$

$$\mathbf{Z} = \mathbf{q} \times \mathbf{A}_\mathbf{q}, \tag{13.77}$$

the contribution of the magnetic field to the Lagrangian is given by

$$\begin{aligned} \mathbf{B}_\mathbf{q}^* \cdot \mathbf{B}_\mathbf{q} &= \left(i\mathbf{q} \times \mathbf{A}_\mathbf{q}^*\right) \cdot (-i\mathbf{q} \times \mathbf{A}_\mathbf{q}) = \\ &= \mathbf{q} \cdot \left[\mathbf{A}_\mathbf{q}^* \times \left(\mathbf{q} \times \mathbf{A}_\mathbf{q}\right)\right]. \end{aligned} \tag{13.78}$$

Applying the cyclic property of the scalar triple product,

$$\mathbf{X} \cdot (\mathbf{Y} \times \mathbf{Z}) = \mathbf{Y} \cdot (\mathbf{Z} \times \mathbf{X}), \tag{13.79}$$

with the same definitions of **X**, **Y**, and **Z** as above, one obtains

$$\mathbf{B}_\mathbf{q}^* \cdot \mathbf{B}_\mathbf{q} = \mathbf{A}_\mathbf{q}^* \cdot \left[\left(\mathbf{q} \times \mathbf{A}_\mathbf{q}\right) \times \mathbf{q}\right]. \tag{13.80}$$

Then

$$\frac{\partial \mathcal{L}}{\partial \mathbf{A}_\mathbf{q}^*} = -\frac{1}{4\pi}\left[\left(\mathbf{q} \times \mathbf{A}_\mathbf{q}\right) \times \mathbf{q}\right] + \frac{1}{c}\mathbf{J}_\mathbf{q}, \tag{13.81}$$

and Lagrange's equation takes the form

$$\frac{1}{4\pi c}\dot{\mathbf{E}}_{\mathbf{q}} - \frac{1}{4\pi}\left[(\mathbf{q}\times\mathbf{A}_{\mathbf{q}})\times\mathbf{q}\right] + \frac{1}{c}\mathbf{J}_{\mathbf{q}} = 0, \tag{13.82}$$

$$\frac{1}{4\pi c}\dot{\mathbf{E}}_{\mathbf{q}} + \frac{1}{c}\mathbf{J}_{\mathbf{q}} = \frac{1}{4\pi}\frac{1}{(-i)}\mathbf{B}_{\mathbf{q}}\times\mathbf{q}, \tag{13.83}$$

$$\frac{1}{4\pi c}\dot{\mathbf{E}}_{\mathbf{q}} + \frac{1}{c}\mathbf{J}_{\mathbf{q}} = -i\frac{1}{4\pi}\mathbf{q}\times\mathbf{B}_{\mathbf{q}}, \tag{13.84}$$

and by multiplying by 4π, we obtain Maxwell's fourth equation in 13.42. With Maxwell's first and fourth equations derived, the Lagrangian 13.70 is related to a system of charged particles in an electromagnetic field.

We now proceed to show that the Lagrangian 13.71 can be made independent of $\mathbf{A}_{\mathbf{q}}^{(L)}$ and $\varphi_{\mathbf{q}}$. From Maxwell's first equation in 13.42, writing $\mathbf{E}_{\mathbf{q}}$ as a function of potentials, we get

$$i\mathbf{q}\cdot\left(-\frac{1}{c}\dot{\mathbf{A}}_{\mathbf{q}} + i\varphi_{\mathbf{q}}\mathbf{q}\right) = -4\pi\rho_{\mathbf{q}}, \tag{13.85}$$

$$q^2\varphi_{\mathbf{q}} = 4\pi\rho_{\mathbf{q}} - i\frac{\mathbf{q}\cdot\dot{\mathbf{A}}_{\mathbf{q}}}{c}, \tag{13.86}$$

that is, the scalar potential is then written as a function of the vector potential as

$$\varphi_{\mathbf{q}} = \frac{4\pi\rho_{\mathbf{q}}}{q^2} - i\dot{A}_{\mathbf{q}}^{(L)}\frac{1}{cq}. \tag{13.87}$$

By substituting into the expression for the electric field, we obtain

$$\begin{aligned}
\mathbf{E}_{\mathbf{q}} &= -\frac{1}{c}\left(\dot{\mathbf{A}}_{\mathbf{q}}^{(L)} + \dot{\mathbf{A}}_{\mathbf{q}}^{(T)}\right) + i\left(\frac{4\pi\rho_{\mathbf{q}}}{q^2} - i\dot{A}_{\mathbf{q}}^{(L)}\frac{1}{cq}\right)\mathbf{q} = \\
&= -\frac{1}{c}\dot{\mathbf{A}}_{\mathbf{q}}^{(L)} - \frac{1}{c}\dot{\mathbf{A}}_{\mathbf{q}}^{(T)} + i\frac{4\pi\rho_{\mathbf{q}}}{q^2}\mathbf{q} + \frac{1}{cq}\dot{A}_{\mathbf{q}}^{(L)}\mathbf{q} = \\
&= -\frac{1}{c}\dot{\mathbf{A}}_{\mathbf{q}}^{(T)} + i\frac{4\pi\rho_{\mathbf{q}}}{q^2}\mathbf{q}.
\end{aligned} \tag{13.88}$$

In particular, the term that appears in the Lagrangian becomes

$$\mathbf{E}_{\mathbf{q}}^{*}\cdot\mathbf{E}_{\mathbf{q}} = \frac{1}{c^2}\dot{\mathbf{A}}_{\mathbf{q}}^{(T)*}\dot{\mathbf{A}}_{\mathbf{q}}^{(T)} + \frac{(4\pi)^2\rho_{\mathbf{q}}^{*}\rho_{\mathbf{q}}}{q^2}. \tag{13.89}$$

Let us now consider the term in the Lagrangian that is dependent on the magnetic field. From Eq. 13.21, we write the magnetic field term as follows

$$\mathbf{B}_{\mathbf{q}}^{*} \cdot \mathbf{B}_{\mathbf{q}} = q^{2}\mathbf{A}_{\mathbf{q}}^{*} \cdot \mathbf{A}_{\mathbf{q}}^{(T)} = \\ = q^{2}\mathbf{A}_{\mathbf{q}}^{(T)*} \cdot \mathbf{A}_{\mathbf{q}}^{(T)}. \tag{13.90}$$

Upon substituting into the Lagrangian, we obtain

$$\begin{aligned}\mathcal{L} = &\frac{1}{2}\sum_{\alpha} m_{\alpha}\mathbf{v}_{\alpha}^{2} + \frac{1}{8\pi}\int \frac{d\mathbf{q}}{(2\pi)^{3}}\left(\frac{1}{c^{2}}\dot{\mathbf{A}}_{\mathbf{q}}^{(T)*} \cdot \dot{\mathbf{A}}_{\mathbf{q}}^{(T)} - q^{2}\mathbf{A}_{\mathbf{q}}^{(T)*} \cdot \mathbf{A}_{\mathbf{q}}^{(T)}\right) + \\ &+ \frac{1}{2}\int \frac{d\mathbf{q}}{(2\pi)^{3}} \frac{4\pi\rho_{\mathbf{q}}^{*}\rho_{\mathbf{q}}}{q^{2}} + \frac{1}{c}\int \frac{d\mathbf{q}}{(2\pi)^{3}}\left(\mathbf{J}_{\mathbf{q}}^{*} \cdot \mathbf{A}_{\mathbf{q}} - c\frac{4\pi\rho_{\mathbf{q}}^{*}\rho_{\mathbf{q}}}{q^{2}} + \frac{i}{q}\dot{A}_{\mathbf{q}}^{(L)}\rho_{\mathbf{q}}\right) = \\ = &\frac{1}{2}\sum_{\alpha} m_{\alpha}\mathbf{v}_{\alpha}^{2} + \frac{1}{8\pi}\int \frac{d\mathbf{q}}{(2\pi)^{3}}\left(\frac{1}{c^{2}}\dot{\mathbf{A}}_{\mathbf{q}}^{(T)*} \cdot \dot{\mathbf{A}}_{\mathbf{q}}^{(T)} - q^{2}\mathbf{A}_{\mathbf{q}}^{(T)*} \cdot \mathbf{A}_{\mathbf{q}}^{(T)}\right) + \\ &- \frac{1}{2}\int \frac{d\mathbf{q}}{(2\pi)^{3}} \frac{4\pi\rho_{\mathbf{q}}^{*}\rho_{\mathbf{q}}}{q^{2}} + \frac{1}{c}\int \frac{d\mathbf{q}}{(2\pi)^{3}}\left(\mathbf{J}_{\mathbf{q}}^{*} \cdot \mathbf{A}_{\mathbf{q}} + \frac{i}{q}\dot{A}_{\mathbf{q}}^{(L)}\rho_{\mathbf{q}}\right).\end{aligned} \tag{13.91}$$

Now we analyze the term $\mathbf{J}_{\mathbf{q}}^{*} \cdot \mathbf{A}_{\mathbf{q}} = J_{\mathbf{q}}^{(L)*}A_{\mathbf{q}}^{(L)} + J_{\mathbf{q}}^{(T)*}A_{\mathbf{q}}^{(T)}$. From Eq. 13.43, we get

$$J_{\mathbf{q}}^{(L)} = -i\frac{\dot{\rho}_{\mathbf{q}}}{q}, \tag{13.92}$$

$$J_{\mathbf{q}}^{(L)*} = i\frac{\dot{\rho}_{\mathbf{q}}^{*}}{q}, \tag{13.93}$$

and thus

$$\mathbf{J}_{\mathbf{q}}^{*} \cdot \mathbf{A}_{\mathbf{q}} = i\frac{\dot{\rho}_{\mathbf{q}}^{*}}{q}A_{\mathbf{q}}^{(L)} + J_{\mathbf{q}}^{(T)*}A_{\mathbf{q}}^{(T)}. \tag{13.94}$$

In the last term of the Lagrangian, the following expression appears

$$i\frac{\dot{\rho}_{\mathbf{q}}^{*}}{q}A_{\mathbf{q}}^{(L)} + J_{\mathbf{q}}^{(T)*}A_{\mathbf{q}}^{(T)} + \frac{i}{q}\dot{A}_{\mathbf{q}}^{(L)}\rho_{\mathbf{q}} = J_{\mathbf{q}}^{(T)*}A_{\mathbf{q}}^{(T)} + \frac{d}{dt}\left(i\frac{\rho_{\mathbf{q}}}{q}A_{\mathbf{q}}^{(L)}\right). \tag{13.95}$$

The Lagrangian depends on $\mathbf{A}^{(L)}$ solely through a total time derivative. It is well known that two Lagrangians differing only by a total time derivative lead to the same equations of motion. This result is significant because the longitudinal component of the vector potential is gauge-dependent, whereas the transverse component remains invariant. Indeed, the physical dynamics of the electromagnetic field ultimately depends only on the transverse component of the vector potential, while the longitudinal component contributes solely to the instantaneous Coulomb potential. We adopt the Coulomb gauge, in which the last term in Eq. 13.95 vanishes identically. Consequently, we can express the Lagrangian as

$$\mathcal{L} = \frac{1}{2}\sum_{\alpha} m_\alpha \mathbf{v}_\alpha^2 + \frac{1}{8\pi}\int \frac{d\mathbf{q}}{(2\pi)^3}\left(\frac{1}{c^2}\dot{\mathbf{A}}_{\mathbf{q}}^{(T)*}\cdot\dot{\mathbf{A}}_{\mathbf{q}}^{(T)} - q^2\mathbf{A}_{\mathbf{q}}^{(T)*}\cdot\mathbf{A}_{\mathbf{q}}^{(T)}\right) + \\ -\frac{1}{2}\int \frac{d\mathbf{q}}{(2\pi)^3}\frac{4\pi\rho_{\mathbf{q}}^*\rho_{\mathbf{q}}}{q^2} + \frac{1}{c}\int \frac{d\mathbf{q}}{(2\pi)^3}\mathbf{J}_{\mathbf{q}}^{(T)*}\cdot\mathbf{A}_{\mathbf{q}}^{(T)}. \tag{13.96}$$

13.3 Hamiltonian of a System of Charged Particles in an Electromagnetic Field

After deriving the Lagrangian for a system of charged particles interacting with an electromagnetic field (see Eq. 13.96), we now proceed to obtain the corresponding Hamiltonian. Since the Lagrangian depends only on the transverse component of the vector potential, we omit the superscript $^{(T)}$ in what follows for notational simplicity, then

$$\mathcal{L} = \frac{1}{2}\sum_{\alpha} m_\alpha \mathbf{v}_\alpha^2 + \frac{1}{8\pi}\int \frac{d\mathbf{q}}{(2\pi)^3}\left(\frac{1}{c^2}\dot{\mathbf{A}}_{\mathbf{q}}^{*}\cdot\dot{\mathbf{A}}_{\mathbf{q}} - q^2\mathbf{A}_{\mathbf{q}}^{*}\cdot\mathbf{A}_{\mathbf{q}}\right) + \\ -\frac{1}{2}\int \frac{d\mathbf{q}}{(2\pi)^3}\frac{4\pi\rho_{\mathbf{q}}^*\rho_{\mathbf{q}}}{q^2} + \frac{1}{c}\int \frac{d\mathbf{q}}{(2\pi)^3}\mathbf{J}_{\mathbf{q}}^{*}\cdot\mathbf{A}_{\mathbf{q}}. \tag{13.97}$$

Let us define two orthogonal vectors $\mathbf{e}_{\mathbf{q},1}$, $\mathbf{e}_{\mathbf{q},2}$ such that

$$\begin{cases} \mathbf{e}_{\mathbf{q},s}\cdot\mathbf{q} = 0 \\ \mathbf{e}_{-\mathbf{q},s} = \mathbf{e}_{\mathbf{q},s} \end{cases}, \quad s\in\{1,2\},\ \forall\mathbf{q}, \tag{13.98}$$

where s is the polarization index. We then write the transverse field component of the vector potential as

$$\mathbf{A}_{\mathbf{q}}^{(T)} = A_{\mathbf{q},1}\mathbf{e}_{\mathbf{q},1} + A_{\mathbf{q},2}\mathbf{e}_{\mathbf{q},2}. \tag{13.99}$$

Given this expression for $\mathbf{A}_{\mathbf{q}}^{(T)}$, the second term of the Lagrangian 13.97 takes the form

$$\frac{1}{c^2}\left(\dot{\mathbf{A}}_{\mathbf{q}}^{*}\cdot\dot{\mathbf{A}}_{\mathbf{q}} - c^2q^2\mathbf{A}_{\mathbf{q}}^{*}\cdot\mathbf{A}_{\mathbf{q}}\right) = \frac{1}{c^2}\left(\sum_{s=1}^{2}\dot{A}_{\mathbf{q},s}^{*}\dot{A}_{\mathbf{q},s} - c^2q^2A_{\mathbf{q},s}^{*}A_{\mathbf{q},s}\right). \tag{13.100}$$

The canonical momentum conjugate to the Lagrangian variable $A^*_{\mathbf{q},s}$ is defined—by direct analogy with Eq. 4.70—as

$$\begin{aligned}\Pi_{\mathbf{q},s} &= \frac{\partial \mathcal{L}}{\partial \dot{A}^*_{\mathbf{q},s}} = \\ &= \frac{1}{4\pi c^2}\dot{A}_{\mathbf{q},s}.\end{aligned} \tag{13.101}$$

Since the transverse component of the electric field, from Eq. 13.32, one finds

$$\begin{aligned}\mathbf{E_q} &= -\frac{1}{c}\dot{\mathbf{A}}_\mathbf{q} = \\ &= -4\pi c\mathbf{\Pi_q},\end{aligned} \tag{13.102}$$

then

$$\begin{aligned}\mathbf{\Pi_q} &= -\frac{\mathbf{E_q}}{4\pi c} = \\ &= \frac{1}{4\pi c^2}\dot{\mathbf{A}}_\mathbf{q}.\end{aligned} \tag{13.103}$$

From Eq. 13.48, one finds

$$\begin{aligned}\mathbf{\Pi}^*_\mathbf{q} &= \frac{1}{4\pi c^2}\dot{\mathbf{A}}^*{}_\mathbf{q} = \\ &= \frac{1}{4\pi c^2}\dot{\mathbf{A}}_{-\mathbf{q}} = \\ &= \mathbf{\Pi}_{-\mathbf{q}}.\end{aligned} \tag{13.104}$$

Now, focusing on the second term of the Lagrangian given in Eq. 13.97, which corresponds to the free electromagnetic field, we obtain the following relations

$$\frac{\partial \mathcal{L}}{\partial A^*_{\mathbf{q},s}} = -\frac{q^2}{4\pi}A_{\mathbf{q},s}, \tag{13.105}$$

$$\dot{\Pi}_{\mathbf{q},s} = \frac{1}{4\pi c^2}\ddot{A}_{\mathbf{q},s}, \tag{13.106}$$

then

$$\ddot{A}_{\mathbf{q},s} = -\omega^2_\mathbf{q} A_{\mathbf{q},s}, \tag{13.107}$$

where $\omega_{\mathbf{q}} = c^2q^2$. This implies that the Lagrangian for free electromagnetic field corresponds to a system of infinitely many—i.e., a continuous set of $\mathbf{q}$-modes—independent harmonic oscillators (see Sect. 1.3), each with frequency $\omega_{\mathbf{q}}$.

We now turn to the computation of the Hamiltonian of the system. In general, given a Lagrangian $\mathcal{L}(q_h, \dot{q}_h)$, the corresponding Hamiltonian is defined as $\mathcal{H} = \sum_h p_h\dot{q}_h(q_h, p_h) - \mathcal{L}(q_h, \dot{q}_h(q_h, p_h))$, where the generalized velocities $\dot{q}_h$ are expressed in terms of the canonical variables (q_h, p_h). In our case, the Lagrangian coordinates are $\mathbf{r}_\alpha$ and $\mathbf{A}_{\mathbf{q}}$. Consequently, one needs to express the quantities $\dot{\mathbf{r}}_\alpha$ and $\dot{\mathbf{A}}^*_{\mathbf{q}}$ as functions of the canonical variables. Recall that the last term of the Lagrangian depends on $\dot{\mathbf{r}}_\alpha$, indeed

$$\begin{aligned}\frac{1}{c}\int \frac{d\mathbf{q}}{(2\pi)^3}\mathbf{J}^*_{\mathbf{q}} \cdot \mathbf{A}_{\mathbf{q}} &= \frac{1}{c}\int d\mathbf{r}\mathbf{J}(\mathbf{r}) \cdot \mathbf{A}(\mathbf{r}) = \\ &= \frac{1}{c}\sum_\alpha q_\alpha \dot{\mathbf{r}}_\alpha \cdot \mathbf{A}(\mathbf{r}_\alpha), \end{aligned} \tag{13.108}$$

then

$$\begin{aligned} p_{\alpha,i} &= \frac{\partial \mathcal{L}}{\partial \dot{r}_{\alpha,i}} = \\ &= m_\alpha \dot{r}_{\alpha,i} + \frac{Q_\alpha}{c}A_i(\mathbf{r}_\alpha), \end{aligned} \tag{13.109}$$

$$\dot{\mathbf{r}}_\alpha = \frac{\mathbf{p}_\alpha}{m_\alpha} - \frac{Q_\alpha}{m_\alpha c}\mathbf{A}(\mathbf{r}_\alpha). \tag{13.110}$$

Using the relation between the Lagrangian and the Hamiltonian, we find

$$\mathcal{H} = \sum_\alpha \mathbf{p}_\alpha \cdot \dot{\mathbf{r}}_\alpha + \int \frac{d\mathbf{q}}{(2\pi)^3}\mathbf{\Pi}_{\mathbf{q}} \cdot \dot{\mathbf{A}}^*_{\mathbf{q}} - \mathcal{L}. \tag{13.111}$$

The contribution associated with the particles in the electromagnetic field is obtained by combining $\sum_\alpha \mathbf{p}_\alpha \cdot \dot{\mathbf{r}}_\alpha$ with the first and the last terms of the Lagrangian, namely those involving the current density. In this way, we obtain

$$\begin{aligned} &\sum_\alpha \mathbf{p}_\alpha \cdot \dot{\mathbf{r}}_\alpha - \frac{1}{2}\sum_\alpha m_\alpha \dot{\mathbf{r}}^2_\alpha - \frac{1}{c}\sum_\alpha Q_\alpha \dot{\mathbf{r}}_\alpha \cdot \mathbf{A}(\mathbf{r}_\alpha) = \\ &= \sum_\alpha \left(\mathbf{p}_\alpha - \frac{Q_\alpha}{c}\mathbf{A}(\mathbf{r}_\alpha)\right) \cdot \dot{\mathbf{r}}_\alpha - \frac{1}{2}\sum_\alpha m_\alpha \dot{\mathbf{r}}^2_\alpha = \\ &= \sum_\alpha \left(\mathbf{p}_\alpha - \frac{Q_\alpha}{c}\mathbf{A}(\mathbf{r}_\alpha)\right) \cdot \left(\frac{\mathbf{p}_\alpha}{m_\alpha} - \frac{Q_\alpha}{m_\alpha c}\mathbf{A}(\mathbf{r}_\alpha)\right) + \end{aligned}$$

$$
-\frac{1}{2}\sum_{\alpha} m_{\alpha}\left(\frac{\mathbf{p}_{\alpha}}{m_{\alpha}}-\frac{Q_{\alpha}}{m_{\alpha}c}\mathbf{A}(\mathbf{r}_{\alpha})\right)^{2}=
$$

$$
=\sum_{\alpha}\frac{1}{2m_{\alpha}}\left(\mathbf{p}_{\alpha}-\frac{Q_{\alpha}}{c}\mathbf{A}(\mathbf{r}_{\alpha})\right)^{2}. \tag{13.112}
$$

Substituting $\dot{\mathbf{A}}_{\mathbf{q}}$ as a function of $\mathbf{\Pi}_{\mathbf{q}}$, we finally get

$$
\mathcal{H}=\sum_{\alpha}\frac{1}{2m_{\alpha}}\left(\mathbf{p}_{\alpha}-\frac{Q_{\alpha}}{c}\mathbf{A}_{\alpha}(\mathbf{r}_{\alpha})\right)^{2}+\frac{1}{2}\int\frac{d\mathbf{q}}{(2\pi)^{3}}\frac{4\pi}{q^{2}}\rho_{\mathbf{q}}^{*}\rho_{\mathbf{q}}+
$$

$$
+\int\frac{d\mathbf{q}}{(2\pi)^{3}}\frac{1}{2}\left[4\pi c^{2}\mathbf{\Pi}_{\mathbf{q}}^{*}\cdot\mathbf{\Pi}_{\mathbf{q}}+\frac{q^{2}}{4\pi}\mathbf{A}_{\mathbf{q}}^{*}\cdot\mathbf{A}_{\mathbf{q}}\right]. \tag{13.113}
$$

13.4 Quantization of the Hamiltonian of a System of Charged Particles in an Electromagnetic Field

After deriving the Hamiltonian for a system of charges in an electromagnetic field (see Eq. 13.113), we can proceed to its quantization. In the quantized theory, the classical field amplitudes $A_{\mathbf{q},s}$ and their conjugate momenta $\Pi_{\mathbf{q},s}$ are promoted to operators. From Eqs. 13.48 and 13.104, one finds

$$
\hat{A}_{\mathbf{q},s}^{\dagger}=\hat{A}_{-\mathbf{q},s}, \tag{13.114}
$$

$$
\hat{\Pi}_{\mathbf{q},s}^{\dagger}=\hat{\Pi}_{-\mathbf{q},s}, \tag{13.115}
$$

respectively. We now specify the commutation relations they satisfy. In analogy with the phonon case (see Eqs. 4.75–4.78), we postulate that the operators obey the canonical commutation relations

$$
\left[\hat{A}_{\mathbf{q},s}^{\dagger},\hat{\Pi}_{\mathbf{q}',s'}\right]=i\hbar\delta_{s,s'}\delta\left(\mathbf{q}-\mathbf{q}'\right), \tag{13.116}
$$

or equivalently

$$
\left[\hat{A}_{\mathbf{q},s},\hat{\Pi}_{\mathbf{q}',s'}\right]=i\hbar\delta_{s,s'}\delta\left(\mathbf{q}+\mathbf{q}'\right), \tag{13.117}
$$

together with

$$
\left[\hat{A}_{\mathbf{q},s},\hat{A}_{\mathbf{q}',s'}\right]=0,\ \forall\,\mathbf{q},\mathbf{q}',s,s', \tag{13.118}
$$

$$\left[\hat{\Pi}_{\mathbf{q},s}, \hat{\Pi}_{\mathbf{q}',s'}\right] = 0, \ \forall \, \mathbf{q}, \mathbf{q}', s, s'. \tag{13.119}$$

Note that into Eq. 13.117, compared to Eq. 4.76, the Kronecker delta is replaced by a Dirac delta, since in the case of the electromagnetic field the treatment here is semiclassical and $\mathbf{q}$ is a continuous variable. Moreover, in analogy with Eqs. 4.79 and 4.80, we introduce

$$\hat{A}_{\mathbf{q},s} = \sqrt{\frac{2\pi\hbar c^2}{\omega_{\mathbf{q}}}} \left(a_{\mathbf{q},s} + a^{\dagger}_{-\mathbf{q},s}\right), \tag{13.120}$$

$$\hat{\Pi}_{\mathbf{q},s} = i\sqrt{\frac{\hbar\omega_{\mathbf{q}}}{8\pi c^2}} \left(a^{\dagger}_{\mathbf{q},s} - a_{-\mathbf{q},s}\right), \tag{13.121}$$

which can be inverted, yielding

$$a_{\mathbf{q},s} = \frac{1}{2}\left(\sqrt{\frac{2\tilde{m}\omega_{\mathbf{q}}}{\hbar}}\hat{A}_{\mathbf{q},s} + i\sqrt{\frac{2}{\hbar\tilde{m}\omega_{\mathbf{q}}}}\hat{\Pi}_{-\mathbf{q},s}\right), \tag{13.122}$$

$$a^{\dagger}_{\mathbf{q},s} = \frac{1}{2}\left(\sqrt{\frac{2\tilde{m}\omega_{\mathbf{q}}}{\hbar}}\hat{A}_{-\mathbf{q},s} - i\sqrt{\frac{2}{\hbar\tilde{m}\omega_{\mathbf{q}}}}\hat{\Pi}_{\mathbf{q},s}\right), \tag{13.123}$$

where

$$\tilde{m} = \frac{1}{4\pi c^2} \tag{13.124}$$

plays the role of an effective mass in the field Hamiltonian. It can be easily verified that these operators are bosonic and satisfy algebraic relations analogous to Eqs. 4.92–4.94, with the Kronecker delta replaced by a Dirac delta. Finally, in perfect analogy with Theorem 4.3, one finds

$$\hat{\mathcal{H}} = \sum_{\alpha} \frac{1}{2m_\alpha}\left(\hat{\mathbf{p}}_\alpha - \frac{Q_\alpha}{c}\hat{\mathbf{A}}_\alpha(\hat{\mathbf{r}}_\alpha)\right)^2 + \frac{1}{2}\int \frac{d\mathbf{q}}{(2\pi)^3}\frac{4\pi}{q^2}\rho^*_{\mathbf{q}}\rho_{\mathbf{q}} + \\ + \int \frac{d\mathbf{q}}{(2\pi)^3}\sum_{s=1}^{2}\hbar\omega_{\mathbf{q}}\left(a^{\dagger}_{\mathbf{q},s}a_{\mathbf{q},s} + \frac{1}{2}\right), \tag{13.125}$$

which is the Hamiltonian describing a system of non-relativistic charged particles interacting with the quantized electromagnetic field. The first term embodies the minimal coupling between the charged particles and the quantized transverse vector potential $\hat{\mathbf{A}}$, thereby governing the interaction of the charges with the radiation field. The second term accounts for the instantaneous Coulomb interaction among the charges, expressed in Fourier space via the charge density operator $\rho_{\mathbf{q}}$. It should be emphasized that $\rho_{\mathbf{q}}$ denotes here the Fourier component of the total charge distribution and must not be confused with the Fourier transform of the particle density operator $\hat{\rho}_{\mathbf{q}}$ (see Eq. 2.262). The third term corresponds to the Hamiltonian of the free electromagnetic field in the Coulomb gauge, which is quantized as a collection of independent harmonic oscillators, each associated with a wavevector $\mathbf{q}$ and a polarization index s. The bosonic excitations associated with the transverse modes of the electromagnetic field are referred to as photons.

Now, we write the quantized vector potential $\hat{\mathbf{A}}(\mathbf{r}, t)$ and the fields $\hat{\mathbf{E}}(\mathbf{r}, t)$, $\hat{\mathbf{B}}(\mathbf{r}, t)$ as functions of the creation and annihilation operators. From

$$\hat{\mathbf{A}}_{\mathbf{q}} = \sqrt{\frac{2\pi\hbar c^2}{\omega_{\mathbf{q}}}} \sum_{s=1}^{2} \left(a_{\mathbf{q},s} + a^{\dagger}_{-\mathbf{q},s} \right) \mathbf{e}_{\mathbf{q},s}, \tag{13.126}$$

we obtain in the Heisenberg picture (see Eq. 1.102)

$$\begin{aligned}\hat{\mathbf{A}}_{\mathbf{q}}(t) &= \sqrt{\frac{2\pi\hbar c^2}{\omega_{\mathbf{q}}}} \sum_{s=1}^{2} \mathbf{e}_{\mathbf{q},s} \left(a_{\mathbf{q},s} e^{-i\omega_{\mathbf{q}}t} + a^{\dagger}_{-\mathbf{q},s} e^{i\omega_{\mathbf{q}}t} \right) = \\ &= \sqrt{\frac{2\pi\hbar c^2}{\omega_{\mathbf{q}}}} \sum_{s=1}^{2} \mathbf{e}_{\mathbf{q},s} \left(a_{\mathbf{q},s}(t) + a^{\dagger}_{-\mathbf{q},s}(t) \right),\end{aligned} \tag{13.127}$$

then

$$\begin{aligned}\hat{\mathbf{A}}(\mathbf{r}, t) &= \int \frac{d\mathbf{q}}{(2\pi)^3} e^{i\mathbf{q}\cdot\mathbf{r}} \hat{\mathbf{A}}_{\mathbf{q}}(t) = \\ &= \int \frac{d\mathbf{q}}{(2\pi)^3} e^{i\mathbf{q}\cdot\mathbf{r}} \sqrt{\frac{2\pi\hbar c^2}{\omega_{\mathbf{q}}}} \sum_{s=1}^{2} \mathbf{e}_{\mathbf{q},s} \left(a_{\mathbf{q},s}(t) + a^{\dagger}_{-\mathbf{q},s}(t) \right).\end{aligned} \tag{13.128}$$

The vector potential is parallel to $\mathbf{e}_{\mathbf{q},s}$, then it is transverse. Its time evolution exhibits the same factor that characterizes the evolution of an operator according to the Hamiltonian of a harmonic oscillator. We also note that the vector potential can be expressed as a superposition of progressive and regressive plane waves, whose amplitudes are quantized since they are represented by operators. Concerning the electromagnetic field, from Eq. 13.32, we have

$$\hat{\mathbf{E}}^{(T)}(\mathbf{r}, t) = \int \frac{d\mathbf{q}}{(2\pi)^3} e^{i\mathbf{q}\cdot\mathbf{r}} i\sqrt{2\pi\hbar\omega_{\mathbf{q}}} \sum_{s=1}^{2} \mathbf{e}_{\mathbf{q},s} \left(a_{\mathbf{q},s}(t) + a^{\dagger}_{-\mathbf{q},s}(t) \right), \tag{13.129}$$

and from Eq. 13.51, we have

$$\hat{\mathbf{B}}(\mathbf{r},t) = \int \frac{d\mathbf{q}}{(2\pi)^3} e^{i\mathbf{q}\cdot\mathbf{r}} \sqrt{\frac{2\pi\hbar c^2}{\omega_{\mathbf{q}}}} \sum_{s=1}^{2} \left(i\mathbf{q} \times \mathbf{e}_{\mathbf{q},s}\right) \left(a_{\mathbf{q},s}(t) + a^{\dagger}_{-\mathbf{q},s}(t)\right). \tag{13.130}$$

Recall that in classical mechanics the momentum of the electromagnetic field is defined as

$$\mathbf{p} = \int \frac{d\mathbf{r}}{4\pi c} \mathbf{E} \times \mathbf{B}, \tag{13.131}$$

then

$$\hat{\mathbf{p}}^{(T)} = \left(2\pi\hbar c i^2\right) \int \frac{d\mathbf{r}}{4\pi c} \int \frac{d\mathbf{q}}{(2\pi)^3} \int \frac{d\mathbf{q}'}{(2\pi)^3} \sum_{s=1}^{2} \sum_{s'=1}^{2}$$
$$e^{i(\mathbf{q}+\mathbf{q}')\cdot\mathbf{r}} \left(\mathbf{e}_{\mathbf{q},s} \times \left(\mathbf{q}' \times \mathbf{e}_{\mathbf{q}',s'}\right)\right) \left(a_{\mathbf{q},s}(t) + a^{\dagger}_{-\mathbf{q},s}(t)\right) \left(a_{\mathbf{q}',s'}(t) + a^{\dagger}_{-\mathbf{q}',s'}(t)\right). \tag{13.132}$$

The integration over the spatial coordinates yields $\delta(\mathbf{q}+\mathbf{q}')$, then

$$\hat{\mathbf{p}}^{(T)} = \frac{1}{2} \int \frac{d\mathbf{q}}{(2\pi)^3} \sum_{s=1}^{2} \sum_{s'=1}^{2} \left(\mathbf{e}_{\mathbf{q},s} \times \left(\mathbf{q} \times \mathbf{e}_{-\mathbf{q},s'}\right)\right) \hbar \left(a_{\mathbf{q},s}(t) + a^{\dagger}_{-\mathbf{q},s}(t)\right)$$
$$\times \left(a_{-\mathbf{q},s'}(t) + a^{\dagger}_{\mathbf{q},s'}(t)\right) =$$
$$= \frac{1}{2} \int \frac{d\mathbf{q}}{(2\pi)^3} \sum_{s=1}^{2} \sum_{s'=1}^{2} \hbar \delta_{s,s'} \mathbf{q} \left(a_{\mathbf{q},s}(t) + a^{\dagger}_{-\mathbf{q},s}(t)\right) \left(a_{-\mathbf{q},s'}(t) + a^{\dagger}_{\mathbf{q},s'}(t)\right) =$$
$$= \int \frac{d\mathbf{q}}{(2\pi)^3} \sum_{s=1}^{2} \hbar \mathbf{q} a^{\dagger}_{\mathbf{q},s} a_{\mathbf{q},s}, \tag{13.133}$$

where we used the vector triple product identity 13.17, and the transversality of the polarization vectors, $\mathbf{e}_{\mathbf{q},s} \cdot \mathbf{q} = 0$, to reduce $\mathbf{e}_{\mathbf{q},s} \times (\mathbf{q} \times \mathbf{e}_{-\mathbf{q},s'})$ to $\delta_{s,s'}\mathbf{q}$. The integration over the spatial coordinates produced a Dirac delta, $\delta(\mathbf{q}+\mathbf{q}')$, allowing us to set $\mathbf{q}' = -\mathbf{q}$. Finally, we used the properties of the bosonic operators to rewrite $(a_{\mathbf{q},s} + a^{\dagger}_{-\mathbf{q},s})(a_{-\mathbf{q},s'} + a^{\dagger}_{\mathbf{q},s'})$ as $a^{\dagger}_{\mathbf{q},s} a_{\mathbf{q},s}$, since the terms aa and $a^{\dagger}a^{\dagger}$ do not contribute to the total momentum. In this way, the total transverse momentum is simply the sum of $\hbar\mathbf{q}$ for each photon in the corresponding mode. Consequently, the momentum $\hat{\mathbf{p}}^{(T)}$ is the sum of $\hbar\mathbf{q}$ as many times as the number of photons with momentum $\mathbf{q}$ and for each of the two polarizations s.

To summarize, we have shown that the electromagnetic field can be described as a field of bosonic particles, the photons, each carrying energy $\hbar\omega_{\mathbf{q}}$ and momentum $\hbar\mathbf{q}$. The quantization of the photon number follows from the fact that the number operator has a discrete spectrum. Moreover, recalling the relativistic expression for the energy of a particle, i.e., $U = \sqrt{p^2c^2 + m^2c^4}$, and noting that each photon has energy $\hbar\omega_{\mathbf{q}} = \hbar|\mathbf{q}|c$, we conclude that the rest mass of the photon is zero. It is important to emphasize that photons are associated solely with the transverse component of the vector potential. To illustrate this explicitly, consider the linear momentum corresponding to the longitudinal component of the electric field. From the first of Maxwell's equations in 13.42, it follows

$$\begin{aligned}
\hat{\mathbf{p}}^{(L)} &= \int \frac{d\mathbf{r}}{4\pi c} \int \frac{d\mathbf{q}}{(2\pi)^3} \int \frac{d\mathbf{q}'}{(2\pi)^3} e^{-i\mathbf{q}\cdot\mathbf{r}} e^{i\mathbf{q}'\cdot\mathbf{r}} \rho_{\mathbf{q}} \frac{4\pi i^2}{q^2} \mathbf{q} \times \left(\mathbf{q}' \times \hat{\mathbf{A}}_{-\mathbf{q}'}\right) = \\
&= \frac{1}{c} \int \frac{d\mathbf{q}}{(2\pi)^3} \frac{\rho_{\mathbf{q}}}{q^2} \mathbf{q} \times \left(-\mathbf{q} \times \hat{\mathbf{A}}_{-\mathbf{q}}\right) = \\
&= \frac{1}{c} \int \frac{d\mathbf{q}}{(2\pi)^3} \rho_{\mathbf{q}} \hat{\mathbf{A}}^*_{\mathbf{q}} = \\
&= \frac{1}{c} \int d\mathbf{r} \rho(\mathbf{r}) \hat{\mathbf{A}}(\mathbf{r}, t) = \\
&= \frac{1}{c} \sum_{\alpha} Q_\alpha \hat{\mathbf{A}}(\mathbf{r}_\alpha, t),
\end{aligned} \tag{13.134}$$

which is equal to the second term in the expression for the total momentum of particles in the electromagnetic field (see Eq. 13.125). This clearly shows that the longitudinal component of the field is entirely determined by the presence of sources and does not contribute independent quantum excitations. By contrast, the transverse components of the field correspond to free oscillations and are responsible for the quantum excitations of the electromagnetic field, that is, the photons. These considerations provide a direct link between the classical description of the electromagnetic field and its quantized counterpart, emphasizing the fundamental role of transverse modes in quantum electrodynamics.

13.4.1 Photon Coherent States

Let us now focus on the transverse electric field at zero temperature, i.e., in the photon ground state. In this case, the expectation value of the electric field vanishes, since it depends linearly on the annihilation and creation operators, a and $a^\dagger$. To draw a comparison between a quantum harmonic oscillator and its classical counterpart, one must consider the oscillator in a coherent state (see Sect. 1.3), which provides the closest quantum analog to a classical state. From Eq. 1.162, we define the vacuum state of the electromagnetic field, denoted by $|0\rangle$, i.e.,

$$a_{\mathbf{q},s}|0\rangle = 0, \ \forall \mathbf{q}, s. \tag{13.135}$$

The vacuum state is the ground state of the quantized electromagnetic field, and all expectation values in the vacuum correspond to the zero-point fluctuations of the field. Let $|\psi(\theta)\rangle$ denote the coherent state of the mode $(\mathbf{q}, s)$, parametrized by the complex number θ that sets both its amplitude and phase. Recall that this state can be constructed as a displacement of the vacuum, as in Eq. 1.213, with

$$S(\theta) = \theta a^{\dagger}_{-\mathbf{q},s} - \theta^{*} a_{\mathbf{q},s}, \tag{13.136}$$

so that the displacement operator reads

$$D(\theta) = e^{\theta a^{\dagger}_{-\mathbf{q},s} - \theta^{*} a_{\mathbf{q},s}}, \tag{13.137}$$

which defines the coherent state as

$$|\psi(\theta)\rangle = D(\theta)|0\rangle. \tag{13.138}$$

Using the conjugation formula (see Eq. H.5), we have

$$D^{\dagger}(\theta) a_{\mathbf{q},s} D(\theta) = a_{\mathbf{q},s} + \theta, \tag{13.139}$$

$$D^{\dagger}(\theta) a^{\dagger}_{-\mathbf{q},s} D(\theta) = a^{\dagger}_{-\mathbf{q},s} + \theta^{*}. \tag{13.140}$$

Taking expectation values on the vacuum gives

$$\langle\psi(\theta)|a_{\mathbf{q},s}|\psi(\theta)\rangle = \theta, \tag{13.141}$$

$$\langle\psi(\theta)|a^{\dagger}_{\mathbf{q},s}|\psi(\theta)\rangle = \theta^{*}. \tag{13.142}$$

The electric field operator for a mode $(\mathbf{q}, s)$ can be expressed as

$$\hat{\mathbf{E}}_{\mathbf{q},s}(\mathbf{r}, t) \propto \mathbf{e}_{\mathbf{q},s} \left(a_{\mathbf{q},s} e^{i(\mathbf{q}\cdot\mathbf{r} - \omega_{\mathbf{q}} t)} + a^{\dagger}_{-\mathbf{q},s} e^{-i(\mathbf{q}\cdot\mathbf{r} - \omega_{\mathbf{q}} t)} \right). \tag{13.143}$$

Finally, taking the expectation value with respect to the coherent state results in

$$\left\langle \hat{\mathbf{E}}_{\mathbf{q},s}(\mathbf{r}, t) \right\rangle \propto \mathbf{e}_{\mathbf{q},s} \left(\theta e^{i(\mathbf{q}\cdot\mathbf{r} - \omega_{\mathbf{q}} t)} + \theta^{*} e^{-i(\mathbf{q}\cdot\mathbf{r} - \omega_{\mathbf{q}} t)} \right). \tag{13.144}$$

From Eq. 1.221, we can set

$$\theta = \sqrt{\left\langle \hat{N} \right\rangle} e^{i\varphi}, \tag{13.145}$$

and we obtain

$$\left\langle \hat{\mathbf{E}}_{\mathbf{q},s}(\mathbf{r},t) \right\rangle \propto \mathbf{e}_{\mathbf{q},s}\, e^{i\mathbf{q}\cdot\mathbf{r}} \sqrt{\left\langle \hat{N} \right\rangle} e^{i\varphi} e^{-i\omega_{\mathbf{q}}t} + \text{c.c.}, \tag{13.146}$$

in other words, the electric field is described by a plane wave whose intensity, given by the squared modulus of the amplitude, is proportional to $\left\langle \hat{N} \right\rangle$, the average number of photons. Varying the intensity of the electromagnetic wave corresponds to changing its average photon number.

Remark 13.2 (Blackbody Radiation) The quantization of the electromagnetic field leads naturally to the derivation of the blackbody radiation spectrum. Consider a free electromagnetic field confined within a cavity. The quantized field can be described by an Hamiltonian of the form 4.102, where $a^\dagger_{\mathbf{q},s}$ and $a_{\mathbf{q},s}$ denote the creation and annihilation operators for photons with wavevector $\mathbf{q}$ and polarization state s, and $\mathcal{E}_{\mathbf{q},s} = \hbar\omega_{\mathbf{q}}$ is the energy of each mode. At thermal equilibrium at temperature T, the average number of photons in each mode is given by the Bose-Einstein distribution. Therefore, the average value of the Hamiltonian becomes (see Eq. 2.251)

$$\begin{aligned} \left\langle \hat{\mathcal{H}} \right\rangle &= \sum_{\mathbf{q},s} \hbar\omega_{\mathbf{q}} \left\langle \hat{N}_{\mathbf{q},s} \right\rangle = \\ &= \sum_{\mathbf{q},s} \frac{\hbar\omega_{\mathbf{q}}}{e^{\beta\hbar\omega_{\mathbf{q}}} - 1}. \end{aligned} \tag{13.147}$$

By transforming the discrete sum into an integral over the density of states in momentum space, and considering the two possible transverse polarizations of each photon, one obtains the energy density per unit frequency

$$u(\omega) = \frac{\hbar\omega^3}{\pi^2 c^3} \frac{1}{e^{\beta\hbar\omega} - 1}, \tag{13.148}$$

which is Planck's law of blackbody radiation. This result not only explains the observed spectrum of blackbody radiation but also resolves the ultraviolet catastrophe predicted by classical physics.

13.4.2 Absorption, Stimulated Emission and Spontaneous Emission

Here, we see that the quantization of the electromagnetic field naturally gives rise to stimulated emission and absorption processes, and, in particular, provides a fundamental explanation for spontaneous emission, i.e., the intrinsic instability of excited states.

The Hamiltonian of a hydrogen-like atom can be obtained by adapting Eq. 13.125 to a single electron and adding an interaction term

$$\hat{\mathcal{H}}_I = \frac{q_e}{m_e c}\hat{\mathbf{p}} \cdot \hat{\mathbf{A}}. \tag{13.149}$$

The Hamiltonian of interest is then

$$\hat{\mathcal{H}} = \frac{\hat{\mathbf{p}}^2}{2m_e} - \frac{q_e^2}{r} + \sum_{\mathbf{q},s} \hbar\omega_{\mathbf{q}} a^{\dagger}_{\mathbf{q},s} a_{\mathbf{q},s} + \frac{q_e}{m_e c}\hat{\mathbf{p}} \cdot \hat{\mathbf{A}}, \tag{13.150}$$

where we neglected the term proportional to $\hat{\mathbf{A}}^2$. In other words, we consider the Hamiltonian as the sum of the Hamiltonian of the unperturbed system, which describes the system of the electron of the hydrogenoid atom and the e.m. field, i.e., free photons, and the perturbation Hamiltonian $\hat{\mathcal{H}}_I$. The eigenstates of the free electromagnetic field are expressed as in Eq. 2.142. Applying Fermi's golden rule, the transition probability is given by

$$W = \frac{2\pi}{\hbar}\left|\langle i|\hat{\mathcal{H}}_I|f\rangle\right|^2 \delta(U_f - U_i), \tag{13.151}$$

and consider an initial and a final state characterized respectively by the energies

$$U_i = U_{N_1} + \left(\hbar\omega_{\mathbf{q}_1} N_{\mathbf{q}_1,s_1} + \ldots\right), \tag{13.152}$$

$$U_f = U_{N_2} + \left(\hbar\omega_{\mathbf{q}_1} N_{\mathbf{q}_1,s_1} + \ldots\right). \tag{13.153}$$

The interaction term depends on the vector potential $\hat{\mathbf{A}}$, which in turn depends on $(a^{\dagger} + a)$, consequently the only permissible transitions are those such that the final and initial states differ by the energy of a single photon, i.e.,

$$U_{n_2} = U_{n_1} \pm \hbar\omega_{\mathbf{q}}. \tag{13.154}$$

Consequently, considering only the one-photon interaction term we find that radiation allows transitions between electronic states that differ by the energy equal to the energy of a photon. This gives rise to the processes of stimulated absorption and stimulated emission.

Let us now focus on the phenomenon of spontaneous emission. The vector potential contains a term involving the operator $a^{\dagger}$, so even if we start from an initial state with zero photons, the system can evolve into a state with one photon. Consequently, the interaction between the electron and the electromagnetic field, described by the term $\hat{\mathbf{p}} \cdot \hat{\mathbf{A}}$, which depends on $a^{\dagger}$, allows the creation of a photon and enables a transition from a higher to a lower electronic energy level. For this reason, excited electronic states are unstable, as they can decay spontaneously, i.e., without the presence of initial photons, via spontaneous emission processes.

Part V
Toolkit of Group Theory and Symmetries in Physics

Chapter 14
Foundations of Group Theory

Abstract Group theory is a foundational framework in theoretical physics, particularly in quantum mechanics, where it enables the classification of quantum states, the derivation of selection rules, and the prediction of spectral degeneracies through symmetry analysis. This chapter introduces the algebraic definition of groups and then progresses to Lie groups, continuous groups endowed with differentiable structures, highlighting their role in describing fundamental symmetries such as rotations, translations, and gauge transformations. To make the material accessible, the presentation emphasizes key properties without requiring advanced knowledge of geometry or topology. The chapter also discusses Lie algebras, which provide a local, infinitesimal description of Lie groups via a commutator operation. These algebras are instrumental in both the classification and computation of symmetry transformations. Special focus is given to SO(3) and SU(2), which are central to quantum mechanics. SO(3) describes real-space rotations associated with angular momentum conservation. SU(2), a double cover of SO(3) with the same Lie algebra but distinct global topology, is essential for modelling spin-½ systems. The representation theory of SU(2) forms the mathematical basis for quantum angular momentum, illustrating the deep connection between symmetry, group structure, and physical observables in quantum theory.

14.1 Lie Group

This section recalls fundamental algebraic structures and their key properties, which provide the foundation for the study of Lie groups. Lie groups generalize basic concepts, such as groups, by introducing a smooth dependence on continuous parameters. The resulting combination of algebraic and analytic features is central in mathematics and physics, where Lie groups play a crucial role in describing continuous symmetries, conservation laws, and the behavior of physical systems under transformations.

Definition 14.1 (Semigroup) Let S be a set and let $\cdot : S \times S \to S$ be a binary operation, i.e., given two elements s and s' of S, the product $ss' := s \cdot s'$ is equal

F. Tafuri et al., *Understanding Quantum Many-Body Theory*, UNITEXT for Physics, https://doi.org/10.1007/978-3-032-17849-7_14

to another element, say s'', in the set. The operation is also called multiplication or composition. The couple $(S, \cdot)$ is said to be a semigroup if it satisfies the following requirement:

- Associativity. The composition is associative, i.e.,

$$\left(ss'\right)s'' = s\left(s's''\right), \quad \forall\, s, s', s'' \in S. \tag{14.1}$$

A semigroup is the most basic structure where a set is closed under an associative operation. It does not require an identity element or inverses. In physics, examples include time-evolution operators restricted to forward times.

Definition 14.2 (Group) Let G be a set and let $\cdot : G \times G \to G$ be a binary operation, i.e., given two elements g and g' of G, the product $gg' := g \cdot g'$ is equal to another element, say g'', in the set. The operation is also called multiplication or composition. The couple $(G,\cdot)$ is said to be a group if it satisfies the following requirements:

- Associativity. The composition is associative, i.e.,

$$\left(gg'\right)g'' = g\left(g'g''\right), \; \forall\, g, g', g'' \in G. \tag{14.2}$$

- Identity. There exists and is unique an element e in G such that

$$eg = ge = e, \; \forall\, g \in G. \tag{14.3}$$

- Inverse. For each g in G there exists and is unique an element, which we denote by g^{-1}, such that

$$gg^{-1} = g^{-1}g = e. \tag{14.4}$$

A group extends the concept of a semigroup by requiring an identity element and inverses for every element. Groups are fundamental in physics, as they capture symmetries and conservation laws, as we will see in the following. If there is no danger of confusion, the composition dot symbol is usually omitted.

Definition 14.3 (Lie Group) Lie groups are groups (see Definition 14.2) whose elements are analytic functions depending on n real variables. Given a Lie group G, the vector $\bar{a} = (a_1, \ldots, a_n)$, with $a_i \in \mathbb{R}$, identifies the element $g(\bar{a}) \in G$. The integer n is called the dimension of the Lie group.

Let G be an n-dimensional Lie group. Without loss of generality, the identity element $e \in G$ can be identified with the origin of $\mathbb{R}^n$, i.e., with the n-dimensional null vector. Since the group elements depend analytically on n real parameters, both the group multiplication and the inverse operation can be locally expressed through Taylor expansions around the identity element.

Definition 14.4 (Group of Invertible Linear Operators on a Vector Space) For a finite-dimensional vector space V, the general linear group $GL(V)$ is the group of all invertible linear operators $V \to V$, equipped with the composition of maps.

Before proceeding, we introduce a few basic definitions which will be used in the following sections.

Definition 14.5 (Homomorphism Between Groups) Given two groups G and G' (see Definition 14.2), a map

$$f : G \to G' \tag{14.5}$$

is a homomorphism if it preserves the group operation, that is,

$$f(g_1 g_2) = f(g_1) f(g_2). \tag{14.6}$$

Two groups are said to be homomorphic if there exists a homomorphism between them. If the homomorphism is bijective, it is called an isomorphism, and the two groups are said to be isomorphic. By definition, the following results hold.

Theorem 14.1 *Let G and G' be isomorphic groups and let e and e' be the identities in G and in G' respectively, then*

$$f(e) = e', \tag{14.7}$$

that is, the identity $e \in G$ is mapped into the identity $e' \in G'$ by the isomorphism.

Proof From

$$\begin{aligned} g' &= f(g) = \\ &= f(eg) = \\ &= f(e) f(g) = \\ &= f(e) g', \end{aligned} \tag{14.8}$$

and

$$\begin{aligned} g' &= f(g) = \\ &= f(ge) = \\ &= f(g) f(e) = \\ &= g' f(e), \end{aligned} \tag{14.9}$$

the assertion follows. □

Theorem 14.2 *Let $f : G \to G'$ be a group homomorphism (see Definition 14.5) and let $g \in G$. Then*

$$f(g^{-1}) = f(g)^{-1}. \tag{14.10}$$

Proof Since f is a homomorphism, one has

$$\begin{aligned} f(g^{-1})f(g) &= f(g^{-1}g) = \\ &= f(e) = \\ &= e', \end{aligned} \tag{14.11}$$

where e' denotes the identity element of G'. Similarly,

$$\begin{aligned} f(g)f(g^{-1}) &= f(gg^{-1}) = \\ &= f(e) = \\ &= e'. \end{aligned} \tag{14.12}$$

Therefore, $f(g^{-1})$ is the inverse of $f(g)$, and the result follows. □

Definition 14.6 (Representation of a Lie Group) Given a Lie group G (see Definition 14.3), let $GL(V)$ denote the group of invertible linear operators on V (see Definition 14.4). If the map

$$T : G \to GL(V) \tag{14.13}$$

is a homomorphism, T is called a representation of the group G.

Representations allow abstract group elements to be realized concretely as linear operators on vector spaces. In physics, representations describe how physical systems transform under symmetry operations, as we will see in the following.

Definition 14.7 (Projective Representation) Given a Lie group G (see Definition 14.3), let $GL(V)$ denote the group of invertible linear operators on V (see Definition 14.4). A projective representation of G on V is a map

$$S : G \longrightarrow \mathrm{GL}(V) \tag{14.14}$$

such that for all $g_1, g_2 \in G$ one has

$$S(g_1)\, S(g_2) = \omega(g_1, g_2)\, S(g_1 g_2), \tag{14.15}$$

where $\omega(g_1, g_2)$ is a nonzero complex scalar, called a multiplier or factor system. If $\omega(g_1, g_2) = 1$ for all $g_1, g_2 \in G$, the representation is an ordinary representation.

Definition 14.8 (Faithful Representation) Given a Lie group G (see Definition 14.3), let $GL(V)$ denote the group of invertible linear operators on V (see Definition 14.4). A representation

$$S : G \longrightarrow \mathrm{GL}(V) \tag{14.16}$$

is called faithful if it is injective, that is, if

$$S(g) = \mathrm{Id}_V \ \Rightarrow \ g = e, \tag{14.17}$$

where e denotes the identity element of G.

Definition 14.9 (Fully Reducible Representation) Given a Lie group G (see Definition 14.3), let $GL(V)$ denote the group of invertible linear operators on V (see Definition 14.4). A representation

$$S : G \longrightarrow \mathrm{GL}(V) \tag{14.18}$$

is called fully reducible if V can be written as a direct sum of invariant subspaces,

$$V = \bigoplus_{\alpha} V_\alpha, \tag{14.19}$$

where each V_α is invariant under the action of G, and the restriction of S to each V_α is irreducible.

14.1.1 Some Lie Groups in Physics: $T(3)$, $SO(3)$, $E(3)$, and $SU(2)$

Here, we introduce some of the most fundamental Lie groups (see Definition 14.3) in physics, together with their construction and basic properties. We begin with spatial translations $T(3)$ and rotations $SO(3)$, which naturally lead to the Euclidean group $E(3)$, and then proceed to the special unitary group $SU(2)$, whose close connection with $SO(3)$ plays a central role in quantum mechanics and the description of spin.

- Spatial translation group $T(3)$.
  Let $\mathbf{r}$ and $\mathbf{r}'$ be two points in 3-dimensional space $\mathbb{R}^3$, which are connected by a spatial translation of a vector $\mathbf{a}$, i.e.,

$$\mathbf{r}' = \mathbf{r} + \mathbf{a}. \tag{14.20}$$

Given three points $\mathbf{r}_1$, $\mathbf{r}_2$ and $\mathbf{r}_3$ connected as follows

$$\mathbf{r}_1 = \mathbf{r}_2 + \mathbf{a}_{12}, \tag{14.21}$$

$$\mathbf{r}_2 = \mathbf{r}_3 + \mathbf{a}_{23}, \tag{14.22}$$

we have

$$\begin{aligned} \mathbf{r}_1 &= \mathbf{r}_2 + \mathbf{a}_{12} = \\ &= \mathbf{r}_3 + \mathbf{a}_{23} + \mathbf{a}_{12} = \\ &= \mathbf{r}_3 + \mathbf{a}_{13}, \end{aligned} \tag{14.23}$$

with

$$\mathbf{a}_{13} = \mathbf{a}_{12} + \mathbf{a}_{23}. \tag{14.24}$$

Consequently, the pair $(\mathbb{R}^3, +)$ defines a group, denoted $T(3)$, which is known as the group of spatial translations. Let $g_1, g_2 \in T(3)$, with $g_1 = \mathbf{a}_1$ and $g_2 = \mathbf{a}_2$. Then, for any point $\mathbf{r} \in \mathbb{R}^3$, we have

$$\begin{aligned} g_1 g_2 \mathbf{r} &= g_1 \big(g_2 \mathbf{r}\big) = \\ &= g_1 \big(\mathbf{r} + \mathbf{a}_2\big) = \\ &= \mathbf{r} + \mathbf{a}_2 + \mathbf{a}_1 = \\ &= \mathbf{r} + (\mathbf{a}_1 + \mathbf{a}_2) = \\ &= g\mathbf{r}, \end{aligned} \tag{14.25}$$

then the composition law of two spatial translations is simply given by

$$g_1 \cdot g_2 = \mathbf{a}_1 + \mathbf{a}_2. \tag{14.26}$$

The identity element of the group $T(3)$ is the null translation, corresponding to the zero vector $\mathbf{0} \in \mathbb{R}^3$, since for any point $\mathbf{r} \in \mathbb{R}^3$

$$e\,\mathbf{r} = \mathbf{r} + \mathbf{0} = \mathbf{r}. \tag{14.27}$$

For any group element $g \in T(3)$, represented by a translation vector $\mathbf{a}$, the inverse element g^{-1} is the translation by $-\mathbf{a}$, since

$$g^{-1}(g\mathbf{r}) = (\mathbf{r} + \mathbf{a}) - \mathbf{a} = \mathbf{r}, \tag{14.28}$$

for all $\mathbf{r} \in \mathbb{R}^3$.

- The rotation group $SO(3)$.
  Let $\mathbf{r}$ and $\mathbf{r}'$ be two points in 3-dimensional space connected by a rotation matrix R, i.e.,

$$\mathbf{r}' = R\mathbf{r}, \tag{14.29}$$

with $R^T = R^{-1}$ and $\det R = 1$. Given three points $\mathbf{r}_1$, $\mathbf{r}_2$ and $\mathbf{r}_3$ connected as follows

$$\mathbf{r}_1 = R_{12}\mathbf{r}_2, \tag{14.30}$$

$$\mathbf{r}_2 = R_{23}\mathbf{r}_3, \tag{14.31}$$

we get

$$\begin{aligned} \mathbf{r}_1 &= R_{12}\mathbf{r}_2 = \\ &= R_{12}\left(R_{23}\mathbf{r}_3\right) = \\ &= R_{12}R_{23}\mathbf{r}_3 = \\ &= R_{13}\mathbf{r}_3, \end{aligned} \tag{14.32}$$

with

$$R_{13} = R_{12}R_{23}. \tag{14.33}$$

Note that

$$\begin{aligned} R_{13}^T &= \left(R_{12}R_{23}\right)^T = \\ &= R_{23}^T R_{12}^T = \\ &= R_{23}^{-1} R_{12}^{-1} = \\ &= R_{13}^{-1}, \end{aligned} \tag{14.34}$$

and

$$\begin{aligned} \det R_{13} &= \det\left(R_{12}R_{23}\right) = \\ &= \det R_{12} \det R_{23} = \\ &= 1, \end{aligned} \tag{14.35}$$

i.e., R_{13} is a rotation matrix. Moreover, for any $R \in SO(3)$ the inverse element R^{-1} is given by the transpose matrix R^T, and the identity element of the group is the identity matrix $\mathbb{1}$.

Each rotation matrix can be parameterized, for instance, by Euler angles, which are three real numbers, then the set of the rotation matrices together with the composition of the multiplication of matrices form a Lie group, denoted $SO(3)$ and known as the rotation group,

$$SO(3) = \left\{ R \in M_3(\mathbb{R}) \mid RR^T = R^T R = \mathbb{1},\ \det R = 1 \right\}, \tag{14.36}$$

where $M_3(\mathbb{R})$ denotes the set of all real 3×3 matrices. Note that the group $SO(3)$ is also known as the special orthogonal group, since the orthogonal group, denoted $O(3)$, is the group of the 3×3 real matrices A which satisfy

$$O(3) = \left\{ A \in M_3(\mathbb{R}) \mid AA^T = A^T A = \mathbb{1},\ \det A = \pm 1 \right\}. \tag{14.37}$$

- The Euclidean group $E(3)$.
  Let $\mathbf{r}$ and $\mathbf{r}'$ be two points in 3-dimensional space related by a rototranslation, i.e.,

$$\mathbf{r}' = R\mathbf{r} + \mathbf{a}, \tag{14.38}$$

where $\mathbf{a}$ is the spatial translation vector and R is a rotation matrix. Although rototranslations are not linear transformations in $\mathbb{R}^3$, they can be represented as linear transformations by extending the space to homogeneous coordinates. It is immediate to verify that any rototranslation can be written as

$$v_i' = \sum_{j=1}^{4} A_{ij} v_j, \tag{14.39}$$

with

$$A = \begin{pmatrix} R_{11} & R_{12} & R_{13} & a_1 \\ R_{21} & R_{22} & R_{23} & a_2 \\ R_{31} & R_{32} & R_{33} & a_3 \\ 0 & 0 & 0 & 1 \end{pmatrix}. \tag{14.40}$$

Indeed, given two rototranslation matrices $A^{(1)}$, $A^{(2)}$, the product

$$A^{(2)} A^{(1)} = \begin{pmatrix} R_{11}^{(2)} & R_{12}^{(2)} & R_{13}^{(2)} & a_1^{(2)} \\ R_{21}^{(2)} & R_{22}^{(2)} & R_{23}^{(2)} & a_2^{(2)} \\ R_{31}^{(2)} & R_{32}^{(2)} & R_{33}^{(2)} & a_3^{(2)} \\ 0 & 0 & 0 & 1 \end{pmatrix} \begin{pmatrix} R_{11}^{(1)} & R_{12}^{(1)} & R_{13}^{(1)} & a_1^{(1)} \\ R_{21}^{(1)} & R_{22}^{(1)} & R_{23}^{(1)} & a_2^{(1)} \\ R_{31}^{(1)} & R_{32}^{(1)} & R_{33}^{(1)} & a_3^{(1)} \\ 0 & 0 & 0 & 1 \end{pmatrix} =$$

$$= \begin{pmatrix} \sum_{m=1}^{3} R_{1m}^{(2)} R_{m1}^{(1)} & \sum_{m=1}^{3} R_{1m}^{(2)} R_{m2}^{(1)} & \sum_{m=1}^{3} R_{1m}^{(2)} R_{m3}^{(1)} & \sum_{m=1}^{3} R_{1m} a_m^{(1)} + a_1^{(2)} \\ \sum_{m=1}^{3} R_{2m}^{(2)} R_{m1}^{(1)} & \sum_{m=1}^{3} R_{2m}^{(2)} R_{m2}^{(1)} & \sum_{m=1}^{3} R_{2m}^{(2)} R_{m3}^{(1)} & \sum_{m=1}^{3} R_{2m} a_m^{(1)} + a_2^{(2)} \\ \sum_{m=1}^{3} R_{3m}^{(2)} R_{m1}^{(1)} & \sum_{m=1}^{3} R_{3m}^{(2)} R_{m2}^{(1)} & \sum_{m=1}^{3} R_{3m}^{(2)} R_{m3}^{(1)} & \sum_{m=1}^{3} R_{3m} a_m^{(1)} + a_3^{(2)} \\ 0 & 0 & 0 & 1 \end{pmatrix} \tag{14.41}$$

is still a rototranslation matrix, where the rotation matrix element of the resulting matrix $A^{(2)}A^{(1)}$ is given by

$$R_{ij} = \sum_{m=1}^{3} R_{im}^{(2)} R_{mj}^{(1)}, \tag{14.42}$$

and the component of the translation vector of $A^{(2)}A^{(1)}$ is given by

$$a_i = \sum_{m=1}^{3} R_{im}^{(2)} a_m^{(1)} + a_i^{(2)}. \tag{14.43}$$

Consequently, rototranslations together with the matrix multiplication form a group, which is denoted $E(3)$ and known as the Euclidean group. A generical element of the Euclidean group is a couple of the form $g = (R, \mathbf{a}) \in E(3)$, where R is a rotation matrix parameterized by 3 variables and $\mathbf{a}$ is a translation vector, that is, the Euclidean group has dimension 6. The multiplication law in $E(3)$ corresponds to the composition of two rototranslations: for $g_1 = (R_1, \mathbf{a}_1)$ and $g_2 = (R_2, \mathbf{a}_2)$, the product $g_2 g_1 \in E(3)$ acts on a point $\mathbf{r}$ as follows

$$\begin{aligned} g_2 g_1 \mathbf{r} &= g_2(g_1 \mathbf{r}) = \\ &= g_2(R_1 \mathbf{r} + \mathbf{a}_1) = \\ &= R_2(R_1 \mathbf{r} + \mathbf{a}_1) + \mathbf{a}_2 = \\ &= R_2 R_1 \mathbf{r} + R_2 \mathbf{a}_1 + \mathbf{a}_2 = \\ &= g\mathbf{r}, \end{aligned} \tag{14.44}$$

therefore, the composition law of two rototranslations is simply given by

$$g_2 g_1 = (R_2 R_1, R_2 \mathbf{a}_1 + \mathbf{a}_2), \tag{14.45}$$

consistently with the composition law derived earlier from the explicit matrix form (see Eqs. 14.41–14.43). Obviously, spatial translations and rotations with respect to an axis are subgroups of the Euclidean group. Indeed, the group of spatial translations can be seen as a rototranslation of the vector **a** and a null rotation $\mathbb{1}$, that is, it is characterized by the multiplication

$$(\mathbb{1}, \mathbf{a})(\mathbb{1}, \mathbf{a}') = (\mathbb{1}, \mathbf{a} + \mathbf{a}'). \tag{14.46}$$

On the other hand, the group of rotations with respect to an axis can be seen as a rototranslation with translation vector $\mathbf{a} = \mathbf{0}$ and a rotation R, that is, it is characterized by the multiplication

$$(R, \mathbf{0})(R', \mathbf{0}) = (RR', \mathbf{0}). \tag{14.47}$$

- Special unitary group $SU(2)$.
  The special unitary group of degree n, denoted $SU(n)$, consists of all $n \times n$ unitary matrices (see Eqs. I.2 and I.3) with unit determinant. In this book, we focus on the case $n = 2$, i.e.,

$$SU(2) = \left\{ U \in M_2(\mathbb{C}) \mid UU^\dagger = U^\dagger U = \mathbb{1},\ \det U = 1 \right\}, \tag{14.48}$$

where $M_2(\mathbb{C})$ denotes the set of complex 2×2 matrices. In particular, we present a useful characterization of $SU(2)$.

Theorem 14.3 *Any element $M \in SU(2)$ (see Eq. 14.48) can be written as*

$$M = a\mathbb{1} + ib\sigma_x + ic\sigma_y + id\sigma_z \tag{14.49}$$

where σ_x, σ_y and σ_z denote the Pauli matrices given in Eqs. 1.53–1.55, and the real coefficients a, b, c, and d satisfy

$$a^2 + b^2 + c^2 + d^2 = 1. \tag{14.50}$$

Proof Given a 2×2 matrix M, that is of the form given in B.1, its inverse matrix is

$$\begin{aligned} M^{-1} &= \frac{1}{\det M} \begin{pmatrix} M_{22} & -M_{12} \\ -M_{21} & M_{11} \end{pmatrix} = \\ &= \begin{pmatrix} M_{22} & -M_{12} \\ -M_{21} & M_{11} \end{pmatrix}, \end{aligned} \tag{14.51}$$

and from $M^{-1} = M^\dagger$, we get

$$M_{22} = M_{11}^*, \tag{14.52}$$

$$M_{21} = -M_{12}^*. \tag{14.53}$$

Consequently, if $M \in SU(2)$, then

$$M = \begin{pmatrix} M_{11} & M_{12} \\ -M_{12}^* & M_{11}^* \end{pmatrix}. \tag{14.54}$$

Since the set of the identity matrix and the Pauli matrices is a basis for the 2×2 complex matrix space (see Theorem B.1), then M satisfies 14.49, i.e.,

$$\begin{pmatrix} a & 0 \\ 0 & a \end{pmatrix} + \begin{pmatrix} 0 & ib \\ ib & 0 \end{pmatrix} + \begin{pmatrix} 0 & c \\ -c & 0 \end{pmatrix} + \begin{pmatrix} id & 0 \\ 0 & id \end{pmatrix} = \begin{pmatrix} a+id & c+ib \\ -c+ib & a-id \end{pmatrix} \tag{14.55}$$

if and only if Eq. 14.50 holds. □

14.1.2 Homomorphism 2:1 from $SU(2)$ on $SO(3)$

A fundamental connection between the groups $SU(2)$ and $SO(3)$ arises in many areas of physics, particularly in quantum mechanics and the theory of angular momentum. The following result establishes this correspondence precisely.

Theorem 14.4 *There exists a* $2:1$ *homomorphism (see Definition 14.5)* R *from* $SU(2)$ *(see Eq. 14.48) on* $SO(3)$ *(see Eq. 14.36), i.e.,*

$$R : U, -U \in SU(2) \to R(U) = R(-U) \in SO(3). \tag{14.56}$$

Proof The proof is divided into the following steps:

1. Definition of R: there exists R that maps $U \in SU(2)$ to a real 3×3 matrix $R(U)$ such that

$$R(U_1)R(U_2) = R(U_1U_2). \tag{14.57}$$

2. $R(U) \in SO(3)$: the map R is in $SO(3)$, that is, $R^T(U)R(U) = \mathbb{1}$, $\det(R(U)) = +1$.
3. Surjectivity: the map R is on $SO(3)$, i.e., it is a homomorphism.
4. Kernel of R: the homomorphism R is $2:1$, that is, if $R(U_1) = R(U_2)$, then $U_1 = \pm U_2$.

1. Let $\sigma_1, \sigma_2, \sigma_3$ be the Pauli matrices (see Eqs. 1.53–1.55), in particular $\sigma_i^\dagger = \sigma_i$, $\operatorname{Tr}\sigma_i = 0$. The set of three such matrices together with the identical matrix forms a basis for the space of 2×2 complex matrices m (see Eq. B.2). In particular, if M is a hermitian null-trace matrix, we have $\alpha_0 = 0, \alpha_i = \alpha_i^\dagger$ for $i = 1, 2, 3$, i.e., any 2×2 hermitian null-trace matrix is a linear combination of real coefficients of the matrices σ_i. Now, given $\sigma_i' = U\sigma_i U^\dagger$, $U \in SU(2)$, since $\operatorname{Tr}\sigma_i' = 0, \sigma_i'^\dagger = \sigma_i'$,

the matrix σ_i' can be written as a linear combination of the Pauli matrices, i.e.,

$$U\sigma_i U^\dagger = \sum_{j=1}^{3} R_{ji}(U)\sigma_j. \tag{14.58}$$

The 9 real coefficients $R_{ji}(U)$ are uniquely defined by U due to the linear independence of the set $\{\sigma_i\}$. Consequently,

$$R : U \to R(U) \tag{14.59}$$

is a map that for each U associates a real 3×3 matrix $R(U)$. Given $V \in SU(2)$, we have

$$VU\sigma_i U^\dagger V^\dagger = (VU)\sigma_i (VU)^\dagger, \tag{14.60}$$

$$V\left(\sum_{k=1}^{3} R_{ki}(U)\sigma_k\right)V^\dagger = \sum_{j=1}^{3} R_{ji}(VU)\sigma_j, \tag{14.61}$$

$$\sum_{j=1}^{3}\left(\sum_{k=1}^{3} R_{ki}(U)R_{jk}(V)\right)\sigma_j = \sum_{j=1}^{3} R_{ji}(VU)\sigma_j, \tag{14.62}$$

$$\sum_{k=1}^{3} R_{jk}(V)R_{ki}(U) = R_{ji}(VU),\ \ j = 1, 2, 3 \tag{14.63}$$

$$R(V)R(U) = R(VU). \tag{14.64}$$

2. From Eq. 1.58, we have

$$\begin{aligned}
U\sigma_i\sigma_j U^\dagger &= U\sigma_i U^\dagger U\sigma_j U^\dagger = \\
&= \sum_{l=1}^{3} R_{li}(U)\sigma_l \sum_{m=1}^{3} R_{mj}(U)\sigma_m = \\
&= \sum_{l=1}^{3}\sum_{m=1}^{3} R_{li}(U)R_{mj}(U)\sigma_l\sigma_m = \\
&= \sum_{l=1}^{3}\sum_{m=1}^{3} R_{li}(U)R_{mj}(U)\left(\delta_{lm}\mathbb{1} + i\sum_{n=1}^{3}\epsilon_{lmn}\sigma_n\right) =
\end{aligned}$$

$$=\sum_{l=1}^{3}\sum_{m=1}^{3}\delta_{lm}R_{li}(U)R_{mj}(U)\mathbb{1}+i\sum_{l=1}^{3}\sum_{m=1}^{3}\sum_{n=1}^{3}R_{li}(U)R_{mj}(U)\epsilon_{lmn}\sigma_n=$$

$$=\sum_{m=1}^{3}R_{mi}(U)R_{mj}(U)\mathbb{1}+i\sum_{l=1}^{3}\sum_{m=1}^{3}\sum_{n=1}^{3}R_{li}(U)R_{mj}(U)\epsilon_{lmn}\sigma_n=$$

$$=\sum_{m=1}^{3}R^T_{im}(U)R_{mj}(U)\mathbb{1}+i\sum_{l=1}^{3}\sum_{m=1}^{3}\sum_{n=1}^{3}R_{li}(U)R_{mj}(U)\epsilon_{lmn}\sigma_n=$$

$$=\left(R^TR\right)_{ij}\mathbb{1}+i\sum_{l=1}^{3}\sum_{m=1}^{3}\sum_{n=1}^{3}R_{li}(U)R_{mj}(U)\epsilon_{lmn}\sigma_n. \tag{14.65}$$

On the other hand, we have

$$U\sigma_i\sigma_jU^\dagger=\delta_{ij}\mathbb{1}+i\sum_{k=1}^{3}\epsilon_{ijk}R_{nk}(U)\sigma_k, \tag{14.66}$$

which, by comparison, implies $R^TR=\mathbb{1}$, that is $R(U)\in O(3)$, and

$$\sum_{l=1}^{3}\sum_{m=1}^{3}R_{li}(U)R_{mj}(U)\epsilon_{lmn}=\sum_{k=1}^{3}\epsilon_{ijk}R_{nk}(U). \tag{14.67}$$

Multiplying both members of the above equation by $R_{ns}(U)$ and using $R^TR=\mathbb{1}$, we have

$$\sum_{l=1}^{3}\sum_{m=1}^{3}\sum_{n=1}^{3}R_{li}(U)R_{mj}(U)R_{ns}(U)\epsilon_{lmn}=\epsilon_{ijs}, \tag{14.68}$$

$$\det\left(R(U)\right)=1, \tag{14.69}$$

that implies $R(U)\in SO(3)$.

3. Given $S\in SO(3)$, which we parameterize by means of Euler angles as $S=S_3(\gamma)S_2(\beta)S_3(\alpha)$, we show that there exists $U_3(\alpha)$ such that $R\left(U_3(\alpha)\right)=S_3(\alpha)$, similarly for $U_2(\beta)$ and $U_3(\gamma)$. With the composition of the row-by-column product of the matrices, it follows that $R\left(U_3(\gamma)U_2(\beta)U_3(\alpha)\right)=S$, that is, R is a homomorphism on $SO(3)$. Recall

$$S_3(\alpha)=\begin{pmatrix}\cos(\alpha) & \sin(\alpha) & 0\\ -\sin(\alpha) & \cos(\alpha) & 0\\ 0 & 0 & 1\end{pmatrix}. \tag{14.70}$$

From

$$U_3(\alpha)\sigma_i U_3^\dagger(\alpha) = \sum_{j=1}^{3} S_3(\alpha)_{ji}\sigma_j, \tag{14.71}$$

one finds

$$U_3(\alpha)\sigma_1 U_3^\dagger(\alpha) = \cos(\alpha)\sigma_1 - \sin(\alpha)\sigma_2, \tag{14.72}$$

$$U_3(\alpha)\sigma_2 U_3^\dagger(\alpha) = \sin(\alpha)\sigma_1 + \cos(\alpha)\sigma_2, \tag{14.73}$$

$$U_3(\alpha)\sigma_3 U_3^\dagger(\alpha) = \sigma_3. \tag{14.74}$$

Now, it is easily verified that a solution is given by

$$\begin{aligned} U_3(\alpha) = e^{i\frac{\alpha}{2}\sigma_3} &= \\ &= \cos\left(\frac{\alpha}{2}\right) + i\sigma_3 \sin\left(\frac{\alpha}{2}\right) = \\ &= \begin{pmatrix} e^{i\frac{\alpha}{2}} & 0 \\ 0 & e^{-i\frac{\alpha}{2}} \end{pmatrix} \in SU(2), \end{aligned} \tag{14.75}$$

then

$$\begin{aligned} U_3(\alpha)\sigma_1 U_3^\dagger(\alpha) &= \left[\cos\left(\frac{\alpha}{2}\right) + i\sigma_3 \sin\left(\frac{\alpha}{2}\right)\right]\sigma_1\left[\cos\left(\frac{\alpha}{2}\right) - i\sigma_3 \sin\left(\frac{\alpha}{2}\right)\right] = \\ &= \cos^2\left(\frac{\alpha}{2}\right)\sigma_1 - i\sigma_1\sigma_3 \sin\left(\frac{\alpha}{2}\right)\cos\left(\frac{\alpha}{2}\right) \\ &\quad + i\sigma_3\sigma_1 \sin\left(\frac{\alpha}{2}\right)\cos\left(\frac{\alpha}{2}\right) + \sigma_3\sigma_1\sigma_3 \sin^2\left(\frac{\alpha}{2}\right) = \\ &= \cos^2\left(\frac{\alpha}{2}\right)\sigma_1 - i\sigma_1\sigma_3 \sin\left(\frac{\alpha}{2}\right)\cos\left(\frac{\alpha}{2}\right) \\ &\quad - i\sigma_1\sigma_3 \sin\left(\frac{\alpha}{2}\right)\cos\left(\frac{\alpha}{2}\right) + \sigma_3\sigma_1\sigma_3 \sin^2\left(\frac{\alpha}{2}\right) = \\ &= \cos^2\left(\frac{\alpha}{2}\right)\sigma_1 - i\sigma_1\sigma_3 2\sin\left(\frac{\alpha}{2}\right)\cos\left(\frac{\alpha}{2}\right) + \sigma_3\sigma_1\sigma_3 \sin^2\left(\frac{\alpha}{2}\right) = \\ &= \cos^2\left(\frac{\alpha}{2}\right)\sigma_1 + \sigma_3\sigma_1\sigma_3 \sin^2\left(\frac{\alpha}{2}\right) - i\sigma_1\sigma_3 2\sin\left(\frac{\alpha}{2}\right)\cos\left(\frac{\alpha}{2}\right) = \\ &= \cos^2\left(\frac{\alpha}{2}\right)\sigma_1 + \sigma_3\sigma_1\sigma_3 \sin^2\left(\frac{\alpha}{2}\right) - i\sigma_1\sigma_3 \sin(\alpha) = \\ &= \cos^2\left(\frac{\alpha}{2}\right)\sigma_1 + \sigma_3\sigma_1\sigma_3 \sin^2\left(\frac{\alpha}{2}\right) - i\left(i\epsilon_{132}\sigma_2\right)\sin(\alpha) = \end{aligned}$$

$$= \cos^2\left(\frac{\alpha}{2}\right)\sigma_1 + \sigma_3\sigma_1\sigma_3 \sin^2\left(\frac{\alpha}{2}\right) + (-\sigma_2)\sin(\alpha) =$$

$$= \cos^2\left(\frac{\alpha}{2}\right)\sigma_1 - \sin^2\left(\frac{\alpha}{2}\right)\sigma_1 - \sin(\alpha)\sigma_2 =$$

$$= \cos(\alpha)\sigma_1 - \sin(\alpha)\sigma_2, \tag{14.76}$$

and similarly for $U_2(\beta)$ and $U_3(\gamma)$.

4. Let $U, V \in SU(2)$ be such that

$$U\sigma_i U^\dagger = V\sigma_i V^\dagger, \tag{14.77}$$

then $U = \pm V$ and

$$V^\dagger U\sigma_i U^\dagger U = V^\dagger V\sigma_i V^\dagger U, \tag{14.78}$$

$$V^\dagger U\sigma_i = \sigma_i V^\dagger U, \tag{14.79}$$

$$\left[V^\dagger U, \sigma_i\right] = 0, \tag{14.80}$$

so, for every hermitian null-trace matrix M, we have

$$\left[V^\dagger U, M\right] = 0, \tag{14.81}$$

consequently,

$$V^\dagger U = \alpha \mathbb{1}. \tag{14.82}$$

If we compute the determinants of both members of the above equation, we have $\alpha^2 = 1$, then $\alpha = \pm 1$ and U and $-U$ map to the same matrix $R(U)$.

□

14.1.3 Germ of a Lie Group

Definition 14.10 (Germ of a Lie Group) Let G be a Lie group (see Definition 14.3) and let $\bar{a} \in \mathbb{R}^n$ denote local coordinates such that the identity element $e \in G$ corresponds to $\bar{a} = 0$. The germ of G at the identity is defined as

$$U_\epsilon = \{g(\bar{a}) \in G : |\bar{a}| < \epsilon, \ \forall \epsilon > 0\}, \tag{14.83}$$

where $|\bar{a}|$ denotes the Euclidean norm in $\mathbb{R}^n$.

Equivalently, U_ϵ consists of all elements of G sufficiently close to the identity. Furthermore, the germ is closed under inversion, i.e., if $g \in U_\epsilon$, then $g^{-1} \in U_\epsilon$. Importantly, one finds

Theorem 14.5 *The elements of a Lie group (see Definition 14.3) in a neighborhood of the identity can be locally parameterized by the germ of the Lie group at the identity (see Definition 14.10).*

Proof Consider the sets of the form

$$U_\epsilon^n = \{g \in G \,:\, g = g_1 \cdots g_n, \text{ with } g_i \in U_\epsilon \text{ for all } i = 1, \ldots, n\}, \tag{14.84}$$

and define

$$G_0 = \bigcup_{n \in \mathbb{N}} U_\epsilon^n. \tag{14.85}$$

We now show that G_0 is a subgroup of G.

- Closure under multiplication. Let $g, g' \in G_0$. Then there exist $m, m' \in \mathbb{N}$ such that

$$g = g_1 \cdots g_m, \quad g_i \in U_\epsilon, \tag{14.86}$$

$$g' = g_1' \cdots g_{m'}', \quad g_j' \in U_\epsilon. \tag{14.87}$$

It follows that

$$gg' = g_1 \cdots g_m g_1' \cdots g_{m'}' \in U_\epsilon^{m+m'} \subset G_0, \tag{14.88}$$

so G_0 is closed under multiplication.

- Closure under inversion. Let $g \in G_0$. Then there exists $m \in \mathbb{N}$ such that

$$g = g_1 \cdots g_m, \quad g_i \in U_\epsilon. \tag{14.89}$$

Since U_ϵ is symmetric (i.e., closed under inversion), we have $g_i^{-1} \in U_\epsilon$ for each i. Therefore,

$$g^{-1} = g_m^{-1} \cdots g_2^{-1} g_1^{-1} \in U_\epsilon^m \subset G_0. \tag{14.90}$$

- Identity element. The identity e belongs to $U_\epsilon^1 \subset G_0$.

□

Given a group G and an element g in a neighborhood of identity, i.e.,

$$g(\epsilon\bar{v}) = \mathbb{1} + \epsilon \sum_{\alpha=1}^{n} v_\alpha T_\alpha, \tag{14.91}$$

any element of the connected component of the identity of the Lie group can be obtained as a product of such infinitesimal elements. Indeed, given

$$\epsilon = \frac{1}{N} > 0, \tag{14.92}$$

we have

$$\begin{aligned} g(\epsilon\bar{v})\dots g(\epsilon\bar{v}) = g^N(\epsilon\bar{v}) = \\ = \left(\mathbb{1} + \frac{1}{N}\sum_{\alpha=1}^{n} v_\alpha T_\alpha\right)^N, \end{aligned} \tag{14.93}$$

and we finally get

$$\lim_{N\to\infty}\left(\mathbb{1} + \frac{1}{N}\sum_{\alpha=1}^{n} v_\alpha T_\alpha\right)^N = e^{\sum_{\alpha=1}^{n} v_\alpha T_\alpha}. \tag{14.94}$$

The group G_0 that is composed by union of the powers of the germs is the whole group G if G is connected, otherwise it constitutes only its connected component. For example in $O(3)$ the germ reconstructs $SO(3)$ but not $O(3)\setminus SO(3) = PSO(3)$ since the elements of $PSO(3)$ are not connected with continuity to the identity.

14.2 Lie Algebra

The local structure of a Lie group (see Definition 14.3) is naturally captured by its associated Lie algebra. This algebra encodes the infinitesimal behavior of the group near the identity element.

Definition 14.11 (Lie Algebra) A Lie algebra is a real vector space L of finite dimension n, equipped with a bilinear map

$$[\cdot,\cdot] : L \times L \to L, \tag{14.95}$$

called the Lie bracket, satisfying the following properties for all $x, y, z \in L$:

- bilinearity: the bracket is linear in both arguments;
- antisymmetry: $[x, y] = -[y, x]$;
- Jacobi identity: see Eqs. F.1 and F.4.

A standard example of a Lie bracket is the commutator.

Definition 14.12 (Representation of a Lie Algebra) Let W be a finite-dimensional vector space, $\dim(W) = k$. A map

$$T : L \to GL(W) \tag{14.96}$$

that assigns to each element of the Lie algebra L an (see Definition 14.11) invertible linear operator on W, and that is a homomorphism (see Definition 14.5) with respect to the Lie bracket, i.e.,

$$T([x, y]) = [T(x), T(y)],\ \forall\, x, y \in L, \tag{14.97}$$

is called a representation of the Lie algebra L.

The invertible linear operators acting on W that are images of the Lie algebra L through the homomorphism T form a closed vector space with respect to the Lie bracket.

Definition 14.13 (Homomorphism Between Lie Algebras) Two Lie algebras L and L' (see Definition 14.11) are said to be homomorphic if there exists a linear map

$$T : L \to L', \tag{14.98}$$

that preserves the Lie bracket, i.e.,

$$T : [x, y] \to [T(x), T(y)] = T\left([x, y]\right). \tag{14.99}$$

14.2.1 Structure Constants

The structure of a Lie algebra can be fully characterized in terms of its basis elements and the corresponding structure constants.

Definition 14.14 (Structure Constants of a Lie Algebra) Let L be a Lie algebra (see Definition 14.11) with basis $\{e_\alpha\}$. Since $[e_\alpha, e_\beta] \in L$, one can express the bracket in terms of the basis as

$$[e_\alpha, e_\beta] = \sum_\gamma C_{\alpha\beta\gamma} e_\gamma, \tag{14.100}$$

where $C_{\alpha\beta\gamma} \in \mathbb{R}$ are called the structure constants of the Lie algebra. These constants satisfy

$$C_{\alpha\beta\gamma} = -C_{\beta\alpha\gamma}, \tag{14.101}$$

due to the antisymmetry of the Lie bracket.

In addition, we have

Theorem 14.6 *The structure constants (see Definition 14.14) satisfy*

$$\sum_{\delta}\sum_{\epsilon}\left(C_{\alpha\beta\delta}C_{\delta\gamma\epsilon}+C_{\gamma\alpha\delta}C_{\delta\beta\epsilon}+C_{\beta\gamma\delta}C_{\delta\alpha\epsilon}\right)e_{\epsilon}=0. \tag{14.102}$$

Proof From Jacobi identity F.4, we get

$$\begin{aligned}0&=\left[[e_\alpha,e_\beta],e_\gamma\right]+\left[[e_\gamma,e_\alpha],e_\beta\right]+\left[[e_\beta,e_\gamma],e_\alpha\right]=\\&=\left[\sum_\delta C_{\alpha\beta\delta}e_\delta,e_\gamma\right]+\left[\sum_\delta C_{\gamma\alpha\delta}e_\delta,e_\beta\right]+\left[\sum_\delta C_{\beta\gamma\delta}e_\delta,e_\alpha\right],\end{aligned} \tag{14.103}$$

then

$$\begin{aligned}\left[\sum_\delta C_{\alpha\beta\delta}e_\delta,e_\gamma\right]&=\sum_\delta C_{\alpha\beta\delta}\left[e_\delta,e_\gamma\right]=\\&=\sum_\delta\sum_\epsilon C_{\alpha\beta\delta}C_{\delta\gamma\epsilon}e_\epsilon,\end{aligned} \tag{14.104}$$

$$\begin{aligned}\left[\sum_\delta C_{\gamma\alpha\delta}e_\delta,e_\beta\right]&=\sum_\delta C_{\gamma\alpha\delta}\left[e_\delta,e_\beta\right]=\\&=\sum_\delta\sum_\epsilon C_{\gamma\alpha\delta}C_{\delta\beta\epsilon}e_\epsilon,\end{aligned} \tag{14.105}$$

$$\begin{aligned}\left[\sum_\delta C_{\beta\gamma\delta}e_\delta,e_\alpha\right]&=\sum_\delta C_{\beta\gamma\delta}\left[e_\delta,e_\alpha\right]=\\&=\sum_\delta\sum_\epsilon C_{\beta\gamma\delta}C_{\delta\alpha\epsilon}e_\epsilon,\end{aligned} \tag{14.106}$$

from which Eq. 14.102 follows. □

Given any two elements of a Lie algebra, its structure constants allow us to calculate the action of Lie brackets as a combination of the basis of the Lie algebra.

Theorem 14.7 *Given a Lie algebra L (see Definition 14.11) and two bases* $\{e_i\}$, $\{e'_j\}$, *the structure constants (see Definition 14.14) with respect to these bases are related by*

$$C'_{\alpha\beta\tau} = \sum_{\gamma,\delta,\epsilon} A_{\alpha\gamma} A_{\beta\delta} (A)^{-1}_{\epsilon\tau} C_{\gamma\delta\epsilon}. \tag{14.107}$$

Proof Since the bases $\{e_\beta\}$ and $\{e'_\alpha\}$ are related by a linear transformation A as follows

$$e'_\alpha = \sum_\beta A_{\alpha\beta} e_\beta, \tag{14.108}$$

$$e_\beta = \sum_\alpha (A)^{-1}_{\beta\alpha} e'_\alpha, \tag{14.109}$$

and the Lie bracket provides

$$\left[e'_\alpha, e'_\beta\right] = \sum_\tau C'_{\alpha\beta\tau} e'_\tau, \tag{14.110}$$

we have

$$\begin{aligned} \left[e'_\alpha, e'_\beta\right] &= \left[\sum_\gamma A_{\alpha\gamma} e_\gamma, \sum_\delta A_{\beta\delta} e_\delta\right] = \\ &= \sum_{\gamma,\delta} A_{\alpha\gamma} A_{\beta\delta} \left[e_\gamma, e_\delta\right] = \\ &= \sum_{\gamma,\delta,\epsilon} A_{\alpha\gamma} A_{\beta\delta} C_{\gamma\delta\epsilon} e_\epsilon = \\ &= \sum_{\gamma,\delta,\epsilon,\tau} A_{\alpha\gamma} A_{\beta\delta} (A)^{-1}_{\epsilon\tau} C_{\gamma\delta\epsilon}\, e'_\tau, \end{aligned} \tag{14.111}$$

from which Eq. 14.107 follows. □

14.2.2 *The Lie Algebra of the Lie Group of Dimension n of Invertible Matrices of Dimension k*

A central result in the theory of Lie groups is that to every Lie group one can associate a Lie algebra, which captures the local structure of the group around the identity element. For any Lie group, the associated Lie algebra can be identified with the tangent space at the identity element; in the case of matrix Lie groups, the Lie bracket is given by the matrix commutator. The following result formalizes this fact for a Lie group of dimension n consisting of invertible $k \times k$ matrices.

Theorem 14.8 (The Lie Algebra of the Lie Group of Dimension n of Invertible Matrices of Dimension k) *Let G be the Lie group (see Definition 14.3) of dimension n of invertible matrices of dimension k, then the span of the infinitesimal generators of G together with the Lie brackets of the commutator between matrices is a Lie algebra.*

Proof By definition, the elements $g \equiv A \in G$ are invertible matrices of dimension k that depend on n real parameters with respect to which they are analytic functions, so a generic matrix element is written as $A_{ij}(\bar{a}), i, j \in (1, \ldots, k), \bar{a} = (a_1, \ldots, a_n)$, $a_i \in \mathbb{R}$. Consider the a neighborhood of the identity matrix. Since the identity is mapped by the null vector $\bar{0}$, i.e., $A_{ij}(\bar{0}) = \delta_{ij}$, the elements belonging to the neighborhood of the identity are such that the parameters a_α are infinitesimal. Since the elements $g \in G$ are analytic functions with respect to the parameterization vector, an element can be approximated by a first-order Taylor polynomial. Given an infinitesimal parameterization of the form $\bar{a}_\epsilon = \epsilon\bar{v}$, $a_{\epsilon,\alpha} = \epsilon v_\alpha$, $0 < \epsilon < 1$, $\alpha \in \{1, \ldots, n\}$, we have

$$
\begin{aligned}
A_{ij}(\epsilon\bar{v}) &\simeq A_{ij}(\bar{0}) + \epsilon \sum_{\alpha=1}^{n} v_\alpha \left.\frac{\partial A_{ij}}{\partial a_\alpha}\right|_{\bar{a}=\bar{0}} = \\
&= \delta_{ij} + \epsilon \sum_{\alpha=1}^{n} v_\alpha \left.\frac{\partial A_{ij}}{\partial a_\alpha}\right|_{\bar{a}=\bar{0}} = \\
&= \delta_{ij} + \epsilon \sum_{\alpha=1}^{n} v_\alpha (T_\alpha)_{ij},
\end{aligned} \tag{14.112}
$$

where we defined the matrices

$$
(T_\alpha)_{ij} = \left.\frac{\partial A_{ij}}{\partial a_\alpha}\right|_{\bar{a}=\bar{0}}. \tag{14.113}
$$

We set

$$
\begin{aligned}
T_{ij} &= \sum_{\alpha=1}^{n} v_\alpha (T_\alpha)_{ij} = \\
&= \sum_{\alpha=1}^{n} v_\alpha \left.\frac{\partial A_{ij}}{\partial a_\alpha}\right|_{\bar{a}=\bar{0}},
\end{aligned} \tag{14.114}
$$

then

$$
A_{ij}(\epsilon\bar{v}) = \delta_{ij} + \epsilon T_{ij}. \tag{14.115}
$$

By definition, the matrices $\{T_\alpha\}$ are n matrices $k \times k$ and form a linearly independent system. Otherwise, there would exist at least one nonzero λ_β for which

$A_{ij} = \sum_{\alpha=1}^{n} \lambda_\alpha (T_\alpha)_{ij} = \delta_{ij}$, which contradicts the hypothesis of biunivocal correspondence between A_{ij} matrices and the parameters $(a_1, \ldots, a_n)$. Then the system of matrices T_α is a system of generators. Let L be the vector space generated by these matrices, that is, $L = \text{Span}\{T_\alpha\}$. We show that L together with the Lie bracket of the matrix commutator is a Lie algebra. The matrices $\{T_\alpha\}$ of the group G are called infinitesimal generators of the Lie algebra L. Previously, we show that two different systems of generators span the same space L. Let $\bar{b} = (b_1, \ldots, b_n)$ be a new parameterization of the group G, and let it be writable as a function of the old parameterization $\bar{a}$, i.e., $\bar{b} = \bar{b}(\bar{a})$, $\bar{b}(\bar{0}) = \bar{0}$. Let $\bar{b}_\epsilon = \bar{w}$, $b_{\epsilon,\alpha} = \epsilon w_\alpha$, $0 < \epsilon < 1$, $\alpha \in \{1, \ldots, n\}$ a new infinitesimal parameterization, then

$$
\begin{aligned}
A_{ij}(\epsilon\bar{w}) &\simeq A_{ij}(\bar{0}) + \epsilon \sum_{\alpha=1}^{n} v_\alpha w_\alpha \left.\frac{\partial A_{ij}}{\partial b_\alpha}\right|_{\bar{b}=\bar{0}} = \\
&= \delta_{ij} + \epsilon \sum_{\alpha=1}^{n} w_\alpha \left.\frac{\partial A_{ij}}{\partial b_\alpha}\right|_{\bar{b}=\bar{0}} = \\
&= \delta_{ij} + \epsilon \sum_{\alpha=1}^{n} v_\alpha w_\alpha \left(\left.\frac{\partial a_\gamma}{\partial b_\alpha}\right|_{\bar{b}=\bar{0}} \right) \left(\left.\frac{\partial A_{ij}}{\partial a_\gamma}\right|_{\bar{a}=\bar{0}} \right).
\end{aligned}
\tag{14.116}
$$

On the other hand, in the new parameterization must be

$$
A_{ij}(\epsilon\bar{w}) = \delta_{ij} + \epsilon \sum_{\alpha=1}^{n} v_\alpha w_\alpha (T'_\alpha)_{ij}, \tag{14.117}
$$

then

$$
(T'_\alpha)_{ij} = \left(\left.\frac{\partial a_\gamma}{\partial b_\alpha}\right|_{\bar{b}=\bar{0}} \right) (T_\gamma)_{ij}, \tag{14.118}
$$

that is, the infinitesimal generators with respect to the parameterization $\bar{b}$ are linear combination of the generators of the parameterization $\bar{a}$, i.e., they belong to the same space L. Now, we show that L together with the matrix commutator is a Lie algebra. Consider two distinct elements in a neighborhood of the identity, say $g_1 = A_{ij}(\epsilon\bar{v})$, $g_2 = A_{ij}(\epsilon'\bar{w})$. We set

$$
B(\epsilon, \epsilon') = g_2^{-1}(\epsilon')\, g_1^{-1}(\epsilon)\, g_2(\epsilon')\, g_1(\epsilon) \tag{14.119}
$$

and we perform a Taylor expansion of B_{ij} in ϵ and ϵ'. The leading-order term is given by

$$
B_{ij}(\epsilon, \epsilon') \simeq \delta_{ij} + o(\epsilon\epsilon'). \tag{14.120}
$$

To understand why the first nonzero term is a function of the product $\epsilon\epsilon'$, note that

$$B(\epsilon, 0) = B(0, \epsilon') = \mathbb{1}, \ \forall \ \epsilon, \epsilon', \tag{14.121}$$

then, if we set one between ϵ and ϵ' equal to zero, the expansion of B with respect to the other parameter must return zero terms. Now, since

$$g_1 = \mathbb{1} + \epsilon \sum_{\delta=1}^{n} v_\delta T_\delta + o(\epsilon^2), \tag{14.122}$$

$$g_2 = \mathbb{1} + \epsilon' \sum_{\gamma=1}^{n} w_\gamma T_\gamma + o(\epsilon'^2), \tag{14.123}$$

$$g_1^{-1} = \mathbb{1} - \epsilon \sum_{\beta=1}^{n} v_\beta T_\beta + o(\epsilon^2), \tag{14.124}$$

$$g_2^{-1} = \mathbb{1} - \epsilon' \sum_{\alpha=1}^{n} w_\alpha T_\alpha + o(\epsilon'^2), \tag{14.125}$$

we have

$$B(\epsilon, \epsilon') = \left(\mathbb{1} - \epsilon' \sum_{\alpha=1}^{n} w_\alpha T_\alpha\right)\left(\mathbb{1} - \epsilon \sum_{\beta=1}^{n} v_\beta T_\beta\right)\left(\mathbb{1} + \epsilon' \sum_{\gamma=1}^{n} w_\gamma T_\gamma\right)$$
$$\left(\mathbb{1} + \epsilon \sum_{\delta=1}^{n} v_\delta T_\delta\right). \tag{14.126}$$

We compute the products in $B(\epsilon, \epsilon')$ and ignore terms of higher order than the product $\epsilon\epsilon'$, i.e.,

$$B(\epsilon, \epsilon') = \mathbb{1} + \epsilon\epsilon' \sum_{\alpha=1}^{n} w_\alpha T_\alpha \sum_{\beta=1}^{n} v_\beta T_\beta - \epsilon\epsilon' \sum_{\alpha=1}^{n} w_\alpha T_\alpha \sum_{\delta=1}^{n} v_\delta T_\delta +$$
$$- \epsilon\epsilon' \sum_{\beta=1}^{n} v_\beta T_\beta \sum_{\gamma=1}^{n} w_\gamma T_\gamma + \epsilon\epsilon' \sum_{\gamma=1}^{n} w_\gamma T_\gamma \sum_{\delta=1}^{n} v_\delta T_\delta. \tag{14.127}$$

The second and third terms cancel each other, hence

$$B(\epsilon, \epsilon') = \mathbb{1} - \epsilon\epsilon' \sum_{\beta=1}^{n} v_\beta T_\beta \sum_{\gamma=1}^{n} w_\gamma T_\gamma + \epsilon\epsilon' \sum_{\gamma=1}^{n} w_\gamma T_\gamma \sum_{\delta=1}^{n} v_\delta T_\delta =$$

$$= \mathbb{1} + \epsilon\epsilon' \sum_{\gamma=1}^{n} \sum_{\beta=1}^{n} w_\gamma v_\beta \left[T_\gamma, T_\beta\right], \tag{14.128}$$

where we renamed the dummy index $\delta \rightarrow \beta$. If ϵ and ϵ' are infinitesimal, the quantity $\epsilon\epsilon' \sum_{\gamma=1}^{n} \sum_{\beta=1}^{n} w_\gamma v_\beta \left[T_\gamma, T_\beta\right]$ belongs to a neighborhood of the identity, so it belongs to the group, consequently, this object can be written as a linear combination of the matrices T_τ, so there must exist coefficients $C_\tau(\bar{v}, \bar{w})$ such that

$$\sum_{\gamma=1}^{n} \sum_{\beta=1}^{n} w_\gamma v_\beta \left[T_\gamma, T_\beta\right] = \sum_{\tau=1}^{n} C_\tau(\bar{v}, \bar{w}) T_\tau, \ \forall\, \bar{v}, \bar{w}. \tag{14.129}$$

Since the vectors $\bar{v}$ and $\bar{w}$ are arbitrary, we set their components equal to Kronecker deltas, i.e., $w_{\gamma,\alpha} = \delta_{\gamma\alpha}$, $v_{\beta,\delta} = \delta_{\beta\delta}$, then

$$\sum_{\tau=1}^{n} C_{\alpha\beta\tau} T_\tau = \sum_{\gamma=1}^{n} \sum_{\beta=1}^{n} \delta_{\gamma\alpha} \delta_{\beta\delta} \left[T_\gamma, T_\delta\right] =$$

$$= \left[T_\alpha, T_\beta\right]. \tag{14.130}$$

We have shown that the commutator of the generators $\{T_\alpha\}$ of L is itself a generator, then the vector space L together with the application of the commutator of matrices is a Lie algebra, in particular, it is the Lie algebra of the Lie group of dimension n of invertible matrices of dimension k. □

Theorem 14.8 shows that any Lie group of invertible $k \times k$ matrices admits a Lie algebra defined by its infinitesimal generators. We now illustrate this construction explicitly for some physically relevant groups. In particular, in the following sections we derive the Lie algebras of the rotation group $SO(3)$, of the Euclidean group $E(3)$, and of the special unitary group $SU(2)$ (see Sect. 14.1.1).

14.2.3 Lie Algebra of $SO(3)$

Let $\delta\omega$ be an infinitesimal angle, then an infinitesimal rotation between two bases $\{\mathbf{e}_i\}$ e $\{\mathbf{e}'_i\}$ can be written as

$$\mathbf{e}'_i = \mathbf{e}_i + \delta\boldsymbol{\omega} \times \mathbf{e}_i =$$

$$= \mathbf{e}_i + \sum_{k=1}^{3} \delta\omega_k \mathbf{e}_k \times \mathbf{e}_i =$$

$$
\begin{aligned}
&= \mathbf{e}_i + \sum_{k=1}^{3}\sum_{l=1}^{3} \delta\omega_k \epsilon_{kil} \mathbf{e}_l = \\
&= \sum_{l=1}^{3} \delta_{il} \mathbf{e}_l + \sum_{k=1}^{3}\sum_{l=1}^{3} \delta\omega_k \epsilon_{kil} \mathbf{e}_l = \\
&= \sum_{l=1}^{3} \left(\delta_{il} + \sum_{k=1}^{3} \epsilon_{kil} \delta\omega_k \right) \mathbf{e}_l = \\
&= \sum_{l=1}^{3} R_{il} \mathbf{e}_l,
\end{aligned} \tag{14.131}
$$

where

$$
R_{il} = \delta_{il} + \sum_{k=1}^{3} \epsilon_{kil} \delta\omega_k \tag{14.132}
$$

is the matrix element of an infinitesimal rotation and ϵ_{kil} is the Levi-Civita symbol. Note that from Eq. 14.132 it follows that the matrix element of the inverse of an infinitesimal rotation matrix is given by

$$
(R)^{-1}_{li} = \delta_{li} - \sum_{k=1}^{3} \epsilon_{kli} \delta\omega_k . \tag{14.133}
$$

By defining the infinitesimal generators in terms of the inverse rotation, we have

$$
\begin{aligned}
(T_k)_{li} &= \left. \frac{\partial (R)^{-1}_{li}}{\partial w_k} \right|_{\mathbf{w}=\mathbf{0}} = \\
&= -\epsilon_{kli},
\end{aligned} \tag{14.134}
$$

and from $k, l, i \in (1, 2, 3)$, the rotation group admits three infinitesimal generators, represented by 3×3 matrices. With this definition of the generators in terms of the inverse rotation, the Lie algebra of $SO(3)$ takes the form

$$
[T_k, T_l] = \sum_{m=1}^{3} \epsilon_{klm} T_m . \tag{14.135}
$$

The Lie algebra of $SO(3)$ is called $L_{SO(3)}$ or $so(3)$.

14.2.4 Lie Algebra of $E(3)$

Consider the elements $A \in E(3)$ infinitely close to the identity, i.e.,

$$A = \mathbb{1} + \sum_{\alpha=1}^{3} \delta w_\alpha \widetilde{I}_\alpha + \sum_{\alpha=1}^{3} \delta a_\alpha \widetilde{\pi}_\alpha , \tag{14.136}$$

where δw_α and δa_α are the components of infinitesimal three-dimensional vectors of rotations and translations, respectively. The objects

$$\widetilde{I}_\alpha = \begin{pmatrix} I_\alpha & 0 \\ 0 & 0 \end{pmatrix} \tag{14.137}$$

denote the 4×4 matrices that include the infinitesimal generators I_α of three-dimensional rotations, as well as the objects

$$(\widetilde{\pi}_\alpha)_{ij} = \delta_{i\alpha}\, \delta_{j4}, \quad i, j = 1, \ldots, 4 \tag{14.138}$$

denote the 4×4 matrices representing the infinitesimal generators of translations, where only the element in the α-th row and 4-th column is nonzero.

It is easily verified that such a generators satisfy

$$\begin{cases} \left[\widetilde{I}_\alpha, \widetilde{I}_\beta\right] = \epsilon_{\alpha\beta\gamma} \widetilde{I}_\gamma \\ \left[\widetilde{\pi}_\alpha, \widetilde{\pi}_\beta\right] = 0 \\ \left[\widetilde{I}_\alpha, \widetilde{\pi}_\beta\right] = \epsilon_{\alpha\beta\gamma} \widetilde{\pi}_\gamma \end{cases} . \tag{14.139}$$

14.2.5 Lie Algebra of $SU(2)$

Consider an element $U \in SU(2)$ in a neighborhood of the identity, which can be written as

$$U = \mathbb{1} + \epsilon T, \tag{14.140}$$

$$U^\dagger = \mathbb{1} + \epsilon T^\dagger, \tag{14.141}$$

where ϵ is a small parameter. We now determine the conditions that the matrix T must satisfy in order for U to belong to $SU(2)$. We now exploit the unitarity condition given in Eq. I.2 and we get

$$
\begin{aligned}
\mathbb{1} &= UU^\dagger = \\
&= (\mathbb{1} + \epsilon T)(\mathbb{1} + \epsilon T^\dagger) \simeq \\
&\simeq \mathbb{1} + \epsilon(T + T^\dagger),
\end{aligned} \tag{14.142}
$$

it follows $\epsilon(T + T^\dagger) = 0$, then $T = -T^\dagger$, that is, the unitary condition for the U matrix implies that the T matrix must be antihermitian, i.e.,

$$
U = \begin{pmatrix} 1 + \epsilon T_{11} & \epsilon T_{12} \\ \epsilon T_{21} & 1 + \epsilon T_{22} \end{pmatrix}. \tag{14.143}
$$

Let us compute the determinant of U and stop at the first order with respect to ϵ, i.e.,

$$
\begin{aligned}
\det U &= 1 + \epsilon T_{11} + \epsilon T_{22} = \\
&= 1 + \epsilon \operatorname{Tr} T,
\end{aligned} \tag{14.144}
$$

then $\operatorname{Tr} T = 0$. The object T is a traceless anti-Hermitian 2×2 matrix. Hence, the matrix T can be expressed as a linear combination of the Pauli matrices σ_α (see Theorem B.1) as follows

$$
T = \frac{i}{2} \sum_{\alpha=1}^{3} v_\alpha \sigma_\alpha, \tag{14.145}
$$

where the coefficients v_α are real numbers, and the factor i ensures that T is anti-Hermitian. The factor $\frac{1}{2}$ is conventionally introduced in the physics literature for normalization purposes. We have

$$
\begin{aligned}
U &= \mathbb{1} + \epsilon \sum_{\alpha}^{3} v_\alpha \left(\frac{i}{2} \sigma_\alpha \right) = \\
&= \mathbb{1} + \epsilon \sum_{\alpha=1}^{3} v_\alpha T_\alpha,
\end{aligned} \tag{14.146}
$$

with

$$
T_\alpha = \frac{i}{2} \sigma_\alpha, \tag{14.147}
$$

that is, the infinitesimal generators are proportional to the Pauli matrices, then they satisfy (see Eq. 1.56)

$$
[T_\alpha, T_\beta] = -\sum_{\gamma=1}^{3} \epsilon_{\alpha\beta\gamma} T_\gamma, \tag{14.148}
$$

which is the Lie algebra of the group $SU(2)$, and is called $L_{SU(2)}$ or $su(2)$.

14.2.6 *The Algebras of $SO(3)$ and $SU(2)$ are Isomorphic*

It has been shown that there is a $2:1$ homomorphism between $SU(2)$ and $SO(3)$ (see Theorem 14.4). Locally there is a $1:1$ correspondence between the elements of $SU(2)$ and $SO(3)$ ones. Since Lie algebras are derived from the germ of the identity and the germs of $SU(2)$ and $SO(3)$ are equivalent, then the two Lie algebras are equal. Locally the two groups have the same properties, globally, their topology is different.

14.3 From a Lie Algebra to a Lie Group

Given a Lie algebra L of dimension n, it is natural to investigate the existence of a Lie group G whose associated Lie algebra is L, with the elements of L interpreted as the infinitesimal generators of the group. By definition, the object $T \in L$ can be written as a linear combination of the basis of the infinitesimal generators, i.e., $T = \sum_{\alpha}^{n} \lambda_\alpha T_\alpha$. From the matrix exponential (see Appendix H), we introduce

Definition 14.15 (Exponential Map of a Lie Algebra) Let $\{T_\alpha\}_{\alpha=1}^{n}$ be a basis of a Lie algebra L (see Definition 14.11). The exponential map is the mapping

$$\exp:\ \bar{\lambda} = (\lambda_1, \ldots, \lambda_n) \in \mathbb{R}^n \to \exp\left(\sum_{\alpha}^{n} \lambda_\alpha T_\alpha\right), \tag{14.149}$$

which assigns to each element of the Lie algebra an element of the associated Lie group G (see Definition 14.3) via the matrix exponential.

We can now show

Theorem 14.9 (Local Lie Group of a Lie Algebra) *If $|\bar{\lambda}| < \epsilon$, the application 14.149 maps the elements of L to a Lie group and it is a $1:1$ application.*

Proof Consider the set

$$U_\epsilon = \left\{ e^{\sum_{\alpha}^{n} \lambda_\alpha T_\alpha} : |\bar{\lambda}| < \epsilon \right\}, \tag{14.150}$$

where $|\bar{\lambda}|$ denotes the Euclidean norm. First note that the identity e is in U_ϵ; second, if $g \in U_\epsilon$, then from $g^{-1} = e^{-\sum_{\alpha}^{n} \lambda_\alpha T_\alpha}$ and $|-\bar{\lambda}| = |\bar{\lambda}| < \epsilon$ it follows $g^{-1} \in U_\epsilon$, i.e., U_ϵ is a neighborhood of the identity in the local Lie group generated by L, i.e., it is a germ. We call G the Lie group of such a germ. Consequently, from any set of matrices with well-defined structure constants, by means of the exponential map, we can construct a group that has T_α as generators. Given $g_1, g_2 \in U_\epsilon$, with $g_1 = g(\bar{\lambda_1})$, $g_2 = g(\bar{\lambda_2})$, the product $g = g(\bar{\lambda}) = g_1 g_2$ may not be in U_ϵ, however, for continuity, it must be

$$\exists\, \delta < \epsilon,\ U_\delta \subset U_\epsilon : g_1, g_2 \in U_\delta \Rightarrow g_2 g_1 \in U_\epsilon. \tag{14.151}$$

Since $g_2 g_1 \in U_\epsilon$, its canonical coordinates are unique, so there exists $\phi_\alpha = \phi_\alpha(\bar{\lambda}_1, \bar{\lambda}_2)$ such that $e^{\sum_\alpha^n \phi_\alpha T_\alpha} = g_2 g_1$. Since G is a Lie group, the functions $\phi^\alpha(\bar{\lambda}_1, \bar{\lambda}_2)$ must be analytic. Now, we show that the Taylor's expansion of such functions is uniquely determined by the structure constants $C_{\alpha\beta\gamma}$ of the Lie algebra L. Let T_1 and T_2 be the matrices related to g_1 and g_2, respectively, then let us compute the element $T \in L$ such that

$$\begin{aligned} g_2 g_1 &= e^{T_2} e^{T_1} = \\ &= e^T. \end{aligned} \tag{14.152}$$

We multiply both members to the right by e^{-T} as follows

$$e^{T_2} e^{T_1} e^{-T} = \mathbb{1}. \tag{14.153}$$

We parameterize g_1 and g_2 as follows

$$g_1(s) = e^{sT_1(s)}, \tag{14.154}$$

$$g_2(s) = e^{sT_2(s)}, \tag{14.155}$$

with $s \in [0, 1]$. We set an Ansatz

$$\tilde{T}(s) = s(T_2 + T_1) + T(s), \tag{14.156}$$

then

$$e^{sT_2} e^{sT_1} e^{-s(T_2+T_1)-T(s)} = \mathbb{1},\ \forall s \in [0, 1]. \tag{14.157}$$

A differential equation for $T(s)$ can be derived, i.e.,

$$\begin{aligned} 0 &= e^{sT_2} T_2 e^{sT_1} e^{-s(T_1+T_2)-T(s)} + e^{sT_2} e^{sT_1} T_1 e^{-s(T_1+T_2)-T(s)} \\ &\quad + e^{sT_2} e^{sT_1} \left(-T_2 - T_1 - \frac{dT(s)}{ds} \right) e^{-s(T_2+T_1)-T(s)} = \\ &= e^{sT_2} T_2 e^{sT_1} e^{-s(T_1+T_2)-T(s)} - e^{sT_2} e^{sT_1} \left(T_2 + \frac{dT(s)}{ds} \right) e^{-s(T_2+T_1)-T(s)} = \end{aligned}$$

$$= e^{sT_2}e^{sT_1}\left(e^{-sT_1}T_2e^{sT_1} - T_2 - \frac{dT(s)}{ds}\right)e^{-s(T_2+T_1)-T(s)}, \tag{14.158}$$

where in the last step we manipulated the term T_2 by means of the identity $e^{-sT_1}e^{sT_1}$. It must be

$$\begin{aligned}\frac{dT(s)}{ds} &= e^{-sT_1}T_2e^{sT_1} - T_2 = \\ &= U(s) - T_2.\end{aligned} \tag{14.159}$$

Let us calculate $U(s)$, we differentiate $U(s)$ with respect to s as follows

$$\frac{dU(s)}{ds} = [U(s), T_1], \tag{14.160}$$

then, note that $U(0) = T_2$, which is the initial condition to the Cauchy problem

$$\begin{cases} U(s) = T_2 + \int_0^s [U(\tau), T_1]\, d\tau \\ U(0) = T_2 \end{cases}. \tag{14.161}$$

Consider the function $U(s)$ as a series expansion with respect to s and proceed iteratively, i.e.,

$$U(0) = T_2, \tag{14.162}$$

$$U(1) = T_2 + s[T_2, T_1], \tag{14.163}$$

and we insert $U(1)$ into $U(s)$, then integrate. By performing an iteration, we get

$$U(s) = \sum_{n=1}^{\infty} \frac{s^n}{n!}\, T_1^n\{T_2\}, \tag{14.164}$$

with

$$T_1^n\{T_2\} = \underbrace{[[[\ldots}_{\text{n times}} T_2, T_1], T_1], \ldots, T_1]. \tag{14.165}$$

Since $U(s)$ is an iterated commutator, it belongs to the Lie algebra, that is, $U(s) \in L$. Now, we can solve the differential equation as follows

$$\begin{aligned}\frac{dT(s)}{ds} &= U(s) - T_2 = \\ &= T_2 + \sum_{n=1}^{\infty} \frac{s^n}{n!}\, T_1^n\{T_2\} - T_2 = \\ &= \sum_{n=1}^{\infty} \frac{s^n}{n!}\, T_1^n\{T_2\},\end{aligned} \tag{14.166}$$

which implies

$$T(s) = \sum_{n=1}^{\infty} \frac{s^{n+1}}{(n+1)!}\, T_1^n\{T_2\}, \tag{14.167}$$

with

$$T(0) = 0, \tag{14.168}$$

then the object $\tilde{T}$ can be written as

$$\begin{aligned}\tilde{T} &= \hat{T}(s=1) = \\ &= T_1 + T_2 + \sum_{n=1}^{\infty} \frac{1}{(n+1)!}\, T_1^n\{T_2\}.\end{aligned} \tag{14.169}$$

Considering only the first term of the series, we have

$$\begin{aligned}\tilde{T} &= \sum_{\alpha} \left(\lambda_{1,\alpha} T_\alpha + \lambda_{2,\alpha} T_\alpha\right) + \frac{1}{2}[T_2, T_1] + o(|\overline{\lambda}|^3) = \\ &= \sum_{\alpha} \left(\lambda_{1,\alpha} T_\alpha + \lambda_{2,\alpha} T_\alpha\right) + \frac{1}{2}\left[\sum_{\beta} \lambda_{2,\beta} T_\beta, \sum_{\gamma} \lambda_{1,\gamma} T_\gamma\right] + o(|\overline{\lambda}|^3) = \\ &= \sum_{\alpha} \left(\lambda_{1,\alpha} T_\alpha + \lambda_{2,\alpha} T_\alpha\right) + \frac{1}{2} \sum_{\beta,\gamma,\alpha} \lambda_{2,\beta} \lambda_{1,\gamma} C_{\beta\gamma\alpha} T_\alpha + o(|\overline{\lambda}|^3).\end{aligned} \tag{14.170}$$

Regarding the second order term, we have

$$\begin{aligned}\frac{1}{6}\left[[T_2, T_1], T_1\right] &= \frac{1}{6}\left[\left[\sum_{\epsilon} \lambda_{2,\epsilon} T_\epsilon, \sum_{\tau} \lambda_{1,\tau} T_\tau\right], \sum_{\gamma} \lambda_{1,\gamma} T_\gamma\right] = \\ &= \frac{1}{6} \sum_{\epsilon,\tau,\gamma} \lambda_{2,\epsilon} \lambda_{1,\tau} \lambda_{1,\gamma} \left[[T_\epsilon, T_\tau], T_\gamma\right] =\end{aligned}$$

$$= \frac{1}{6} \sum_{\epsilon,\tau,\gamma} \lambda_{2,\epsilon} \lambda_{1,\tau} \lambda_{1,\gamma} \left[\sum_{\xi} C_{\epsilon\tau\xi} T_{\xi}, T_{\gamma} \right] =$$

$$= \frac{1}{6} \sum_{\epsilon,\tau,\gamma,\xi} \lambda_{2,\epsilon} \lambda_{1,\tau} \lambda_{1,\gamma} C_{\epsilon\tau\xi} \left[T_{\xi}, T_{\gamma} \right] =$$

$$= \frac{1}{6} \sum_{\epsilon,\tau,\gamma,\xi,\alpha} \lambda_{2,\epsilon} \lambda_{1,\tau} \lambda_{1,\gamma} C_{\epsilon\tau\xi} C_{\xi\gamma\alpha} T_{\alpha}, \qquad (14.171)$$

which turns out to be uniquely determined by the structure constants of the Lie algebra. The same occurs for every order of the series, then the local $1:1$ correspondence between Lie algebra L and Lie group G is proved. □

We have thus shown that, in a neighborhood of the identity, the Lie algebra L uniquely determines the local structure of a Lie group G. The exponential map establishes a one-to-one correspondence between elements of the algebra and elements of the group, and the local group multiplication law is completely encoded by the structure constants of L, as illustrated by the series expansion in iterated commutators. This demonstrates the existence of a local Lie group associated with any given Lie algebra, whose analytic group law is fully consistent with the algebraic structure of L.

14.3.1 *Connectivity of a Lie Group*

Here we discuss the global properties of the exponential map associated with a basis of a Lie algebra. It is well known that the Lie groups $SU(2)$ and $SO(3)$ are locally isomorphic; that is, they share the same Lie algebra and hence identical structure constants. However, they differ at the global level, as their group manifolds and matrix representations are not isomorphic. The fundamental distinction between these groups lies in their topological connectivity properties.

Let G be a Lie group, which can be viewed as a smooth geometric object with a group structure. Consider a fixed point in G, typically the identity element, and closed curves (loops) that start and end at this point. Two such curves are said to be equivalent if one can be continuously deformed into the other without breaking or detaching from the base point; otherwise, they are not equivalent. This notion partitions all curves into equivalence classes. We define the connectivity of the Lie group as the number of equivalence classes of closed curves based at the identity. A common geometric visualization of $SO(3)$ is provided by the axis–angle representation, in which rotations are parametrized by vectors whose direction determines the rotation axis and whose magnitude corresponds to the rotation angle. In this representation, rotations by an angle π lie on a sphere, and antipodal points on this sphere represent the same rotation. This description is purely illustrative and does not endow $SO(3)$ with a manifold structure. We define a continuous deformation as a family of curves parameterized continuously with respect to t, where $t = 0$ corresponds to the identity element and $t = 1$ to the curve itself. To make the classification of loops more intuitive, we introduce the notion of *jump curves* as a visual tool to distinguish between topological classes of loops in $SO(3)$. This terminology is informal, but it captures the essential idea of curves crossing the boundary where antipodal points are identified. We distinguish between 0-jump curves (C_0) and one-jump curves (C_1), where a jump corresponds to a crossing of the spherical surface associated with the identification of antipodal points. More precisely, a curve is said to have a jump each time it passes through the spherical surface, which identifies two antipodal points as the same rotation. An example of a C_0 curve is one that starts from the identity and returns to it without crossing the spherical surface. In contrast, a C_1 curve starts at the identity, reaches the spherical surface (crossing it once), and then continues from the identified antipodal point to return to the origin. It can be shown that curves with an even number of jumps belong to the equivalence class C_0, while curves with an odd number of jumps belong to C_1. Therefore, there are exactly two equivalence classes of closed curves in $SO(3)$, implying that

$$\text{conn}[SO(3)] = 2, \tag{14.172}$$

on the other hand, it can be shown that

$$\text{conn}[SU(2)] = 1. \tag{14.173}$$

14.3.2 Universal Covering Group

Before introducing the notion of a universal covering group, we recall two basic definitions that will be used throughout this section.

Definition 14.16 (Kernel of a Group Homomorphism) If

$$T : G \longrightarrow H \tag{14.174}$$

is a group homomorphism (see Definition 14.5), then the kernel of T is given by

$$\ker(T) = \{\, g \in G \mid T(g) = e_H \,\}, \tag{14.175}$$

where e_H denotes the identity element of H.

Definition 14.17 (Left Coset) Let G be a group (see Definition 14.2) and let $H \subset G$ be a subgroup. For any element $g \in G$, the left coset of H with respect to g is defined as

$$gH = \{gh \mid h \in H\}. \tag{14.176}$$

Definition 14.18 (Universal Covering Group) Given two groups (see Definition 14.2) $\overline{G}$ and G, we say that $\overline{G}$ is the *universal covering group* of G if:

- $\overline{G}$ is simply connected, i.e., $\text{conn}[\overline{G}] = 1$;
- there exists a group homomorphism

$$T : \overline{G} \to G \tag{14.177}$$

  such that $\ker(T)$ is a discrete subgroup with $h_1, \ldots, h_k$ distinct elements, and

$$k = \text{conn}[G]. \tag{14.178}$$

Any Lie group (see Definition 14.3) that is not simply connected admits a universal covering group. The universal covering group of a Lie group is itself a Lie group and is unique up to isomorphism. Given a Lie algebra (see Definition 14.11), there exists a unique simply connected Lie group whose Lie algebra is generated by it. To summarize, the process of constructing Lie groups from a given Lie algebra proceeds as follows:

(a) Apply the exponential map to the generators of the Lie algebra L, yielding a simply connected Lie group $\overline{G}$:

$$\exp : L \to \overline{G}; \tag{14.179}$$

(b) Construct group homomorphisms $T_i : \overline{G} \to G_i$ such that $\ker(T_i)$ is a discrete normal subgroup of $\overline{G}$;
(c) Define the quotient group

$$G_i = \overline{G}/\ker(T_i), \tag{14.180}$$

which is a Lie group with the same Lie algebra as $\overline{G}$.

In this way, we identify all possible Lie groups sharing a given Lie algebra. Among these, the simply connected one is unique (up to isomorphism). All such groups share the same universal covering group.

We now prove point (c).

- Normal subgroup. Let $f : \overline{G} \to G$ be a group homomorphism. The kernel of f, denoted $\ker(f)$, is the set of elements in $\overline{G}$ that are mapped to the identity element $e \in G$. That is,

$$\ker(f) = \{\overline{g} \in \overline{G} \mid f(\overline{g}) = e\}. \tag{14.181}$$

  We now show that:

(i) $\ker(f)$ is a subgroup of $\overline{G}$, that is,

$$\overline{e} \in \ker(f), \quad \overline{h}_1, \overline{h}_2 \in \ker(f) \Rightarrow \overline{h}_1\overline{h}_2 \in \ker(f); \tag{14.182}$$

(ii) $\ker(f)$ is a normal (invariant) subgroup of $\overline{G}$, i.e., for all $\overline{g} \in \overline{G}$ and $\overline{h} \in \ker(f)$,

$$\overline{g}\,\overline{h}\,\overline{g}^{-1} \in \ker(f). \tag{14.183}$$

Indeed, since f is a homomorphism, we get

$$\begin{aligned} f(\overline{g}\,\overline{h}\,\overline{g}^{-1}) &= f(\overline{g})f(\overline{h})f(\overline{g}^{-1}) = \\ &= f(\overline{g})ef(\overline{g})^{-1} = \\ &= f(\overline{g})f(\overline{g})^{-1} = \\ &= f(\overline{e}) = \\ &= e, \end{aligned} \tag{14.184}$$

then the conjugate element lies in the kernel.

- Quotient group. The kernel $\ker(f)$ induces an equivalence relation on $\overline{G}$: two elements $\overline{g}_1, \overline{g}_2 \in \overline{G}$ are equivalent if and only if

$$\overline{g}_1\overline{g}_2^{-1} \in \ker(f), \tag{14.185}$$

i.e.,

$$\overline{g}_1 \sim \overline{g}_2 \quad \Leftrightarrow \quad \exists\, \overline{h} \in \ker(f) \text{ such that } \overline{g}_1 = \overline{h}\overline{g}_2. \tag{14.186}$$

This partitions $\overline{G}$ into left cosets (see Definition 14.17) of $\ker(f)$. The set of all such forms the quotient group

$$Q = \overline{G}/\ker(f), \tag{14.187}$$

whose elements are the equivalence classes in $\overline{G}$ under this relation. To verify that Q is a group, consider two cosets $g_1 = \overline{h}_1\overline{g}_1$, $g_2 = \overline{h}_2\overline{g}_2$, with $\overline{h}_1, \overline{h}_2 \in \ker(f)$. Then, the product of the two cosets is

$$\begin{aligned} g_1g_2 &= (\overline{h}_1\overline{g}_1)(\overline{h}_2\overline{g}_2) = \\ &= \overline{h}_1(\overline{g}_1\overline{h}_2\overline{g}_1^{-1})(\overline{g}_1\overline{g}_2), \end{aligned} \tag{14.188}$$

where we used the associativity of the group product. Since $\ker(f)$ is normal, $\overline{g}_1\overline{h}_2\overline{g}_1^{-1} \in \ker(f)$, and the product lies in the same coset class:

$$g_1 g_2 \in Q. \tag{14.189}$$

If $\ker(f)$ is discrete and $\overline{G}$ is simply connected, then the quotient group $G = \overline{G}/\ker(f)$ is a Lie group with the same Lie algebra as $\overline{G}$.

The group $SU(2)$ is simply connected and is the universal covering group of $SO(3)$. Its only nontrivial discrete normal subgroup is

$$\overline{H} = \{\mathbb{1}, -\mathbb{1}\}, \tag{14.190}$$

and the corresponding quotient is

$$SU(2)/\overline{H} = SO(3). \tag{14.191}$$

There are no other discrete normal subgroups of $SU(2)$, so the only Lie groups with the Lie algebra of $SU(2)$ are $SU(2)$ and $SO(3)$.

14.4 From Lie Group Representations to Lie Algebra Representations

Let G' be the representation of the Lie group G by the homomorphism S (see Definition 14.6), and in particular G' is a group of invertible linear operators on a vector space V (see Definition 14.4). If V has finite dimension, we can take G' as a group of invertible matrices; if V has infinite dimension, then G' is a group of invertible linear operators on V. We have

Theorem 14.10 *Given a Lie group (see Definition 14.3) and its representation (see Definition 14.6), G and G' respectively, let L and L' be the respective Lie algebras (see Definition 14.11), then a representation of G induces a representation of L (see Definition 14.12).*

Proof Let G be a group of matrices and $g \in G$ an element close to the identity, i.e.,

$$g = \mathbb{1} + \epsilon T, \ T \in L. \tag{14.192}$$

Now, the representation is a homomorphism, the identity of G is mapped to the identity of G', and the homomorphism is continuous, so elements of G near the identity e go into elements of G' near the identity $e' = S(e)$

$$S(g) = \mathbb{1}' + \epsilon T', \ T' \in L'. \tag{14.193}$$

Consequently

$$L \ni T \to T' \in L' \tag{14.194}$$

is a map. We show that such a map:

(i) is linear;
(ii) is a homomorphism between Lie algebras.

(i) Given

$$g_1 = \mathbb{1} + \epsilon_1 T_1, \tag{14.195}$$

$$g_2 = \mathbb{1} + \epsilon_2 T_2, \tag{14.196}$$

$$g_1' = \mathbb{1}' + \epsilon_1 T_1', \tag{14.197}$$

$$g_2' = \mathbb{1}' + \epsilon_2 T_2', \tag{14.198}$$

S is a homomorphism, then

$$S(g_1 g_2) = S(g_1) S(g_2). \tag{14.199}$$

The product of elements near the identity is, to first order in ϵ_i,

$$g_1 g_2 \simeq \mathbb{1} + \epsilon_1 T_1 + \epsilon_2 T_2, \tag{14.200}$$

which under the representation S maps to

$$g_1' g_2' \simeq \mathbb{1}' + \epsilon_1 T_1' + \epsilon_2 T_2', \tag{14.201}$$

then the map

$$s : \epsilon_1 T_1 + \epsilon_2 T_2 \rightarrow \epsilon_1 T_1' + \epsilon_2 T_2' \tag{14.202}$$

is a linear map.

(ii) We show that s is a homomorphism and in particular that it preserves commutators. We introduce the quantities

$$B(\epsilon_1, \epsilon_2) = g_1^{-1} g_2^{-1} g_1 g_2, \tag{14.203}$$

$$B'(\epsilon_1, \epsilon_2) = g_1'^{-1} g_2'^{-1} g_1' g_2'. \tag{14.204}$$

At the first order in ϵ_1, ϵ_2, i.e., neglecting terms of order $O(\epsilon_i^3)$, we have

$$B(\epsilon_1, \epsilon_2) = \mathbb{1} + \epsilon_1 \epsilon_2 \left(T_1 T_2 - T_2 T_1\right), \tag{14.205}$$

$$B'(\epsilon_1, \epsilon_2) = \mathbb{1}' + \epsilon_1 \epsilon_2 \left(T_1' T_2' - T_2' T_1'\right), \tag{14.206}$$

then

$$S : B \to B' \tag{14.207}$$

implies that

$$s : [T_1, T_2] \to [T_1', T_2'] \tag{14.208}$$

is a homomorphism of Lie algebras.

□

Chapter 15
Symmetry Principles in Quantum Physics

Abstract This chapter examines the fundamental role of symmetries in quantum mechanics. It begins with a rigorous proof of Wigner's theorem, which shows that any symmetry transformation preserving quantum probabilities must be represented by a unitary or antiunitary operator. This establishes a deep connection between physical symmetries and their mathematical representations. The chapter then introduces the operators associated with key spatial symmetries: translations and rotations. These operators act on both quantum states and physical observables, such as position, linear momentum, and angular momentum, transforming them according to precise rules. Using the group theory framework, the link between these symmetries and their infinitesimal generators (the linear and angular momentum operators) becomes evident, emphasizing their role as conserved quantities. Finally, the chapter presents applications, such as the conservation of linear momentum from translational invariance and the implications of time-reversal symmetry for quantum dynamics and spectra. Together, these concepts highlight how symmetries constrain and shape the physical behaviour of quantum systems.

15.1 Symmetry Transformation and Wigner's Theorem

Given two observers O and O', their measuring instruments are said to be identical if they yield the same measurement outcomes, and their physical descriptions are considered equivalent. Mathematically, this corresponds to the existence of a symmetry transformation relating the two measurement processes. Such a transformation leaves invariant a property of the physical system, regardless of the observer. Let A, B be observables acting on the Hilbert space H associated with observer O, and A', B' be observables acting on the Hilbert space H' associated with observer O'. Assume these observables have nondegenerate spectra, then

$$A\,|\varphi_r\rangle = \lambda_r\,|\varphi_r\rangle, \tag{15.1}$$

$$B\,|\psi_s\rangle = \mu_s\,|\psi_s\rangle, \tag{15.2}$$

F. Tafuri et al., *Understanding Quantum Many-Body Theory*, UNITEXT for Physics, https://doi.org/10.1007/978-3-032-17849-7_15

$$A' \, |\varphi_r'\rangle = \lambda_r \, |\varphi_r'\rangle, \tag{15.3}$$

$$B' \, |\psi_s'\rangle = \mu_s \, |\psi_s'\rangle. \tag{15.4}$$

In quantum mechanics, the act of measurement is related not only to the numerical values of the spectrum but also to the measurement probabilities associated with the eigenstates. Specifically, if observer O first measures the observable A and immediately afterward measures the observable B, the probability of obtaining the eigenvalue μ_s following the eigenvalue λ_r is given by the transition (or superposition) probability $|\langle\psi_s|\varphi_r\rangle|^2$. If the two physical descriptions are equivalent, this probability must be preserved under the transformation, so that

$$|\langle\psi_s|\varphi_r\rangle|^2 = |\langle\psi_s'|\varphi_r'\rangle|^2. \tag{15.5}$$

Since state vectors are defined only up to an overall phase factor, it is natural to consider equivalence classes of vectors of fixed norm $\|\psi\|^2 = \langle\psi|\psi\rangle$ rather than the vectors themselves. We denote by $r = \|\psi\|^2$ the norm squared (or radius) of the equivalence class of vectors of the form $e^{i\alpha}|\psi\rangle$, with $\alpha \in \mathbb{R}$, and call any vector $|\psi\rangle$ a representative of its equivalence class of radius r. Consequently, a symmetry transformation establishes a one-to-one correspondence between the equivalence classes (radii) of H and those of H' such that Eq. 15.5 holds. We now proceed to characterize such symmetry transformations. We prove the result in the finite-dimensional case. The statement, known as Wigner's theorem, remains valid in arbitrary Hilbert spaces, although its proof requires additional technical assumptions.

Theorem 15.1 (Wigner's Theorem) *Given a symmetry transformation, that is, a map between rays of Hilbert spaces H and H' of dimension n that preserves the overlap function, i.e.,*

$$|\langle\psi|\varphi\rangle|^2 = \left|\langle\psi'|\varphi'\rangle\right|^2, \tag{15.6}$$

such a map can be represented by an operator

$$\hat{\Omega} : H \to H', \tag{15.7}$$

either linear and unitary (see Eqs. I.1–I.3), or antilinear and antiunitary (see Eqs. I.4 and I.5).

Proof We divide the proof into several steps.

1. Let $\{|i\rangle\}$ be a basis in H and let $\{|i'\rangle\}$ be the vectors transformed through a symmetry transformation. We show that the symmetry transforms complete orthonormal systems into complete orthonormal systems, i.e., the superscripted system is a basis in H'. From the conservation of the overlap function it follows that

$$|\langle i'|j'\rangle|^2 = |\langle i|j\rangle|^2 = \\ = \delta_{ij}, \tag{15.8}$$

then the orthonormality is preserved. Now, suppose that $\{|i'\rangle\}$ is not complete, then there exists a vector $|a'\rangle$ orthogonal to all superscripted versors

$$0 = |\langle a'|i'\rangle|^2, \tag{15.9}$$

and the transformed of $|a'\rangle$, i.e., $|a\rangle$, should satisfy the equality

$$0 = |\langle a|i\rangle|^2, \tag{15.10}$$

which is an absurdity because by hypothesis $\{|i\rangle\}$ is complete. So, the system $\{|i'\rangle\}$ is complete and orthonormal, i.e., it is a basis.

2. We need to establish a convention for fixing the phase of vectors in H'. Specifically, we choose, for each basis vector in H', a representative that satisfies

$$|1\rangle + |j\rangle \longrightarrow |1'\rangle + |j'\rangle, \ \forall\, j. \tag{15.11}$$

Consider

$$|\psi_j\rangle = \frac{1}{\sqrt{2}}\left[|1\rangle + |j\rangle\right], \ j \neq 1, \tag{15.12}$$

then we expand the transformed $|\psi_j'\rangle$ respect to the basis $\{|i'\rangle\}$, i.e.,

$$|\psi_j'\rangle = \sum_l c_{jl}|l'\rangle. \tag{15.13}$$

From the conservation of the overlap function, it follows

$$|c_{jj}|^2 = |c_{j1}|^2 = \\ = \frac{1}{\sqrt{2}}, \tag{15.14}$$

$$|c_{jl}|^2 = 0, \ j \neq l, \ j \neq 1, \tag{15.15}$$

we can choose the representatives so that

$$c_{jj} = c_{jl} = 1, \tag{15.16}$$

fixing the relative phases consistently. Then

$$|1\rangle + |j\rangle \longrightarrow |1'\rangle + |j'\rangle, \ \forall\, j. \tag{15.17}$$

3. Let us see how any state with real components is transformed under the symmetry transformation. Given

$$|u\rangle = \sum_j c_j |j\rangle, \tag{15.18}$$

and its transformed

$$|u'\rangle = \sum_{j'} c'_j |j'\rangle, \tag{15.19}$$

we get

$$|\langle \psi_j | u\rangle|^2 = |c_1 + c_j|^2, \tag{15.20}$$

$$|\langle \psi'_j | u'\rangle|^2 = |c'_1 + c'_j|^2, \tag{15.21}$$

which together with

$$\begin{aligned} |\langle j | u\rangle|^2 &= |c_j|^2 = \\ &= |c'_{j'}|^2 = \\ &= |\langle j' | u'\rangle|^2, \ \forall\, j \end{aligned} \tag{15.22}$$

implies

$$\begin{cases} |c_j|^2 = |c'_{j'}|^2, \ \forall\, j \\ |c_1|^2 = |c'_{1'}|^2, \\ |c_1 + c_j|^2 = |c'_{1'} + c'_{j'}|^2, \ \forall\, j \neq 1 \end{cases} . \tag{15.23}$$

If we chose the representative of $|u'\rangle$ such that

$$c'_{1'} = c_1, \tag{15.24}$$

$$c'_{j'} = c_j, \tag{15.25}$$

the last equality of the system is valid if and only if all the coefficients are real. Note that in the case of real components, the operator acts as a unitary transformation (up to a global phase).

4. We generalize step 3 to a symmetry transformation of states with any complex components. Given j, k, consider the states

$$|j\rangle + |j+1\rangle + \ldots + |j+k\rangle \longrightarrow |j'\rangle + |j'+1\rangle + \ldots + |j'+k'\rangle, \tag{15.26}$$

where the images are written by choosing each representative with real components. Consider the scalar product of $|j\rangle + |j+1\rangle + \ldots + |j+k\rangle$ by

$$|u\rangle = \sum_l c_l |l\rangle, \tag{15.27}$$

then

$$\langle u| \left(|j\rangle + |j+1\rangle + \ldots + |j+k\rangle\right) = \sum_{s=j}^{j+l} c_s, \tag{15.28}$$

and similarly

$$\langle u'| \left(|j'\rangle + |j'+1\rangle + \ldots + |j'+k'\rangle\right) = \sum_{s'=j'}^{j'+l'} c_{s'}. \tag{15.29}$$

We represent the two summations as broken lines Γ and Γ' in the complex plane. Each segment of Γ, labeled by a given index, corresponds to a segment of Γ' with the same modulus, owing to the correspondence between the radii. By probability conservation, i.e., the conservation of transition probabilities, the total length of Γ and Γ' must therefore have the same modulus. Such condition can be satisfied in only two possible ways:

- The coefficients c_i and $c'_{i'}$ differ by a global phase rotation

$$c'_{i'} = e^{i\alpha} c_i, \tag{15.30}$$

  which, of course, preserves the squared moduli and hence the length of the broken line.
- The broken line Γ' is obtained from Γ by a combination of rotation and reflection in the complex plane. As a result, each coefficient transforms as

$$c'_{i'} = e^{i\beta} c_i^*, \tag{15.31}$$

  which also preserves the moduli and the total length.

Since the coefficients are defined up to a phase, we can choose the representatives such that $\alpha = \beta = 0$ so there are only two cases, i.e.,

$$c'_{i'} = c_i, \tag{15.32}$$

and

$$c'_{i'} = c_i^*. \tag{15.33}$$

It remains to be shown that the two possibilities are mutually exclusive, that is, the choice of a transformation for the coefficients of a state involves all vectors of the H space.

5. Given

$$|j\rangle + e^{i\alpha}|k\rangle, \quad \alpha \neq n\pi, \tag{15.34}$$

let us assume that this vector is transformed through a linear transformation, i.e., $c'_j = c_j$, $c'_k = c_k$. According to Step 2, using the convention of fixing the phases of the representatives for these vectors, we have

$$|j\rangle + e^{i\alpha}|k\rangle \longrightarrow |j'\rangle + e^{i\alpha}|k'\rangle. \tag{15.35}$$

Given a vector $|u\rangle$, assume that it transforms through an antilinear transformation, and consider the scalar product

$$\left(\langle j| + e^{-i\alpha}\langle k|\right)|u\rangle = c_j + e^{-i\alpha}c_k, \tag{15.36}$$

$$\left(\langle j'| + e^{-i\alpha}\langle k'|\right)|u'\rangle = c'_{j'} + e^{-i\alpha}c'_{k'} = c_j^* + e^{-i\alpha}c_k^*. \tag{15.37}$$

where in the last step we used the antilinearity condition. From the conservation of the transition probability we have

$$|c_j + e^{-i\alpha}c_k|^2 = |c_j^* + e^{-i\alpha}c_k^*|^2, \tag{15.38}$$

$$c_j c_k^* e^{i\alpha} + c_j^* c_k e^{-i\alpha} = c_j^* c_k e^{i\alpha} + c_j c_k^* e^{-i\alpha}, \tag{15.39}$$

and let γ be the phase between c_j and c_k, meaning that

$$\cos(\gamma + \alpha) = \cos(\gamma - \alpha). \tag{15.40}$$

We note that it must be $\gamma \neq m\pi$, with $m \in \mathbb{Z}$, since otherwise it would be $c_i = c_i^*$ and we are assuming instead that the vectors have components with non-zero imaginary part. In addition, the condition

$$\cos(\gamma + \alpha) = \cos(\gamma - \alpha) \tag{15.41}$$

implies

$$\begin{cases} \gamma + \alpha = \alpha - \gamma + 2m\pi \\ \gamma + \alpha = \gamma - \alpha + 2m\pi \end{cases}, \quad m \in \mathbb{Z}, \tag{15.42}$$

$$\begin{cases} \gamma = m\pi \\ \alpha = m\pi \end{cases}, \quad m \in \mathbb{Z}, \tag{15.43}$$

this contradiction shows that mixed linear/antilinear actions are impossible, so $|u\rangle$ must transform consistently with the linear case.

□

15.2 Wigner's Operator for Rotations

Given two reference systems at rest, rotated with respect to each other by an arbitrary angle, and assuming the measurement equivalence principle for such rotations, Wigner's theorem 15.1 ensures the existence of an operator $\hat{\Omega}$ relating the corresponding measurement states. Consider now three observers at rest, denoted O, O', and O'', rotated with respect to each other by arbitrary angles around a common axis. Let R_1 map O to O' and R_2 map O' to O''. By Wigner's theorem 15.1, each rotation corresponds to a unitary or antiunitary operator, denoted $\hat{\Omega}_1$ and $\hat{\Omega}_2$, respectively. Since O and O'' are also related by a rotation, and rotations form a group (see Sect. 14.1.1), the transformation from O to O'' is given by the composition $R_2 R_1$. Therefore, there exists an operator $\hat{\Omega}^*$ relating the Hilbert spaces of O and O''. Given a state $|\psi\rangle$ in the reference frame of O, the action of $\hat{\Omega}_1$ yields the state $|\psi'\rangle$ in the frame of O', and subsequently, the action of $\hat{\Omega}_2$ on $|\psi'\rangle$ produces the state $|\psi''\rangle$ in the frame of O''. Consequently, we have

$$\begin{aligned} |\psi''\rangle &= \hat{\Omega}_2|\psi'\rangle = \\ &= \hat{\Omega}_2\hat{\Omega}_1|\psi\rangle, \end{aligned} \tag{15.44}$$

$$\hat{\Omega}_2\hat{\Omega}_1 = e^{i\gamma}\hat{\Omega}^*. \tag{15.45}$$

On the other hand,

$$\hat{\Omega}(R_2)\hat{\Omega}(R_1) = e^{i\gamma}\hat{\Omega}(R_2 R_1), \tag{15.46}$$

that is, the operators $\hat{\Omega}$ provide a projective representation (see Eq. 14.15) of the rotation group $SO(3)$ through the homomorphism

$$R \to \hat{\Omega}(R). \tag{15.47}$$

Theorem 15.2 *The $\hat{\Omega}(R)$ representation of $SO(3)$ (see Eq. 14.36) is unitary (see Eqs. I.2 and I.3).*

Proof An antiunitary matrix cannot become the identity matrix with continuity. The rotation group is a connected group: any rotation can gradually recur to the identity with continuity, and since the trivial rotation, i.e., the identity, must be matched by $\hat{\Omega}(0) = e^{i\gamma}\mathbb{1}$, $\hat{\Omega}(R)$ must be a unitary operator. □

15.3 Mathematical Derivation of the Angular Momentum Operator

Here, the angular momentum operator will be deduced from the general theory of rotation symmetry. Consider a Hilbert space H, on which a local representation of the rotations is provided by a unitary operator $\hat{R}$, with $\hat{R} = \hat{\Omega}(R)$ (see Theorem 15.2). Given an infinitesimal rotation 14.132, the infinitesimal rotation operator $\hat{R}$ has the form

$$\hat{R} = \mathbb{1} + \sum_{\alpha=1}^{3} \delta w_\alpha \hat{I}_\alpha, \tag{15.48}$$

where $\hat{I}_\alpha$ denote the infinitesimal generators of the representation of the rotation group. Given a representation of the group $SO(3)$ (see Sect. 14.1.1), the associated infinitesimal generators define a representation of its Lie algebra $L_{SO(3)}$ (see Sect. 14.2.3). As a consequence, such operators satisfy the corresponding commutation relations (see Eq. 14.135)

$$\left[\hat{I}_\alpha, \hat{I}_\beta\right] = \sum_{\gamma=1}^{3} \epsilon_{\alpha\beta\gamma} \hat{I}_\gamma. \tag{15.49}$$

Since $\hat{R}$ is unitary (see Eqs. I.2 and I.3), it follows

$$\begin{aligned}
\mathbb{1} &= \left(\mathbb{1} + \sum_{\alpha=1}^{3} \delta w_\alpha \hat{I}_\alpha\right)\left(\mathbb{1} + \sum_{\alpha=1}^{3} \delta w_\alpha \hat{I}_\alpha^\dagger\right) = \\
&= \mathbb{1} + \sum_{\alpha=1}^{3} \delta w_\alpha \hat{I}_\alpha^\dagger + \sum_{\alpha=1}^{3} \delta w_\alpha \hat{I}_\alpha + \sum_{\alpha=1}^{3}\sum_{\beta=1}^{3} \delta w_\alpha \delta w_\beta \hat{I}_\alpha \hat{I}_\beta^\dagger \simeq \\
&\simeq \mathbb{1} + \sum_{\alpha=1}^{3} \delta w_\alpha \left(\hat{I}_\alpha + \hat{I}_\alpha^\dagger\right),
\end{aligned} \tag{15.50}$$

i.e., if and only if

$$\hat{I}_\alpha = -\hat{I}_\alpha^\dagger, \tag{15.51}$$

that is, if and only if the operators $\hat{I}_\alpha$ are anti-Hermitian. Note that from Eq. 15.48, it follows

$$\begin{aligned}\hat{R}^\dagger &= \mathbb{1} + \sum_{\alpha=1}^{3} \delta w_\alpha \hat{I}_\alpha^\dagger = \\ &= \mathbb{1} - \sum_{\alpha=1}^{3} \delta w_\alpha \hat{I}_\alpha,\end{aligned} \tag{15.52}$$

where we used Eq. 15.51. Now, we define the hermitian operators $\hat{J}_\alpha$ as

$$\hat{I}_\alpha = -\frac{i}{\hbar}\hat{J}_\alpha. \tag{15.53}$$

Such operators satisfy

$$-\frac{1}{\hbar^2}\left[\hat{J}_\alpha, \hat{J}_\beta\right] = -\frac{i}{\hbar}\sum_{\gamma=1}^{3} \epsilon_{\alpha\beta\gamma}\hat{J}_\gamma, \tag{15.54}$$

$$\frac{1}{\hbar^2}\left[\hat{J}_\alpha, \hat{J}_\beta\right] = \frac{i}{\hbar}\sum_{\gamma=1}^{3} \epsilon_{\alpha\beta\gamma}\hat{J}_\gamma, \tag{15.55}$$

$$\left[\hat{J}_\alpha, \hat{J}_\beta\right] = i\hbar\sum_{\gamma=1}^{3} \epsilon_{\alpha\beta\gamma}\hat{J}_\gamma. \tag{15.56}$$

By requiring only that the operators $\hat{J}_\alpha$ are Hermitian and satisfy the commutation relations of the rotation group, we obtain the general definition of angular momentum, as given in Eq. 15.56. Note that Eqs. 15.48 and 15.52, become

$$\hat{R} = \mathbb{1} - \frac{i}{\hbar}\sum_{\alpha=1}^{3} \delta w_\alpha \hat{J}_\alpha, \tag{15.57}$$

$$\hat{R}^\dagger = \mathbb{1} + \frac{i}{\hbar}\sum_{\alpha=1}^{3} \delta w_\alpha \hat{J}_\alpha, \tag{15.58}$$

respectively. Locally, any rotation can be expressed as the exponential of a linear combination of the generators of the Lie algebra $L_{SO(3)}$ associated with the $SO(3)$ group (see Sect. 14.3), i.e.,

$$R_{\mathbf{u}}(\delta\varphi) = e^{\delta\varphi \mathbf{u}\cdot\mathbf{I}}, \tag{15.59}$$

then, such a rotation is represented by the unitary operator

$$\hat{R}_{\mathbf{u}}(\delta\varphi) = e^{-\frac{i}{\hbar}\delta\varphi\mathbf{u}\cdot\hat{\mathbf{J}}}. \tag{15.60}$$

For finite rotations, we have

$$R_{\mathbf{u}}(\varphi) = \lim_{N\to\infty}\left[R_{\mathbf{u}}\left(\frac{\varphi}{N}\right)\right]^N. \tag{15.61}$$

Recall, however, that the exponentiation of the generators of a Lie algebra, in this case $L_{SO(3)}$, does not produce a representation of the group $SO(3)$ itself, but rather a representation of its universal covering group, $SU(2)$ (see Sect. 14.3.2).

15.4 Law of Transformations of Observables Under Finite and Infinitesimal Rotations

Let A and A' be a generic observable measured by O and O' respectively, the principle of equivalence of measures implies

$$\begin{aligned}|\langle\psi|A|\psi\rangle|^2 &= |\langle\psi'|A'|\psi'\rangle|^2 = \\ &= |\langle\psi|\hat{R}^\dagger A'\hat{R}|\psi\rangle|^2, \quad \forall\,|\psi\rangle,\end{aligned} \tag{15.62}$$

then

$$A = \hat{R}^\dagger A'\hat{R}, \tag{15.63}$$

$$A' = \hat{R}A\hat{R}^\dagger. \tag{15.64}$$

15.4.1 Scalar Operators

Let S be the operator associated with an observable that is a scalar quantity, that is, $S = S'$. Then

$$S = \hat{R}S\hat{R}^\dagger, \tag{15.65}$$

from which multiplying to the right by $\hat{R}$, we have for scalar observables the following characterization

$$S\hat{R} = \hat{R}S, \tag{15.66}$$

then

$$\left[\hat{R}, S\right] = 0. \tag{15.67}$$

15.4.2 *Vector Operators*

Here, we emphasize that the transformation of a vector under a rotation, aimed at characterizing the vector transformation laws, can be described in two equivalent ways: either by comparing two physically equivalent rotated systems (active view), or by expressing the same system with respect to two rotated spatial reference frames (passive view).

Definition 15.1 (Active and Passive Rotations)

- Active rotations. Given a vector $\mathbf{V}$, the rotated vector is the vector $\mathbf{V}'$ whose components are obtained by applying the rotation matrix R as follows

$$V_i' = \sum_{j=1}^{3} R_{ij} V_j, \tag{15.68}$$

  or equivalently

$$\mathbf{V}' = R\mathbf{V}, \tag{15.69}$$

  where $\mathbf{V}'$ and $\mathbf{V}$ are column vectors.
- Passive rotations. Given the components of a vector $\mathbf{V}$ with respect to two rotated spatial reference systems, i.e.,

$$\mathbf{e}_i' = \sum_{j=1}^{3} R_{ij}\mathbf{e}_j, \tag{15.70}$$

$$\mathbf{e}_j = \sum_{j=1}^{3} R_{ji}^{-1}\mathbf{e}_j', \tag{15.71}$$

$$\mathbf{V} = \sum_{j=1}^{3} V_j\mathbf{e}_j, \tag{15.72}$$

$$\mathbf{V} = \sum_{i=1}^{3} V_i'\mathbf{e}_i'. \tag{15.73}$$

The components V_j and V_i' must satisfy

$$\sum_{i=1}^{3} V_i'\mathbf{e}_i' = \sum_{j=1}^{3} V_j\mathbf{e}_j, \tag{15.74}$$

then

$$\sum_{i=1}^{3} V_i'\mathbf{e}_i' = \sum_{j=1}^{3} V_j \sum_{i=1}^{3} R_{ji}^{-1}\mathbf{e}_i', \tag{15.75}$$

$$V_i' = \sum_{j=1}^{3} V_j R_{ji}^{-1}, \tag{15.76}$$

or equivalently

$$\mathbf{V}'^T = \mathbf{V}^T R^{-1}. \tag{15.77}$$

where $\mathbf{V}'^T$ and $\mathbf{V}^T$ are row vectors.

Given the position vector operator $\hat{\mathbf{r}}$, we postulate that its coordinates transform under rotation analogously to the associated classical dynamical quantities, according to the interpretation of passive rotations, as both observers measure the same state in different references, rotated with respect to each other. If an observer O measures the state

$$\hat{r}_i|\mathbf{r}\rangle = r_i|\mathbf{r}\rangle,\ i \in \{1, 2, 3\}, \tag{15.78}$$

then the observer O' measures

$$\begin{aligned} \hat{r}_i'|\mathbf{r}'\rangle = r_i'|\mathbf{r}'\rangle &= \\ &= \sum_{j=1}^{3} r_j R_{ji}^{-1}|\mathbf{r}'\rangle = \\ &= \sum_{j=1}^{3} R_{ji}^{-1} r_j|\mathbf{r}'\rangle = \\ &= \sum_{j=1}^{3} R_{ji}^{-1}\hat{r}_j|\mathbf{r}'\rangle. \end{aligned} \tag{15.79}$$

Since the previous relation must hold for every state $|\mathbf{r}'\rangle$, it follows that it is an operator identity, i.e.,

$$\hat{r}_i' = \sum_{j=1}^{3} R_{ji}^{-1} \hat{r}_j, \tag{15.80}$$

$$\hat{R}\hat{r}_i\hat{R}^\dagger = \sum_{j=1}^{3} R_{ji}^{-1} \hat{r}_j. \tag{15.81}$$

We take the above relation as the defining property for the transformation of a vector operator with components $\{K_i\}$ under finite rotations. Accordingly,

$$\hat{R}K_i\hat{R}^\dagger = \sum_{j=1}^{3} R_{ji}^{-1} K_j. \tag{15.82}$$

Let us now derive the transformation rule for a vector operator under an infinitesimal rotation. According to Eq. 15.82, from Eqs. 14.133, 15.57 and 15.58, we get

$$\sum_{j=1}^{3}\left(\delta_{ji} - \sum_{\alpha=1}^{3} \delta w_\alpha \epsilon_{\alpha ji}\right) K_j = \left(\mathbb{1} - \frac{i}{\hbar}\sum_{\alpha=1}^{3} \delta w_\alpha \hat{J}_\alpha\right) K_i \left(\mathbb{1} + \frac{i}{\hbar}\sum_{\alpha=1}^{3} \delta w_\alpha \hat{J}_\alpha\right), \tag{15.83}$$

$$\sum_{j=1}^{3} \delta_{ji} K_j - \sum_{j=1}^{3}\sum_{\alpha=1}^{3} \delta w_\alpha \epsilon_{\alpha ji} K_j = \left(K_i - \frac{i}{\hbar}\sum_{\alpha=1}^{3} \delta w_\alpha \hat{J}_\alpha K_i\right)\left(\mathbb{1} + \frac{i}{\hbar}\sum_{\alpha=1}^{3} \delta w_\alpha \hat{J}_\alpha\right), \tag{15.84}$$

$$K_i - \sum_{j=1}^{3}\sum_{\alpha=1}^{3} \delta w_\alpha \epsilon_{\alpha ji} K_j = K_i - \frac{i}{\hbar}\sum_{\alpha=1}^{3} \delta w_\alpha \hat{J}_\alpha K_i + \frac{i}{\hbar}\sum_{\alpha=1}^{3} \delta w_\alpha K_i \hat{J}_\alpha$$
$$+ \frac{1}{\hbar^2}\sum_{\alpha=1}^{3}\sum_{\beta=1}^{3} \delta w_\alpha \delta w_\beta \hat{J}_\alpha K_i \hat{J}_\beta, \tag{15.85}$$

$$K_i - \sum_{j=1}^{3}\sum_{\alpha=1}^{3} \delta w_\alpha \epsilon_{\alpha ji} K_j \simeq K_i - \frac{i}{\hbar}\sum_{\alpha=1}^{3} \delta w_\alpha \hat{J}_\alpha K_i + \frac{i}{\hbar}\sum_{\alpha=1}^{3} \delta w_\alpha K_i \hat{J}_\alpha, \tag{15.86}$$

$$-\sum_{j=1}^{3}\sum_{\alpha=1}^{3} \delta w_\alpha \epsilon_{\alpha ji} K_j = -\frac{i}{\hbar}\sum_{\alpha=1}^{3} \delta w_\alpha \hat{J}_\alpha K_i + \frac{i}{\hbar}\sum_{\alpha=1}^{3} \delta w_\alpha K_i \hat{J}_\alpha, \tag{15.87}$$

$$-\sum_{j=1}^{3}\sum_{\alpha=1}^{3} \delta w_\alpha \epsilon_{\alpha ji} K_j = -\frac{i}{\hbar}\sum_{\alpha=1}^{3} \delta w_\alpha \hat{J}_\alpha K_i + \frac{i}{\hbar}\sum_{\alpha=1}^{3} \delta w_\alpha K_i \hat{J}_\alpha. \tag{15.88}$$

Ultimately, since infinitesimal increments are arbitrary, it must be

$$-\sum_{j=1}^{3} \epsilon_{\alpha ji} K_j = \frac{i}{\hbar}\left(K_i \hat{J}_\alpha - \hat{J}_\alpha K_i\right), \tag{15.89}$$

$$-\sum_{j=1}^{3} \epsilon_{\alpha ji} K_j = \frac{i}{\hbar}\left[K_i, \hat{J}_\alpha\right], \tag{15.90}$$

$$i\hbar \sum_{j=1}^{3} \epsilon_{\alpha ji} K_j = \left[K_i, \hat{J}_\alpha\right], \tag{15.91}$$

$$-i\hbar \sum_{j=1}^{3} \epsilon_{\alpha ij} K_j = \left[K_i, \hat{J}_\alpha\right], \tag{15.92}$$

$$i\hbar \sum_{j=1}^{3} \epsilon_{\alpha ij} K_j = \left[\hat{J}_\alpha, K_i\right]. \tag{15.93}$$

Since the quantity $\sum_{j=1}^{3} \epsilon_{\alpha ij} K_j$ is nonzero only when the indices α, i, and j are all distinct, it follows that

$$\left[\hat{J}_\alpha, K_i\right] = i\hbar \sum_{j=1}^{3} \epsilon_{\alpha ij} K_j \tag{15.94}$$

is the transformation of a vector operator under an infinitesimal rotation. In particular, for the position operator $\hat{r}_\beta$, Eq. 15.94 reads

$$\left[\hat{J}_\alpha, \hat{r}_\beta\right] = i\hbar \sum_{\gamma=1}^{3} \epsilon_{\alpha\beta\gamma} \hat{r}_\gamma. \tag{15.95}$$

15.5 Angular Momentum Operator

Here, we focus on irreducible sets of observables, which form the minimal sets of operators that fully characterize the physical degrees of freedom of a system and whose transformation properties under symmetry operations, such as rotations, determine the behavior of the system's fundamental observables.

15.5.1 Spinless Particle

For a spinless particle we refer to the irreducible system of observables $(\hat{r_\alpha}, \hat{p}_\alpha)$, $\alpha = 1, 2, 3$ on the space $\mathcal{L}^2(\mathbb{R}^3)$, i.e., the space of square-integrable complex-valued functions over $\mathbb{R}^3$. Since the momentum is a vector operator, in the case of finite or infinitesimal rotation it must transform as the position operator does (see Eqs. 15.82 and 15.94). For instance, under an infinitesimal rotation, we have

$$\left[\hat{J}_\alpha, \hat{p}_\beta\right] = i\hbar \sum_{\gamma=1}^{3} \epsilon_{\alpha\beta\gamma} \hat{p}_\gamma. \tag{15.96}$$

Due to irreducibility of the system $(\hat{r_\alpha}, \hat{p}_\alpha)$, from the commutators $\left[\hat{J}_\alpha, \hat{r}_\beta\right]$, $\left[\hat{J}_\alpha, \hat{p}_\beta\right]$, the difference of any two $\hat{\mathbf{J}}$, $\hat{\mathbf{J}}'$ can be calculated. Indeed, suppose that $\hat{\mathbf{J}}$ and $\hat{\mathbf{J}}'$ satisfy the same commutator, then it must be

$$\left[\hat{J}_\alpha - \hat{J}'_\alpha, \hat{r}_\beta\right] = 0, \tag{15.97}$$

$$\left[\hat{J}_\alpha - \hat{J}'_\alpha, \hat{p}_\beta\right] = 0, \tag{15.98}$$

and Schur's lemma implies

$$\hat{J}_\alpha - \hat{J}'_\alpha = c_\alpha \mathbb{1}, \tag{15.99}$$

$$\hat{J}'_\alpha = \hat{J}_\alpha - c_\alpha \mathbb{1}. \tag{15.100}$$

Consequently, the finite rotation operator $\hat{R}$ satisfies

$$\begin{aligned} \hat{R}_{\mathbf{u}}(\varphi) &= e^{-\frac{i}{\hbar}\varphi \hat{\mathbf{J}}' \cdot \mathbf{u}} = \\ &= e^{-\frac{i}{\hbar}\varphi\left(\hat{\mathbf{J}} - \mathbf{c}\mathbb{1}\right)\cdot \mathbf{u}} = \\ &= e^{\frac{i}{\hbar}\varphi \mathbf{c}\cdot\mathbf{u}} e^{-\frac{i}{\hbar}\varphi \hat{\mathbf{J}}\cdot\mathbf{u}}, \end{aligned} \tag{15.101}$$

i.e., the operators of the rotations obtained by exponentiating the generators $\hat{\mathbf{J}}'$ differ from the operators associated with the generators $\hat{\mathbf{J}}$ by a phase factor. In general, the unitary operators representing the action of a rotation provide projective representations (see Definition 14.7) of rotation group, that is, they are defined up to a phase factor, so $\hat{\mathbf{J}}$ and $\hat{\mathbf{J}}'$ are physically equivalent. It is easy to verify that a solution $\hat{\mathbf{J}}$ is provided by the definition of classical orbital angular momentum

$$\hat{\mathbf{J}} = \hat{\mathbf{L}} = \hat{\mathbf{r}} \times \hat{\mathbf{p}}. \tag{15.102}$$

For a spinless particle, the generators of the infinitesimal rotations are the components of the particle's orbital angular momentum.

15.5.2 Particle with Spin

For a particle with fixed spin, the irreducible set of observables is given by $(\hat{r}_\alpha, \hat{p}_\alpha, \hat{S}_\alpha)$, $\alpha = 1, 2, 3$. The spin operators transform as vector operators under infinitesimal rotations (see Eq. 15.94)

$$\left[\hat{J}_\alpha, \hat{S}_\beta\right] = i\hbar \sum_{\gamma=1}^{3} \epsilon_{\alpha\beta\gamma} \hat{S}_\gamma. \tag{15.103}$$

For particles with spin, the Hilbert space factorizes as in Eq. 1.1, and the total angular momentum operator is given by

$$\hat{J}_\alpha = \hat{L}_\alpha + \hat{S}_\alpha, \tag{15.104}$$

or equivalently

$$\hat{\mathbf{J}} = \hat{\mathbf{L}} + \hat{\mathbf{S}}, \tag{15.105}$$

with the understanding that $\hat{L}_\alpha$ and $\hat{S}_\alpha$ act on different factors of the Hilbert space.

15.6 Action of Rotations on the State Vector

Here, we examine the action of the rotation operator on a generic state vector. In general, exponentiating the infinitesimal generators of the Lie algebra $so(3)$ yields elements of the universal covering group (see Definition 14.18) of $SO(3)$, namely $SU(2)$. Our goal is to determine under which conditions this exponential map induces a faithful representation (see Definition 14.8) of the rotation group $SO(3)$ itself.

15.6.1 Spinless Particle

Under an infinitesimal rotation R, and from the perspective of a passive rotation (see Definition 15.1), a vector $\mathbf{V}$ transforms according to

$$\begin{aligned}\mathbf{V}' &= R^{-1}\mathbf{V} = \\ &= \mathbf{V} - \delta\boldsymbol{\omega} \times \mathbf{V} = \\ &= \mathbf{V} - \delta\varphi\ \mathbf{u} \times \mathbf{V},\end{aligned} \tag{15.106}$$

where $\delta\boldsymbol{\omega} = \delta\varphi\,\mathbf{u}$ is the infinitesimal rotation vector. Consequently, the operator representing an infinitesimal rotation about the axis $\mathbf{u}$ by an angle $\delta\varphi$ can be written as

$$\begin{aligned}\hat{R}_{\mathbf{u}}(\delta\varphi) &= e^{-\frac{i}{\hbar}\delta\varphi\mathbf{u}\cdot\hat{\mathbf{J}}} \simeq \\ &\simeq \mathbb{1} - \frac{i}{\hbar}\delta\varphi\mathbf{u}\cdot\hat{\mathbf{J}} \equiv \\ &\equiv \mathbb{1} - \frac{i}{\hbar}\delta\varphi\mathbf{u}\cdot\hat{\mathbf{L}},\end{aligned} \tag{15.107}$$

where in the last step the total angular momentum operator has been replaced by the orbital angular momentum operator, since the particle has no spin. We now consider the action of this rotation operator on the state vector $\psi(\mathbf{r})$, which gives

$$\begin{aligned}\left(\hat{R}_{\mathbf{u}}(\delta\varphi)\psi\right)(\mathbf{r}) &= \left(\left(\mathbb{1} - \frac{i}{\hbar}\delta\varphi\ \mathbf{u}\cdot\hat{\mathbf{L}}\right)\psi\right)(\mathbf{r}) = \\ &= \left(\left(\mathbb{1} - \frac{i}{\hbar}\delta\varphi\ \mathbf{u}\cdot\left(\mathbf{r}\times\frac{\hbar}{i}\nabla\right)\right)\psi\right)(\mathbf{r}) = \\ &= ((\mathbb{1} - \delta\varphi\ \mathbf{u}\cdot(\mathbf{r}\times\nabla))\,\psi)\,(\mathbf{r}),\end{aligned} \tag{15.108}$$

which corresponds to the first-order Taylor expansion of the function $\psi(\mathbf{r}-\delta\varphi\ \mathbf{u}\times\mathbf{r})$, as

$$\begin{aligned}\psi(\mathbf{r} - \delta\varphi\ \mathbf{u}\times\mathbf{r}) &= \psi(\mathbf{r}) - \delta\varphi\,(\mathbf{u}\times\mathbf{r})\cdot\nabla\psi(\mathbf{r}) + o(\delta\varphi^2) = \\ &= \psi(\mathbf{r}) - \delta\varphi\,(\mathbf{u}\cdot(\mathbf{r}\times\nabla\psi(\mathbf{r}))) + o(\delta\varphi^2),\end{aligned} \tag{15.109}$$

i.e., the wave function evaluated at the position vector transformed according to the inverse of the rotation matrix R, that is,

$$\begin{aligned}\left(\hat{R}_{\mathbf{u}}(\delta\varphi)\psi\right)(\mathbf{r}) &= \psi(\mathbf{r} - \delta\varphi\ \mathbf{u}\times\mathbf{r}) = \\ &= \psi\left(R_{\mathbf{u}}^{-1}(\delta\varphi)\mathbf{r}\right).\end{aligned} \tag{15.110}$$

From Eq. 15.61, the operator of a finite rotation is

$$\hat{R}_{\mathbf{u}}(\varphi) = \lim_{N\to\infty}\left[\mathbb{1} - \frac{i}{\hbar}\frac{\varphi}{N}\mathbf{u}\cdot\hat{\mathbf{J}}\right]^N = \\ = e^{-\frac{i}{\hbar}\varphi\mathbf{u}\cdot\hat{\mathbf{J}}}. \tag{15.111}$$

Finally, the action of a finite rotation on the state vector is given by

$$\left(\hat{R}_{\mathbf{u}}(\varphi)\psi\right)(\mathbf{r}) = e^{-\frac{i}{\hbar}\varphi\mathbf{u}\cdot\hat{\mathbf{J}}}\psi(\mathbf{r}) = \\ = e^{-\frac{i}{\hbar}\varphi\mathbf{u}\cdot\hat{\mathbf{L}}}\psi(\mathbf{r}) = \\ = \left(\lim_{N\to+\infty}\left(\mathbb{1} - \frac{i}{\hbar}\frac{\varphi}{N}\mathbf{u}\cdot\hat{\mathbf{L}}\right)^N\psi\right)(\mathbf{r}) = \\ = \lim_{N\to+\infty}\psi\left(R_{\mathbf{u}}^{-1}\left(\left(\frac{\varphi}{N}\right)^N\right)\mathbf{r}\right) = \\ = \psi\left(R_{\mathbf{u}}^{-1}(\varphi)\,\mathbf{r}\right). \tag{15.112}$$

Equation 15.112 shows that, under a finite rotation, the wave function of a spinless particle behaves as a scalar field, remaining invariant under coordinate transformations.

15.6.1.1 Fully Reducible Faithful Unitary Representation

We start to investigate whether the rotation operators for a spinless particle constitute a projective or a faithful representation (see Definitions 14.7 and 14.8). To this end, we consider the transformation law given in Eq. 15.112 and evaluate it for a complete rotation, $\varphi = 2\pi$, that is,

$$\left(\hat{R}_{\mathbf{u}}(2\pi)\psi\right)(\mathbf{r}) = \psi\left(R_{\mathbf{u}}^{-1}(2\pi)\,\mathbf{r}\right) = \\ = \psi(\mathbf{r}), \tag{15.113}$$

showing that for a spinless particle, the operator $\hat{R}$ provides a faithful unitary representation of the group $SO(3)$.

Next, we examine the reducibility of this representation (see Definition 14.9). Specifically, we investigate whether the matrices representing rotations can, in an appropriate basis, be brought into block-diagonal form. The representation turns out to be completely reducible: each block corresponds to a definite eigenvalue $\hbar^2L(L+1)$ of the total angular momentum operator $\hat{\mathbf{L}}^2$. This structure follows from the fact that $\hat{\mathbf{L}}^2$ commutes with all rotation operators. Indeed, from

$$\left[\hat{\mathbf{L}}^2, \hat{L}_i\right] = 0 \tag{15.114}$$

it follows that

$$\left[\hat{\mathbf{L}}^2, \hat{R}_{\mathbf{u}}(\varphi)\right] = 0, \tag{15.115}$$

so the action of a rotation preserves the eigenspaces of $\hat{\mathbf{L}}^2$. Consequently, the Hilbert space can be decomposed as a direct sum of invariant subspaces corresponding to different eigenvalues L of $\hat{\mathbf{L}}^2$ as follows

$$H = \bigoplus_L H_L, \tag{15.116}$$

where each subspace H_L is invariant under all rotations and has dimension $2L + 1$.

Therefore, the representation provided by $\hat{R}_{\mathbf{u}}(\varphi)$ is faithful, unitary, and completely reducible.

15.6.2 Particle with Spin

In Schrödinger's representation, the dynamical evolution of a particle with spin is described by 1.2, where ψ_o is the wavefunction of the orbital (spatial) degrees of freedom, ψ_s describes the spin degrees of freedom. Its total angular momentum is given by Eq. 15.105, where $\hat{\mathbf{L}}$ and $\hat{\mathbf{S}}$ act on the orbital and spin spaces of the wave function, respectively. Since $[\hat{\mathbf{L}}, \hat{\mathbf{S}}] = 0$, the rotation operator $\hat{R}_{\mathbf{u}}(\varphi)$ can be manipulated as follows

$$\begin{aligned}\hat{R}_{\mathbf{u}}(\varphi) &= e^{-\frac{i}{\hbar}\varphi\mathbf{u}\cdot\left(\hat{\mathbf{L}}+\hat{\mathbf{S}}\right)} = \\ &= e^{-\frac{i}{\hbar}\varphi\mathbf{u}\cdot\hat{\mathbf{L}}}e^{-\frac{i}{\hbar}\varphi\mathbf{u}\cdot\hat{\mathbf{S}}},\end{aligned} \tag{15.117}$$

then

$$\begin{aligned}\hat{R}_{\mathbf{u}}(\varphi)\psi &= \hat{R}_{\mathbf{u}}(\varphi)(\psi_o \otimes \psi_s) = \\ &= \left(e^{-\frac{i}{\hbar}\varphi\mathbf{u}\cdot\hat{\mathbf{L}}}\,\psi_o\right) \otimes \left(e^{-\frac{i}{\hbar}\varphi\mathbf{u}\cdot\hat{\mathbf{S}}}\,\psi_s\right).\end{aligned} \tag{15.118}$$

Since the rotation of the orbital part ψ_o has already been computed (see Sect. 15.6.1), we now focus on the spin part. That is, we proceed to compute $e^{-\frac{i}{\hbar}\varphi\mathbf{u}\cdot\hat{\mathbf{S}}}\,\psi_s$. We will analyze the case $S = \frac{1}{2}$. Let us compute the matrix elements of the exponentials of the spin operators, i.e.,

$$\left\langle i \left| e^{-\frac{i}{\hbar}\varphi \mathbf{u}\cdot\hat{\mathbf{S}}} \right| j \right\rangle = \left\langle i \left| e^{-\frac{i}{\hbar}\varphi \mathbf{u}\cdot\hat{\mathbf{S}}} \right| j \right\rangle =$$
$$= \left(e^{-\frac{i}{2}\varphi \mathbf{u}\cdot\boldsymbol{\sigma}} \right)_{ij}, \tag{15.119}$$

and we expand the exponential in a series of powers as follows

$$e^{-\frac{i}{2}\varphi \mathbf{u}\cdot\boldsymbol{\sigma}} = \sum_{n=0}^{+\infty} \frac{1}{n!} \left(-\frac{i\varphi}{2} \right)^n (\mathbf{u}\cdot\boldsymbol{\sigma})^n. \tag{15.120}$$

From the Pauli identity 1.59, setting $\mathbf{v}_1 = \mathbf{v}_2 = \mathbf{u}$, we have

$$(\mathbf{u}\cdot\boldsymbol{\sigma})^2 = \mathbb{1}, \tag{15.121}$$

accordingly, the exponential series can be separated into its even and odd components, yielding

$$\begin{aligned}
e^{-\frac{i}{2}\varphi \mathbf{u}\cdot\boldsymbol{\sigma}} &= \sum_{n=0}^{+\infty} \frac{1}{n!} \left(-\frac{i\varphi}{2} \right)^n (\mathbf{u}\cdot\boldsymbol{\sigma})^n = \\
&= \sum_{n=0}^{+\infty} \left[\frac{1}{(2n+1)!} \left(-\frac{i\varphi}{2} \right)^{2n+1} (\mathbf{u}\cdot\boldsymbol{\sigma})^{2n+1} + \frac{1}{(2n)!} \left(-\frac{i\varphi}{2} \right)^{2n} (\mathbf{u}\cdot\boldsymbol{\sigma})^{2n} \right] = \\
&= \sum_{n=0}^{+\infty} \left[\frac{1}{(2n+1)!} \left(-\frac{i\varphi}{2} \right)^{2n+1} (\mathbf{u}\cdot\boldsymbol{\sigma})^{2n} (\mathbf{u}\cdot\boldsymbol{\sigma}) + \frac{1}{(2n)!} \left(-\frac{i\varphi}{2} \right)^{2n} \mathbb{1} \right] = \\
&= \sum_{n=0}^{+\infty} \left[\frac{1}{(2n+1)!} \left(-\frac{i\varphi}{2} \right)^{2n+1} (\mathbf{u}\cdot\boldsymbol{\sigma}) + \frac{1}{(2n)!} \left(-\frac{i\varphi}{2} \right)^{2n} \mathbb{1} \right] = \\
&= \cos\left(\frac{\varphi}{2}\right)\mathbb{1} - i \sin\left(\frac{\varphi}{2}\right)(\mathbf{u}\cdot\boldsymbol{\sigma}).
\end{aligned} \tag{15.122}$$

Equation 15.122 represents the rotation of the spin $S = \frac{1}{2}$ by a finite angle φ around the direction $\mathbf{u}$.

15.6.2.1 Projective Unitary Representation

Here, we follow the same procedure as in the spinless case to examine whether the rotation operators for a particle with spin furnish a projective or a faithful representation (see Definitions 14.7 and 14.8). Again, we consider the transformation law given in Eq. 15.112 and we evaluate it for a complete rotation, $\varphi = 2\pi$, that is,

$$\hat{R}_{\mathbf{u}}(2\pi)(\psi_o \otimes \psi_s) = \psi_o \otimes \left(e^{-\frac{i}{\hbar}2\pi\mathbf{u}\cdot\hat{\mathbf{S}}}\,\psi_s\right), \tag{15.123}$$

where $\mathbf{u} \cdot \hat{\mathbf{S}}$ is the projection of the spin $\hat{\mathbf{S}}$ along the direction $\mathbf{u}$. Since the spin component in the irreducible set $(\hat{r_\alpha}, \hat{p}_\alpha, \hat{S}_\alpha)$ is arbitrary, we can arrange the axes of the spatial reference system such that

$$\hat{S}_z = \mathbf{u} \cdot \hat{\mathbf{S}}, \tag{15.124}$$

and we use the eigenfunctions of $\hat{S}_z$, denoted by $\chi_\sigma(s)$ (see Sect. 1.1), and the corresponding eigenvalue equation (see Eqs. 1.33 and 1.27), as follows

$$\begin{aligned} e^{-\frac{i}{\hbar}2\pi\hat{S}_z}|\psi_s\rangle &= \sum_{\sigma=-S}^{S} e^{-\frac{i}{\hbar}2\pi\hat{S}_z} c_\sigma \chi_\sigma(s) = \\ &= \sum_{\sigma=-S}^{S} e^{-\frac{i}{\hbar}2\pi\hbar\sigma} c_\sigma \chi_\sigma(s) = \\ &= \sum_{\sigma=-S}^{S} e^{-i2\pi\sigma} c_\sigma \chi_\sigma(s). \end{aligned} \tag{15.125}$$

If spin S is integer, σ is integer; if S is semi-integer, σ is semi-integer, so

$$e^{-\frac{i}{\hbar}2\pi\hat{S}_z}|\psi_s\rangle = \begin{cases} +|\psi_s\rangle, & \text{if } S \text{ is integer} \\ -|\psi_s\rangle, & \text{if } S \text{ is half-integer} \end{cases}, \tag{15.126}$$

$$\begin{cases} S \text{ integer} \Rightarrow \hat{R}_{\mathbf{u}}(2\pi) = \mathbb{1} \otimes \mathbb{1} \Rightarrow \text{faithful representation} \\ S \text{ half-integer} \Rightarrow \hat{R}_{\mathbf{u}}(2\pi) = -\mathbb{1} \otimes \mathbb{1} \Rightarrow \text{double-valued representation} \end{cases}. \tag{15.127}$$

This result is a direct consequence of the fact that $SU(2)$ is a double covering of $SO(3)$ (see Sect. 14.3.2), so that half-integer-spin representations are single-valued representations of $SU(2)$ but appear as double-valued (or projective) representations of $SO(3)$.

Remark 15.1 (Spin and Physical Equivalence of Unitary Representations) Unitary operators are defined up to a phase factor, so the operators $\mathbb{1}$ and $-\mathbb{1}$ are physically equivalent, since measurements depend only on the squared modulus of the wavefunction. In contrast, if the wavefunctions were classical fields, only faithful representations of spin would have physical meaning. In that case, a correspondence with classical state vectors would exist only for integer spin values S.

15.7 The Angular Momentum and Momentum Operators as Infinitesimal Generators of $E(3)$

Let O and O' be two spatial reference frames related by a roto-translation. For a state $|\psi\rangle$ defined in the reference frame O, the unitary operator $\hat{\Omega}(R_{\mathbf{u}}(\varphi), \mathbf{a})$, associated with the element $g = (R, \mathbf{a}) \in E(3)$ (see Sect. 14.1.1), maps $|\psi\rangle$ to the corresponding state in the frame O' that represents the same physical situation. Analogously to rotations, the operator $\hat{\Omega}$ must be unitary also for translations. Given an infinitesimal roto-translation 14.136, the corresponding infinitesimal transformation operator $\hat{\Omega}$ takes the form

$$\hat{\Omega} = \mathbb{1} + \sum_{\alpha=1}^{3} \delta w_\alpha \hat{\tilde{I}}_\alpha + \sum_{\alpha=1}^{3} \delta a_\alpha \hat{\tilde{\pi}}_\alpha. \tag{15.128}$$

Since $\hat{\Omega}$ is unitary (see Eqs. I.2 and 1.3), one finds the following constraints on the generators

$$\hat{\tilde{I}}_\alpha = -\hat{\tilde{I}}^\dagger_\alpha, \tag{15.129}$$

$$\hat{\pi}_\alpha = -\hat{\tilde{\pi}}^\dagger_\alpha. \tag{15.130}$$

As a consequence, the Hermitian conjugate of $\hat{\Omega}$ reads

$$\hat{\Omega}^\dagger = \mathbb{1} - \sum_{\alpha=1}^{3} \delta w_\alpha \hat{\tilde{I}}_\alpha - \sum_{\alpha=1}^{3} \delta a_\alpha \hat{\tilde{\pi}}_\alpha. \tag{15.131}$$

We define the two hermitian operators by means of

$$\hat{\tilde{I}}_\alpha = -\frac{i}{\hbar} \hat{J}_\alpha, \tag{15.132}$$

$$\hat{\tilde{\pi}}_\alpha = -\frac{i}{\hbar} \hat{P}_\alpha, \tag{15.133}$$

which satisfy the algebras

$$\begin{cases} \left[\hat{J}_\alpha, \hat{J}_\beta\right] = i\hbar \sum_{\gamma=1}^{3} \epsilon_{\alpha\beta\gamma} \hat{J}_\gamma \\ \left[\hat{P}_\alpha, \hat{P}_\beta\right] = 0 \\ \left[\hat{J}_\alpha, \hat{P}_\beta\right] = i\hbar \sum_{\gamma=1}^{3} \epsilon_{\alpha\beta\gamma} \hat{P}_\gamma \end{cases}. \tag{15.134}$$

Independently of a physical system, we define the momentum $\hat{\mathbf{P}}$ and the angular momentum $\hat{\mathbf{J}}$ operators as the infinitesimal generators of translations and rotations, respectively. Note that Eqs. 15.128 and 15.131 become

$$\hat{\Omega} = \mathbb{1} - \frac{i}{\hbar}\sum_{\alpha=1}^{3}\delta w_\alpha \hat{J}_\alpha - \frac{i}{\hbar}\sum_{\alpha=1}^{3}\delta a_\alpha \hat{P}_\alpha, \tag{15.135}$$

$$\hat{\Omega}^\dagger = \mathbb{1} + \frac{i}{\hbar}\sum_{\alpha=1}^{3}\delta w_\alpha \hat{J}_\alpha + \frac{i}{\hbar}\sum_{\alpha=1}^{3}\delta a_\alpha \hat{P}_\alpha, \tag{15.136}$$

respectively. The action of infinite infinitesimal rotations and infinite infinitesimal translations lead to the operator $\hat{\Omega}\left(R_{\hat{\mathbf{u}}}(\varphi), \mathbf{a}\right)$ of a finite rototranslation, i.e.,

$$\hat{\Omega}(R_{\mathbf{u}}(\varphi), \mathbf{a}) = e^{-\frac{i}{\hbar}\varphi\mathbf{u}\cdot\hat{\mathbf{J}} - \frac{i}{\hbar}\mathbf{a}\cdot\hat{\mathbf{P}}}. \tag{15.137}$$

15.8 Angular Momentum and Momentum of a Spinless Particle

Let $(\hat{r}_i, \hat{p}_j)$ denote an irreducible set of operators acting on the Hilbert space H of a quantum system. To understand how the total angular momentum $\hat{\mathbf{J}}$ and the momentum $\hat{\mathbf{P}}$ depend on these operators, it is necessary to analyze their transformation properties under roto-translations. In this discussion, we adopt the passive interpretation of rotations (see Definition 15.1). Consider two spatial reference frames, O and O', related by a roto-translation. Their respective orthonormal bases are related by a rotation matrix R, so that each vector of the rotated basis can be expressed as a linear combination of the original basis vectors, $e_i' = \sum_{j=1}^{3} R_{ij} e_j$, and conversely, each original can be written in terms of the rotated basis, $e_j = \sum_{i=1}^{3} R_{ji}^{-1} e_i'$. Accordingly, the coordinates of a generic point P in the two reference frames are related as follows

$$\begin{aligned}\sum_{i=1}^{3} r_i' e_i' = O'P = \\ &= OP - OO' = \\ &= \sum_{j=1}^{3} r_j e_j - \sum_{j=1}^{3} a_j e_j = \\ &= \sum_{j=1}^{3} (r_j - a_j) e_j = \\ &= \sum_{j=1}^{3}\sum_{i=1}^{3} (r_j - a_j) R_{ji}^{-1} e_i',\end{aligned} \tag{15.138}$$

from which we immediately obtain

$$r'_i = \sum_{j=1}^{3} (r_j - a_j) R^{-1}_{ji}. \tag{15.139}$$

We postulate that the position operator undergoes a rototranslation in direct analogy with the transformation of its classical counterpart, namely

$$\begin{aligned}
\hat{r}'_i |\mathbf{r}'\rangle &= r'_i |\mathbf{r}'\rangle = \\
&= \sum_{j=1}^{3} (r_j - a_j) R^{-1}_{ji} |\mathbf{r}'\rangle = \\
&= \sum_{j=1}^{3} R^{-1}_{ji} (r_j - a_j) |\mathbf{r}'\rangle = \\
&= \sum_{j=1}^{3} R^{-1}_{ji} \left(\hat{r}_j - a_j \mathbb{1}\right) |\mathbf{r}'\rangle,
\end{aligned} \tag{15.140}$$

then

$$\hat{r}'_i = \sum_{j=1}^{3} R^{-1}_{ji} \left(\hat{r}_j - a_j \mathbb{1}\right), \tag{15.141}$$

$$\hat{\Omega} \hat{r}_i \hat{\Omega}^\dagger = \sum_{j=1}^{3} R^{-1}_{ji} \left(\hat{r}_j - a_j \mathbb{1}\right). \tag{15.142}$$

In classical physics, the momentum of a particle does not vary under a translation, so we postulate that the action of a rototranslation on $\hat{\mathbf{p}}$ reduces to a rotation as follows (see Eq. 15.82)

$$\hat{p}'_i = \sum_{j=1}^{3} R^{-1}_{ji} \hat{p}_j, \tag{15.143}$$

$$\hat{\Omega} \hat{p}_i \hat{\Omega}^\dagger = \sum_{j=1}^{3} R^{-1}_{ji} \hat{p}_j. \tag{15.144}$$

The infinitesimal generators $\hat{\mathbf{J}}$ and $\hat{\mathbf{P}}$ can now be calculated from the system given by 15.142 and 15.144, where R^{-1}_{ji}, $\hat{\Omega}$ and $\hat{\Omega}^\dagger$, are given by 14.133, 15.135 and

15.136, respectively, and the translation vector **a** is replaced by the infinitesimal translation vector $\delta\mathbf{a}$. From Eq. 15.144, one recovers Eq. 15.96. In addition, since the translations leave the momentum unchanged, we have

$$\left[\hat{p}_\alpha, \hat{P}_\beta\right] = 0, \ \forall\, \alpha, \beta. \tag{15.145}$$

From

$$\hat{\Omega}\hat{r}_i\hat{\Omega}^\dagger = \sum_{j=1}^{3} R^{-1}_{ji}\left(\hat{r}_j - \delta a_j \mathbb{1}\right), \tag{15.146}$$

we have

$$\left(\mathbb{1} - \frac{i}{\hbar}\sum_{\alpha=1}^{3}\delta w_\alpha \hat{J}_\alpha - \frac{i}{\hbar}\sum_{\alpha=1}^{3}\delta a_\alpha \hat{P}_\alpha\right)\hat{r}_i\left(\mathbb{1} + \frac{i}{\hbar}\sum_{\alpha=1}^{3}\delta w_\alpha \hat{J}_\alpha + \frac{i}{\hbar}\sum_{\alpha=1}^{3}\delta a_\alpha \hat{P}_\alpha\right) = \\ = \sum_{j=1}^{3}\left(\delta_{ji} - \sum_{\alpha=1}^{3}\delta w_\alpha \epsilon_{\alpha j i}\right)\left(\hat{r}_j - \delta a_j \mathbb{1}\right), \tag{15.147}$$

$$\left(\hat{r}_i - \frac{i}{\hbar}\sum_{\alpha=1}^{3}\delta w_\alpha \hat{J}_\alpha \hat{r}_i - \frac{i}{\hbar}\sum_{\alpha=1}^{3}\delta a_\alpha \hat{P}_\alpha \hat{r}_i\right)\left(\mathbb{1} + \frac{i}{\hbar}\sum_{\alpha=1}^{3}\delta w_\alpha \hat{J}_\alpha + \frac{i}{\hbar}\sum_{\alpha=1}^{3}\delta a_\alpha \hat{P}_\alpha\right) = \\ = \sum_{j=1}^{3}\delta_{ji}\hat{r}_j - \sum_{j=1}^{3}\delta_{ji}\delta a_j \mathbb{1} - \sum_{j=1}^{3}\sum_{\alpha=1}^{3}\delta w_\alpha \epsilon_{\alpha j i}\hat{r}_j + \sum_{j=1}^{3}\sum_{\alpha=1}^{3}\delta w_\alpha \delta a_j \epsilon_{\alpha j i}, \tag{15.148}$$

$$\hat{r}_i + \frac{i}{\hbar}\sum_{\alpha=1}^{3}\delta w_\alpha \hat{r}_i \hat{J}_\alpha + \frac{i}{\hbar}\sum_{\alpha=1}^{3}\delta a_\alpha \hat{r}_i \hat{P}_\alpha - \frac{i}{\hbar}\sum_{\alpha=1}^{3}\delta w_\alpha \hat{J}_\alpha \hat{r}_i + \frac{1}{\hbar^2}\sum_{\alpha,\beta=1}^{3}\delta w_\alpha \delta w_\beta \hat{J}_\alpha \hat{r}_i \hat{J}_\beta \\ + \frac{1}{\hbar^2}\sum_{\alpha,\beta=1}^{3}\delta w_\alpha \delta a_\beta \hat{J}_\alpha \hat{r}_i \hat{P}_\beta - \frac{i}{\hbar}\sum_{\alpha=1}^{3}\delta a_\alpha \hat{P}_\alpha \hat{r}_i + \frac{1}{\hbar^2}\sum_{\alpha,\beta=1}^{3}\delta a_\alpha \delta w_\beta \hat{P}_\alpha \hat{r}_i \hat{J}_\beta + \\ + \frac{1}{\hbar^2}\sum_{\alpha,\beta=1}^{3}\delta a_\alpha \delta a_\beta \hat{P}_\alpha \hat{r}_i \hat{P}_\beta = \hat{r}_i - \delta a_i \mathbb{1} \\ - \sum_{j=1}^{3}\sum_{\alpha=1}^{3}\delta w_\alpha \epsilon_{\alpha j i}\hat{r}_j + \sum_{j=1}^{3}\sum_{\alpha=1}^{3}\delta w_\alpha \delta a_j \epsilon_{\alpha j i}, \tag{15.149}$$

$$\hat{r}_i + \frac{i}{\hbar}\sum_{\alpha=1}^{3}\delta w_\alpha \hat{r}_i \hat{J}_\alpha + \frac{i}{\hbar}\sum_{\alpha=1}^{3}\delta a_\alpha \hat{r}_i \hat{P}_\alpha - \frac{i}{\hbar}\sum_{\alpha=1}^{3}\delta w_\alpha \hat{J}_\alpha \hat{r}_i$$
$$- \frac{i}{\hbar}\sum_{\alpha=1}^{3}\delta a_\alpha \hat{P}_\alpha \hat{r}_i = \hat{r}_i - \delta a_i \mathbb{1} - \sum_{j=1}^{3}\sum_{\alpha=1}^{3}\delta w_\alpha \epsilon_{\alpha j i}\hat{r}_j, \tag{15.150}$$

where in the last step we neglected second-order terms with respect to infinitesimal increments. Then

$$\frac{i}{\hbar}\sum_{\alpha=1}^{3}\delta w_\alpha \hat{r}_i \hat{J}_\alpha + \frac{i}{\hbar}\sum_{\alpha=1}^{3}\delta a_\alpha \hat{r}_i \hat{P}_\alpha - \frac{i}{\hbar}\sum_{\alpha=1}^{3}\delta w_\alpha \hat{J}_\alpha \hat{r}_i - \frac{i}{\hbar}\sum_{\alpha=1}^{3}\delta a_\alpha \hat{P}_\alpha \hat{r}_i$$
$$= -\delta a_i \mathbb{1} - \sum_{j=1}^{3}\sum_{\alpha=1}^{3}\delta w_\alpha \epsilon_{\alpha j i}\hat{r}_j, \tag{15.151}$$

$$\frac{i}{\hbar}\sum_{\alpha=1}^{3}\delta w_\alpha \left[\hat{r}_i, \hat{J}_\alpha\right] + \frac{i}{\hbar}\sum_{\alpha=1}^{3}\delta a_\alpha \left[\hat{r}_i, \hat{P}_\alpha\right] = -\sum_{j=1}^{3}\sum_{\alpha=1}^{3}\delta w_\alpha \epsilon_{\alpha j i}\hat{r}_j - \delta a_i \mathbb{1}. \tag{15.152}$$

By comparison we get

$$\frac{i}{\hbar}\left[\hat{r}_i, \hat{J}_\alpha\right] = -\sum_{j=1}^{3}\epsilon_{\alpha j i}\hat{r}_j, \tag{15.153}$$

$$\frac{i}{\hbar}\sum_{\alpha=1}^{3}\delta a_\alpha \left[\hat{r}_i, \hat{P}_\alpha\right] = -\delta a_i \mathbb{1}. \tag{15.154}$$

Regarding the first equation, in analogy with the treatment of the angular momentum, we obtain

$$\left[\hat{r}_i, \hat{J}_\alpha\right] = i\hbar \sum_{j=1}^{3}\epsilon_{\alpha j i}\hat{r}_j, \tag{15.155}$$

$$\left[\hat{r}_i, \hat{J}_\alpha\right] = -i\hbar \sum_{j=1}^{3}\epsilon_{\alpha i j}\hat{r}_j, \tag{15.156}$$

$$-\left[\hat{J}_\alpha, \hat{r}_i\right] = -i\hbar \sum_{j=1}^{3}\epsilon_{\alpha i j}\hat{r}_j, \tag{15.157}$$

that is, one recovers Eq. 15.95. Concerning the second equation of the system, we note that, in order for it to be satisfied for an arbitrary choice of increments δa_α, the commutator $\left[\hat{r}_i, \hat{P}_\alpha\right]$ must vanish for $i \neq \alpha$. Hence,

$$\left[\hat{r}_i, \hat{P}_\alpha\right] = i\hbar\delta_{i,\alpha}. \tag{15.158}$$

Ultimately, the solutions satisfy

$$\begin{cases} \left[\hat{J}_\alpha, \hat{r}_i\right] = i\hbar \sum_{j=1}^{3} \epsilon_{\alpha i j}\hat{r}_j \\ \left[\hat{r}_i, \hat{P}_\alpha\right] = i\hbar\delta_{i,\alpha} \\ \left[\hat{p}_i, \hat{J}_\alpha\right] = i\hbar \sum_{j=1}^{3} \epsilon_{i\alpha j}\hat{p}_j \\ \left[\hat{p}_\alpha, \hat{P}_\beta\right] = 0 \end{cases} . \tag{15.159}$$

i.e., for a particle with no spin and subject to infinitesimal rototranslation, we derive in a general way the identifications

$$\begin{aligned} \hat{J}_\alpha &= \hat{L}_\alpha = \\ &= \sum_{\gamma=1}^{3} \epsilon_{\alpha\beta\gamma}\hat{r}_\beta\hat{p}_\gamma, \end{aligned} \tag{15.160}$$

and

$$\hat{P}_\alpha = \hat{p}_\alpha. \tag{15.161}$$

15.9 Translation Operator

Based on Eq. 15.137, we define the spatial translation operator $\hat{T}(\mathbf{a})$ as a pure translation, i.e., the roto-translation operator corresponding to the identity rotation matrix (no rotation) and a translation by the vector **a**, that is,

$$\begin{aligned} \hat{T}(\mathbf{a}) &\equiv \hat{\Omega}(\mathbb{1}, \mathbf{a}) = \\ &= e^{-\frac{i}{\hbar}\mathbf{a}\cdot\hat{\mathbf{P}}} = \\ &= e^{-\frac{i}{\hbar}\mathbf{a}\cdot\hat{\mathbf{p}}}, \end{aligned} \tag{15.162}$$

where in the last step, we have replaced the translation generator $\hat{\mathbf{P}}$ with the momentum operator $\hat{\mathbf{p}}$, according to Eq. 15.161. The definition of the translation operator leads to a fundamental result in quantum mechanics: the theorem of invariance under spatial translations. Although this theorem arises from the formal structure of quantum theory, it is particularly valuable in quantum many-body physics, as it has been extensively employed throughout this text enabling the simplification of calculations and the systematic analysis.

Theorem 15.3 (Invariance Under Spatial Translations) *Spatial translation invariance of the Hamiltonian implies the conservation of the momentum operator.*

Proof Let $\mathcal{H}$ be the Hamiltonian of the system, then the symmetry under spatial translations implies

$$\left[\hat{T}(\mathbf{a}), \mathcal{H}\right] = 0, \ \forall \mathbf{a}, \tag{15.163}$$

$$\hat{T}^\dagger(\mathbf{a})\mathcal{H}\hat{T}(\mathbf{a}) = \mathcal{H}. \tag{15.164}$$

We differentiate both sides with respect to $\mathbf{a}$, and evaluate at $\mathbf{a} = \mathbf{0}$, that is

$$\begin{aligned}\nabla_{\mathbf{a}}\left(\hat{T}^\dagger(\mathbf{a})\mathcal{H}\hat{T}(\mathbf{a})\right)\Big|_{\mathbf{a}=\mathbf{0}} &= \left(\nabla_{\mathbf{a}}\hat{T}(\mathbf{a})\mathcal{H} + \mathcal{H}\nabla_{\mathbf{a}}\hat{T}(\mathbf{a})\right)\Big|_{\mathbf{a}=\mathbf{0}} = \\ &= \frac{i}{\hbar}\hat{\mathbf{p}}\mathcal{H} - \frac{i}{\hbar}\mathcal{H}\hat{\mathbf{p}} = \\ &= \frac{i}{\hbar}\left[\hat{\mathbf{p}}, \mathcal{H}\right].\end{aligned} \tag{15.165}$$

Since the derivative of the left-hand side vanishes, because it equals the derivative of $\mathcal{H}$, a constant operator, we get $\left[\hat{\mathbf{p}}, \mathcal{H}\right] = 0$. From the Heisenberg equation of motion (see Eq. 1.104), it follows

$$\frac{d\hat{\mathbf{p}}}{dt} = 0, \tag{15.166}$$

and the statement follows. □

Ultimately, we show that a state vector transforms under a spatial translation as does a scalar field, indeed

$$\begin{aligned}\left(\hat{T}(\mathbf{a})\psi\right)(\mathbf{r}) &= e^{-\frac{i}{\hbar}\mathbf{a}\cdot\hat{\mathbf{p}}}\psi(\mathbf{r}) = \\ &= \left(\lim_{N\to+\infty}\left(\mathbb{1} - \frac{i}{\hbar}\frac{\mathbf{a}}{N}\cdot\hat{\mathbf{p}}\right)^N\psi\right)(\mathbf{r}) = \\ &= \left(\lim_{N\to+\infty}\left(\mathbb{1} - \frac{\mathbf{a}}{N}\cdot\nabla\right)^N\psi\right)(\mathbf{r}) = \\ &= \lim_{N\to+\infty}\psi\left(\left(\mathbb{1} - \frac{\mathbf{a}}{N}\cdot\nabla\right)^N\mathbf{r}\right) = \\ &= \psi(\mathbf{r} - \mathbf{a}).\end{aligned} \tag{15.167}$$

15.10 Ordinary $SO(3)$ Matrix Representations of Operators

Let $\{A_1, \dots, A_n\}$ be a set of linearly independent operators that transform, under rotations, into linear combinations of the set itself by means of $n \times n$ matrices $M(R)$, i.e.,

$$\hat{\Omega}(R) A_i \hat{\Omega}^\dagger(R) = \sum_{j=1}^{n} A_j M_{ji}(R), \tag{15.168}$$

that is, any linear combination with complex coefficients of the $\{A_i\}$ will transform under rotation into an element of the n-dimensional space $\text{Span}\{A_1, \dots, A_n\}$. We have

Theorem 15.4 *If $\hat{\Omega}(R)$ is a projective representation of the rotation group (see Eq. 15.46), then the matrices $M(R)$ define an ordinary representation (see Definition 14.7) of $SO(3)$ acting on the operators, i.e.,*

$$M(R_2)M(R_1) = M(R_2 R_1). \tag{15.169}$$

Proof The rotation $R_2 R_1$ satisfies

$$\hat{\Omega}(R_2 R_1) A_i \hat{\Omega}^\dagger(R_2 R_1) = \sum_{k=1}^{n} A_k M_{ki}(R_2 R_1), \tag{15.170}$$

and since $\hat{\Omega}(R)$ is a projective representation, we have

$$\begin{aligned} &e^{i\varphi(R_2,R_1)} \hat{\Omega}(R_2) \hat{\Omega}(R_1) A_i e^{-i\varphi(R_2,R_1)} \hat{\Omega}^\dagger(R_1) \hat{\Omega}^\dagger(R_2) \\ &= \hat{\Omega}(R_2) \left(\hat{\Omega}(R_1) A_i \hat{\Omega}^\dagger(R_1) \right) \hat{\Omega}^\dagger(R_2) = \\ &= \hat{\Omega}(R_2) \left(\sum_{j=1}^{n} A_j M_{ji}(R_1) \right) \hat{\Omega}^\dagger(R_2) = \\ &= \sum_{j=1}^{n} \sum_{k=1}^{n} A_k M_{kj}(R_2) M_{ji}(R_1), \end{aligned} \tag{15.171}$$

then

$$\sum_{j=1}^{n} \sum_{k=1}^{n} A_k M_{kj}(R_2) M_{ji}(R_1) = \sum_{k=1}^{n} A_k M_{ki}(R_2 R_1), \tag{15.172}$$

and since $\{A_i\}$ is a linearly independent set, the assertion follows. □

Given a vector operator $\mathbf{K}$, it transforms under rotations as

$$\begin{aligned}\hat{\Omega}(R)K_i\hat{\Omega}^\dagger(R) &= \sum_{j=1}^{3} R^{-1}_{ji}K_j = \\ &= \sum_{j=1}^{3} R^{T}_{ji}K_j = \\ &= \sum_{j=1}^{3} R_{ij}K_j. \end{aligned} \tag{15.173}$$

The matrices $M(R)$ form a faithful representation (see Definition 14.8) of the three-dimensional rotation matrices $R \in SO(3)$, and in this case they coincide with them, i.e.,

$$M_{ij}(R) \equiv R_{ij}, \ i, j = 1, 2, 3. \tag{15.174}$$

15.11 Time Reversal Symmetry

Time-reversal symmetry is a fundamental property of many physical systems, reflecting the invariance of their dynamics under the inversion of the time coordinate. In both classical and quantum mechanics, this symmetry provides important constraints on the evolution of states and the transformation properties of observables.

15.11.1 Time Reversal in Classical Mechanics

Consider a single particle moving in a conservative potential, whose Hamiltonian is given by $\mathcal{H} = \frac{\mathbf{p}^2}{2m} + V(\mathbf{r})$, where $V(\mathbf{r})$ is the potential associated with a conservative force field, the Hamiltonian is an even function of the conjugate momentum, that is, $\mathcal{H}(\mathbf{r}, \mathbf{p}) = \mathcal{H}(\mathbf{r}, -\mathbf{p})$. Indeed, under the substitution $\mathbf{p} \to -\mathbf{p}$, both the potential term $V(\mathbf{r})$ and the kinetic energy remain unchanged. Consequently, if $\mathbf{r}(t)$, $\mathbf{p}(t)$ are solutions of the Hamilton equations of motion,

$$\frac{d\mathbf{r}(t)}{dt} = \frac{\partial \mathcal{H}}{\partial \mathbf{p}}, \tag{15.175}$$

$$\frac{d\mathbf{p}(t)}{dt} = -\frac{\partial \mathcal{H}}{\partial \mathbf{r}}, \tag{15.176}$$

then also the functions

$$\mathbf{r}'(t) = \mathbf{r}(-t), \tag{15.177}$$

$$\mathbf{p}'(t) = -\mathbf{p}(-t) \tag{15.178}$$

satisfy the same equations of motion. In other words, $\mathbf{r}'(t)$ and $\mathbf{p}'(t)$ describe a valid physical trajectory obtained by time reversal: the particle retraces its path in reverse, with the momentum inverted at each point. Indeed,

$$\begin{aligned}\frac{d\mathbf{p}'(t)}{dt} &= -\frac{d\mathbf{p}(-t)}{dt} = \\ &= \frac{d\mathbf{p}(t)}{dt} = \\ &= -\frac{\partial \mathcal{H}}{\partial \mathbf{r}} = \\ &= -\frac{\partial \mathcal{H}}{\partial \mathbf{r}'}\frac{\partial \mathbf{r}'_i}{\partial \mathbf{r}} = \\ &= -\frac{\partial \mathcal{H}}{\partial \mathbf{r}'},\end{aligned} \tag{15.179}$$

and

$$\begin{aligned}\frac{d\mathbf{r}'(t)}{dt} &= \frac{d\mathbf{r}(-t)}{dt} = \\ &= -\frac{d\mathbf{r}(t)}{dt} = \\ &= -\frac{\partial \mathcal{H}}{\partial \mathbf{p}} = \\ &= \frac{\partial \mathcal{H}}{\partial (-\mathbf{p})},\end{aligned} \tag{15.180}$$

where the last step is due to the fact that $\mathcal{H}$ is even with respect to $\mathbf{p}$, then

$$\frac{d\mathbf{r}'(t)}{dt} = \frac{\partial \mathcal{H}}{\partial \mathbf{p}'}. \tag{15.181}$$

The solutions $\mathbf{r}'(t), \mathbf{p}'(t)$ have considerable physical significance. If a particle moves from A at time $t - t_0$ to B at time t_0, its dynamics is described by the solutions $\mathbf{r}(t)$, $\mathbf{p}(t)$, that is,

$$\begin{aligned} A \quad &\underset{+t}{\rightarrow} \quad B \\ (\mathbf{r}(t_0 - t), \mathbf{p}(t_0 - t)) \quad &\underset{+t}{\rightarrow} \quad (\mathbf{r}(t_0), \mathbf{p}(t_0)) \,. \end{aligned} \tag{15.182}$$

The functions $\mathbf{r}'(t)$, $\mathbf{p}'(t)$ are solutions of the particle dynamics. In particular, given Eqs. 15.177 and 15.178, if the motion starts at the time $-t_0$ (note that we reverse the sign of time) and we set

$$B \equiv (\mathbf{r}(t_0), \mathbf{p}(t_0)) \equiv (\mathbf{r}(-t_0), -\mathbf{p}(t_0)) \tag{15.183}$$

since the functions are equal at each instant, the evolution described by $\mathbf{r}'(t)$, $\mathbf{p}'(t)$ predicts that the particle moves from B to A, in the time interval t, with reversed momentum, i.e.,

$$\begin{aligned} B \quad &\underset{+\mathrm{t}}{\rightarrow} \quad A \\ (\mathbf{r}(-t_0), \mathbf{p}(-t_0)) \quad &\underset{+\mathrm{t}}{\rightarrow} \quad (\mathbf{r}(t-t_0), \mathbf{p}(t-t_0))\,. \end{aligned} \tag{15.184}$$

Therefore,

$$A \equiv (\mathbf{r}(t_0 - t), \mathbf{p}(t_0 - t)) \equiv (\mathbf{r}(t - t_0), -\mathbf{p}(t - t_0))\,, \tag{15.185}$$

$$B \equiv (\mathbf{r}(t_0), \mathbf{p}(t_0)) \equiv (\mathbf{x}(-t_0), -\mathbf{p}(-t_0))\,. \tag{15.186}$$

Due to the inversion of the sign of time, the functions $\mathbf{r}'(t)$ and $\mathbf{p}'(t)$ are referred to as the time-reversed (or temporal inverses) of $\mathbf{r}(t)$ and $\mathbf{p}(t)$, respectively. In fact, the initial point, the final point, and the direction along the trajectory are all reversed under this transformation.

15.11.2 Time Reversal in Quantum Mechanics

Here, we will study a formal theory of symmetry by time reversal in quantum mechanics from which we will also derive the properties of the time reversal operator. For this purpose, it is sufficient to restrict the discussion to the Hilbert space of a spinless particle. Let $\hat{\Theta}$ be the time reversal operator

$$\hat{\Theta}: \; |\alpha\rangle \longrightarrow \hat{\Theta}|\alpha\rangle. \tag{15.187}$$

Now, given a state $|\alpha\rangle$ at time $t = 0$, its time evolution in an infinitesimal time interval δt is given by

$$|\alpha, \delta t\rangle = \left(\mathbb{1} - i\frac{\mathcal{H}}{\hbar}\delta t\right)|\alpha\rangle. \tag{15.188}$$

On the other hand, if $\hat{\Theta}$ acts on the state $|\alpha\rangle$ at $t = 0$, after a time δt, that state becomes

$$\hat{\Theta}|\alpha\rangle \longrightarrow \left(\mathbb{1} - i\frac{\mathcal{H}}{\hbar}\delta t\right)\hat{\Theta}|\alpha\rangle. \tag{15.189}$$

If the system is invariant under time reversal, then the previous state is equal to the action of the time reversal operator $\hat{\Theta}$ on the state evolved from time 0 to time $-\delta t$, that is, $\hat{\Theta}|\alpha, -\delta t\rangle$. Therefore,

$$\left(\mathbb{1} - i\frac{\mathcal{H}}{\hbar}\delta t\right)\hat{\Theta}|\alpha\rangle = \hat{\Theta}\left(\mathbb{1} - i\frac{\mathcal{H}}{\hbar}(-\delta t)\right)|\alpha\rangle, \tag{15.190}$$

$$-i\mathcal{H}\hat{\Theta}|\alpha\rangle = \hat{\Theta}i\mathcal{H}|\alpha\rangle. \tag{15.191}$$

According to Wigner's theorem 15.1, the time reversal operator is either unitary or antiunitary. If we assume that it is unitary, and thus linear, then the presence of the imaginary unit i in the previous equation becomes irrelevant, and we obtain

$$-\mathcal{H}\hat{\Theta}|\alpha\rangle = \hat{\Theta}\mathcal{H}|\alpha\rangle, \tag{15.192}$$

which, however, leads to physically incorrect conclusions. Indeed, if $\mathcal{E}_\alpha$ is the eigenvalue associated with the state $|\alpha\rangle$, the time-reversed state satisfies

$$\begin{aligned}\mathcal{H}\hat{\Theta}|\alpha\rangle &= -\hat{\Theta}\mathcal{H}|\alpha\rangle = \\ &= -\mathcal{E}_\alpha\hat{\Theta}|\alpha\rangle,\end{aligned} \tag{15.193}$$

that is, the opposite of a state is associated with the opposite of energy, which makes no sense. Thus, the time reversal operator is antiunitary, in particular is antilinear

$$-i\mathcal{H}\hat{\Theta}|\alpha\rangle = \hat{\Theta}i\mathcal{H}|\alpha\rangle, \tag{15.194}$$

$$-i\mathcal{H}\hat{\Theta}|\alpha\rangle = -i\hat{\Theta}\mathcal{H}|\alpha\rangle, \tag{15.195}$$

$$\left[\mathcal{H}, \hat{\Theta}\right] = 0. \tag{15.196}$$

15.11.3 Transformation of Operators Under Time Reversal

Now, we examine the transformation properties of physical observables under the action of the time-reversal operator. To this end, we specialize Eq. I.22 to the present context, yielding

$$\langle\beta|A|\alpha\rangle = \langle\tilde{\alpha}|\hat{\Theta}A\hat{\Theta}^{-1}|\tilde{\beta}\rangle \tag{15.197}$$

where A is a physical observable and $\hat{\Theta}$ is an antiunitary operator. The observable A is classified as even or odd under time reversal depending on whether it transforms into itself or into its negative, respectively, i.e.,

$$\hat{\Theta}A\hat{\Theta}^{-1} = \pm A. \tag{15.198}$$

Equations 15.197 and 15.198 impose a constraint on the phase factors of the matrix elements of the operator A taken between time-reversed state vectors. Indeed,

$$\begin{aligned}\langle\beta|A|\alpha\rangle &= \langle\tilde{\alpha}|\hat{\Theta}A\hat{\Theta}^{-1}|\tilde{\beta}\rangle = \\ &= \pm\langle\tilde{\alpha}|A|\tilde{\beta}\rangle = \\ &= \pm\langle\tilde{\beta}|A|\tilde{\alpha}\rangle^*,\end{aligned} \tag{15.199}$$

and in particular, the expectation values of the observable satisfy

$$\langle\alpha|A|\alpha\rangle = \pm\langle\tilde{\alpha}|A|\tilde{\alpha}\rangle^*. \tag{15.200}$$

Equation 15.200 expresses the condition that an observable operator must satisfy under time reversal. For the momentum operator $\hat{\mathbf{p}}$, it follows from the classical analysis that its expectation value in the time-reversed state is the negative of that in the original state, i.e., it changes sign,

$$\langle\alpha|\hat{\mathbf{p}}|\alpha\rangle = -\langle\tilde{\alpha}|\hat{\mathbf{p}}|\tilde{\alpha}\rangle^*, \tag{15.201}$$

that is, the momentum operator is an odd operator, i.e.,

$$\hat{\Theta}\hat{\mathbf{p}}\hat{\Theta}^{-1} = -\hat{\mathbf{p}}, \tag{15.202}$$

then

$$\begin{aligned}\hat{\mathbf{p}}\,\hat{\Theta}|\mathbf{p}'\rangle &= \left(-\hat{\Theta}\hat{\mathbf{p}}\hat{\Theta}^{-1}\right)\hat{\Theta}|\mathbf{p}'\rangle = \\ &= -\hat{\Theta}\hat{\mathbf{p}}|\mathbf{p}'\rangle = \\ &= -\hat{\Theta}\mathbf{p}'|\mathbf{p}'\rangle = \\ &= -\mathbf{p}'\,\hat{\Theta}|\mathbf{p}'\rangle,\end{aligned} \tag{15.203}$$

This confirms that $\hat{\Theta}|\mathbf{p}'\rangle$ is an eigenstate of the transformed momentum operator with eigenvalue $-\ \mathbf{p}'$. For the position operator, instead, we have

$$\langle\alpha|\hat{\mathbf{r}}|\alpha\rangle = \langle\tilde{\alpha}|\hat{\mathbf{r}}|\tilde{\alpha}\rangle^*, \tag{15.204}$$

$$\hat{\Theta}\hat{\mathbf{r}}\hat{\Theta}^{-1} = \hat{\mathbf{r}}, \tag{15.205}$$

and

$$
\begin{aligned}
\hat{\mathbf{r}}\,\hat{\Theta}|\mathbf{r}'\rangle &= \left(\hat{\Theta}\,\hat{\mathbf{r}}\,\hat{\Theta}^{-1}\right)\hat{\Theta}|\mathbf{r}'\rangle = \\
&= \hat{\Theta}\,\hat{\mathbf{r}}|\mathbf{r}'\rangle = \\
&= \hat{\Theta}\,\mathbf{r}'|\mathbf{r}'\rangle = \\
&= \mathbf{r}'\,\hat{\Theta}|\mathbf{r}'\rangle.
\end{aligned}
\tag{15.206}
$$

Let us now consider a useful application of the time-reversal operator. Specifically, we examine a Hamiltonian that is quadratic in both the position and momentum operators. Such a Hamiltonian is invariant under time-reversal symmetry. Under the action of the time-reversal operator, the momentum operator transforms as

$$
\hat{\Theta}\hat{\mathbf{k}}\hat{\Theta}^{-1} = -\hat{\mathbf{k}}. \tag{15.207}
$$

Then, the relation

$$
\hat{\Theta}\mathcal{H}(\mathbf{k})\hat{\Theta}^{-1} = \mathcal{H}(-\mathbf{k}) \tag{15.208}
$$

implies that the eigenvalues of the Hamiltonian must satisfy the symmetry

$$
\mathcal{E}_{\mathbf{k}} = \mathcal{E}_{-\mathbf{k}}. \tag{15.209}
$$

Indeed, if $|\psi_{\mathbf{k}}\rangle$ is an eigenstate of $\mathcal{H}(\mathbf{k})$ with eigenvalue $\mathcal{E}_{\mathbf{k}}$, then

$$
\mathcal{H}(\mathbf{k})|\psi_{\mathbf{k}}\rangle = \mathcal{E}_{\mathbf{k}}|\psi_{\mathbf{k}}\rangle. \tag{15.210}
$$

Applying the time-reversal operator $\hat{\Theta}$ and using the transformation property of the Hamiltonian, one finds

$$
\begin{aligned}
\mathcal{H}(-\mathbf{k})\left(\hat{\Theta}|\psi_{\mathbf{k}}\rangle\right) &= \hat{\Theta}\mathcal{H}(\mathbf{k})|\psi_{\mathbf{k}}\rangle = \\
&= \mathcal{E}_{\mathbf{k}}\left(\hat{\Theta}|\psi_{\mathbf{k}}\rangle\right).
\end{aligned}
\tag{15.211}
$$

Therefore, $\hat{\Theta}|\psi_{\mathbf{k}}\rangle$ is an eigenstate of $\mathcal{H}(-\mathbf{k})$ with the same eigenvalue $\mathcal{E}_{\mathbf{k}}$. This demonstrates that the spectrum of $\mathcal{H}$ is symmetric under the transformation $\mathbf{k} \to -\mathbf{k}$. Physically, the time-reversal operator inverts the direction of momentum while leaving the energy, a scalar quantity, unchanged. Consequently, the energy associated with a state of momentum $\mathbf{k}$ must coincide with that of the state with inverted momentum $-\mathbf{k}$. In particular, this symmetry also applies to the corresponding frequencies. Denoting the frequency associated with the state $|\psi_{\mathbf{k}}\rangle$ by $\omega_{\mathbf{k}}$, we have

$$
\omega_{\mathbf{k}} = \omega_{-\mathbf{k}}. \tag{15.212}
$$

Thus, both energies and frequencies exhibit the same symmetry under momentum inversion.

Appendix A
Operational Aspects of Tensor Products

The tensor product provides a natural framework for describing composite systems. It allows us to combine vector spaces or operators acting on separate subsystems into a single, higher-dimensional space that captures all correlations between them. Here, we introduce the tensor product between vectors and show how to construct a basis for the combined space. We then extend the concept to matrices, providing a clear procedure for forming their tensor product and understanding its structure. This serves as a practical tool for becoming familiar with calculations involving tensor products, useful in many-body systems and operations on composite spaces.

Definition A.1 (Tensor Product Between Vectors) Let V and W be finite-dimensional vector spaces over a field, with dimensions n and m, respectively. The tensor product $V \otimes W$ of two vector spaces V and W (over the same field) is a vector space to which is associated a bilinear map $V \times W \to V \otimes W$ that maps a pair (v, w), $v \in V$, $w \in W$ to an element of $V \otimes W$ denoted $v \otimes w$. Let $e_1, \ldots, e_n, f_1, \ldots, f_m$, be bases for V and W respectively. Then the tensor product basis $\left\{e_i \otimes f_j\right\}_{i,j}^{n,m}$ of $V \otimes W$ can be constructed as follows

$$
\begin{aligned}
\left\{e_i \otimes f_j\right\}_{i,j=1}^{n,m} &= \begin{pmatrix} e_1 \\ \vdots \\ e_n \end{pmatrix} \begin{pmatrix} f_1 \ldots f_m \end{pmatrix} = \\
&= \begin{pmatrix} e_1 f_1 & e_1 f_2 & \ldots & e_1 f_m \\ e_2 f_1 & e_2 f_2 & \ldots & e_2 f_m \\ \vdots & \vdots & \ddots & \vdots \\ e_n f_1 & e_n f_2 & \ldots & e_n f_m \end{pmatrix},
\end{aligned}
\tag{A.1}
$$

with this definition we have that $\dim(V \otimes W) = nm$.

F. Tafuri et al., *Understanding Quantum Many-Body Theory*, UNITEXT for Physics, https://doi.org/10.1007/978-3-032-17849-7

The basis elements are conventionally ordered in lexicographic order: the index i (corresponding to the basis of V) varies more slowly than j (corresponding to W). Consequently, the first basis vector is $e_1 \otimes f_1$, followed by $e_1 \otimes f_2, \ldots,$ up to $e_n \otimes f_m$. When tensor products are represented as matrices, this ordering implies that the row and column indices correspond to pairs (i, l) and (j, k), respectively.

Definition A.2 (Tensor Product Between Matrices) Let A and B be matrices of sizes $n \times m$ and $p \times q$, with entries

$$A = (a_{ij})_{1 \le i \le n,\, 1 \le j \le m}, \tag{A.2}$$

$$B = (b_{lk})_{1 \le l \le p,\, 1 \le k \le q}. \tag{A.3}$$

The tensor product can be constructed as follows

$$\begin{aligned}
A \otimes B &= \begin{pmatrix} a_{11} & \ldots & a_{1m} \\ \vdots & \ddots & \vdots \\ a_{n1} & \ldots & a_{nm} \end{pmatrix} \otimes \begin{pmatrix} b_{11} & \ldots & b_{1q} \\ \vdots & \ddots & \vdots \\ b_{p1} & \ldots & b_{pq} \end{pmatrix} = \\
&= \begin{pmatrix} a_{11} \begin{pmatrix} b_{11} & \ldots & b_{1q} \\ \vdots & \ddots & \vdots \\ b_{p1} & \ldots & b_{pq} \end{pmatrix} & \ldots & a_{1m} \begin{pmatrix} b_{11} & \ldots & b_{1q} \\ \vdots & \ddots & \vdots \\ b_{p1} & \ldots & b_{pq} \end{pmatrix} \\ \vdots & \ddots & \vdots \\ a_{n1} \begin{pmatrix} b_{11} & \ldots & b_{1q} \\ \vdots & \ddots & \vdots \\ b_{p1} & \ldots & b_{pq} \end{pmatrix} & \ldots & a_{nm} \begin{pmatrix} b_{11} & \ldots & b_{1q} \\ \vdots & \ddots & \vdots \\ b_{p1} & \ldots & b_{pq} \end{pmatrix} \end{pmatrix} = \\
&= \begin{pmatrix}
a_{11}b_{11} & a_{11}b_{12} & \ldots & a_{11}b_{1q} & \ldots\ldots & a_{1m}b_{11} & a_{1m}b_{12} & \ldots & a_{1m}b_{1q} \\
a_{11}b_{21} & a_{11}b_{22} & \ldots & a_{11}b_{2q} & \ldots\ldots & a_{1m}b_{21} & a_{1m}b_{22} & \ldots & a_{1m}b_{2q} \\
\vdots & \vdots & \ddots & \vdots & & \vdots & \vdots & \ddots & \vdots \\
a_{11}b_{p1} & a_{11}b_{p2} & \ldots & a_{11}b_{pq} & \ldots\ldots & a_{1m}b_{p1} & a_{1m}b_{p2} & \ldots & a_{1m}b_{pq} \\
\vdots & \vdots & \ddots & \vdots & \ddots & \vdots & \vdots & \ddots & \vdots \\
\vdots & \vdots & \ddots & \vdots & \ddots & \vdots & \vdots & \ddots & \vdots \\
a_{n1}b_{11} & a_{n1}b_{12} & \ldots & a_{n1}b_{1q} & \ldots\ldots & a_{nm}b_{11} & a_{nm}b_{12} & \ldots & a_{nm}b_{1q} \\
a_{n1}b_{21} & a_{n1}b_{22} & \ldots & a_{n1}b_{2q} & \ldots\ldots & a_{nm}b_{21} & a_{nm}b_{22} & \ldots & a_{nm}b_{2q} \\
\vdots & \vdots & \ddots & \vdots & & \vdots & \vdots & \ddots & \vdots \\
a_{n1}b_{p1} & a_{n1}b_{p2} & \ldots & a_{n1}b_{pq} & \ldots\ldots & a_{nm}b_{p1} & a_{nm}b_{p2} & \ldots & a_{nm}b_{pq}
\end{pmatrix},
\end{aligned} \tag{A.4}$$

which is a matrix of dimension $np \times mq$. Explicitly, each block $a_{ij}B$ is a scalar multiple of the entire matrix B. Thus, the resulting matrix has size $np \times mq$, where each scalar a_{ij} multiplies every entry of B in its corresponding block. Accordingly, if $T = A \otimes B$, the generic entry of T can be labeled as

$$\left\{t_{ij,lk}\right\}, \ 1 \le i \le n, \ 1 \le j \le m, \ 1 \le l \le p, \ 1 \le k \le q, \tag{A.5}$$

i.e., the row and column indices are ordered lexicographically according to the structure of the tensor product.

Appendix B
Complex 2 x 2 Matrices

Complex 2×2 matrices are fundamental in linear algebra and its applications, from group representations to quantum mechanics. We denote by $M_2(\mathbb{C})$ the vector space of all 2×2 complex matrices. Being finite-dimensional, it admits a basis that can be explicitly constructed for practical use.

Theorem B.1 (A Basis of $M_2(\mathbb{C})$) *Given the vector space $M_2(\mathbb{C})$ with elements*

$$\begin{aligned} M &= \begin{pmatrix} M_{11} & M_{12} \\ M_{21} & M_{22} \end{pmatrix} = \\ &= M_{11}\begin{pmatrix} 1 & 0 \\ 0 & 0 \end{pmatrix} + M_{12}\begin{pmatrix} 0 & 1 \\ 0 & 0 \end{pmatrix} + M_{21}\begin{pmatrix} 0 & 0 \\ 1 & 0 \end{pmatrix} + M_{22}\begin{pmatrix} 0 & 0 \\ 0 & 1 \end{pmatrix}, \end{aligned} \tag{B.1}$$

a basis is given by the set formed by the 2×2 identity matrix $\mathbb{1}$ and the Pauli matrices $\{\sigma_i\}_{i=1,2,3}$ (see Eqs. 1.53–1.55).

Proof The set $\{\mathbb{1}, \sigma_1, \sigma_2, \sigma_3\}$ is a basis of $M_2(\mathbb{C})$ if and only if

1. the set $\{\mathbb{1}, \sigma_1, \sigma_2, \sigma_3\}$ is linearly independent;
2. each element $M \in M_2(\mathbb{C})$ can be written as

$$M = \alpha_0\mathbb{1} + \alpha_1\sigma_1 + \alpha_2\sigma_2 + \alpha_3\sigma_3, \ \{\alpha_j\}_{j=0}^3 \in \mathbb{C}. \tag{B.2}$$

1. Let $\{\alpha_j\}_{j=0}^3 \in \mathbb{C}$ such that

$$\alpha_0\mathbb{1} + \alpha_1\sigma_1 + \alpha_2\sigma_2 + \alpha_3\sigma_3 = 0, \tag{B.3}$$

then

F. Tafuri et al., *Understanding Quantum Many-Body Theory*, UNITEXT for Physics, https://doi.org/10.1007/978-3-032-17849-7

$$\begin{pmatrix}\alpha_0 & 0\\ 0 & \alpha_0\end{pmatrix}+\begin{pmatrix}0 & \alpha_1\\ \alpha_1 & 0\end{pmatrix}+\begin{pmatrix}0 & -i\alpha_2\\ i\alpha_2 & 0\end{pmatrix}+\begin{pmatrix}\alpha_3 & 0\\ 0 & -\alpha_3\end{pmatrix}=\begin{pmatrix}0 & 0\\ 0 & 0\end{pmatrix}, \tag{B.4}$$

$$\begin{pmatrix}\alpha_0+\alpha_3 & \alpha_1-i\alpha_2\\ \alpha_1+i\alpha_2 & \alpha_0-\alpha_3\end{pmatrix}=\begin{pmatrix}0 & 0\\ 0 & 0\end{pmatrix}, \tag{B.5}$$

$$\alpha_0=\alpha_1=\alpha_2=\alpha_3=0. \tag{B.6}$$

2. From

$$\begin{pmatrix}\alpha_0+\alpha_3 & \alpha_1-i\alpha_2\\ \alpha_1+i\alpha_2 & \alpha_0-\alpha_3\end{pmatrix}=\begin{pmatrix}M_{11} & M_{12}\\ M_{21} & M_{22}\end{pmatrix}, \tag{B.7}$$

we get

$$\begin{cases}\alpha_0+\alpha_3=M_{11}\\ \alpha_1-i\alpha_2=M_{12}\\ \alpha_1+i\alpha_2=M_{21}\\ \alpha_0-\alpha_3=M_{22}\end{cases}, \tag{B.8}$$

which has solutions given by

$$\alpha_0=\frac{1}{2}\left(M_{11}+M_{22}\right), \tag{B.9}$$

$$\alpha_1=\frac{1}{2}\left(M_{12}+M_{21}\right), \tag{B.10}$$

$$\alpha_2=\frac{i}{2}\left(M_{12}-M_{21}\right), \tag{B.11}$$

$$\alpha_3=\frac{1}{2}\left(M_{11}-M_{22}\right). \tag{B.12}$$

Having established that any complex 2×2 matrix can be expressed as a linear combination of the 2×2 identity matrix and the Pauli matrices, we can now focus on a special class of matrices: traceless Hermitian matrices. In the following result, we explicitly show that such matrices are precisely linear combinations of the Pauli matrices with real coefficients.

Theorem B.2 (Characterization of Traceless Hermitian 2×2 Matrices) *Let $M \in M_2(\mathbb{C})$ be a traceless and hermitian matrix, then*

$$M = \sum_{i=1}^{3} \alpha_i \sigma_i, \ \alpha_i \in \mathbb{R}. \tag{B.13}$$

Proof Since the matrix is hermitian, i.e., $M = M^\dagger$, it satisfies

$$\begin{pmatrix} M_{11} & M_{12} \\ M_{21} & M_{22} \end{pmatrix} = \begin{pmatrix} M_{11}^* & M_{21}^* \\ M_{12}^* & M_{22}^* \end{pmatrix}, \tag{B.14}$$

then

$$M = \begin{pmatrix} M_{11} & M_{12} \\ M_{12}^* & M_{22} \end{pmatrix}, \ M_{11}, M_{22} \in \mathbb{R}, \ M_{12} \in \mathbb{C}. \tag{B.15}$$

On the other hand, the null trace condition implies

$$M_{11} + M_{22} = 0. \tag{B.16}$$

From Eqs. B.9–B.16, one finds

$$\alpha_0 = 0, \tag{B.17}$$

$$\begin{aligned} \alpha_1 &= \frac{1}{2}\left(M_{12} + M_{21}\right) = \\ &= \frac{1}{2}\left(M_{12} + M_{12}^*\right) = \\ &= \frac{2\,\mathrm{Re}(M_{12})}{2} = \\ &= \mathrm{Re}(M_{12}) \in \mathbb{R}, \end{aligned} \tag{B.18}$$

$$\begin{aligned} \alpha_2 &= \frac{i}{2}\left(M_{12} - M_{21}\right) = \\ &= \frac{i}{2}\left(M_{12} - M_{12}^*\right) = \\ &= \frac{i}{2}\left(-2i\,\mathrm{Im}(M_{12})\right) = \\ &= \mathrm{Im}(M_{12}) \in \mathbb{R}, \end{aligned} \tag{B.19}$$

$$\alpha_3 = \frac{1}{2}\left(M_{11} - M_{22}\right) \in \mathbb{R}, \tag{B.20}$$

from which the statement follows. □

Hence, the traceless Hermitian 2×2 matrices form a 3-dimensional real vector space spanned by the Pauli matrices.

Appendix C
The Thermodynamic Limit

In statistical physics and condensed matter theory, many-body systems are often described in terms of discrete modes arising from finite, periodic volumes. However, for macroscopic systems, it is convenient to replace these discrete sums by continuous integrals, corresponding to the limit of infinitely large volume and particle number, known as the thermodynamic limit. This procedure allows for a smooth transition from finite to extended systems and provides the physical foundation for the use of Fourier transforms in describing spatial or time distributions.

In the following, we discuss its physical interpretation and show how it motivates the transition from discrete sums to continuous integrals in the thermodynamic limit.

Theorem C.1 (Fourier Series Expansion) *Let $f(x)$ be a square summable function defined on the interval $\left[-\frac{L}{2}, \frac{L}{2}\right]$ and periodic of period L. Such a function can be computed at any point on the real axis by means of its periodic extension, which is given by its Fourier series, i.e., $f(x)$ can be written as follows*

$$f(x) = \sum_{n=-\infty}^{\infty} c_n \frac{e^{-ik_n x}}{\sqrt{L}}, \tag{C.1}$$

with

$$k = \frac{2\pi}{L}, \tag{C.2}$$

$$\begin{aligned} k_n &= nk = \\ &= \frac{2\pi n}{L}, \end{aligned} \tag{C.3}$$

F. Tafuri et al., *Understanding Quantum Many-Body Theory*, UNITEXT for Physics, https://doi.org/10.1007/978-3-032-17849-7

$$c_n = \int_{-\frac{L}{2}}^{+\frac{L}{2}} f(x) \frac{e^{ik_n x}}{\sqrt{L}}\, dx, \ \forall n \in \mathbb{Z}. \tag{C.4}$$

Remark C.1 (Normalization Conventions in Fourier Series) The choice of normalization in a Fourier series, such as the factors $\frac{1}{\sqrt{L}}$ in the coefficients and the 2π in the wave numbers, is purely conventional. Different communities and textbooks adopt different conventions. For example, one may use $\frac{1}{L}$ instead of $\frac{1}{\sqrt{L}}$, or distribute the 2π factors symmetrically between the forward and inverse transforms. These choices do not change the underlying mathematics; they only rescale the Fourier coefficients and the corresponding basis functions. Therefore, all such conventions are equivalent, and one may select the convention that is most convenient for the physical or mathematical context at hand.

From a physical point of view, Fourier series is the expansion of a periodic function in the spatial plane wave basis

$$\left\{ \frac{e^{ik_n x}}{\sqrt{L}} \right\}, \tag{C.5}$$

which satisfy

$$e^{ik_n x} = e^{ik_n (x+L)}. \tag{C.6}$$

Similarly, the same interpretation applies to periodic functions with time period T. In this case, Theorem C.1 is adapted by substituting k and k_n with

$$\omega = \frac{2\pi}{T}, \tag{C.7}$$

$$\omega_n = \frac{2\pi n}{T}, \tag{C.8}$$

respectively. Consequently, a function periodic in time is expanded in the basis of time plane waves

$$\left\{ \frac{e^{i\omega_n t}}{\sqrt{T}} \right\}, \tag{C.9}$$

which satisfy

$$e^{i\omega_n t} = e^{i\omega_n (t+T)}. \tag{C.10}$$

Now we are able to introduce the thermodynamic limit, i.e.,

$$\sum_{\mathbf{k}} \cdots \to \frac{V}{(2\pi)^3} \int d^3\mathbf{k} \ldots \tag{C.11}$$

which is valid at the limits $N \to \infty$, $V \to \infty$. Indeed, consider a sum with respect to the wave vector $\mathbf{k}$ and recall that

$$k_i = \frac{n_i}{2\pi}, \; n_i \in \mathbb{Z}, \; i = x, y, z. \tag{C.12}$$

Now, if the sum can be decomposed into three sums with respect to the coordinate axes, i.e.,

$$\sum_{\mathbf{k}} \cdots = \sum_{k_x} \sum_{k_y} \sum_{k_z} \cdots \tag{C.13}$$

and the limits $N \to \infty$, $V \to \infty$ are valid, each of these sums can be approximated by an integral, as in Theorem C.1. Then, we multiply and divide by the volume $V = L_x L_y L_z$ of the physical system as follows

$$V \left(\frac{1}{L_x} \frac{1}{L_y} \frac{1}{L_z} \sum_{k_x} \sum_{k_y} \sum_{k_z} \cdots \right). \tag{C.14}$$

Consequently, the discrete quantity $\mathbf{k}$ is replaced by the measure $d^3\mathbf{k}$, the sum gives way to an integral extended over $\mathbb{R}^3$, and a factor $(2\pi)^{-1}$ arises for each coordinate axis. Eq. C.11 has been proved.

Ultimately, note that the factor $(2\pi)^{-3}$ in the three-dimensional integral corresponds to the density of discrete wave vectors in momentum space, consistent with the choice k_n in the Fourier series.

Appendix D
Creation, Annihilation, and Number Operators: Bosonic and Fermionic Cases

In quantum statistical mechanics, particles are classified into two fundamental categories according to their statistical behavior: fermions and bosons. A fundamental way to distinguish between bosons and fermions is by examining the algebra satisfied by their respective annihilation and creation operators.

Definition D.1 (Bosonic and Fermionic Operator Algebra) Let a_α and $a_\alpha^\dagger$ denote the annihilation and creation operators associated with the one-particle state α (see Eqs. 1.44 and 1.45). The algebraic relations obeyed by these operators are determined by the statistical nature of the particles. We have

- Bosons. The annihilation and creation operators obey the commutation relation: the commutator between an annihilation and a creation operator is

$$\left[a_\alpha, a_\beta^\dagger\right] = \delta_{\alpha,\beta}\mathbb{1}, \tag{D.1}$$

  while commutators between two annihilation operators or two creation operators vanish, i.e.,

$$\left[a_\alpha, a_\beta\right] = 0, \ \forall\, \alpha, \beta, \tag{D.2}$$

$$\left[a_\alpha^\dagger, a_\beta^\dagger\right] = 0, \ \forall\, \alpha, \beta. \tag{D.3}$$

- Fermions. The annihilation and creation operators obey the anticommutation relation: the anticommutator between an annihilation and a creation operator is

$$\left\{a_\alpha, a_\beta^\dagger\right\} = \delta_{\alpha,\beta}\mathbb{1}, \tag{D.4}$$

  while the anticommutators between two annihilation operators or two creation operators vanish, i.e.,

F. Tafuri et al., *Understanding Quantum Many-Body Theory*, UNITEXT for Physics, https://doi.org/10.1007/978-3-032-17849-7

$$\left\{a_\alpha, a_\beta\right\} = 0, \ \forall\, \alpha, \beta, \tag{D.5}$$

$$\left\{a_\alpha^\dagger, a_\beta^\dagger\right\} = 0, \ \forall\, \alpha, \beta. \tag{D.6}$$

Thanks to the parentheses 2.102, we can write the bosonic and fermionic algebras, that is, D.1 and D.4, respectively, in a compact form as follows

$$a_\alpha a_\beta^\dagger - \varepsilon a_\beta^\dagger a_\alpha = \delta_{\alpha,\beta}\mathbb{1}, \tag{D.7}$$

which will often be useful in the form

$$a_\alpha a_\beta^\dagger = \delta_{\alpha,\beta}\mathbb{1} + \varepsilon a_\beta^\dagger a_\alpha. \tag{D.8}$$

In particular, for a creation and annihilation operators, bosonic or fermionic, referring to different labels, the commutation or anticommutation rules are both described in compact form by

$$a_\alpha a_\beta^\dagger = \varepsilon a_\beta^\dagger a_\alpha, \ \alpha \neq \beta. \tag{D.9}$$

Furthermore, by means of the statistical index 2.52, Eqs. D.2 and D.5, as well as Eqs. D.3 and D.6, can also be written in a unified form as follows

$$a_\alpha a_\beta = \varepsilon a_\beta a_\alpha, \ \forall\, \alpha, \beta, \tag{D.10}$$

$$a_\alpha^\dagger a_\beta^\dagger = \varepsilon a_\beta^\dagger a_\alpha^\dagger, \ \forall\, \alpha, \beta. \tag{D.11}$$

Remark D.1 (Notation Convention) In many applications, the identity operator $\mathbb{1}$ appearing in commutation or anticommutation relations is understood and thus often omitted for brevity. This simplification is particularly common when the algebraic structure is the main focus, and the presence of the identity operator does not affect the physical interpretation. Whenever omitted, its presence should be considered implicit.

Now, given a one-particle state labeled by α, the number operator associated with this state, denoted by $\hat{N}_\alpha$, is adapted from 1.139 as follows

$$\hat{N}_\alpha = a_\alpha^\dagger a_\alpha. \tag{D.12}$$

We have

Theorem D.1 *Regardless of the statistical nature of creation and annihilation operators, the number operator D.12 satisfies the following commutators*

$$\left[a_\alpha, \hat{N}_\beta\right] = \delta_{\alpha,\beta} a_\beta, \tag{D.13}$$

$$\left[a^\dagger_\alpha, \hat{N}_\beta\right] = -\delta_{\alpha,\beta} a^\dagger_\beta. \tag{D.14}$$

Proof For both computations, we will make use of the unified algebraic relation D.8. In addition, Eq. D.10 will be used in the first computation, while Eq. D.11 will be used in the second. One finds

$$\begin{aligned}
\left[a_\alpha, \hat{N}_\beta\right] &= a_\alpha \hat{N}_\beta - \hat{N}_\beta a_\alpha = \\
&= a_\alpha a^\dagger_\beta a_\beta - a^\dagger_\beta a_\beta a_\alpha = \\
&= \left(\delta_{\alpha,\beta} a_\beta + \varepsilon a^\dagger_\beta a_\alpha\right) a_\beta - a^\dagger_\beta a_\beta a_\alpha = \\
&= \delta_{\alpha,\beta} a_\beta + \varepsilon a^\dagger_\beta a_\alpha a_\beta - a^\dagger_\beta a_\beta a_\alpha = \\
&= \delta_{\alpha,\beta} a_\beta + \varepsilon^2 a^\dagger_\beta a_\beta a_\alpha - a^\dagger_\beta a_\beta a_\alpha = \\
&= \delta_{\alpha,\beta} a_\beta.
\end{aligned} \tag{D.15}$$

Similarly, for the second commutator, we have

$$\begin{aligned}
\left[a^\dagger_\alpha, \hat{N}_\beta\right] &= a^\dagger_\alpha \hat{N}_\beta - \hat{N}_\beta a^\dagger_\alpha = \\
&= a^\dagger_\alpha a^\dagger_\beta a_\beta - a^\dagger_\beta a_\beta a^\dagger_\alpha = \\
&= a^\dagger_\alpha a^\dagger_\beta a_\beta - a^\dagger_\beta \left(\delta_{\beta,\alpha} + \varepsilon a^\dagger_\alpha a_\beta\right) = \\
&= a^\dagger_\alpha a^\dagger_\beta a_\beta - \delta_{\beta,\alpha} a^\dagger_\beta - \varepsilon a^\dagger_\beta a^\dagger_\alpha a_\beta = \\
&= a^\dagger_\alpha a^\dagger_\beta a_\beta - \delta_{\alpha,\beta} a^\dagger_\beta - \varepsilon^2 a^\dagger_\alpha a^\dagger_\beta a_\beta = \\
&= -\delta_{\alpha,\beta} a^\dagger_\beta.
\end{aligned} \tag{D.16}$$

It should be noted that Eqs. 1.141 and 1.142, derived for the quantum harmonic oscillator are, of course, a special instance of the more general commutation relations. In particular, Eqs. D.13 and D.14, will be particularly useful when dealing with Hamiltonians written in second quantization, such as those involving interaction terms or mean-field approximations. We now extend this result to more general bilinear forms of creation and annihilation operators.

Theorem D.2 *Regardless of the statistical nature of creation and annihilation operators, they satisfy the following commutators*

$$\left[a_\alpha, a^\dagger_\beta a_\gamma\right] = \delta_{\alpha,\beta} a_\gamma, \tag{D.17}$$

$$\left[a^\dagger_\alpha, a^\dagger_\beta a_\gamma\right] = -\delta_{\alpha,\gamma} a^\dagger_\beta. \tag{D.18}$$

Proof As in the previous case, we will make use of the unified algebraic relation for both computations, along with the same additional identities respectively used in the first and second parts of the earlier derivation. One finds

$$\begin{aligned}\left[a_\alpha, a_\beta^\dagger a_\gamma\right] &= a_\alpha a_\beta^\dagger a_\gamma - a_\beta^\dagger a_\gamma a_\alpha = \\ &= \left(\delta_{\alpha,\beta} + \varepsilon a_\beta^\dagger a_\alpha\right) a_\gamma - a_\beta^\dagger a_\gamma a_\alpha = \\ &= \delta_{\alpha,\beta} a_\gamma + \varepsilon a_\beta^\dagger a_\alpha a_\gamma - a_\beta^\dagger a_\gamma a_\alpha = \\ &= \delta_{\alpha,\beta} a_\gamma + \varepsilon^2 a_\beta^\dagger a_\gamma a_\alpha - a_\beta^\dagger a_\gamma a_\alpha = \\ &= \delta_{\alpha,\beta} a_\gamma .\end{aligned} \tag{D.19}$$

Similarly, for the second commutator, we have

$$\begin{aligned}\left[a_\alpha^\dagger, a_\beta^\dagger a_\gamma\right] &= a_\alpha^\dagger a_\beta^\dagger a_\gamma - a_\beta^\dagger a_\gamma a_\alpha^\dagger = \\ &= a_\alpha^\dagger a_\beta^\dagger a_\gamma - a_\beta^\dagger \left(\delta_{\gamma,\alpha} + \varepsilon a_\alpha^\dagger a_\gamma\right) = \\ &= a_\alpha^\dagger a_\beta^\dagger a_\gamma - \delta_{\alpha,\gamma} a_\beta^\dagger - \varepsilon a_\beta^\dagger a_\alpha^\dagger a_\gamma = \\ &= a_\alpha^\dagger a_\beta^\dagger a_\gamma - \delta_{\alpha,\gamma} a_\beta^\dagger - \varepsilon^2 a_\alpha^\dagger a_\beta^\dagger a_\gamma = \\ &= -\delta_{\alpha,\gamma} a_\beta^\dagger .\end{aligned} \tag{D.20}$$

The above identities can be regarded as a direct generalization of the commutators with the number operator, obtained by replacing N_β with an arbitrary bilinear operator. We next proceed to the most general bilinear-bilinear commutator.

Theorem D.3 *Regardless of the statistical nature of creation and annihilation operators, they satisfy the following commutators*

$$\left[a_\alpha^\dagger a_\beta,\, a_\gamma^\dagger a_\eta\right] = \delta_{\beta,\gamma}\, a_\alpha^\dagger a_\eta - \delta_{\alpha,\eta}\, a_\gamma^\dagger a_\beta . \tag{D.21}$$

Proof Starting from the definition of the commutator,

$$\left[a_\alpha^\dagger a_\beta,\, a_\gamma^\dagger a_\eta\right] = a_\alpha^\dagger a_\beta a_\gamma^\dagger a_\eta - a_\gamma^\dagger a_\eta a_\alpha^\dagger a_\beta . \tag{D.22}$$

We use the unified algebra D.8 as follows

$$a_\beta a_\gamma^\dagger = \delta_{\beta\gamma} + \varepsilon a_\gamma^\dagger a_\beta , \tag{D.23}$$

$$a_\eta a_\alpha^\dagger = \delta_{\eta\alpha} + \varepsilon a_\alpha^\dagger a_\eta . \tag{D.24}$$

Applying the first identity to the first product:

$$\begin{aligned} a_\alpha^\dagger a_\beta a_\gamma^\dagger a_\eta &= a_\alpha^\dagger \big(\delta_{\beta,\gamma} + \varepsilon\, a_\gamma^\dagger a_\beta\big) a_\eta = \\ &= \delta_{\beta,\gamma}\, a_\alpha^\dagger a_\eta + \varepsilon\, a_\alpha^\dagger a_\gamma^\dagger a_\beta a_\eta . \end{aligned} \tag{D.25}$$

Applying the second identity to the second product:

$$\begin{aligned} a_\gamma^\dagger a_\eta a_\alpha^\dagger a_\beta &= a_\gamma^\dagger \big(\delta_{\eta,\alpha} + \varepsilon\, a_\alpha^\dagger a_\eta\big) a_\beta = \\ &= \delta_{\eta,\alpha}\, a_\gamma^\dagger a_\beta + \varepsilon\, a_\gamma^\dagger a_\alpha^\dagger a_\eta a_\beta . \end{aligned} \tag{D.26}$$

Subtracting the two contributions, we find

$$\Big[a_\alpha^\dagger a_\beta,\, a_\gamma^\dagger a_\eta\Big] = \delta_{\beta,\gamma}\, a_\alpha^\dagger a_\eta - \delta_{\eta,\alpha}\, a_\gamma^\dagger a_\beta + \varepsilon\left(a_\alpha^\dagger a_\gamma^\dagger a_\beta a_\eta - a_\gamma^\dagger a_\alpha^\dagger a_\eta a_\beta\right). \tag{D.27}$$

The quartic term vanishes in both cases. For fermions, from the anticommutative relations, Eqs. D.6 and D.5, we have $a_\alpha^\dagger a_\gamma^\dagger = -\, a_\gamma^\dagger a_\alpha^\dagger$, $a_\beta a_\eta = -\, a_\eta a_\beta$, so that the two terms cancel exactly. For bosons, the commutation relations relations, Eqs. D.3 and D.2, yield $a_\alpha^\dagger a_\gamma^\dagger = a_\gamma^\dagger a_\alpha^\dagger$, $a_\beta a_\eta = a_\eta a_\beta$, and again the two terms cancel. Consequently, only the bilinear contributions remain, and the commutator reduces to the stated expression. □

This result completes the hierarchy, showing that the commutators of arbitrary bilinear combinations of creation and annihilation operators remain closed within the same algebraic structure.

Appendix E
Bose-Einstein and Fermi-Dirac Statistics

Fermions obey the Pauli exclusion principle, which implies that each quantum state can be occupied by at most one particle. In contrast, bosons are not subject to this restriction, and multiple particles can occupy the same quantum state simultaneously. Within the framework of the grand canonical ensemble, these intrinsic properties are reflected in the statistical distributions that determine the average occupation number of a quantum state.

Theorem E.1 (Average Occupation Number in a Non-Interacting System: Statistical Mechanics Formulation) *Consider a system of non-interacting particles in thermal and chemical equilibrium with a reservoir of energy and particles, described within the framework of the grand canonical ensemble. Let* $\{\mathcal{E}_\alpha\}$ *denote the one-particle energy levels and* N_α *the occupation number of state* α*. For fermions no restriction applies, whereas for bosons the chemical potential must satisfy*

$$\mu < \inf_\alpha \mathcal{E}_\alpha \equiv \mathcal{E}_0. \tag{E.1}$$

Then, the thermal average (see Eq. 1.379) of the occupation number of a one-particle state α *is given by*

$$\begin{aligned} \left\langle \hat{N}_\alpha \right\rangle &= \frac{1}{e^{\beta(\mathcal{E}_\alpha - \mu)} \pm 1} = \\ &= n_\varepsilon(\mathcal{E}_\alpha), \end{aligned} \tag{E.2}$$

where ε *is given in Eq. 2.52, corresponding to Bose-Einstein statistics*

$$n_{+1}(\mathcal{E}_\alpha) = \frac{1}{e^{\beta(\mathcal{E}_\alpha - \mu)} - 1}, \tag{E.3}$$

or Fermi-Dirac statistics

F. Tafuri et al., *Understanding Quantum Many-Body Theory*, UNITEXT for Physics, https://doi.org/10.1007/978-3-032-17849-7

$$n_{-1}(\mathcal{E}_\alpha) = \frac{1}{e^{\beta(\mathcal{E}_\alpha - \mu)} + 1}. \tag{E.4}$$

Proof In the grand canonical ensemble, the system can exchange both energy and particles with a reservoir. The grand partition function is defined as

$$Z = \sum_{\{N_\alpha\}} e^{-\beta \sum_i N_\alpha(\mathcal{E}_\alpha - \mu)}, \tag{E.5}$$

where the sum $\sum_{\{N_\alpha\}}$ runs over all possible configurations allowed by the particle statistics, i.e.,

- Fermions: $N_\alpha = 0, 1$ (Pauli exclusion principle),
- Bosons: $N_\alpha = 0, 1, 2, \ldots$.

Since the levels are independent, the grand partition function factorizes as follows

$$\begin{aligned} Z = \prod_\alpha \sum_{N_\alpha} e^{-\beta N_\alpha(\mathcal{E}_\alpha - \mu)} \equiv \\ \equiv \prod_\alpha Z_\alpha, \end{aligned} \tag{E.6}$$

with

$$Z_\alpha = \sum_{N_\alpha} e^{-\beta N_\alpha(\mathcal{E}_\alpha - \mu)}. \tag{E.7}$$

- Fermions.

$$\begin{aligned} Z_\alpha = \sum_{N_\alpha = 0,1} e^{-\beta N_\alpha(\mathcal{E}_\alpha - \mu)} = \\ = 1 + e^{-\beta(\mathcal{E}_\alpha - \mu)}, \end{aligned} \tag{E.8}$$

  so that

$$Z^{(f)} = \prod_\alpha \left(1 + e^{-\beta(\mathcal{E}_\alpha - \mu)}\right). \tag{E.9}$$

- Bosons. From

$$Z_\alpha = \sum_{N_\alpha = 0}^{+\infty} e^{-\beta N_\alpha(\mathcal{E}_\alpha - \mu)}, \tag{E.10}$$

setting $n = N_\alpha$ and introducing $q \equiv e^{-\beta(\mathcal{E}_\alpha-\mu)}$, the sum can be recognized as the well-known geometric series

$$\begin{aligned} S(q) &= \sum_{n=0}^{\infty} q^n = \\ &= \frac{1}{1-q}. \end{aligned} \tag{E.11}$$

The assumption in Eq. E.1 guarantees that $\mathcal{E}_\alpha - \mu > 0$, for all α, and therefore $e^{-\beta(\mathcal{E}_\alpha-\mu)} < 1$, ensuring the convergence of the geometric series defining $Z_\alpha^{(b)}$, i.e.,

$$\begin{aligned} Z_\alpha &= \sum_{N_\alpha=0}^{\infty} e^{-\beta N_\alpha(\mathcal{E}_\alpha-\mu)} = \\ &= \frac{1}{1-e^{-\beta(\mathcal{E}_\alpha-\mu)}}, \end{aligned} \tag{E.12}$$

so that

$$Z^{(b)} = \prod_\alpha \frac{1}{1-e^{-\beta(\mathcal{E}_\alpha-\mu)}}. \tag{E.13}$$

According to statistical mechanics, the average occupation number of the one-particle level α is given by

$$\left\langle \hat{N}_\alpha \right\rangle = -\frac{1}{\beta}\frac{\partial \ln Z}{\partial \mathcal{E}_\alpha}. \tag{E.14}$$

- Fermions:

$$\ln Z^{(f)} = \sum_\alpha \ln\left(1 + e^{-\beta(\mathcal{E}_\alpha-\mu)}\right), \tag{E.15}$$

$$\begin{aligned} \left\langle \hat{N}_\alpha \right\rangle &= -\frac{1}{\beta}\frac{\partial}{\partial \mathcal{E}_\alpha}\left[\sum_{\alpha'} \ln\left(1 + e^{-\beta(\mathcal{E}_{\alpha'}-\mu)}\right)\right] = \\ &= -\frac{1}{\beta}\frac{\partial}{\partial \mathcal{E}_\alpha} \ln\left(1 + e^{-\beta(\mathcal{E}_\alpha-\mu)}\right) = \\ &= \frac{1}{e^{\beta(\mathcal{E}_\alpha-\mu)}+1}, \end{aligned} \tag{E.16}$$

which is the Fermi-Dirac distribution.

- Bosons:

$$\ln Z^{(b)} = -\sum_{\alpha} \ln\left(1 - e^{-\beta(\mathcal{E}_\alpha - \mu)}\right), \tag{E.17}$$

$$\begin{aligned}
\left\langle \hat{N}_\alpha \right\rangle &= -\frac{1}{\beta}\frac{\partial}{\partial \mathcal{E}_\alpha}\left[-\sum_{\alpha'} \ln\left(1 - e^{-\beta(\mathcal{E}_{\alpha'} - \mu)}\right)\right] = \\
&= -\frac{1}{\beta}\frac{\partial}{\partial \mathcal{E}_\alpha}\left[-\ln\left(1 - e^{-\beta(\mathcal{E}_\alpha - \mu)}\right)\right] = \\
&= \frac{1}{e^{\beta(\mathcal{E}_\alpha - \mu)} - 1},
\end{aligned} \tag{E.18}$$

which is the Bose-Einstein distribution.

Collecting both cases, the unified expression E.2 follows. □

In the grand canonical ensemble, the total number of particles is not fixed, but fluctuates around an average value. The chemical potential μ acts as a Lagrange multiplier that enforces the constraint on the average particle number. It is precisely this role that allows us to derive the Fermi-Dirac and Bose-Einstein distributions: μ adjusts the occupation probabilities of one-particle states so that the mean total particle number equals the physical value of the system, without needing to fix N exactly. However, these distributions remain valid under certain conditions even in the canonical ensemble, where N is fixed. In particular, they emerge in the thermodynamic limit (see Chap. C), as we now show, when the number of particles becomes very large and fluctuations relative to N become negligible.

Theorem E.2 (Fermi-Dirac and Bose-Einstein Distributions from the Canonical Ensemble in the Thermodynamic Limit) *Consider a system of N non-interacting particles with one-particle energy levels $\{\mathcal{E}_\alpha\}$ in the canonical ensemble. Let $g_\alpha(N_\alpha)$ denote the number of configurations for occupation number N_α, depending on fermionic or bosonic statistics. Then, in the thermodynamic limit $N \to \infty$, the average occupation number of level α approaches the Bose-Einstein (see Eq. E.3) or Fermi-Dirac (see Eq. E.4) distributions.*

Proof The canonical partition function reads

$$Z_N = \sum_{N_\alpha} \delta\left(\sum_{\alpha} N_\alpha - N\right) \prod_{\alpha} g_\alpha(N_\alpha) e^{-\beta \mathcal{E}_\alpha N_\alpha}, \tag{E.19}$$

and the mean occupation of level α is

$$\langle N_\alpha \rangle = \frac{1}{Z_N} \sum_{N_\alpha} N_\alpha\, \delta\left(\sum_{\alpha} N_\alpha - N\right) \prod_{\alpha} g_\alpha(N_\alpha) e^{-\beta \mathcal{E}_\alpha N_\alpha}. \tag{E.20}$$

The total-number constraint couples all levels, so the sum cannot be factorized directly. The delta function can be represented as a Fourier integral:

$$\delta\left(\sum_\alpha N_\alpha - N\right) = \frac{1}{2\pi}\int_0^{2\pi} d\phi\, e^{i\phi(\sum_\alpha N_\alpha - N)}. \tag{E.21}$$

Substituting this into Z_N gives

$$Z_N = \frac{1}{2\pi}\int_0^{2\pi} d\phi\, e^{-i\phi N}\prod_\alpha\sum_{N_\alpha} g_\alpha(N_\alpha)e^{-(\beta\mathcal{E}_\alpha - i\phi)N_\alpha}. \tag{E.22}$$

Now each level appears independently, but with a complex parameter $i\phi$ controlling the total sum. For large N, the saddle-point method estimates the integral over ϕ. The stationary point ϕ_0 is determined by the condition

$$\sum_\alpha \langle N_\alpha\rangle_{\phi_0} = N, \tag{E.23}$$

with

$$\langle N_\alpha\rangle_{\phi_0} = \frac{\sum_{N_\alpha} N_\alpha g_\alpha(N_\alpha)e^{-(\beta\mathcal{E}_\alpha - i\phi_0)N_\alpha}}{\sum_{N_\alpha} g_\alpha(N_\alpha)e^{-(\beta\mathcal{E}_\alpha - i\phi_0)N_\alpha}}. \tag{E.24}$$

Defining

$$\mu \equiv \frac{i\phi_0}{\beta}, \tag{E.25}$$

the average occupation number takes the grand-canonical-like form:

$$\langle N_\alpha\rangle \simeq \frac{\sum_{N_\alpha} N_\alpha g_\alpha(N_\alpha)e^{-\beta(\mathcal{E}_\alpha - \mu)N_\alpha}}{\sum_{N_\alpha} g_\alpha(N_\alpha)e^{-\beta(\mathcal{E}_\alpha - \mu)N_\alpha}}. \tag{E.26}$$

Since $g_\alpha(N_\alpha)$ takes the value 1 for $N_\alpha = 0, 1$ in the case of fermions and 1 for all $N_\alpha \geq 0$ in the case of bosons, Eq. E.26 reduces exactly to the two cases treated in Theorem E.1, namely the Bose-Einstein distribution for bosons and the Fermi-Dirac distribution for fermions. □

As a consequence, the grand-canonical results for the average occupation numbers remain accurate in the canonical ensemble in the thermodynamic limit. The thermodynamic limit will be widely used throughout this text, as we are dealing with systems containing very large numbers of particles.

Appendix F
Jacobi Identity

The Jacobi identity is a fundamental property of commutators that governs how nested operations behave under permutation of the operands. It provides a systematic way to expand and manipulate commutators of products of operators, ensuring consistency in algebraic calculations.

Theorem F.1 (Jacobi Identity) *Given the operators* A, B *and* C, *let* $[\cdot, \cdot]$ *the commutator parentheses, then*

$$[A, BC] = [A, B]\, C + B\, [A, C]\,. \tag{F.1}$$

Proof We exploit the commutator as follows

$$[A, BC] = ABC - BCA, \tag{F.2}$$

and we add and substract the operator BAC and obtain

$$\begin{aligned}[A, BC] &= ABC - BAC + BAC - BCA = \\ &= (ABC - BAC) + (BAC - BCA) = \\ &= (AB - BA)\, C + B\,(AC - CA)\,,\end{aligned} \tag{F.3}$$

from which the thesis follows. □

Note that Eq. F.1 can be equivalently rewritten as

$$[A, [B, C]] + [B, [C, A]] + [C, [A, B]] = 0. \tag{F.4}$$

F. Tafuri et al., *Understanding Quantum Many-Body Theory*, UNITEXT for Physics, https://doi.org/10.1007/978-3-032-17849-7

Moreover, although the identity is a purely algebraic property valid for arbitrary operators in any associative algebra, its significance becomes particularly evident in quantum many-body theory. In this context, one often deals with bosonic operators, which obey commutation relations.

Appendix G
The Commutator-Anticommutator Identity

The commutator-anticommutator identity plays a role analogous to the Jacobi identity, but is particularly convenient when dealing with fermionic systems. It provides a direct algebraic relation between the commutator of an operator with a product and the corresponding anticommutators, allowing for systematic simplification of operator expressions.

Theorem G.1 (The Commutator-Anticommutator Identity) *For any operators X, Y, and Z, the commutator of X with the product YZ can be expressed in terms of anticommutators as*

$$[X, YZ] = \{X, Y\}Z - Y\{X, Z\}. \tag{G.1}$$

Proof We start from the definitions of the commutator and the anticommutator:

$$[X, YZ] = XYZ - YZX, \tag{G.2}$$

$$\{X, Y\} = XY + YX. \tag{G.3}$$

Using the relation $XY = \{X, Y\} - YX$, we obtain

$$\begin{aligned} XYZ &= (\{X, Y\} - YX)Z = \\ &= \{X, Y\}Z - YXZ. \end{aligned} \tag{G.4}$$

Therefore,

$$\begin{aligned} [X, YZ] &= XYZ - YZX = \\ &= \{X, Y\}Z - YXZ - YZX = \\ &= \{X, Y\}Z - Y(XZ + ZX). \end{aligned} \tag{G.5}$$

F. Tafuri et al., *Understanding Quantum Many-Body Theory*, UNITEXT for Physics, https://doi.org/10.1007/978-3-032-17849-7

Since $XZ + ZX = \{X, Z\}$, the thesis follows. □

Moreover, while the identity is a purely algebraic property valid for arbitrary operators within any associative algebra, its relevance becomes especially clear in the context of quantum many-body theory. Similarly to the Jacobi identity (see Eq. F.1), which governs the commutation relations of bosonic operators and ensures the consistency of their algebraic structure, the identity G.1 acquires particular significance for fermionic operators, whose algebra is defined by anticommutation relations. In this case, it greatly simplifies operator manipulations, as many anticommutators either vanish or reduce to Kronecker delta terms.

Appendix H
Matrix Exponential

Given a square matrix $A \in \mathbb{C}^{n\times n}$, the matrix exponential e^A generalizes the scalar exponential function to matrices, allowing one to describe continuous transformations, solve systems of linear differential equations, and study the structure of Lie groups and Lie algebras. Here, we introduce the definition of the matrix exponential via a power series, discuss its convergence, and explore key properties such as the Baker-Campbell-Hausdorff formula, the conjugation formula, the Glauber factorization, and the behavior under differentiation. These properties not only provide practical tools for computation but also reveal the deep algebraic structure underlying non-commutative operators, which is central in quantum mechanics.

Definition H.1 Let $A \in \mathbb{C}^{n\times n}$ be a square matrix. The matrix exponential e^A is defined via the power series, i.e.,

$$\begin{aligned} e^A &= \sum_{k=0}^{\infty} \frac{1}{k!} A^k = \\ &= \mathbb{1} + A + \frac{1}{2!}A^2 + \frac{1}{3!}A^3 + \cdots \end{aligned} \tag{H.1}$$

Note that this series converges absolutely for any matrix $A \in \mathbb{C}^{n\times n}$, since the operator norm of each term is bounded by the corresponding term in the scalar exponential series:

$$\left\| \frac{1}{k!} A^k \right\| \leq \frac{1}{k!} \|A\|^k . \tag{H.2}$$

Therefore, by comparison with the convergent series

F. Tafuri et al., *Understanding Quantum Many-Body Theory*, UNITEXT for Physics, https://doi.org/10.1007/978-3-032-17849-7

$$\sum_{k=0}^{\infty} \frac{1}{k!} \|A\|^k = e^{\|A\|}, \tag{H.3}$$

the matrix exponential is well defined for all square matrices. The matrix exponential plays a fundamental role in many areas of mathematics and physics, such as in the representation theory of Lie groups (see Chap. 14). It satisfies several important properties, among which we highlight some that will be used in this text.

- Baker-Campbell-Hausdorff (BCH) formula. For any two square matrices (or operators) A and B, the following relation holds

$$e^A e^B = \exp\left(A + B + \frac{1}{2}[A, B] + \frac{1}{12}[A, [A, B]] - \frac{1}{12}[B, [A, B]] + \cdots \right), \tag{H.4}$$

  where the series continues with higher-order nested commutators and converges under appropriate conditions. This formula allows expressing the product of exponentials as the exponential of a single matrix, capturing the non-commutativity of A and B.
- The conjugation formula

$$e^X Y e^{-X} = Y + [X, Y] + \frac{1}{2!}[X, [X, Y]] + \frac{1}{3!}[X, [X, [X, Y]]] + \ldots \tag{H.5}$$

  can be seen as a special case of the BCH formula. Start from the BCH formula for two operators A and B, we set $A = X$ and $B = tY$, where t is a real parameter, and expand in powers of t, that is,

$$\exp\left(X + tY + \frac{1}{2}[X, tY] + \ldots\right) \approx e^X \left(1 + tY + t[X, Y] + \frac{t^2}{2!}[X, [X, Y]] + \ldots\right). \tag{H.6}$$

  Multiplying from the right by e^{-X} and taking the limit $t \to 1$, we obtain exactly the conjugation formula H.5. Thus, the conjugation formula is essentially the BCH formula applied to $e^X e^{tY}$, linearized in t. The following special case is widely used in quantum optics and quantum many-body theory.

Theorem H.1 (Truncated Conjugation Formula) *If the nested commutator satisfies*

$$[[A, B], B] = 0, \tag{H.7}$$

then the conjugation formula H.5 truncates after the first term and yields

$$\left[A, e^B\right] = [A, B]\, e^B. \tag{H.8}$$

Proof We first prove the elementary identity

$$[A, B^n] = \sum_{j=0}^{n-1} B^j \, [A, B] \, B^{n-1-j}, \qquad n \geq 1, \tag{H.9}$$

by induction on n. For $n = 1$ the statement is trivial. Assume Eq. H.9 holds for n. Then

$$\begin{aligned}\left[A, B^{n+1}\right] &= [A, B^n B] = \\ &= [A, B^n]B + B^n[A, B].\end{aligned} \tag{H.10}$$

Applying the inductive hypothesis to $[A, B^n]$ gives

$$\begin{aligned}\left[A, B^{n+1}\right] &= \left(\sum_{j=0}^{n-1} B^j[A, B]B^{n-1-j}\right) B + B^n[A, B] = \\ &= \sum_{j=0}^{n} B^j[A, B]B^{n-j},\end{aligned} \tag{H.11}$$

which proves the formula for $n + 1$. Then Eq. H.9 holds for all $n \geq 1$. Under the additional hypothesis $[[A, B], B] = 0$, the commutator $[A, B]$ commutes with B, therefore each term in the sum given in Eq. H.9 can be reordered as

$$B^j[A, B]B^{n-1-j} = B^{n-1}[A, B], \tag{H.12}$$

so that

$$\begin{aligned}[A, B^n] &= \sum_{j=0}^{n-1} B^{n-1}[A, B] = \\ &= n \, [A, B] \, B^{n-1}.\end{aligned} \tag{H.13}$$

Finally, expanding the exponential $e^B = \sum_{n=0}^{\infty} \frac{1}{n!} B^n$ and using the previous result, we get

$$\begin{aligned}\left[A, e^B\right] &= \sum_{n=1}^{\infty} \frac{1}{n!}[A, B^n] = \\ &= \sum_{n=1}^{\infty} \frac{1}{n!} n \, [A, B] \, B^{n-1} =\end{aligned}$$

$$= [A, B] \sum_{n=1}^{\infty} \frac{1}{(n-1)!} B^{n-1} =$$
$$= [A, B]\, e^{B}, \tag{H.14}$$

which is the claimed truncated conjugation formula. □

From Theorem H.1 we can deduce the following fundamental result concerning operators and their exponential.

Theorem H.2 (Commutation of an Operator with Its Exponential) *Let A be a linear operator on a Hilbert space. Then, for any $t \in \mathbb{R}$, we have*

$$\left[A,\, e^{At}\right] = 0, \tag{H.15}$$

that is, A commutes with its exponential.

Proof Applying Theorem H.1 with $B = At$, we observe that

$$\begin{aligned}[A, B] &= [A, At] = \\ &= t\,[A, A] = \\ &= 0.\end{aligned} \tag{H.16}$$

Therefore, the condition on the nested commutator is satisfied, and the truncated formula immediately gives

$$\begin{aligned}\left[A, e^{At}\right] &= [A, At]e^{At} = \\ &= 0.\end{aligned} \tag{H.17}$$

- Glauber formula (special case of BCH). For two operators A and B whose commutator $[A, B]$ is a c-number or it is proportional to the identity operator, and such that all higher nested commutators vanish, the exponential of their sum can be factorized as

$$e^{A+B} = e^{-\frac{1}{2}[A,B]} e^{A} e^{B}. \tag{H.18}$$

  This formula is a special case of the BCH formula and applies, for example, to bosonic creation and annihilation operators $a^{\dagger}$ and a (see Eq. D.1).

Theorem H.3 (Additivity of the Exponential for Commuting Operators) *Let A and B be two operators such that $[A, B] = 0$. Then the exponential of their sum factorizes as follows*

$$e^{A+B} = e^A e^B. \tag{H.19}$$

Proof This is an immediate consequence of the Glauber formula H.18, since the exponential $e^{-\frac{1}{2}[A,B]}$ reduces to the identity operator. □

Theorem H.4 (Commutation of Exponentials of Commuting Operators) *Let A and B be two operators such that $[A, B] = 0$. Then their exponentials commute:*

$$\left[e^A, e^B\right] = 0. \tag{H.20}$$

Proof Since $[A, B] = 0$, by the Theorem H.3 we have

$$e^{A+B} = e^A e^B, \tag{H.21}$$

$$e^{B+A} = e^B e^A, \tag{H.22}$$

but $e^{A+B} = e^{B+A}$, so it follows immediately that

$$e^A e^B = e^B e^A, \tag{H.23}$$

which proves the claim.

- Differentiation rule.

□

Theorem H.5 (Derivative of the Matrix Exponential) *Let A be a constant square matrix. Then, for any real parameter t,*

$$\frac{d}{dt} e^{tA} = Ae^{tA} = e^{tA} A.$$

Proof Start from the power series definition of the matrix exponential:

$$e^{tA} = \sum_{n=0}^{\infty} \frac{t^n}{n!} A^n.$$

Due to the absolute convergence of the series, we can differentiate it term by term, and we get

$$\begin{aligned}\frac{d}{dt} e^{tA} &= \sum_{n=1}^{\infty} \frac{nt^{n-1}}{n!} A^n = \\ &= \sum_{n=1}^{\infty} \frac{t^{n-1}}{(n-1)!} A^n =\end{aligned}$$

$$
\begin{aligned}
&= A \sum_{m=0}^{\infty} \frac{t^m}{m!} A^m = \\
&= A e^{tA}. \qquad \text{(H.24)}
\end{aligned}
$$

Finally, from Eq. H.15, we also have $e^{tA} A = A e^{tA}$. □

Appendix I
Unitary and Antiunitary Operators

Unitary and antiunitary operators are fundamental tools in quantum mechanics to describe symmetries and reversible transformations. Unitary operators are linear and preserve inner products, representing reversible evolutions, while antiunitary operators are antilinear and preserve norms up to complex conjugation, encoding symmetries such as time-reversal. Here, we present precise definitions of these operators, explore their key properties, and provide a useful characterization of antiunitary operators in terms of unitary operators and complex conjugation.

Definition I.1 (Linear, Unitary, Antilinear, and Antiunitary Operators)

- Linear operator. An operator U acting on a Hilbert space is linear if for all vectors $|\alpha\rangle$, $|\beta\rangle$ and all scalars $c_1, c_2 \in \mathbb{C}$:

$$U\left(c_1|\alpha\rangle + c_2|\beta\rangle\right) = c_1 U|\alpha\rangle + c_2 U|\beta\rangle. \tag{I.1}$$

- Unitary operator. A linear operator U is unitary if it preserves inner products and satisfies

$$UU^\dagger = U^\dagger U = \mathbb{1} \Leftrightarrow U^\dagger = U^{-1}, \tag{I.2}$$

$$\langle U\alpha|U\beta\rangle = \langle\alpha|\beta\rangle. \tag{I.3}$$

  Unitary operators describe reversible transformations that preserve norms and scalar products.
- Antilinear operator. An operator Θ is antilinear if for all vectors $|\alpha\rangle$, $|\beta\rangle$ and scalars $c_1, c_2 \in \mathbb{C}$:

$$\Theta\left(c_1|\alpha\rangle + c_2|\beta\rangle\right) = c_1^*\Theta|\alpha\rangle + c_2^*\Theta|\beta\rangle, \tag{I.4}$$

F. Tafuri et al., *Understanding Quantum Many-Body Theory*, UNITEXT for Physics, https://doi.org/10.1007/978-3-032-17849-7

where c^* denotes complex conjugation. Unlike linear operators, constants and antilinear operators do not commute.

- Antiunitary operator. An antilinear operator Θ is antiunitary if it preserves inner products up to complex conjugation:

$$\langle \Theta\alpha | \Theta\beta \rangle = \langle \beta | \alpha \rangle . \tag{I.5}$$

To exploit the properties of antilinearity, recall that every bra can be seen as a linear application between the space of kets and the complex field, so the bra $\langle \alpha | \Theta$ will act as an antilinear application on a generic ket $|\beta\rangle$. In the scalar product we consider the action of Θ on the ket and apply a complex conjugate as follows

$$(\langle \alpha | \, \Theta) \, |\beta\rangle = [\langle \alpha | (\Theta \, |\beta\rangle)\rangle]^* . \tag{I.6}$$

An antilinear operator Θ is said to be antiunitary if it satisfies $\Theta\Theta^\dagger = \Theta^\dagger\Theta = \mathbb{1}$ which coincides with the first property of unitary operators and depends on the fact that the product of two antiunitary operators is a unitary operator. If an antiunitary operator acts on a bra and its ket, the scalar product becomes

$$\begin{aligned} \left(\langle \alpha | \, \Theta^\dagger\right) (\Theta \, |\beta\rangle) &= \left[\left\langle \alpha \middle| \Theta^\dagger \Theta \middle| \beta \right\rangle\right]^* = \\ &= \langle \beta | \alpha \rangle . \end{aligned} \tag{I.7}$$

that is, only the square modulus of the scalar product is preserved.

Now, we want to show that an antiunitary operator Θ is equivalent to the nonunique product of a unitary operator U and an operator K of complex conjugation, defined as

$$K c|\alpha\rangle = c^* K |\alpha\rangle . \tag{I.8}$$

We now explicate the action of K on the state vectors. Given the basis $|a\rangle$ and the expansion

$$\begin{aligned} |\alpha\rangle &= \sum_a |a\rangle \, \langle a|\alpha\rangle = \\ &= \sum_a \langle a|\alpha\rangle \, |a\rangle , \end{aligned} \tag{I.9}$$

the transformed vector is

$$\begin{aligned} K \, |\alpha\rangle &= K \sum_a \langle a|\alpha\rangle \, |a\rangle = \\ &= \sum_a K \, \langle a|\alpha\rangle \, |a\rangle = \end{aligned}$$

$$
\begin{aligned}
&= \sum_a \langle a|\alpha\rangle^* \, K \, |a\rangle = \\
&= \sum_a \langle a|\alpha\rangle^* \, |a\rangle \,, \qquad \text{(I.10)}
\end{aligned}
$$

where we chose the standard basis $|a\rangle$ on which of course K acts as the identity operator. If the basis is changed, the action of K on a state vector changes. Consequently, the unitary operator U in the product $\Theta = UK$ is not unique but depends on the basis. Now, we verify that the factorization $\Theta = UK$ is an antiunitary operator. First,

$$
\begin{aligned}
\Theta\left(c_1 |\alpha\rangle + c_2 |\beta\rangle\right) &= UK\left(c_1 |\alpha\rangle + c_2 |\beta\rangle\right) = \\
&= c_1^* UK |\alpha\rangle + c_2^* UK |\beta\rangle = \\
&= c_1^* \Theta |\alpha\rangle + c_2^* \Theta |\beta\rangle \,. \qquad \text{(I.11)}
\end{aligned}
$$

To prove Eq. I.7, we follow an approach frequent in the scientific literature, i.e., we consider the action of Θ on the kets and calculate the changes of the bras by complex conjugation of the kets as follows

$$
\begin{aligned}
|\tilde{\alpha}\rangle &= \Theta |\alpha\rangle = \\
&= \sum_{a_1} \langle a_1|\alpha\rangle^* \, U \, |a_1\rangle = \\
&= \sum_{a_1} \langle \alpha|a_1\rangle \, U \, |a_1\rangle \,, \qquad \text{(I.12)}
\end{aligned}
$$

$$
\begin{aligned}
\left|\tilde{\beta}\right\rangle &= \Theta |\beta\rangle = \\
&= \sum_{a_2} \langle \beta|a_2\rangle \, U \, |a_2\rangle \,, \qquad \text{(I.13)} \\
\left\langle\tilde{\beta}\right| &= \sum_{a_2} \langle a_2|\beta\rangle \, \langle a_2| \, U^\dagger, \qquad \text{(I.14)}
\end{aligned}
$$

$$
\begin{aligned}
\left\langle\tilde{\beta}\middle|\tilde{\alpha}\right\rangle &= \sum_{a_1,a_2} \langle a_2|\beta\rangle \left\langle a_2\middle|U^\dagger U\middle|a_1\right\rangle \langle \alpha|a_1\rangle = \\
&= \sum_{a_1,a_2} \langle a_2|\beta\rangle \, \langle a_2|a_1\rangle \, \langle \alpha|a_1\rangle = \\
&= \sum_{a_1} \langle a_1|\beta\rangle \, \langle \alpha|a_1\rangle =
\end{aligned}
$$

$$
\begin{aligned}
&= \sum_{a_1} \langle \alpha | a_1 \rangle \langle a_1 | \beta \rangle = \\
&= \langle \alpha | \beta \rangle = \\
&= \langle \beta | \alpha \rangle^* .
\end{aligned} \tag{I.15}
$$

Theorem I.1 (A Useful Characterization of Anti-Unitary Operators) *Let O be a linear operator, then*

$$
\langle \beta | O | \alpha \rangle = \left\langle \tilde{\alpha} \middle| \Theta O^\dagger \Theta^{-1} \middle| \tilde{\beta} \right\rangle , \tag{I.16}
$$

where Θ is an antiunitary operator.

Proof We define

$$
|\tilde{\alpha}\rangle = \Theta |\alpha\rangle , \tag{I.17}
$$

$$
\left|\tilde{\beta}\right\rangle = \Theta |\beta\rangle , \tag{I.18}
$$

$$
|\gamma\rangle = O^\dagger |\beta\rangle , \tag{I.19}
$$

$$
|\tilde{\gamma}\rangle = \Theta O^\dagger |\beta\rangle , \tag{I.20}
$$

then

$$
\begin{aligned}
\langle \beta | \, O \, | \alpha \rangle &= \langle \gamma | \, \alpha \rangle = \\
&= \langle \tilde{\alpha} | \, \tilde{\gamma} \rangle = \\
&= \langle \tilde{\alpha} | \, \Theta O^\dagger \, | \beta \rangle = \\
&= \langle \tilde{\alpha} | \, \Theta O^\dagger \Theta^{-1} \Theta \, | \beta \rangle = \\
&= \langle \tilde{\alpha} | \, \Theta O^\dagger \, | \beta \rangle = \\
&= \langle \tilde{\alpha} | \Theta O^\dagger \Theta^{-1} | \tilde{\beta} \rangle .
\end{aligned} \tag{I.21}
$$

In the physically relevant case of observables, where $A = A^\dagger$, the above equation takes the following form

$$
\langle \beta | \, A \, | \alpha \rangle = \langle \tilde{\alpha} | \Theta A \Theta^{-1} | \tilde{\beta} \rangle . \tag{I.22}
$$

Appendix J
Trace of an Operator

The trace of a linear operator provides a basis-independent measure of an operator, capturing the sum of its diagonal elements in any orthonormal basis. The trace plays a key role in the study of operator properties, including cyclicity, projections, and the characterization of trace-class operators in infinite-dimensional Hilbert spaces. Here, we define the trace, prove its main properties, and discuss its extensions to projections and infinite-dimensional settings.

Definition J.1 (Trace of a Linear Operator on a Finite-Dimensional Hilbert Space) Let A be a linear operator acting on the finite-dimensional complex Hilbert space H, $\dim(H) = r$, and $\{\varphi_i\}$ a complete orthonormal system, the trace of an operator A is defined as

$$\mathrm{Tr}(A) = \sum_{i=1}^{r} \langle \varphi_i | A | \varphi_i \rangle. \tag{J.1}$$

It follows from the definition that the trace is a linear functional on operators, being a finite sum of diagonal elements with respect to the basis φ_i. For a Hermitian operator, the related trace is a real number.

Theorem J.1 *The trace of an operator is invariant under the choice of basis.*

Proof Given a basis $\{\varphi_i'\}$, we have

$$\begin{aligned}\mathrm{Tr}(A) &= \sum_{i=1}^{r} \langle \varphi_i' | A | \varphi_i' \rangle = \\ &= \sum_i \sum_j \sum_k \langle \varphi_i' | \varphi_j \rangle \langle \varphi_j | A | \varphi_k \rangle \langle \varphi_k | \varphi_i' \rangle =\end{aligned}$$

F. Tafuri et al., *Understanding Quantum Many-Body Theory*, UNITEXT for Physics, https://doi.org/10.1007/978-3-032-17849-7

$$= \sum_i \sum_j \sum_k \delta_{ij}\delta_{ki}\langle\varphi_j|A|\varphi_k\rangle =$$

$$= \sum_j \langle\varphi_j|A|\varphi_j\rangle. \tag{J.2}$$

Theorem J.2 (Cyclic Property of the Trace) *Given a set of any n operators* $\{A_1, \ldots, A_n\}$, *the trace satisfies*

$$\mathrm{Tr}\,(A_1A_2\ldots A_n) = \mathrm{Tr}\,(A_nA_1\ldots A_{n-1}) = \ldots = \mathrm{Tr}\,(A_2\ldots A_nA_1)\,. \tag{J.3}$$

which is known as the cyclic property of the trace operator.

Proof Since the proof does not depend on the number of the operators, for simplicity, consider the case $n = 3$. By definition, the trace of the product $A_1A_2A_3$ is given in the bracket notation by

$$\mathrm{Tr}\,(ABC) = \sum_n \langle n|ABC|n\rangle, \tag{J.4}$$

then we insert the identities $\sum_m |m\rangle\langle m|$, $\sum_p |p\rangle\langle p|$ (see Eq. 1.14), as follows

$$\mathrm{Tr}\,(ABC) = \sum_n \sum_m \sum_p \langle n|A|m\rangle\langle m|B|p\rangle\langle p|C|n\rangle \equiv$$

$$\equiv \sum_n \sum_m \sum_p A_{nm}B_{mp}C_{pn}. \tag{J.5}$$

Note that row-by-column products between operators can be arranged in several equivalent ways, i.e.,

$$\sum_n \sum_m \sum_p A_{nm}B_{mp}C_{pn} = \sum_p \sum_m \sum_n C_{pn}A_{nm}B_{mp} =$$

$$= \sum_m \sum_n \sum_p B_{mp}C_{pn}A_{nm}. \tag{J.6}$$

Since the summation indices are dummy variables, their permutation does not change the sum, which establishes cyclicity. □

Theorem J.3 *Given a projection operator on the space* $H_g \subset H$, *with* $\dim(H) = r$ *and* $\dim(H_g) = g < r$

$$P^g = \sum_{j=1}^{g} |\varphi_j\rangle\langle\varphi_j| \tag{J.7}$$

we have

$$\mathrm{Tr}\left(P^{g}\right) = g. \tag{J.8}$$

Proof From definition it follows

$$\begin{aligned}
\mathrm{Tr}(P^{g}) &= \sum_{i=1}^{g} \langle \varphi_i | P^{g} | \varphi_i \rangle = \\
&= \sum_{i} \sum_{j} \langle \varphi_i | \varphi_j \rangle \langle \varphi_j | \varphi_i \rangle = \\
&= \sum_{i} \sum_{j} \langle \varphi_j | \varphi_i \rangle^{*} \langle \varphi_j | \varphi_i \rangle = \\
&= \sum_{j=1}^{g} |\langle \varphi_j | \varphi_j \rangle|^{2} = \\
&= g.
\end{aligned} \tag{J.9}$$

Finally, for infinitely dimensional Hilbert spaces, the sum defining the trace becomes an infinite series, so the definition is adapted as follows.

Definition J.2 (A Trace Class Operator) The trace class operators are the linear operators defined on infinite-dimensional Hilbert spaces for which the series defined by the trace is absolutely convergent.

Appendix K
Residue Theorem

The residue theorem, supported by tools such as Jordan's lemma, provides an elegant and powerful method for evaluating complex and real integrals, especially those involving functions with isolated singularities. Its power lies in the fact that, instead of integrating directly along the contour, it suffices to know the residues of the function at the interior singular points.

Definition K.1 (Simple Pole) An isolated singularity z_0 of a complex function $f(z)$ is called a simple pole if, in a neighborhood of z_0, the function can be written in the form

$$f(z) = \frac{g(z)}{z - z_0}, \tag{K.1}$$

where $g(z)$ is complex differentiable in a neighborhood of z_0 and $g(z_0) \neq 0$.

Definition K.2 (Residue at a Simple Pole) Let $f(z)$ be a complex function with an isolated singularity at z_0. If z_0 is a simple pole of f, then in a neighborhood of z_0 the function can be expressed by the Laurent series expansion

$$f(z) = \frac{a_{-1}}{z - z_0} + a_0 + a_1(z - z_0) + \cdots , \tag{K.2}$$

where the coefficient a_{-1} of the $(z - z_0)^{-1}$ term is nonzero. The residue of f at z_0 is precisely this coefficient a_{-1}, which can be computed conveniently by the formula

$$\mathrm{Res}_{z_0}(f) = \lim_{z \to z_0} (z - z_0) f(z). \tag{K.3}$$

In practice, multiplying $f(z)$ by $(z - z_0)$ and taking the limit as $z \to z_0$ yields a finite number that captures the most significant part of the singularity of f at z_0. This value is fundamental for computing complex integrals via the residue theorem, as we show.

F. Tafuri et al., *Understanding Quantum Many-Body Theory*, UNITEXT for Physics, https://doi.org/10.1007/978-3-032-17849-7

Theorem K.1 (Residue Theorem) *Let $\Omega \subset \mathbb{C}$ be an open set, and let $z_1, \ldots, z_n$ be isolated singularities of a complex-valued function f. Let γ be a positively oriented, simple closed curve lying in $\Omega \setminus \{z_1, \ldots, z_n\}$, and suppose that each z_k lies in the interior of the region enclosed by Γ. Assume that the function $f : \Omega \setminus \{z_1, \ldots, z_n\} \to \mathbb{C}$ is complex-differentiable at every point of $\Omega \setminus \{z_1, \ldots, z_n\}$. Then the integral of f around Γ is given by*

$$\oint_{\Gamma} f(z)\, dz = 2\pi i \sum_{k=1}^{n} I_{z_k}(\Gamma)\, \mathrm{Res}_{z_k}(f), \tag{K.4}$$

where $\mathrm{Res}_{z_k}(f)$ denotes the residue of f at the singularity z_k, and $I_{z_k}(\Gamma)$ is the winding number of the curve γ around z_k.

In the special case where γ is positively oriented, that is, traversed in the counterclockwise direction so that the interior region lies on the left while moving along the curve, and encloses each z_k exactly once, we have $I_{z_k}(\Gamma) = 1$, and the formula simplifies to

$$\oint_{\Gamma} f(z)\, dz = 2\pi i \sum_{k=1}^{n} \mathrm{Res}_{z_k}(f). \tag{K.5}$$

Lemma K.1 *Let $f(z)$ be continuous on the upper semicircular arc C_R^+ of radius R centered at the origin, and let $a > 0$. Assume there exists a constant M such that*

$$|f(z)| \leq \frac{M}{|z|} \quad \text{for } |z| = R. \tag{K.6}$$

Then, for the arc C_R^+ traversed counterclockwise, we get

$$\lim_{R\to\infty} \int_{C_R^+} f(z) e^{iaz}\, dz = 0. \tag{K.7}$$

Appendix L
Kramers-Krönig Relations

The Kramers-Krönig relations provide a fundamental connection between the real and imaginary parts of a complex function that is analytic in the upper half-plane and vanishes at infinity. These relations are widely used in physics to ensure causality and consistency in response functions, such as dielectric permittivity, refractive index, or susceptibility, where the real and imaginary parts correspond to dispersion and absorption, respectively. Here, we introduce the Cauchy principal value to handle integrals with singularities, and then derive the Kramers-Krönig relations rigorously using contour integration and Cauchy's theorem, highlighting their role in linking the dispersive and absorptive behavior of physical systems.

Definition L.1 (Cauchy Principal Value) Let $f(x)$ be a function with a singularity at a point $c \in (a, b)$. The Cauchy principal value of the integral

$$\int_a^b f(x)\,dx \tag{L.1}$$

is defined as

$$\text{P.V.} \int_a^b f(x)\,dx = \lim_{\epsilon \to 0^+} \left(\int_a^{c-\epsilon} f(x)\,dx + \int_{c+\epsilon}^b f(x)\,dx \right), \tag{L.2}$$

and if this limit exists, the integral is said to admit a principal value.

Thanks to Definition L.1, we can now prove a useful relation in theoretical physics. We have

Theorem L.1 (Kramers–Krönig Relations) *Let $k(z)$ be an analytic function in the upper half-plane and such that*

$$\lim_{|z| \to \infty} k(z) = 0, \tag{L.3}$$

F. Tafuri et al., *Understanding Quantum Many-Body Theory*, UNITEXT for Physics, https://doi.org/10.1007/978-3-032-17849-7

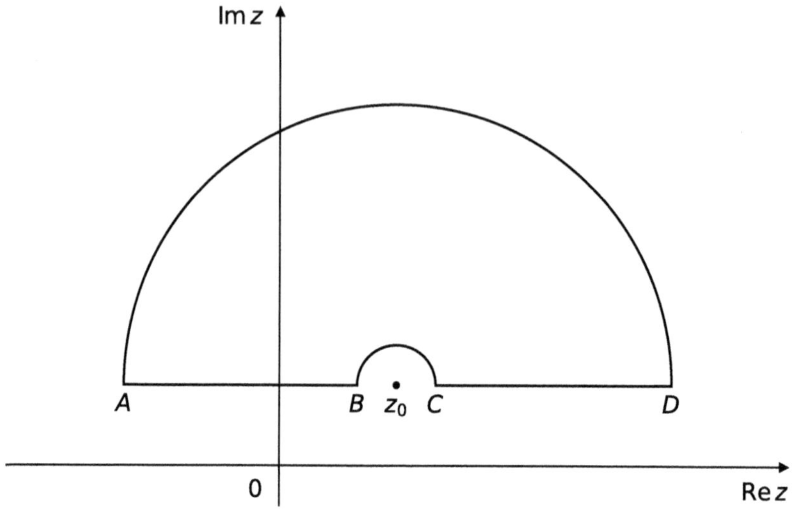

Fig. L.1 The integration contour Γ appearing in Eq. L.6, employed in the derivation of the Kramers-Krönig relations

then, for every $x_0 \in \mathbb{R}$, the function $k(z)$ satisfies the Kramers-Krönig relations

$$\mathrm{Re}\, k(x_0) = \frac{1}{\pi}\ \mathrm{P.V.} \int_{-\infty}^{+\infty} \frac{\mathrm{Im}\, k(x)}{x - x_0}\, dx, \tag{L.4}$$

$$\mathrm{Im}\, k(x_0) = -\frac{1}{\pi}\ \mathrm{P.V.} \int_{-\infty}^{+\infty} \frac{\mathrm{Re}\, k(x)}{x - x_0}\, dx. \tag{L.5}$$

Proof Let $z_0 = x_0 + iy_0$ with $y_0 > 0$. Let us consider the function

$$F(z) = \frac{k(z)}{z - z_0}. \tag{L.6}$$

Since $k(z)$ is analytic for $\mathrm{Im}\, z > 0$ and tends to zero as $|z| \to \infty$, the function $F(z)$ is analytic in the upper half-plane except for a simple pole at $z = z_0$. We define a closed contour Γ (see Fig. L.1) given by the union of

- two horizontal segments parallel to the real axis at height y_0, i.e.,

$$AB : x \in [x_0 - R, x_0 - \varepsilon], \tag{L.7}$$

$$CD : x \in [x_0 + \varepsilon, x_0 + R], \tag{L.8}$$

  with $R \to +\infty$ and $\varepsilon \to 0^+$,
- two semicircles in the upper half-plane, i.e.,

$$\text{small semicircle of radius } \varepsilon \text{ around } z_0, \tag{L.9}$$

$$\text{big semicircle of radius } R \text{ closed in the upper half-plane.} \tag{L.10}$$

Since F is analytic inside Γ except for the pole at z_0, by Cauchy's theorem we have

$$\oint_\Gamma F(z)\, dz = 0. \tag{L.11}$$

Let us compute the individual contributions.

- Integrals on the segments

$$\int_{x_0-R}^{x_0-\varepsilon} \frac{k(x+iy_0)}{x-x_0}\, dx + \int_{x_0+\varepsilon}^{x_0+R} \frac{k(x+iy_0)}{x-x_0}\, dx = \text{P.V.} \int_{-\infty}^{+\infty} \frac{k(x+iy_0)}{x-x_0}\, dx. \tag{L.12}$$

- Integral on the semicircle of radius ε Using the parametrization $z = z_0 + \varepsilon e^{i\theta}$, $\theta \in [\pi, 0]$,

$$\int_\pi^0 ik(z_0 + \varepsilon e^{i\theta})\, d\theta \xrightarrow[\varepsilon \to 0^+]{} -i\pi k(z_0). \tag{L.13}$$

- Integral on the semicircle of radius R Using the parametrization $z = z_0 + Re^{i\theta}$, $\theta \in [0, \pi]$,

$$\int_0^\pi ik(z_0 + Re^{i\theta})\, d\theta \xrightarrow[R \to \infty]{} 0, \tag{L.14}$$

by the hypothesis $\lim_{|z|\to\infty} k(z) = 0$.

Putting together the contributions and taking the limits $R \to \infty$, $\varepsilon \to 0^+$, we get

$$\text{P.V.} \int_{-\infty}^{+\infty} \frac{k(x+iy_0)}{x-x_0}\, dx = i\pi k(x_0 + iy_0), \tag{L.15}$$

from which, separating the real and imaginary parts of k, we obtain the Kramers-Krönig relations. □

Appendix M
Euler Gamma Function

The Euler gamma function is a fundamental extension of the factorial function to complex numbers. It plays a central role in mathematical analysis, probability theory, and statistical physics, providing a continuous interpolation of the factorial and serving as a key tool in asymptotic approximations such as Stirling's formula. The Euler gamma function is defined as

$$\Gamma(z) = \int_0^\infty t^{z-1} e^{-t} dt, \ z \in \mathbb{C}, \ \mathrm{Re}(z) > 0. \tag{M.1}$$

Theorem M.1 (Recurrence Relation for the Euler Gamma Function) *The Euler gamma function satisfies the following recurrence relation*

$$z\Gamma(z) = \Gamma(z+1). \tag{M.2}$$

Proof Consider

$$\Gamma(z+1) = \int_0^\infty t^z e^{-t} \, dt. \tag{M.3}$$

We apply integration by parts by choosing

$$u = t^z, \tag{M.4}$$

$$du = z t^{z-1} dt, \tag{M.5}$$

$$dv = e^{-t} dt, \tag{M.6}$$

$$v = -e^{-t}. \tag{M.7}$$

F. Tafuri et al., *Understanding Quantum Many-Body Theory*, UNITEXT for Physics, https://doi.org/10.1007/978-3-032-17849-7

Then,

$$\Gamma(z+1) = \int_0^\infty t^z e^{-t} dt = \\ = \left[-t^z e^{-t}\right]_0^\infty + \int_0^\infty z t^{z-1} e^{-t}\, dt. \tag{M.8}$$

We now examine the boundary term. For $\mathrm{Re}(z) > 0$,

$$\lim_{t\to\infty} t^z e^{-t} = 0, \tag{M.9}$$

$$\lim_{t\to 0^+} t^z e^{-t} = 0, \tag{M.10}$$

since the exponential decay dominates at infinity and $t^z \to 0$ as $t \to 0^+$. Hence, the boundary term in Eq. M.8 vanishes, and we obtain

$$\Gamma(z+1) = z \int_0^\infty t^{z-1} e^{-t} dt, \tag{M.11}$$

which yields the recurrence in Eq. M.2. □

In the particular case $z = n \in \mathbb{N}$, Eq. M.2 gives

$$\Gamma(n+1) = n\Gamma(n) = \\ = n(n-1)\Gamma(n-1) = \\ \vdots \\ = n!\Gamma(1), \tag{M.12}$$

with

$$\Gamma(1) = \int_0^\infty t^0\, e^{-t}\, dt = \\ = \int_0^\infty e^{-t}\, dt = \\ = \left[-e^{-t}\right]_0^\infty = \\ = 1. \tag{M.13}$$

In other words, the Euler gamma function is a generalization of the factorial function, that is

$$\Gamma(n+1) = n!,\ \forall\, n \in \mathbb{N}. \tag{M.14}$$

Consequently, for large values of $z \in \mathbb{N}$ we can use the Stirling approximation, i.e.,

$$\begin{aligned}\Gamma(z+1) &= z! \simeq \\ &\simeq z^z\, e^{-z}\, \sqrt{2\pi z} = \\ &= z^{z+\frac{1}{2}}\, e^{-z}\, \sqrt{2\pi},\end{aligned} \tag{M.15}$$

with

$$z \in \mathbb{N},\ z \to \infty. \tag{M.16}$$

It can be shown that this approximation is valid even for complex numbers, i.e.,

$$\Gamma(z+1) \simeq z^{z-\frac{1}{2}}\, e^{-z}\, \sqrt{2\pi} \tag{M.17}$$

with

$$z \in \mathbb{C},\ \mathrm{Re}(z) > 0,\ \mathrm{Re}(z) \to \infty. \tag{M.18}$$

In statistical physics, a widely used equation derived from the Stirling approximation is given by

$$\begin{aligned}\ln\Gamma(z+1) &\simeq \ln\left(z^{z+\frac{1}{2}}\, e^{-z}\, \sqrt{2\pi}\right) = \\ &= \left(z+\frac{1}{2}\right)\ln z - z + \ln\sqrt{2\pi} = \\ &= z\ln z + \frac{1}{2}\ln z - z + \ln\sqrt{2\pi} \simeq \\ &\simeq z\ln z - z,\, z \in \mathbb{R}^+,\ z \to \infty.\end{aligned} \tag{M.19}$$

Appendix N
The Volume of a Hypersphere

Here, we compute the volume of a hypersphere in an arbitrary number of dimensions. In statistical physics, both classical and quantum, the dimension of the space often corresponds to the number of degrees of freedom of the system. For a system of N particles in three-dimensional space, the momentum phase space is $3N$-dimensional, with each degree of freedom associated with a Cartesian component of a particle's momentum. We focus on a hypersphere in this momentum subspace, whose radius is determined by the maximal total kinetic energy of the system. The computation of its volume is a fundamental step in evaluating phase-space integrals and understanding the statistical properties of many-particle systems. To derive a general formula, we will employ Gaussian integrals, hyperspherical coordinates, and the Euler Gamma function.

Theorem N.1 *Let $B^{3N}(R) = \left\{x \in \mathbb{R}^{3N} : \|x\| \leq R\right\}$ be the hypersphere of radius $R = \sqrt{2m\mathcal{E}}$ in the momentum phase space of N particles in 3 dimensions. Here, the radius R is set by the maximal total kinetic energy $\mathcal{E}$ of the system, assuming each particle has mass m. Then, the volume of this hypersphere is*

$$V_{3N} = \frac{\pi^{\frac{3N}{2}}}{\Gamma\left(\frac{3N}{2}+1\right)} R^{3N}. \tag{N.1}$$

Proof By dimensional analysis and rotational invariance, the volume must scale as a constant B_{3N}, depending on the space dimension, multiplied by a power of the radius R dictated by the dimensionality of the space, namely,

$$V_{3N} = B_{3N} R^{3N}. \tag{N.2}$$

The hypersphere is defined in momentum space, then

F. Tafuri et al., *Understanding Quantum Many-Body Theory*, UNITEXT for Physics, https://doi.org/10.1007/978-3-032-17849-7

$$0 \leq \sum_{i=1}^{3N} \frac{p_i^2}{2m} \leq \mathcal{E}, \tag{N.3}$$

$$0 \leq \sum_{i=1}^{3N} p_i^2 \leq 2m\mathcal{E}. \tag{N.4}$$

Since the phase space is constructed from orthogonal coordinates, this condition defines a ball in $\mathbb{R}^{3N}$ endowed with the Euclidean norm. Accordingly, let us consider a Gaussian integral in Cartesian coordinates, i.e.,

$$\begin{aligned} I &= \int_{0 \leq \sum_{i=1}^{3N} p_i^2 \leq 2m\mathcal{E}} e^{-\sum_{i=1}^{3N} p_i^2} d^{3N} p = \\ &= \int_{-\sqrt{2m\mathcal{E}}}^{+\sqrt{2m\mathcal{E}}} e^{-p_1^2} dp_1 \int_{-\sqrt{2m\mathcal{E}}}^{+\sqrt{2m\mathcal{E}}} e^{-p_2^2} dp_2 \ldots \int_{-\sqrt{2m\mathcal{E}}}^{+\sqrt{2m\mathcal{E}}} e^{-p_{3N}^2} dp_{3N}. \end{aligned} \tag{N.5}$$

In the thermodynamic limit, the contribution from the boundary of the integration domain is negligible, allowing the extension of the integration limits to infinity, i.e.,

$$\begin{aligned} I &\simeq \int_{-\infty}^{+\infty} e^{-p_1^2} dp_1 \int_{-\infty}^{+\infty} e^{-p_2^2} dp_2 \ldots \int_{-\infty}^{+\infty} e^{-p_{3N}^2} dp_{3N} = \\ &= \left[\sqrt{\pi}\right]^{3N} = \\ &= \pi^{\frac{3N}{2}}. \end{aligned} \tag{N.6}$$

Let us express this Gaussian integral in a hyperspherical coordinate system, with one radial dimension p and $3N - 1$ angular dimensions. Note that in $\mathbb{R}^3$ the surface (V_2) and volume (V_3) elements satisfy

$$dV_2 = 2\pi R dR, \tag{N.7}$$

$$dV_3 = 4\pi R^2 dR, \tag{N.8}$$

and since the $(3N - 1)$-dimensional hypersurface is obtained by differentiating the $3N$-dimensional volume with respect to the radius,

$$V_{d-1} = \frac{d}{dR} V_d, \tag{N.9}$$

we have in particular

$$V_{3N-1} = 3N B_{3N} R^{3N-1}. \tag{N.10}$$

It follows that the hypervolume element in hyperspherical coordinates is

$$dV_{3N} = 3NB_{3N}R^{3N-1}dR, \tag{N.11}$$

after integration over $3N-1$ angular variables. Consequently, from Cartesian to hyperspherical coordinates, the hypervolume element is transformed as

$$dV_{3N} = \prod_{i=1}^{3N} dp_i, \tag{N.12}$$

$$dV_{3N} = 3NB_{3N}p^{3N-1}dp, \tag{N.13}$$

and the Gaussian integral is written as

$$\begin{aligned} I &= \pi^{\frac{3N}{2}} = \\ &= 3NB_{3N}\int_0^{\sqrt{2m\mathcal{E}}} e^{-p^2}p^{3N-1}dp \simeq \\ &\simeq 3NB_{3N}\int_0^{\infty} e^{-p^2}p^{3N-1}dp. \end{aligned} \tag{N.14}$$

Let us compute B_{3N}. We have

$$\begin{aligned} \pi^{\frac{3N}{2}} &= 3NB_{3N}\int_0^{\infty} e^{-p^2}p^{3N-1}dp = \\ &= 3NB_{3N}\int_0^{\infty} \frac{2p}{2p}e^{-p^2}p^{3N-1}dp, \end{aligned} \tag{N.15}$$

and from

$$t = p^2, \tag{N.16}$$

$$dt = 2pdp, \tag{N.17}$$

it follows

$$\begin{aligned} \pi^{\frac{3N}{2}} &= 3NB_{3N}\int_0^{\infty} \frac{2p}{2p}e^{-p^2}p^{3N-1}dp = \\ &= \frac{3N}{2}B_{3N}\int_0^{\infty} e^{-p^2}\frac{p^{3N-1}}{p}2pdp = \\ &= \frac{3N}{2}B_{3N}\int_0^{\infty} e^{-p^2}p^{3N-2}2pdp = \end{aligned}$$

$$
\begin{aligned}
&= \frac{3N}{2} B_{3N} \int_0^\infty e^{-p^2} p^{2\left(\frac{3N}{2}-1\right)} 2p dp = \\
&= \frac{3N}{2} B_{3N} \int_0^\infty e^{-t}\, t^{\left(\frac{3N}{2}-1\right)}\, dt,
\end{aligned}
\tag{N.18}
$$

and recalling the definition of the Euler Gamma function $\Gamma(z)$ from Eq. M.1, one finds

$$
\pi^{\frac{3N}{2}} = \frac{3N}{2} B_{3N}\, \Gamma\left(\frac{3N}{2}\right), \tag{N.19}
$$

$$
B_{3N} = \frac{\pi^{\frac{3N}{2}}}{\frac{3N}{2}\, \Gamma\left(\frac{3N}{2}\right)}. \tag{N.20}
$$

From the property M.2 of Gamma function, one finds

$$
B_{3N} = \frac{\pi^{\frac{3N}{2}}}{\Gamma\left(\frac{3N}{2}+1\right)}, \tag{N.21}
$$

from which the thesis follows. □

Appendix O
The Fermi Momentum

The Fermi momentum k_F is a fundamental quantity in the description of electrons in a solid, especially at zero temperature. It represents the maximum magnitude of the wavevector of electrons occupied in their ground state, corresponding to the Fermi energy $\mathcal{E}_F$: all states with momentum less than or equal to k_F are occupied, while those with higher momentum are empty. The following theorem provides a general relation, valid in an arbitrary d-dimensional space, between the particle number density ρ_0 of fermionic particles and the Fermi momentum k_F, where the subscript 0 emphasizes that we are considering non-interacting fermions.

Theorem O.1 (Particle Density of Non-Interacting Particles as a Function of Fermi Momentum in d-Dimensions) *Consider a system of non-interacting fermions, with spin degeneracy g, confined in a volume $V \subset \mathbb{R}^d$, at zero temperature. The particle number density $\rho_0 = \frac{N}{V}$ is related to the magnitude of the Fermi momentum $\mathbf{k}_F$ by the relation*

$$\rho_0 = g\frac{1}{(2\pi)^d}\frac{\pi^{\frac{d}{2}}}{\Gamma\left(\frac{d}{2}+1\right)}k_F^d. \tag{O.1}$$

Proof At zero temperature, all states with energy less than or equal to the Fermi energy are occupied. Since the energy depends only on the magnitude of the wavevector $\mathbf{k}$, the occupied states correspond to all vectors $\mathbf{k} \in \mathbb{R}^d$ such that $k \leq k_F$. In the thermodynamic limit, momentum space becomes continuous, and the number of states in a volume element $d^d k$ in momentum space is

$$dN = g\frac{V}{(2\pi)^d}d^d k, \tag{O.2}$$

where the factor $\frac{V}{(2\pi)^d}$ comes from the periodic quantization in a volume V, and g is the number of spin states accessible for each $\mathbf{k}$. The total number of particles is

F. Tafuri et al., *Understanding Quantum Many-Body Theory*, UNITEXT for Physics, https://doi.org/10.1007/978-3-032-17849-7

obtained by integrating over all states inside the Fermi sphere, that is

$$N = g\frac{V}{(2\pi)^d}\int_{k\leq k_F} d^d k. \tag{O.3}$$

The volume of the hypersphere in $\mathbb{R}^d$ is given by Eq. N.1 with $d = 3N$ and radius $R = k_F$, i.e.,

$$\int_{k\leq k_F} d^d k = \frac{\pi^{\frac{d}{2}}}{\Gamma\left(\frac{d}{2}+1\right)}k_F^d, \tag{O.4}$$

then we obtain

$$N = g\frac{V}{(2\pi)^d}\frac{\pi^{\frac{d}{2}}}{\Gamma\left(\frac{d}{2}+1\right)}k_F^d, \tag{O.5}$$

and dividing both sides by the volume V, the thesis follows. □

Such relation is exact for free electron models, such as the ideal Fermi gas (an ideal gas that obeys Fermi statistics) and the jellium model, where the complexity of electron interactions and the crystal's periodic potentials are neglected, allowing electrons to be treated as free particles confined within a finite volume. In particular, from Theorem O.1, for the physically relevant cases $d = 1, 2, 3$, we have the following

Theorem O.2 (Particle Density of Non-Interacting Particles as a Function of Fermi Momentum in 1D, 2D, and 3D Dimensions) *The relation between the particle density ρ_0 and the magnitude of the Fermi momentum k_F for non-interacting fermions with spin degeneracy g in spatial dimensions $d = 1, 2, 3$ is*

$$\rho_0 = \frac{g}{2\pi}k_F, \tag{O.6}$$

$$\rho_0 = \frac{g}{4\pi}k_F^2, \tag{O.7}$$

$$\rho_0 = \frac{g}{6\pi^2}k_F^3. \tag{O.8}$$

Proof These formulas are obtained as special cases of Eq. O.1, by substituting $d = 1, 2, 3$ and evaluating the corresponding Gamma function given in Eq. M.1. □

Appendix P
Breit-Wigner Formula

The Breit-Wigner formula provides a link between a Lorentzian function and the Dirac delta distribution. It is widely used in physics to model resonances, spectral line shapes, and decay processes, where a sharply peaked function can be approximated by a delta function in the limit of vanishing width. In this chapter, we present the precise sense in which the Lorentzian converges to the delta function as a distribution and provide a step-by-step demonstration of this limiting behavior.

Theorem P.1 *The following result holds*

$$\lim_{y\to 0^+} \frac{1}{\pi}\frac{y}{x^2+y^2} = \delta(x) \tag{P.1}$$

in the sense of distributions, for every continuous and bounded test function $\varphi(x)$ *we have*

$$\lim_{y\to 0^+} \int_{-\infty}^{+\infty} \frac{1}{\pi}\frac{y}{x^2+y^2}\varphi(x)\,dx = \varphi(0). \tag{P.2}$$

Proof We will proceed step by step.

1. Normalization. We verify that the function

$$f_y(x) = \frac{1}{\pi}\frac{y}{x^2+y^2} \tag{P.3}$$

is normalized, then

$$\int_{-\infty}^{+\infty} f_y(x)\,dx = \int_{-\infty}^{+\infty} \frac{1}{\pi}\frac{y}{x^2+y^2}\,dx. \tag{P.4}$$

F. Tafuri et al., *Understanding Quantum Many-Body Theory*, UNITEXT for Physics, https://doi.org/10.1007/978-3-032-17849-7

We make the change of variable $t = \frac{x}{y}$, from which it follows that $dx = y\,dt$, then

$$\int_{-\infty}^{+\infty} f_y(x)\,dx = \int_{-\infty}^{+\infty} \frac{1}{\pi}\frac{y}{y^2t^2 + y^2} y\,dt =$$
$$= \int_{-\infty}^{+\infty} \frac{1}{\pi}\frac{1}{t^2+1}\,dt =$$
$$= 1. \tag{P.5}$$

2. Distributional limit. Let us consider the integral

$$I_y = \int_{-\infty}^{+\infty} f_y(x)\varphi(x)\,dx =$$
$$= \int_{-\infty}^{+\infty} \frac{1}{\pi}\frac{y}{x^2 + y^2}\varphi(x)\,dx. \tag{P.6}$$

We make again the change of variable $t = \frac{x}{y}$ and obtain

$$I_y = \int_{-\infty}^{+\infty} \frac{1}{\pi}\frac{y}{y^2t^2 + y^2}\varphi(yt)y\,dt =$$
$$= \int_{-\infty}^{+\infty} \frac{1}{\pi}\frac{1}{t^2+1}\varphi(yt)\,dt. \tag{P.7}$$

3. The limit. Since φ is a continuous function, if $y \to 0^+$, we have

$$\varphi(yt) \to \varphi(0), \tag{P.8}$$

and by applying the dominated convergence theorem, we obtain

$$\lim_{y\to 0^+} I_y = \varphi(0) \int_{-\infty}^{+\infty} \frac{1}{\pi}\frac{1}{t^2+1}\,dt =$$
$$= \varphi(0), \tag{P.9}$$

which is the thesis.

□

Glossary

Absorption, Stimulated Emission, and Spontaneous Emission Fundamental light-matter interaction processes: absorption transfers energy from a photon to a quantum system; stimulated emission occurs when an incoming photon induces emission from an excited state; spontaneous emission is the decay of an excited state emitting a photon without external stimulation.

Acoustic and Optical Branches Two types of phonon dispersion relations occur in a crystal: acoustic branches correspond to vibrations in which the atoms in the basis move in phase, leading, in the long-wavelength limit, to low-frequency sound modes, while optical branches correspond to out-of-phase motions of atoms in the basis and remain at finite frequency in the same limit.

Active and Passive Rotations Two equivalent ways of describing rotations in space: active rotations rotate the physical object while keeping the coordinate system fixed, whereas passive rotations rotate the coordinate system while keeping the object fixed; both lead to the same transformed vectors or operators.

Adiabatic Describing a process in which a quantum system evolves slowly enough that it remains in an instantaneous eigenstate of its Hamiltonian, often used in the study of quantum dynamics and perturbations.

Algebra A set equipped with operations such as addition and multiplication that satisfy specific axioms, providing the framework for operators, commutators, and anticommutators; in quantum many-body theory, the algebraic relations often determine the statistics and type of particles (bosons or fermions).

Anderson Single-Impurity Model A quantum model describing a localized impurity level hybridized with a conduction electron band, including local Coulomb repulsion.

F. Tafuri et al., *Understanding Quantum Many-Body Theory*, UNITEXT for Physics, https://doi.org/10.1007/978-3-032-17849-7

Annihilation and Creation Operators Operators that remove or add a particle in a given quantum state, forming the foundation of second quantization and enabling the description of many-body systems and their excitations.

Bloch Functions Solutions of the Schrödinger equation for electrons in a periodic potential, expressed as a plane wave modulated by a function with the periodicity of the lattice, forming the basis for electronic band structure in crystalline solids.

Bloch Problem The quantum problem of electrons moving in a periodic potential, leading to Bloch's theorem and the description of electronic band structure in crystalline solids.

Bogoljubov Transformation A canonical transformation used to diagonalize quadratic Hamiltonians, commonly applied in the study of superconductivity and Bose-Einstein condensates.

Boltzmann Constant A fundamental physical constant relating temperature to energy, appearing in statistical mechanics, thermodynamics, and distributions such as Boltzmann, Fermi-Dirac, and Bose-Einstein ones.

Bose-Einstein Condensation A phase transition in which a macroscopic number of bosons occupy the lowest quantum state at low temperature, leading to quantum coherence on a macroscopic scale and phenomena such as superfluidity and collective excitations.

Bose-Einstein Distribution A statistical distribution describing the occupation numbers of indistinguishable bosons in thermal equilibrium, allowing multiple particles to occupy the same quantum state.

Bosons and Fermions Two classes of indistinguishable quantum particles distinguished by their spin and exchange statistics: bosons have integer spin and obey Bose-Einstein statistics, allowing multiple occupancy of a single state, while fermions have half-integer spin and obey Fermi-Dirac statistics, subject to the Pauli exclusion principle.

Boson and Fermion Matsubara Frequencies In finite-temperature quantum many-body theory, the formulation is commonly carried out in imaginary time, defined over a finite interval. This naturally leads to a discrete set of frequency components. Bosonic fields are periodic in imaginary time, so that only modes with an even number of phase windings are allowed, giving rise to frequencies that are integer multiples of a fundamental unit. Fermionic fields, by contrast, are antiperiodic, which enforces an odd number of phase windings and shifts the allowed frequencies by half a unit.

Bravais Lattice An infinite array of discrete points in space generated by integer linear combinations of primitive vectors, defining the translational symmetry of a crystalline solid.

Bubble Polarization A diagrammatic contribution to the polarization propagator consisting of a particle-hole loop (*bubble*), representing the simplest electronic response to an external perturbation and forming the basis for screening and dielectric function calculations.

Cauchy Principal Value A prescription for assigning a finite value to certain improper integrals with singularities by symmetrically excluding infinitesimal regions around the singular points, commonly used in contour integrals.

Chemical Potential The parameter controlling the exchange of particles between a system and a reservoir, determining the average particle number in statistical ensembles and appearing in Fermi-Dirac and Bose-Einstein distributions.

Coherent State A quantum state of a harmonic oscillator or bosonic mode that most closely resembles classical behavior, being an eigenstate of the annihilation operator, exhibiting minimal uncertainty, and maintaining its form under time evolution.

Connected and Disconnected Feynman Diagrams Feynman diagrams that, respectively, cannot or can be separated into independent pieces without cutting propagator lines; connected diagrams contribute to physical quantities such as the free energy or self-energy, while disconnected ones factorize and typically encode normalization or combinatorial contributions.

Coulomb Interaction The electrostatic interaction between charged particles, described by a potential proportional to the product of the charges divided by their separation, playing a central role in electron correlations, collective excitations, and screening effects in many-body systems.

Debye Temperature A characteristic temperature in the Debye model of solids, associated with an effective cutoff frequency introduced to account for the finite number of lattice vibrational modes. It sets the energy scale for phonon excitations and is commonly used to estimate the specific heat and other thermal properties of solids.

Density Operator An operator representing the statistical state of a quantum system, capable of describing both pure and mixed states, and used to compute expectation values in many-body systems.

Density of States The function describing the number of available quantum states per unit energy (and often per unit volume).

Determinant A scalar value associated with a square matrix, encoding properties such as invertibility, volume scaling, and the sign change under row exchanges, commonly used to construct wavefunctions like Slater determinants in many-body systems.

Dielectric Function A response function that quantifies how an interacting many-electron system screens external electric fields, encoding collective excitations, screening strength, and the modification of effective interactions.

Dirac Delta A distribution $\delta(x)$ that is zero everywhere except at $x = 0$, where it is singular, and integrates to 1, used to represent continuous orthonormality and point-like interactions in quantum mechanics.

Dyson Algebraic Equation An equation relating the full Green's function G to the non-interacting Green's function G_0 and the self-energy Σ, $G = G_0 + G_0\Sigma G$, providing a compact algebraic formulation of the Dyson series for systematic inclusion of interaction effects.

Dyson Expansion (or Dyson Series) An expansion of the time-evolution operator in terms of time-ordered integrals of the interaction Hamiltonian, which provides the foundation for diagrammatic techniques and the systematic construction of Green's functions in quantum many-body physics.

Dyson Time-Ordering Operator An operator that arranges products of operators according to their time arguments, used to systematically expand the time-evolution operator in perturbation theory and Feynman diagram calculations.

EELS (Electron Energy-Loss Spectroscopy) A spectroscopy technique in which fast electrons lose energy by exciting collective and one-particle modes of a material; the measured loss spectrum probes the dielectric function, revealing plasmons, interband transitions, and other electronic excitations.

Einstein Temperature A characteristic temperature in the Einstein model of a solid that sets the energy scale of lattice vibrations by assuming all atoms oscillate with the same frequency. It provides a simple estimate of the contribution of phonons to the specific heat and other thermal properties.

Eigenvalue It is a scalar that characterizes how a linear operator acts on one of its eigenvectors or eigenfunctions. It specifies the factor by which the corresponding state is scaled when the operator is applied.

Eigenvector/Eigenfunction An eigenvector—or eigenfunction when represented in a functional basis—is a nonzero element of a Hilbert space that is mapped to a scalar multiple of itself under the action of a linear operator. The associated scalar is the eigenvalue.

Electron-Phonon Coupling The interaction between electrons and quanta of lattice vibrations (phonons), responsible for phenomena such as electrical resistivity, superconductivity, and renormalization of electronic properties in solids.

Entropy A measure of uncertainty or disorder in a quantum system, quantifying the degree of mixedness of a state through the density matrix, often used to characterize correlations and entanglement in many-body systems.

Fermi-Dirac Distribution A statistical distribution describing the occupation numbers of indistinguishable fermions in thermal equilibrium, constrained by the Pauli exclusion principle, where each quantum state can be occupied by at most one particle.

Fermi Momentum The magnitude of the momentum of the highest occupied state at zero temperature in a system of non-interacting fermions, defining the Fermi surface and setting the scale for many electronic properties.

Feynman Diagrammatics A graphical formalism that represents terms in Dyson expansion of Green's functions or scattering amplitudes as diagrams built from propagators and interaction vertices, providing an intuitive and systematic tool for computing many-body and quantum-field-theoretic processes.

Field Operator An operator that depends on position (and possibly time) and creates or annihilates particles at a given point in space, forming the basic building blocks of second quantization and quantum field theory.

First Brillouin Zone The Wigner-Seitz cell of the reciprocal lattice, representing the set of inequivalent wavevectors in a periodic system and forming the fundamental domain for the electronic band structure.

First Quantization A formulation of quantum mechanics in which classical dynamical variables, such as position and momentum, are promoted to operators acting on wavefunctions, describing one-particle quantum states.

Fixed-ion Electronic Problem The quantum problem of determining the electronic structure and properties of a system while treating the nuclei as fixed classical points, forming the basis of the Born-Oppenheimer approximation and many electronic structure calculations.

Fock Space A Hilbert space constructed as the direct sum of N-particle spaces, allowing the description of states with variable particle numbers using creation and annihilation operators in second quantization.

Force Constants Parameters quantifying the strength of interatomic interactions in a crystal, determining the frequencies of lattice vibrations (phonons) and governing the dynamical stability and elastic properties of the solid.

Fugacity An effective chemical potential-like parameter used to characterize particle number in a grand-canonical ensemble, particularly convenient in the treatment of ideal quantum gases and virial expansions.

Gamma Function A generalization of the factorial to complex numbers, widely used in statistical mechanics.

Green's Function A mathematical object describing the propagation of a particle or excitation in a quantum many-body system, encoding information about energy spectra, correlations, and response functions.

Group A set of elements equipped with a binary operation that is associative, has an identity element, and for which every element has an inverse; fundamental for describing symmetries in physics and mathematics.

Hamiltonian The operator representing the total energy of a quantum system, including kinetic and potential terms, whose eigenvalues determine the allowed energy levels and dynamics of many-body states.

Hartree-Fock Equations A set of self-consistent equations determining one-particle orbitals in a many-electron system, where each particle moves in an effective mean field generated by the average distribution of all other electrons, including exchange interactions.

Hartree-Fock, Thomas-Fermi, and RPA Approximations (for Dielectric Functions) Approximate schemes for the dielectric response of electrons: Hartree-Fock includes exchange at the one-particle level; Thomas-Fermi provides a static, semiclassical screening; RPA sums bubble diagrams to capture dynamical screening and collective excitations.

Heaviside Step Function (Θ) A function defined as $\Theta(x) = 0$ for $x < 0$ and $\Theta(x) = 1$ for $x \geq 0$, commonly used to represent abrupt changes, switch-like behavior, and as a convenient tool in integrals and Green's function formulations.

Helmholtz Decomposition A mathematical theorem stating that any sufficiently smooth vector field can be uniquely decomposed into a divergence-free (solenoidal) component and a curl-free (irrotational) component, widely used in electrodynamics.

Hermitian and Anti-Hermitian Hermitian operators satisfy $A^\dagger = A$, ensuring real eigenvalues and representing observables, while anti-Hermitian operators satisfy $A^\dagger = -A$, often appearing in generators of unitary and anti-unitary transformations.

Hilbert Space A complete vector space equipped with an inner product, providing the framework for quantum states and operators in quantum mechanics.

Homomorphism A structure-preserving map between two groups that respects the group operation, mapping the product of elements in the first group to the product of their images in the second.

Hopping The quantum-mechanical process by which a particle, typically an electron, moves between neighboring sites of a lattice, described by off-diagonal terms in tight-binding and related many-body Hamiltonians.

Hubbard Model A lattice model describing interacting electrons, combining a hopping term that allows electrons to move between neighboring sites and an on-site Coulomb repulsion term, widely used to study magnetism, metal-insulator transitions, and correlated electron phenomena.

Identical and Indistinguishable Particles Particles that have the same intrinsic properties and cannot be distinguished experimentally, so that exchanging two particles leaves the physical state unchanged up to a phase, giving rise to Bose-Einstein or Fermi-Dirac statistics.

Identity The operator or matrix $\mathbb{1}$ that leaves all Hilbert space vectors unchanged ($\mathbb{1}v = v$), acting as the neutral element under operator composition.

Independent Boson Model (IBM) A model describing a quantum system in which a localized electronic state is linearly coupled to a bath of non-interacting bosons.

Infinitesimal Generators Operators associated with symmetry transformations that generate finite transformations via exponentiation, forming the basis of the corresponding Lie algebra and determining the system's conserved quantities.

Inverse Temperature The inverse temperature parameter in statistical mechanics, defined as $\beta = (k_B T)^{-1}$, appearing in the Maxwell-Boltzmann, Fermi-Dirac, and Bose-Einstein distributions and governing thermal occupation probabilities.

Jacobi Identity A fundamental property of the Lie bracket in a Lie algebra, ensuring the consistency of the algebra's structure and the closure of commutation relations.

Jellium Model A simplified model of a many-electron system in which the positive ionic background is treated as a uniform charge density; despite its simplicity, it provides a fundamental framework for studying electron correlations, collective excitations, and screening effects, and serves as the starting point for many approximations and models in metals and plasmas.

Jellium (or Constant) Background Term A uniform positive charge distribution that naturally arises in the calculations of the jellium model, simplifying the treatment of electron-electron interactions and screening.

Kramers-Kronig Relations Integral relations connecting the real and imaginary parts of a response function or Green's function, ensuring causality and analyticity, and widely used to extract complete spectral information from measurable quantities.

Kronecker Delta A function δ_{ij} of two discrete indices that equals 1 if $i = j$ and 0 otherwise, commonly used to express orthonormality of basis vectors in discrete spaces.

Kubo Formula A fundamental linear-response relation expressing transport and response coefficients in terms of equilibrium correlation functions of the corresponding operators. It provides the bridge between microscopic quantum dynamics and macroscopic response to weak external perturbations.

Lie Algebra A vector space equipped with a bilinear, antisymmetric bracket (the Lie bracket) satisfying the Jacobi identity, associated with a Lie group and describing its infinitesimal generators and continuous symmetries.

Lie Group A group whose elements depend smoothly on n real parameters, forming a continuous (differentiable) manifold, such that group operations (multiplication and inversion) are smooth functions; widely used to describe continuous symmetries in physics, including rotations, translations, and internal symmetries of quantum systems.

Linear and Antilinear Operators Linear operators satisfy additivity and homogeneity with complex scalars ($A(\alpha v + \beta w) = \alpha Av + \beta Aw$), while antilinear operators reverse the complex scalar ($A(\alpha v + \beta w) = \alpha^* Av + \beta^* Aw$), both acting on Hilbert space vectors.

Lorentz and Drude Models Classical models describing the optical response of materials: the Drude model accounts for free-electron absorption and low-frequency conductivity, while the Lorentz model treats bound charges as harmonic oscillators, capturing optical resonances and absorption spectra.

Magnetic Susceptibility A response function quantifying how a system's magnetization responds to an applied magnetic field, encoding information on spin correlations, collective magnetic excitations, and the tendency toward magnetic ordering.

Matrix Exponential A generalization of the exponential function to matrices, used to generate finite transformations from infinitesimal generators and to solve linear systems of differential equations.

Matsubara Formalism A framework for finite-temperature quantum many-body systems in which time is extended to imaginary values, allowing the calculation of thermal Green's functions and facilitating perturbation expansions in the imaginary-time (Matsubara) representation.

Matsubara Sums Sums over discrete imaginary-time (Matsubara) frequencies used to compute thermal Green's functions and thermodynamic quantities at finite temperature, often evaluated using contour integration or analytic continuation techniques.

Maxwell-Boltzmann Distribution A statistical distribution describing the probabilities of particles occupying energy states in a classical ideal gas, where particles are distinguishable and non-interacting, with velocities following an exponential dependence on kinetic energy.

Mixed State A quantum state described by a statistical ensemble of pure states, represented by a density matrix that is not a projector, reflecting partial knowledge or classical uncertainty about the system.

Number Operator An operator that counts the number of particles in a given quantum state, defined as the product of the corresponding creation and annihilation operators, central to the second quantization formalism.

One-Body, Two-Body, ..., Operators These denote operators acting respectively on a one particle, on pairs of particles, or more generally on subsets of several particles. One-body operators describe one-particle properties, whereas two-body and higher-body operators encode interactions or correlations among multiple particles.

Operator A mathematical object acting on elements of a Hilbert space, representing physical observables, transformations, or interactions in quantum mechanics.

Optical Conductivity A response function describing how a system conducts electric current in response to an oscillating electric field, providing information on charge transport, excitations, and energy dissipation in materials.

Pair Correlation Function A function that measures the probability of finding two particles at specified positions simultaneously relative to an uncorrelated distribution, used to characterize spatial correlations, ordering, and interactions in many-body systems.

Parseval Identity An identity stating that the integral of the squared modulus of a function is equal to the integral of the squared modulus of its Fourier transform.

Particle Density (Fluctuation) Operator An operator that measures the local particle number at a given position, with its deviation from the mean defining density fluctuations, used to study correlations, collective excitations, and response functions in many-body systems.

Partition Function A fundamental quantity in statistical mechanics that assigns statistical weights to the possible states of a system in thermal equilibrium and serves as the basis for the description of its thermodynamic behavior.

Pauli Matrices A set of three 2×2 Hermitian and unitary matrices $(\sigma_x, \sigma_y, \sigma_z)$ representing spin-$\frac{1}{2}$ operators, forming the basis for describing spin dynamics and interactions in quantum systems.

Phonon Propagator The Green's function describing the propagation of phonons in space and time, encoding information about phonon energies, lifetimes, and interactions with electrons or other excitations.

Phonons Quanta of lattice vibrations in a crystalline solid, representing collective excitations of atoms around their equilibrium positions and playing a central role in thermal properties, electron-phonon interactions, and superconductivity.

Photons Quanta of the electromagnetic field, massless bosons with spin 1, mediating electromagnetic interactions and exhibiting both particle-like and wave-like properties, fundamental to light-matter interactions, cavity quantum electrodynamics, and quantum optics.

Permanent A scalar function of a square matrix similar to the determinant but with all positive signs in the expansion, often appearing in the construction of bosonic many-body wavefunctions.

Permutation A rearrangement of a set of elements or indices, used in quantum mechanics to describe particle exchanges and to define symmetrization (bosons) or antisymmetrization (fermions) of many-body wavefunctions.

Plane Waves Solutions of the Schrödinger equation with constant amplitude and a spatially periodic phase, characterized by a well-defined wavevector and momentum, commonly used as basis states in many-body and solid-state systems.

Plasmons Collective oscillations of the electron density in a conducting system, representing quantized charge-density waves, playing a central role in screening, optical properties, and electron energy-loss spectroscopy.

Polylogarithmic Function A special function that naturally appears in the calculation of thermodynamic properties of ideal quantum gases, such as particle number and energy, in terms of fugacity and temperature.

Pure State A quantum state that can be represented by a single state vector in Hilbert space, exhibiting maximal knowledge of the system and yielding a density matrix that is a projector onto that vector.

Quantum Pictures Equivalent formulations of quantum dynamics—such as the Schrödinger, Heisenberg, and interaction pictures—differing in the distribution of time dependence between states and operators, used to simplify the description of evolving quantum systems.

Quasi-Particle An emergent excitation in a quantum many-body system that behaves like a particle with modified properties (e.g., effective mass, charge), representing collective effects.

Reduced Planck Constant The reduced Planck constant $\hbar = \frac{h}{2\pi}$, where h is the Planck constant, is a fundamental physical constant relating the energy of a quantum to its frequency. It appears in the quantization of energy levels, in the Schrödinger equation, and in virtually all expressions of quantum mechanics.

Representation of a Group A mapping of the elements of a group to linear operators or matrices acting on a vector space, preserving the group multiplication, used to implement group symmetries in quantum mechanics and other areas of physics.

Residue Theorem and Jordan's Lemma Fundamental tools of complex analysis: the residue theorem allows evaluation of contour integrals via the sum of residues of singularities inside the contour, while Jordan's lemma provides conditions for the vanishing of integrals over large semicircular arcs.

Second Quantization A formalism in which fields or particle creation and annihilation operators replace one-particle wavefunctions, providing a natural framework to describe many-body quantum systems and particle statistics.

Self-Energy (Proper and Improper) The self-energy Σ describes the interaction-induced modification of a particle's propagator; proper self-energy diagrams cannot be split into two disconnected parts by cutting a single propagator line, while improper diagrams can, and only proper diagrams contribute directly to the Dyson equation.

State Vector or Wavefunction A vector in Hilbert space (or its representation as a function) that fully describes the quantum state of a system, encoding all probabilistic information about observables and allowing computation of expectation values.

Spatial Translation Operator An operator that shifts a quantum state by an arbitrary spatial vector, including lattice vectors in a periodic system, leaving the Hamiltonian invariant.

Spectral Function A function describing the distribution of energy and momentum of excitations in a quantum system, related to the imaginary part of the Green's function and used to analyze quasiparticle properties, lifetimes, and density of states.

Spin An intrinsic form of angular momentum carried by quantum particles, characterized by discrete values and governing magnetic and exchange interactions in many-body systems.

Spin Volume Density Local operator or field representing the spin angular momentum per unit volume. It characterizes the spatial structure of magnetization and provides the basic variable for the theoretical description of spin-dependent responses and magnetic excitations.

Structure Constants Constants defining the commutation relations of the basis elements of a Lie algebra, encoding the algebra's structure and symmetry properties.

Sum Rules for the Moments of the Spectral Function Constraints that relate the moments of a spectral function to expectation values of commutators or powers of the Hamiltonian, providing exact relationships useful for testing approximations and understanding the distribution of spectral weight in quantum many-body systems.

Symmetry Transformation An operation on a physical system that leaves the observable properties or the Hamiltonian invariant, often represented by unitary or antiunitary operators in quantum mechanics.

Temperature A measure of the average kinetic energy of particles in a system, governing thermal fluctuations, population of energy states, and appearing in statistical distributions through the inverse temperature $\beta = (k_B T)^{-1}$.

Thermal Average The expectation value of an observable computed over a statistical ensemble at finite temperature, typically using the density matrix $\rho = \frac{e^{-\beta H}}{Z}$ to weight contributions from different energy states.

Thermal Wick's Theorem A theorem that expresses time- or normal-ordered products of many operators as sums of fully contracted terms, allowing the systematic evaluation of correlation functions and forming the foundation of diagrammatic expansions in quantum many-body theory.

Thermodynamic Limit The limit in which the number of particles N and the system volume V both go to infinity while keeping the density $\frac{N}{V}$ constant, allowing the emergence of well-defined macroscopic properties and the application of statistical mechanics.

Tight-Binding Model A quantum model describing electrons as strongly localized around atomic sites, with hopping terms connecting neighboring sites, used to calculate electronic band structures and study transport properties in crystalline solids.

Unit Cell (Primitive Cell) The smallest volume of space that, when repeated through all lattice translations, reproduces the entire crystal lattice, uniquely defining the lattice structure.

Unitary and Antiunitary Unitary operators preserve inner products and norms in Hilbert space ($A^\dagger A = \mathbb{1}$), representing reversible quantum transformations, while antiunitary operators combine antilinearity with norm preservation, often describing time-reversal symmetry.

Wannier Functions A set of orthonormal functions localized in real space, obtained from Bloch functions through a unitary transformation, used to describe electrons in crystalline solids and to construct tight-binding models.

Wick Time-Ordering Operator An operator that arranges products of time-dependent operators such that their times increase from right to left, introducing a sign factor for fermionic operators according to their exchange statistics, and enabling systematic evaluation of correlation functions and perturbation expansions in quantum field theory and many-body systems.

Wigner Theorem A fundamental result stating that any symmetry transformation of quantum states that preserves transition probabilities is represented by either a unitary or antiunitary operator on the Hilbert space.

Wigner-Seitz Cell The region of space around a lattice point that is closer to that point than to any other lattice point, defining a primitive unit cell that reflects the local symmetry of the lattice.

Yukawa Interaction A screened Coulomb interaction characterized by an exponentially decaying strength with increasing distance. It provides an effective description of forces with finite range and is widely used in plasma physics, nuclear physics, and other systems where screening effects suppress long-range interactions.

GPSR Compliance

The European Union's (EU) General Product Safety Regulation (GPSR) is a set of rules that requires consumer products to be safe and our obligations to ensure this.

If you have any concerns about our products, you can contact us on ProductSafety@springernature.com

In case Publisher is established outside the EU, the EU authorized representative is:

Springer Nature Customer Service Center GmbH
Europaplatz 3
69115 Heidelberg, Germany

Batch number: 10505346

Printed by Printforce, the Netherlands